Contemporary Logic Design

Second Edition

Randy H. Katz
University of California, Berkeley

Gaetano Borriello
University of Washington

PEARSON
Prentice Hall

Upper Saddle River, New Jersey 07458

Original edition, entitled CONTEMPORARY LOGIC DESIGN 2nd Edition, by KATZ, RANDY H.; BORRIELLO, GAETANO, published by Pearson Education, Inc, publishing as Prentice Hall, Copyright © 2005.

ISBN 978-89-450-0148-1
ISBN 89-450-0148-4

Dedicated to our students, who motivate,
challenge, and make us proud every day.

Contents

Preface

A Second Edition

In the decade since the first edition of this book was published, the technologies of digital design have continued to evolve. The evolution has run along two closely related tracks: the underlying physical technology and the software tools that facilitate the application of the new devices. The trends identified in the first edition have continued stronger than ever and promise to continue for some time to come. Specifically, programmable logic has become virtually the norm for digital designers and the art of digital design now absolutely requires the software skills to deal with hardware description languages.

No longer do we see the familiar yellow cover of the TTL Data Book on every designer's bookshelf. In fact, for many application areas, even small programmable logic devices (PLDs), the mainstays of the 1970s and early 1980s, are rapidly disappearing. The burgeoning market for smaller, lower power, and more portable devices has driven high levels of integration into almost every product. This also has changed the nature of optimization; the focus is now on what goes into each chip rather than on the collection of individual gates needed to realize the design. The optimizations of today are more and more often made at the architecture level rather than in the switches.

Hardware designers now spend the majority of their time dealing with software. Specifically, the tools needed to efficiently map digital designs onto the emerging programmable devices that are growing ever more sophisticated. They capture their design specifications in software with description languages appropriate for describing the parallelism of hardware; they use software tools to simulate their designs and then to synthesize it into the implementation technology of choice. Design time is reduced radically as market pressures require products to be introduced quickly, at the right price and performance.

Although the evergrowing complexity of designs necessitates more powerful abstractions, the fundamentals haven't changed. In fact, the contemporary digital designer must have a broader understanding of the discipline of computation than ever before, including both hardware and software. In this second edition, we provide this broader perspective.

Changes from the First Edition

There are many changes from the first edition that can be grouped into four rough categories. First, we updated the hardware technologies discussed in the book. Second, we added a more complete, if nevertheless introductory, treatment of the software tools that are now so commonplace in the designer's tool kit. Third, we responded to the comments and suggestions received over the years by the many faculty and practitioners who have used the book. Finally, we rationalized the organization of the text so that concepts, technologies, tools, and practical matters were more clearly defined.

New Introduction

The introduction has been changed from one that focused on the process of design to one that introduces the concepts of computation, encoding, and sequencing. This sets out a better road map to the rest of the book and provides a rationale for its organization. Rather than discussing the design process in the abstract, we now include many more case studies to help the student gain that understanding by seeing the process in action.

Repartitioning of Material

Each of the two major sections on combinational and sequential logic was divided into a set of chapters. These first cover the fundamental concepts, then describe the principles of manipulating the logic into different forms, followed by a discussion of the optimizations and tools that are available, and concludes with an overview of the technologies available to build logic circuits. Each is capped by a set of comprehensive design case studies that make each of the issues concrete.

More Emphasis on Programmable Logic

We have added new material on the latest programmable logic technologies that have quickly become the dominant style for realizing digital designs. We do not attempt to provide all the information needed to work with any one technology. Those used will vary dramatically from institution to institution. Therefore, the book needs to be supplemented with a laboratory guide that covers the specifics of a particular installation. In this text, we focused on the underlying concepts. We expect laboratory guides to be available in the form of web-based materials that can be easily customized to the variety already out there and updated as new technologies emerge.

Inclusion of Hardware Description Languages

HDLs are now given a more central role to reflect their total acceptance by the design community over the past 10 years. We describe only the basics of one of the dominant languages, namely Verilog, focusing on describing behavior, as well as covering the basics of HDL simulation models. We highlight the power of the languages in making designs more parameterizable and customizable and designers more efficient.

New Design Case Studies

Nothing helps students learn design as much as designing for themselves. The next best thing is to provide a large collection of examples where the intuitions and rules of thumb are discussed explicitly. The hope is that this will help bootstrap new digital designers into the world of practical applications rather than the drill problems that were the norm in simpler times. There are many new and extensive design examples sprinkled throughout the text and in two large case study chapters focusing on combinational and sequential logic.

Elimination of Chapters on Datapath, Control, and Register-Transfer

We decided to remove the last two chapters of the first edition, that focused on datapath and register-transfer design, and a simple processor as an in-depth design case study of the interaction of control and datapath. While these topics are without a doubt important, on reflection we felt they are better left for a more extended study of digital design than could be included within the page limit of this edition. Instead, we chose more intensive coverage of programmable logic and HDLs, with extensive but smaller design examples spread throughout the text. We plan to make supplementary materials on the eliminated topics available on the web.

Navigating the Book

The book is organized into 10 chapters and three appendices. Chapter 1 is an overall introduction to the field. Chapters 2 through 5 cover combinational logic. Chapters 6 through 10 cover sequential logic. The three appendices provide some potentially useful background material that may have been part of other courses in a computer or electrical engineering curriculum.

Chapter 1 is an ambitious attempt to introduce many of the concepts of digital design through a short history of the evolution of digital hardware and two simple examples. Many may find that it introduces too many concepts too quickly for students to grasp their importance. However, this was not the intent. We fully expect students to be somewhat overwhelmed by the number of new concepts that come up in the discussion of the example. The purpose of the chapter is to provide an aerial view of the field so that students find it easier to see how the pieces they will see, in much greater detail and depth in later chapters, fit together coherently. It is certainly possible to replace this chapter with a more traditional introduction.

The next four chapters lay out the concepts of combinational logic design, closing with a set of comprehensive examples.

Chapter 2 covers the basics of combinational logic from simple gates to their time behavior. It lays out the concepts of two-level and multilevel logic and motivates why we would want to simplify logic. Some of the basic machinery for manipulating logic is presented with an emphasis on pencil-and-paper methods.

Chapter 3 delves into methods for working with combinational logic. It begins by describing the algorithms inside of today's CAD tools and ends with an overview of hardware description languages and uses Verilog to demonstrate key elements. Included is a discussion of the discrete simulation concepts that help to clarify the language constructs. There is probably not enough detail to make this book the sole resource for laboratory work with CAD tools. We wanted to keep the book focused on key concepts rather than on details of particular tools. It will need to be supplemented with appropriate manuals for the particular tools students will find in their own laboratories. This chapter also covers timing issues in more detail, including hazards and hazard-elimination strategies.

Chapter 4 presents the full range of implementation technologies available to the logic designer for combinational logic. It is paired with Chapter 9 that does the same for sequential logic. Chapter 4 starts with basic logic gates (as in the traditional TTL-based courses), but quickly progresses to programmable logic (PLDs and two-level forms) and then to field-programmable gate arrays. We also discuss other types of logic constructs such as tri-state and open-collector logic. Basic electronics to support this discussion are in Appendix B.

Chapter 5 culminates the combinational logic section of the book with seven examples of increasing complexity. We emphasize problem solving from the initial specification and have provided considerable discussion of how to transform an initial informal description of the problem into precise logical statements while keeping track of the assumptions that are being made. Our goal in this chapter is to show the range of logic design and how to judge design tradeoffs and take advantage of optimization opportunities.

The remaining five chapters do the same for sequential logic what the Chapters 2 through 5 did for combinational logic.

Chapter 6 begins this section by introducing the idea of circuits with feedback and how they can be analyzed. We develop the basic elements of sequential logic, latches, and flip-flops by recapitulating their evolution. This is coupled with a discussion of the timing methodologies that make it practical to build large sequential logic systems. These methodologies are illustrated with simple sequential systems of shift registers. The chapter concludes with a continuation of the exposition of hardware description languages started in Chapter 3 and extends it to basic sequential logic elements.

Chapter 7 covers the central concept of finite state machines. It begins by using counters as a simple form of FSM and then moves on to the basic Moore and Mealy models for organizing sequential behavior. Like Chapter 2, it concludes by motivating the various optimization opportunities.

Chapter 8 extends the basic ideas of Chapter 7 and expands on the details of FSM optimization by treating state minimization, state encoding, and FSM partitioning, in turn. Each of these is illustrated with examples that highlight the tradeoffs at each stage of optimization. An additional section at the end of the chapter provides some guidelines for structuring FSM descriptions in HDLs.

Chapter 9 concludes the discussion of implementation technologies. It recapitulates all the technologies used for combinational logic introduced in Chapter 4 but focusing on their sequential logic elements.

Chapter 10 is a large chapter with six comprehensive design examples that bring to practice all the concepts in the text. It begins with the sequential logic example from Chapter 1, now discussed in full detail, to tie back to the start of the text and ends with the serial transmission of characters from a keypad to display. The latter examples focus on the partitioning of design problems into communicating pieces along two dimensions: parallel state machines and partitioning into data-path and control.

The three appendices cover number systems, basic electronics, and flip-flop types. The first two cover concepts that students are likely to have already seen in mathematics, physics, electrical engineering, or computer science introductory courses. They are not intended to be extensive treatment of these topics but only provide the background most directly connected to the main topics of this text. The appendix on flip-flop types is provided for historical completeness.

The Complete Teaching Package

The material in this book easily fills a quarter-long course and can be comfortably covered in a semester-long course. In fact, it is likely that supplemental topics, governed by the place in the curriculum the semester-long course occupies, can and should be included. These could be: more in-depth discussion of CAD algorithms including their data structures, efficiency, and implementation; further discussion of design tradeoffs in a particular implementation technology such as FPGAs; a larger design problem that can serve as a term project to highlight issues of scale and debugging; and topics from computer organization emphasizing partitioning into data-path and control and optimizations of both. Of course, individual instructors may also find that re-ordering some of the material makes more sense in their environments. For example, it is certainly possible to proceed by following the two sections in parallel rather than serially. Chapter 2 plus 6 and 7 can be paired, followed by 3 and 8, then 4 and 9, with the larger design examples of 5 and 10 together at the end. Many topics can also be skipped altogether. For example, CAD tools and their algorithms may be relegated to another course. Similarly, HDLs do not need to be included if the design environment focuses on schematic-level design. In the technology dimension, FPGAs can be skipped as they may be included in a later course on more advanced design methods. Our goal in organizing the book was to make it easier to make these customizations.

Finally, we are making a wealth of supplementary material available to course instructors and students. Our publisher's web site includes:

- A set of CAD tools that supports all the concepts presented in this text;
- A comprehensive set of lecture slides;
- Samples of possible laboratory assignments and projects;
- Solutions to all the problems in the text; and
- Supplementary material on computer organization for those that include that material in their introductory logic design classes.

We hope you will agree with us that this second edition is a worthy successor to the first.

Randy H. Katz
Gaetano Borriello

CHAPTER 1

Introduction

Introduction

Computer hardware, in its short 50-year history, has experienced the most dramatic improvement in capabilities and costs ever experienced by humankind. Computing devices are now ubiquitous in our daily lives. They've come a long way since they were machines used exclusively by a few scientists. They are used now by an ever-growing portion of the world's population. From their birth as automatic machines for arithmetic calculations, computers now are part of everyday appliances such as microwave ovens, anti-lock brakes, media players, and pocket calendars. We no longer think about computers explicitly when we use the devices that they make possible. CD and DVD players let us listen to music and watch movies—but they are, in fact, *computing* devices in that they include several microprocessors as well as other digital hardware (see Figure 1.1). Just think how the telephone has become a highly integrated communication system with many components. A cellular phone allows you to make and answer calls from just about anywhere in the world. As electronics shrink in size, we also benefit from lower power requirements and smaller batteries that last longer. A music player has become portable, holds thousands of songs, and can play them for hours without recharging. Your home phone or cable line now also lets your computers access an unprecedented wealth of information and services over global networks.

Logic design is one of the disciplines that has enabled the digital revolution which has dramatically altered our economies, communication systems, and, consequently, our lives. Not only have the hardware components evolved dramatically, but the tools of the logic designers' trade are also quite different that what they were even a short 10 years ago.

Any introduction to this subject must necessarily begin by defining what we mean by design and logic design, in particular. We will then provide some historical perspective by quickly reviewing the evolution of the underlying technology that makes our digital world possible. We'll conclude the chapter with some examples of logic design that will serve as a preview to the remainder of the text.

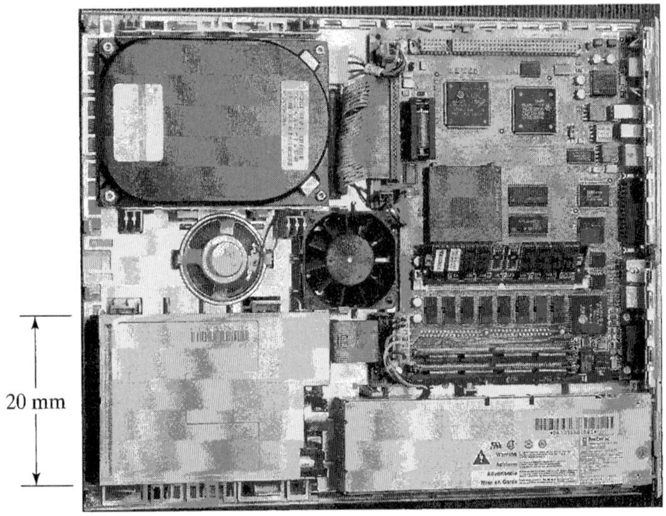

Figure 1.1 **The internal electronics of a digital video disk (DVD) player consist of several computing devices that process the light levels from a reflected laser beam and transform them into an audio stream for speakers or headphones and video stream for a display.**

1.1 Dissecting the Title

1.1.1 Design

Design is the process of coming up with a solution to a problem. To do this, we not only have to understand precisely what the problem is, but also the constraints our solution must meet. We can't come up with solutions that require an unlimited amount of money or carelessly waste resources and pollute our environment. Constraints can arise from physical limitations or even from aesthetic and subjective criteria. For example, if we are designing a building, the problem may be to create 50,000 square feet of space that will make office workers very productive. Physical constraints on the building design could include limits on its height, how far it can go underground, the number of offices with windows, the number and type of common spaces such as conference rooms, kitchens, and atria, and all the service utilities the building will need. The building's external design also will have to fit in to the surrounding neighborhood and be welcomed by its neighbors. These are, of course, only a small part of the constraints that architects must manage when designing a solution. The architect must also ensure that the building is completed within its cost budget, that it can be heated and cooled efficiently, and that noise levels generated by ventilation systems are within tolerable limits for the people that will occupy the building.

We recognize building design as a very complex process. To handle this complexity, we've developed an important strategy based on the tried and true divide-and-conquer approach. We do this so that we can divide our problem into smaller sub-problems. Human beings are only able to keep so many details and their interrelationships in their heads at one time. By dividing the building into its constituent parts, and using a team of designers (or in this case, contractors and their sub-contractors) each working on their own parts of the whole, we are better able to manage a large design task. Each component is now a design problem for its respective designer who may in turn choose to divide it down further. In our building example, these will include office layouts, window frames, ventilation systems, etc. Note that the decomposition of the design tasks is often functional and not spatial. For example, the elevator design may be further decomposed into the design of the shafts, mechanical and electronic controls, as well as their relationship to fire safety rules. The building's chief architect is responsible for bringing these pieces together to form the complete building. Although, the architect may have some influence over the design of these pieces, it is likely that, due to cost considerations, the choice may be limited to pre-designed solutions available only in specific pre-fabricated configurations. Good communication between all the contractors involved in the construction of the building is essential in guaranteeing that inconsistencies and errors are caught as early as possible when there is more time to make adjustments. It is difficult to widen an elevator shaft, for example, after the building's skeleton is already completed.

Most of the objects we design are themselves going to consist of components. In fact, the design process turns out to be very similar in a wide range of disciplines. This may not be obvious at first. Clearly, a civil engineer designing a new highway interchange has a different set of problems than a computer engineer building a new holographic display. But it turns out they are much more similar than they first appear. Each has to breakdown the problem into smaller pieces, each has to deal with constraints beyond their control, each has to put all the pieces together to solve the bigger problem, each has to be part of a larger team and/or manage a part of the effort, etc.

For example, designing a new software application is not really all that different from our building example. When we pick a particular computing platform (a processor and its memory system) it imposes limitations on our software. Some programs may need more memory resources or not run fast enough on the platform we chose. User interface concerns also have an influence on our software's design. Our software designer will be trying to be efficient and will want to reuse utilities available from the computer's operating system or parts of software packages that were previously written and with which she is already familiar. Thus, the software will have to be structured in such a way to take advantage of the pieces that may already be available. Finally, the software required may be too large for one individual, group, or even company, and pieces will have to be sub-contracted to others to complete. Our designers will have to make sure they communicate regularly and clearly so that they don't make any mistakes that will cost them time later when they try to pull all the pieces together.

Complex systems, such as our building or personal computer software, require us to adopt a design methodology if we are going to be able to manage the process efficiently and effectively. A design methodology is a domain-specific formalization of the design process into a set of well understood steps. Architects have developed a design methodology that involves contractors and builders, materials suppliers, community and municipal agencies, and the eventual occupants of the building. An automobile designer also has a methodology but it is likely to be quite different from that of the building architect and include much more concern with safety, reusing parts from existing car models, and streamlining the assembly line.

Design has three important facets. It is a creative process of coming up with a vision of the solution. It is an engineering process of evaluating tradeoffs and making decisions among many alternatives. It is an optimization process of choosing the best combination of components to realize the vision. In this text we'll be working on all three facets.

1.1.2 Logic Design

The components that the digital designer has at the ready are made from digital electronics. Combinations of switches, built from semiconductor transistors, form the basis of all of today's digital hardware. We generally refer to an interconnected collection of switches as a *circuit*. Individual switches are not the only building blocks available. There are also higher-level circuit modules such as logic gates and memories that use switches already pre-arranged in an efficient and flexible way. The logic designer's job is to choose the right components to solve a logic design problem. Constraints in logic design are often related to some combination of size, cost, performance, and power consumption. Cost and size are very closely related. A component's complexity is determined not only from all the switches it contains but more importantly from all the wires used to connect the switches together. A component's size usually has a direct relationship to the cost of manufacturing the component. Performance and power consumption are determined by the particular arrangement of switches and wires, the underlying materials from which they are constructed, their size, and how fast they are switched on and off. Interestingly, as integrated circuit technology further shrinks the size of transistors and the width of wires, it is the wires that are coming to dominate designers' concerns as they already add more to the size of the circuit and to its performance concerns than do the transistors.

All digital components have a set of input wires and a set of output wires both of which carry digital logic values. In other words, wires that are set to some voltages we have agreed to will represent two different values, that is, 0 and 1. Of course, in the real world, we have continuous phenomena and the wire can in fact be set to any voltage. In the case of digital electronic circuits, any voltage below some level, say 1 volt, is interpreted as a logical 0 any voltage above some other level, say 2 volts, is viewed as a logical 1.

Arbitrary information can be represented using this digital abstraction. Binary notation is used to represent integers as well as floating-point or even complex numbers through the use of a collection of wires each of which is a binary digit and together form the number. More interestingly, the color values for each pixel of a digital image or the volume of the sound created by a speaker can also be represented digitally. The digital abstraction is extremely powerful because it is more tolerant to variations in voltages which are difficult to make precisely identical in every single copy of a circuit. Furthermore, 0s and 1s can be used to represent analog values to arbitrary precision and have led to a powerful convergence of information that lets us store, transfer, and display information ranging from telephone calls to digitized maps to movies to e-mail messages using the same basic concept, namely, binary voltage levels on wires.

The transistors inside a digital component react to the voltage levels on the input wires. Changes in voltages cause transistors to change their conductivity, that is, switch their state, and thereby cause a change in some of the voltage levels on the output wires of the component. We often refer to the inputs *causing* a value to be applied to the outputs. For example, we may have a circuit with two inputs and one output where the output is set to one if both inputs are also one. This is referred to as an *AND logic gate* because both the first input and the second input must be at voltage levels corresponding to a 1. We'll later see a wide range of logic gates. These types of circuits, where inputs directly influence the value on outputs, are referred to as combinational logic circuits. They form the basis of all our computational elements including components that can add, subtract, or even multiply.

Other types of digital circuits are referred to as sequential logic circuits. Their outputs not only react to the current values on the input wires but also to the past history of values on those same input wires. Thus, sequential circuits have memory in that they will remember the past inputs and react to the current inputs while taking their history into account. A simple example of a sequential logic circuit is a memory component that has two inputs and one output; the output is set to the same value as one of the input wires when the second input wire signals it to do so. The input value *sampled* in this manner is then held indefinitely until the circuit is signaled again to store a new value. These types of circuits form the basis of memory devices that allow our computers to store and recall data as well as keep track of the step in a sequence of instructions, embodied in a computer program, they are to execute next.

Logic design is a set of abstractions and methodologies that let us devise, understand, and manipulate large collections of digital circuits. We've seen one abstraction already, namely, using 0 and 1 to represent all sorts of data. We'll see several others in this chapter and many others in the remainder of the text. Design methodologies are important procedures for ensuring a principled and effective design process. We'll be developing several of these for different types of digital logic. Finally, we'll also develop ways of transforming and optimizing our circuits so that they can have a better chance of meeting design constraints and be efficiently implemented using the particular components we may have at our disposal.

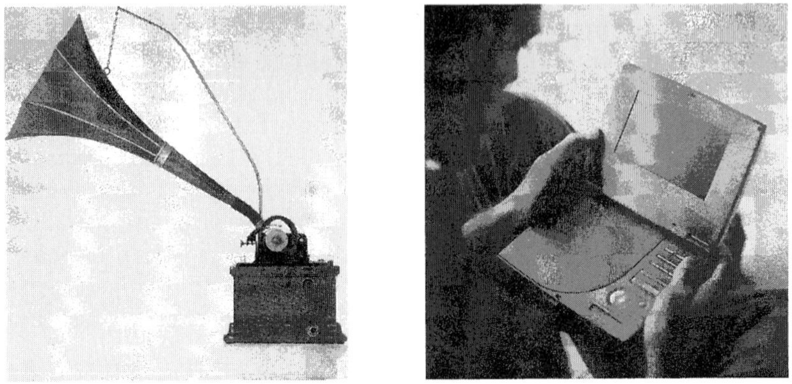

Figure 1.2 **Photographs of a late 19th century phonograph and a late 20th century portable digital video disc (DVD) player (not to scale).**

1.1.3 Contemporary Logic Design

Logic design has been around for at least 150 years. In that time, there has been continuous evolution of the abstractions, the basic building blocks, and design methodologies, the rules for putting the building blocks together. We can extend our analogy to building design as it also has evolved, using different materials, and different methods, for example, new materials as well as new earthquake and fire safety codes. In digital design, the changes have been quite dramatic and there are several important trends that are still at work.

First, our systems are becoming ever more complex as we integrate more functions into a device and perform computations on ever larger quantities of data. Just compare the complexity of a turn-of-the-19th-century phonograph and today's DVD players (see Figure 1.2). The older device is mostly mechanical and dealt with continuous values in both time and amplitude (loudness of the sound), namely, the vibration of a needle on a vinyl record and the amplification of that vibration into a vibration of the speakers. The DVD player reads digital data representing the sounds using a laser that doesn't even touch the spinning plastic disk. The information read from the disk can be in any form including audio, video, and text. These binary numbers are translated into either vibrations of a speaker through a digital-to-analog converter that translates a number into a proportional voltage level, or into images by turning pixels on and off on a screen. Furthermore, enough of these numbers have to be read and translated every second to make the sound seem continuous to the human ear rather than choppy. The phonograph consisted of a handful of moving parts. A DVD player includes several microprocessors operating on a digital representation of images and sound giving us access to any part of a film instantaneously with no degradation over time.

Second, the design of today's digital systems is happening in a much faster time frame as the demands of the consumer market place inexorable pressure on products to have a wide range of features appropriate for different uses and situations. New models of phonographs were introduced every few years. In contrast, there are dozens or, perhaps, even hundreds of new models of DVD players produced each year. Each model solves the problem of getting images and audio from the DVD disc to the screen a bit differently and thus provides for choices in cost, form-factor, user interfaces, and advanced features.

Third, and finally, the cost of digital hardware has become so low and its performance so high that we no longer need to be concerned with engineering the absolutely lowest cost solution. It is becoming more important to design rapidly and getting it right the first time. Getting new products out to market quickly is what generates the largest profits.

These three trends have led to a radical change in the methodologies of logic design over the past 10 to 15 years. First among these is automatic generation of logic circuits using software tools. We can now specify the functions we want our circuits to perform using a high-level specification language and have a *logic compiler* refine and transform that specification into a set of components. Second, we have created digital components that do not have fixed functionality but can be used to perform a wide range of functions based on a configuration performed after they are manufactured. This provides immense economies of scale, as we no longer need to carry large inventories of different digital components. Third, the emphasis has now shifted from the crafting of the implementation (i.e., the arrangement of switches into circuits) to the crafting of the specification. Designers focus more on getting the high-level specification right, to meet all the functional requirements of their product, rather than on arrangements of transistors. They rely on the compilation tools to determine the best set of components to use and how they will be configured.

Contemporary logic design now faces many of the same problems as software design. Designers want to work using specification languages at ever higher-levels of abstraction. They can be much more productive at higher levels, as there are fewer components to consider. However, they also want to ensure that the resulting design will meet the design constraints. Hardware designers are interested in ensuring that an appropriate collection of components is used that costs no more than necessary and will perform their computations quickly enough. Similarly, software designers are also concerned with performance and the memory requirements of their applications. Both types of designers also need to be able to visualize their specifications and debug them when their artefacts do not behave as intended. In logic design, simulation tools, that mimic the behavior of the real physical components but allow the designer to easily peek inside, are an essential part of the designer's arsenal serving similar functions as the debugger does for software designers. Another important similarity between hardware and software exists in their respective methodologies for design. Both hardware and software designers understand the need to re-use as many portions of designs as possible. It is still an engineering art to devise components that can be used in many different contexts. This is crucial to making the design process more economically efficient.

1.2 A Brief History of Logic Design

We will begin the history of logic design in 1854 when George Boole invented an algebraic system for manipulating logical propositions. Boolean algebra is now the mathematical foundation of logic design. It forms the basis for the optimization of digital logic much in the same way we use arithmetic algebra to transform expressions on variables into equivalent ones that have fewer operations. For example, by using the distributive law to perform one less multiplication in an algebraic expression ($a * x + b * x = [a + b] * x$).

Claude Shannon's seminal paper in 1938 established a link between Boolean algebra and the switches used in the relay circuits of the day. This was an important step in moving Boolean algebra from the realm of abstract mathematical logic to physical devices that actually computed a logical expression with voltages.

The first general-purpose digital electronic computer, the electronic numerical integrator and computer (ENIAC), was designed and build by J. Presper Eckert and John W. Mauchly at the University of Pennsylvania between 1943 and 1946. It was the first machine to have all the classical elements we now consider part of a modern computer. This was a big advance from relays because vacuum tubes were much smaller and could switch much more quickly than relays. These machines could perform several hundred multiplications per minute with 18,000 vacuum tubes—a huge advance from the hand-operated calculating machines, which were then the mainstay of scientific computation at the time. However, it did weigh 30 tons, consumed 200 KW of power, and occupied 1500 square feet. Even with these staggering dimensions, its principal disadvantage was that vacuum tubes were highly unreliable and over the course of a day an average of 50 tubes had to be replaced.

The invention of the transistor in 1947 heralded the dawn of the integrated circuit age. Highly reliable semiconductor switches replaced vacuum tubes very quickly. The first commercially available transistor computer was the Ferranti Mark I in 1951. Soon thereafter, scientists were able to manufacture multiple transistors simultaneously using the technique of photolithography, invented in 1957, to create patterns of semiconductor materials using light-sensitive materials and chemical etchants. This was a revolutionary change as circuit elements no longer needed to be wired together individually but could be manufactured as an already wired-up circuit. Thus began the era of integrated circuits, which led to the first microprocessor: the 4-bit Intel 4004, in 1971, was first used in the Busicom electronic calculator.

By the end of the 1960s, logic designers had available a large catalog of logic components (such as those described in the then ubiquitous Texas Instruments TTL data book, the preeminent catalog of available parts). Arbitrary logic circuits could be built from these basic primitives which were mass produced in great quantities and so beginning the inexorable advance toward cheaper electronic circuits with higher reliability that continues to this day.

Programmable logic arrays, collections of switches in regular arrangements that could be configured by the logic designer to implement any one of a huge quantity of possible functions, soon arose to increase levels of integration and to make it easier for designers to change the wiring pattern between logic functions. These devices started to see wide use with the introduction of Monolithic Memories' Programmable Array Logic (PAL) line of components in 1975.

With the increased levels of integration in digital circuits and the need for designers to program their reconfigurable logic components came the development of logic synthesis tools. Starting from Boolean algebra expressions, these software packages could determine the precise configuration of logic arrays to implement the specified function and therefore free designers from the drudgery of dealing with every single switch and allowing them to focus on higher level design. One can think of these early tools as the assemblers of their day, translating from assembly language to machine code. Their successors today are much closer to compilers for high-level languages (e.g., C, Java) and the software development environments that go with them (e.g., Visual Studio).

Today, we see the continued development of programmable logic, in the form of field-programmable gate arrays (introduced by Xilinx in 1984), which can now be reconfigured over and over again. This makes possible logic circuits that can be altered over time, field upgraded after a product has been purchased, or even re-programmed quickly enough from one use of the product to the next. Synthesis tools have followed closely with the appropriate compilation technology to configure these new types of components. Finally, the level of integration has continued to increase. We can now consider using many transistors just to give our circuits more flexibility rather than stingily allocating each individual transistor as in the early days of integrated circuits.

1.3 Computation

Up to now, for most of you, computation has been an abstract process. You may have specified the steps of a calculation for a computer to execute but the details of how that computation is actually accomplished have probably been somewhat of a mystery. You may know that digital computers operate on binary digits (also called bits), 0s and 1s, and you may know how they represent integers and characters using strings of bits, usually 32- and 8-bits long, respectively. But what role do these strings of bits play in the execution of complex programs?

This text is about de-mystifying computation. It will guide you through the first steps in understanding how computers work. In the coming chapters, we'll see how to implement all the common programming constructs, including variable assignment, arithmetic operations, conditional and iterative statements, and subprocedures. We will learn how to construct circuits that perform all these functions using a few simple primitive elements. In addition, we'll also see that circuits can perform functions in parallel, not just sequentially, and operate on arbitrary data types, not just bits and integers. After all, we can have many circuits working in parallel on different inputs. At a small scale, arithmetic circuits provide an excellent motivation for parallelism since we can work on the different parts of the strings of bits in parallel. At a larger scale we'll see how simple computers, working independently, can go about communicating and coordinating their activities.

1.3.1 Switches, Relays, and Circuits

Switches are the basic building blocks of digital computers. The proper arrangement is the physical embodiment of abstract computations in digital circuits.

Let's review how switches work. In Figure 1.3, we show a simple switching network that is found in every home, probably in every room. A switch is used to disconnect a light bulb from its power source. If the switch is open, current does not flow through the circuit and the light bulb is off. If the switch is closed, then the light bulb is turned on as the battery's current can now travel through the light bulb and return to the battery—a completed circuit.

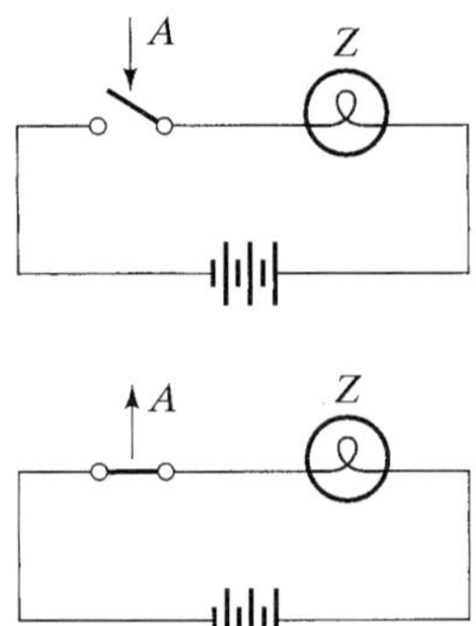

Figure 1.3 A simple switching circuit found in every home shown with its switch in open and closed positions, respectively.

If we represent the state of the switch by using a Boolean variable, say A, that we set to 0 if the switch is open and 1 if it is closed, and represent the state of the light bulb using another Boolean variable, say Z, that we set to 0 if the bulb is off and 1 if it is on, then we can write:

$$Z = A$$

as a Boolean expression that represents the functionality of this circuit. If A is 0, then Z is 0 (switch open, light off) and if A is 1, then Z is 1 (switch closed, light on). We can also use the symbol in Figure 1.4 to indicate this circuit in a schematic drawing. Note the arrow shape indicating that the output Z is affected by the input A.

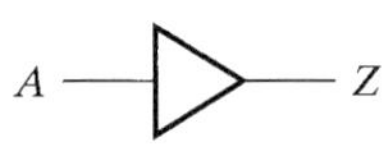

Figure 1.4 Schematic symbol for the simple circuit of Figure 1.3.

We can make our simple circuit a bit more interesting if we now add a second switch in series with our original switch represented by A (see Figure 1.5). We will represent this new switch with the Boolean variable B. Now the functionality of our circuit is such that both switches have to be closed for the bulb to turn on. We can write this expression as:

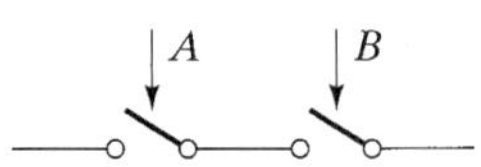

Figure 1.5 Two switches in series will close a circuit if both *A* <u>and</u> *B* are closed.

$$Z = A \text{ and } B$$

to indicate that both A and B have to be 1. A circuit like this is commonly found in automobiles where the key activates one switch and the windshield wiper wand activates another. Both have to be *on*, that is, both switches have to be closed for the windshield wipers to work. The schematic symbol for AND is shown in Figure 1.6. Note that it also "points" in the direction of the output.

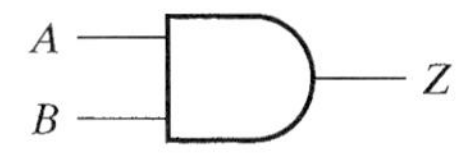

Figure 1.6 The schematic symbol for the AND function.

At this point, it is clear that we are thinking of 1 as having a meaning of *true* and 0 as *false*. This is, in fact, the common convention, and it makes sense in everyday English. However, the choice is really arbitrary. We'll see later on that it really doesn't matter what value represents true or false, just as long as we are consistent.

But let's continue on to another parallel arrangement for the two switches (see Figure 1.7). In this case, either A or B being closed will turn on the bulb. The current can flow through either path or both around the circuit. Our expression for this circuit is

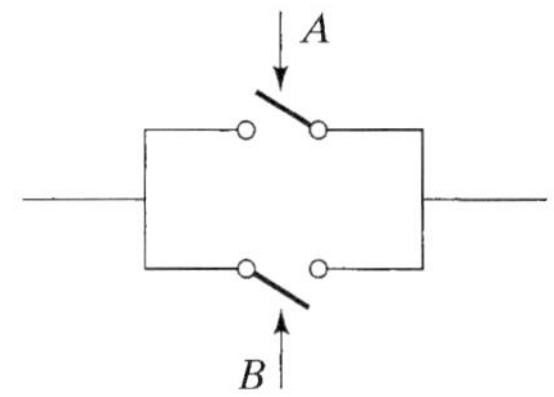

Figure 1.7 Two switches in parallel will close a circuit if either A or B is closed.

$$Z = A \text{ or } B$$

to indicate that either A or B or both have to be 1 for the bulb to be turned on. An example of this is also found in cars where the dome light will turn on whether the driver's or the passenger's door is opened. The schematic symbol for OR is shown in Figure 1.8.

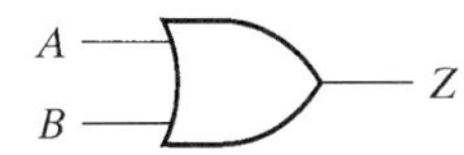

Figure 1.8 The schematic symbol for the OR function.

Switch settings determine whether a complete circuit, a conducting path, exists to light the bulb. We've assumed so far that someone sets the switches to be open or closed. To build larger and more interesting circuits, we have to be able to combine the basic ones we've just discussed. We need a way to get the state of a light bulb to control a switch on another circuit. In the early days of digital circuits, this was accomplished using a special device called a *relay* (see Figure 1.9). Its name is derived from the fact that it serves as a connection point between two otherwise independent circuits by passing a value from one circuit to the other in a way loosely analogous to the passing of a baton in a relay race.

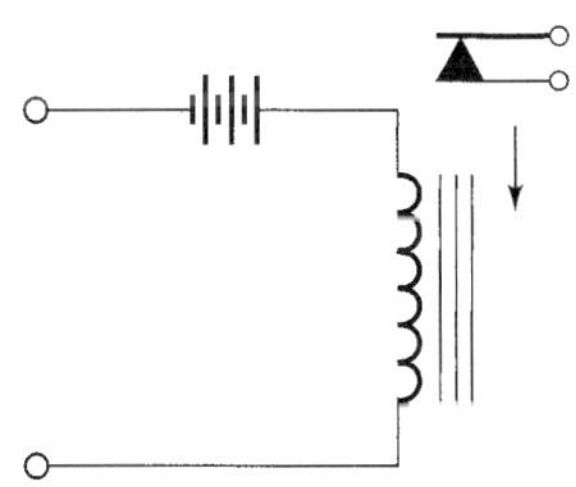

Figure 1.9 A simple magnetic relay. The switch at the top is opened when a current passes through the electromagnetic material directly below it creating an electromagnet.

A relay operates much in the same way as a light bulb. The difference is that instead of creating a glowing light, the current that passes through the device creates an electromagnet. A special switch is constructed from a ferric material so that if the magnet is on, it opens the switch by attracting the flexible half of it away from the stationary half as shown by the arrow in Figure 1.9. If the magnet is off, the switch's flexible half snaps back into position and closes the switch. This is an example of a switch that normally is closed and opened when current flows through the nearby electromagnet. The magnet and switch, together, form the relay.

Relay circuits were large and slow and could never have been used to construct large-scale computing machines. Magnets take time to charge and mechanical switches are slow to move as they have inertia directly proportional to their mass. It took the invention of electronic devices such as vacuum tubes to make it possible to build larger systems. However, vacuum tubes were still quite large and often unreliable. The invention of the metal-oxide-semiconductor, or MOS transistor, was needed to open the doors to both immensely useful and easily affordable computing devices.

1.3.2 Transistors

In today's integrated circuit technologies we have two types of switches. They are referred to as *n*-type and *p*-type transistors and are named after the type of semiconductor material from which they are constructed. The switch can be opened or closed by applying a control voltage. The *n*-type devices are switches just like the ones we discussed above; they are open when a low voltage is applied to their *gate,* or controlling terminal, and they act as an open switch; when a high voltage is applied they act as a closed switch completing the connection across their other two terminals (labeled *S* and *D* in Figure 1.10). The *p*-type devices are exactly the opposite in that they normally are closed and become open when a high voltage is applied to their *gate.* A small circle at the gate of the transistor is used to indicate a *p*-type device. Hence, the name CMOS technology, which stands for *complementary MOS technology* referring to its two complementary types of switches (see Figure 1.10).

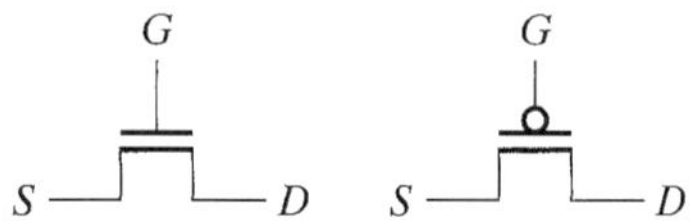

Figure 1.10 The symbols for *n*-type and *p*-type CMOS transistors. The terminals are labeled *G* (gate), *S* (source), and *D* (drain).

A simple CMOS network is shown in Figure 1.11. It consists of one switch of each type. The *p*-type device is used to connect the output, *Y*, to a high voltage while the *n*-type device is used to connect the output to a low voltage. Both transistors are controlled by the same voltage, as their gates are connected to the same wire. If the input, *X*, is 0, then the *p*-type switch will be closed and the *n*-type switch will be open, causing the output to be connected to a high voltage (a digital 1 and logical true). The reverse is true if the input voltage is high, then the output is connected to a low voltage (a digital 0 and logical false). This is a very simple logic device called an *inverter* because its output is the digital opposite of its input. It is often referred to as a NOT device or gate (gate is an historical term that, by an unfortunate coincidence, is the same word used for the controlling terminal of a transistor). The symbol for the NOT gate is shown in Figure 1.12. The bubble at the end of the arrow shape is used to indicate the inversion property of the circuit. This is the same notation as in the *p*-type transistor, whose gate input worked in the opposite sense as the *n*-type transistor.

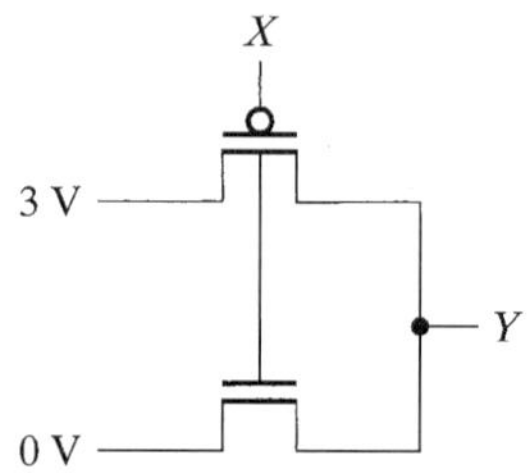

Figure 1.11 A simple CMOS network utilizing two transistors to connect the output *Y* to either 3 V or 0 V based on the value of *X*.

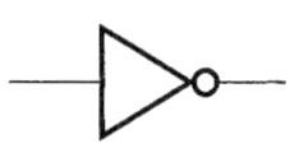

Figure 1.12 The schematic symbol for the NOT circuit of Figure 1.11.

You should now be asking yourself: why this particular arrangement of switches? The motivation comes from the fact that our transistors are not idealized switches. The *p*-type devices do a good job of connecting to high voltages but are not very good at connecting to low voltages. What happens is that the output voltage may not get quite low enough and could not be used appropriately to control a transistor in another circuit. Similarly, the *n*-type device is good at connecting to low voltages but not to high ones. Fortunately, we have both types and can arrange the two on the same integrated circuit as needed. Because the transistors are imperfect switches, and we use a 0 input to connect to 1 and a 1 to connect to 0, all of our gates have an inversion at their output.

More generally, CMOS transistor networks are constructed by creating a *p*-type switch network for the cases when we want the output to be high and an *n*-type switch network for the remainder of the cases where the output is to be low. The next most basic CMOS networks demonstrate this concept and are shown in Figure 1.13. They represent serial and parallel arrangements of the transistors. You'll note that when the *p*-type devices are serially connected, the *n*-type devices are in parallel and vice versa.

Let's step through the operation of the device on the top of Figure 1.13. When either input is low, the output will be connected to a high voltage. Only when both inputs are high, is the output connected, through the two serially connected *n*-type devices, to a low voltage. This is starting to sound like an AND arrangement except that the output is low when both inputs are high. Thus, we refer to this circuit as a NAND gate (for a NOT-AND gate: an AND gate with its output the opposite of what is should be). Its dual is the switching network on the bottom which is referred to as a NOR gate for NOT-OR. The schematic symbols for NAND and NOR are shown in Figure 1.14. Note, again, the use of the bubbles at the output to indicate inversion. All of the switching networks that we build with CMOS technology will have this inversion property because of the fundamental properties of the transistors we use.

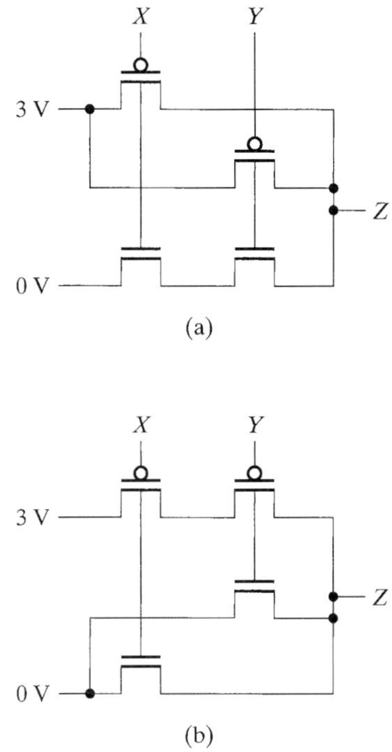

Figure 1.13 **Two CMOS transistors networks that demonstrate the arrangement of switches for a NAND operation (a) and a NOR operation (b).**

Transistor networks are much faster than relays. The switching time is determined by the flow of electronics in the semiconductor materials that make up the transistors. A water faucet provides a useful analogy (this idea is not really that far-fetched, since electrical effects ultimately are due to the flow of electrons, which act in a somewhat analogous way to water). Theoretically, a water faucet is either on, with water flowing, or off, with no flow. However, if you observed the action of a faucet being shut off, you would see the stream of water change from a strong flow, to a dribbling weak flow, to a few drips, and finally to no flow at all. The same thing happens in electrical devices. They start out by moving electrons rather quickly, but eventually the flow slows down to a trickle and finally stops. To complicate matters further, transistors are leaky faucets and never quite stop the flow of electrons completely.

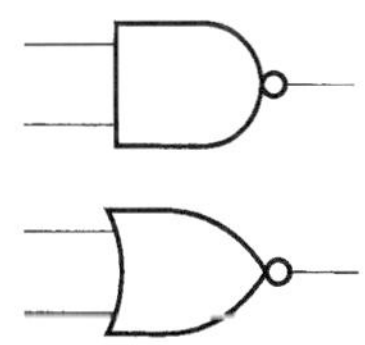

Figure 1.14 **Schematic symbols for the NAND and NOR circuits of Figure 1.13, respectively.**

As transistors have shrunk, and continue to shrink, in size, their switching speeds are increasing. There is less water in the pipes and it can be moved more quickly. This is what has fueled the rapid advances in computing technology of the past 50 years. Of course, we may soon see the day when we do reach fundamental limits and our circuits will depend on the movement of so few electrons that other physical effects will come into play.

1.3.3 Digital Representations

Real-world electronics are quite complex. The digital abstraction, interpreting all voltages as falling either into the *high* or *low* category, is a crucial step in being able to build digital circuits from imperfect switches such as transistors. Just imagine for a moment how difficult it might be to realize a circuit that depended upon precise and continuous voltages. We would have to be concerned that every slight variation in a transistor, due to the process involved in making it and printing it on to an integrated circuit chip, could alter the operating voltages and render the circuit useless. We would have to be sure our circuit could deal with a wide range of manufacturing variations, making the job of the designer impossibly difficult. Transistors, as all electronic components, exhibit continuous, or analog, behavior. They do so because output transitions are not instantaneous, they require electrons to flow and the charging or discharging of wires analogous to a water faucet or drain filling or emptying a tank.

Digital logic eliminates these problems by not taking on the difficult task of recognizing a single voltage value as logic 1 or 0. Digital logic must be able to deal with degraded or imperfect signals. Since digital circuits recognize any analog value above a specified voltage as logic 1, they can use degraded inputs to still generate correct output voltage levels. This is why we refer to digital logic as "restoring." Degraded values are restored as they affect the arrangement of switches. As an example, assume that *on* or logic 1 is represented by +3 volts. *Off* or logic 0 is represented by 0 volts. But we will interpret any voltage above 2 volts as a 1 and any voltage below 1 volt as a 0. Figure 1.15 shows a plot, or waveform, of an output switching from on/1 to off/0. You might observe a waveform like this on an oscilloscope. The transition in the figure certainly is not instantaneous. Other values, between 3 and 0 volts, are visible even if only for a relatively brief instant in time. These are interpreted by digital circuits as either 0 or 1.

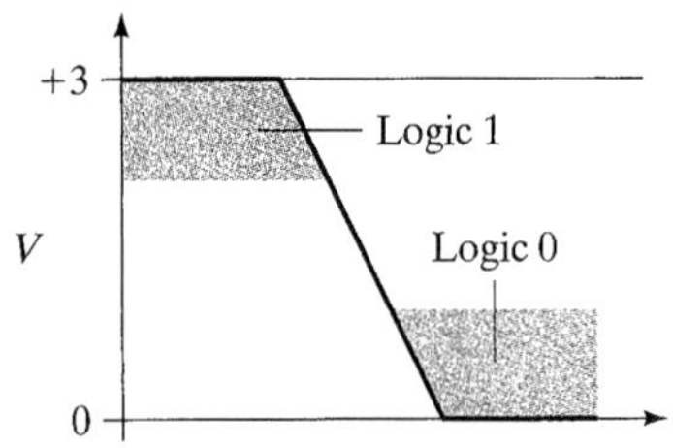

Figure 1.15 Digital transition versus time.

Digital circuits also do not output perfect voltages (0 and 3 volts). Their transistors, again due to imperfections in the manufacturing process, will most likely output voltages somewhat greater and less than the nominal 0 and 3 volts. Furthermore, ambient conditions such as humidity, temperature, and radio waves may also affect the behavior of devices. We cannot count on the outputs of our circuits to be perfect. This variation is often called *noise*. The *noise margin* of our circuits is the tolerance in voltage values, that is, the range of input voltages that will be interpreted properly.

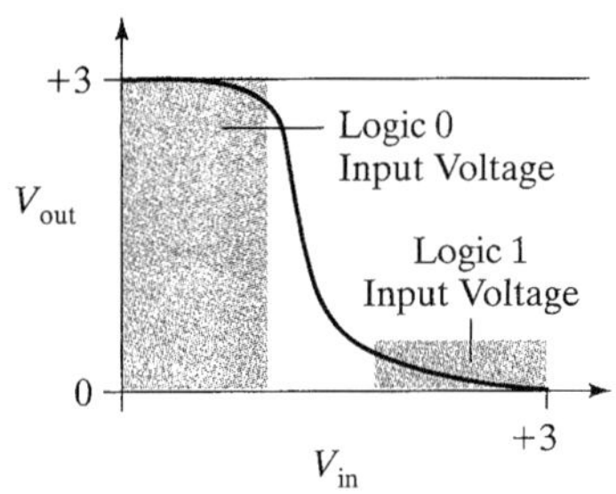

Figure 1.16 Inverter transfer characteristic.

Let's look at the transfer characteristic of the inverter. Figure 1.16 is a plot of the inverter's output voltage given all possible values of its input voltage. You'll note as the input goes higher, the output goes lower. However, it doesn't do so linearly. For input voltages between 0 and 1 volts, the output is very close to 3 volts. Similarly, for input voltages from 2 to 3 volts, the output is very close to 0 volts. Intermediate values of input voltage, between 1 and 2 volts, cause the output to be further away from either of the two nominal values of 0 and 3 volts. This one-volt range for logic 0 and logic 1 inputs is called the noise margin. Inputs and outputs can differ by as much as 1 volt from the nominal value when there is a 1 volt noise margin. Clearly, we need to avoid voltages in that intermediate range between 1 and 2 volts. We can do so by using transistors properly and not making wires too long.

1.3.4 Encoding

The manipulation of digital data is at the heart of computing devices. Digital representations exist for everything from numbers and characters to music and images. A digital representation is simply a string of 0s and 1s with an agreed upon interpretation. Of course, the more people, organizations, and software applications agree on the interpretation, the better.

A simple example of encoding is the audio stored on a CD or DVD. Small pits in the plastic of the disc cause a fine laser beam to reflect differently than if the surface patch did not have a pit. Collections of these bits make up a sound sample, or the volume to be reproduced on a speaker at the particular rate of playback. If the sample is interpreted correctly, that is, the bits are in the right order to represent the volume level, and they are sent to the speaker's digital-to-analog converter at the right rate, then the recorded sound will be faithfully reproduced.

Another example of encoding led us through the Y2K issues at the close of the millennium. Early programmers in the mid-20th century used only two digits to represent the year. This enabled them to encode the year of a database record in only two 8-bit bytes or sometimes just 8-bits altogether (two 4-bit numbers for the last two digits of the year). Of course, when the year 2000 finally arrived, 1900 was indistinguishable from 2000 and records were in danger of wrongly being sorted and correlated.

We also use encoding when we write software programs. The variables we use are often encoded as 32 or 64-bit binary numbers. The characters we use to represent text are represented as 8- or 16-bit values. We'll encounter encoding issues over and over again in this book. In each case, we'll have some freedom to decide how many bits will be used and what the order of the bits will mean. For example, a traffic light could have its interface consist of three signals, one for each bulb, or only two, as there are only three possible settings for the light and these can be encoded in a 2-bit binary number, say 10 for green, 11 for yellow, and 01 for red. We must take care in deciding on our encoding that we've taken into account all of the possible values that we may need to represent so that we do not repeat the problems of Y2K.

1.4 Examples

It is now time to turn to two examples to illustrate many of the concepts discussed so far. We'll use the examples to bring up some more of the terminology of logic design and introduce the primitive logic elements we'll be using in the rest of the text. Try not to be overwhelmed by the detail in these examples. The objective is to simply preview some of the topics we'll go into much more detail in later chapters in the book. The discussion of these two examples is used simply to provide some context for the topics that are to follow. Just try to get a general sense of the issues and approaches rather than feeling the need to understand every detail—that and more will come soon enough by the end of this book.

EXAMPLE 1.1 CALENDAR

Our first example is a small part of a larger design we'll build as we progress through the next few chapters. Its function is to decide, based on the month of a year and whether that year is a leap year, how many days are in that month. This will be used as part of a calendar display that could be part of a wristwatch.

Let's begin by thinking about the specification we've just been given. Our simple system has two inputs: a month (January to December) and leap-year indicator. It also has to have some outputs that tell us whether the month has 28, 29, 30, or 31 days. We could write a simple procedure that performs the function of our circuit. We'll use the C language, as it is quite common and relatively easy to read.

```
integer number_of_days (month, leap_year_flag) {
     switch (month) {
          case 'january': return (31);
          case 'february': if (leap_year_flag == 1) then
            return (29) else return (28);
          case 'march': return (31);
          case 'april': return (30);
          case 'may': return (31);
          case 'june': return (30);
          case 'july': return (31);
          case 'august': return (31);
          case 'september': return (30);
          case 'october': return (31);
          case 'november': return (30);
          case 'december': return (31);
          default: return (0);
     }
}
```

The procedure is quite simple. It has the two inputs as parameters and the output as the return value. The body of the procedure is a large switch statement that branches based on the value of the parameter for the month and returns the appropriate value. The only complication is February, which has a conditional statement that checks the leap year flag.

In implementing this program in digital logic, we are already faced with some encoding problems. How do we represent the month? In software, it is an enumerated type that will likely be implemented as an integer; but there are only 12 months and we are unlikely to add any more. We do not need all of the values possible with a 32-bit integer. To represent twelve possibilities, we need a minimum of 4 bits if we use a *binary* encoding. But 4 bits have 16 possible values, and we'll only need 12. Four of them will be unused. We term these *don't care* combinations because they should never occur and we will not be concerned with how our circuit reacts. Of course, should they occur, we might be in trouble. Later on in the text, we'll see how we can use don't cares to help us make smaller and faster circuits. Using 4 bits is not our only option. We could use 12 wires, one for each month. This is called a *one-hot* encoding. In this type of encoding, we have one and only one wire ever carry logic 1: the wire corresponding to that month. This is a very simple encoding but usually requires many more wires: in this case, 12 instead of 4.

The leap-year flag is straightforward, as it is a simple Boolean value. We'll need only a single wire that will be a 1 if the year is leap and 0 if it is not. We have to make sure to decide on that encoding and not its opposite. This decision better make its way into our documentation so that whoever builds the rest of our calendar system will be well informed of the choice we made.

Finally, we have to encode our output. The value will be as high as 31. This would require 5 bits in a binary encoding. On the other hand, there are only four possible results. We could use four wires and a one-hot encoding, namely, no more than one of the outputs will be true at any one time.

Let's use the minimum number of wires and get on to designing our logic. We'll choose a 4-bit binary encoding for the month, a single wire for the leap-year flag, and a 4-bit one-hot encoding for the result. Figure 1.17 shows the inputs and outputs of our circuits, schematically.

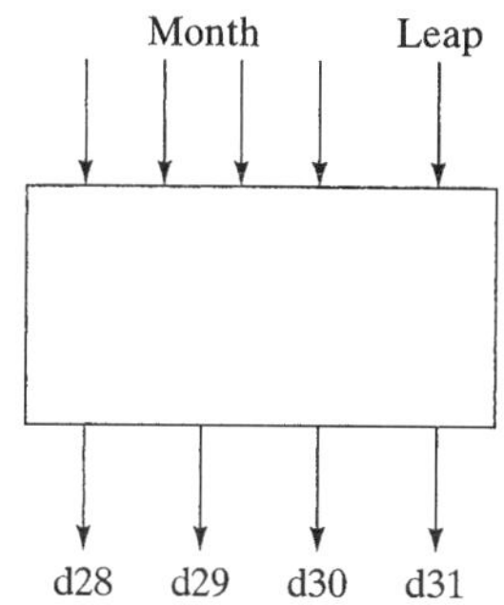

Figure 1.17 Inputs and outputs of the calendar subsystem.

Our next step is to figure out what the output should be for each combination of inputs. For example, if the month is February or 0010, and the leap-year flag is 1, then the result is 29. This would be encoded as 0100 if the 4 bits are 28, 29, 30, and 31 days, from left to right. We've already done something very similar to this in our program code. The completed table is shown in Figure 1.18. It has columns for all the inputs on the left side of the vertical line and a column for each output to the right. There is one row for each possible input combination. This type of table is referred to as a *truth table* because it specifies when a particular output should be true (1, high voltage). It could have just as easily been called a *falsity table,* but truth table sounds a whole lot better.

Month	Leap	d28	d29	d30	d31
0000	–	–	–	–	–
0001	–	0	0	0	1
0010	0	1	0	0	0
0010	1	0	1	0	0
0011	–	0	0	0	1
0100	–	0	0	1	0
0101	–	0	0	0	1
0110	–	0	0	1	0
0111	–	0	0	0	1
1000	–	0	0	0	1
1001	–	0	0	1	0
1010	–	0	0	0	1
1011	–	0	0	1	0
1100	–	0	0	0	1
1101	–	–	–	–	–
111–	–	–	–	–	–

Figure 1.18 Table of output values for the number of days in a month.

You should notice that the table is not complete in that every single input combination is not shown explicitly. We took some shortcuts. For example, for March (0011) we do not care about the value of the leap-year flag, March always has 31 days. Rather than having two rows as for February, we only have one with a dash in the column for the leap-year flag indicating an *input don't care*. Don't cares also appear in the output columns. This occurs because of those four unused values of our 4-bit month number (0, 13, 14, 15). Note also the use of the don't care to merge the rows for 14 and 15. Truth tables should be complete. This means that we should make sure to specify what we want the output to be for every single input combination. If we had left any out, then we would have their outputs open to interpretation and a likely error when we start combining our piece with other components. Output don't cares let us say explicitly that a designer can choose either a 0 or 1 without worry. In our program, we chose 0 for all the output don't cares in the table (the default return value). Here, we are saying its "ok" to choose 0 or 1 for any of those outputs. Of course, we had better be sure we really don't care, if we do, we should be explicit about the output value we want. Note that "don't care" is different than "don't know." Not knowing a value does not mean we don't care what it is.

The truth table can now be used to derive Boolean logic expressions for each of our outputs. For example, in the case of the output for 28 days, we have a very simple expression that says the month must be

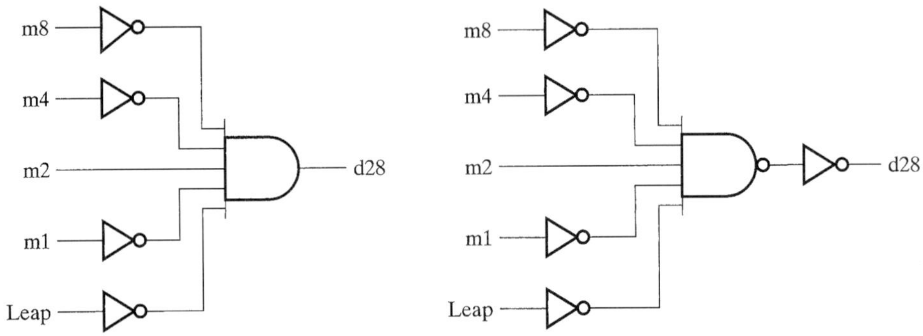

Figure 1.19 **Logic gate diagrams for the circuit of the d28 output of the calendar.**

February and the leap-year flag must be 0. Let's give names to the four bits for the month and the four bits for the number of days. We'll use m8, m4, m2, m1 for the month, chosen so that the names correspond to the weight of the bit in the binary representation, and we'll use d28, d29, d30, and d31 for the number of days. Our Boolean expression for d28 can then be refined as follows:

d28 = "February" AND (leap == 0)
d28 = (m8 == 0) AND (m4 == 0) AND (m2 == 1) AND (m1 == 0)
AND (leap == 0)
d28 = m8' AND m4' AND m2 AND m1' AND leap'

The last line uses a symbol for negation. m2 stands for m2 being a 1 (m2 == 1) while m1' stands for m1 being a 0 (m1 == 0). We can implement the quote or negation symbol (') using an inverter or NOT gate.

We can use AND and NOT circuits to implement our logic for computing the value of d28. On the left side of Figure 1.19 we used four inverters (for m8, m4, m1, and leap) and one 5-input AND gate. These gates are only logical constructs that help us draw a diagram for the Boolean expression above. Our next step is to find physical gates that implement these same functions. Now, we know how to implement a NOT gate, but how does one build an AND gate? In CMOS technology, we've seen that it's easy to get a NAND function. Thus, we'll use a 5-input NAND gate and then invert its output with a fifth inverter as shown on the right side of Figure 1.19. You may be wondering how we can construct a 5-input gate. Quite simply, we can increase the number of inputs for an AND gate by putting more switches in parallel (and in series on the complementary side of the gate).

Let's move on to another output. d31 is true in seven cases (January or 0001, March or 0011, . . . , December or 1100). Our expression refinement is as follows:

d31 = "January" or "March" or "May" or "July" or
"August" or "October" or "December"
d31 = (m8' AND m4' AND m2' AND m1) OR (m8' AND m4' AND
m2 AND m1) OR . . . OR (m8 AND m4 AND m2' AND m1')

This output will require the use of a seven input OR gate as shown on the top in Figure 1.20. Does such a gate exist? Can we increase the number of inputs arbitrarily? We know we can do the same as before and invert the output of a NOR gate to get an OR function by using a NOR gate followed by a NOT gate. But what will we do about the number of inputs or *fan-in* to the gate? Recall that the transistors we use for building our switching circuits are not perfect. Those imperfections also make putting too many transistors in series impractical because the circuit gets too slow and may cease to work entirely. We'll limit our design to 4-input gates and replace the 7-input OR gate we need for the d31 output with the circuit shown on the bottom in Figure 1.20. Similarly, we could have split the 5-input AND gate in the circuit for d28 with smaller gates. We'll see later how to convert these circuits to NOR and NAND gates.

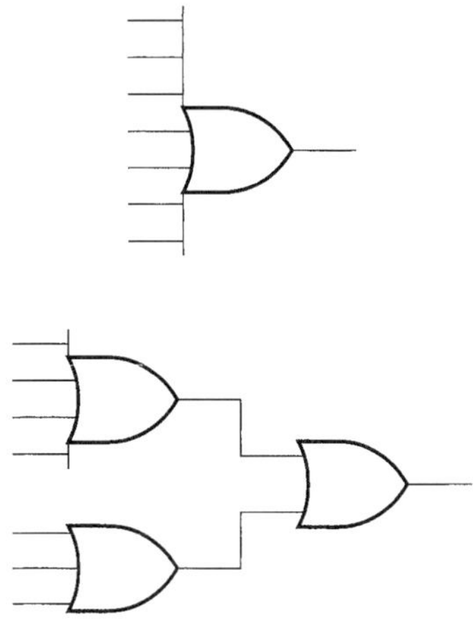

Figure 1.20 Alternate realizations for a 7-input OR function.

Note that for this example, when we completed our truth table, we simply looked at each possible combination of input values in isolation from the others. This is why we refer to this type of circuit as combinational logic. Our next example introduces sequential logic, where the order in which different input values appear (their sequence) is important as well.

EXAMPLE 1.2 COMBINATION LOCK

A simple door combination lock might consist of punching in a sequence of three specific keys on a small keyboard. The lock opens if the sequence is correct and stays locked if it is incorrect. Let's see how we would go about implementing such a lock.

Again, let's begin by writing a software program that has the functionality of our lock. The procedure has no explicit inputs and assumes there is another procedure that can tell when a new key has been pressed (`new_value`) and another procedure that can read the value of the key that was pressed (`read value`). You should note that the program has the combination for opening the door explicitly encoded in some static variables arranged into an array of three elements. It then uses conditional expressions to check the key pressed against the stored combination. A `while` statement is used so that the procedure can keep checking if a new key press has been made. After it detects a key press and there is a match it continues on to wait for the next key press. If there is a mismatch, it also continues on, but first sets an error flag. Finally, after three keys are pressed, it either opens the lock or not depending on the error flag.

```
integer combination_lock ( ) {
    integer v1, v2, v3;
    integer error = 0;
    static integer c[3] = 3, 4, 2;

    while (!new_value( ));
    v1 = read_value( );
    if (v1 != c[1]) then error = 1;
```

```
        while (!new_value( ));
        v2 = read_value( );
        if (v2 != c[2]) then error = 1;

        while (!new_value( ));
        v3 = read_value( );
        if (v3 != c[3]) then error = 1;

        if (error == 1) then return(0); else return (1);
    }
```

Another assumption we can make is that there is a reset button on our lock that, no matter at which step we are in our program, will cause our little procedure to restart from its first statement. Think of this as a reboot button on our lock.

Our logic design can now begin by determining the precise nature of the inputs and outputs of our system. First, we have the reset button (that is not visible in the code above). It's a simple Boolean valued input. Second, we need to decide how many keys to have and how they should be encoded. We'll choose 4-bit values in a binary encoding giving us 16 keys on the keypad and a total of 16*16*16 or 4096 different combinations for the lock. The output is a simple Boolean value that either opens the lock or keeps it closed (as does the return value above). But where in the program are the inputs to our system? They are encapsulated in the calls to `new_value` and `read_value`. These procedures are the ones that actually connect to the keypad of the lock. Before we can tackle how to design logic for `new_value` and `read_value` there are a few other issues we need to resolve.

You may have noticed that we need to be able to tell how far along we are in entering a combination into the lock. Are we looking for the first key press, the second, or the third? To do this, we need to introduce the notion of sequence and with it a time component to the behavior of our circuit. In the program, this was done using the sequential nature of our computers (in that they execute one step at a time) and the fact that our computers have what is called a *program counter* that points to the statement of the program to be executed next. Conditional statements, such as the while statements in our code, alter the program counter so it can execute instructions that do not follow each other in linear order in the program.

To accomplish the same thing in our digital circuit, we'll use a special signal that exists in every computer; a signal called a *clock*. Its name is quite descriptive indicating that it regularly "ticks," that is, it alternates between logic 1 and logic 0 at a regular rate. The period of this oscillation (the time it takes to complete an entire cycle of being set to 0 and 1) is the inverse of the clock's frequency. This is the parameter commonly used to discuss the performance of computers. The clock's tick lets us advance from one step to the next and also tells our circuit when to sample inputs and change outputs. We'll also need some memory to keep track of where we at any given time in completing a combination sequence. This is called the *state* of our

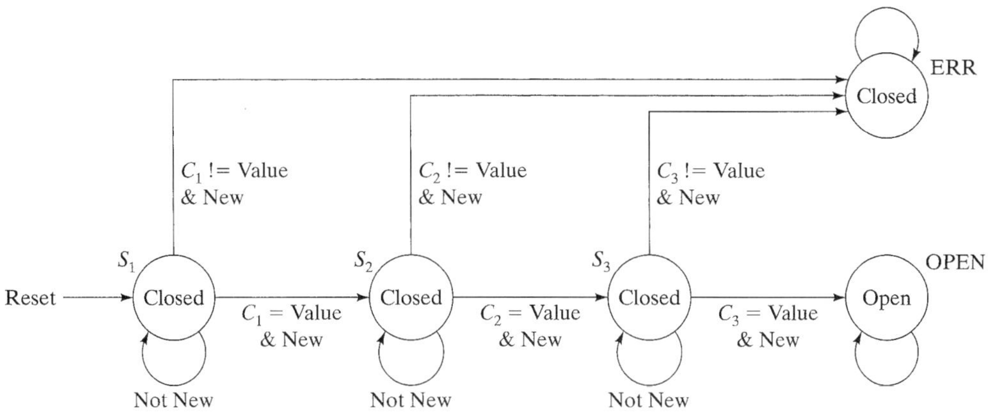

Figure 1.21 State diagram for the combination lock.

system. For example, a program counter is an important part of computer's state.

We can now begin describing how the state of our system changes over time. This is accomplished through a *state diagram* (see Figure 1.21). Each possible state is represented as a bubble. Arcs drawn from one state to another indicate under what conditions the system's state will change as indicated. The change will not occur until there is another clock tick. That way there is time for our logic circuitry to make a decision, based on sampled inputs and its current state, as to which state to go to next. If the clock is too fast, the decision may not be complete and our system will go to the wrong state. If its too slow, then we'll waste time waiting for the clock tick with the decision already made. Designers of high-performance computers spend a lot of time balancing this tradeoff so that their computers can run with the fastest possible clocks.

Clearly, there should be a different state for each step of the sequence so that we can tell the steps apart. We'll also have two states to represent whether we open the lock or if there has been an error. These five states are shown in the state diagram above. The conditions on the arcs coming out of a state represent the decisions that need to be made when the system is in that state. For example, when waiting for the first key press in state S_1, our circuit is checking to see that this is a newly pressed key (the signal *new* indicates this) and that it matches the first element of the combination (C_1 = Value) or not. If there is no new key pressed, then the system's state does not change (the arc coming back to the same state S_1). If there is a new key pressed, then depending on whether it matches, the state either changes to the error state, *ERR*, or to state S_2 so that the circuit can wait for the second key press.

How do we know when a key is pressed? Key presses may last a while. We only want to get a signal that there was a new key pressed once per key entered. We wouldn't want to view a single, long key press as two or three separate key presses. We'll defer the details of this to a later chapter. For now, let's assume that every time a new key is pressed the *new* signal is set to 1 for exactly 1 clock period and no more so that a 1 is sampled only once. This is another example of a constraint now imposed on whoever will design the keypad circuitry and it certainly belongs in our documentation.

Next on our list is the reset button. What does it do exactly? The state diagram provides a clear explanation. If the reset signal is ever 1, then no matter the present system state, it will change to state S_1 at the next clock tick. This is shown in the diagram of Figure 1.21 in shorthand by using an arrow that just points into state S_1. This implies an arc from every other state to S_1 that is traversed whenever reset is true. Of course, this also implies that the condition on every other arc also includes reset being false (this typically is not shown in the state transition diagram for the sake of brevity and a less cluttered diagram).

Finally, we have our system's inputs and outputs. There is a signal wire for when a new key is pressed (*new*), there are four wires for the value of the key (which only have significance if *new* is true), and there is a *reset* wire. In addition, there is the clock signal that can also be viewed as another input but is distinguished with a small triangle symbol. The output is a single wire that controls whether the lock is open or not. Figure 1.22 shows the inputs and outputs of our system and the fact that is has internal state.

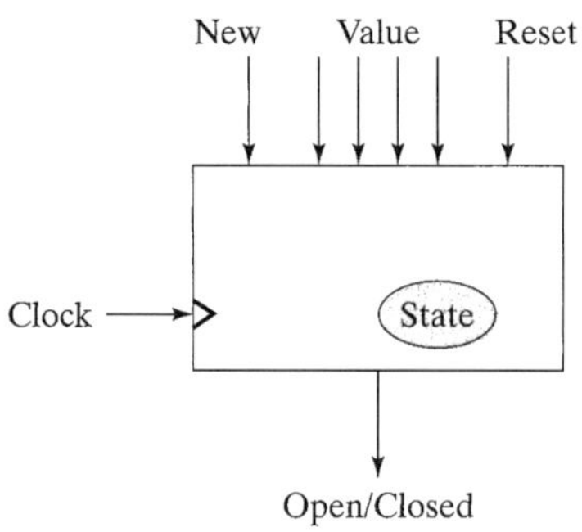

Figure 1.22 **Inputs, outputs, and internal state of the combinational lock.**

We can now turn to the internal structure of our lock system. We'll begin by separating the portions of the circuit that operate on the key values from those that concern themselves with the proper sequence of comparisons and their result. These are referred to as the *data path* and the *controller* of the circuit. Elements of the data path operate on their inputs the same way no matter what the values of those inputs. For example, it always compares the new key pressed with an element of the combination no matter what their values are. The controller is responsible for the sequence of steps (or states) our circuit will take. The result of the comparison is used by the controller to decide the next step to take: go and wait for the next input, open the lock, or keep the door locked in the error state.

The internal structure of our circuit is shown in Figure 1.23. The data path consists of three memory components that store the combination

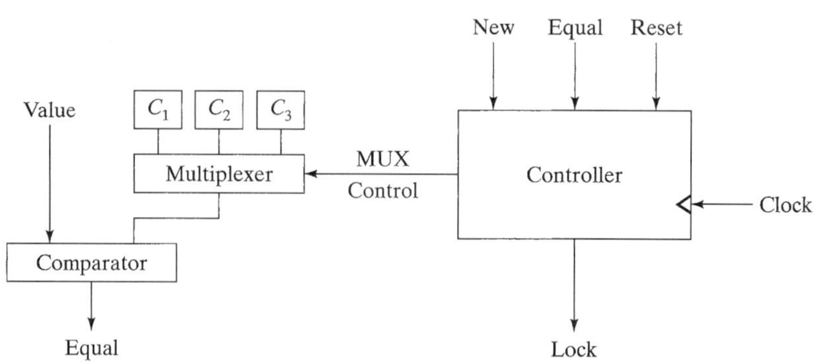

Figure 1.23 **Internal structure of the combination lock showing data path and controller.**

(we'll see in a later chapter how these values are set in memory), a multiplexer, and a comparator. All of these circuits operate on 4-bit quantities. A multiplexer is simply a way to choose between inputs. Its output is set to the same value as one of its multiple inputs. Control inputs determine which input's values will be used for the output. We'll see later how to build a multiplexer from combinational logic gates. The comparator is also combinational logic that outputs a single bit to indicate if the values it compared were equal. Data-path elements are either combinational logic such as multiplexers and comparators or they are memory elements, usually called *registers.*

The controller is a *finite state machine,* a concept that we will refine quite a bit further in later chapters. It will have five possible internal states (corresponding to S_1, S_2, S_3, *ERR,* and *OPEN* in the state diagram) and inputs for *reset* and *new* as well as *equal,* the result of the comparator. The finite state machine, of course, also requires a clock input to be used to advance from one state to the next. Its outputs are the controlling wire for the lock as well as the control for the multiplexer in the data path so that the correct element of the combination is used by the comparator.

To implement our finite state machine we need to revisit the state diagram and turn it into a state table. The latter is very much like a truth table (see Figure 1.24). A row corresponds to a different combination of inputs and internal state values. The outputs for each row are not only the output signals but also what the next state of the machine should be. There is no difference in the information content of a state diagram and a state table. Let's take a look at one of the rows. The second row of the table states that in state S_1, if there is not a new key press and *reset* is also 0, then it doesn't matter what the value of *equal* is, the next state will be S_1, the multiplexer will be set to channel the first number of the combination to the comparator, and the lock will be closed. The third row of the table says that if a new key is pressed and result of the comparison is false, then the next state is the error state and the lock is kept closed.

Reset	New	Equal	State	Next State	MUX	Lock
1	–	–	–	S_1	C_1	Closed
0	0	–	S_1	S_1	C_1	Closed
0	1	0	S_1	ERR	–	Closed
0	1	1	S_1	S_2	C_2	Closed
0	0	–	S_2	S_2	C_2	Closed
0	1	0	S_2	ERR	–	Closed
0	1	1	S_2	S_3	C_3	Closed
0	0	–	S_3	S_3	C_3	Closed
0	1	0	S_3	ERR	–	Closed
0	1	1	S_3	OPEN	–	Open
0	–	–	OPEN	OPEN	–	Open
0	–	–	ERR	ERR	–	Closed

Figure 1.24 State table for combination lock.

What is different about the state table versus a truth table is that it has symbolic names for the states rather than 0s and 1s. Our next step is to derive an encoded state table where we assign a unique binary code to each and every state so that the circuit can tell them apart. We also have to assign codes to the multiplexer control wires and the lock output. We have many choices in coming up with these encodings.

For the five states, we can use anywhere from three to five bits (some examples are: binary 000, 001, 010, 011, 100; to one-hot encoding, 00001, 00010, 00100, 01000, 10000). Conceptually, it doesn't matter which we use as long as each state has a unique identifying code. Practically, one code may lead to a much smaller circuit than another one would. We'll revisit this topic in later chapters as well.

The encoding of the multiplexer control signals depends on how we designed the multiplexer or the particular multiplexer we chose from a catalog. This is an example of interrelated component design. One can't be designed completely without the other being completed—a chicken-and-egg problem. The solution is not to charge ahead on one component's design but rather to iterate and refine both in parallel. In our case, we could use two or three wires to indicate which element of the combination to compare. Finally, the lock output is simple enough that we can use a single wire that is 0 when the door is locked and 1 when it is open.

An encoded state table is shown below in Figure 1.25. Note that the multiplexer uses a 3-bit one-hot encoding for its control signals while the state code was chosen to be 4 bits with 11 of the 16 possible encoding going unused. You'll note however, that the first three bits of the next state are identical to the multiplexer control outputs and the fourth bit of the next state is identical to the lock output. Because we had lots of flexibility to choose five codes out of the 16 available, we were able to choose wisely so that we only need to implement four output circuits rather than eight. In other words, we can use the same wires we'll be using to represent our current state to also control the multiplexer and the lock.

Reset	New	Equal	State	Next State	MUX	Lock
1	–	–	–	0001	001	0
0	0	–	0001	0001	001	0
0	1	0	0001	0000	–	0
0	1	1	0001	0010	010	0
0	0	–	0010	0010	010	0
0	1	0	0010	0000	–	0
0	1	1	0010	0100	100	0
0	0	–	0100	0100	100	0
0	1	0	0100	0000	–	0
0	1	1	0100	1000	–	1
0	–	–	1000	1000	–	1
0	–	–	0000	0000	–	0

Figure 1.25 **Encoded state table for the combination lock.**

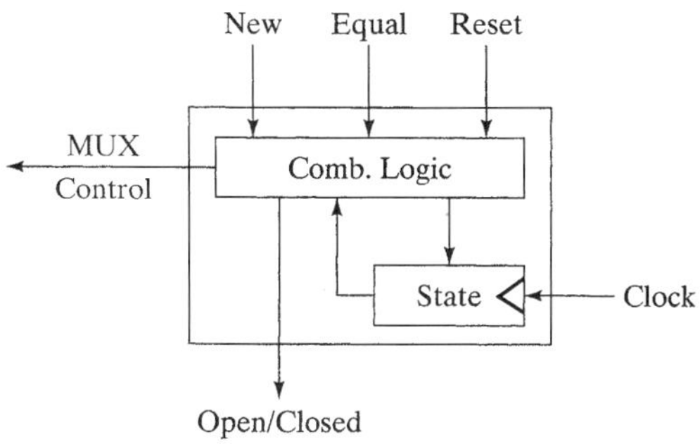

Figure 1.26 **Internal structure of combination lock controller.**

The controller's internal structure is shown in Figure 1.26. It consists of two parts: a state register that keeps track of the current state and is updated every time the clock ticks and a combinational logic module that given the current state (the output of the state register) and the inputs to the controller determine the output and next state. The next state will become the current state at the next clock tick.

This example was a whirlwind tour of sequential-circuit design and many of the issues designers face. Don't be concerned if you didn't understand every detail of these examples. In the following chapters, we'll be revisiting all the issues we breezed through with these two examples and go into a much more extensive discussion. We simply used these examples to set some context for the things to come in the rest of the book and to highlight the difference between combinational and sequential logic, namely, the use of memory to keep track of the internal state of a sequential circuit.

CHAPTER REVIEW

This first chapter has introduced what very well may be an overwhelming number of definitions and concepts. The following chapters hopefully will make these crystal clear by covering them more slowly, in much more depth, and with practice through examples and problem sets. The hope is that this introduction will have served a purpose by giving you, the reader, a sense of where you'll be heading as you make your way through the rest of the book.

The chapter introduced many of the abstractions crucial to making possible the logic design of today's complex digital systems. We introduced transistors and relays as underlying technologies but quickly abstracted their details into switches. Switching networks were abstracted into truth tables. Boolean algebra was introduced as the mathematical foundation for manipulating the logic expressions implemented by the switching networks. Logic gates abstracted away the implementation of the switching networks themselves by leveraging Boolean operators. We then talked about time and sequencing culminating in the design of a sequential circuit that included a finite state machine and a data path. These are the same conceptual components of all digital computing devices. We will revisit all of these abstractions in the chapters ahead.

We also briefly discussed the similarity between hardware and software and hiinted at the parallelism that we can achieve in logic circuits that we can't achieve on our general-purpose computing platforms. Finally, through two examples, we demonstrated how starting from simple primitives such as logic gates and memory registers we can construct such circuits as our days-in-a-month calculator and combination lock. Hierarchy will be an important concept in design that enables us to tackle larger and larger problems (we'll see several large examples in later chapters in the text). As shown in Figure 1.27, everything we'll build will be constructed from simple switching elements. Switches are used to construct our logic gates and registers. These, in turn, are used to construct combinational and sequential logic. Our last example used sequential and combinational logic in both its data path and controller. Combinational logic appeared as a multiplexer, comparator, and the next-state computation of the finite state machine. Sequential logic appeared in the memory elements that stored the numbers for our combination and in the state registers of the finite

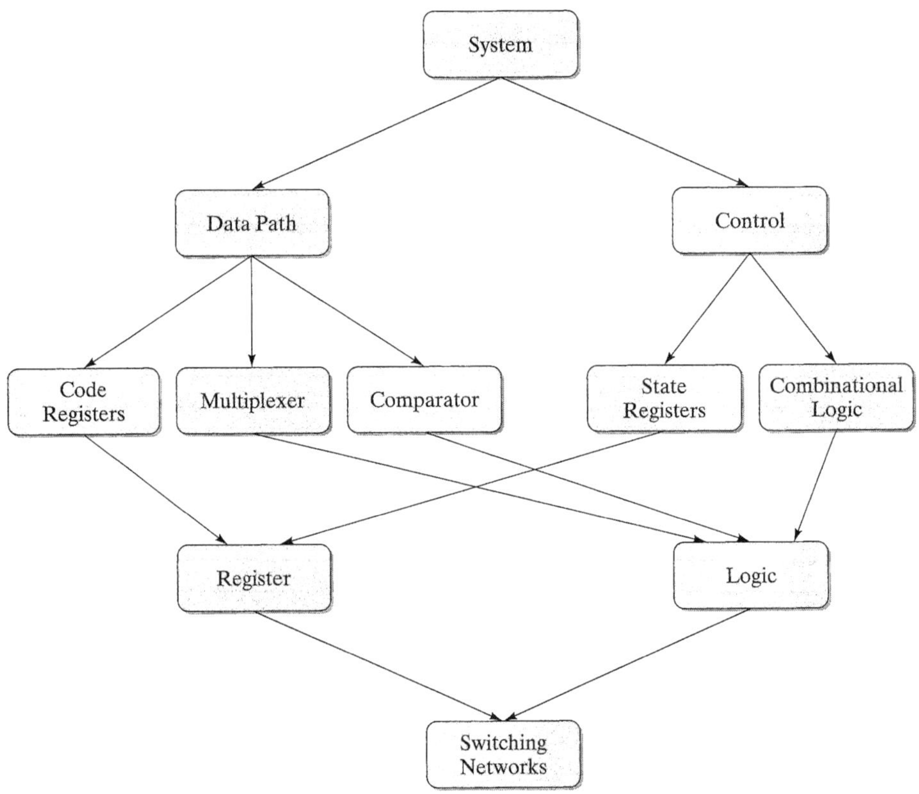

Figure 1.27 Hierarchy in digital design as illustrated in the combination lock example.

state machine. The next layer in the hierarchy divided the logic according to function, whether it was a component of the data path or the controller. We could, of course, continue further and user our combination lock as a component of a larger system.

Our goal it to take you on an interesting and intellectually stimulating journey through the landscape of digital design. Before embarking on any journey, however, we usually collect guidebooks and maps that help us plan our route and read about the places we will visit. This chapter should serve as that guidebook to the rest of the text. We have used it to preview all the topics you will encounter in the following chapters. Necessarily, it only gave you a broad and superficial view. The journey itself, the visits to the later chapters and the direct experience of their material through exercises and laboratory assignments, will provide the depth and details. Combinational logic will be covered in Chapters 2 through 5. Sequential logic will be the subject of Chapters 6 through 10.

FURTHER READING

A very good description of the design process can be found in Chapter 3 of S. Dasgupta's book *Computer Architecture: A Modern Synthesis,* John Wiley, New York, 1989. Other descriptions can be found in most texts on digital design. Especially interesting are J. F. Wakerly's *Digital Design: Principles and Practices, Third Edition,* Prentice Hall, New Jersey, 2000 and D. Gajski's *Principles of Digital Design,* Prentice Hall, New Jersey, 1997. For those not familiar with the basic background concepts of electronics, a gentle introduction can be found in T. M. Frederiksen's work, *Intuitive Digital Computer Basics,* published by McGraw-Hill, New York, in 1988 (the entire "Intuitive" series is quite good). The classic text on digital design for very large-scale integrated circuits is by Carver Mead and Lynn Conway, *Introduction to VLSI Systems,* Addison-Wesley, Reading, MA, 1980. A more recent text by Neil Weste and David Harris, *CMOS VLSI Design: A Circuits and Systems Perspective, Third edition,* Pearson/Addison-Wesley, Reading, MA, 2004.

For the history buffs, the original monograph by George Boole describing what would become Boolean algebra was published as *An Investigation of the Laws of Thought, on Which Are Founded the Mathematical Theories of Logic and Probabilities,* in 1854. It can be found reprinted by Dover Publications, 1973. The work in Clause Shannon's MIT Masters thesis was published as *A Symbolic Analysis of Relay and Switching Circuits,* in the Transactions of the American Institute of Electrical Engineers, 1938. William Schockley, John Bardeen, and Walter Brattain's "transfer resistor," their original name for what would become the transistor, is best described in their Nobel Lectures found in Elsevier's *Nobel Lectures—Physics, 1942–62.* They shared the Nobel Prize in 1956. John von Neumann was first to publish the ideas embodied in the ENIAC in a report on its successor machine the EDVAC in 1945. It was titled *First Draft of a Report on the EDVAC* and can now be found in the IEEE Annals of the History of Computing, 1993.

EXERCISES

1.1 *(Computation)* Construct a switching circuit using simple switches (like that of Figure 1.3) controlled by three control signals. Show the arrangement of the switches if we want the light bulb to light when:

(a) All the control signals are true
(b) Any two of the control signals are true
(c) Any one of the control signals is true

1.2 *(Computation)* Show the arrangement of transistors needed to construct:

(a) A 3-input NAND gate
(b) A 3-input NOR gate
(c) A gate whose output is the inverted value of *A* OR (*B* AND *C*)
(d) A gate whose output is the inverted value of (*A* OR *B*) AND *C*

1.3 *(Encoding)* To digitally manipulate information we must encode it using binary numbers. Consider an encoding for playing cards. There are 52 cards in a deck and they are divided into four suits with values from 1 to 13 (an ace is 1, jack is 11, queen is 12, king is 13). Show two possible encodings for the cards.

1.4 *(Encoding)* Given your encodings for Exercise 1.3, show the logic expression that describes (give names to each of signals you use to represent the card):

(a) A jack of diamonds
(b) A seven of any suit
(c) Any card of the heart suit

1.5 *(Calendar Example)* Derive the equations for d29, d30, and d31.

1.6 *(Calendar Example)* Show how you can derive an equation for d31 in terms of d28, d29, and d30 rather than the inputs m8, m4, m2, and m1.

1.7 *(Calendar Example)* Derive an expression for the months that contain the letter R using the same encoding as for the calendar example.

1.8 *(Calendar Example)* Consider a different encoding for the months of the year that numbers the months from 0 (January) to 11 (December) rather than 1 to 12? Show the effect of this new encoding by rederiving the equations of Exercise 1.5.

1.9 *(Calendar Example)* Consider a different encoding for the number of days in a month if we have a display that uses d28 to turn on the first 28 numbers, d29 to turn on just the 29th, d30 to turn on just the 30th, and d31 to turn on just the 31st. In this case, the output for April would be d28 = 1, d29 = 1, d30 = 1, d31 = 0 and the output for February of a non-leap year would be d28 = 1 with all the others equal to 0. Show the effect of this new encoding by rederiving the equations for the four outputs.

1.10 *(Combination Lock Example)* Change the C program for the combination lock to include two combinations for opening the lock instead of only one. Derive a new state diagram corresponding to this new functionality.

1.11 *(Combination Lock Example)* Instead of using a comparator to check for the combination, we could simply embed the combination into the state diagram. This can be accomplished by having an arc to the next state if the bits of the value entered match a predetermined number and an arc to an error state if they don't. Show how this would be reflected in the state diagram notation.

1.12 *(Combination Lock Example)* Derive expressions for the next-state functions of the combination lock from the encoded state table in Figure 1.25. There are four columns for the next-state and you can label your functions NS_1 through NS_4.

1.13 *(Combination Lock Example)* Try out your own state encoding for the state table of Figure 1.24. Explain the rationale for your choices. Derive equations for the next state functions (NS_1 through NS_4) given your encoding.

1.14 *(Combination Lock Example)* Encode the multiplexer control signals using only two bits. Derive expressions for your two multiplexer control functions (MUX_1 and MUX_2).

1.15 *(Logical Statements)* Write logic statements for the light control variants of Exercise 1.1, using AND, OR, and NOT operators, as described in Section 1.3.

1.16 *(Analog versus Digital)* Consider the inverter transfer characteristic described in Figure 1.16. Suppose two inverter circuits are placed in series so that the output of the first inverter is the input to the second inverter. Assume initially that the input to the first stage is a logic 1 represented by 3 volts. Of course, the output of the second stage will be identical, at least initially. Describe what happens to the outputs of the first and second stages as the first stage input slowly changes from 3 volts to 0 volts. Do this by drawing a graph whose X axis is time and whose Y axis is voltage, showing two curves, one each for (a) the first inverter's output and (b) the second inverter's output.

1.17 *(Truth Tables)* Consider a function that takes as input two 2-bit numbers and produces as output a 3-bit sum. Write the truth table for this function. It should have four input columns, 16 rows, and three output columns.

1.18 *(Truth Tables)* Write truth tables for the three functions of Exercise 1.1.

1.19 *(Boolean Algebra)* Write logic expressions for each of the three functions in the truth tables of Exercise 1.18.

1.20 *(Gates)* Given the Boolean expressions of Exercise 1.19, draw logic schematics using AND, OR, and NOT gates that implement those functions.

1.21 *(Gates)* Show how you could construct the equivalent of a 9-input AND gate from a set of 3-input AND gates. Do this again for a 10-input AND gate.

1.22 *(Hierarchy)* Describe the hierarchical composition of a building, house, or train. Draw a diagram similar to Figure 1.27 for your example.

1.23 *(Combinational versus Sequential Circuits)* Which of the following contain circuits that are likely to be combinational and which contain sequential circuits? Explain your rationale.

- **(a)** A washing machine that sequences through the soak, wash, and spin cycles for preset periods of time.
- **(b)** A circuit that divides two 2-bit numbers to yield a quotient and a remainder.
- **(c)** A machine that takes a dollar bill and gives three quarters, two dimes, and a nickel in change, one at a time through a single coin change slot.
- **(d)** A digital alarm clock that generates an alarm when a preset time has been reached.
- **(e)** A circuit that takes as input two decimal numbers in the range from 0 to 9, outputs a 0 if they are different, and a 1 if they are identical.
- **(f)** A circuit that turns on or off a hall light based on the configuration of two input switches. If both switches are in the same position, the light is off. If they are in different positions the light is on.
- **(g)** A circuit that takes a sequence of bits, one bit at a time, and outputs a 0 or 1 after each bit that indicates if the number of 1s in the sequence seen so far is even or odd, respectively.
- **(h)** A circuit with two binary inputs and four binary outputs that works as follows. The binary input indicates which of the four outputs should be driven to a 1 with the other outputs set to 0.

1.24 *(Design Problem)* Consider a digital system that works as follows. It takes as input a number in the range of 0_{10} to 15_{10} in binary and outputs a function, F_4, that is 1 if the number is a multiple of 4.

- **(a)** Develop the truth table for this function, with four binary inputs (00002 to 11112) and the output as indicated above.
- **(b)** Write down the Boolean equations for the function F_4.
- **(c)** Characterize the complexity of this implementation by counting the number of AND, OR, and NOT gates of various input sizes needed to realize the output (e.g., so many 2-input ANDs, 3-input ANDs, etc.).

1.25 *(Design Problem)* Extend the system of Exercise 1.24 with two or more outputs: F_2 (that is 1 when the input is a multiple of 2) and F_8 (that is 1 when the input is a multiple of 8).

(a) Repeat parts (a) through (c) of Exercise 1.24 for F_2 and F_8.
(b) Can you implement the function F_4 in terms of F_2? How?
(c) Can you implement the function F_8 in terms of F_4? How?

1.26 *(Design Problem)* Consider the Calendar subsystem presented in this chapter. We will change the output specifications slightly while the inputs will remain the same. Directly generate the 5-bit binary number for the number of days in the month: $28 = 11100_2$, $29 = 11101_2$, $30 = 11110_2$, and $31 = 11111_2$.

(a) Develop the truth table for the revised function, with four inputs to represent the month, one input to indicate a leap year, and the five outputs as indicated above.
(b) Write down the Boolean equations for each of the five outputs.
(c) Characterize the complexity of this implementation by counting the number of AND, OR, and NOT gates of various input sizes needed to realize each of the five outputs (e.g., so many 2-input ANDs, 3-input ANDs, etc.).

1.27 *(Design Problem)* Now consider a different way to achieve the same result. Keep the Calendar system exactly as discussed in class. But, add a new component that takes as inputs the four outputs—d28, d29, d30, and d31—and maps these into the five outputs as described in Exercise 1.26.

(a) Develop the truth table for the new portion of the function.
(b) Write down the Boolean equations for each of the five outputs.
(c) Once again, characterize the complexity of the implementation by tabulating the number of gates of various inputs needed to realize the five outputs.
(d) Given the complexity of the original calendar subsystem, and this new subsystem, how does this solution compare with the one you developed for Exercise 1.26? Which is better?

1.28 *(Design Problem)* Can you think how to dramatically reduce the complexity of implementing the function of Exercise 1.27? (*Hint:* Rather than starting with the truth table, think through a simple implementation of direct Boolean functions to implement the mappings between the four inputs—d28, d29, d30, and d31—and the five binary outputs).

(a) Write down your equations for a simplified implementation of Exercise 1.27.
(b) Characterize its complexity in the same way as in part (c) of Exercise 1.26 and Exercise 1.27.

1.29 *(Design Problem)* Now consider the door combination lock example discussed in Section 1.4.2. This implementation enters an error state as soon as an incorrect bit is entered. This could make it an exceedingly easy lock to pick if this were observable at the lock. Change the design so all three combination bits must be entered before an error state is entered.

(a) Draw a revised state diagram for your revised design.
(b) How many states does your design now have? Write down your state table.
(c) Choose a state encoding and describe the rationale behind your choice. Write down your encoded state table.
(d) Consider the output function for Open. What was the Boolean equation for the original implementation as described in class? What is the Boolean equation for your implementation? How does its complexity compare with the original?

1.30 *(Design Problem)* We can make one more improvement to the combinational lock of Section 1.4.2. The lock needs to be reset in order to get back to its starting state. Let's eliminate this requirement and make a state diagram that always returns to the starting state after three keys are pressed. Ensure that the lock is in the state where the output is to open the lock for at least one clock cycle so that the lock can be released.

1.31 *(Design Problem)* The Standard Master Combination Lock (the kind you find on gym lockers) has a dial with the numbers 0 through 39 on it. It works as follows. You reset it by spinning it clockwise past 0 a few times. Then you turn it counterclockwise to the first number in the combination, positioning the dial so that the number is directly under an indicator arrow. Then you spin the dial clockwise again, past 0 once, to the second number of the combination. Finally you spin the dial counterclockwise a second time directly to the last number of the combination. At this point the lock should open.

(a) Define the system's inputs and outputs. Consider signals that indicate clockwise or counterclockwise motion, the number where the dial is currently positioned, etc.
(b) Draw a finite state diagram for this subsystem, showing states, transition arcs, and logical conditions under which the machine moves from one state to the next.

1.32 *(Design Problem)* For the system described in Exercise 1.31, suggest a method for making it possible for the user to program a new combination.

(a) What additional inputs do you need to make the lock combination "programmable?"
(b) Draw a revised state diagram that shows the state sequencing for programming the lock with a new combination.

CHAPTER

2

Combinational Logic

Introduction

This chapter begins our detailed examination of the implementation of digital systems. We start with *combinational logic design,* the design and implementation of logic functions whose outputs depend solely on their inputs. The calendar sub-system described in Section 1.4.1 is just such a circuit.

After starting with the representation of a function as a truth table or a Boolean equation, we will introduce a *canonical,* or standard, representation, called the *sum-of-products two-level form.* We can think of this as a unique way to represent a Boolean function, like a fingerprint. The form expresses the function as ANDed terms (first level of gates) that are then ORed together (second level of gates). An alternative canonical form, the *product-of-sums form,* has ORs at the first level and ANDs at the second level.

You can implement a Boolean function with logic gates in more than one way. It is highly desirable to find the simplest implementation—that is, the one with the smallest number of gates or wires. The process of reducing a Boolean function to its simplest two-level form is called *Boolean minimization.* We will introduce Karnaugh maps as another way of representing Boolean functions that make it easy to see the relationships between terms and can assist in coming up with a minimal implementation. The formal process for Boolean minimization will be described in more detail in Chapter 3.

Just as a complex algebraic expression can be simplified by factoring out common subexpressions, you can implement a Boolean function in fewer gates if you factor it judiciously. This leads to a fundamental trade-off between time (more levels of logic—analogous to fewer levels of parentheses in algebra) and space (fewer gates needed to implement the function—analogous to the number of arithmetic steps). We will introduce the basic ideas of multilevel logic in this chapter.

Logic circuits are more than simply abstract implementations of mathematical equations. They are constructed from physical devices that take some time to compute their functions. An actual implementation of a logic circuit does not determine its output instantly; it takes time for the signals to *propagate* starting from a change in the inputs of a logic gate to a final change on its outputs. In this chapter, we will introduce the concept of *time response in digital networks.*

Hardware description languages (HDLs) are an alternative way to describe logic circuits in a textual rather than graphical form. We will introduce the basic concepts here and will use a simplified form of a particular HDL called Verilog throughout this book.

2.1 Outputs as a Function of Inputs

2.1.1 Combinational Logic Defined

Combinational logic is the kind of digital system whose output behavior depends only on the current inputs. Such a system is *memoryless:* its outputs are independent of the historical sequence of values presented to it as inputs.

For example, a digital system that adds two input bits together to form sum and carry output bits is combinational. Changing the inputs causes the outputs to change, after a small delay they will reach a final value. But input values from the past have no effect on the final value. Only the current input values matter.

Sequential logic, on the other hand, adds the notion of *memory* or *state* to produce systems whose output behavior does depend on the sequence of inputs and not just the last inputs (and hence the name sequential logic). We will examine this kind of digital systems in detail starting with Chapter 6.

A traffic light controller is an example of sequential logic. A traffic light cycles through the sequence green–yellow–red. So when the light changes in response to input changes (perhaps a pedestrian has pushed the crossing button, a car is detected as waiting to cross the intersection, or a timer has gone off), the next light to illuminate depends also on the currently illuminated light.

For now, our interest is combinational logic. There are many ways to describe combinational logic: Boolean algebra expressions, wired up logic gates, truth tables tabulating input and output combinations, graphical maps, and even program statements in a hardware description language. Each of these will be introduced in the following sections.

2.1.2 Examples of Combinational Logic

Let's look at several digital systems that can be implemented as combinational logic. We will begin with a system that detects equivalence among its inputs. Given two binary inputs, X and Y, the Equal output is set to 1 if both inputs have the same value. Either X and Y are both 0 or both 1. Such a system depends only on its current inputs, and not the sequence of previous inputs. Thus, the system is combinational. It is easily described in terms of a truth table, as shown in Figure 2.1.

X	Y	Equal
0	0	1
0	1	0
1	0	0
1	1	1

Figure 2.1 **Truth table for the equivalence circuit.**

Now, consider a "tally" circuit of two binary inputs, X and Y, and three binary outputs, *Zero, One,* and *Two*. This digital system counts the number of ones among its inputs and has the appropriate output set to 1 to indicate the result. The tally circuit makes the output signal *Zero* true if both inputs are 0. The output *One* is true if either input is 1 but not both. The output *Two* is true only if both inputs are 1. Again, the outputs depend only on the current inputs, and so the system is combinational. Its truth table is shown in Figure 2.2.

X	Y	Zero	One	Two
0	0	1	0	0
0	1	0	1	0
1	0	0	1	0
1	1	0	0	1

Figure 2.2 Truth table for the tally circuit.

Notice that X and Y are a binary encoding while *Zero, One,* and *Two* form a one hot encoding.

Suppose that you wanted to design a system that, given a binary number in the range of 000 to 111 (0 to 7), could determine if the number was divisible by 2, 3, or 5 evenly (without a remainder). Such a circuit turns out to be a useful component of a variety of clock and timer systems. A generalization of this circuit could be used, for example, to cause a digital watch to chime differently at 15 minutes before and after the hour (the time is divisible by 3 and 5 but not 2) then at the half hour and hour (divisible by 2, 3, and 5, simultaneously). We can write down the desired behavior as another truth table, as shown in Figure 2.3. Since the outputs only depend on the current inputs, this is yet another combinational system.

X	Y	Z	By2	By3	By5
0	0	0	1	1	1
0	0	1	0	0	0
0	1	0	1	0	0
0	1	1	0	1	0
1	0	0	1	0	0
1	0	1	0	0	1
1	1	0	1	1	0
1	1	1	0	0	0

Figure 2.3 Truth table for divide-by-2, -3, -5.

Imagine we have been asked to design a circuit that accepts a binary digit, 0 or 1, and decodes it into a set of signals to drive a 7-segment display. The display element and its control signals are shown in Figure 2.4. A segment within the display is illuminated if its associated control signal is asserted (we use asserted to indicate that an output is set to 1 and deasserted to indicate a 0). To display a 0, all segments except the center segment are turned on: the control signals C_0 through C_5 are asserted while C_6 is deasserted. A 1 is represented by only illuminating the segments controlled by C_1 and C_2. Since the illuminated segments only depend on the current inputs, the system must be combinational.

EXAMPLE 2.1 BINARY ADDER

Another example of a combinational network is a two-data-input binary adder. This circuit adds together two binary digits. Its result is a one bit sum and a carry. For example, 0 + 0 = 0, 0 + 1 = 1, 1 + 0 = 1 (all with a carry of 0), and 1 + 1 = 0 (with a carry of 1). This system with two inputs (A and B) and two outputs (*Carry* and *Sum*) is referred to as

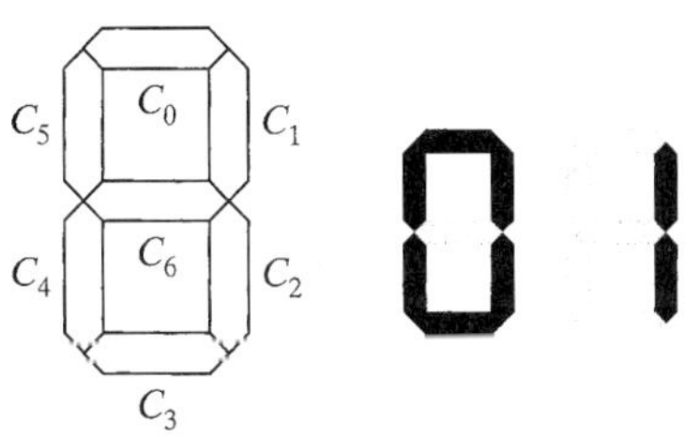

X	C_0	C_1	C_2	C_3	C_4	C_5	C_6
0	1	1	1	1	1	1	0
1	0	1	1	0	0	0	0

Figure 2.4 Binary digit-display combinational system.

A	B	Carry	Sum
0	0	0	0
0	1	0	1
1	0	0	1
1	1	1	0

Figure 2.5 Half-adder truth table.

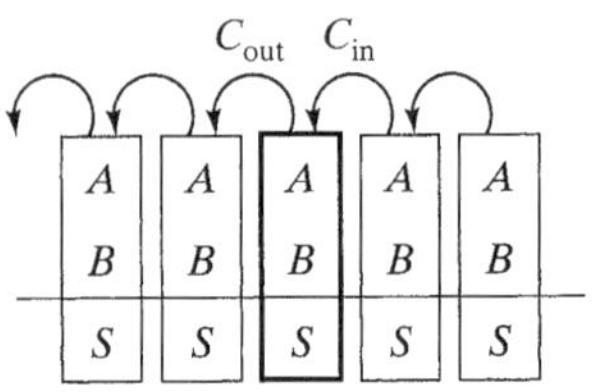

Figure 2.6 Binary addition of multi-bit numbers.

A	B	C_{in}	C_{out}	S
0	0	0	0	0
0	0	1	0	1
0	1	0	0	1
0	1	1	1	0
1	0	0	0	1
1	0	1	1	0
1	1	0	1	0
1	1	1	1	1

Figure 2.7 Full-adder truth table.

a *half adder* because it can't be chained with other adders to add larger binary numbers (it has no way of considering a carry from another column for the addition of numbers that are more than one binary digit). Its truth table is shown in Figure 2.5. Since the network has two inputs, A and B, and two outputs, *Carry* and *Sum*, the truth table has four columns, one for each input and output, and four rows, for each of the four possible combinations (2^2) of input values.

The *full adder* adds a third input that represents the carry-in of a previous addition column. Like the half adder, it also generates a carry-out to the next addition column. Figure 2.6 highlights a column of a binary addition of two multi-digit numbers. The bits to be added in every column are A (a bit from the first number), B (a bit from the second number), and the carry-in (from the previous column), C_{in}, the sum is S, and the carry-out is C_{out} (which becomes the carry-in of the next column). For example, if $A = 1$, $B = 1$, and $C_{in} = 1$, then $S = 1$ and $C_{out} = 1$ (that is, one plus one plus one is three which in binary is 11_2). Figure 2.7 shows the truth table for the full adder. The three inputs have 2^3 unique binary combinations, leading to a truth table with eight rows.

Truth tables are fine for describing functions with a modest number of inputs. But for large numbers of inputs, the truth table grows too large, as the exponential of the number of inputs (2^n rows where n is the number of inputs). An alternative approach, often more compact, is an expression in Boolean algebra rather than a truth table. The operators of the algebra are the logic functions such as AND and OR. We will look at this algebra in the next section.

But before we do, we don't want to leave you with the impression that just about every digital system is combinational. Here are some examples of systems that are *not* combinational. A digital system takes as input a stream of binary digits on a single input signal and asserts is single output whenever it has seen the sequence 1, 0, 0 on its input wire. When such a system sees a 0, it has to remember if it saw another 0 just before it and a 1 just before that, before it can assert its output (set it to 1). Since its behavior depends on its memory of the earlier input sequence, it is a sequential system. Another complication is that the system needs to be able to tell how long a single input lasts so that it can distinguish two 0s in a row from a single 0. A clock is usually used to accomplish this by keeping a steady beat that separates time intervals.

Counters are another example of digital systems that are inherently sequential. A 3-bit counter that advances in a binary sequence, such as 000, 001, 010, 011, 100, 101, 110, 111, and repeats, needs to keep track of the current position in this count sequence before it advances to the next one in the proper order. This implies that the counter must use memory for this purpose. Memory lets the system remember where it is in a sequence, just as it can help remember what it saw on previous inputs.

A combinational lock, such as the one of Section 1.4.2, offers a third example of a sequential system. It must keep the combination to be used in memory and must keep track of what point a user is at in entering a new combination that will be compared against the preset one. Thus, the same input (the user pressing a number twice) in a different context (or memory state) may cause a different output. For example, pressing 5 as the first number of a combination preset to 515 will not cause the door to unlock, but on the third button press, it will. This functionality requires a sequential system.

2.2 Laws and Theorems of Boolean Logic

In Chapter 1, you saw that (at least some) Boolean expressions can be represented by logic gates and vice versa. Actually, all Boolean functions can be implemented in terms of collections of AND, OR, and NOT gates. Because of this close relationship between the laws of Boolean algebra and the behavior of logic gates, the theorems of Boolean algebra can also be used to transform digital logic—usually for the purposes of deriving simpler and/or standardized forms.

Boolean algebra is the mathematical foundation of digital systems. We will see that an algebraic expression provides a convenient shorthand notation for the truth table of a function.

Basic Concept The operations of a Boolean algebra must adhere to certain properties, called *laws* or *axioms*. One of these axioms is that the Boolean operations are commutative: you can reverse the order in which the variables are written without changing the meaning of the Boolean expression. For example, OR is commutative: X OR Y is identical to Y OR X, where X and Y are Boolean variables.

The axioms can be used to prove more general laws about Boolean expressions. You can use them to simplify expressions in the algebra. For example, it can be shown that X AND (Y OR NOT Y) is the same as X, since Y OR NOT Y is always true. The procedures you will learn for optimizing combinational and sequential networks are based on the principles of Boolean algebra, and thus Boolean expressions are often used as input to computer-aided design tools.

Boolean Operations Most designers find it a little cumbersome to keep writing Boolean expressions with AND, OR, and NOT operations, so they have developed a shorthand for the operators. If we use X and Y as the Boolean variables, then we write the complement (inversion, negation) of X as one of X', $\overline{X}$ $!X$, $/X$, or $\backslash X$ (we'll use most of these in this book as do most designers). The OR operation is written as $X + Y$, $X \vee Y$, or $X \mid Y$ (we'll favor the +). The AND operation is written as $X \bullet Y$, $X \wedge Y$, $X \mathbin{\&} Y$, or more simply $X\,Y$ (we'll tend to favor the last two ways of describing AND). Although there are certain analogies between OR and PLUS and between AND and MULTIPLY, the logic operations are not the same as the arithmetic operations.

Complement is always applied first, followed by AND, followed by OR. We say that complement has the highest priority or precedence, followed by AND and then OR (this is similar to negation, multiplication, and addition in algebraic expressions). Parentheses can be used to change the default order of evaluation. The default grouping of operations is illustrated by the following examples:

$$\overline{A} \bullet B + C = ((\overline{A}) \bullet B) + C$$

$$\overline{A} + B \bullet C = (\overline{A}) + (B \bullet C)$$

Equivalence of Boolean Expressions and Truth Tables A Boolean expression can be readily derived from a truth table and vice versa. In fact, Boolean expressions and truth tables convey exactly the same information.

Let's consider the structure of a truth table, with one column for each input variable and a column for the expression's output. Each row in which the output column is a 1 contributes a single ANDed term of the input variables to the Boolean expression. This is called a *product term,* because of the analogy between AND and MULTIPLY. Looking at the row, we see that if the column associated with variable X has a 0 in it, the expression $\overline{X}$ is part of the ANDed term. Otherwise the expression X is part of the term. Each variable in either its non-complemented (X) or complemented ($\overline{X}$) form is called a *literal.*

There is one product term for each row with a 1 in the output column. All these product terms are ORed together to complete the expression. A Boolean expression written in this form is called a *sum of products.*

EXAMPLE 2.2 DERIVING EXPRESSIONS FROM TRUTH TABLES

Let's go back to Figure 2.5 and Figure 2.7, the truth tables for the half adder and the full adder, respectively. Each output column leads to a new Boolean expression, but each of these output expression is defined over the same variables (the input columns). The Boolean expressions for the half adder's *Sum* and *Carry* outputs can be written as:

$$Sum = (A\overline{B}) + (\overline{A}B)$$
$$Carry = AB$$

The half adder *Sum* is 1 in two rows: $A = 1, B = 0$ and $A = 0, B = 1$. The half adder *Carry* is 1 in only one row: $A = 1, B = 1$.

The truth table for the full adder is considerably more complex. Both S and C_{out} have four rows with 1s in the output columns. The two functions are written as:

$$S = (A\overline{B}\overline{C}_{in}) + (\overline{A}B\overline{C}_{in}) + (\overline{A}\overline{B}C_{in}) + (ABC_{in})$$
$$C_{out} = (\overline{A}BC_{in}) + (A\overline{B}C_{in}) + (AB\overline{C}_{in}) + (ABC_{in})$$

As we shall see, we can exploit Boolean algebra to simplify Boolean expressions. By applying some of the simplification theorems of Boolean algebra, we can reduce the expression for the full adder's C_{out} output to the following:

$$C_{out} = (AC_{in}) + (BC_{in}) + (AB)$$

Such simplified forms reduce the amount of gates, transistors, wires, and so on, needed to implement the expression. Simplification is an extremely valuable tool.

You can use a truth table to verify that the simplified expression just obtained is equivalent to the original. Start with a truth table with

A	B	C_{in}	AC_{in}	BC_{in}	AB	C_{out}
0	0	0	0	0	0	0
0	0	1	0	0	0	0
0	1	0	0	0	0	0
0	1	1	0	1	0	1
1	0	0	0	0	0	0
1	0	1	1	0	0	1
1	1	0	0	0	1	1
1	1	1	1	1	1	1

Figure 2.8 Truth table of reduced carry-out expression.

filled-in input columns but empty output columns. Then, find all rows of the truth table for which the product terms are true, and enter a 1 in the associated output column. For example, the term AC_{in} is true wherever $A = 1$ and $C_{in} = 1$, independent of the value of B. We say that AC_{in} *covers* two truth-table rows: $A = 1,\ B = 0,\ C_{in} = 1$ and $A = 1,\ B = 1,\ C_{in} = 1$.

Figure 2.8 shows the filled-in truth table and indicates the rows covered by each of the terms and the original output function (the rows with a 1 in that term's or output's column). Since the resulting truth-table column for C_{out} is the same as that of the original truth table (see Figure 2.7), the two expressions for C_{out} are logically equivalent.

Now, lets turn to the details of Boolean algebra and we will then revisit this example and perform the simplification step by step.

2.2.1 Axioms of Boolean Algebra

A Boolean algebra consists of a set of elements B, together with two binary operations $\{+\}$ and $\{\bullet\}$ and a unary operation $\{'\}$, so that the following hold:

1. The set B contains at least two elements a, b such that a is not equal to b.
2. *Closure:* For every a, b in B,
   - **a.** $a + b$ is in B
   - **b.** $a \bullet b$ is in B
3. *Commutative laws:* For every a, b in B,
   - **a.** $a + b = b + a$
   - **b.** $a \bullet b = b \bullet a$
4. *Associative laws:* For every a, b, c in B,
   - **a.** $(a + b) + c = a + (b + c) = a + b + c$
   - **b.** $(a \bullet b) \bullet c = a \bullet (b \bullet c) = a \bullet b \bullet c$
5. *Identities:*
   - **a.** There exists an identity element with respect to $\{+\}$, designated by 0, such that $a + 0 = a$ for every a in B.
   - **b.** There exists an identity element with respect to $\{\bullet\}$, designated by 1, such that $a \bullet 1 = a$ for every a in B.

6. *Distributive laws:* For every a, b, c in B,
   a. $a + (b \cdot c) = (a + b) \cdot (a + c)$
   b. $a \cdot (b + c) = (a \cdot b) + (a \cdot c)$
7. *Complement:* For each a in B, there exists an element a' in B (the complement of a) such that
   a. $a + a' = 1$
   b. $a \cdot a' = 0$

NOTE: Remember that we'll use A' and $\bar{A}$ interchangeably to represent the complement of a variable.

It is easy to verify that the set $B = \{0, 1\}$ and the logical operations OR, AND, and NOT satisfy all the axioms of a Boolean algebra. Simply substitute 0 and 1 for a and b, OR for +, AND for •, and NOT for ′, and show that the expressions are true. For example, to verify the commutative law for +:

$$0 + 1 = 1 + 0 \qquad 0 \cdot 1 = 1 \cdot 0$$
$$1 = 1 \ \surd \qquad 0 = 0 \ \surd$$

A *Boolean function* uniquely maps some number of inputs over the set {0, 1} into an output set {0, 1} represented by an output variable. Arbitrary functions can be represented by a truth table. A *Boolean expression* is an algebraic statement containing Boolean variables and operators without an assignment to an output variable. A theorem in Boolean algebra states that any Boolean function can be expressed in terms of AND, OR, and NOT operations. In other words, there is a Boolean expression for every Boolean function. For example, as we just saw for the full adder outputs, one way to map a truth table into a Boolean expression in the *sum* (OR) of *products* (AND) form. We looked at each row for which an output was true and ORed together terms formed by ANDing the complemented or uncomplemented forms of the inputs for that row.

In fact, there are many other ways to represent Boolean functions. Some use new logical operations (to be introduced next) that are interesting because they are easier to implement with real transistor switches.

Boolean Operations Revisited Let's review the elementary Boolean operations and how these are represented as gates and truth tables. Figures 2.9, 2.10, and 2.11 summarize the representations for the NOT, AND, and OR operations, respectively.

Take the Boolean expression $Z = \bar{A} \cdot \bar{B} \cdot (C + D)$. A version of the expression with parentheses would be $(\bar{A} \cdot (\bar{B} \cdot (C + D)))$. Each pair of parentheses represents the expression generated by a single gate. Thus,

NOT $\qquad X' \qquad \bar{X} \qquad {\sim}X$

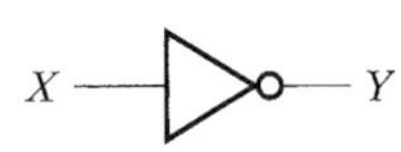

X	Y
0	1
1	0

Figure 2.9 Alternative representations of NOT (also known as COMPLEMENT).

AND $X \cdot Y$ XY $X \wedge Y$

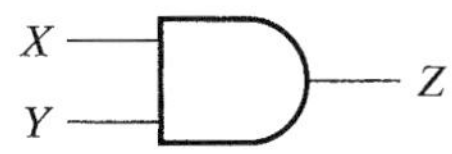

X	Y	Z
0	0	0
0	1	0
1	0	0
1	1	1

Figure 2.10 **Alternative representations of AND.**

OR $X + Y$ $X \vee Y$

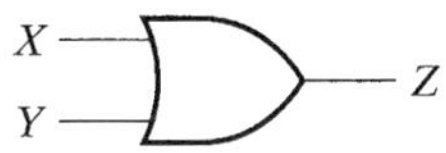

X	Y	Z
0	0	0
0	1	1
1	0	1
1	1	1

Figure 2.11 **Alternative representations of OR.**

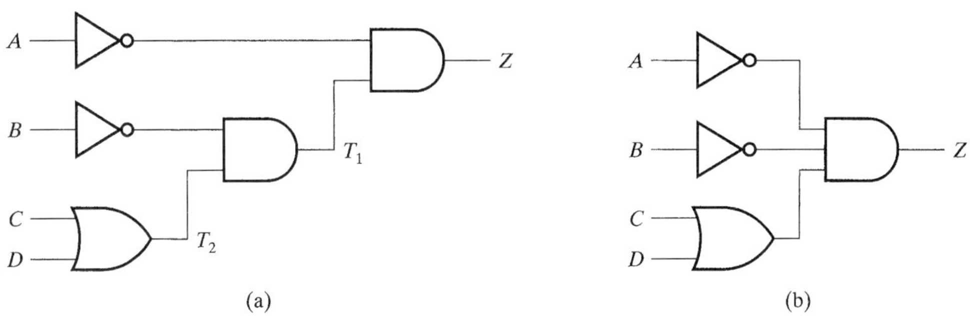

Figure 2.12 **Two equivalent gate-level implementations.**

the circuit is built up through a set of intermediate results, from the inside out:

$$T_2 = (C + D)$$
$$T_1 = (\bar{B} \cdot T_2)$$
$$Z = (\bar{A} \cdot T_1)$$

The gate-level implementation is shown in Figure 2.12(a), using two-input gates. The primitive gates need not be limited to two inputs, however. Figure 2.12(b) shows the same circuit implemented using a three-input AND gate. These implementations are equivalent because of the associative law of Boolean algebra. We could also use the associative law to reorder the expression. For example, switching A and B would create a new, but functionally equivalent, circuit that would have the value of A passing through the gate T_1 in Figure 2.12(a) rather than B.

Recall that each appearance of a variable or its complement in an expression is called a literal. In the preceding expression, we can see that there are four variables and four literals. The following expression has 10 literals but only three variables (A, B, and C): $Z = A\bar{B}C + AB + \bar{A}B\bar{C} + \bar{B}C$. Each literal represents the connection of a variable or its complement to a unique gate input. Later, we will use literals as a rough measure of the complexity of a Boolean functions.

2.2.2 Theorems of Boolean Algebra

Boolean algebra provides the foundation for all of the simplification techniques we shall discuss. Based on the Boolean laws of the previous section we can prove additional theorems that can be used as tools to simplify Boolean expressions. For example, if E_1 and E_2 are two expressions for the same Boolean function, we say that E_2 is simpler than E_1 if it contains fewer literals. This usually (but not always) means that the simpler expression will contain fewer Boolean operations.

Duality Before we provide a tabulation of useful laws and theorems, it is important to describe the concept of *duality*. Every Boolean expression has a *dual*. A dual of an expression is derived from the original expression by replacing AND operations by OR operations and vice versa, and replacing constant logic 0s by logic 1s and vice versa, while leaving the literals unchanged. It is a fundamental theorem of Boolean algebra, which we do not prove here, that any statement that is true about a Boolean expression is also true for its dual. Remember, an expression and its dual are not equivalent. It is just that, once we discover a useful theorem for simplifying a Boolean expression, we obtain a second, dual theorem as a bonus. For example, the *dual* of the Boolean theorem $X + 0 = X$, written $(X + 0 = X)^D$, is the theorem $X \cdot 1 = X$.

Useful Theorems The following is a list of frequently used laws and theorems of Boolean algebra. Some are generalized from Section 2.2.1. The second column shows the duals of the expression in the first column.

Operations with 0 and 1:

1. $X + 0 = X$ **1D.** $X \cdot 1 = X$

2. $X + 1 = 1$ **2D.** $X \cdot 0 = 0$

Idempotent theorem:

3. $X + X = X$ **3D.** $X \cdot X = X$

Involution theorem:

4. $(X')' = X$

Theorem of complementarity:

5. $X + X' = 1$ **5D.** $X \cdot X' = 0$

Commutative law:

6. $X + Y = Y + X$ **6D.** $X \cdot Y = Y \cdot X$

Associative law:

7. $(X + Y) + Z = X + (Y + Z) = X + Y + Z$ **7D.** $(X \cdot Y) \cdot Z = X \cdot (Y \cdot Z) = X \cdot Y \cdot Z$

Distributive law:

8. $X \cdot (Y+Z) = X \cdot Y + X \cdot Z$ **8D.** $X + (Y \cdot Z) = (X+Y) \cdot (X+Z)$

Simplification theorems:

9. $X \cdot Y + X \cdot Y' = X$ **9D.** $(X+Y) \cdot (X+Y') = X$

10. $X + X \cdot Y = X$ **10D.** $X \cdot (X+Y) = X$

11. $(X+Y') \cdot Y = X \cdot Y$ **11D.** $(X \cdot Y') + Y = X + Y$

DeMorgan's law:

12. $(X+Y+Z+\cdots)' = X' \cdot Y' \cdot Z' \cdot \cdots$ **12D.** $(X \cdot Y \cdot Z \cdot \cdots)' = X' + Y' + Z' + \cdots$

General form:

13. $\{f(X_1,X_2,\ldots,X_n,0,1,+,\cdot)\}' = \{f(X_1',X_2',\ldots,X_n',1,0,\cdot,+)\}$

Duality:

14. $(X+Y+Z+\cdots)^D = X \cdot Y \cdot Z \cdot \cdots$ **14D.** $(X \cdot Y \cdot Z \cdot \cdots)^D = X + Y + Z + \cdots$

General form:

15. $\{f(X_1,X_2,\ldots,X_n,0,1,+,\cdot)\}^D = f(X_1,X_2,\ldots,X_n,1,0,\cdot,+)$

Theorem for multiplying and factoring:

16. $(X+Y) \cdot (X'+Z) = X \cdot Z + X' \cdot Y$ **16D.** $X \cdot Y + X' \cdot Z = (X+Z) \cdot (X'+Y)$

Consensus theorem:

17. $X \cdot Y + Y \cdot Z + X' \cdot Z = X \cdot Y + X' \cdot Z$ **17D.** $(X+Y) \cdot (Y+Z) \cdot (X'+Z) = (X+Y) \cdot (X'+Z)$

The notation $f(X_1,X_2,\ldots,X_n,0,1,+,\cdot)$ used in Theorems 13 and 15 represents an expression in terms of the variables $X_1, X_2, \ldots, X_n$, the constants 0, 1, and the Boolean operations + and •. Theorem 13 states succinctly that, in forming the complement of an expression, the variables are replaced by their complements, 0 is replaced by 1 and 1 by 0, and + is replaced by • and • by +.

Since any of the listed theorems can be derived from the original laws shown in Section 2.1, there is no reason to memorize all of them, they can always be re-proven. They are listed here for convenience.

Verifying the Boolean Theorems We can prove the first Simplification theorem (9), sometimes called the *Uniting theorem,* as follows:

$X \cdot Y + X \cdot Y' = X?$	
$X(Y+Y') = X$	Distributive law (8)
$X(1) = X$	Complementarity theorem (5)
$X = X\surd$	Identity (1D)

As another example, let's look at the second Simplification theorem (10):

$$
\begin{aligned}
X + X \cdot Y &= X? \\
X \cdot 1 + X \cdot Y &= X && \text{Identity (1D)} \\
X(1 + Y) &= X && \text{Distributive law (8)} \\
X(1) &= X && \text{Identity (2)} \\
X &= X \surd && \text{Identity (1)}
\end{aligned}
$$

EXAMPLE 2.3 THE BINARY FULL-ADDER CARRY-OUT

We can use the laws and theorems just introduced to verify the simplified expression for the full adder's carry-out function. The original expression, derived from the truth table, is:

$$C_{out} = \bar{A}BC_{in} + A\bar{B}C_{in} + AB\bar{C}_{in} + ABC_{in}$$

The first step uses Theorem 3, the Idempotent theorem, to introduce a copy of the term ABC_{in}. Then we use the commutative law to rearrange the terms:

$$
\begin{aligned}
&= \bar{A}BC_{in} + A\bar{B}C_{in} + AB\bar{C}_{in} + ABC_{in} + ABC_{in} \\
&= \bar{A}BC_{in} + ABC_{in} + A\bar{B}C_{in} + AB\bar{C}_{in} + ABC_{in}
\end{aligned}
$$

We next use the Distributive law to factor out the common literals from the first two terms:

$$= (\bar{A} + A)BC_{in} + A\bar{B}C_{in} + AB\bar{C}_{in} + ABC_{in}$$

We apply the Complementarity law:

$$= (1)BC_{in} + A\bar{B}C_{in} + AB\bar{C}_{in} + ABC_{in}$$

and the Identity law:

$$= BC_{in} + A\bar{B}C_{in} + AB\bar{C}_{in} + ABC_{in}$$

We can repeat the process for the second and third terms. The steps are: (1) Idempotent theorem to introduce a redundant term, (2) Commutative law to rearrange terms, (3) Distributive law to factor out common literals, (4) Complementarity theorem to replace $(\bar{X} + X)$ with 1, and (5) Identity law to replace $1 \cdot X$ by X:

$$
\begin{aligned}
&= BC_{in} + A\bar{B}C_{in} + AB\bar{C}_{in} + ABC_{in} + ABC_{in} \\
&= BC_{in} + A\bar{B}C_{in} + ABC_{in} + AB\bar{C}_{in} + ABC_{in} \\
&= BC_{in} + A(\bar{B} + B)C_{in} + AB\bar{C}_{in} + ABC_{in} \\
&= BC_{in} + A(1)C_{in} + AB\bar{C}_{in} + ABC_{in} \\
&= BC_{in} + AC_{in} + AB\bar{C}_{in} + ABC_{in}
\end{aligned}
$$

The final simplification, using the Distributive theorem, Complementarity theorem, and Identity law, proceeds similarly:

$$= BC_{in} + AC_{in} + AB(\overline{C}_{in} + C_{in})$$
$$= BC_{in} + AC_{in} + AB(1)$$
$$= BC_{in} + AC_{in} + AB$$

This is exactly the reduced form of the expression we derived above. Although it leads to a simpler expression, applying the rules of Boolean algebra in this *ad hoc* fashion does not guarantee you will always get the simplest expression as a result. We will introduce a more systematic approach in Chapter 3.

2.2.3 Duality and DeMorgan's Law

Duality Duality is a very useful property of Boolean algebra. Recall that the dual of a Boolean expression is derived by replacing • by +, + by •, 0 by 1 and 1 by 0, while leaving the Boolean variables unchanged (laws 14 and 15 in the last section). Any theorem of Boolean algebra that is shown to be true implies that its dual is also true. In essence, duality is a *meta-theorem,* a theorem about theorems. While it is not a way to simplify expressions directly, it does allow you to derive new theorems from those you already know to help in the simplification process.

For example, the dual of the Uniting theorem (9), $X \bullet Y + X \bullet Y' = X$, is $(X + Y) \bullet (X + Y') = X$. The proof of the dual follows step-by-step from the original, simply using the duals of the laws used in the original proof. Compare the following to our original proof of its dual:

$(X + Y) \bullet (X + Y') = X?$	
$X + (Y \bullet Y') = X$	Distributive law (8D)
$X + 0 = X$	Complementarity theorem (5D)
$X = X \surd$	Identity (1)

DeMorgan's Law DeMorgan's law (12 and 13) gives a procedure for complementing a complex function. The complemented expression is formed from the original by replacing all literals by their complements; all 1s become 0s and vice versa, and ANDs become ORs and vice versa. This theorem indicates an interesting relationship between AND, OR, and their complements NOT-OR (NOR) and NOT-AND (NAND):

$$\overline{X + Y} = \overline{X} \bullet \overline{Y} \qquad \overline{X \bullet Y} = \overline{X} + \overline{Y}$$

Note that $\overline{X + Y} \neq \overline{X} + \overline{Y}$ and $\overline{X \bullet Y} \neq \overline{X} \bullet \overline{Y}$ because on the left-hand side of each expression we are complementing the entire expression while on the right-hand side we are complementing individual variables. In other words, NOR is the same as AND with complemented

X	Y	X'	Y'	$(X+Y)'$	$X' \cdot Y'$
0	0	1	1	1	1
0	1	1	0	0	0
1	0	0	1	0	0
1	1	0	0	0	0

X	Y	X'	Y'	$(X \cdot Y)'$	$X' + Y'$
0	0	1	1	1	1
0	1	1	0	1	1
1	0	0	1	1	1
1	1	0	0	0	0

Figure 2.13 **DeMorgan's law.**

inputs while NAND is equivalent to OR with complemented inputs! This is easily seen to be true from the truth tables of Figure 2.13.

Let's use DeMorgan's law to find the complement of the following expression:

$$Z = \bar{A}\bar{B}C + \bar{A}BC + A\bar{B}C + AB\bar{C}$$

Step-by-step, the complement is formed as follows:

$$\bar{Z} = \overline{(\bar{A}\bar{B}C + \bar{A}BC + A\bar{B}C + AB\bar{C})}$$

$$\bar{Z} = \overline{\bar{A}\bar{B}C} \cdot \overline{\bar{A}BC} \cdot \overline{A\bar{B}C} \cdot \overline{AB\bar{C}}$$

$$\bar{Z} = (A + B + \bar{C})(A + \bar{B} + \bar{C})(\bar{A} + B + \bar{C})(\bar{A} + \bar{B} + C)$$

Note that duality and DeMorgan's law are *not* the same thing. The procedure for producing the dual is similar, but the literals are not complemented during the process. Thus, the dual of NOR is NAND (and vice versa); the dual of OR is AND (and vice versa). Remember, any theorem that is true for an expression is also true for its dual. When the duality theorem is applied to a function the result is a function that is totally different from the original, when DeMorgan's law is applied to a function, the result is a function that is the complement of the original. For example if we have the expression:

$$Z = \bar{A}B + C\bar{D}$$

then the dual of Z is:

$$Z^D = (\bar{A} + B)(C + \bar{D})$$

while applying DeMorgan's law yields:

$$\bar{Z} = (A + \bar{B})(\bar{C} + D)$$

2.3 Realizing Boolean Formulas

2.3.1 Logic Gates

There are other functions of two Boolean variables besides AND, OR, and NOT. In fact, there are 16 possible functions, the number of different ways you can write down the different choices of 0 and 1 for the four possible truth table rows. A truth table representation of the 16 functions is shown in Figure 2.14. The constant functions 0 and 1 and the functions $X, \bar{X}, Y, \bar{Y}, X \cdot Y$, and $X + Y$ represent only half of the possible functions. We now introduce the remaining Boolean operators.

X	Y	Zero	X AND Y	NOT (X implies Y)	X	NOT (Y implies X)	Y	X XOR Y	X OR Y	NOT (X OR Y)	X = Y	NOT Y	Y implies X	NOT X	X implies Y	NOT (X and Y)	One
0	0	0	0	0	0	0	0	0	0	1	1	1	1	1	1	1	1
0	1	0	0	0	0	1	1	1	1	0	0	0	0	1	1	1	1
1	0	0	0	1	1	0	0	1	1	0	0	1	1	0	0	1	1
1	1	0	1	0	1	0	1	0	1	0	1	0	1	0	1	0	1

(16 Possible Functions (F_0–F_{15}))

Figure 2.14 The 16 functions of two Boolean variables.

NAND

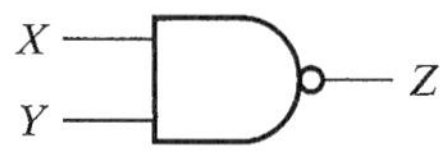

X	Y	Z
0	0	1
0	1	1
1	0	1
1	1	0

Figure 2.15 Gate and truth table representations of NAND.

NOR

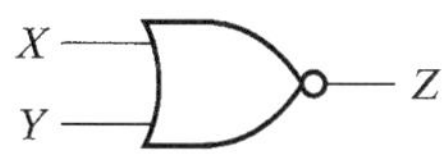

X	Y	Z
0	0	1
0	1	0
1	0	0
1	1	0

Figure 2.16 Gate and truth table representations of NOR.

In addition to their different truth tables, each function has a different number of switches associated with its implementation. For example, F_0 and F_{15}, constants 0 and 1 respectively, require no switches for their implementation at all. These logical values can be obtained by directly connecting to the reference low and high voltage signals. The implementation complexity of the other functions is as follows: X (F_3) and Y (F_5) also require no switches, as they are simply direct-through connections to the appropriate input. $\overline{X}$ (F_{12}) and $\overline{Y}$ (F_{10}) are the complement functions of X and Y, and require two switches to implement an inverter. X NOR Y (F_4) and X NAND Y (F_{14}) are implemented using four switches. X OR Y (F_7) and X AND Y (F_1), on the other hand, require six switches, while $X = Y$ (F_9) and X XNOR Y (F_6) demand 16 switches for their implementations. So while one way of expressing a Boolean function might minimize the number of Boolean operations, it may not minimize the number of switches for its realization.

NAND and NOR Two of the most frequently encountered Boolean operators are NAND (NOT-AND) and NOR (NOT-OR). Their gate and truth table representations are shown in Figure 2.15 and Figure 2.16, respectively. The NAND operation behaves as if an AND is composed with a NOT: it yields a logic 0 in the truth table rows where AND is a 1, and it yields a 1 in the rows where AND is 0. The gate representation is an AND gate with a small circle or "bubble" at its output, denoting negation.

If you take a close look at the truth table representation in Figure 2.15 and compare it to Figure 2.10, you will see that it looks like an AND function with the true and false outputs reversed. NAND is true when either X is 0 or Y is 0. Alternatively, it is false when X and Y are both true.

NOR behaves in a similar fashion, but now with respect to OR. Once again the truth table outputs are complemented, and we draw the NOR gate as an OR gate with a bubble at the output. Both X and Y must be 0 to force the output to be true.

NAND and NOR gates far outnumber AND and OR gates in a typical digital design, even though these functions are less intuitive, for the simple reason that they can be implemented in fewer switches than AND and OR gates.

Since any Boolean expression can be represented in terms of AND, OR, and NOT gates, it is hardly surprising that the same statement can be made about NAND, NOR, and NOT gates. In fact, NOT gates are superfluous: if you carefully examine the truth tables of Figure 2.15 and Figure 2.16, you'll see that NAND and NOR act like NOT when both inputs are both 0 or both 1. We shall see an efficient method for mapping Boolean expressions into NAND and NOR logic in Section 2.4.

XOR/XNOR (Equality) This leaves six functions still unnamed in Figure 2.14. Two of these, frequently of use, are exclusive OR (XOR, also known as the *inequality gate* or *difference function*) and exclusive NOR (XNOR, also known as the *equality gate* or *coincidence function*). Their truth tables and gate representations are given in Figure 2.17. XOR is true when its inputs differ in value. XNOR is true when its inputs coincide in value. The Boolean operator for XOR is ⊕; XNOR is usually represented by ≡ or as the complement of XOR. As with any Boolean function, these can be implemented in terms of AND, OR, and NOT operations:

$$\text{XOR:} \quad X \oplus Y = X\overline{Y} + \overline{X}Y$$

$$\text{XNOR:} \quad X \equiv Y = \overline{X \oplus Y} = \overline{X}\,\overline{Y} + XY$$

XOR $X \oplus Y$

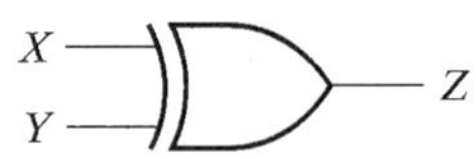

X	Y	Z
0	0	0
0	1	1
1	0	1
1	1	0

XNOR $X \equiv Y$

X	Y	Z
0	0	1
0	1	0
1	0	0
1	1	1

Figure 2.17 Representations of the XOR and XNOR operations.

If you examine the truth table of Figure 2.17(a), you can see that XOR is precisely the function needed to implement the half-adder sum of Figure 2.5 and XNOR directly implements the equivalence function of Figure 2.1.

Implication The remaining four functions are based on a Boolean operator called *implication*. X implies Y (written $X \Rightarrow Y$) is false only when X it true and Y is false, otherwise, it is true. The group of four functions are $X \Rightarrow Y$, $Y \Rightarrow X$, NOT $(X \Rightarrow Y)$, and NOT $(Y \Rightarrow X)$. These commonly are not found as primitives due to their asymmetry. Thus, they are not available readily for realizing digital systems and we won't be making much use of them in this text.

2.3.2 Logic Blocks and Hierarchy

Just as a program can be constructed from simpler subroutines, even the most complex logic function can be constructed from more primitive functions by wiring up logic gates.

In most integrated circuit technologies, libraries of prepackaged functions are made available to the designer. The library components include all of the major logic gates, in the form of 2, 3, and possibly higher numbers of inputs, as well as more complex functions constructed from a collection of basic logic gates. Examples of these are the half adder, full adder, and multi-bit adder functions.

EXAMPLE 2.4 TWO-BIT ADDER

We have already seen the half-adder function with its two binary inputs (A, B), and its two outputs, sum (S) and carry (C). Figure 2.18 represents the half adder (HA) as a block diagram, a black box with inputs and outputs. Next to its block diagram is the truth table that describes the function's input/output behavior. The gate-level implementation of the half adder is on the right in the figure. The half-adder's sum output is implemented by the XOR gate, and the carry by the AND gate.

A multiple-bit adder can be built by up by wiring together single bit adders. The stage that computes the sum of the bits in the first, lowest order position passes its carry-out to the carry-in of the next higher order adder. Unfortunately, the half adder function, since it doesn't have a carry-in input, cannot be used in this way. This requires the full adder.

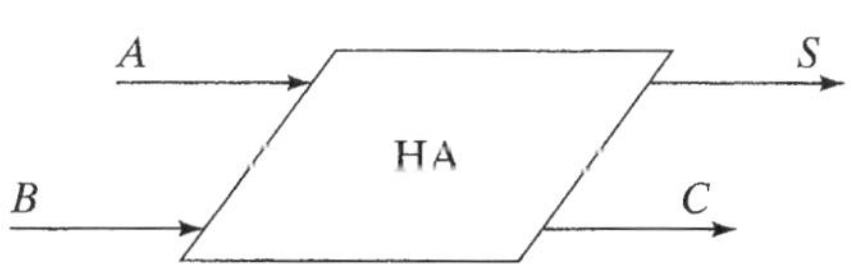

A	B	C	S
0	0	0	0
0	1	0	1
1	0	0	1
1	1	1	0

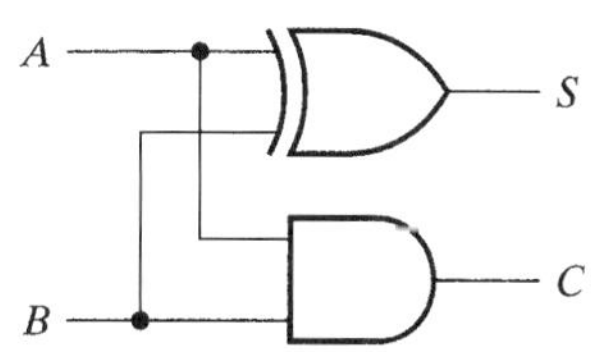

Figure 2.18 Half-adder function implemented via interconnected gates.

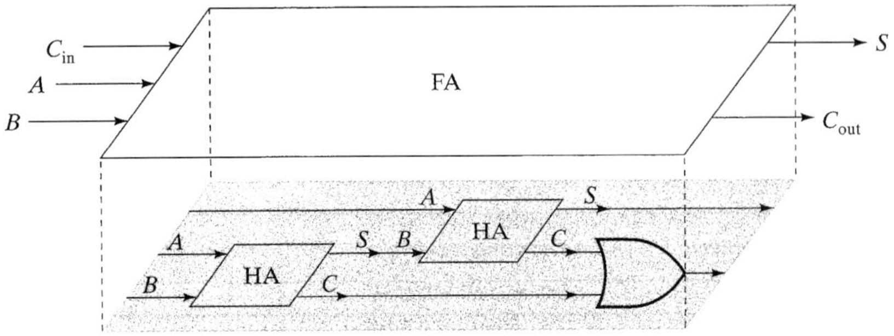

C_{in}	A	B	C_{out}	S
0	0	0	0	0
0	0	1	0	1
0	1	0	0	1
0	1	1	1	0
1	0	0	0	1
1	0	1	1	0
1	1	0	1	0
1	1	1	1	1

Figure 2.19 **Full-adder function implemented as cascaded half adders.**

The full-adder (FA) block diagram is shown in Figure 2.19, it has: three inputs: A, B, and carry-in (C_{in}); and two outputs: sum (S) and carry-out (C_{out}). The full adder functionality is described by the truth table in the figure, and can be implemented by the cascaded half adders, as shown.

It is a fair question as to how this particular combination of half-adder building blocks and an OR gate realizes the full-adder function. It comes from a careful examination of the full-adder truth table. Consider the case when $C_{in} = 0$, basically the top half of the truth table. The function for SUM is exactly as for the HA. When $C_{in} = 1$, the bottom half of the truth table, the SUM output is exactly the complement of the HA's S output. The XOR gate's behavior is such that when one input is 0, the output is the same as the other input. And when one input is 1, the output is the complement of the other input. Connecting one XOR input to the FA's carry-in input, and the other to the sum of A and B computed by a first stage HA, gives us exactly the *Sum* output behavior we need. But this is exactly what the HA does to its two inputs! We are simply adding three bits, two at a time and taking the sum of the first addition of two bits and adding it to the third bit.

Implementing the FA carry-out is also straightforward. We simply OR together both carry's of the two HAs. If either has a carry, then the FA has a carry-out. You can check that this composition implements the full adder by working through the truth table for each of the possible input combinations.

Now the 2-bit adder can be constructed by interconnecting two full adders. The wiring is shown in Figure 2.20. A and B are now 2-bit numbers (A_1, A_0 and B_1, B_0). The carry-in for the low order stage (C_0) is connected to logic 0. Its carry-out is connected to the carry-in of the high-order stage. The carry-out of the second stage (C_2) becomes the carry-out of the 2-bit adder.

2.3.3 Time Behavior and Waveforms

Logic gates do not operate infinitely fast. When the inputs change, it takes some time for the these changes to be reflected in the gate's output, as the electrical signals propagate through the interconnected switches.

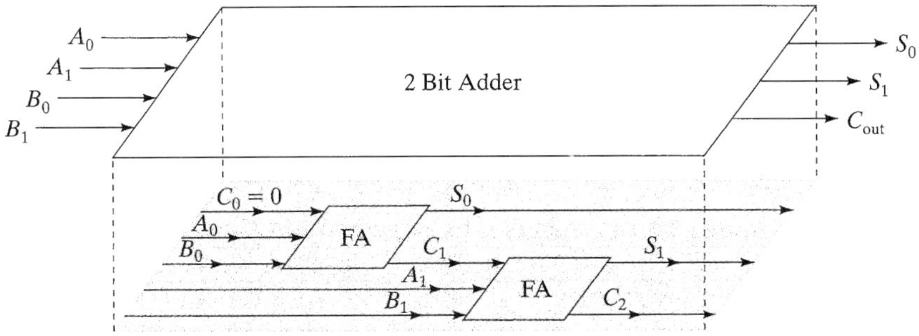

Figure 2.20 Two-bit binary adder composed from two full adders.

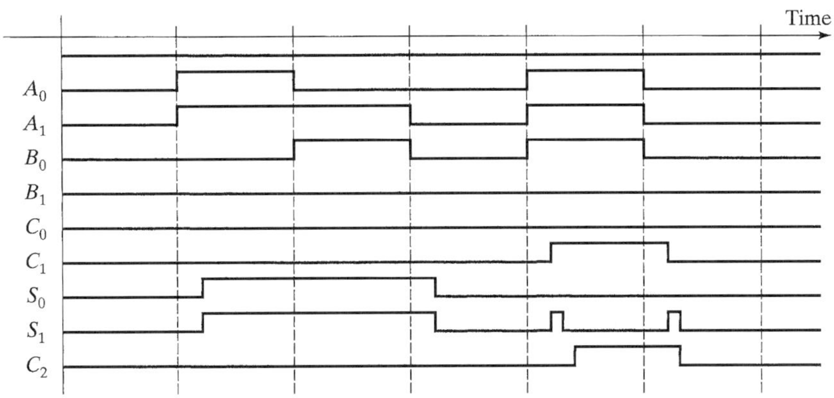

Figure 2.21 Timing waveforms for the 2-bit adder of Figure 2.20 for several different combinations of inputs.

A *waveform* representation is a good way to represent signal propagation over time. The *x*-axis represents the time steps. The *y*-axis shows the logical value of various points in the circuit. These *probe* points can be an output of the function, or the value on any internal wire of the circuit.

For ease of discussion, we consider the delay through any gate as taking exactly one time unit. This is called the *unit delay model,* and it really is a simplifying assumption. Real gates rarely exhibit such uniformly simple performance. For example, an AND gate is usually implemented as a NAND gate composed with a NOT gate. So it makes sense that an AND gate should be slower to change its output in response to changes in its inputs than a simpler, more primitive NAND gate.

Timing waveforms for the full-adder of Figure 2.20 are shown in Figure 2.21. Again, the *x*-axis represents time (there are now units shown as they are unimportant for our purposes) and we use different horizontal traces to show how the value of individual wires changes over time. The inputs are changed every 10 time units (the tick mark on the horizontal axis) to allow sufficient time for propagation through all the gates of the circuit to the output wires. The full adders are constructed using two half adders (recall that these contain an XOR and

AND gate). To simplify matters we have set each gate to have a delay of one unit regardless of how complex they are.

The inputs start out all equal to 0 and so are the outputs for this condition, $0 + 0 = 0$. The inputs then change to $3 + 0$ and this requires two time units for the S outputs to reach their final value at time 12. When the inputs change to $2 + 1$ at time 20 (A_1, $A_0 = 1, 0$ and B_1, $B_0 = 0, 1$) there is no change on the outputs, as expected. At time 30, the inputs change to back to $0 + 0$ and then go to $3 + 1$ at time 40. Note that there is a longer delay to reach the final value for $3 + 1 = 4$ than for $2 + 1 = 3$. The final value is not reached until time 44, four time units after the change in inputs, with 0s on the S outputs and a carry-out of 1 on C_2. C_0 is set to 0 for the entire duration of the waveform.

Also, note the glitches on the S_1 output. Any other system using the output of this full-adder for its own input would have to make sure to wait long enough for all the signals to reach their final value and ignore any of the temporary *glitches* we see in the waveform of Figure 2.21.

2.3.4 Minimizing the Number of Gates and Wires

Logic minimization uses a variety of techniques to obtain the simplest gate-level implementation of a Boolean function. But simplicity depends on the metric we use. We examine these metrics in this subsection.

Time and Space Trade-Offs One way to measure the complexity of a Boolean function is to count the number of literals it contains. Literals are an approximate measure of the number of transistors and the amount of wiring that will be needed to implement a function. For electrical and packaging reasons, gates in a given technology typically have a limited number of inputs. While two-, three-, and four-input gates are common, gates with more than eight or nine inputs are rather rare. Thus, one of the primary reasons for performing logic minimization is to reduce the number of literals in the expression of a function, thus reducing the number of gate inputs.

An alternative metric is the number of gates, this focuses on the area a circuit will occupy. There is a strong correlation between the number of gates in a design and the number of components, whether library modules or integrated circuit packages, needed for its implementation. The simplest design to manufacture is often the one with the fewest gates, not the fewest literals. Of course, wires take up space also, so this measure is lessening in importance as gates are more integrated on a single chip and transistor sizes keep shrinking.

A third metric is the number of cascaded levels of gates. Reducing the number of logic levels reduces overall delay, as there are fewer gate delays on the path from inputs to outputs. However, putting a circuit in a form suitable for minimum delay rarely yields an implementation with the fewest gates or the simplest gates. It is not possible to minimize all three metrics at the same time.

The traditional minimization techniques you will study in this chapter emphasize reducing delay at the expense of adding more gates.

Newer methods, covered in the next chapter, allow a trade-off between increased circuit delay and reduced gate count.

▶ EXAMPLE 2.5 DIFFERENT IMPLEMENTATIONS OF ONE FUNCTION

To illustrate the trade-offs just discussed, consider the following 3-variable Boolean function:

$$Z = \bar{A}\bar{B}C + \bar{A}BC + A\bar{B}C + AB\bar{C}$$

The truth table for this function is shown in Figure 2.22. You would probably implement the function directly from the preceding equation, using three NOT gates, four 3-input AND gates, and a single 4-input OR gate. This is called a *two-level implementation,* with variables and their complements at the zeroth level, AND gates at the first level, and an OR gate at the second level.

You could implement the same truth table with fewer gates. An alternative two-level implementation is given in Figure 2.22 as function Z_1:

$$Z_1 = AB\bar{C} + \bar{A}C + \bar{B}C$$

It uses the same number of inverters and OR gates but only three AND gates and an OR gate with only three inputs instead of four. The

A	B	C	Z
0	0	0	0
0	0	1	1
0	1	0	0
0	1	1	1
1	0	0	0
1	0	1	1
1	1	0	1
1	1	1	0

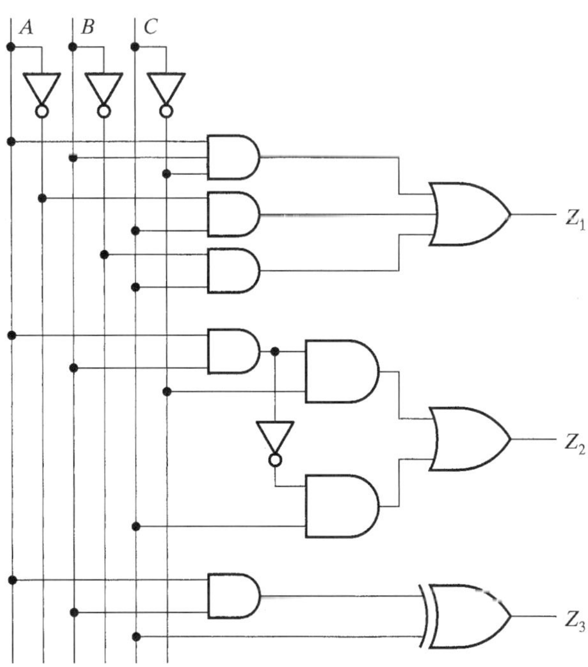

Figure 2.22 Alternative realizations of Z.

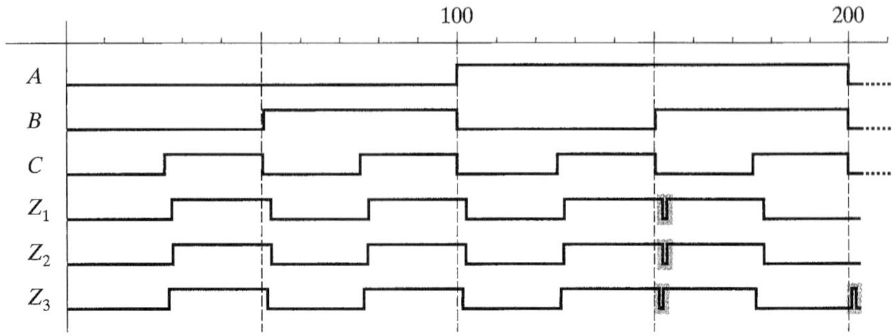

Figure 2.23 Waveform behavior of three implementations of the truth table of Figure 2.22.

original function has 12 literals. This alternative has only seven, thus reducing the wiring complexity.

The implementation labeled Z_2 is called *multilevel:*

$$Z_2 = T\bar{C} + \bar{T}C$$
$$T = AB$$

The longest path from an input to an output passes through four gates. This contrasts with three gate delays in the two-level functions. In terms of gate counts, the circuit uses two rather than three inverters and only 2-input AND and OR gates. Here you can see a trade-off between gate count and performance: Z_2 uses simpler and faster gates, but it has more level than Z_1. We would have to look at the precise delay of each type of gate to decide which is fastest.

Z_3 shows a third realization that uses an XOR gate:

$$Z_3 = (AB) \oplus C$$

XOR is sometimes called a *complex gate,* because you normally implement it by combining several NAND or NOR gates. Although this implementation has the lowest gate count, it is also likely to have the worst delay. An XOR gate tends to be slow compared with the implementations for Z based on simple AND and OR gates.

Figure 2.23 shows the timing waveforms for the three circuit alternatives, assuming a single time unit delay per gate (somewhat unrealistically especially in the case of the XOR gate). All have equivalent behavior, although they exhibit slightly different propagation delays. All three circuits show a glitch on the transition $ABC = 101$ to $ABC = 110$. Glitches can often be a problem as they temporarily cause the output to have a wrong value. We'll see later how we can prevent these from happening.

2.3.5 Case Study: 7-Segment Decoder

We have already introduced the 7-segment display for displaying a binary digit earlier in this chapter and its truth table that tells us which segments to turn on and off (see Figure 2.4). In this subsection, we will generalize the display to handle the decimal digits 0 through 9. The input to the system now consists of four bits in a *binary-coded-decimal* (BCD) form, where decimal 0 is represented by the four binary functions $ABCD = 0000$, 1 is represented by $ABCD = 0001$, 2 by 0010, 3 by 0011, through 9 = 1001. We assume that the input bit patterns 1010, 1011, 1100, 1101, 1110, 1111 are never encountered in practice (we call such inputs *don't cares,* and we will see how to make use of them for simplification in Chapter 3.

Figure 2.24 shows a block diagram of the decoder, the display element, and the way the segments should be driven to represent the appropriate digit. Figure 2.25 shows the truth table for this circuit with the don't-care conditions in a gray box.

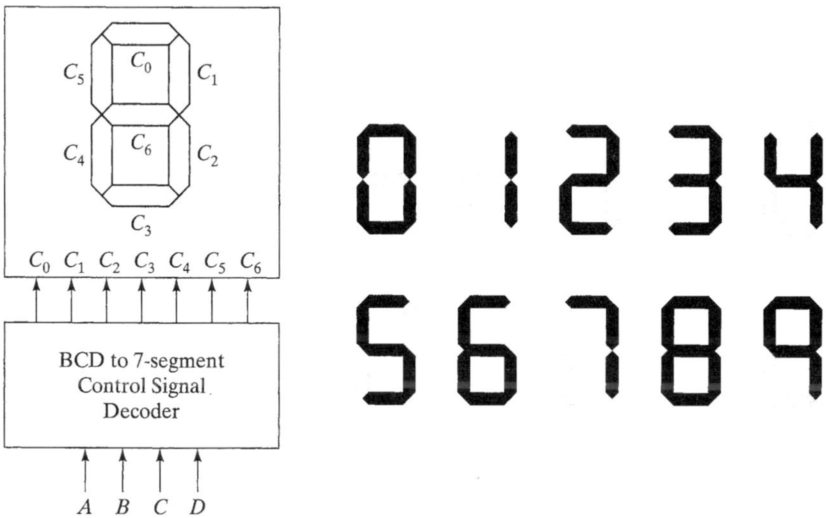

Figure 2.24 **Binary-coded decimal digital-display system.**

A	B	C	D	C_0	C_1	C_2	C_3	C_4	C_5	C_6
0	0	0	0	1	1	1	1	1	1	0
0	0	0	1	0	1	1	0	0	0	0
0	0	1	0	1	1	0	1	1	0	1
0	0	1	1	1	1	1	1	0	0	1
0	1	0	0	0	1	1	0	0	1	1
0	1	0	1	1	0	1	1	0	1	1
0	1	1	0	1	0	1	1	1	1	1
0	1	1	1	1	1	1	0	0	0	0
1	0	0	0	1	1	1	1	1	1	1
1	0	0	1	1	1	1	0	0	1	1
1	0	1	0	0	0	0	0	0	0	0
1	0	1	1	0	0	0	0	0	0	0
1	1	0	0	0	0	0	0	0	0	0
1	1	0	1	0	0	0	0	0	0	0
1	1	1	0	0	0	0	0	0	0	0
1	1	1	1	0	0	0	0	0	0	0

Figure 2.25 **Digital-display system truth table.**

Let's specify the Boolean logic for just one of the decoder's outputs: C_2. This particular segment is illuminated for every valid input combination except for digit 2 (0010). The Boolean expression for C_2 is all of the input combinations that cause that segment to be illuminated ORed together:

$$\begin{aligned} C_2 = {} & \bar{A}\bar{B}\bar{C}\bar{D} + \bar{A}\bar{B}\bar{C}D + \bar{A}\bar{B}CD + \bar{A}B\bar{C}\bar{D} + \bar{A}B\bar{C}D \\ & + \bar{A}BC\bar{D} + \bar{A}BCD + A\bar{B}\bar{C}\bar{D} + A\bar{B}\bar{C}D \end{aligned}$$

This Boolean function is quite complicated: 36 literals, nine 4-input AND gates, and one 9-input OR gate. Fortunately, its complexity, measured in terms of the total number of literals, can be significantly reduced (from 36 to 20) by the right use of simplification theorems. By rearranging terms, introducing redundant terms, and factoring, we can rewrite C_2 as follows:

$$\begin{aligned} C_2 = {} & \bar{A}B(\bar{C}\bar{D} + \bar{C}D + C\bar{D} + CD) + \bar{B}\bar{C}(\bar{A}\bar{D} + \bar{A}D + A\bar{D} + AD) \\ & + \bar{A}D(\bar{B}\bar{C} + \bar{B}C + B\bar{C} + BC) \end{aligned}$$

Successive use of the unifying theorem and the complementarity theorem for the expressions in the parentheses reduces each of them to the constant 1. The resulting simplification is:

$$C_2 = \bar{A}B + \bar{B}\bar{C} + \bar{A}D$$

This expression has only six literals, three 2-input AND gates, and one 3-input OR gate. That is a considerable savings in terms of all of the relevant metrics: less wires, fewer gates, and much simpler gates at that! If we take advantage of the don't care elements of the truth table, we could simplify the expression even further, as we will see in Section 2.5.3.

2.4 Two-Level Logic

You already know that there are many gate-level implementations with the same truth table behavior. In this section, you will learn the methods for deriving a reduced gate-level implementation of a Boolean function in two-level form. This minimizes the levels of logic but usually yields circuits that do not have minimum delay or the smallest gate counts. Two-level forms are still interesting because they are easy to understand and realize as logic.

2.4.1 Canonical Forms

To compare Boolean functions expressed in algebraic terms, it is useful to have a standard form with which to represent the function. This standard term is called a *canonical form,* and it is a unique algebraic signature of the function. You will frequently encounter two alternative forms: *sum-of-products* and *product-of-sums.* We will now introduce these formally.

Sum-of-Products You have already met the *sum-of-products* form in Section 1.4.1. It is also sometimes known as a *disjunctive normal form* or *minterm expansion*. A sum-of-products expression is formed as follows. Each row of the truth table in which the function takes on the value 1 contributes an ANDed term, using the asserted variable if there is a 1 in its column for that row or its complement if there is a 0. These are called *minterms*. Technically, a minterm is defined as an ANDed product of literals in which each variable appears exactly once in either true or complemented form, but not both. The minterms are then ORed to form the expression for the function. The minterm expansion is unique because it is derived deterministically from the truth table. There are is only one way to do it.

A	B	C	F	F'
0	0	0	0	1
0	0	1	1	0
0	1	0	0	1
0	1	1	1	0
1	0	0	0	1
1	0	1	1	0
1	1	0	1	0
1	1	1	1	0

Figure 2.26 Sample truth table.

Figure 2.26 shows a truth table for a function and its complement. The minterm expansions for F and $\bar{F}$ are

$$F = \bar{A}\bar{B}C + \bar{A}BC + A\bar{B}C + AB\bar{C} + ABC$$

$$\bar{F} = \bar{A}\bar{B}\bar{C} + \bar{A}B\bar{C} + A\bar{B}\bar{C}$$

We can write such expressions in a shorthand notation using the binary number system to encode the minterms. Figure 2.27 shows the relationship between the truth table row and the numbering of

A	B	C	Minterms	
0	0	0	$A'B'C'$	m_0
0	0	1	$A'B'C$	m_1
0	1	0	$A'BC'$	m_2
0	1	1	$A'BC$	m_3
1	0	0	$AB'C'$	m_4
1	0	1	$AB'C$	m_5
1	1	0	ABC'	m_6
1	1	1	ABC	m_7

(a)

A	B	C	Minterms	
0	0	0	$A'B'C'$	m_0
0	0	1	$A'B'C$	m_1
0	1	0	$A'BC'$	m_2
0	1	1	$A'BC$	m_3
1	0	0	$AB'C'$	m_4
1	0	1	$AB'C$	m_5
1	1	0	ABC'	m_6
1	1	1	ABC	m_7

(b)

A	B	C	Minterms	
0	0	0	$A'B'C'$	m_0
0	0	1	$A'B'C$	m_1
0	1	0	$A'BC'$	m_2
0	1	1	$A'BC$	m_3
1	0	0	$AB'C'$	m_4
1	0	1	$AB'C$	m_5
1	1	0	ABC'	m_6
1	1	1	ABC	m_7

(c)

A	B	C	Minterms	
0	0	0	$A'B'C'$	m_0
0	0	1	$A'B'C$	m_1
0	1	0	$A'BC'$	m_2
0	1	1	$A'BC$	m_3
1	0	0	$AB'C'$	m_4
1	0	1	$AB'C$	m_5
1	1	0	ABC'	m_6
1	1	1	ABC	m_7

(d)

Figure 2.27 Shorthand notation for minterms of three variables. Note that each minterm covers exactly one row. Minterms are ORed together to form the complete function (note that the figure does not show separate tables for m_6 and m_7). Therefore, a row is shaded in the final table (d) if it is shaded in any of the minterm tables (parts a,b,c, and two others not shown above).

the minterm. Note that the ordering of the Boolean variables is critical in deriving the minterm index. In this case, A determines the most significant bit and C is the least significant bit. You can write the shorthand expression for F and $\bar{F}$ as

$$F(A,B,C) = \Sigma\ m(1,3,5,6,7) = m_1 + m_3 + m_5 + m_6 + m_7$$

$$\bar{F}(A,B,C) = \Sigma\ m(0,2,4) = m_0 + m_2 + m_4$$

where m_i represents the ith minterm. In Figure 2.27, there are four truth-tables. The first shows the minterm for $\bar{A}\bar{B}C$, the second for $\bar{A}BC$, the third for $A\bar{B}C$, and finally, the fourth table shows all five minterms needed to *cover* the function together (the fourth and fifth minterm are not shown separately in the figure). The function has a true output if the combination of input values for inputs A, B, and C correspond to *any* of the minterms, hence, a disjunctive normal form (OR) because the function is true if the inputs fall within the row labeled m_1, *or* the row labeled m_3, *or* . . . .

The indices of the minterms are used as a shorthand notation and generalize nicely for functions of more variables. For example, if F is defined over the variables A, B, C, then m_3 (011_2) is the minterm $\bar{A}BC$. But if F is defined over A, B, C, D, then m_3 (0011_2) is $\bar{A}\bar{B}CD$.

The minterm expansion is not guaranteed to be the simplest form of the function, in terms of the fewest literals or terms, nor is it likely to be. You can further reduce the expression for F by applying Boolean algebra:

$$\begin{aligned} F(A, B, C) &= \bar{A}\bar{B}C + \bar{A}BC + A\bar{B}C + AB\bar{C} + ABC \\ &= (\bar{A}\bar{B} + \bar{A}B + A\bar{B} + AB)C + AB\bar{C} \\ &= ((\bar{A} + A)(\bar{B} + B))C + AB\bar{C} \\ &= C + AB\bar{C} \\ &= AB\bar{C} + C \\ &= AB + C \end{aligned}$$

The one step you may find tricky is the last one, which applies rule 11D, $(X \bullet \bar{Y}) + Y = X + Y$, substituting AB for X and C for Y.

AB and C are called *product terms*: ANDed strings of literals containing a subset of the possible Boolean variables or their complements. For F defined over the variables A, B, and C, $\bar{A}BC$ is a minterm (because it contains one version of every literal—complemented or uncomplemented) as well as a product term, but AB is only a product term.

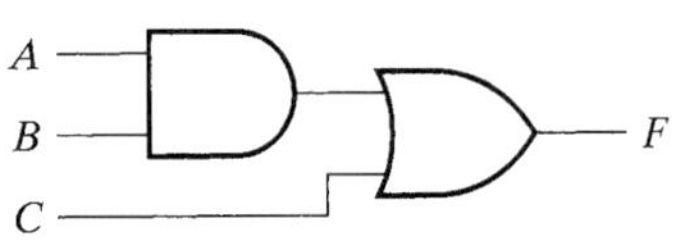

Figure 2.28 **Two-level AND-OR gate-level implementation.**

The minimized gate-level implementation of F is shown in Figure 2.28. Each product term is realized by its own AND gate. The product term A is the degenerate case of a single literal. No AND gate is needed to form this term. The product terms' implementations are then input to a second-level OR gate. The sum-of-products form leads directly to a two-level realization.

We can repeat the simplification process for $\overline{F}$ but DeMorgan's law gives us a good starting point for applying Boolean simplification:

$$\overline{F} = \overline{(AB + C)} = (\overline{A} + \overline{B})\overline{C} = \overline{A}\overline{C} + \overline{B}\overline{C}$$

Although this procedure is not guaranteed to obtain the simplest form of the complement, it does so in this case.

Product-of-Sums The involution theorem states that the complement of a Boolean expression's complement is the expression itself. By using DeMorgan's law twice, we can derive a second canonical form for Boolean equations. This form is called the *product-of-sums* and sometimes the *conjunctive normal form* or *maxterm expansion*.

The procedure for deriving a product-of-sums expression from a truth table is the logical dual of the sum-of-products case. First, find the rows of the truth table where the function is 0. A *maxterm* is defined as an ORed sum of literals in which each variable appears exactly once in either true or complemented form, but not both. We form a maxterm by ORing the uncomplemented variable if there is a 0 in its column for that row, or the complemented variable if there is a 1 there. This is exactly opposite to the way we formed minterms. There is one maxterm for each 0 row of the truth table; these are ANDed together at the second level.

The product-of-sums for the functions F and $\overline{F}$ of Figure 2.26 are

$$F = (A + B + C)(A + \overline{B} + C)(\overline{A} + B + C)$$

$$\overline{F} = (A + B + \overline{C})(A + \overline{B} + \overline{C})(\overline{A} + B + \overline{C})(\overline{A} + \overline{B} + C)(\overline{A} + \overline{B} + \overline{C})$$

Once again, we often use a shorthand notation. Figure 2.29 shows the relationship between maxterms and their shorthand form. We can write F and $\overline{F}$ as

$$F(A, B, C) = \Pi M(0,2,4) = M_0 \bullet M_2 \bullet M_4$$

$$\overline{F}(A, B, C) = \Pi M(1,3,5,6,7) = M_1 \bullet M_3 \bullet M_5 \bullet M_6 \bullet M_7$$

where M_i is the ith maxterm. In Figure 2.29, the maxterms that represent the *cover* of the function of Figure 2.26 are shown in the first three parts of the figure. Each maxterm covers all the rows except for the row corresponding to the maxterm. Therefore, to form the complete function, we AND the maxterms together. All rows will be true except for the ones with a corresponding maxterm (those with a value of 0 for the function). There are three different maxterms in this case, because the output is 0 in exactly three cases. The first corresponds to M_0, the second corresponds to M_2, and the third corresponds to M_4. The function is true if the combination of input values for inputs A, B, and C lies within *all* of the maxterm regions, hence, a conjunctive normal form because the function is true if the inputs fall within the region for M_0 *and* the region for M_2 *and* the region for M_4.

A	B	C	Maxterms	
0	0	0	$A+B+C$	M_0
0	0	1	$A+B+C'$	M_1
0	1	0	$A+B'+C$	M_2
0	1	1	$A+B'+C'$	M_3
1	0	0	$A'+B+C$	M_4
1	0	1	$A'+B+C'$	M_5
1	1	0	$A'+B'+C$	M_6
1	1	1	$A'+B'+C'$	M_7

(a)

A	B	C	Maxterms	
0	0	0	$A+B+C$	M_0
0	0	1	$A+B+C'$	M_1
0	1	0	$A+B'+C$	M_2
0	1	1	$A+B'+C'$	M_3
1	0	0	$A'+B+C$	M_4
1	0	1	$A'+B+C'$	M_5
1	1	0	$A'+B'+C$	M_6
1	1	1	$A'+B'+C'$	M_7

(b)

A	B	C	Maxterms	
0	0	0	$A+B+C$	M_0
0	0	1	$A+B+C'$	M_1
0	1	0	$A+B'+C$	M_2
0	1	1	$A+B'+C'$	M_3
1	0	0	$A'+B+C$	M_4
1	0	1	$A'+B+C'$	M_5
1	1	0	$A'+B'+C$	M_6
1	1	1	$A'+B'+C'$	M_7

(c)

A	B	C	Maxterms	
0	0	0	$A+B+C$	M_0
0	0	1	$A+B+C'$	M_1
0	1	0	$A+B'+C$	M_2
0	1	1	$A+B'+C'$	M_3
1	0	0	$A'+B+C$	M_4
1	0	1	$A'+B+C'$	M_5
1	1	0	$A'+B'+C$	M_6
1	1	1	$A'+B'+C'$	M_7

(d)

Figure 2.29 Maxterm shorthand for a function of three variables. Note that each maxterm covers all but one row. Maxterms are ANDed together to form the complete function. Therefore, a row is shaded in final table (d) if it is shaded in all of the maxterm tables (a,b,c).

Interestingly, the maxterm expansion of F could have been formed directly by applying DeMorgan's law to the minterm expansion of $\bar{F}$:

$$\bar{F} = \bar{A}\bar{B}\bar{C} + \bar{A}B\bar{C} + A\bar{B}\bar{C}$$

$$\bar{\bar{F}} = \overline{\bar{A}\bar{B}\bar{C} + \bar{A}B\bar{C} + A\bar{B}\bar{C}}$$

$$F = (A + B + C)(A + \bar{B} + C)(\bar{A} + B + C)$$

Of course, the same is true for deriving the minterm form of F from the maxterm form of $\bar{F}$

$$\bar{F} = (A + B + \bar{C})(A + \bar{B} + \bar{C})(\bar{A} + B + \bar{C})(\bar{A} + \bar{B} + C)(\bar{A} + \bar{B} + \bar{C})$$

$$\bar{\bar{F}} = \overline{(A + B + \bar{C})(A + \bar{B} + \bar{C})(\bar{A} + B + \bar{C})(\bar{A} + \bar{B} + C)(\bar{A} + \bar{B} + \bar{C})}$$

$$F = \bar{A}\bar{B}C + \bar{A}BC + A\bar{B}C + AB\bar{C} + ABC$$

It is easy to translate a product-of-sums expression into a gate-level realization. The zeroth level forms the complements of the variables if they are needed to realize the function. The first level creates the individual maxterms as outputs of OR gates. The second level is an AND gate that combines the maxterms.

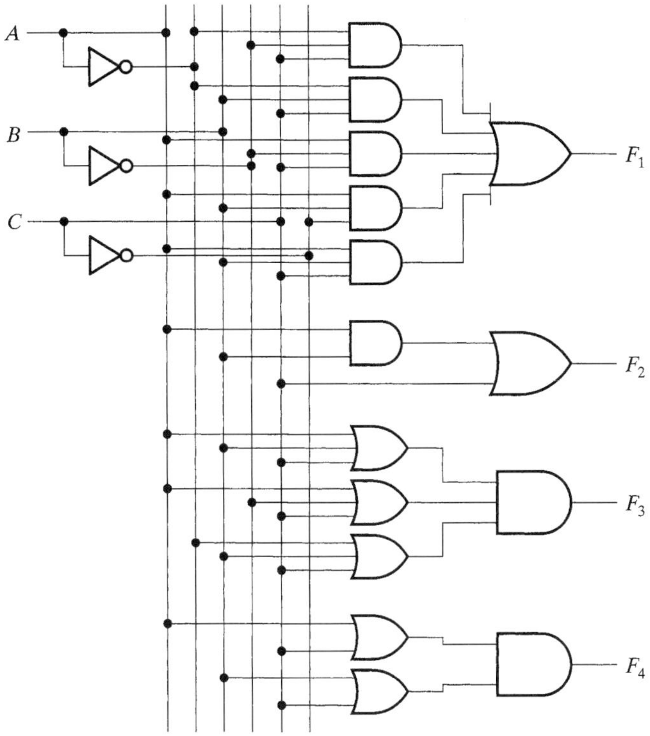

Figure 2.30 **Four implementations of *F*.**

We can find a minimized product-of-sums form by starting with the minimized sum-of-products expression of $\overline{F}$. To complement this expression, we use DeMorgan's law:

$$\overline{F} = \overline{A}\,\overline{C} + \overline{B}\,\overline{C}$$
$$\overline{\overline{F}} = \overline{\overline{A}\,\overline{C} + \overline{B}\,\overline{C}}$$
$$F = (A + C)(B + C)$$

Figure 2.30 shows the four different gate-level implementations for F discussed so far: canonical sum-of-products (F_1), minimized sum-of-products (F_2), canonical product-of-sums (F_3), and minimized product-of-sums (F_4). In terms of gate counts, the product-of-sums canonical form is more economical than the sum-of-products canonical form. But the minimized sum-of-products form uses fewer gates than the minimized product-of-sums form. Depending on the function, one or the other of these forms will be better for implementing the function.

To demonstrate that the implementations are equivalent, Figure 2.31 shows the timing waveforms for the circuits' responses to the same inputs. Except for short-duration glitches in the waveforms, their shapes are identical.

Conversion between Canonical Forms We can place any Boolean function in one of the two canonical forms, sum-of-products or product-of-sums.

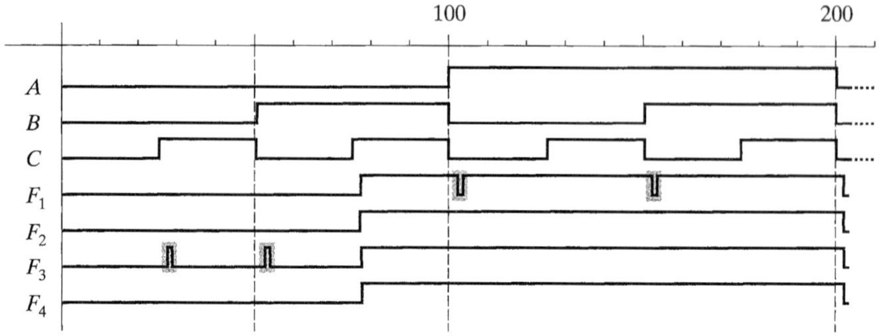

Figure 2.31 **Timing waveforms for the four implementations of *F*.**

It is easy to map an expression in one canonical form into the other. The procedure, using the shorthand notation we already introduced, is summarized here:

1. To convert from the minterm expansion to the maxterm expansion, you rewrite the minterm shorthand notation to maxterm shorthand, replacing the term numbers with those not used in the minterm list. This is equivalent to applying DeMorgan's law to the complement of the function in minterm form.

   Example: $F(A,B,C) = \Sigma m(1,3,5,6,7) = \Pi M(0,2,4)$

2. To convert from the maxterm expansion to the minterm expansion, you rewrite the maxterm shorthand notation to minterm shorthand, replacing term numbers with those not used in the maxterm list. This is equivalent to applying DeMorgan's law to the complement of the function in maxterm form.

   Example: $F(A,B,C) = \Pi M(0,2,4) = \Sigma m(1,3,5,6,7)$

3. To obtain the minterm expansion of the complement, given the minterm expansion of the function, you simply list the minterms not in *F*. The same procedure works for obtaining the maxterm complement of a function expressed in maxterm form.

   Example:

   $F(A,B,C) = \Sigma m(1,3,5,6,7)$ $\quad$ $F(A,B,C) = \Pi M(0,2,4)$

   $\overline{F}(A,B,C) = \Sigma m(0,2,4)$ $\quad$ $\overline{F}(A,B,C) = \Pi M(1,3,5,6,7)$

4. To obtain the maxterm expansion of the complement, given the minterm expansion of the function, you simply use the same maxterm numbers as used in *F*'s minterm expansion. The same procedure applies if a minterm expansion of the complement is to be derived from the maxterm expansion of the function.

   Example:

   $F(A,B,C) = \Sigma m(1,3,5,6,7)$ $\quad$ $F(A,B,C) = \Pi M(0,2,4)$

   $\overline{F}(A,B,C) = \Pi M(1,3,5,6,7)$ $\quad$ $\overline{F}(A,B,C) = \Sigma m(0,2,4)$

2.4.2 Incompletely Specified Functions

We have assumed that we must define an n-input function on all of its 2^n possible input combinations. This is not always the case. Often we have some flexibility in the specification of a function. Making this flexibility explicit permits optimization procedures to take that it into account when trying to find the smallest or fastest circuit to implement a function. We study the case of incompletely specified functions in this subsection.

EXAMPLE 2.6 INCOMPLETELY SPECIFIED FUNCTIONS

Let's consider a logic function that takes as input a binary-coded decimal (BCD) digit. *BCD digits* are decimal digits, in the range 0 through 9, that are represented by four-bit binary numbers, using the combinations 0000_2 (0) through 1001_2 (9). The other combinations, 1010_2 (10) through 1111_2 (15), should never be encountered. It is possible to simplify the Boolean expressions for the function if we assume that we *do not care* about its behavior in these "out of range" cases.

Figure 2.32 shows the truth table for a BCD increment-by-1 circuit. Each BCD number is represented by four Boolean variables, $A\ B\ C\ D$. The output of the incrementer is represented by four Boolean functions, $W\,X\,Y\,Z$.

The output functions have the value "X" for each of the input combinations we should never encounter. When used in a truth table, the value X is often called a *don't care*. Do not confuse this with the value X reported by many logic simulators, where it represents an undefined value or a *don't know*. Any actual implementation of the circuit will generate some output for the don't-care cases. When used in a truth table, an X simply means that we have a choice of assigning a 0 or 1 to the truth table entry. We should choose the value that will lead to the simplest implementation.

A	B	C	D	W	X	Y	Z
0	0	0	0	0	0	0	1
0	0	0	1	0	0	1	0
0	0	1	0	0	0	1	1
0	0	1	1	0	1	0	0
0	1	0	0	0	1	0	1
0	1	0	1	0	1	1	0
0	1	1	0	0	1	1	1
0	1	1	1	1	0	0	0
1	0	0	0	1	0	0	1
1	0	0	1	0	0	0	0
1	0	1	0	X	X	X	X
1	0	1	1	X	X	X	X
1	1	0	0	X	X	X	X
1	1	0	1	X	X	X	X
1	1	1	0	X	X	X	X
1	1	1	1	X	X	X	X

Figure 2.32 Truth table for BCD increment by 1.

To see that don't cares eventually are replaced by some logic value, let's consider the BCD incrementer truth table. The function Z looks as if it could be realized quite simply as the function $\bar{D}$. If we choose to implement Z in this way, the Xs will be replaced by real logic values. Since the inputs 1010_2 through 1111_2 will never be encountered by the operational circuit, it shouldn't matter which values we assign to those truth table rows. We choose an assignment that makes this implementation as simple as possible.

EXAMPLE 2.7 DECIMAL DIGIT TO SEVEN SEGMENT DISPLAY DECODER

Figure 2.33 shows a revised truth table for Seven Segment Display Decoder that takes full advantage of the fact that the input bit patterns 1010 through 1111 should never be presented to the system. Note that the last six rows of Figure 2.25 have been condensed into only two. There are now don't cares in the input columns as well. The first of these two rows tells us we have *output don't cares* for all the outputs when the inputs are $A = 1$, $B = 0$, and $C = 1$ regardless of the value of the input D (an *input don't care*). This handles two cases: when $D = 0$ and when $D = 1$. The last row of this table captures the same meaning as the last 4 rows of the truth table of Figure 2.25.

Don't Cares and the Terminology of Canonical Forms In terms of the standard Σ and Π notations, minterms or maxterms assigned a don't care are written as d_i or D_i, respectively. Thus, the canonical form for Z (from Figure 2.32) is written as:

$$Z = m_0 + m_2 + m_4 + m_6 + m_8 + d_{10} + d_{11} + d_{12} + d_{13} + d_{14} + d_{15}$$
$$Z = M_1 \bullet M_3 \bullet M_5 \bullet M_7 \bullet M_9 \bullet D_{10} \bullet D_{11} \bullet D_{12} \bullet D_{13} \bullet D_{14} \bullet D_{15}$$

Notice that the same don't-care terms appear in both expressions because we can assign 0 or 1 for the don't cares.

A	B	C	D	C_0	C_1	C_2	C_3	C_4	C_5	C_6
0	0	0	0	1	1	1	1	1	1	0
0	0	0	1	0	1	1	0	0	0	0
0	0	1	0	1	1	0	1	1	0	1
0	0	1	1	1	1	1	1	0	0	1
0	1	0	0	0	1	1	0	0	1	1
0	1	0	1	1	0	1	1	0	1	1
0	1	1	0	1	0	1	1	1	1	1
0	1	1	1	1	1	1	0	0	0	0
1	0	0	0	1	1	1	1	1	1	1
1	0	0	1	1	1	1	0	0	1	1
1	0	1	X	X	X	X	X	X	X	X
1	1	X	X	X	X	X	X	X	X	X

Figure 2.33 Seven-segment display controller truth table with don't cares.

Similarly, C_2 (from Figure 2.33) can be written as:

$$\begin{aligned} C_2 &= m_0 + m_1 + m_3 + m_4 + m_5 + m_6 + m_7 + m_8 + m_9 \\ &\quad + d_{10} + d_{11} + d_{12} + d_{13} + d_{14} + d_{15} \\ C_2 &= M_2 \bullet D_{10} \bullet D_{11} \bullet D_{12} \bullet D_{13} \bullet D_{14} \bullet D_{15} \end{aligned}$$

It is important to understand the terminology of Boolean functions. Let's introduce the principle definitions by way of the truth table of Figure 2.32. This function is *multi-output* because it is represented by four output bits defined over the same inputs. It is *incompletely specified* because it contains don't cares in its outputs. For each of the function's output columns, we can define three sets: the on-set, off-set, and don't-care set. The *on-set* contains all input combinations for which the function is 1. The *off-set* and *don't-care set* are defined analogously for 0 and X, respectively. Thus, the on-set for the incrementer's W output is {[0,1,1,1], [1,0,0,0]}; its off-set is {[0,0,0,0], [0,0,0,1], [0,0,1,0], [0,0,1,1], [0,1,0,0], [0,1,0,1], [0,1,1,0], [1,0,0,1]; and its don't-care set is {[1,0,1,0], [1,0,1,1], [1,1,0,0], [1,1,0,1], [1,1,1,0], [1,1,1,1]}. In Figure 2.33 we see not only *output don't cares* but also *input don't cares* that are used to condense the size of the truth tables.

Proper specification of don't cares is critical for the successful operation of many computer-aided design tools that perform minimization of Boolean expressions. It is impossible to extract all don't care information automatically. Designers must consciously include this information in their design descriptions and documentation. Unfortunately, don't cares are not expressible in a schematic diagram. In the next chapter, we'll see how we can use hardware description languages to capture incompletely specified functions.

2.5 Motivation for Two-Level Simplification

We can always use the rules of Boolean algebra from Section 2.1 to simplify an expression, but this method has a few problems. For one thing, there is no algorithm you can use to ensure that you've obtained a minimum solution. When do you stop looking for a Simplification theorem to apply? It might be easy when we have only a few variables, but many large designs involve dozens or hundreds of variables. Another problem is that you often have to make the expression more complex before you can simplify it, for example, replacing $(X + \overline{X})$ for 1 to add more terms and actually make the expression temporarily bigger! We follow this expansion with rearranging of terms to obtain advantageous groupings that help to simplify the expression in a later step. It is against human nature to climb out of such a *local minimum* in the hope of finding a better global solution, the *global minimum*. But this is exactly what we often have to do. And finally, it is just too cumbersome (and error prone) to manipulate Boolean expressions by hand.

Given that computer-based tools have been developed for Boolean simplification, why bother to learn any by-hand method, especially

A	B	F
0	0	0
0	1	0
1	0	1
1	1	1

(a)

A	B	G
0	0	1
0	1	0
1	0	1
1	1	0

(b)

Figure 2.34 Two simple truth tables.

when these break down for problems with many variables? Certainly no by-hand method will be effective for equations involving more than six variables. But you still need knowledge of the basic approach. Observing the symmetries in a circuit's function helps to understand its behavior and to visualize what is going on inside the circuit. As CAD tools become ever more sophisticated, you need a deeper knowledge of the algorithms they apply to use the tools effectively. And don't forget that CAD tools are written by mere mortals and do not always do things right! You must still be able to check the output of the tool.

The Essence of Boolean Simplification Let's look at what is really going on in simplification with the simple truth table of Figure 2.34(a). The function is $F = A\overline{B} + AB$. We can simplify this equation by applying one of the Boolean simplification theorems, called the *Uniting theorem:* $A(\overline{B} + B) = A$.

Notice that the two truth table rows that make up the on-set of F have A asserted, while one row has $B = 0$ and the other has $B = 1$. For the on-set, *A's value stays the same while B's value changes.* This allows us to factor out B using the Uniting theorem.

Now examine Figure 2.34(b). The function is $G = \overline{A}\overline{B} + A\overline{B}$. Applying the Uniting theorem again, we obtain $(\overline{A} + A)\overline{B} = \overline{B}$. Once again, the on-set contains two rows in which B's value does not change (it is equal to 0) and A's does change. Thus, we can factor out A, leaving $\overline{B}$.

The essence of simplification is to repeatedly find two-element subsets of the on-set in which only one variable changes its value while the other variables do not. You can eliminate the single varying variable from the term by applying the Uniting theorem. Wouldn't it be nice if there was a way to arrange the truth table rows so that the entries to which this technique could be applied are obvious? We will introduce two representations that will help us with this task: the Boolean cube and the Karnaugh map. In this section, we will focus on-hand methods, Chapter 3 will present computer algorithms for performing two-level logic minimization that are based on the same concepts.

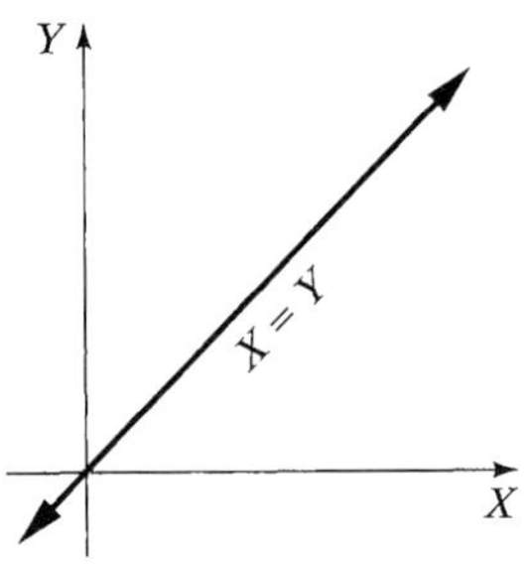

Figure 2.35 Graph of the arithmetic expression $X = Y$.

2.5.1 Graphing Boolean Expressions

We can graphically represent algebraic expressions on multiple axes, one for each variable. For example, the arithmetic expression $X = Y$ can be graphed on X and Y axes as a line as shown in Figure 2.35. The Boolean expression $X = Y$ can be graphed similarly, but with some important differences, namely, X and Y do not take on continuous values along the axes but can only take on the values 0 and 1. Therefore, rather than a line we have two points: one at 0,0 and one at 1,1 as shown in Figure 2.36.

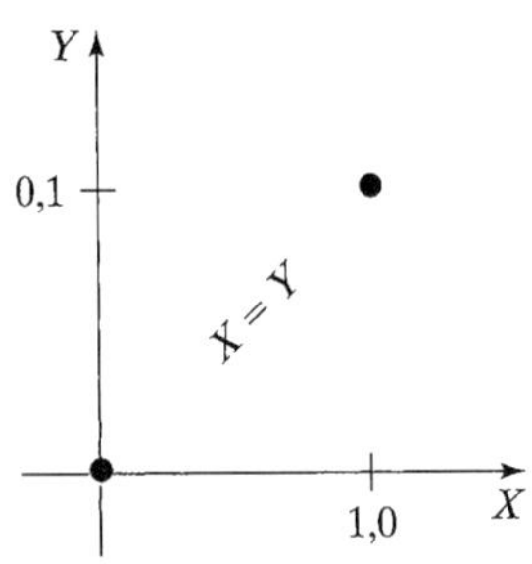

Figure 2.36 Graph of the Boolean expression $X = Y$.

This analogy can continue on to 3-dimensional spaces. In that case, the three axes will define the points of a cube on which our Boolean expressions will take on a true or false value.

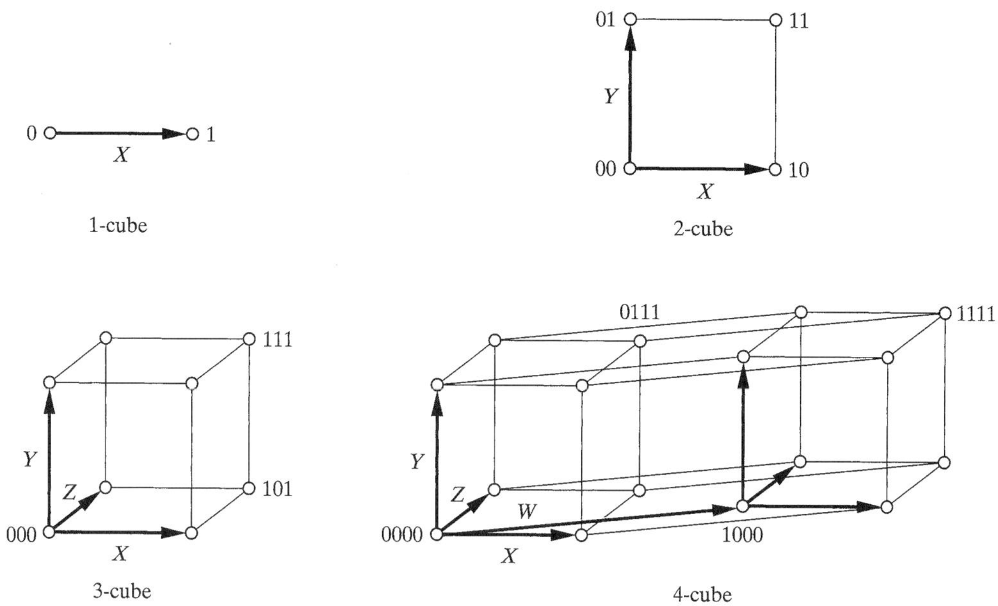

Figure 2.37 **Some n-dimensional cubes.**

2.5.2 Boolean Cubes

A *cube* is usually defined as a solid object bounded by six equal squares. This concept can be generalized to other than three dimensions. A 2-dimensional cube is a square, a 1-dimensional cube is a line, and a 0-dimensional cube is a point.

We can represent the truth table of an n-input Boolean function as a "cube" in an n-dimensional Boolean space. There is one axis for each variable, over which it can take on exactly two values: 0 or 1. For example, the cube of Figure 2.36 is a 2-dimensional cube where the value at the points 0,0 and 1,1 are 1 and the values at 0,1 and 1,0 are 0.

Figure 2.37 shows the form of Boolean 1-, 2-, 3-, and 4-cubes. Each node in the figure is labeled with its coordinates in the n-dimensional space. The structure generalizes beyond four dimensions, but it is rather hard to visualize these complex cubes.

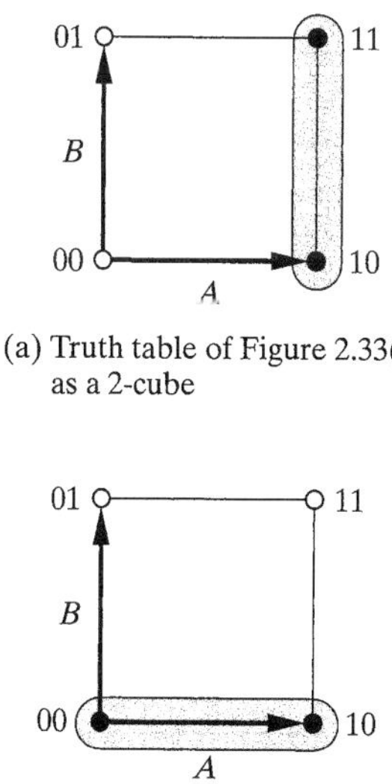

(a) Truth table of Figure 2.33(a) as a 2-cube

(b) Truth table of Figure 2.33(b) as a 2-cube

Figure 2.38 **Examples of mapping truth tables onto cubes.**

EXAMPLE 2.8 MAPPING TRUTH TABLES ONTO CUBES

Now let's examine how to map a truth table onto a cube, using the simple examples of Figure 2.34. The elements of the on-set are represented by black nodes and those of the off-set by white nodes (don't cares are represented by X nodes, although we do not have any in this example).

Figure 2.38 shows the mapping. Observe that the elements of the functions' on-sets are next to each other in the truth table's Boolean cube. This tells you that the Uniting theorem can be used to eliminate a variable. In the figure, we have circled elements of the on-set that are directly adjacent. We call such a circled group of nodes an *adjacency plane*. Each adjacency plane corresponds to a product term. In Figure 2.38(a), the circled nodes yield the term A; in Figure 2.38(b), the term is $\overline{B}$.

You can think about these adjacencies as distances between nodes in the Boolean cube. Two nodes are distance 1 apart if they are connected by an edge in the cube. They are distance 2 apart if they are separated by a path of two connected edges. If this is the case, the two nodes are on the same plane. Two nodes are distance 3 apart if they are separated by a path of three connected edges and no shorter path exists between the two nodes. Then the nodes are in the same 3-dimensional cube.

In the on-set/adjacency plane of Figure 2.38(a), A's value stays 1 while B's varies between 0 and 1. This is the clue that the uniting theorem can reduce the function to the single literal A. Similarly, in Figure 2.38(b) the adjacency plane is circled, and the nodes involved have B retaining its value while A varies. The Uniting theorem reduces the expression to the term $\overline{B}$.

As an example of a 3-variable function and its mapping onto a 3-cube, let's return to the full adder's carry-out function examined in Section 2.2.2. The truth table and its mapping onto a 3-cube are shown in Figure 2.39. The on-set is arranged on three 1-dimensional adjacency planes: the edges containing the nodes 011-111, 101-111, and 110-111. In the first segment, along the top of the back side of the cube, A varies between 0 and 1 while B and C_{in} remain asserted and unvarying. This reduces to the term BC_{in}. For the second segment, the vertical one at the back right, B varies, yielding the term AC_{in}. In the final segment, C_{in} varies, and the resulting term is $A\,B$. The final expression becomes

$$C_{out} = BC_{in} + AC_{in} + AB$$

This method lets us obtain the final expression much more quickly than using Boolean algebra. Note that each adjacency plane contributes one product term to the final expression. Since each plane is an edge in a 3-dimensional cube, the two 3-variable minterms in the plane are reduced to a single 2-variable product term.

A	B	C_{in}	C_{out}
0	0	0	0
0	0	1	0
0	1	0	0
0	1	1	1
1	0	0	0
1	0	1	1
1	1	0	1
1	1	1	1

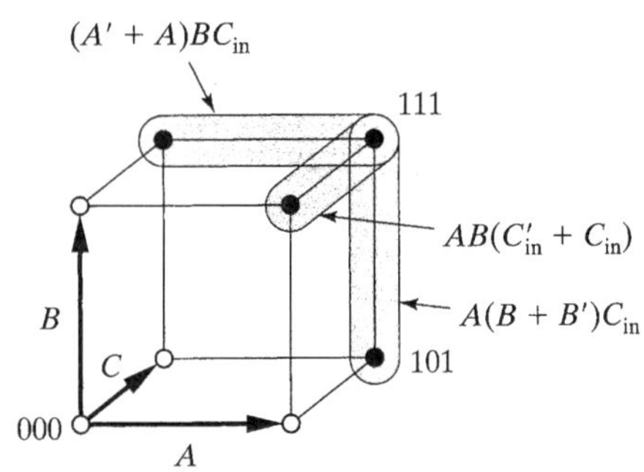

Figure 2.39 Full-adder carry-out truth table in a 3-cube.

EXAMPLE 2.9 ADJACENCIES OF HIGHER DIMENSIONS

What about adjacency planes of higher dimensionality than single edges such as a 2-dimensional plane within a 3-cube? Consider the function depicted in the 3-cube of Figure 2.40. One entire face of the cube is included in the on-set. Intuitively, we should expect this surface to reduce to the single literal A because all other variables vary between 0 and 1 within the surface except A which is true at all four vertices.

Let's verify that this is the case. The four line segments on the surface are denoted by the nodes 110–111, 101–111, 100–101, and 100–110. Applying the uniting theorem to each segment independently yields the terms: AB, AC, $A\bar{C}$ and $A\bar{B}$, respectively. We can continue to apply the Uniting theorem:

$$AB + AC + A\bar{B} + A\bar{C} = A(B + \bar{B}) + A(C + \bar{C}) = A + A = A$$

In the 3-cube, if the on-set completely covers a 2-dimensional plane, that plane contributes a single 1-variable product term to the expression for the function.

For the 3-cube, the relationship between the dimensionality of the adjacency plane and the term it represents is the following:

- A 0-dimensional adjacency plane, a single node, yields a three-literal product term. For example, 101 = $A\bar{B}C$. This is the same as a minterm.
- A 1-dimensional plane, an edge, yields a two-literal product term. For example, 100-101 = $A\bar{B}$.
- A 2-dimensional plane yields a one-literal product term. For example, 100-101-111-110 = A.
- A 3-dimensional plane, the whole cube, yields a term with no literals in it; that is, it reduces to the constant logic 1.

This generalizes to cubes of higher dimensions. An m-dimensional adjacency plane within an n-dimensional cube will reduce to a term with $n - m$ literals.

The fewer planes needed to include all of the 1s in the truth table, the fewer the terms in the function's final expression. Planes of higher dimensionality generate terms with fewer literals in them. Thus, computer-aided design algorithms for minimization attempt to find the smallest number of the highest dimensionality adjacency planes that contain all the nodes of the function's on-set (perhaps including elements of the don't-care set as well). This process is called finding a *minimum cover* of the function.

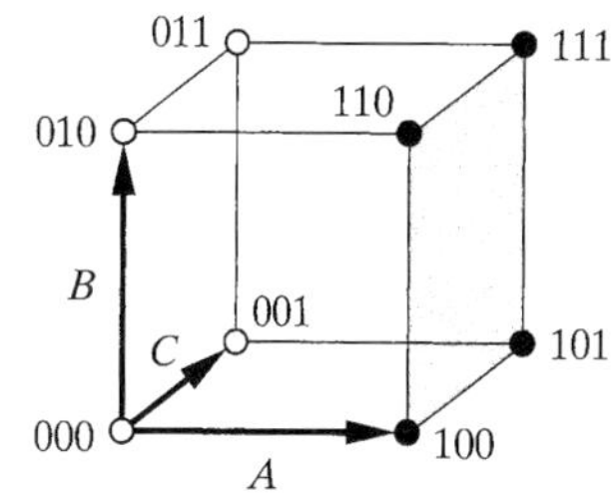

Figure 2.40 Higher dimensions of adjacency.

2.5.3 Karnaugh Maps

The cube notation makes obvious the adjacencies among truth table rows. Adjacencies provide visual clues as to where to apply the uniting theorem to elements of the on-set that are clustered in groups of 2 (a line), 4 (a square), 8 (a cube), etc. within the n-dimensional cube that represents the function. The problem for humans is the difficulty of visualizing adjacencies in more than three dimensions. To circumvent this problem, at least for expressions up to six variables, we introduce an alternative reformulation of the truth table called the Karnaugh map or *K-map*. The problem of projecting multi-dimensional spaces onto a 2-dimensional surface is not new. Cartographers have been dealing with the problem ever since people started using maps. Figure 2.41 shows a globe representing the 3-dimensional Earth. The flat map next to it is a Mercator projection of the surface of the globe onto two dimensions. Notice how features are distorted. However, things that were close on the globe are still close on the map. We do need to consider that the left and right edges of the map wrap around and connect. The top and bottom edges come together at points corresponding to the two poles.

General Concept A K-map for a Boolean function specifies values of the function for every combination of the function's variables. Figure 2.42

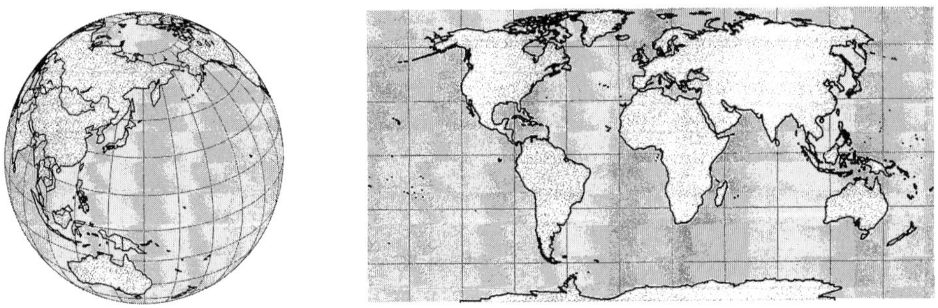

Figure 2.41 Mercator projection of the Earth. Similarly to the K-map, we need to consider its left and right edges connected. Its top and bottom edge come together at two points: the poles.

B \ A	0	1
0	0	2
1	1	3

C \ AB	00	01	11 (A)	10 (A)
0	0	2	6	4
1 (C)	1	3	7	5
		B	B	

CD \ AB	00	01	11 (A)	10 (A)
00	0	4	12	8
01 (D)	1	5	13	9
11 (C, D)	3	7	15	11
10 (C)	2	6	14	10
		B	B	

Figure 2.42 Two-, three-, and four-variable K-map templates.

shows K-map templates for two-, three-, and four-variable Boolean functions. The 2-dimensional K-map is very similar to the 2-dimensional Boolean cube in appearance. Roughly, one can think of the 3-dimensional K-map as an unfolded cube much in the same way as the Mercator projection is an unfolding of the surface of Earth's sphere. We must remember to keep in mind that the cells at one edge are adjacent to cells on the opposite edge. Unlike the Mercator map the top and bottom edges are also adjacent. The entries are labeled with the decimal equivalent of the binary number formed by joining the column with the row indices. For example, entry 3 in the 3-variable K-map is labeled by the column $AB = 01$ and the row $C = 1$ ($ABC = 011_2 = 3$). The labels are included only for your convenience in filling in the K-map from the truth table: they correspond to the row (or minterm) number of the associated truth-table entry.

The only surprising thing is the ordering of the indices: 00, 01, 11, 10. This is called a *Gray code*. It has the property that, when advancing from one index to the next adjacent index, only a single bit changes value. This property is not true for the standard numerically sequential binary sequence: 00, 01, 10, 11.

The structure of the K-map is chosen so that any two adjacent (horizontal or vertical, but not diagonal) elements are distance 1 apart in the equivalent cube representation (they share a common edge). This is shown in Figure 2.43 for a 3-variable K-map and 3-dimensional Boolean cube. Note that K-map square 0 is adjacent to squares 1, 2, and 4. The K-map actually folds back on itself in each dimension. The elements in the rightmost column are adjacent to the elements in the leftmost column; the elements in the top row are adjacent to the elements in the bottom row. Note that in this case there are only three adjacent cells for cell 0 but four directions. Cell 1 is counted twice because it is adjacent to 0 from above and below. Note that for this K-map we did not show the variable values at the top of each column and side of each row. Instead, we used a short-hand notation using the bold bars, one for each variable, that indicate the region of the map where the variable is true. Thus, A is 1 in the right half of the map and 0 in the left. B is 1 in the middle two columns and C is 1 in the bottom half.

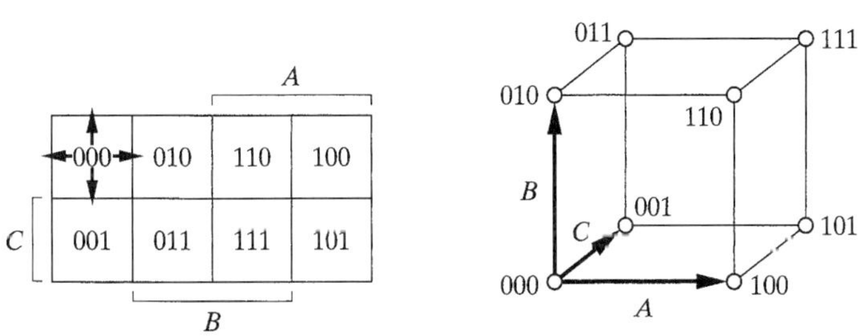

Figure 2.43 K-map adjacencies.

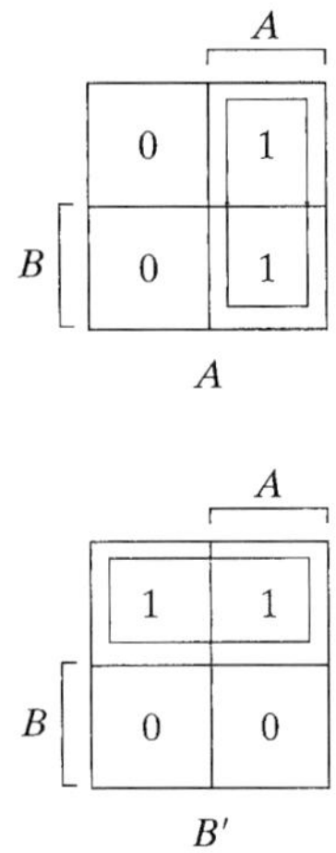

Figure 2.44 **Two-variable maps for the example functions of Figure 2.34.**

EXAMPLE 2.10 TWO-VARIABLE MAPS

Figure 2.44 shows the two-variable maps for the example functions F and G of Figure 2.34. The K-map is filled in directly from the truth table. Each truth table row corresponds to a K-map entry. The values of the variables index into a cell in the map, and the truth table's output value is placed into the K-map's cell with that corresponding index.

On the left in Figure 2.44, the terms of the function are $A\overline{B} + AB$ are denoted by the 1s in the $A = 1$, $B = 0$ and $A = 1$, $B = 1$ map entries. We can apply the Uniting theorem to reduce this to the single literal A. The K-map shows immediately that the two entries are adjacent. The A variable value remains unchanged while the B value varies from 0 to 1. Looking at this group should tell you that the B term can be eliminated, leaving us with A.

The same analysis holds for the right side of Figure 2.44. The function is $\overline{A}\overline{B} + A\overline{B}$, and its on-set is row adjacent in the K-map. This demonstrates the advantage of the K-map over the truth table for recognizing adjacencies. A varies from 0 to 1 while B holds at 0 for this K-map row. We can reduce the function to the single literal $\overline{B}$.

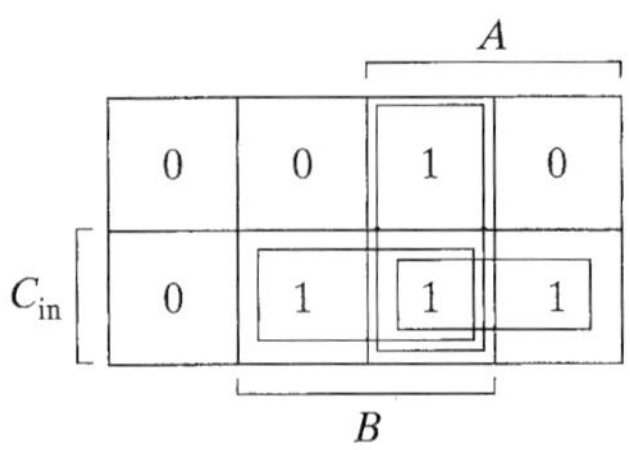

Figure 2.45 **K-map for truth table of Figure 2.39.**

EXAMPLE 2.11 THREE-VARIABLE MAPS

Figure 2.45 shows the three-variable K-map for the full adder carry-out function of Figure 2.39. You can see that three different two-element adjacencies cover the on-set (recall that adjacency is not defined for diagonal entries). The first is the column indexed by $A = 1$, $B = 1$. Since C_{in} varies from 0 to 1, that variable can be eliminated, yielding the term AB. The second is the adjacency indexed by $A = 0$, $B = 1$, $C_{in} = 1$ and $A = 1$, $B = 1$, $C_{in} = 1$. A varies while B and C_{in} remain unchanged, yielding the term BC_{in}.

The final adjacency is indexed by $A = 1$, $B = 1$, $C_{in} = 1$ and $A = 1$, $B = 0$, $C_{in} = 1$. B varies and A and C_{in} remain unchanged, resulting in the term AC_{in}. Once again, the labeled bar at the top of the K-map reminds us that A remains unchanged within the last two columns and $\overline{A}$ in the first two columns. The final expression is $AB + BC_{in} + AC_{in}$. There is one term in the reduced expression for each circled adjacency group in the K-map.

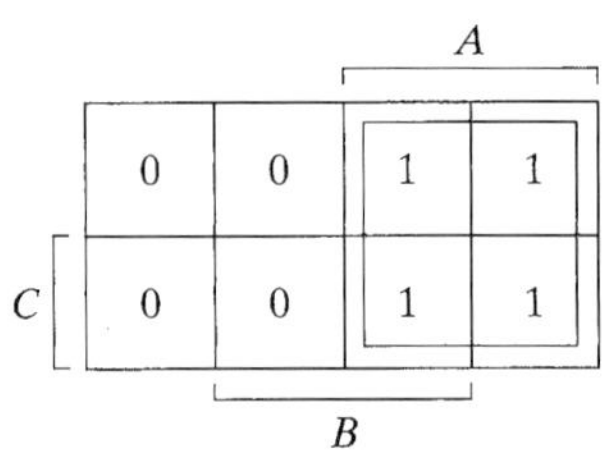

Figure 2.46 **K-map for truth table of Figure 2.40.**

Let's revisit the function of Figure 2.40. Its K-map is given in Figure 2.46. The four elements of the on-set are adjacent, and we can circle them all. Within this grouping, both B and C vary while A remains asserted, reducing to the single literal A.

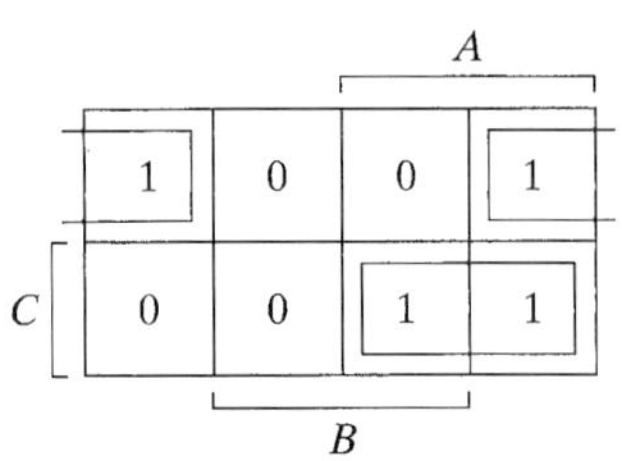

Figure 2.47 **K-map for $f = \Sigma\, m(0,4,5,7)$.**

Another case of adjacency is illustrated by Figure 2.47, which shows the K-map for the function $F(A, B, C) = m_0 + m_4 + m_5 + m_7$. Recall that the leftmost and rightmost columns of the K-map are adjacent. Thus, we can combine $m_0(\overline{A}\overline{B}\overline{C})$ and $m_4(A\overline{B}\overline{C})$ to form the term $\overline{B}\overline{C}$. We also can combine m_5 and m_7 to form $F = AC + \overline{B}\overline{C}$.

You might be tempted to circle the terms m_4 and m_5, as they are also adjacent. But you obtain the most reduced solution only by finding the smallest number of the largest possible adjacency groups that completely cover the on-set. Recall that the number of groups equals the number of product terms, and larger groupings have a smaller number of literals. The term formed from m_4 and m_5 is redundant, because both entries already are covered by other terms. We will become more formal about the process of obtaining the minimum solution a bit later on in this section.

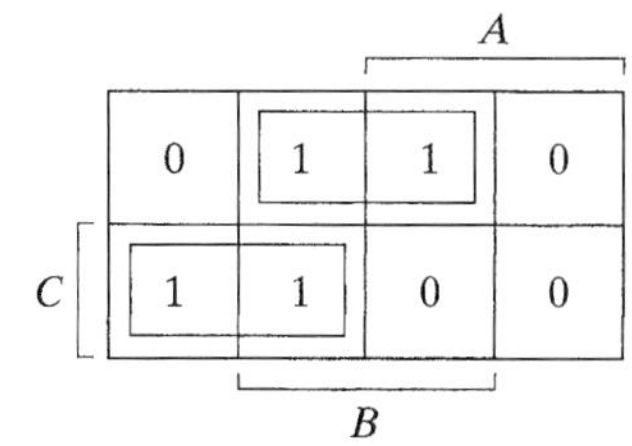

Figure 2.48 Complement of Figure 2.47.

K-maps provide a good mechanism for finding the reduced form of the complement very quickly. Figure 2.48 contains the K-map for the complement of Figure 2.47. All we have done is to replace the 0s with 1s and vice versa. The complement can be read out immediately as $\overline{F} = B\overline{C} + \overline{A}C$. Contrast this method with the method using Boolean algebra:

$F = AC + \overline{B}\,\overline{C}$	
$\overline{F} = \overline{AC + \overline{B}\,\overline{C}}$	
$= \overline{(AC)}\,\overline{(\overline{B}\,\overline{C})}$	DeMorgan's law
$= \overline{(AC)}(\overline{\overline{B}}\,\overline{\overline{C}})$	DeMorgan's law
$= (\overline{A} + \overline{C})(B + C)$	DeMorgan's law
$= \overline{A}B + \overline{A}C + B\overline{C} + C\overline{C}$	Distribution law
$= \overline{A}B + \overline{A}C + B\overline{C}$	Complement and identity
$= \overline{A}B(C + \overline{C}) + \overline{A}C + B\overline{C}$	Complement and identity
$= \overline{A}BC + \overline{A}C + \overline{A}B\overline{C} + B\overline{C}$	Distribution and commutivity
$= \overline{A}C + \overline{A}B\overline{C} + B\overline{C}$	$X + XY = X$, with $X = A\overline{C}$, $Y = B$
$= \overline{A}C + B\overline{C}$ √	$X + XY = X$, with $X = B\overline{C}$, $Y = \overline{A}$

The K-map yields the result much more quickly!

EXAMPLE 2.12 FOUR-VARIABLE MAPS

Now let's consider a 4-variable function $F(A,B,C,D) = \Sigma m(0,2,3,5,6,7,8,10,11,14,15)$. The K-map is shown in Figure 2.49. Remember that the strategy is to cover the on-set with as few groups as possible, so we must try to find large groups of adjacency. Also, don't forget that the number of elements in an adjacency group is always a power of 2 and the grouping does form a cube in some number of dimensions.

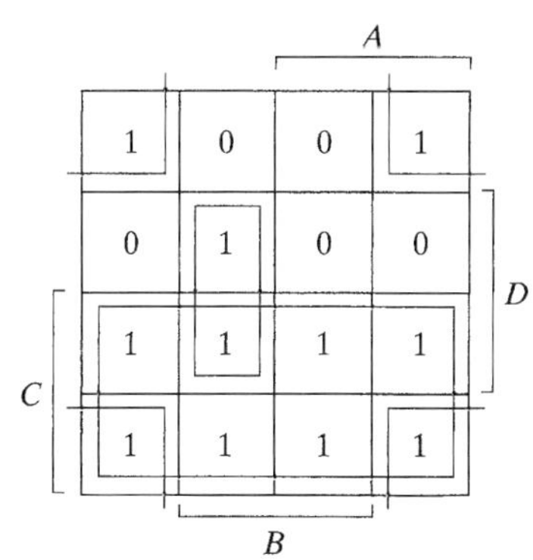

Figure 2.49 K-map of a 4-variable function.

The elements m_2, m_3, m_6, m_7, m_{10}, m_{11}, m_{14}, m_{15} form an adjacency group of eight in the bottom half of the map. This collapses to the single literal C. (Recall that a 3-dimensional plane within a 4-dimensional cube yields a term with $4 - 3 = 1$ literal.) The elements m_5 and m_7 result in the term $\overline{A}BD$ (a 1-dimensional plane in a 4-dimensional space results in a term with $4 - 1 = 3$ literals). The final grouping involves the corner terms: m_0, m_2, m_8, m_{10}. To see this adjacency, you must remember that the map folds around in both dimensions.

Examining Figure 2.49 should make this clearer. In the figure, look for the minterm indices 0000, 0010, 1000, and 1010. These corner elements in the K-map reduce to the term $\overline{B}\overline{D}$ (a 2-dimensional plane within a 4-dimensional space results in a term with $4 - 2 = 2$ literals). Figure 2.50 shows this 2-dimensional sub-cube shaded inside the larger 4-cube. The entire top half of the 4-dimensional cube is the 3-dimensional cube described as C. The 1-dimensional cube consisting of indices 0101 and 0111 completes the cover. The minimized form of the function is

$$F = C + \overline{A}BD + \overline{B}\overline{D}$$

Finding the Minimum Product-of-Sums Form The K-map can also be used to find a function's minimum product of sums expression. In this case we search for elements of the off-set, simply circling the maximal adjacent groups of 0s. We interpret the indices in a fashion complementary to the procedure for finding the minimum sum of products expression. If the variable that is left unchanged in a grouping of 0s has an index of 0, then that variable contributes an asserted literal to the term. If the index is 1, it contributes a complemented literal.

This method works because we begin by solving for the function's complement in sum of products form, by circling the 0s. Then we apply DeMorgan's law to get a product of sums expression by interpreting the indices as complements. Let's look at an example.

Let's reconsider the K-map in Figure 2.49. $\overline{F}$ in minimum sum of products form is found by circling the K-map's 0s rather than its 1s: $\overline{F} = B\overline{C}\overline{D} + A\overline{C}D + \overline{B}\overline{C}D$. By applying DeMorgan's law, we get F in the product-of-sums form:

$$\overline{F} = B\overline{C}\overline{D} + A\overline{C}D + \overline{B}\overline{C}D$$

$$F = (\overline{B} + C + D)(\overline{A} + C + \overline{D})(B + C + \overline{D})$$

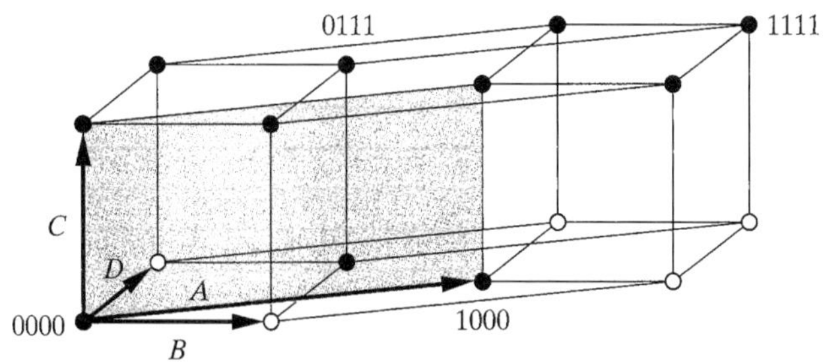

Figure 2.50 **The adjacency of the four corner squares of the K-map when viewed in the 4-cube.**

Figure 2.51 shows the same K-map as Figure 2.49, but this time with the 0s circled. You can see that there are three groups of two 0s each that are needed to cover the function's complement. Since these are 1-dimensional adjacency groups in a 4-dimensional space, there are three literals in each term. The term formed from M_4 and M_{12} is $(\bar{B} + C + D)$: B, C, and D remain unchanged and B's index is 1 while C's and D's are 0. This is just shorthand for applying DeMorgan's law to the $B\bar{C}\bar{D}$ term of the complement. The terms formed from M_1, M_9 and M_9, M_{13} are $(B + C + \bar{D})$ and $(\bar{A} + C + \bar{D})$, respectively. Each term is ANDed to form the final expression (same as the above).

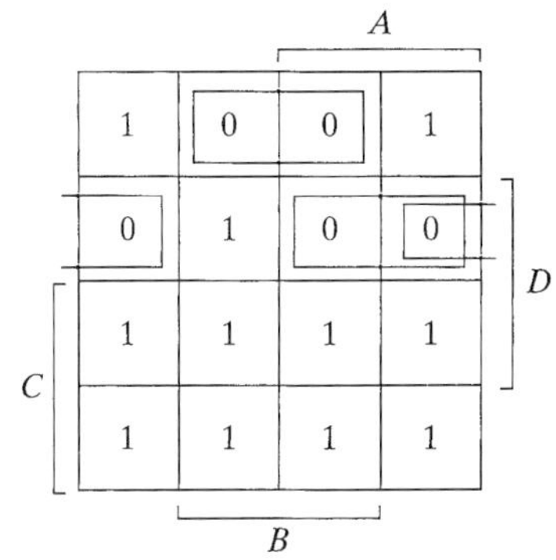

Figure 2.51 **K-map of a 4-variable function.**

Don't Cares in K-Maps The last wrinkle we consider is the use of don't cares within the K-map. Figure 2.52 shows a K-map for the function $F(A,B,C,D) = \Sigma\, m(1,3,5,7,9) + \Sigma\, d(6,12,13)$. The group of four elements reduces to $\bar{A}D$. If we assume that the Xs are all 0, we can cover the remaining member of the on-set with the term $\bar{B}\bar{C}D$. However, if we assume that the element d_{13} is 1 (while d_6 and d_{12} are 0), we can form a larger adjacency group that yields the term $\bar{C}D$. This has fewer literals. Thus, by appropriately choosing the values of the don't cares, we can maximize the size of the adjacency groups. The expression for F becomes $\bar{A}D + \bar{C}D$.

In product-of-sums form, the shorthand expression is written as $F(A,B,C,D) = \Pi\, M(0,2,4,8,10,11,14,15) \bullet \Pi\, D(6,12,13)$. Figure 2.53 shows the groupings we use to find the minimum product-of-sums form. We form a group of eight 0s (remember that the top and bottom rows are adjacent) and one of four 0s by judicious use of don't cares (D_6, $D_{12} = 0$, $D_{13} = 1$). The expression for F becomes $D(\bar{A} + \bar{C})$.

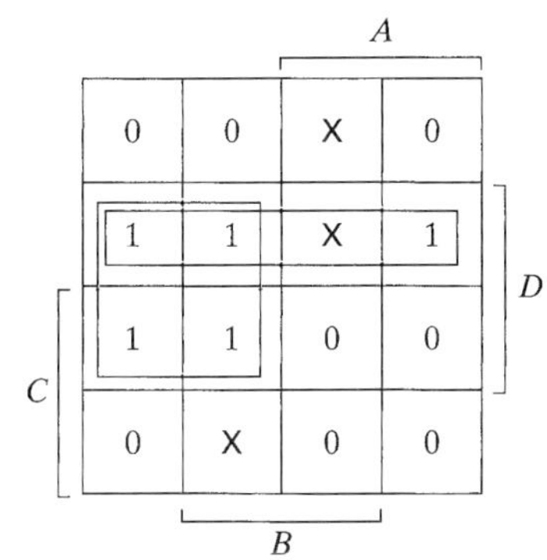

Figure 2.52 **Use of don't-care entries in the K-map.**

Let's now revisit the function C_2 for the BCD display of Section 2.3.5. Recall that the expression for C_2 derived from the truth table of Figure 2.25 was:

$$C_2 = m_0 + m_1 + m_3 + m_4 + m_5 + m_6 + m_7 + m_8 + m_9 + d_{10} + d_{11} + d_{12} + d_{13} + d_{14} + d_{15}$$

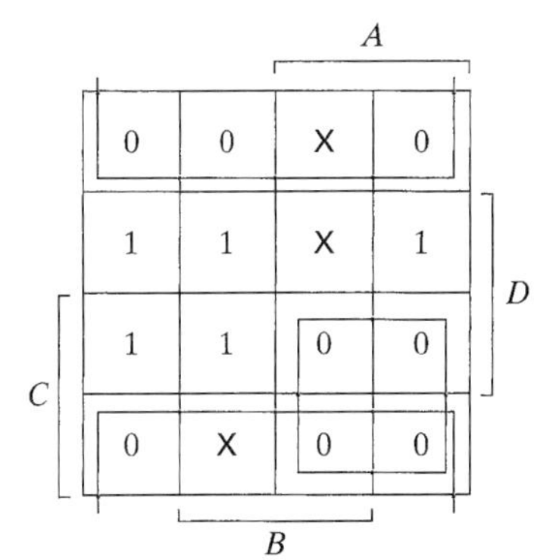

Figure 2.53 **Groupings for product of sums form.**

Through a lot of manipulation of Boolean expressions, we were able to simplify it to the following:

$$C_2 = \bar{A}B + \bar{B}\bar{C} + \bar{A}D$$

However, we did not take advantage of the many don't cares that are available. Let's see if Karnaugh maps can help us achieve the same result and improve upon it. Two K-maps for C_2 are shown in Figure 2.54. The one on the left shows a cover that does not use don't cares. This corresponds exactly to the expression we derived laboriously using Boolean algebra. The K-map on the right shows that we can simplify the expression even further and much more easily (to only three literals instead of six and utilizing only one 3-input OR

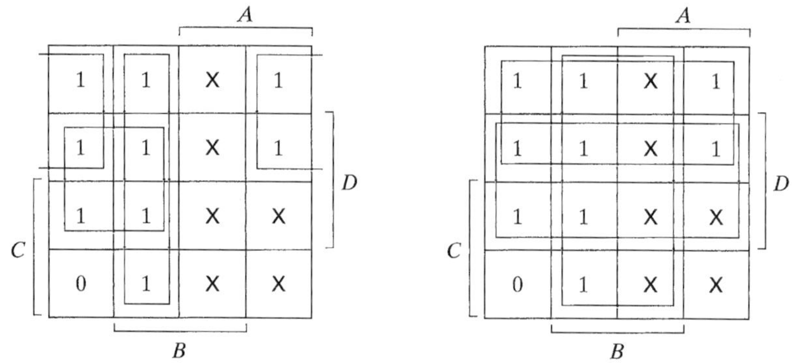

Figure 2.54 K-maps showing covers with and without the use of don't cares for the output C_2 of Figure 2.25.

gate) by exploiting the don't cares available to us. The more minimized expression is:

$$C_2 = B + \bar{C} + D$$

Note that although it would have been possible to derive this much more minimized form by using the don't cares in our Boolean manipulations, it would have been considerably more difficult, especially when the potential end result is not known. K-maps can save a lot of time when we deal with functions of only a few variables. For larger numbers of inputs we will require automated methods.

2.6 Multilevel Logic

At this point, given a Boolean function expressed in minterm or maxterm canonical form, we now have a pretty good idea as to how we can reduce it into a minimal two-level form with the fewest terms and literals. In this section, we'll see that we can often do better than a two-level form. By better, we mean a smaller number of gates with a smaller number of inputs to each gate. We can achieve this improvement if we abandon the restriction of just two levels of logic (AND-OR or OR-AND) and instead consider an arbitrary number of levels. We'll see later that this greatly complicates the process of minimizing functions but it usually leads to smaller and faster implementations.

Let's begin by considering the function $Z(A, B, C, D, E, F, G)$:

$$Z = ADF + AEF + BDF + BEF + CDF + CEF + G$$

It is already in its minimal sum of products form. Its implementation as a two-level network of AND and OR gates requires six 3-input AND gates and one 7-input OR gate, a total of seven gates and 19 literals (see Figure 2.55(a)).

We can do better if we replace the two-level form with a so-called *factored form*. We express the function with common literals factored out from the product terms wherever possible.

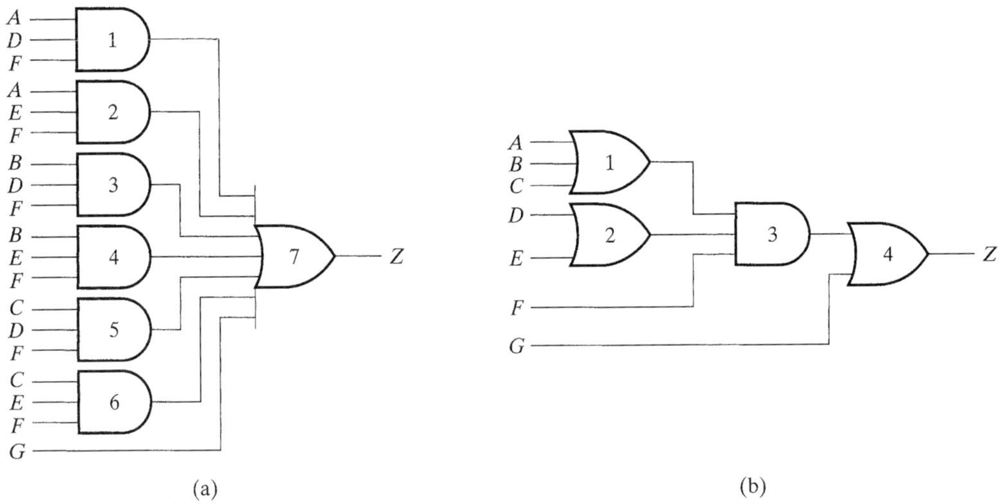

Figure 2.55 Multilevel circuit implementation.

By recursively factoring out common literals, we can express the function Z as:

$$Z = (AD + AE + BD + BE + CD + CE)F + G$$
$$Z = [(A + B + C)D + (A + B + C)E]F + G$$
$$Z = (A + B + C)(D + E)F + G$$

Expressed as a series of expressions, each in the two-level form, Z becomes

$$Z = XYF + G$$
$$X = A + B + C$$
$$Y = D + E$$

When written this way, the function requires one 3-input OR gate, two 2-input OR gates, and a 3-input AND gate for a total of four gates and nine literals. The intermediate functions X and Y count as literals in the final expression for Z.

The implementation of the factored form is shown in Figure 2.55(b). The number of wires and gates needed to implement the function is significantly reduced, but this implementation may have worse delay because of the increased levels of logic. On the other hand, the 7-input OR gate of the two-level implementation is likely to be quite slow. In general, it is difficult to tell a priori whether a two-level or multilevel implementation will be faster or smaller. However, multilevel implementations tend to use gates with fewer inputs (because expressions are factored), which tend to be faster; however, their multiple levels of logic can add up to a larger combined delay than a two-level implementation with two levels of larger and slower gates.

EXAMPLE 2.13 FULL ADDER

We have seen multilevel logic already in this chapter in our discussion of the full-adder implementation in Section 2.2. The full adder of Figure 2.19 is implemented using two half adders and an OR gate. If we show the gate logic for this circuit by looking into the half-adder boxes, the circuit would be as shown in Figure 2.56(a). If we remove the XOR gates and use only AND and OR gates, we obtain the circuit of Figure 2.56(b), which we will use as the reference multilevel design for the remainder of the discussion. Note that there are five levels of logic (in the worst case and not counting inverters at the inputs) from the inputs to the outputs. This is the case for the *Sum* output, C_{out} has four levels. Figure 2.57 shows the two-level implementation for the full adder taken directly from the equations of Section 2.2:

$$Sum = (A\bar{B}\bar{C}_{in}) + (\bar{A}B\bar{C}_{in}) + (\bar{A}\bar{B}C_{in}) + (ABC_{in})$$
$$C_{out} = (AC_{in}) + (BC_{in}) + (AB)$$

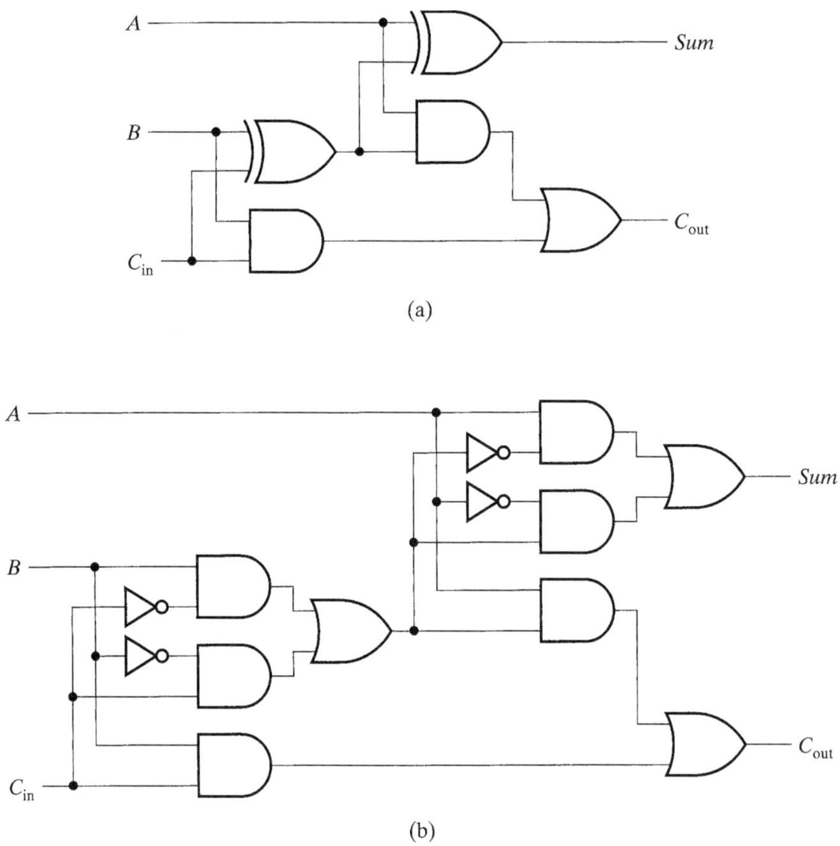

Figure 2.56 **Multilevel circuit implementation of the full adder: (a) using XOR gates, and (b) not using XOR gates.**

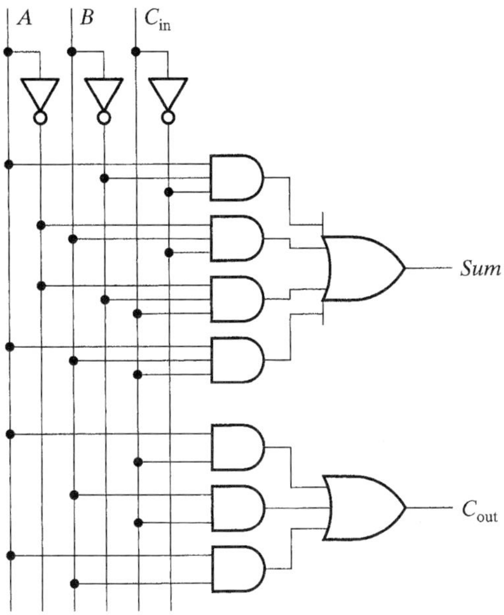

Figure 2.57 Two-level circuit implementation of the full adder.

There is quite a bit difference between the two implementations. A count of gates shows that the multilevel implementation uses three 2-input OR gates, six 2-input AND gates, and four inverters for a total of 13. The two-level implementation uses one 4-input OR gates, one 3-input OR gates, four 3-input AND gates, three 2-input gates, and three inverters, for a total of 12. If we do not count inverters used at the inputs, then the multilevel implementation uses 10 gates (one of which is an inverter), while the two-level implementation uses nine gates. In terms of wires, the multilevel implementation has 22 wires while the two-level implementation has 28. The number of wires is computed by simply counting the inputs to every gate in each circuit (inverters included).

But gate and wire counts don't tell the entire story. Note that all the gates in the multilevel implementation have (at most) 2 inputs, while in the multilevel implementation we have many 3-input gates as well as one 4-input OR gate. The larger fan-in gates will be slower and their propagation delay increases more than linearly with the number of inputs. Therefore, it is a good rule-of-thumb to not use gates with more than four inputs. Remember that this is only a rule-of-thumb, and in the real circuit, there are many other factors that affect the speed of a gate.

Let's revisit our circuits and see which may be faster than the other. The multilevel implementation has a worst-case delay for its *Sum* output through four 2-input gates and an inverter (again, we don't count input inverters as it could easily be the case that the complement of input signals are readily available and we may not, in fact, need the input inverters). We can say that the delay will be proportional to four basic gate delays plus one inverter delay. For the two-level implementation, the worst-case delay (also for the Sum output) involves a 3-input gate and one 4-input gate. Typically, a 3-input gate is at least twice as slow as a 2-input gate, and a 4-input gate is at least another factor of two slower than that (approximately two and four basic gate delays).

In summary, for this example, the multilevel implementation seems to win in terms of delay and number of wires while the two-level implementation uses less gates. When designers have to make a choice, these are some of the comparisons they consider, however, there are others such as the parts that are actually available for constructing the circuits. We'll discuss the technologies and parts available in Chapter 4.

2.7 Motivation for Multilevel Minimization

We can obtain minimized two-level logic networks from the canonical sum of products or product of sums form, by applying Boolean algebra, as well as other reduction methods we will introduce in the next chapter. Signal propagations can be fast, because no signal has to travel through more than two gate levels (not counting zeroeth-level inversions). As we have seen from the full-adder example of the previous section, the drawback is the potential for large gate fan-ins, which can reduce gate performance and increase circuit area in some technologies. In many technologies, gate-level building blocks with more than four inputs are rare. Most only have two or three. Large fan-in gates are found in programmable two-level logic but in a limited way, namely, to form product terms of many inputs. Field-programmable gate arrays, larger and more general programmable logic components, also limit the number of inputs to a logic function for two common reasons: to make the programmable logic blocks smaller and to use less wiring resources to interconnect signals between logic blocks. Thus, in virtually all practical gate-level realizations of a logic network, large fan-in gates must be replaced by a multiple-level network of smaller fan-in gates. This motivates much of the interest in automatic multilevel logic optimization.

An optimal two-level network uses the smallest number of product terms and literals to realize a given truth table. It is not so easy to define optimality for multilevel networks. Is the network optimal if it has the smallest number of gates? Or is it more critical to have the fewest literals in the resulting expression (corresponding to number of wires)? Or is the number of inputs per gate a major consideration in the technology being used?

To get a taste for how to simplify multilevel expressions, we will discuss a useful way to represent such expressions, and the kinds of operations that can be applied to them to achieve their simplification. The actual methods used by modern computer-aided design tools will be described in the next chapter.

2.7.1 Factored Forms

The standard way to represent a multilevel equation is in a *factored form*. Simply stated, this is an expression that alternates between AND and OR

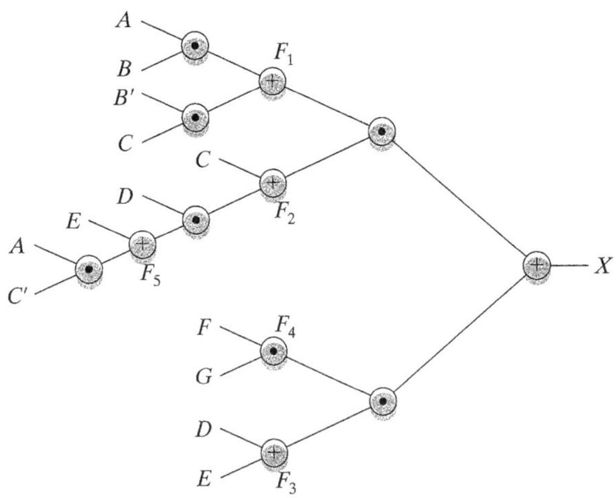

Figure 2.58 Graphical representation of a factored form with simple nodes.

operations, a kind of "sum-of-products of sums-of-products . . . ." The following function is in factored form:

$$X = (AB + \bar{B}C)[C + D(E + A\bar{C})] + (D + E)(FG)$$

Examining the expression, we see that no further subexpressions can be factored out. For the purpose of counting literals, the factored form for X can be rewritten as a sequence of two-level expressions:

$$\begin{aligned} X &= F_1F_2 + F_3F_4 & F_3 &= D + E \\ F_1 &= (AB + \bar{B}C) & F_4 &= FG \\ F_2 &= C + DF_5 & F_5 &= E + A\bar{C} \end{aligned}$$

The structure of the expression is a little clearer if it is represented in the form of a graph or tree, where the "leaves" represent literals and the internal nodes represent either an AND or OR operation. This graphical representation is shown in Figure 2.58. For the most part, the ANDs and ORs alternate between adjacent nodes of the tree.

2.7.2 Criteria for Multilevel Simplification

In modern logic families, designers have observed that gates (internal nodes of the graph of Figure 2.58) require relatively little circuit area but connections (edges of the graph) use significant area. Stated in a different way, the implementation complexity is strongly related to the number of wires used to construct a circuit. Because the number of internal connections scale with the number of literals, it is a reasonable strategy to attempt to minimize the number of literals. This can be accomplished, in part, by having more complex expressions at each of the nodes in the graph and choosing them judiciously so that they are

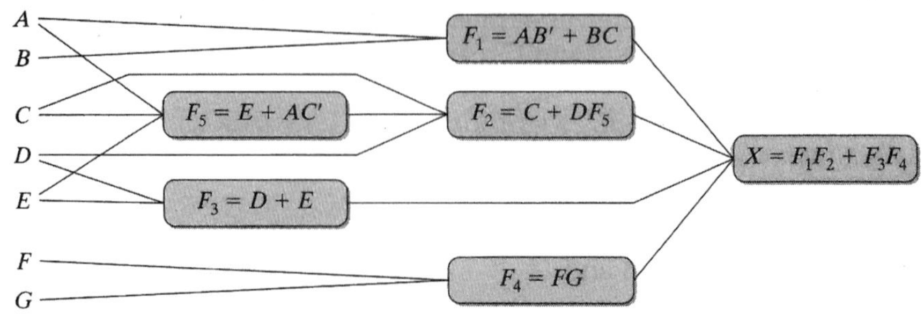

Figure 2.59 **Graphical representation of a factored form with complex nodes.**

likely to be reused. Figure 2.59 shows the corresponding complex graph for the particular intermediate functions chosen previously.

To count the number of literals in a multilevel expression, simply add up the number of literals found in its equivalent implementation in terms of two-level expressions. For the function X, as depicted in Figure 2.59, the literal count is 18. Note that when they are referenced in expressions, the intermediate functions F_1, F_2, F_3, F_4, and F_5 count as literals. One of the important choices to be made in multilevel optimization is how to group the nodes into intermediate expressions that can be reused multiple times.

Why is this factored form useful? It has 18 literals instead of our original 13. Won't the realization of the factored form have more wires? In answering these questions we have to keep one important thing in mind: the technology we are using to realize our circuit. Our original expression may have had fewer literals but there may not be building blocks available that can build it directly. We may be forced to use expressions with a larger total number of literals just to make it possible to build the circuit. This is referred to as the *technology-dependent* optimization step and it can have a major impact on the number of gates, wiring complexity, and delay of our circuit.

In the next chapter, we will provide an overview of an approach to multilevel logic minimization. There are five basic *technology-independent* operations for manipulating multilevel networks: (1) decomposition, (2) extraction, (3) factoring, (4) substitution, and (5) collapsing. We will describe each of these operations and illustrate how each alters the structure of a multilevel implementation. By using the right sequence of transformations, we strive to transform the design into something that better meets our constraints of speed, size, or parts used. Unfortunately, unlike for two-level logic minimization, there is no procedure that will guarantee we will find the minimal expression. Heuristics (rules-of-thumb) are used that are known to do well most of the time—but with no guarantees!

EXAMPLE 2.14 FULL ADDER

You might be wondering if the transformations on multilevel logic can help us transform the two-level implementation of the full adder in

Figure 2.57 to the multilevel implementation in Figure 2.56. Our two-level expressions for *Sum* and C_{out} are:

$$Sum = (A\bar{B}\bar{C}_{in}) + (\bar{A}B\bar{C}_{in}) + (\bar{A}\bar{B}C_{in}) + (ABC_{in})$$
$$C_{out} = (AC_{in}) + (BC_{in}) + (AB)$$

Looking at *Sum* we can see that it can be easily factored into two expressions based on A and $\bar{A}$.

$$Sum = A(\bar{B}\bar{C}_{in} + BC_{in}) + \bar{A}(B\bar{C}_{in} + \bar{B}C_{in})$$

The next step is to realize that we have two subexpressions that are related in that they are complements of each other. Multilevel logic optimizations systems looks for opportunities involving terms that are complements of each other. If we give this subexpression a new name, X, then we can rewrite *Sum* as:

$$Sum = A\bar{X} + \bar{A}X$$
$$X = B\bar{C}_{in} + \bar{B}C_{in}$$

Note that *Sum* is now clearly the XOR of A and X and X is in turn the XOR of B and C_{in}. Turning to C_{out}, we see that it, too, can be factored although less completely. Since we chose to factor based on A for *Sum*, we'll choose to use A again as this will increase the likelihood of finding common subexpressions:

$$C_{out} = A(C_{in} + B) + (BC_{in})$$

At this point, we can step back and change the expression for $C_{in} + B$ into canonical two-level form and re-factor the expression yielding:

$$C_{out} = A(\bar{B}C_{in} + B\bar{C}_{in} + BC_{in}) + (BC_{in})$$
$$C_{out} = A(\bar{B}C_{in} + B\bar{C}_{in}) + BC_{in}(1 + A)$$

Note that now we have another use of X:

$$C_{out} = AX + BC_{in}$$

We have completed the transformation we set out to accomplish and can verify that our final equation for *Sum* and C_{out} correspond exactly to the multilevel circuitry of Figure 2.56.

CHAPTER REVIEW

In this chapter, we have introduced a variety of primitive logic building blocks: the NOT, AND, OR, NAND, NOR, XOR, and XNOR gates, with which we can implement any Boolean function. We have also presented the two primary canonical forms for describing a Boolean function: sum-of-products and product-of-sums. A function may have many equivalent Boolean expressions but only one representation in a canonical form. Logic minimization, to be discussed in more detail in the next chapter, seeks to find the equivalent expression with the minimum number of terms and the fewest literals per term. Don't-care conditions on the inputs can be used to simplify the expression substantially. Reduced expressions lead to realizations of circuits with the fewest gates and the fewest inputs, at least if the target is a two-level circuit realization.

The final section dealt with the concept of multilevel logic and introduced its advantages in terms of reduced literal counts and simplified wiring complexity. Multilevel logic introduces the possibility of increased circuit delay by placing additional levels of gates between inputs and outputs. In general, two-level logic tends to yield the fastest implementations and multilevel logic results in circuits with fewer wires requiring less area for implementation. However, the precise details of the technology we use, the area and speed of the different logic building blocks we have available, can make a big difference in any decision.

FURTHER READING

George Boole's original work defining Boolean algebra was already referenced at the end of Chapter 1. Obviously, he did not have computer hardware in mind at the time. Instead, he attempted to develop a mathematical basis for logic. The set of basic axioms presented in Section 2.1 are called Huntington's axioms. These were published by E. V. Huntington in a paper entitled "Sets of Independent Postulates for the Algebra of Logic," in *Transactions of the American Mathematical Society* (Volume 5) in 1904. C. E. Shannon was the first to show how Boolean algebra could be applied to digital design. The K-map method was described in an article by M. Karnaugh in 1953 ("A Map Method for Synthesis of Combinational Logic Circuits," *Transactions of the AIEE, Communications and Electronics,* 72, I, pp. 593–599, November 1953). Interestingly, despite the fact that Karnaugh is given the credit, the original idea is from E. W. Veitch ("A Chart Method for Simplifying Boolean Functions," *Proceedings of the ACM,* May 1952, pp. 127–133). The major difference is in how the boxes of the map are labeled: Karnaugh used the familiar Gray-code scheme and Veitch used an alternative "distance 1" code, which is somewhat harder to remember and did not gain as much popularity.

All digital-design textbooks describe Boolean simplification in one form or another. For alternative explanations, see J. F. Wakerly, *Digital Design Principles and Practices, Third edition,* Prentice Hall, New Jersey, 2000; M. M. Mano and C. R. Kime, *Logic and Computer Design Fundamentals, Third edition,* Prentice Hall, New Jersey, 2000; and D. D. Gajski, *Principles of Digital Design,* Prentice Hall, New Jersey, 1997.

EXERCISES

2.1 *(Truth Tables and Gate Logic)* Derive a gate-logic implementation for the three functions in the truth table of Figure 2.3. Try to come up with as small an implementation as you can in terms of gate count.

2.2 *(Gate Logic)* Draw schematics for the following functions in terms of AND, OR, and NOT gates.

(a) $X(Y + Z)$
(b) $XY + XZ$
(c) $\overline{X(Y + Z)}$
(d) $\bar{X} + \bar{Y}\bar{Z}$
(e) $W(X + YZ)$

2.3 *(Gate Logic)* Draw the schematics for the following functions using NOR gates and inverters only:

(a) $\overline{[\bar{X} + \overline{(Y + Z)}]}$

(b) $\overline{[\overline{(\bar{X} + \bar{Y})} + \overline{(\bar{X} + \bar{Z})}]}$

2.4 *(Gate Logic)* Draw the schematics for the following functions using NAND gates and inverters only:

(a) $\overline{[X(\bar{Y}\bar{Z})]}$
(b) $XY + XZ$

2.5 *(Gate Logic)* Design a hall light circuit to the following specification. There is a switch at either end of a hall that controls a single light. If the light is off, changing the position of either switch causes the light to turn on. Similarly, if the light is on, changing the position of either switch causes the light to turn off. Write your assumptions, derive a truth table, and describe how to implement this function in terms of logic gates.

2.6 *(Laws and Theorems of Boolean Algebra)* Prove the following simplification theorems using the first eight laws of Boolean algebra:

(a) $(X + Y)(X + \bar{Y}) = X$
(b) $X(X + Y) = X$
(c) $(X + \bar{Y})Y = XY$
(d) $(X + Y)(\bar{X} + Z) = XZ + \bar{X}Y$

2.7 *(Laws and Theorems of Boolean Algebra)* Verify that

(a) OR and AND are duals of each other
(b) NOR and NAND are duals of each other
(c) XNOR and XOR are duals of each other
(d) XNOR is the complement of XOR:

$$\overline{(X\bar{Y} + \bar{X}Y)} = \bar{X}\bar{Y} + XY$$

2.8 *(Laws and Theorems of Boolean Algebra)* Prove, using truth tables, that:

(a) $XY + YZ + X\bar{Z} = YZ + X\bar{Z}$
(b) $(A + \bar{B})B = AB$
(c) $(A + B)(\bar{A} + C) = AC + \bar{A}B$
(d) $ABC + \bar{A}BC + \bar{A}\bar{B}C + \bar{A}B\bar{C} + \bar{A}\bar{B}\bar{C} = BC + \bar{A}\bar{B} + \bar{A}\bar{C}$

2.9 *(Laws and Theorems of Boolean Algebra)* Prove, using the theorems of Boolean algebra, that $BC + \bar{A}\bar{B} + \bar{A}\bar{C} = ABC + \bar{A}$.

2.10 *(Laws and Theorems of Boolean Algebra)* Use DeMorgan's law to compute the complement of the following Boolean expressions:

(a) $A(B + CD)$
(b) $ABC + B(\bar{C} + \bar{D})$
(c) $\bar{X} + \bar{Y}$
(d) $X + Y\bar{Z}$
(e) $(X + Y)Z$
(f) $X + \overline{(YZ)}$
(g) $X(Y + Z\bar{W} + \bar{V}S)$

2.11 *(Laws and Theorems of Boolean Algebra)* Form the complement of the following functions:

(a) $f(A,B,C,D) = [A + \overline{BCD}][\overline{AD} + B(\bar{C} + A)]$
(b) $f(A,B,C,D) = \bar{A}BC + (\bar{A} + B + D)(AB\bar{D} + \bar{B})$

2.12 *(Laws and Theorems of Boolean Algebra)* Using Boolean algebra, verify that the schematic of Figure Ex. 2.12 implements an XOR function.

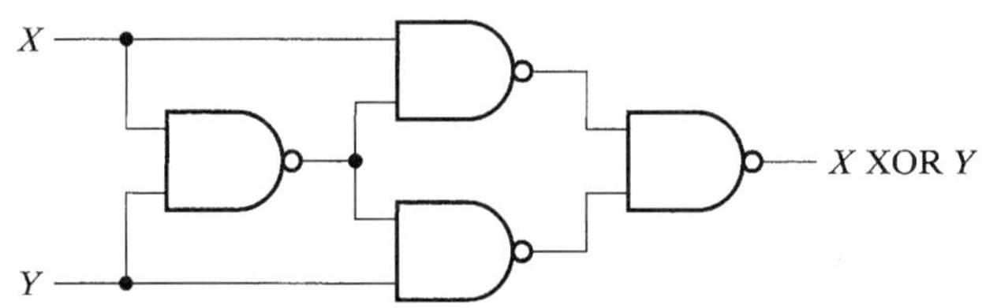

Figure Ex. 2.12 **XOR implemented by NAND gates.**

2.13 *(Laws and Theorems of Boolean Algebra)* Demonstrate that a 2-input NOR gate is a universal logic element. You can do this by showing how they can be used to make: NOT, AND, OR, and XOR gates. Remember that each input of the NOR gates must be used, it can not be left unconnected. Is an XOR gate a universal logic element? Why or why not? What about a 2-input NAND gate?

2.14 *(Block Diagrams)* Given the truth table for the half adder, show that the composition of two half adders and an OR gate, as in Figure 2.19, yield the same truth table as the full adder.

2.15 *(Waveform Verification)* This chapter has described two different gate-level implementations for a full-adder circuit: direct implementation from the Boolean equations derived in Section 2.2 and hierarchical implementation via cascaded half adders, as in Figure 2.19. Would you expect the waveform behaviors of these implementations to be identical? Justify your answer.

2.16 *(Block Diagrams)* Using the formulas for the two full-adder outputs, derive expressions for the three outputs of the 2-bit adder of Figure 2.20 in terms of the 2-bit inputs A and B.

2.17 *(Boolean Simplification)* Simplify the following functions using the theorems of Boolean algebra. Write the particular law or theorem you are using in each step. For each function, by how many literals did you reduce its representation?

(a) $f(X,Y) = XY + X\overline{Y}$
(b) $f(X,Y) = (X + Y)(X + \overline{Y})$
(c) $f(X,Y,Z) = \overline{Y}Z + \overline{X}YZ + XYZ$
(d) $f(X,Y,Z) = (X + Y)(\overline{X} + Y + Z)(\overline{X} + Y + \overline{Z})$
(e) $f(W,X,Y,Z) = X + XYZ + \overline{X}YZ + \overline{X}Y + WX + W\overline{X}$

2.18 *(Boolean Simplification)* Consider the function:

$$f(A,B,C,D) = (AD + \overline{A}C)[\overline{B}(C + B\overline{D})].$$

(a) Draw its schematic using AND, OR, and NOT gates.
(b) Using Boolean algebra, put the function into its minimized form and draw the resulting schematic.

2.19 *(Canonical Forms)* Consider the function:

$$f(A,B,C,D) = \Sigma\, m(0,1,2,7,8,9,10,15).$$

(a) Write this as a Boolean expression in canonical minterm form.
(b) Rewrite the expression in canonical maxterm form.
(c) Write the complement of f in "little m" notation and as a canonical minterm expression.
(d) Write the complement of f in "big M" notation and as a canonical maxterm expression.

2.20 *(Canonical Forms)* Consider the function:

$$f(A,B,C,D) = \Sigma\, m(1,\ 2,\ 3,\ 5,\ 8,\ 13).$$

(a) Write this as a Boolean expression in canonical minterm form.
(b) Rewrite the expression in canonical maxterm form.
(c) Write the complement of f in "little m" notation and as a canonical minterm expression.
(d) Write the complement of f in "big M" notation and as a canonical maxterm expression.

2.21 *(Canonical Forms)* Consider the function:

$$f(A,B,C) = AB + \overline{B}\overline{C} + A\overline{C}$$

(a) Express the function in canonical sum-of-products form. Use "little m" notation.
(b) Express the complement of the function in canonical product-of-sums form. Use "big M" notation.

2.22 *(Canonical Forms)* Given the following function in sum-of-products form (not necessarily minimized):

$$F(A,B,C,D) = \bar{A}BC + AD + AC$$

Re-express the function in:

(a) Canonical product-of-sums form. Use ΠM notation.
(b) Minimized product-of-sums form.
(c) $\bar{F}$ in minimized product-of-sums form.
(d) $\bar{F}$ in minimized sum-of-products form.
(e) Implement F and $\bar{F}$ using NAND gates only. You may assume that literals and their complements are available.
(f) Implement F and $\bar{F}$ using NOR gates only. You may assume that literals and their complements are available.

2.23 *(Canonical Forms and Boolean Simplification)* Given the following function in product-of-sums form, not necessarily minimized:

$$F(W,X,Y,Z) = (W + \bar{X} + \bar{Y})(\bar{W} + \bar{Z})(W + Y)$$

(a) Express the function in the canonical sum-of-products form. Use "little m" notation.
(b) Re-express the function in minimized sum-of-products form.
(c) Express $\bar{F}$ in minimized sum-of-products form.
(d) Re-express $\bar{F}$ in minimized product-of-sums form.

2.24 *(Boolean Simplification)* Using K-maps, find the following:

(a) Minimum sum-of-products form for the function and its complement given in Exercise 2.19.
(b) Minimum product-of-sums form for the function and its complement given in Exercise 2.19.

2.25 *(Boolean Simplification)* Use Karnaugh maps (K-maps) to simplify the following functions in sum-of-products form. How many literals appear in your minimized solutions?

(a) $f(X,Y,Z) = \Pi\, M(0,1,6,7)$
(b) $f(W,X,Y,Z) = \Pi\, M(1,3,7,9,11,15)$
(c) $f(A,B,C,D) = \Sigma\, m(0,2,4,6)$

2.26 *(Boolean Simplification)* Determine the minimized realization of the following functions in the sum-of-products form:

(a) $f(W,X,Y,Z) = \Sigma\, m(0,2,8,9) + \Sigma\, d(1,3)$
(b) $f(W,X,Y,Z) = \Sigma\, m(1,7,11,13) + \Sigma\, d(0,5,10,15)$

(c) $f(A,B,C,D) = \Sigma\ m(1,2,11,13,14,15) + \Sigma\ d(0,3,6,10)$.
(d) $f(A,B,C,D) = \Pi\ M(2,5,6,8,9,10) * \Pi\ D(4,11,12)$.

2.27 *(Boolean Simplification)* Use the K-map method to find the minimized product-of-sums expressions for the following Boolean functions:

(a) $f(A,B,C) = A \oplus B \oplus C$
(b) $f(A,B,C) = AB + BC + AC$
(c) $f(A,B,C,D) = \Sigma\ m(1,3,5,7,9) + \Sigma\ d(6,12,13)$
(d) $f(A,B,C,D) = \Pi\ M(0,1,6,7)$
(e) $f(A,B,C,D) = \Sigma\ m(0,2,4,6)$

2.28 *(Boolean Simplification)* Simplify the following expressions using the laws and theorems of Boolean algebra:

(a) $S(A,B,C) = \bar{A}\bar{B}C + \bar{A}B\bar{C} + A\bar{B}\bar{C} + ABC$
(b) $F(A,B,C) = \bar{A}\bar{B}\bar{C} + \bar{A}\bar{B}C + A\bar{B}\bar{C} + A\bar{B}C + AB\bar{C} + ABC$
(c) $G(A,B,C,D) = \bar{A}\bar{B}\bar{C}\bar{D} + \bar{A}\bar{B}C\bar{D} + A\bar{B}\bar{C}\bar{D} + A\bar{B}CD + AB\bar{C}\bar{D} + ABCD$

2.29 *(Boolean Simplification)* Use K-maps on the expressions of Problem 2.28. Show your work in K-map form. From 2.28 (a) through (c):

(a) Find the minimized sum-of-products form.
(b) Find the minimized product-of-sums form.
(c) Find the minimized sum-of-products form of the function's complement.
(d) Find the minimized product-of-sums form of the function's complement.

2.30 *(Laws and Theorems of Boolean Algebra)* Simplify the following expressions using the laws and theorems of Boolean algebra:

(a) $W(A,B,C) = \bar{A}B\bar{C} + \bar{A}BC + A\bar{B}\bar{C} + A\bar{B}C$
(b) $X(A,B,C) = \bar{A}\bar{B}\bar{C} + \bar{A}BC + A\bar{B}\bar{C} + ABC$
(c) $Y(A,B,C,D) = \bar{A}\bar{B}\bar{C}\bar{D} + \bar{A}\bar{B}\bar{C}D + \bar{A}\bar{B}C\bar{D} + \bar{A}\bar{B}CD + A\bar{B}\bar{C}\bar{D} + A\bar{B}C\bar{D}$

2.31 *(Karnaugh Map Method)* Use K-maps on the expressions of Exercise 2.30. Show your work in K-map form.

(a) Find the minimized sum-of-products form.
(b) Find the minimized product-of-sums form.
(c) Find the minimized sum-of-products form of the function's complement.
(d) Find the minimized product-of-sums form of the function's complement.

2.32 *(Karnaugh Map Method)* There may be more than one true minimum equivalent form for a given Boolean expression. Demonstrate this by drawing a 4-variable K-map that has two different minimized forms for the same Boolean expression, each with the same number of terms and literals.

2.33 *(Karnaugh Map Method)* Using a 4-variable K-map, fill it with 1s and 0s to find a function that illustrates the following points. Write the expressions for each of the requested forms and count the number of terms and literals for each one:

(a) The minimized sum-of-products and product-of-sums forms have the same number of terms and literals.
(b) The minimized sum-of-products form has fewer terms and literals than the minimized product-of-sums form.
(c) The minimized product-of-sums form has fewer terms and literals than the minimized sum-of-products form.

2.34 *(Incompletely Specified Functions)* Use K-maps to derive minimal expressions for the functions for C_0, C_1, C_3, C_4, C_5, and C_6 in Figure 2.33.

2.35 *(Incompletely Specified Functions)* Use K-maps to derive minimal sum-of-product expressions for d30 and d31 from the calendar example of Section 1.4.1. Make sure to take advantage of don't care information.

2.36 *(Encoding)* Use K-maps to derive minimal sum-of-product expressions for d30 and d31 from the calendar example of Section 1.4.1 but with a different encoding for the months. In this case, have the months start with January = 0000 and December = 1011. Make sure to take advantage of don't care information. Is this encoding better or worse than the original one?

2.37 *(Multilevel Logic)* Factor the following sum-of-products expressions:

(a) $ABCD + ABDE$
(b) $ACD + BC + ABE + BD$
(c) $AC + ADE + BC + BDE$
(d) $AD + AE + BD + BE + CD + CE + AF$
(e) $ACE + ACF + ADE + ADF + BCE + BCF + BDE + BDF$

2.38 *(Multilevel Logic)* Write down the function represented by the circuit network in Figure Ex. 2.38 in a multilevel factored form using AND, OR, and NOT operations only—that is, no NAND or NOR operations.

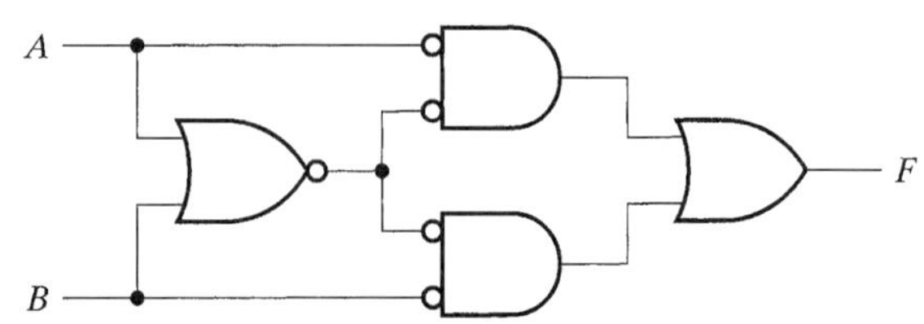

Figure Ex. 2.38

Derive the simplest Boolean expression you can (minimum number of literals and fewest gates) for the function represented by this schematic. You may use any kind of logic operators described in this chapter.

2.39 *(Multilevel Logic)* Derive the multilevel logic realization of the 2-bit adder of Figure 2.20. Compare it to a two-level logic realization of the same function. List the different types of gates (function and fanin) for each of the two implementations. Which has fewer gates? Which has fewer wires? Which is likely to be faster (do this by showing the relative delays along the worst-case paths from an input to an output)? This is similar to the comparison for the full adder at the end of Section 2.7.

2.40 *(Combinational Logic Design)* Write truth tables for the three functions described by the following specifications:

(a) A 2-bit-wide *shifter* takes two input signals, i_0 and i_1, and shifts them to two outputs, o_0 and o_1, under the control of a shift signal. If this signal *SHIFT* is false, then the outputs are equal to their corresponding inputs. If *SHIFT* is true, then o_1 is equal to i_0, and o_0 is set a 0.

(b) A 1-bit *demultiplexer* takes an input signal *IN* and "routes" it to one of two outputs, o_0 and o_1, under the control of a single *SELECT* signal. If *SELECT* is 0, then o_0 has the value of *IN* and o_1 is a 0. If *SELECT* is 1, then o_1 has the value of *IN* and o_0 is a 0.

(c) A 2-bit *multiplexer* takes two input signals, i_0 and i_1, and "routes" one of them to the single output *OUT* under the control of a 1-bit select signal. If the *SELECT* signal is false, then *OUT* is equal to i_0. If *SELECT* is true, then *OUT* is equal to i_1.

2.41 *(Boolean Simplification)* Write sum-of-products expressions for the truth tables of Exercise 2.40. Minimize them using K-maps.

2.42 *(Gates)* Given the Boolean expressions of Exercise 2.41, draw logic schematics using AND, OR, and INVERT gates that implement those functions.

2.43 *(Combinational Logic Design)* Consider a 4-input Boolean function that is asserted whenever exactly two of its inputs are asserted.

(a) Construct its truth table.

(b) What is the function in sum-of-products form, using "little *m*" notation?

(c) What is the function in product-of-sums form, using "big *M*" notation?

(d) Use the Karnaugh map method to simplify the function in sum of products form.

2.44 *(Combinational Logic Design)* In this chapter, we've examined the BCD increment-by-1 function (see Figure 2.32). Now consider a binary increment-by-1 function defined over the 4-bit binary numbers 0000 through 1111.

(a) Fill in the truth table for the function.

(b) Fill in the four 4-variable K-maps, and find the minimized sum-of-products for each output function.

(c) Repeat the process for the minimized product-of-sums form. Which leads to the simpler implementation in terms of the number of literals?

2.45 *(Combinational Logic Design)* Consider a four-input function that outputs a 1 whenever an odd number of its inputs are 1.

(a) Fill in the truth table for the function.

(b) Fill in the K-map to find the minimum sum-of-products expression for the function. What is it? Can the function be minimized using the K-map method?

(c) Can you think of a more economical way to implement this function if XOR gates are allowed? (*Warning:* It will be very tedious to try to simplify this function using Boolean algebra, so think about the question first!)

2.46 *(Combinational Logic Design)* In this chapter, we've examined a 2-bit binary adder circuit. Now consider a 2-bit binary subtractor, defined as follows. The inputs A, B and C, D form the two 2-bit numbers N_1 and N_2. The circuit will form the difference $N_1 - N_2$ on the output bits F (most significant) and G (least significant). Assume that the circuit never sees an input combination in which N_1 is less than N_2. The output bits are don't cares in these cases.

(a) Fill in the 4-variable truth table for F and G.

(b) Fill in the K-map for the minimum sum-of-products expression for the functions F and G.

(c) Repeat to find the minimum product-of-sums expression for F and G.

CHAPTER

3

Working with Combinational Logic

Introduction

Now that we've learned about two-level logic and had a short introduction to multilevel logic, it should be becoming obvious that we have many choices in how we go about realizing a Boolean function. We will need tools to aid us in transforming our logic expressions and gate-level circuits so that we can explore all these possibilities. When we can, we will also want to automate these tools so that they can help us obtain optimized designs (with the criteria of number of logic gates, number of literals or wires, and delay) for large and complex designs that would be too cumbersome to do by hand.

In this chapter, we will look at the underlying approaches to tools for two-level and multilevel logic minimization. We will also see how the resulting logic formulas can be mapped to NAND and NOR gates, the most common gates available in our implementation technologies.

The chapter will conclude with a more detailed discussion of time behavior in logic networks and how we can use hardware description languages (HDLs) to specify the implementation and, alternatively, the behavior of our combinational circuits. HDLs have become the method of choice not only for describing and documenting, but also for simulating and automatically synthesizing logic circuits. They are the primary interface to the automatic tools designers use today.

3.1 Two-Level Simplification

You now have the foundation to learn the practical methods for reducing a Boolean expression to its simplest form when we use only two levels of logic gates to realize it. The result is an expression with the fewest terms and thus less gates, and the fewest literals and thus less wires in the final gate-level implementation.

Our description of two-level simplification actually began in Chapter 2 with Section 2.5. We defined the essential concept of Boolean cubes that provide us with a way to visualize truth tables so that we can see easily see when two terms closely are related and can be combined into a single term. We also introduced Karnaugh maps as a way of drawing Boolean cubes on 2-dimensional paper. K-maps are an effective paper-and-pencil approach for identifying Boolean cubes for functions with a modest number of variables. The method rearranges the truth table rows into a special tabular structure that places entries that might be grouped right next to each other, so that the uniting possibilities are easier to spot by humans.

Let's review K-maps and Boolean simplification with some examples that are a bit more substantial.

EXAMPLE 3.1 TWO-BIT COMPARATOR

The goal is to design a circuit that takes as input two 2-bit numbers for comparison, N_1 and N_2, and generates three outputs: $N_1 = N_2$, $N_1 < N_2$, and $N_1 > N_2$. We denote the numbers N_1 and N_2 by the single-bit inputs A, B and C, D, respectively. A and C are most significant bits.

The first step in tackling any problem like this is to understand fully the behavior of the circuit being designed. You can do this best by drawing a block diagram showing inputs and outputs and creating a truth table for each output as a function of the inputs.

These are shown in Figure 3.1. It is fairly straightforward to fill in the table. For example, the first row compares the N_1 input 00 to the N_2 input 00. The F_{eq} function (=) is true, while F_{lt} (<) and F_{gt} (>) are false. In the second row, 00 < 01, so F_{eq} and F_{gt} are false while F_{lt} is true. The rest of the table is filled in a similar way.

The next step is to prepare K-maps for each of the outputs. This is shown in Figure 3.2. Let's start with the K-map for F_{eq}. There are no

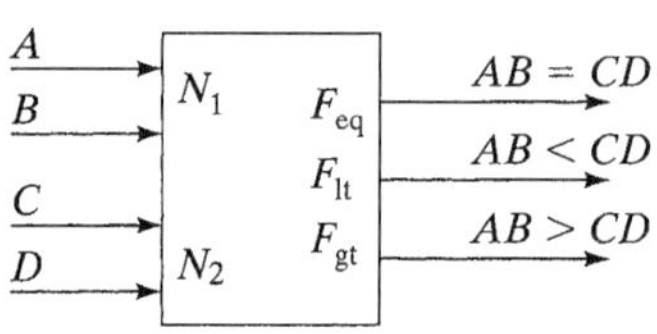

A	B	C	D	F_{eq}	F_{lt}	F_{gt}
0	0	0	0	1	0	0
		0	1	0	1	0
		1	0	0	1	0
		1	1	0	1	0
0	1	0	0	0	0	1
		0	1	1	0	0
		1	0	0	1	0
		1	1	0	1	0
1	0	0	0	0	0	1
		0	1	0	0	1
		1	0	1	0	0
		1	1	0	1	0
1	1	0	0	0	0	1
		0	1	0	0	1
		1	0	0	0	1
		1	1	1	0	0

Figure 3.1 **Block diagram and truth table of 2-bit comparator.**

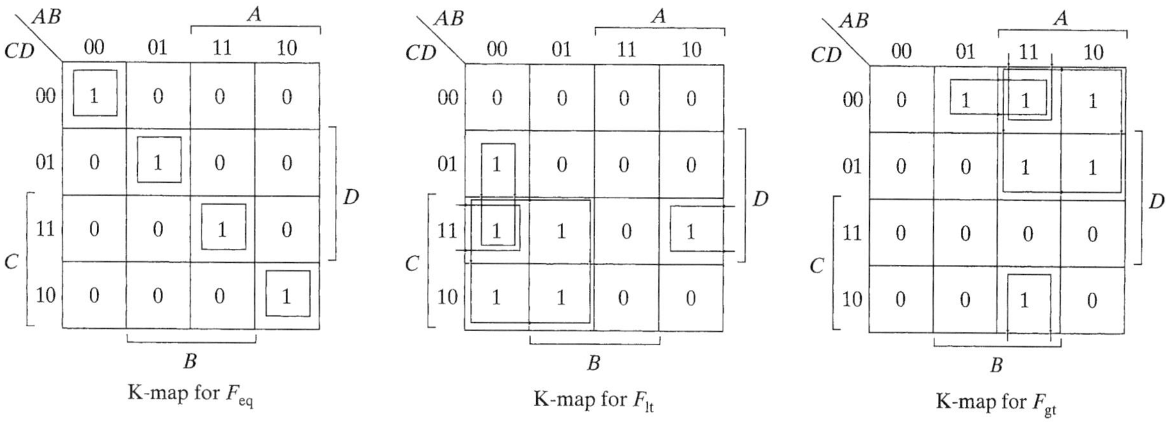

Figure 3.2 **K-maps for the 2-bit comparator.**

adjacencies in this K-map! Each of the four elements of the on-set contributes a 4-variable term to the expression for F_{eq}:

$$F_{eq} = \bar{A}\bar{B}\bar{C}\bar{D} + \bar{A}B\bar{C}D + A\bar{B}C\bar{D} + ABCD$$

Note the first term corresponds to both input values being 0, the next term to them both being 1, the third term for both 2, and the last one for both 3. This is the minimized sum-of-products form for the function. However, by using Boolean algebra, we can simplify this expression some more:

$$F_{eq} = \bar{A}\bar{C}(\bar{B}\bar{D} + BD) + AC(\bar{B}\bar{D} + BD)$$
$$F_{eq} = (\bar{A}\bar{C} + AC)(\bar{B}\bar{D} + BD)$$
$$F_{eq} = \overline{(A \oplus C)}\overline{(B \oplus D)}$$
$$F_{eq} = (A \equiv C)(B \equiv D)$$

Our final expression is $(A$ XNOR $C)$ $(B$ XNOR $D)$. However, this is not a sum of products form as it uses XNORs and not just ANDs and ORs. 1s on K-map diagonals provide a good hint that the function can be expressed more economically in terms of XOR or XNOR operations.

The K-map for F_{lt} has three adjacency groups, two with two elements and one with four elements. The former yield-product terms with three literals, $\bar{A}\bar{B}D$ and $\bar{B}CD$; the latter is a term with two literals, $\bar{A}C$. The minimum sum-of-products expression for F_{lt} is:

$$F_{lt} = \bar{A}\bar{B}D + \bar{B}CD + \bar{A}C$$

The K-map for F_{gt} is very similar as we would expect from the symmetry of the less-than and greater-than operators. It also consists of two groups of two elements each (three literals) and one group of four elements (two literals). The minimum sum-of-products expression for F_{gt} is:

$$F_{gt} = A\bar{C} + AB\bar{D} + B\bar{C}\bar{D}$$

EXAMPLE 3.2 TWO-BIT BINARY ADDER

The next function we examine is a 2-bit binary adder. It takes as input the same two 2-bit binary numbers, N_1 and N_2, and produces a 3-bit binary number, N_3, as the result. The block diagram and truth table for these functions are shown in Figure 3.3. As in the previous example, N_1 is represented by the inputs A and B, N_2 by C and D, while N_3 is composed of the Boolean functions X, Y, and Z. X represents the most significant bit.

The K-maps for the outputs are shown in Figure 3.4. The maps for the X and Z outputs are more straightforward than for Y, and we will start with these. The function for X reduces to two 2-element groups (three literals each) and one 4-element group (two literals):

$$X = AC + BCD + ABD$$

Z exhibits two 4-element groups (two literals each) and reduces to the expression:

$$Z = B\bar{D} + \bar{B}D = B \oplus D$$

By careful examination of the K-map, we can often spot opportunities for reduction using XOR and XNOR operators. We will show this by reducing the literal count for the function Y by making good use of XOR/XNOR.

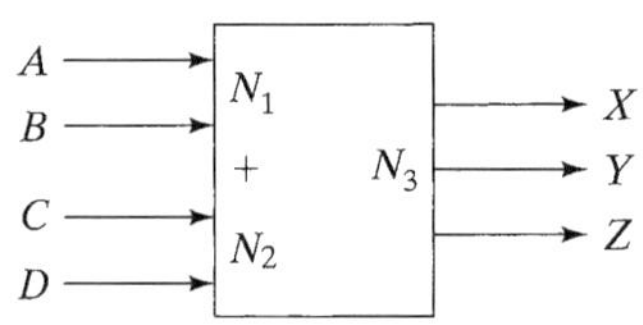

A	B	C	D	X	Y	Z
0	0	0	0	0	0	0
		0	1	0	0	1
		1	0	0	1	0
		1	1	0	1	1
0	1	0	0	0	0	1
		0	1	0	1	0
		1	0	0	1	1
		1	1	1	0	0
1	0	0	0	0	1	0
		0	1	0	1	1
		1	0	1	0	0
		1	1	1	0	1
1	1	0	0	0	1	1
		0	1	1	0	0
		1	0	1	0	1
		1	1	1	1	0

Figure 3.3 Block diagram and truth table of 2-bit binary adder.

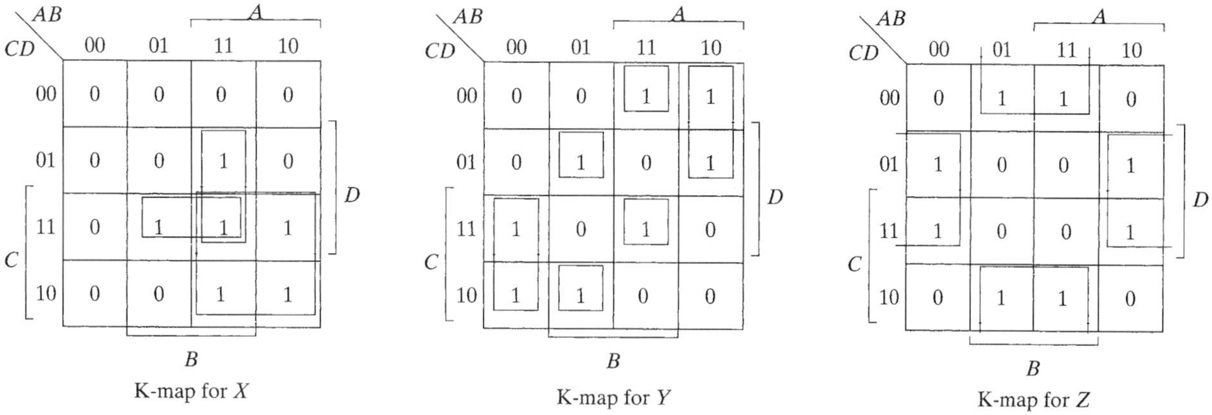

Figure 3.4 K-maps for the 2-bit binary adder.

Two straightforward 2-element terms of Y are $\bar{A}\bar{B}C$ and $A\bar{B}\bar{C}$. The remaining four single-element groups yield the terms: $\bar{A}B\bar{C}D$, $\bar{A}BC\bar{D}$, $AB\bar{C}\bar{D}$, and $ABCD$. We can further reduce $AB\bar{C}\bar{D}$ and $\bar{A}BC\bar{D}$ (to $A\bar{C}\bar{D}$ and $\bar{A}C\bar{D}$) but for the moment we will not do this. Factoring yields the three expressions (which represent three parts of Y):

$$\bar{A}\bar{B}C + A\bar{B}\bar{C} = \bar{B}(\bar{A}C + A\bar{C}) = \bar{B}(A \oplus C)$$
$$\bar{A}B\bar{C}D + \bar{A}BC\bar{D} = \bar{A}B(\bar{C}D + C\bar{D}) = \bar{A}B(C \oplus D)$$
$$AB\bar{C}\bar{D} + ABCD = AB(\bar{C}\bar{D} + CD) = AB\overline{(C \oplus D)}$$

We can factor the latter two expressions after combining them:

$$\bar{A}B(C \oplus D) + AB\overline{(C \oplus D)} = B(\bar{A}(C \oplus D) + A\overline{(C \oplus D)})$$
$$\bar{A}B(C \oplus D) + AB\overline{(C \oplus D)} = B(A \oplus C \oplus D)$$

Then, by combining them with the first part of Y, we get:

$$Y = \bar{B}(A \oplus C) + B(A \oplus C \oplus D)$$

This expression has just seven literals. Compare it to the reduced form, assuming only AND, OR, and NOT gates are allowed:

$$Y = \bar{A}\bar{B}C + A\bar{B}\bar{C} + A\bar{C}\bar{D} + \bar{A}C\bar{D} + \bar{A}B\bar{C}D + ABCD$$

This expression requires two 4-input AND gates, four 3-input AND gates, and a 6-input OR gate, for a total of seven gates and 20 literals. The first multilevel expression we derived requires a 2-input OR gate, two 2-input XOR gates, and two 2-input AND gates, a total of only five gates and seven literals to implement the function. However, the XOR gates are more complex gates, but in balance, all the gates of the multilevel implementation have only two inputs. The two alternative implementations are shown in Figure 3.5.

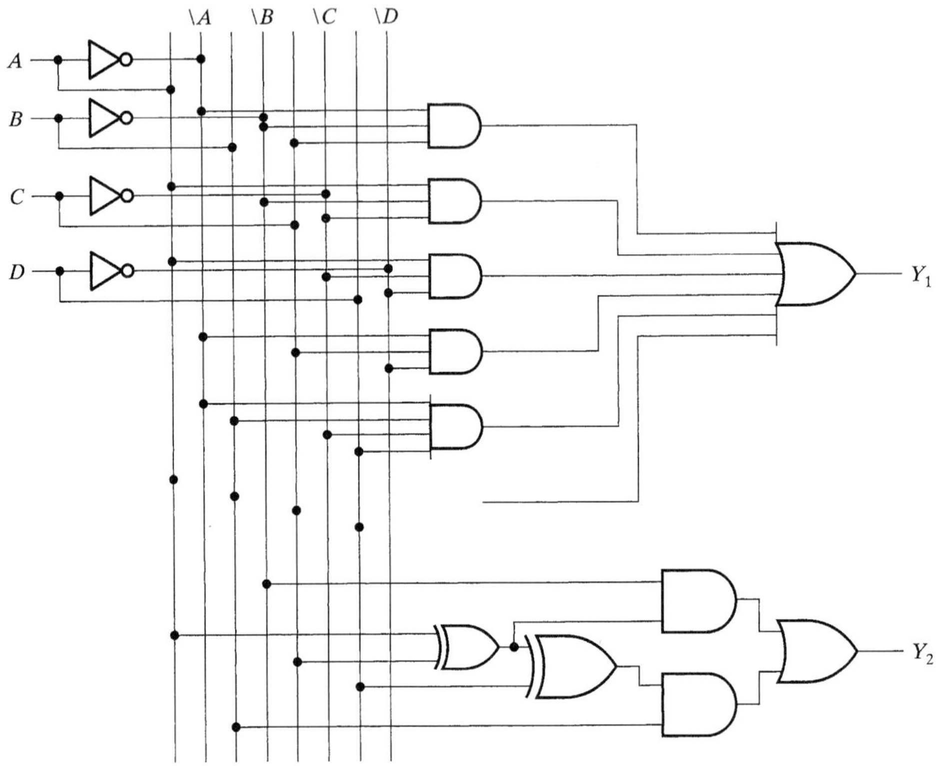

Figure 3.5 **Two alternative implementations of Y.**

EXAMPLE 3.3 BCD INCREMENT-BY-1 FUNCTION

We introduced the BCD increment-by-1 function in Section 2.4.2 as an example of a function with don't cares. The truth table of Figure 2.32 yields the four 4-variable K-maps of Figure 3.6.

We attempt to form the largest adjacency groups we can, taking advantage of don't cares to expand the group wherever possible. The function W can be implemented by two terms: $W = BCD + A\bar{D}$. These are formed from adjacency groups of two elements and four elements, respectively. Notice how we have taken advantage of adjacencies that wrap from the top of the K-map to the bottom-most row. The don't cares help us make a larger grouping by setting them to be 1. Remember that don't cares are set to 0 when we do not cover them. For W, we set four of the don't cares to 1 and two to 0.

The function X is implemented by three terms: $X = B\bar{D} + B\bar{C} + \bar{B}CD$. Once again, we have attempted to take advantage of adjacencies that wrap from top to bottom or left to right in the K-map. Four don't cares were also set to 1 while two were set to 0.

The function Y is implemented by two terms: $Y = \bar{A}\bar{C}D + C\bar{D}$. This is derived from groups of two and four entries, respectively. We set two don't cares to 1 to get this simplified expression for Y. The final

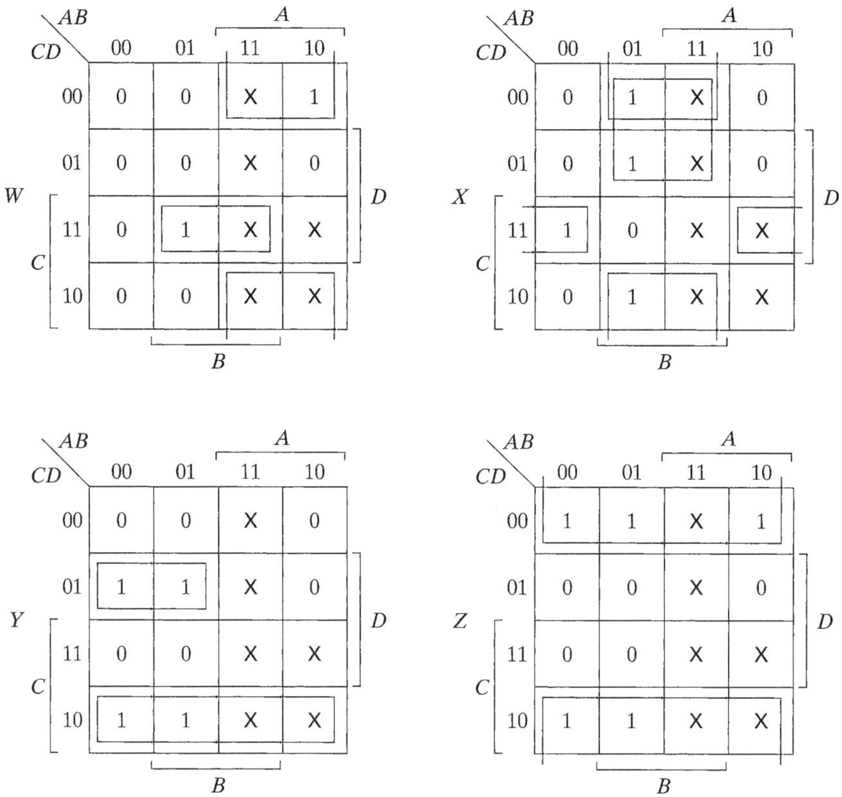

Figure 3.6 **BCD increment-by-1 K-maps.**

function Z is implemented by a group of eight nodes, which reduces to the single literal $\bar{D}$ by using three don't cares.

Once again, notice that adjacency groups are always formed by groups of 1 (4 literals), 2 (3 literals), 4 (2 literals), 8 (1 literal), or 16 (0 literals, a constant 0 or 1) elements, always a power of 2. Also notice how the adjacencies are formed: above, below, to the left, to the right of an element of the on-set, including those that wrap around the edges of the K-map and taking advantage of available don't cares. Keep in mind that a K-map is just a flattened multi-dimensional Boolean cube and the adjacencies we are searching for are sub-cubes.

3.1.1 Formalizing the Process of Boolean Minimization

We are now ready to be more precise about the process for obtaining a minimized expression. An *implicant* of a function F is a single element of the on-set or any group of elements that can be combined together in a K-map. We have been calling these adjacency groups or sub-cubes up to this point. A *prime implicant* is an implicant that cannot be combined with another one to eliminate a literal. In other words, you grow implicants to cover as much of the on-set as possible and a prime implicant is an implicant with as few literals as possible. Each prime

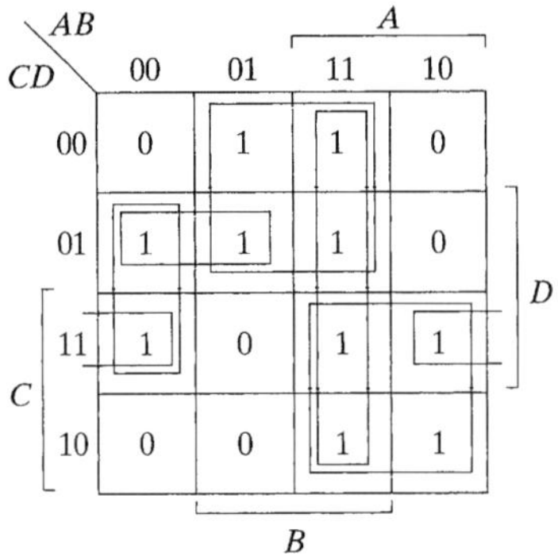

Figure 3.7 **Prime implicants.**

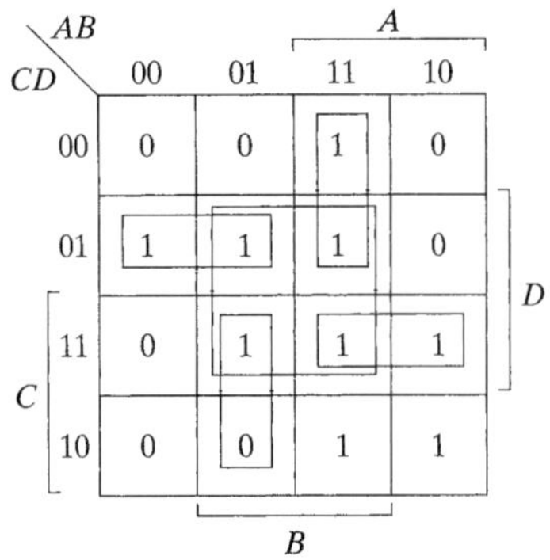

Figure 3.8 **Essential primes.**

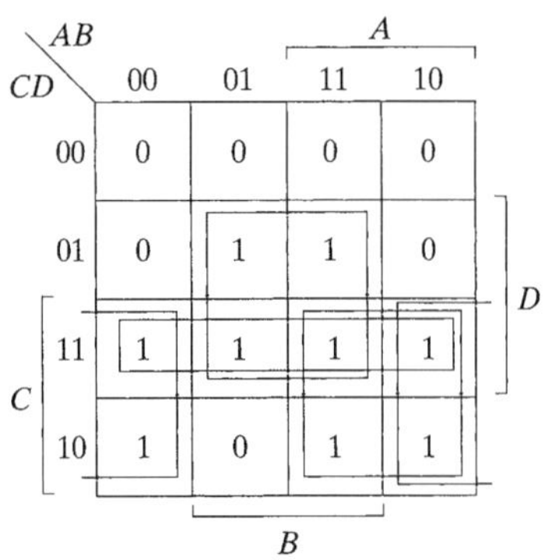

Figure 3.9 **More prime implicants.**

implicant corresponds to a product term in the minimum sum-of-products expression for the function. The objective is to find the smallest set of prime implicants that together cover all the elements of the on-set (and, optionally, some don't cares). If a particular element of the on-set is covered by only one prime implicant, then that implicant is called an *essential prime implicant.* All essential primes must be part of the minimized expression as they are needed for any and all covers.

Illustrating the Definitions Let's look at some examples to make these concepts more concrete. The 4-variable K-map of Figure 3.7 contains six prime implicants: $\overline{A}\overline{B}D$, $B\overline{C}$, AC, $\overline{A}\overline{C}D$, AB, and $\overline{B}CD$. Of these, only AC and $B\overline{C}$ are essential. Adding the additional implicant $\overline{A}\overline{B}D$ covers the remainder of the on-set. The other three prime implicants are *redundant* and are not needed in the expression, so we leave them out. Thus, the minimized expression for the function becomes

$$F = \overline{A}\overline{B}D + B\overline{C} + AC$$

As another example, consider the function whose K-map is given in Figure 3.8. It contains five prime implicants: BD, $AB\overline{C}$, ACD, $\overline{A}BC$, and $\overline{A}\overline{C}D$. All but the first implicant are essential. Interestingly, it is the largest prime implicant that is redundant. The minimized form is

$$F = AB\overline{C} + ACD + \overline{A}BC + \overline{A}\overline{C}D$$

As a final example, consider the K-map of Figure 3.9. It contains four prime implicants: BD, CD, AC, and $\overline{B}C$. The implicant CD is not needed, since the 1s it covers are covered already by the remaining implicants, all of which are essential. The minimized function is

$$F = BD + AC + \overline{B}C$$

Deriving a Minimized Expression from a K-Map A procedure for finding a minimum sum-of-products expression from the K-map is the following:

STEP 1 Choose an element from the on-set. Find all of the "maximal" groups of 1s and Xs adjacent to that element. Check for adjacency in the horizontal and vertical directions. Remember that the K-map wraps from top row to bottom and left-most column to right-most. The prime implicants (adjacency groups) thus formed always contain a number of elements that is a power of 2 (a sub-cube of the cube of the K-map). Repeat Step 1 for each element of the on-set to find all prime implicants.

STEP 2 Visit an element of the on-set. If it is covered by a single prime implicant, it is essential and will contribute a term to the final sum-of-products expression. The other 1s covered by the essential implicant do not need to be visited again as they definitely will be covered. Repeat Step 2 until all essential prime implicants have been found.

STEP 3 If there remain 1s uncovered by essential prime implicants, select a minimum number of prime implicants that cover them. Try several alternative covers to find the one with the fewest possible implicants.

EXAMPLE 3.4 APPLICATION OF THE STEP-BY-STEP ALGORITHM

Figure 3.10 shows the algorithm applied to a complete example. The function represented in the K-map is $F(A,B,C,D) = \Sigma\, m(4,5,6,8,9,10,13) + d(0,7,15)$.

Figure 3.10(a) gives the starting configuration. We scan down the K-map's columns, top to bottom and left to right, skipping 0s and Xs, searching for a 1. The first 1 we encounter is term $m_4(\bar{A}B\bar{C}\bar{D})$. We expand it in all directions, combining adjacent 1s and Xs into the largest implicant groups we can find. Two such groupings are possible, represented by the terms $\bar{A}B$ and $\bar{A}\bar{C}\bar{D}$. These are circled in Figure 3.10(b).

Continuing down the column, we next come to $m_5(\bar{A}B\bar{C}D)$. At this point, we should add only new implicants that are not contained

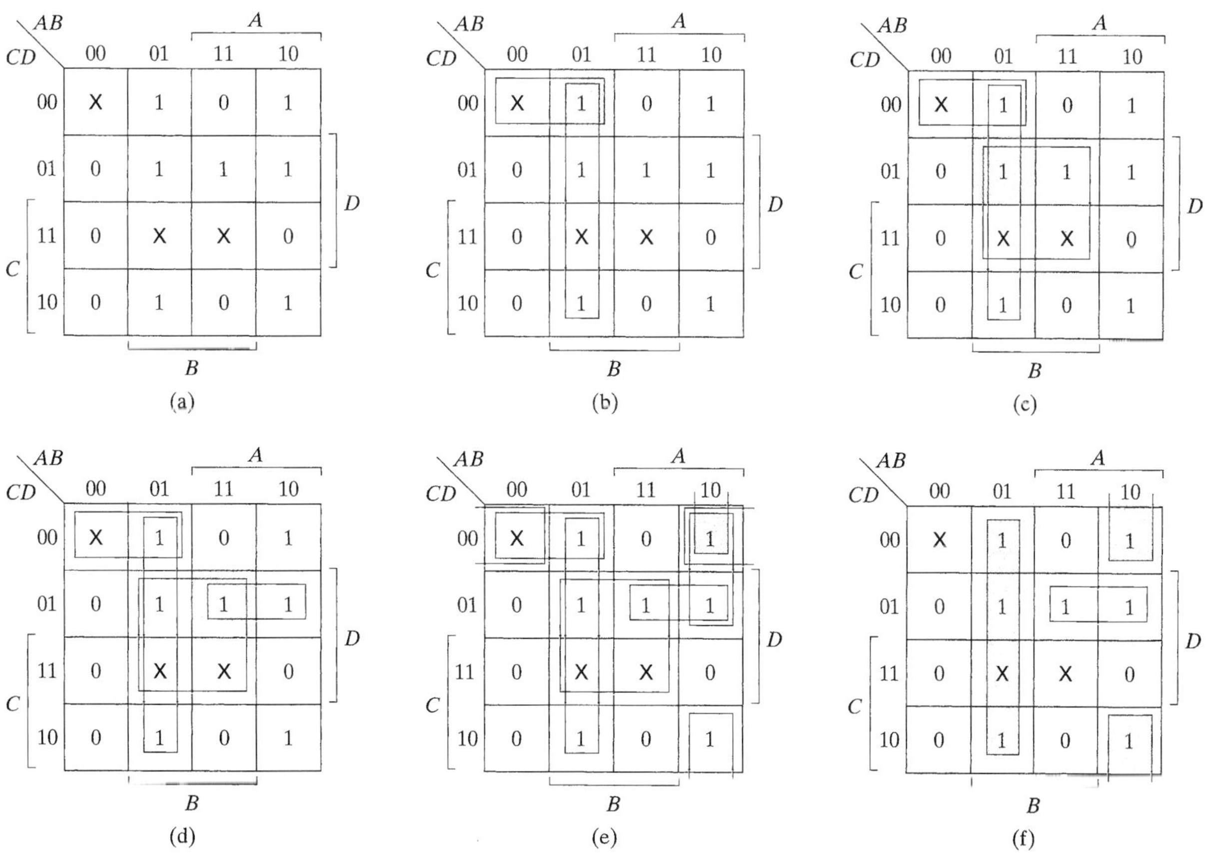

Figure 3.10 **Finding prime implicants, step by step.**

already within an implicant found so far. BD is the only new implicant we can add under this rule, as shown in Figure 3.10(c).

The next element of the on-set is $m_6(\bar{A}BC\bar{D})$, but no new implicant can be added because the set of implicants already contains $\bar{A}B$. So we continue searching for the next element of the on-set, which is $m_{13}(AB\bar{C}D)$. We now add the implicant $A\bar{C}D$. Note that the implicant ABD is not prime as it is already covered by the prime implicant BD. The state of the process to this point is shown in Figure 3.10(d).

The next 1 is m_8, which contributes three additional prime implicants, $A\bar{B}\bar{C}$, $A\bar{B}\bar{D}$, and $\bar{B}\bar{C}\bar{D}$. This is shown in Figure 3.10(e). The process continues with m_9 and m_{10}, but these add no new prime implicants.

All the elements of the on-set are now covered, and we have obtained all the prime implicants. The next step is to identify the essential prime implicants. The highlighted prime implicants of Figure 3.10(f), $\bar{A}B$ and $A\bar{B}\bar{D}$, are the essential primes because they exclusively cover m_6 and m_{10}, respectively.

The last step is to cover the remaining 1s not already covered by the essential primes. This is accomplished by the single prime implicant, $A\bar{C}D$. The process of enumerating prime implicants found six of them, yet three were sufficient to cover the entire on-set (two of which were essential). The final minimized function is:

$$F(A,B,C,D) = \bar{A}B + A\bar{B}\bar{D} + A\bar{C}D$$

3.1.2 K-Maps Revisited: Five- and Six-Variable Functions

The K-map method is not limited to four variables or less, although using it to visualize functions with more than four variables becomes more challenging as does trying to map a 5- or 6-dimensional cube to a 2-dimensional paper. It is important to remember that within an n-variable map we must check for adjacencies in n directions. Fortunately, we can handle the adjacencies for up to six variables, simply by stacking 4-variable K-maps.

EXAMPLE 3.5 FIVE-VARIABLE K-MAPS

A 5-variable map is shown in Figure 3.11(a). Let's consider the following Boolean function:

$$F(A,B,C,D,E) = \Sigma m(2,5,7,8,10,13,15,17,19,21,23,24,29,31)$$

The filled in K-map is shown in Figure 3.11(b). We have omitted the 0 entries to reduce the clutter in the figure. When searching for adjacencies, besides looking in the four horizontal and vertical squares as we did in the 4-variable K-map, we must also look either up or down. The example's on-set is covered by the four prime implicants CE (group of 8),

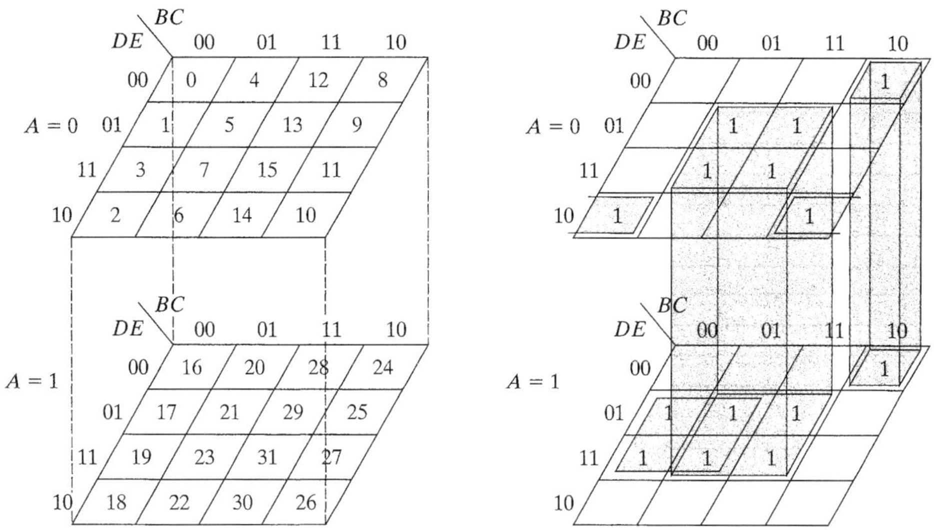

(a) Five-variable K-map (b) Example

Figure 3.11 Five-variable K-map and example.

$A\bar{B}E$ (group of 4), $B\bar{C}\bar{D}\bar{E}$ (group of 2), and $\bar{A}\bar{C}D\bar{E}$ (group of 2). This is hard to visualize and not something we want to do by hand very often.

EXAMPLE 3.6 SIX-VARIABLE K-MAP

The six-variable K-map is to the 5-variable map as the 5-variable is to the 4-variable: the number of 4-variable planes is increased from two to four. This is shown in Figure 3.12(a).

An example 6-variable K-map is shown in Figure 3.12(b). The function is

$$F(A,B,C,D,E,F) = \Sigma m(2,8,10,18,24,26,34,37,42,45,50,53,58,61)$$

In addition to horizontal and vertical adjacencies, the planes immediately above and below the element being examined must be checked. The top plane also wraps around onto the bottom plane. In the figure, the on-set is covered by three prime implicants: $\bar{D}E\bar{F}$ (a group of eight), $AD\bar{E}F$ (a group of four), and $\bar{A}C\bar{D}\bar{F}$ (a group of four).

3.2 Automating Two-Level Simplification

The algorithm we presented in the previous section for extracting essential prime implicants from a K-map could form the basis for computer-based tools. In this section, we examine computer-based algorithms for two-level simplification in more detail. We begin with the Quine-McCluskey method, the first complete algorithm for simplification. We complete this section by covering the methods used in *espresso,* a popular tool for two-level minimization. While *not guaranteed* to find the

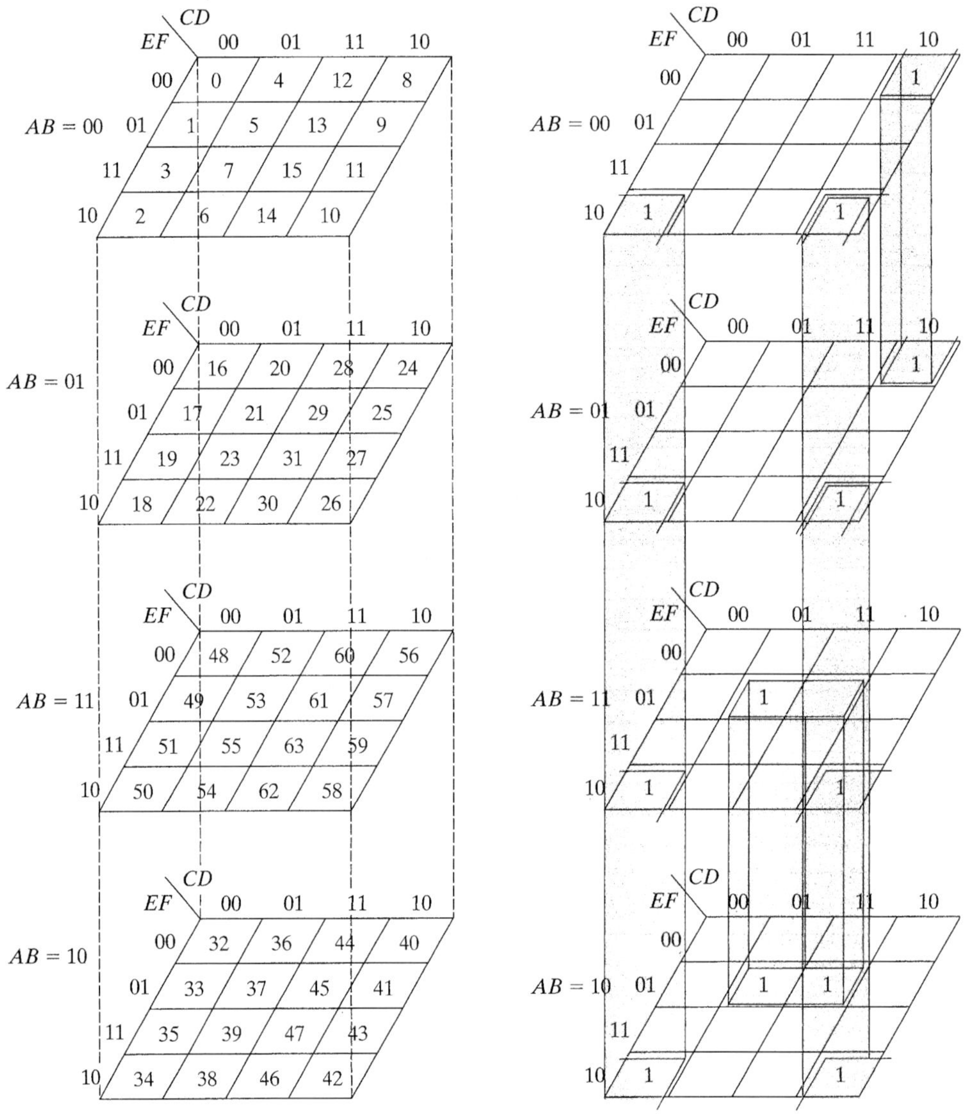

(a) Six-variable K-map

(b) Example

Figure 3.12 **Six-variable K-map and example.**

best two-level expression, *espresso* uses several tricks to find a good solution as fast as possible.

3.2.1 Quine-McCluskey Method

Except in special cases or for particularly sparse truth tables, the K-map method simply breaks down beyond six variables. We need a more systematic algorithm. The Quine-McCluskey method, developed in the mid-1950s, finds the minimized representation of any Boolean expression. It provides a systematic procedure for generating all prime

TABLE 3.1 Quine-McCluskey Implication Table

Column I	*Column II*	*Column III*
0000 √	0-00 *	01-- *
	-000 *	
0100 √		-1-1 *
1000 √	010- √	
	01-0 √	
0101 √	100- *	
0110 √	10-0 *	
1001 √		
1010 √	01-1 √	
	-101 √	
0111 √	011- √	
1101 √	1-01 *	
1111 √	-111 √	
	11-1 √	

√ implicant
* prime implicant

implicants and then extracting a minimum set of primes covering the on-set.

The method finds the prime implicants by repeatedly applying the Uniting theorem, just as we did earlier in this section. The contribution of Quine-McCluskey is to provide a tabular method that ensures that all prime implicants are found. To understand how it works, let's use the same example as in Figure 3.10: $F = \Sigma\, m(4,5,6,8,9,10,13) + d(0,7,15)$.

Finding Prime Implicants The first step is to list all elements of the on-set and don't-care set in terms of their minterm indices, represented as a binary number. The elements are grouped according to the number of 1s in the binary representation. This will make it easier to compare pairs systematically.

Table 3.1 shows the structure of a Quine-McCluskey implicant table. The first column contains the minterms of the on-set and don't-care-set, that is, single points in the Boolean space. In the example, each of these represents a 4-variable product term (a minterm). As a result of applying the method, the second column will contain implicants that form edges in the Boolean space: 3-variable product terms. After another iteration of the method, the third column will contain larger implicants that represent planes in the Boolean space: 2-variable terms. A third iteration will find implicant 3-dimensional cubes in the space: 1-variable terms.

We begin the method by filling in the first column of the table as follows. Each group of the minterm indices of the on-set and don't-care-set is separated by a blank line. The first group has no 1s in the indices, the second has one 1 in each index, the third has two 1s in each index, and so on.

To apply the Uniting theorem systematically, compare the elements in the first group against each element in the second. If they differ by a single bit, it means that the minterms the numbers represent are adjacent in n-dimensional Boolean space. For example, 0000 = $\bar{A}\bar{B}\bar{C}\bar{D}$ and 0100 = $\bar{A}B\bar{C}\bar{D}$ can be combined into the implicant $\bar{A}\bar{C}\bar{D}$ according to the Uniting theorem. The latter term is represented symbolically by 0-00. Every time a new implicant is formed, it is placed in the next column. Since each group differs in its 1s count by one, it is sufficient to restrict comparisons to adjacent groups to detect when the uniting theorem can be applied.

Let's apply the Quine-McCluskey algorithm to the whole first column. We begin with the first group (no 1s) against the second group (one 1). 0000 is compared against 0100 and 1000, yielding terms for the second column of 0-00 and -000, respectively. Every time a term contributes to a new implicant, it receives a check mark (√). This is how we remind ourselves that the implicant is not prime: it was combined with some other element to form a larger implicant.

We repeat for the second group against the third group. 0100 combines with 0101 and 0110, giving 010- and 01-0 in the second column. 1000 combines with 1001 and 1010, resulting in 100- and 10-0.

Now we try the third group against the fourth group. 0101 combines with 0111 and 1101 to give 01-1 and -101. 0110 combines with 0111 to yield 011-. 1001 combines with 1101 to give 1-01. 1010 does not combine with any element of the fourth group.

When we compare the fourth to the fifth group, two additional terms are added: -111 and 11-1.

The procedure is repeated for the groups in column II. For the elements to be combined, they must differ by a single bit and must have their "-", the variables eliminated by the uniting theorem, in the same bit position. This corresponds to taking 1-dimensional cubes and trying to combine them into 2-dimensional cubes. The elements of the first group do not combine with any elements of the second group. We mark these two elements of the first group in column II with asterisks (*) because they are prime implicants: we have expanded them as much as possible.

In the second and third groups, 010- can be combined with 011-, yielding 01-- for the third column. 01-0 and 01-1 are combined to derive the same condensed term, so we only list it once. 100- and 10-0 cannot be combined further and are prime implicants.

Between the third and fourth groups, only the additional term -1-1 is added to the third column, derived from the combinations of -101 and -111, and 01-1 and 11-1.

The elements of the third column cannot be reduced further. Both are marked as prime implicants. Since no new implicants are added, we have found all prime implicants, and the first phase of the algorithm can now terminate.

The algorithm has found the following prime implicants:

$$0\text{-}00 = \bar{A}\bar{C}\bar{D} \qquad \text{-}000 = \bar{B}\bar{C}\bar{D}$$
$$100\text{-} = A\bar{B}\bar{C} \qquad 10\text{-}0 = A\bar{B}\bar{D}$$
$$1\text{-}01 = A\bar{C}D \qquad 01\text{-}\text{-} = \bar{A}B$$
$$\text{-}1\text{-}1 = BD$$

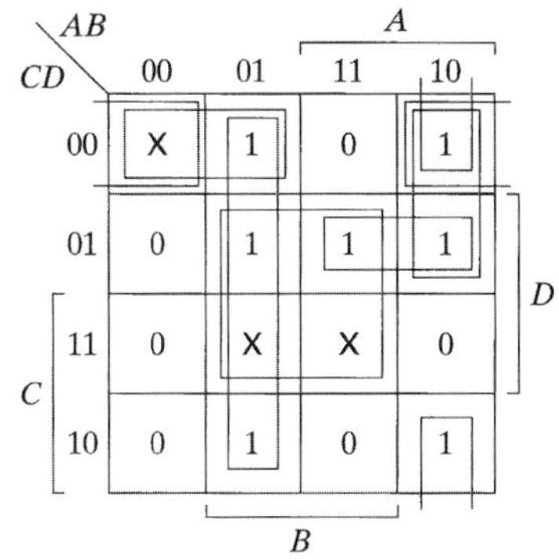

Figure 3.13 Prime implicants found by the Quine-McCluskey method.

These are shown circled in the K-map of Figure 3.13. They are exactly the same as the prime implicants we found in the previous section (see the identical Figure 3.10(e)). Note that, as before, we treated the don't cares as though they were 1s to get the largest prime implicants possible.

Finding the Minimum Cover The second step of the method is to find the smallest collection of prime implicants that cover the complete on-set of the function. This is accomplished through the *prime implicant chart* (as opposed to the implication chart used in the first phase), as shown in Figure 3.14.

The prime implicant chart is organized as follows. The columns are labeled with the minterm indices of the on-set. The rows are labeled with the minterms covered by a given prime implicant. This is

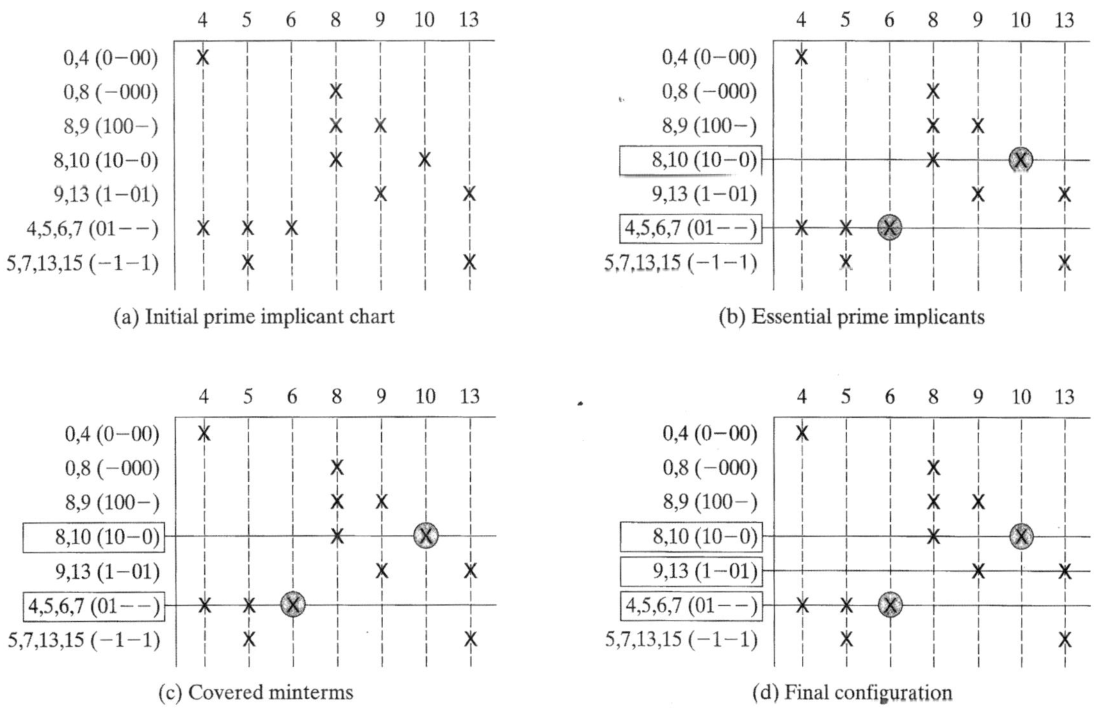

Figure 3.14 The prime implicant chart.

done by taking the indices of the prime implicant representation and replacing each "-" by all possible combinations of 1s and 0s. For example, -1-1 becomes 0101, 0111, 1101, 1111, which are the indices of the minterms m_5, m_7, m_{13}, and m_{15}. An X is placed in the (row, column) location if the minterm represented by the column is covered by the prime implicant associated with the row. The initial configuration is given in Figure 3.14(a). Note that don't cares are not included in the columns as we do not need to cover them.

Next, we look for essential prime implicants. These are immediately apparent whenever there is a single X in any column. This means that there is a minterm that is covered by one and only one prime implicant. These essential prime implicants must participate in the final cover. We place a line through the column and row in which the essential prime implicant has been found and place a box around the prime. This is shown in Figure 3.14(b).

The essential prime implicants usually cover additional minterms. We cross out any columns that have an X in a row associated with an essential prime. These minterms are already covered by the essential primes and we do need to be concerned with them any longer in finding a complete cover. This is shown in Figure 3.14(c).

In the example, two minterms are still uncovered, represented by the columns 9 and 13. The final step is to find as few primes as possible that cover the remaining minterms. In our example, the single prime implicant 1-01 covers both of these. In contrast, we could have used 100- and -1-1, but this would have added two terms to our solution instead of one. Adding 1-01 to the two essential prime implicants completes the cover. This is shown in Figure 3.14(d). The solution found here is identical to the one we found earlier in Figure 3.10(f).

3.2.2 *Espresso* Method

Unfortunately, the number of prime implicants grows very quickly as the number of inputs increases. It can be shown that the upper bound on the number of prime implicants is $3^n/n$. Finding a minimum set cover is also known to be a very difficult problem, a so-called "NP-complete" problem. This means that there are not likely to be any efficient algorithms for solving it and we can only guarantee that we've found the best solution if we look at all possible solutions.

Thus, much of the work in logic minimization has concentrated on heuristic methods to perform these two steps more quickly, finding a good solution rapidly rather than guaranteeing a minimum solution will be found. The primary technique avoids generating all prime implicants by judiciously computing a subset of primes that still cover the on-set. In this subsection, we will examine the algorithms and techniques used in *espresso*.

Algorithms used in Espresso *Espresso* is a program for two-level Boolean function minimization, developed at the University of California at Berkeley, and now a common subroutine for many logic minimization tools. It combines many of the best heuristic techniques developed in earlier programs, such as *mini* and *presto*. Although a detailed explanation of the operation of *espresso* is beyond the scope of this book (see Brayton, et al., in the suggestions for further reading at the end of this chapter), the basic ideas employed by the program are not difficult to understand. They are as follows:

STEP 1 Rather than start by generating all implicants and then finding those that are prime, *espresso* expands implicants into their maximum size. Implicants that are covered by an expanded implicant are removed from further consideration. This process is called EXPAND. How well this works depends critically on the order and direction in which implicants are expanded. Much of the power of *espresso* lies in its methods for directing and ordering the expansion.

STEP 2 An irredundant cover (that is, one for which no proper subset is also a cover) is extracted from the expanded implicants. The method is essentially the same as the Quine-McCluskey prime-implicant chart method. This step is called IRREDUNDANT COVER.

STEP 3 At this point, the solution is usually quite good, but under certain conditions it can still be improved. There might be another cover with fewer terms or the same number of terms but fewer literals. To try to find a better cover, *espresso* first shrinks the prime implicants to the smallest size that still covers the logic function. This process is called REDUCE.

STEP 4 Since reduction yields a cover that no longer consists of only primes, the EXPAND, and IRREDUNDANT COVER steps are repeated in such a fashion that alternative prime implicants are derived. *Espresso* will continue repeating these steps as long as it generates a cover that improves on the last solution found.

STEP 5 A number of other strategies are used to improve the result or to compute it more quickly. These include: (a) early identification and extraction of essential prime implicants, so they need not be revisited during Step 4; (b) using the function's complement to check efficiently whether an EXPAND step actually increases the coverage of the function (the minterms covered by an expanded implicant may already be covered by another expanded implicant, so the newly expanded implicant should not be placed in the cover); and (c) a special last step that guarantees no single prime implicant can be added to the cover in such a way that two primes can then be eliminated.

0100	1
0101	1
0110	1
1000	1
1001	1
1010	1
1101	1
0000	–
0111	–
1111	–

Figure 3.15 ***Espresso* input table.**

Input to tools that use *Espresso* is usually provided in the form of an encoded truth table using a similar notation to that of the Quine-McCluskey implication table. For the example of Figure 3.10, the input is shown in Figure 3.15. There are 10 terms specified: seven on-set terms and three don't-care terms. Each term is listed by its minterm

```
1-01  1
10-0  1
01--  1
```

Figure 3.16 ***Espresso* output table.**

index, encoded in binary, followed by a 1 or a "-", indicating a don't care. Off-set elements can also be specified and are represented with a 0 after the minterm index. The output minimized truth table is shown in Figure 3.16. The encoding for the minimum cover is identical to that used in the Quine-McCluskey method. The three terms 1-01, 10-0, and 01-- correspond to $A\bar{C}D$, $A\bar{B}\bar{D}$, and $\bar{A}B$, respectively.

EXAMPLE 3.7 USE OF ESPRESSO

To see how the iteration of REDUCE, EXPAND, IRREDUNDANT COVER can improve the cover, consider the four-variable K-map of Figure 3.17. Figure 3.17(a) shows the primes as found by *espresso* after executing Steps 1 and 2 for the first time. It has four prime implicants and is an irredundant cover, but this is not the minimum cover possible.

The results of the REDUCE step are shown in Figure 3.17(b). The prime implicant $\bar{C}D$ has been reduced to the implicant (no longer prime) $A\bar{C}D$, and $C\bar{D}$ has been reduced to $\bar{A}C\bar{D}$ (also no longer prime). The particular choice of reductions depends on heuristics and *espresso*'s order of execution. The result of the second iteration of EXPAND is shown in Figure 3.17(c). *Espresso* retains the last irredundant cover,

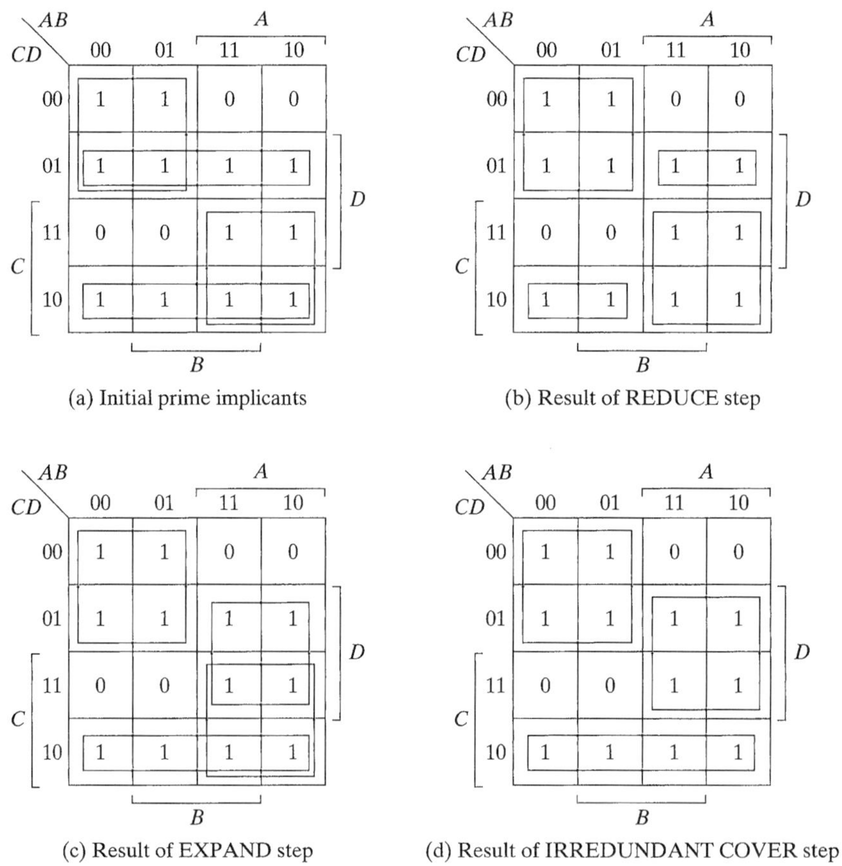

Figure 3.17 **Four-variable K-maps.**

and its expansion algorithms guarantee that it never generates the same cover twice. The IRREDUNDANT COVER extracted by *espresso* is given in Figure 3.17(d). This three-product-term cover is indeed an improvement on the original result in Figure 3.17(a).

3.2.3 Realizing S-o-P and P-o-S Logic Networks

In this section, we will be concerned with how to express logic networks solely in terms of NAND and NOR gates. The canonical forms you have studied so far are expressed in terms of AND and OR gates, but you rarely will encounter these in digital systems. The underlying technologies are more efficient at implementing NAND and NOR gates. In fact, AND and OR gates are most commonly realized by following a NAND or NOR gate with an inverter. In addition, NAND and NOR functions are *complete;* that is, a function expressed in terms of AND, OR, and NOT operations can be implemented solely in terms of NAND or NOR operations. Frequently, you will be confronted with the task of mapping a network with an arbitrary number of levels of AND and OR gates into one that consists only of NAND or NOR gates. We will begin the discussion with two-level networks and extend it to multilevel networks later in the chapter.

Visualization: DeMorgan's Law and Pushing Bubbles The conversion process depends critically on DeMorgan's law. Recall that

$$\overline{AB} = (\bar{A} + \bar{B}) \qquad \overline{A + B} = (\bar{A}\bar{B})$$

and that

$$AB = (\overline{\bar{A} + \bar{B}}) \qquad A + B = (\overline{\bar{A}\bar{B}})$$

In essence, the first expression above states that a NAND function can be implemented just as well by an OR gate with its inputs complemented. Similarly, a NOR function can be implemented by an AND gate with its inputs complemented. The conversion from one form to the other is often called "pushing the bubble." This is simply a way to remember DeMorgan's law As the bubble on the output of an AND gate "pushes through" the AND shape from the output toward the inputs, it changes the gate to an OR shape with bubbles on the inputs and no bubble on the output. Similarly, pushing the bubble through an OR shape transforms it to an AND shape with bubbles on the inputs.

Figure 3.18 summarizes the relationship between OR and NAND gates. An OR gate is logically equivalent to a NAND gate with inputs complemented and a NAND gate is equivalent to an OR gate with inputs complemented. There is the same kind of relationship between AND and NOR, an AND gate is equivalent to a NOR gate with inputs complemented and a NOR gate is equivalent to an AND gate with inputs complemented. These last two relationships shown in Figure 3.19. The corresponding schematic symbols can be freely exchanged without altering the logic function.

A	$\bar{A}$	B	$\bar{B}$	$A+B$	$\overline{\bar{A}\cdot\bar{B}}$	$\bar{A}+\bar{B}$	$\overline{A\cdot B}$
0	1	0	1	0	0	1	1
0	1	1	0	1	1	1	1
1	0	0	1	1	1	1	1
1	0	1	0	1	1	0	0

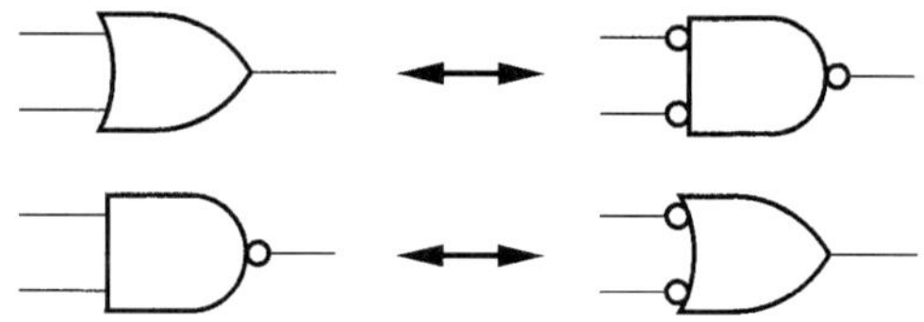

Figure 3.18 OR/NAND equivalences.

A	$\bar{A}$	B	$\bar{B}$	$A\cdot B$	$\overline{\bar{A}+\bar{B}}$	$\bar{A}\cdot\bar{B}$	$\overline{A+B}$
0	1	0	1	0	0	1	1
0	1	1	0	0	0	0	0
1	0	0	1	0	0	0	0
1	0	1	0	1	1	0	0

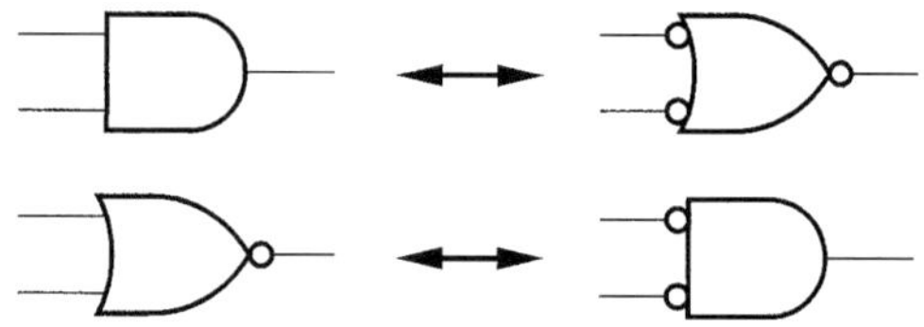

Figure 3.19 AND/NOR equivalence.

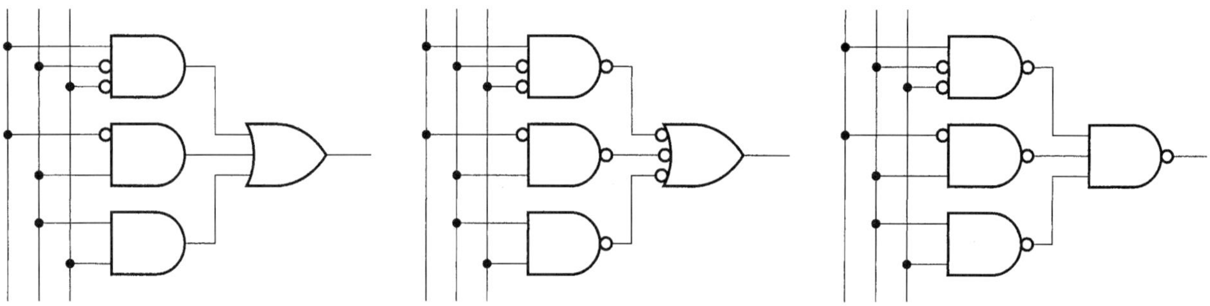

(a) (b) (c)

Figure 3.20 AND/OR to NAND/NAND.

EXAMPLE 3.8 AND/OR CONVERSION TO NAND/NAND NETWORKS

Consider the AND-OR network in Figure 3.20(a). The bubbles on the first-stage gate inputs indicate that these should be complemented. Let's replace the first-level AND gates by their NAND equivalents and complement the inputs to the OR gate so that we do not change the circuit's logic function. The equivalent circuit is shown in Figure 3.20(b). We can replace the second level OR gate with complemented inputs by its more conventional, alternative form: the NAND gate. The result is shown in Figure 3.20(c).

EXAMPLE 3.9 AND/OR CONVERSION TO NOR/NOR NETWORKS

Suppose you are now asked to map an AND/OR network into a NOR-only network. We can start by complementing the inputs to the first-level AND gates to create NOR-gate equivalents. Of course, every time a new inversion is introduced, it must be balanced by a complementary inversion. We call this "conserving bubbles." To accomplish this, we introduce additional inverters at the inputs. We

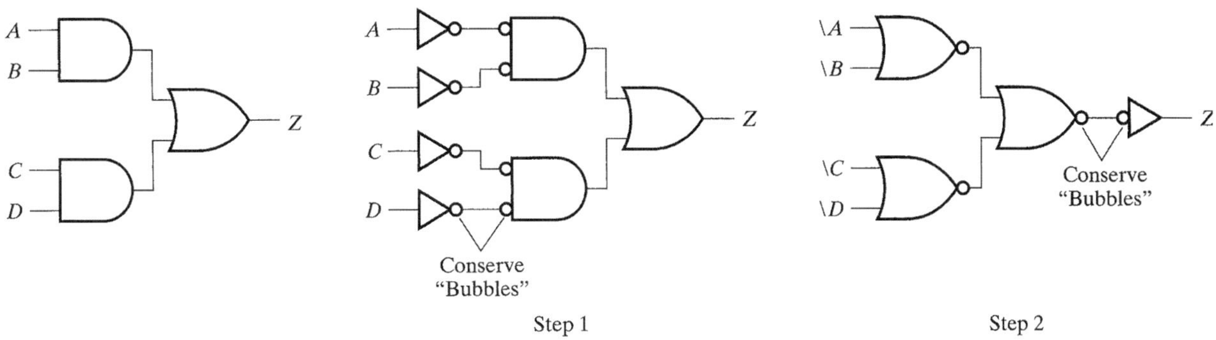

Figure 3.21 **AND/OR conversion to NOR/NOR.**

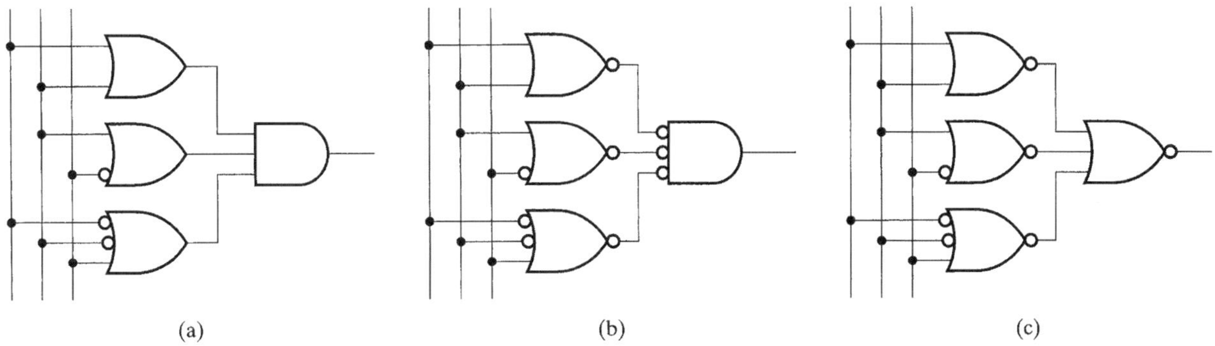

Figure 3.22 **OR/AND conversion to NOR/NOR.**

can then complement the output of the second-level OR gate to make a NOR after introducing a matching inverter on the output to keep the logic constant. This two-step process is illustrated in Figure 3.21. Note that the two AND gates with complemented inputs created in Step 1 are replaced by NOR gates (their equivalent form) in Step 2.

To keep track of the need for inverted inputs to the first-level gates, we use the notation: \A, \B, \C, and \D. Since it is common to have both a Boolean variable and its complement available as circuit inputs, the conversion may not necessarily lead to additional inverters. To eliminate the inverter at the output, we may be able to use the complement of the function, \Z, wherever in the rest of the circuit Z was needed or we may need to add an extra inverter (we do so in this case).

EXAMPLE 3.10 OR/AND CONVERSION TO NOR/NOR NETWORK

Now let's consider a gate implementation for a simple expression in product of sums form. We can map this expression into a NOR/NOR network simply by replacing the OR gates with NOR gates and the AND gate with a NOR gate (an AND gate with inverted inputs). You can see in Figure 3.22 that the inversions are appropriately conserved.

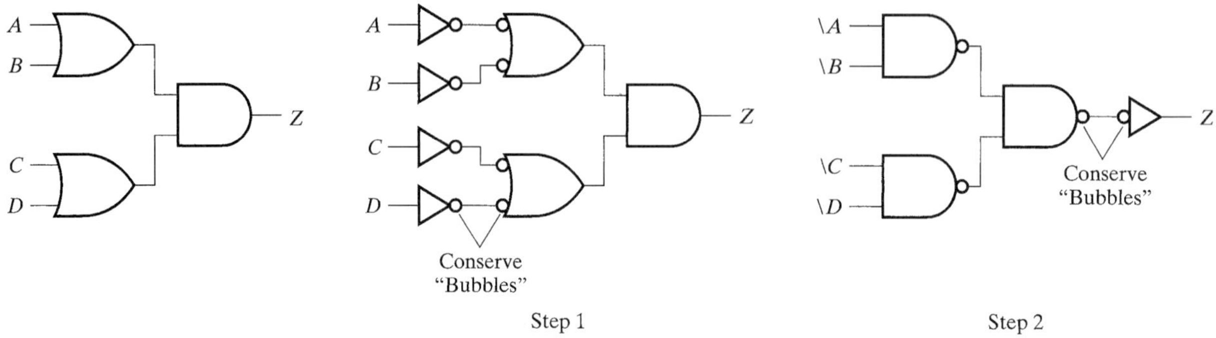

Figure 3.23 OR/AND conversion to NAND/NAND.

EXAMPLE 3.11 OR/AND CONVERSION TO NAND/NAND NETWORKS

Implementing the expression using NAND-only logic introduces exactly the same problems we have already seen in Figure 3.21. The correct transformation replaces the OR gates with NAND gates (OR gates with inverted inputs) and the AND gate with a NAND gate. To maintain equivalence with the original function, we place inversions at the inputs (in the form of an annotation) and at the output (in the form of an inverter). This is shown in Figure 3.23.

This section illustrated some of the practical issues in how to transform logic functions so that they are more easily mapped to available gates. This general problem is referred to as *technology mapping*. We will discuss this problem further as we discuss simplification of multilevel logic and in more detail in Chapter 4 where we will introduce a wide array of implementation technologies.

3.3 Multilevel Simplification

We can obtain minimized two-level logic networks from the canonical sum of products or product of sums form, by applying the appropriate reduction methods (K-maps, Quine-McCluskey, a two-level logic minimizer like *espresso*, etc.). Signal propagations can be fast, because no signal has to travel through more than two gate levels. The drawback is the potential for large gate fan-ins, which can reduce gate performance and increase circuit area in some technologies. In many technologies, gates with more than four inputs are rare, if they exist at all. In a practical gate-level realization of a network, the large fan-in gates must be replaced by a multiple-level network of smaller fan-in gates. This has motivated much recent interest in multilevel logic synthesis.

An optimal two-level network is one that uses the smallest number of product terms and literals to realize a given truth table. It is not so

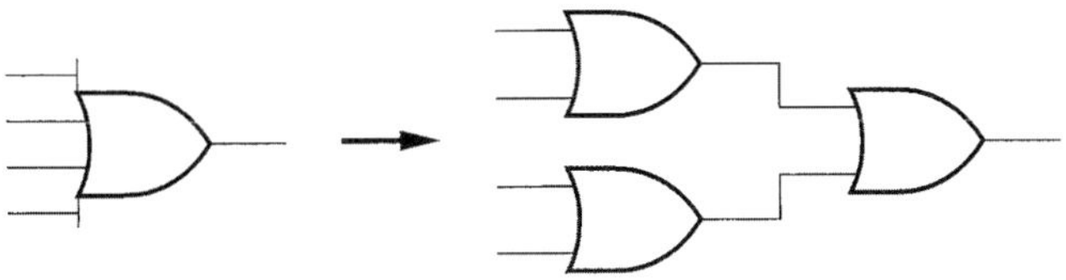

Figure 3.24 **Effects of technology mapping.**

easy to define optimality for multilevel networks. Is the network optimal if it has the smallest number of gates? Or is it more critical to have the fewest literals in the resulting expression? Hence, the process with multilevel systems is called *synthesis* rather than *optimization*. The goal is to create or "synthesize" a reasonable multilevel implementation, without any pretense that the result is the best that is possible.

The synthesis process involves two steps. The first, *technology-independent* stage factors out common sub-logic to reduce gate fan-ins, increasing the number of gate levels as a side effect. This step is independent of the kinds of gates that eventually will be used to implement the network. It works by exploiting basic mathematical properties of Boolean expressions.

The second, *technology-dependent* stage maps the resulting factored Boolean equations into a particular implementation using a library of available gates. For example, if only 2-input OR gates are found in a particular library, then a 4-input OR gate would have to be mapped into three 2-input OR gates (see Figure 3.24). Such reorganizations of the logic network can introduce additional levels of logic in the function's implementation, possibly affecting the delay through the circuit. We'll save further discussions of this issue after we introduce implementation technologies in the next chapter.

Unfortunately, multilevel simplification is a much more difficult problem than two-level simplification. The approach taken by researchers in multilevel simplification, which seems to work reasonably well in most situations, is to provide basic transformations that designers can apply to multilevel formulas so that they can guide the process to the kind of realization they are most interested in. We'll see that automated methods take the most successful of these strategies and provide them to all designers so that each person does not have to re-derive them (a process that can be very time consuming and only gets better with experience). There are strategies, or *scripts*, for the fewest literals, smallest delay, lowest fan-in gates, etc.

There are five basic operations for manipulating multilevel networks: factoring, decomposition, extraction, substitution, and collapsing. In the following subsections, we will describe each of these operations and illustrate how each alters the multilevel expression with the simple example we ended with in Section 2.7 and is re-drawn in Figure 3.25.

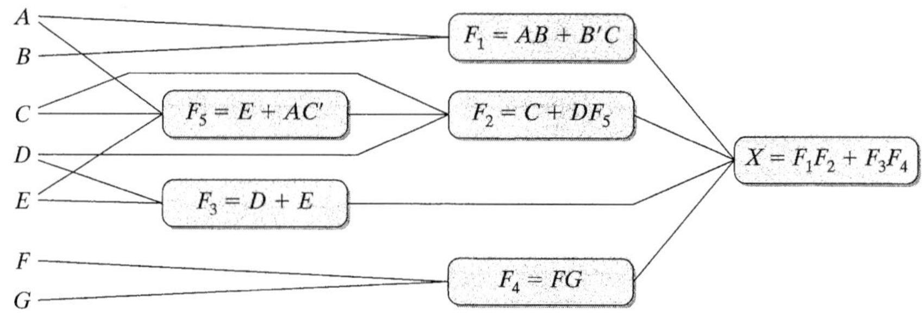

Figure 3.25 **Graphical representation of a factored form with complex nodes.**

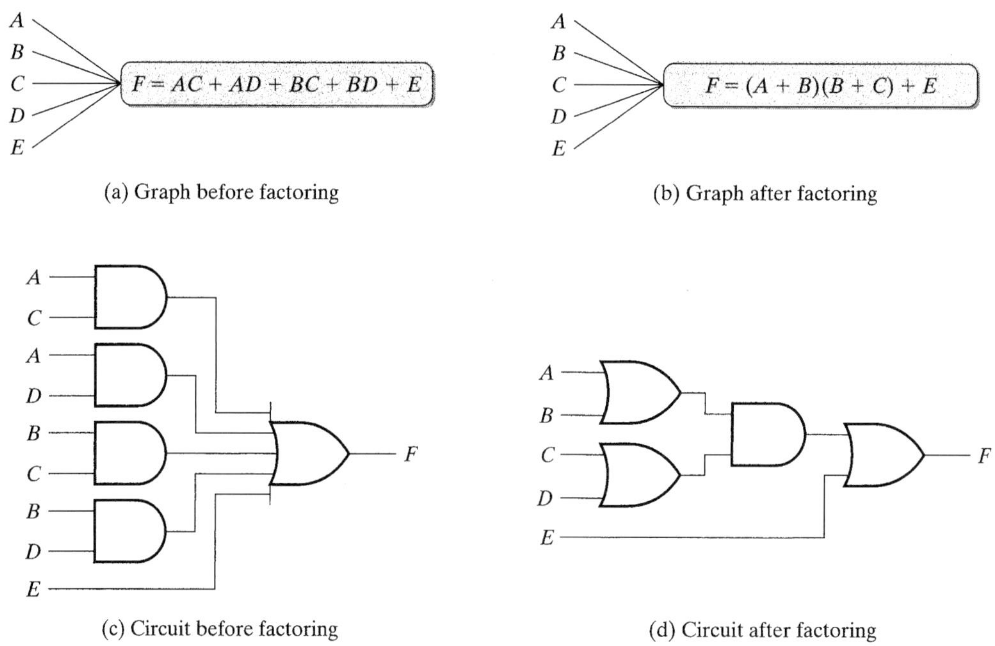

Figure 3.26 **Effects of factoring operation.**

Factoring Factoring takes an expression in two-level form and re-expresses it as a multilevel function without introducing any intermediate subfunctions. Thus, factoring simply rewrites the expression within a node rather than changing the structure of the graph (see Figure 3.26(a,b)). It is used just before decomposition and/or extraction to identify potential common subexpressions.

As an example, let's consider the following function in sum-of-products form. It has nine literals and can be implemented with five gates (Figure 3.26(c)):

$$F = AC + AD + BC + BD + E$$

After factoring, the number of literals is reduced to five:

$$F = (A + B)(C + D) + E$$

This can be implemented with four gates (Figure 3.26(d)). Note that the graph representation for F does not change. It is still a single node with five inputs. Factoring only changes the representation of a function within a node.

Decomposition Decomposition takes a single Boolean expression and replaces it by a collection of new expressions. It is applied to functions that have already been factored. Consider the function

$$F = ABC + ABD + \bar{A}\bar{C}\bar{D} + \bar{B}\bar{C}\bar{D}$$

This expression is in reduced sum of products form and has 12 literals. It requires nine gates, counting inverters, for implementation (see Figure 3.27(c)). It would be represented by a single node in our complex graph representation. However, the expression can first be factored into:

$$F = (AB)(C + D) + (\bar{A} + \bar{B})(\bar{C}\bar{D}) = (AB)(C + D) + \overline{(AB)}\,\overline{(C + D)}$$

and then decomposed individually into three much simpler functions:

$$F = XY + \bar{X}\bar{Y} \qquad X = AB \qquad Y = C + D$$

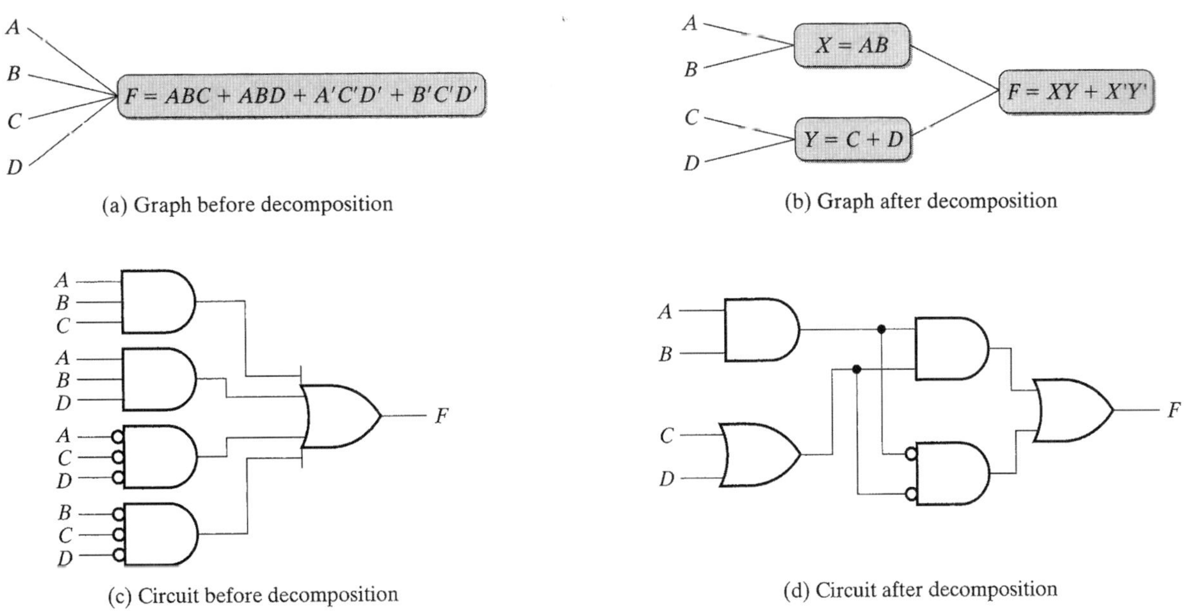

(a) Graph before decomposition

(b) Graph after decomposition

(c) Circuit before decomposition

(d) Circuit after decomposition

Figure 3.27 **Effects of decomposition operation.**

This resulting set of functions has eight literals and requires seven gates for implementation (see Figure 3.27(d))—nothing is free, though, and the number of gate levels has increased from two to three. These functions now require three nodes to be represented in our graph rather than one (see Figure 3.27(a,b)). As you would expect from the name, decomposition breaks a node into smaller, simpler pieces.

Extraction Whereas you apply decomposition to a single function, you apply extraction to a collection of functions. This operation identifies common subexpressions in the collection of functions to which it is applied. Performing extraction requires that the functions be expressed in terms of their factors, and then the common factors must be "pulled out."

Let's look at an example of extraction. We start with three functions represented as three nodes

$$F = (A + B)CD + E$$
$$G = (A + B)\overline{E}$$
$$H = CDE$$

Note that the extraction operation does not require the functions to be in a two-level form.

In this example, the extraction operation discovers that the subexpressions, $X = (A + B)$ and $Y = (CD)$, are common to F, G, and F, H, respectively. These sub-expressions are called "primary divisors" or, more technically, *kernels* and *cubes*. We have seen cubes already, they are simply sub-cubes of the larger Boolean cube for the function. Kernels are not cubes but disjunctions of cubes (cubes ORed together to form a factor that combines more than a single product term). Re-express in these terms, the functions can be rewritten as:

$$F = XY + E$$
$$G = X\overline{E}$$
$$H = YE$$
$$X = A + B$$
$$Y = CD$$

The original collection of functions contains 11 literals and requires eight gates for implementation. (In Figure 3.28(c), the bubble at the input of the gate that computes G counts as one inverter gate because it is internal to the circuit and not at the primary inputs.) The revised set of functions, after extraction, still contains 11 literals but now needs only seven gates for its implementation. The graph now has five nodes, the modified versions of the three original ones plus the two new nodes to represent the new subexpressions (see Figure 3.28(a,b)). You can see in Figure 3.28(d) that the single-level implementation for H has been replaced by a two-level implementation after extraction. The number of gates is reduced, but the function H may incur worse delay than in the original version.

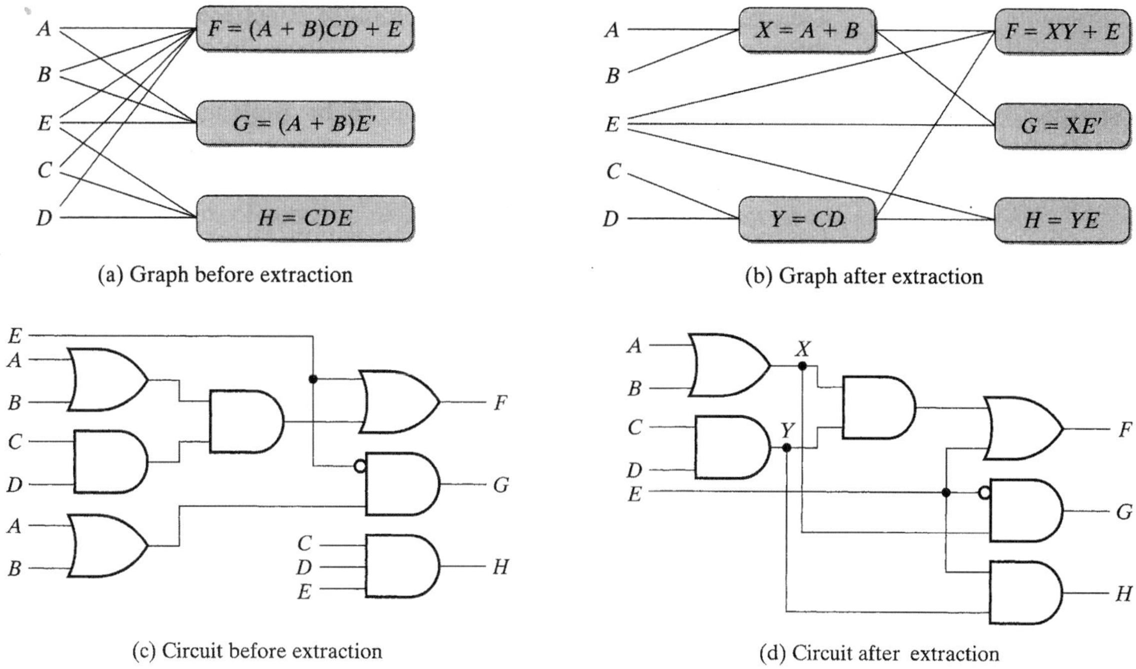

Figure 3.28 **Effects of extraction operation.**

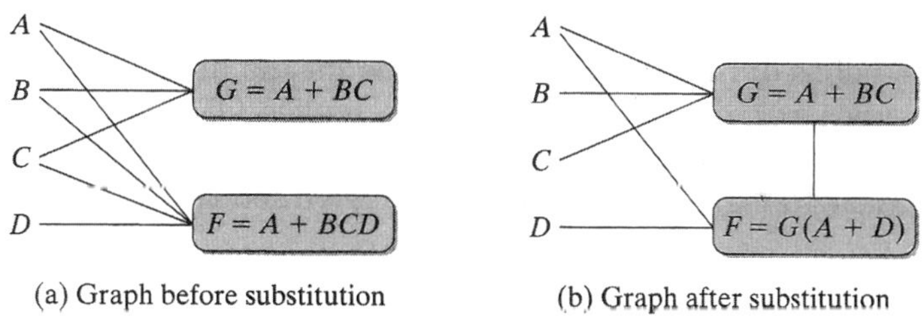

Figure 3.29 **Effects of substitution operation on circuit graph.**

Substitution Substituting a function G into a function F re-expresses F in terms of G. For example, if $F = A + BCD$, and $G = A + BC$, then F can be rewritten in terms of G as follows:

$$F = A + BCD = G(A + D)$$

Once common subexpressions have been identified, substitution can be used to reexpress functions as factored forms over the subexpressions. This operation also changes the structure of the graph by adding an arc from the node for G to the node for F and changing the definition of F. The number of literals for F is reduced from four to three. This is advantageous if we need G elsewhere as well, otherwise the three literal cost for G is not amortized. A new connection from G to F is created in the graph as shown in Figure 3.29.

Collapsing Collapsing is the reverse of substitution. It might be used to reduce the number of levels of logic to meet a timing constraint. As an example, we can collapse G back into F:

$$F = G(A + D)$$
$$F = (A + BC)(A + D)$$
$$F = AA + AD + ABC + BCD$$
$$F = A + BCD$$

If G is not needed elsewhere in our circuit, then we have reduced the cost of implementing F from six literals (three for F and three for G) to only four. The graph is also modified in the opposite way as substitution (going from the graph of Figure 3.29(b) to (a)) in that an arc that existed between nodes G and F is now removed.

Clean-up operations on a graph may be necessary as the by-products of multiple transformations accumulate. For example, the node for G in the collapsing step described above may be removed if it is no longer used by any other expression. Also, small nodes that have been greatly simplified may be collapsed into the nodes that use their values. This is especially the case for nodes that end up being simple identity functions or simple inversions.

Polynomial Division and Multilevel Factoring All of the multilevel operations have strong analogies with the multiplication and division of polynomials. The strategy is to rewrite the expression for a function F in terms of the subexpressions P, Q, and R, which represent the divisor, quotient, and remainder, respectively. In generic terms, F is written as

$$F = PQ + R$$

As a more concrete example, given the two expressions:

$$X = AC + AD + BC + BD + E$$
$$Y = A + B$$

We could write X "divided" by Y as follows:

$$X = Y(C + D) + E$$

The divisor is $Y = (A + B)$, the quotient is $(C + D)$, and the remainder is E. Expanding the expression with the distributive law would yield the original equation for X.

Finding divisors is a considerably more difficult problem in Boolean algebra are considered. Multiplying Boolean expressions can yield results that are very different from what we would expect from polynomial arithmetic because of the variety of simplification theorems you can apply. For example, consider the functions F and G:

$$F = AD + BCD + E$$
$$G = A + B$$

Under the normal rules of algebra, G does not divide into F. The *algebraic divisors* of F are D and $(A + BC)$. That is, F can be divided by D, leaving the quotient $A + BC$ with remainder E. It can also be divided by $A + BC$, leaving D as the quotient and E as the remainder.

However, if we apply the rules of Boolean algebra, then G does divide into F (with E as a remainder). We call G a *Boolean divisor* of F and can write the quotient of F divided by G as

$$F/G = (A + C)D$$

This is because F can be rewritten as

$$F = [G(A + C)D] + E$$
$$F = [(A + B)(A + C)D] + E$$
$$F = (AA + AC + AB + BC)D + E$$
$$F = (A + BC)D + E$$
$$F = AD + BCD + E$$

The principal difference is our ability to use the absorption theorem of Boolean algebra which is has no analog in polynomial arithmetic. The existence of Boolean divisors greatly increases the number of potential factorings of a set of expressions. It is not uncommon to restrict the search to the much easier-to-find algebraic factors.

It should be clear from this discussion that the challenge in multilevel logic simplification is to find good divisors. These lead to factored expressions with the greatest number of common subexpressions. By factoring these out, we minimize the number of literals needed to express a set of functions. The five graph operations discussed above must be applied in an order that guides the transformations of the circuit toward a result that meets our optimization criteria: smallest size, smallest gates, fewest levels, etc.

Another complication to multilevel simplification is the handling of "don't care" information. With a circuit graph consisting of many interconnected nodes, it is not as straightforward to express don't cares for each of them. Input and output "don't care" conditions must be propagated throughout the network. In addition, other forms of don't cares arise due to the multilevel structure of the logic itself. These are called *structural don't cares*. An example is shown in Figure 3.30. Notice that a, b, and x can't possibly be all equal to 1 at the same time. The larger node can take advantage of this fact in its simplification. The details of don't-care manipulation in multilevel circuits are quite elaborate and outside the scope of this text.

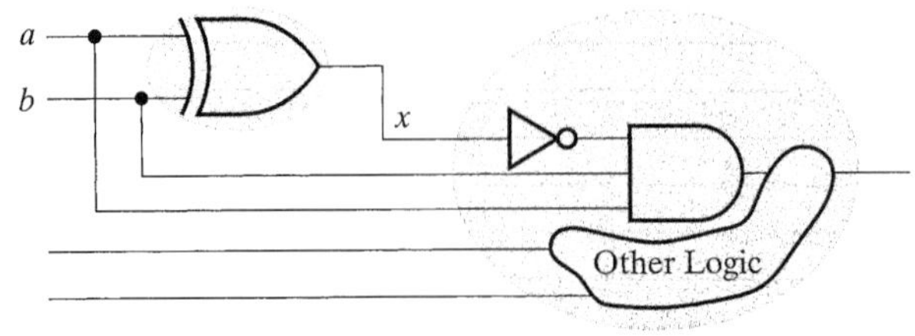

Figure 3.30 **An example of structural don't cares.**

3.4 Automating Multilevel Simplification

Good scripts to guide multilevel logic transformations are valued highly. They usually represent the accumulated experience of many designers and the sequence of transformations that they have found to work well for a large class of circuits. Often scripts are specialized to particular types of circuits (e.g., large numbers of inputs with low-complexity functions) or for particular optimization criteria (e.g., faster implementations that sacrifice size). Highly evolved scripts are invaluable to novice or less-experienced designers as they can be used to obtain the same results as the more experienced designers that developed them originally.

One of the first systems to take the script approach to multilevel simplification was the Multilevel Implementation System (MIS) tool developed at the University of California at Berkeley. The complete range of operations it supports is much richer than the basic transformations we described in the previous section. You can learn much more about MIS by looking at some of the suggested readings at the end of this chapter. For now, let's look at the sequence of transformations that is typical of most multilevel logic optimization scripts.

3.4.1 Multilevel Logic Optimization Scripts

A typical optimization script begins by reading in a set of Boolean expressions corresponding to the nodes of a multilevel network. The first thing it does is to look over the nodes and see if there are any that are simple, for example, a single literal that is best collapsed into its fan-out nodes. The reason these nodes can exist is that the logic specification may have come from a computer-aided design tool that generates logic specification from a hardware description language (we'll see an example of such a language later in this chapter).

Next, each node is simplified using a two-level logic optimization program such as *espresso*. This is done so that the nodes are expressed in a more succinct form than the way they may have been originally generated. The elimination step may be repeated just in case some nodes are reduced to a very simple form by the simplification step.

At this point, nodes are typically factored. This is a two-step process that involves first extracting good common divisors (both cubes and kernels). After factoring, common sub-expressions are identified and substituted so as to remove redundant logic. This process is repeated starting with large divisors and then considers smaller and smaller ones. The idea is to take advantage of large common sub-expressions first.

The technology-independent part of the script usually ends with a clean-up step that again looks to collapse nodes that end up being too small (one or two literals).

Finally, a technology-mapping step is performed using a library of available logic gates. Nodes are matched to the parts and may need to be broken up into smaller pieces. This process may introduce new intermediate nodes and, because of this, usually includes some collapsing and clean-up steps as well.

EXAMPLE 3.12 A SESSION WITH AN OPTIMIZATION SCRIPT

In the following sequence of operations, we will (1) start with Boolean equations that describe the combinational network to be manipulated (it is also possible to describe the network as truth tables, using *espresso* input format for each node rather than a Boolean expression), (2) initially perform a two-level simplification (in this case, we'll use *espresso* as a subroutine), (3) decompose the resulting functions into a multilevel network, and (4) map the decomposed network onto the gates of a given gate library.

We'll use the equations of our now very familiar example, the full adder already expressed in two-level canonical form:

$$\begin{aligned} Sum &= (A\bar{B}\bar{C}_{in}) + (\bar{A}B\bar{C}_{in}) + (\bar{A}\bar{B}C_{in}) + (ABC_{in}) \\ C_{out} &= (ABC_{in}) + (AB\bar{C}_{in}) + (A\bar{B}C_{in}) + (\bar{A}BC_{in}) \end{aligned}$$

These can be simplified using *espresso* to yield

$$\begin{aligned} Sum &= (A\bar{B}\bar{C}_{in}) + (\bar{A}B\bar{C}_{in}) + (\bar{A}\bar{B}C_{in}) + (ABC_{in}) \\ C_{out} &= (AB) + (AC_{in}) + (BC_{in}) \end{aligned}$$

And then factored into

$$\begin{aligned} Sum &= \bar{C}_{in}(A\bar{B} + \bar{A}B) + C_{in}(\bar{A}\bar{B} + AB) \\ C_{out} &= (AB) + C_{in}(A + B) \end{aligned}$$

After a decomposition we obtain two sub-expressions as separate nodes:

$$\begin{aligned} Sum &= \bar{C}_{in}\bar{X} + C_{in}X \\ C_{out} &= (AB) + C_{in}Y \\ X &= \bar{A}\bar{B} + AB \\ Y &= A + B \end{aligned}$$

Finally, we map our nodes to available components. In this case, our library includes an XOR gate, an inverter, and a complex gate that implements the function $\overline{((F + G)H)}$. The resulting mapping is

$$\begin{aligned}
Sum &= C_{in} \oplus \overline{X} \\
C_{out} &= \overline{((F1 + G1)H1)} = \overline{F1}\,\overline{G1} + \overline{H1} \\
\overline{X} &= A \oplus B \\
F1 &= \overline{A} \\
G1 &= \overline{B}_{in} \\
H1 &= \overline{((F2 + G2)H2)} = \overline{F2}\,\overline{G2} + \overline{H2} \\
F2 &= A \\
G2 &= B_{in} \\
H2 &= C_{in}
\end{aligned}$$

Note that two complex gates and two inverters are used to implement C_{out} ($Y3$ is the same complex-gate equation after applying DeMorgan's law), while two XOR gates are used to implement *Sum*. The schematic for the corresponding circuit is shown in Figure 3.31.

At this point, we could also compute the size, cost, and delay of our circuit using analysis commands found in most optimization tools. The library includes this information for each type of gate. Typical values for area may be 40 versus 16 for the relative size of an XOR and inverter; 3 versus 1 for cost; and 4 versus 1 for delay. Delay estimates not only include the effects of large fan-in (an intrinsic delay of the gate available from the circuit library) but also fan-out, that is, when a gate output is used by many other gates as an input, its delay increases (this is an extrinsic property of CMOS gates that must be computed during the analysis of the circuit and can not be included in a library).

If our library contains only simple gates, our mapping result will be quite different. A library containing only simple AND, OR, and NOT gates may yield:

$$\begin{aligned}
Sum &= A\overline{C}_{out} + B\overline{C}_{out} + C_{in}\overline{C}_{out} + ABC_{in} \\
C_{out} &= BC_{in} + AC_{in} + AB
\end{aligned}$$

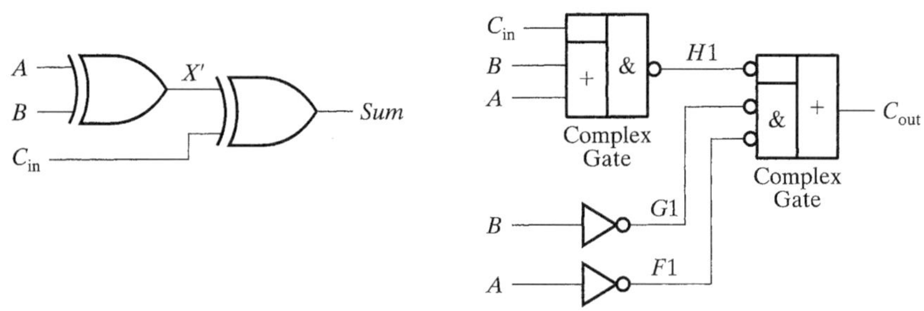

Figure 3.31 Implementation of full adder using a generic library of parts (note the two complex gates are duals of the same part).

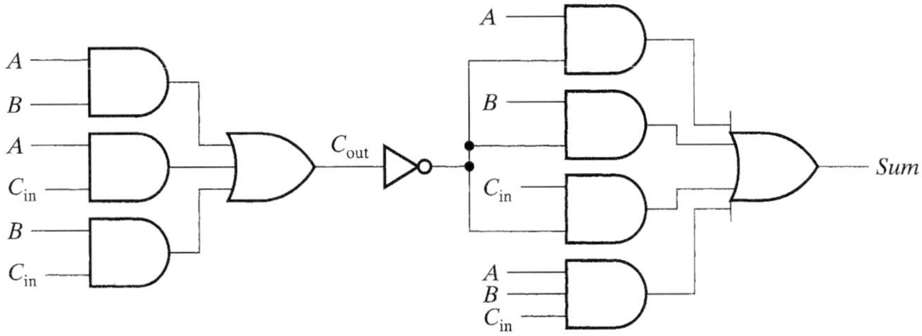

Figure 3.32 Multilevel implementation of full adder.

While the expression for C_{out} is identical to what we have seen before as well to the one derived by *espresso,* the expression for *Sum* is a bit surprising. Let's look at it in more detail.

Recall that the full adder *Sum* is $A \oplus B \oplus C_{in}$. However, our library cannot make use of XOR gates. Expressed in sum of products form, *Sum* becomes $Sum = (A\bar{B}\bar{C}_{in}) + (\bar{A}B\bar{C}_{in}) + (\bar{A}\bar{B}C_{in}) + (ABC_{in})$. Our new expression for *Sum,* by using $\bar{C}_{out}$, effectively reduces the number of literals from 12 to 9. However, there is another aspect to this solution. The two-level expression for *Sum* in the standard sum-of-products form has been replaced by a five-level expression (see Figure 3.32). It is likely this probably will have a large performance penalty—a larger delay. Clearly, our script is focusing on wiring complexity by emphasizing literal count with little or no regard for delay. This is still a useful script, however, as many parts of our designs are not performance critical and can be minimized for size as much as possible.

EXAMPLE 3.13 TWO-BIT BINARY ADDER

Let's look at a second example of multilevel minimization, the 2-bit binary adder. Recall that the inputs are two 2-bit binary numbers to be summed, represented by the inputs *A*, *B* and *C*, *D*, respectively. The output is a 3-bit binary number *X*, *Y*, *Z*. The result is as follows:

$$Z = \bar{B}D + B\bar{D} = B \oplus D\backslash$$
$$Y = \bar{W}\bar{A}C + \bar{W}A\bar{C} + WAC + W\bar{A}\bar{C} = W \oplus (A \oplus C)$$
$$X = AC + W(A + C)$$
$$W = BD$$

You can see that there are four functions, even though the function has only three outputs. A new intermediate result, denoted by *W*, has been introduced. Outputs *X* and *Y* have been expressed as functions of the new intermediate function *W*.

This solution represents considerable savings of literals compared to the solution we found in Section 3.1. In sum-of-products form, *X* required eight literals, *Y* used 20, and *Z* was expressed in four, a total of 32. The multilevel implementation described above uses 23 literals without using XOR gates, a 28% savings. Of course, there are performance implications of the multilevel implementation. The worst-case delay in the sum-of-products form is, as always, two gate levels (with potentially large fan-in gates and not counting inverted input literals). In the multilevel form, it is four gate levels (first level forms W, then a second-level inverter to create $\overline{W}$, then two more levels to form the sum-of-products for Y).

3.4.2 Realizing Multilevel Logic Networks

As was the case for two-level logic, we also need to be able to implement our multilevel logic in terms of common gates. In this section, we will be concerned with how to express multilevel logic networks solely in terms of NAND and NOR gates.

Generalization to Multilevel Circuits We can extend the transformation techniques we developed for two-level networks to multi-level networks. Consider the function

$$F = A(B + CD) + B\overline{C}$$

Its implementation in AND/OR form is shown in Figure 3.33(a). You can see how we have arranged the logic into alternating levels of AND and OR gates. This makes it easier to observe the places where the conversion to NAND/NAND gates can take place. You simply replace each AND with a NAND and each OR with a NAND in its "alternative" form (OR with inverted inputs).

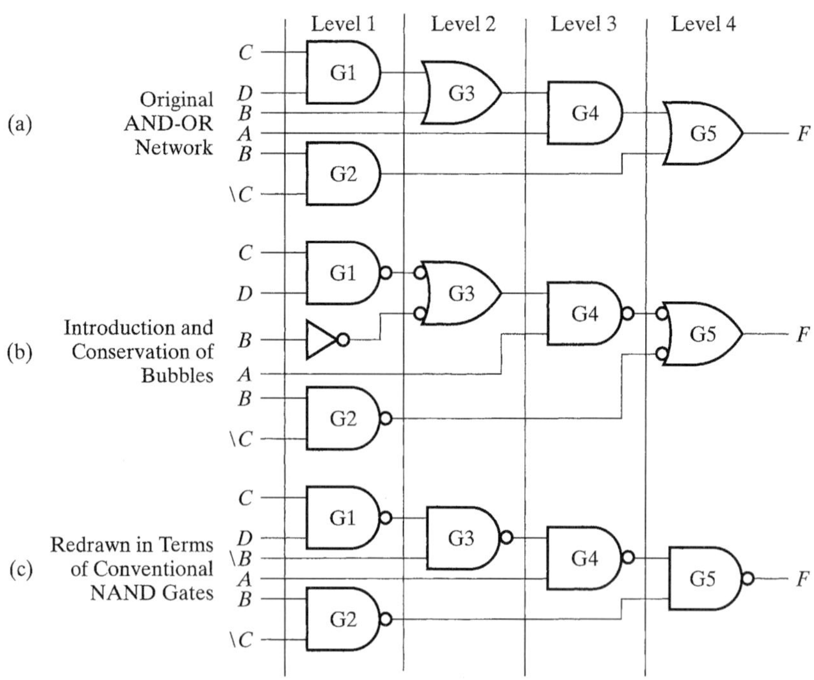

Figure 3.33 Multilevel conversion to NAND gates.

The application of this procedure is shown in Figure 3.33(b). We have grouped levels 1, 2 and 3, 4 into AND/OR circuits. These can be replaced by equivalent NAND/NAND networks directly.

Note that the literal *B* input to gate G3 must be inverted to preserve the original sense of the signal wire. Always remember to introduce logic inversions in pairs to maintain logic consistency. Any internal signal wires that undergo an odd number of inversions must have an additional inverter inserted in the path.

The final NAND-only network is shown in Figure 3.33(c). We have eliminated an inverter by replacing the *B* input to G3 with a connection to its complemented literal.

Suppose your target is a NOR-only network. You can take the same approach when the initial network is expressed as alternating OR and AND levels. You should place OR gates at the odd levels and AND gates at the even levels. You can immediately replace these by NOR gates. Any unmatched input bubbles should be corrected by inserting inverters or using the complemented literal where necessary.

It is just a little more complicated when transforming alternating AND/OR networks (OR/AND networks) into NOR-only circuits (NAND-only circuits). Nevertheless, you can still apply the same basic techniques.

For example, suppose you want to map the AND/OR/AND/OR network of Figure 3.34(a) into NOR gates. You should invert the inputs to

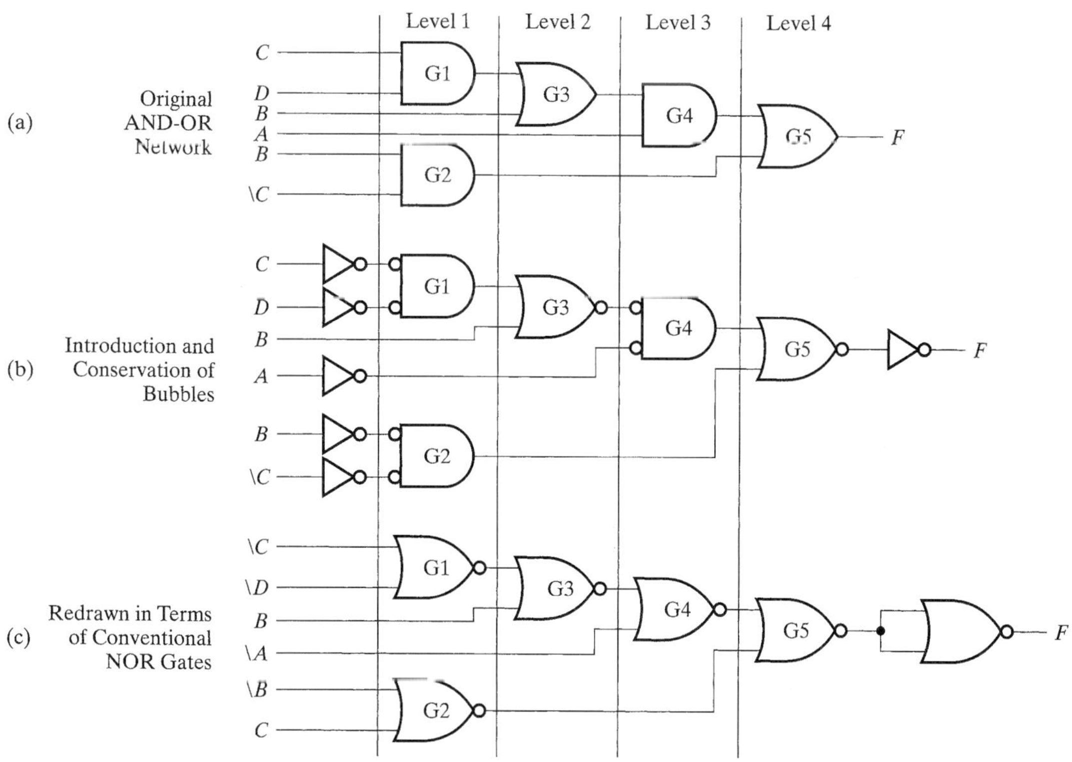

Figure 3.34 Multilevel conversion to NOR gates.

the odd levels while inserting an extra inversion at the output of the even levels. The extra inversions between adjacent even and odd levels can be saved if they cancel each other. This is shown in Figure 3.34(b) between levels 2 and 3, for gates G3 and G4. The final NOR-only circuit is shown in Figure 3.34(c). All but one of the literals have been inverted and an extra inversion has been inserted at the output. You can implement this last inversion by a NOR gate with both inputs tied to the same signal.

EXAMPLE 3.14 NON-ALTERNATING NAND/NOR MULTILEVEL NETWORKS

Figure 3.35 shows an example in which the circuit cannot be placed into a form that alternates between AND and OR gates. The multilevel function is:

$$F = AX + \overline{X} + D$$
$$X = BC$$

Figure 3.35(a) shows the initial AND/OR network. We begin by introducing double inversions at the noncomplemented inputs to the last-stage OR gate (see Figure 3.35(b)). We propagate these back to the outputs of the top AND gate and the input D (Figure 3.35(c)). With all of its inputs complemented, the OR gate is now equivalent to a NAND gate. Note how the connection to D has been replaced by a connection to its complement (denoted by \\D) to compensate for the bubble on the OR gate's input.

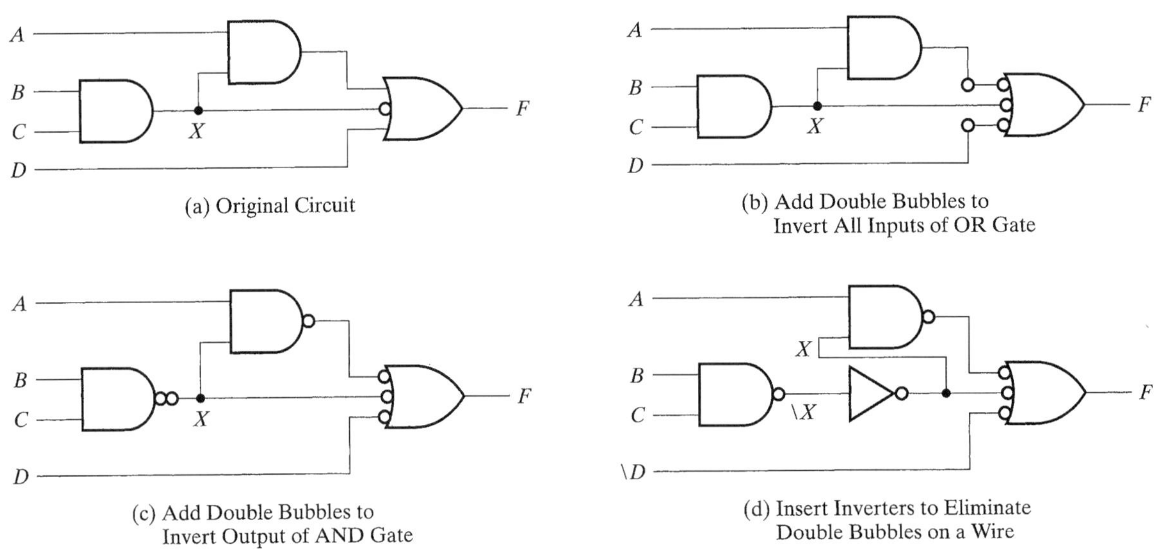

Figure 3.35 **Another multilevel conversion example.**

We now have two NAND gates, but there is still an AND gate left with input B and C. The next step is to add two bubbles to its output. Unfortunately, there is no place to propagate the second bubble. We don't want it at the input to two NAND gates. To accommodate this bubble, we add a new inverter to the circuit. Conversions between forms can sometimes introduce extra inverters, but only inverters. The final converted circuit is shown in Figure 3.35(d).

Our purpose in this section was to give you a flavor of the approaches to multilevel logic synthesis. The key idea is to identify common Boolean subexpressions across a collection of equations. If we can factor these out and share them among several functions, we can reduce the total number of literals needed to realize the functions. Fewer literals mean fewer wires, an important criterion in determining the complexity of a circuit.

We introduced the basic operations for manipulating multilevel networks: decomposition, extraction, factoring, substitution, and collapsing. We showed how an script created by a design expert it could perform a sequence of operations to optimize circuits such as the full adder and 2-bit adder. The key is to reduce literal counts, although this usually has the effect of increasing the number of levels of gates and can have a negative effect on delay. Other scripts can be designed to minimize delay at the expense of size or strike a balance between the two.

3.5 Time Response in Combinational Networks

Our analysis of circuits so far has concentrated on the static behavior of combinational networks. The analysis adequately describes a circuit in a steady state, but it is not enough to tell us about a circuit's dynamic behavior. Remember that the propagation of signals through a network is not instantaneous. This characteristic can be useful, for example, when creating circuits that output signals that assume a value for a limited duration. But it causes problems if the momentary changes of signals at the outputs lead to logical errors. Such transient output changes are called glitches. A logic circuit is said to have a hazard if it has the potential for these glitches.

As a hardware designer, it is extremely important to be able to visualize the behavior of a circuit as a function of time—that is, to be able to look at a circuit and see how signals move through it and recognize asymmetric delays along paths that can lead to transitory behavior at the outputs. This is not an easy skill to acquire, even after extensive design experience. Fortunately, simulation tools can offer great assistance in visualizing the time-based behavior of circuits.

3.5.1 Gate Delays

Outputs in combinational logic are functions of the inputs and some delay. As we stated in Chapter 2, gate delay is the amount of time it takes for a change at the gate input to cause a change at its output. Most circuit families define delays in terms of minimum (best case), typical (average), and maximum (worst case) times. A corollary to Murphy's law, well known to experienced digital designers, is that if a circuit can run at its worst-case delay, it will. Never assume that the parts you purchase will all run with minimum delay. In fact, interesting effects often arise when one part runs slowly and another of the same type is much faster. We can make very few assumptions about component delays and it is always prudent to check that all possible variations the circuit will still have the performance and function intended. The various families of logic gates exhibit trade-offs between delay and power.

In the simplest case, what would happen if you depended on a portion of your design running with minimum delay? If its delay is longer than you designed for, you may examine its output too soon, incorrectly computing the final output of your overall system. However, choosing a component that is faster generally means that it will consume more power. Designers often have to balance delay requirements against power consumption.

3.5.2 Timing Waveforms

Let's now consider the circuit shown in Figure 3.36. An input signal A passes through three inversions, leaving it in its inverted state, which is then ANDed with the original input. This appears to implement a rather useless function: $A \bullet \bar{A} = 0$. However, the timing diagram of Figure 3.37 tells us a different story. After the input A goes high, the output waveform goes high for a short time before going low. Such a circuit is called a *pulse shaper* because a change at its input causes a short-duration pulse at the output.

The circuit operates as follows. Assume that the initial state has $A = 0$, $B = 1$, $C = 0$, $D = 1$, and $F = 0$, as shown in Figure 3.37 at a time step of 0. Further, we assume that each gate has a propagation delay of 10 time units. When input A changes from 0 to 1 at time 10, it takes 10 time units, a gate delay, before B changes from 1 to 0 (time step 20). After a second gate delay, C changes from 0 to 1 (time step 30). D changes from 1 to 0 after a third gate delay (time step 40). However, between time 10 and time 40, both A and D are logic 1. If the AND gate also has a 10-unit gate delay, the output F will be high between time steps 20 and 50. This is exactly what is shown in the timing diagram. In effect, the three inverters stretch the time during which A and D are both logic 1 after A changes from 0 to 1. Eventually, the change in A

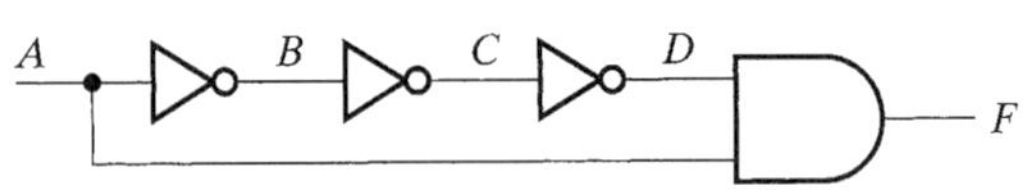

Figure 3.36 **Circuit that exploits delay to provide a useful function.**

propagates to D as a 0, causing F to fall after one more gate delay. It is no surprise that the pulse is exactly three inverter-delays wide. If we increased the number of inverters to five, the width of the pulse would be five gate delays instead.

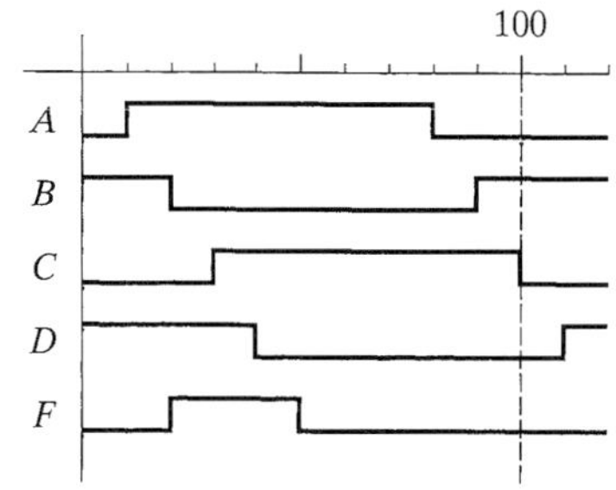

Figure 3.37 Pulse-shaper waveform.

3.5.3 Analysis of a Pulse-Shaper Circuit

In this section, we will analyze the operation of another pulse shaper circuit shown in Figure 3.38. A pulse-shaper circuit exploits the propagation asymmetries in signal paths with the explicit purpose of creating short-duration changes at the output. In this case, it generates a periodic waveform that could be used, for example, as a clock in a digital system. It operates much like a stopwatch. With its switch in one position, the circuit does nothing. In the other position, the circuit generates a periodic sequence of pulses.

The circuit has a single input A that can be connected to a logic 1 or to a logic 0, depending on the position of a switch. We will assume that the propagation delay of all gates is 10 time units. The timing behavior of a typical use of the circuit is summarized in the timing diagram of Figure 3.39.

Suppose that at time-step 0, the switch has just been connected to 0 (ground). We begin by determining the initial value for each of the circuit's wires. A goes to 0 instantly. Since a NAND gate will output a 1 whenever one of its inputs is 0, B goes to 1, but only after a gate delay of 10 time units. So we say that B goes to 1 at time-step 10. Since we do

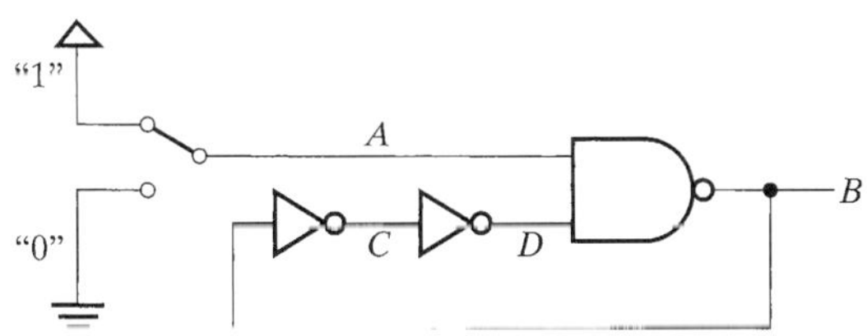

Figure 3.38 Pulse-shaper example.

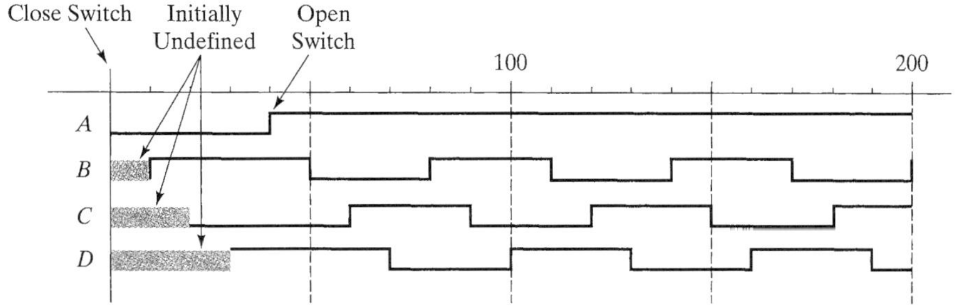

Figure 3.39 Timing waveform for pulse-shaping circuit.

not know the value of B, C, and D at the start of our waveform, we shade the waveform to indicate an unknown value.

C is the complement of B, but once again only after an inverter propagation delay. Thus C goes to 0 at time-step 20. D becomes the complement of C after another inverter delay. So it goes to 1 at time-step 30. Since A is 0 and D is 1, the output of the NAND gate stays at 0. The circuit is said to be in a *steady state*.

What happens if the switch is flipped and connected to 1 (power) at time-step 40? The input A immediately goes to 1. Now both inputs to the NAND gate are 1, so after a gate propagation of 10 time units, B will go low. This happens at time-step 50.

The change in B propagates to C after another inverter delay. Thus, at time-step 60, C goes to 1. In a similar fashion, D goes to 0 at time-step 70. Now the NAND gate has one of its inputs at 0, so at time-step 80, B will go to 1.

Note that B first goes low at time-step 50 and goes high at time-step 80—a difference of 30 time units. This is exactly three gate delays: the delay through the NAND gate and the two inverter gates on the path from signal B to C to D and back to B.

Now that B is at 1, C will go to 0 at time-step 90, D will go to 1 at time-step 100, and B will return to 0 at time-step 110. The circuit is no longer in a steady state. It now oscillates with B, C, and D varying between 1 and 0, staying at each value for three gate delays (30 time units).

3.5.4 Hazards and Glitches

While we have been looking at some circuits that take advantage of the delay of digital circuits, there are times when these effects are unwanted. A *glitch* is an unwanted pulse at the output of a combinational logic network—a momentary change in an output value that should have remained unchanged. A circuit with the potential for a glitch is said to have a *hazard*. In other words, a hazard is something intrinsic about a circuit; a circuit with a hazard may or may not glitch, depending on the input patterns and the electrical characteristics of the circuit. In this section, we will develop a procedure that leads to hazard-free circuits.

Hazards are a problem for digital systems in two cases. In the first case, we may use the value of a signal without waiting for it to settle to its final value. The problem with this time-sensitive logic can be solved by measuring an appropriately long interval between the time when inputs first begin to change and the time when the outputs are examined by the decision-making logic. Clock signals are typically used for this purpose, for example, you can lengthen the waiting interval by increasing your system's clock period. We will discuss the topic of clocking methodologies in considerably more detail when we discuss sequential logic in later chapters.

In the second case, a hazardous output is connected to a component with asynchronous inputs. *Asynchronous inputs* take effect as soon as they change, rather than when sampled with a standard reference clock. Of course, we can try to avoid the situation of having an asynchronous signal sampled by a clock, but this is not always possible. For example, when a signal forms a communication link to an entirely different subsystem with its own independent clock. This occurs most typically with computers connected over a network.

There are very useful components that, unavoidably, have asynchronous inputs. For example, many components that implement counting functions or storage elements have asynchronous inputs to reset them or initiate their operation. Therefore, if you ever design circuits that interface to such components (and you will later in this text), you should understand how to design hazard-free logic.

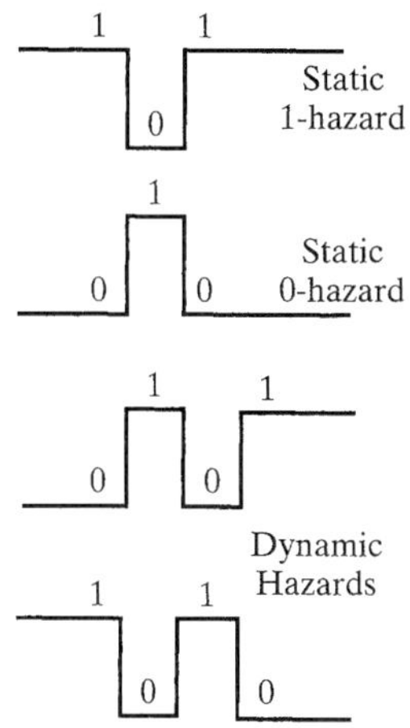

Figure 3.40 **Kinds of hazards on an output function.**

Static and Dynamic Hazards Methods for eliminating hazards always depend on the assumption that the unexpected changes in the outputs are in response to *single-bit changes* in the inputs. This assumption is equivalent to moving along an edge in the Boolean cube that describes the function's truth table. The techniques simply do not apply when more than one input bit changes at the same time. Hazards caused by simultaneous multiple-input changes are unavoidable (we'll see why shortly).

The various kinds of hazards are summarized in Figure 3.40. A *static hazard* occurs when it is possible for an output to undergo a momentary transition when it is expected to remain unchanged. A *static 1-hazard* occurs when the output momentarily goes to 0 when it should remain at 1. Similarly, a *static 0-hazard* occurs when the output should remain at 0 but momentarily changes to 1. We will develop techniques that can eliminate static hazards from two-level and multilevel circuits.

Dynamic hazards occur when the output signal has the potential to change more than once when it is expected to make a single transition from 0 to 1 or 1 to 0. Dynamic hazards cause glitches in multilevel circuits, where there are multiple paths with different delays from the inputs to the outputs. Unfortunately, it is quite difficult to eliminate dynamic hazards, in general. The best approach to dealing with this problem is to transform a multilevel circuit with a dynamic hazard into a static hazard-free two-level circuit.

3.5.5 Hazard Detection and Elimination in Two-Level Networks

Consider the 4-variable function: $F(A,B,C,D) = \Sigma\, m(1,3,5,7,8,9,12,13)$. Its K-map is shown in Figure 3.41. The minimum sum-of-products form for the function is $A\bar{C} + \bar{A}D$.

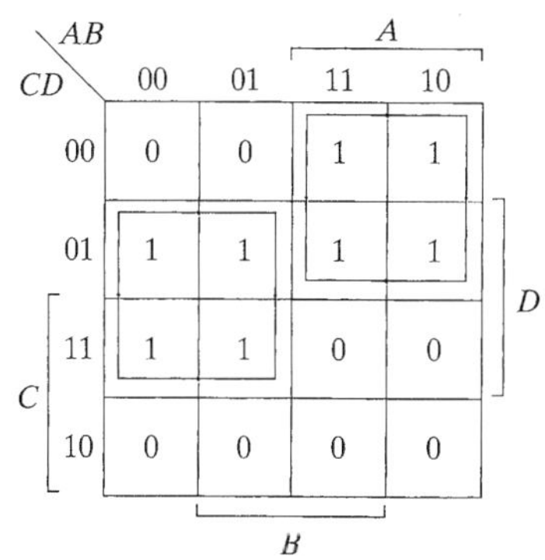

Figure 3.41 **K-map for example circuit.**

The gate-level implementation of F is given in Figure 3.42. Let's examine what happens when the inputs change from $ABCD = 1100$ to 1101. When the inputs are 1100, the output of gate G1 is 1 while G2's output is 0. Thus, the output from G3 is 1. When the input changes by a single bit to 1101, the outputs of the gates remain unchanged.

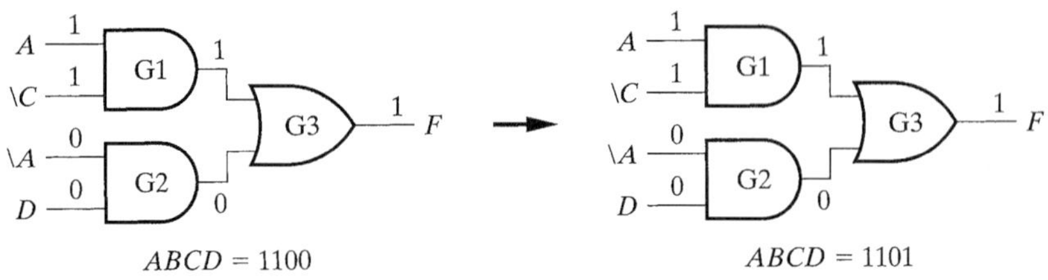

Figure 3.42 Effect of input change from 1100 to 1101.

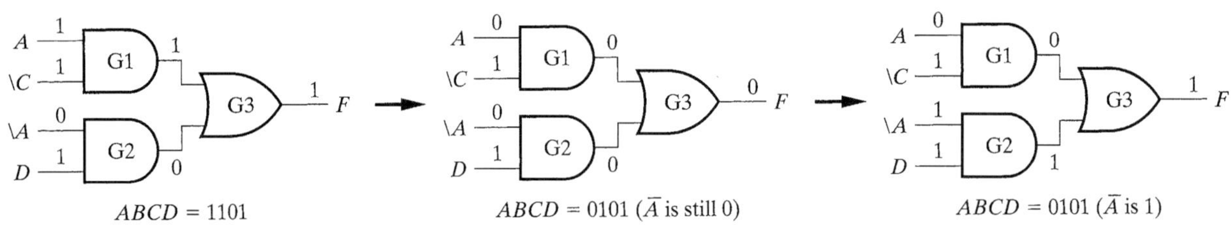

Figure 3.43 Effect of input change from 1101 to 0101.

G1 implements the prime implicant that covers both of the input configurations we considered; it remains asserted despite the input changes. A glitch cannot happen in this case.

Now consider an input change from 1101 to 0101, another single-bit change in the inputs. When A goes low, $\bar{A}$ goes high, but only after a gate delay. For a short time, both A and $\bar{A}$ are low. This allows the outputs from G1 and G2 to be low at the same time, and thus F goes low. When $\bar{A}$ finally does go high, G2 will go high and F will return to 1. A glitch has happened! The step-by-step process is shown in Figure 3.43. This can also happen if G2 is slightly slower than G1.

A close examination of the K-map of Figure 3.41 suggests what caused the problem. When the initial and final inputs are covered by the same prime implicant, no glitch is possible. But when the input change spans prime implicants, a glitch can happen. We move from under the cover of one prime implicant to the cover of another and are momentarily "exposed." Of course, if G1 is much slower to change than G2, you might not ever see a glitch on F. The hazard is always there, however, as that is a property of the logic itself. Whether you actually see a glitch depends on the timings of the individual gates.

The fundamental strategy for eliminating a hazard is to add redundant prime implicants to guarantee that all single-bit input changes are covered by one such implicant. This is a key insight. Suppose we add the implicant $\bar{C}D$ to the implementation for F. This strategy does not change the function's truth table. By adding the term $\bar{C}D$, F remains asserted for the inputs 1101 and 0101, independently of the change to input A.

You can now understand why we can only eliminate hazards for single-output changes. When two inputs change simultaneously we move "diagonally" in the K-map and there is no way we could always put a cover across to diagonal elements of the on-set. However, the single-input change assumption isn't really as bad as it sounds. In reality, inputs do change in some order, even when more than one will change. This reduces to the single-input change problem.

Let's return to eliminating hazards. Our method eliminates the static 1-hazard, but what about static 0-hazards? First, reexpress the function F in minimum product of sums form:

$$F = (\bar{A} + \bar{C})(A + D)$$

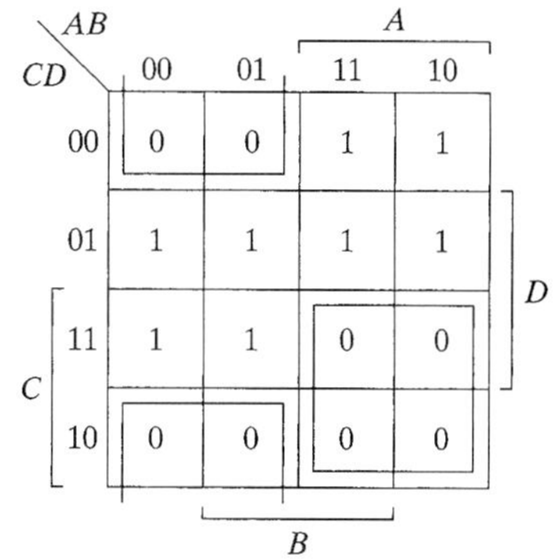

Figure 3.44 K-map for example circuit.

The K-map in Figure 3.44 clearly indicates that a static 0-hazard exists when the input changes from 1110 to 0110.

The solution is to add the redundant prime implicant $(\bar{C} + D)$ to the product of sums expression for F. The resulting expression is equivalent to the sum of products form that eliminates the static 1-hazard:

$$F = (\bar{A} + \bar{C})(A + D)(\bar{C} + D)$$
$$F = (\bar{C} + \bar{A}D)(A + D)$$
$$F = A\bar{C} + A\bar{A}D + \bar{C}D + \bar{A}D$$
$$F = A\bar{C} + \bar{A}D + \bar{C}D$$

Alternatively, we can use a shortcut to analyze the function for 0-hazards. We start with the expression that is free of static 1-hazards and work with its complement. We can then superimpose the analysis on the original K-map, looking at the zeros of the original function. A static 0-hazard exists if the implicants of the complement do not cover all adjacent pairs of 0s.

The revised expression for F is $A\bar{C} + \bar{A}D + \bar{C}D$. Working with its complement, we get the following:

$$\bar{F} = \overline{A\bar{C} + \bar{A}D + \bar{C}D}$$
$$\bar{F} = (\bar{A} + C)(A + \bar{D})(C + \bar{D})$$
$$\bar{F} = \bar{A}C\bar{D} + \bar{A}\bar{D} + AC + AC\bar{D} + C\bar{D}$$
$$\bar{F} = \bar{A}\bar{D} + AC + C\bar{D}$$

This collection of terms does indeed cover all adjacent 0s in the K-map for the revised F. This expression is free of both static 1-hazards and 0-hazards.

General Strategy for Static Hazard Elimination The preceding example leads to a general strategy for eliminating static hazards in two-level networks. Let's consider static 1-hazards first. Starting with the K-map, we examine it to make sure that all adjacent elements of the on-set are covered by a prime implicant. If they are not, we add redundant prime implicants until all adjacent elements of the on-set are covered by a prime implicant.

We follow a similar procedure to eliminate static 0-hazards. Given the sum of products form for the function that eliminates the static 1-hazards, we write it in product of sums form using Boolean algebra. Then we verify that adjacent elements of the off-set are covered by a common prime implicant in the product of sums form. If necessary, we add more prime implicants to cover any uncovered adjacencies.

3.5.6 Static Hazards in Multilevel Networks

We can generalize the techniques described in the previous section for multilevel circuits. We begin by mapping the multilevel function into a two-level form called the *transient output function*. In forming the transient output function, we treat a variable and its complement as independent variables. This means that we can no longer make use of the Boolean laws that state that $X \bullet \bar{X} = 0$ and $X + \bar{X} = 1$. The former introduces static 0-hazards, while the latter leads to static 1-hazards. Furthermore, we can no longer use any of the simplification theorems derived from these Boolean laws, such as simplification theorems 9, 10, and 11 and the Consensus theorem (#17) of Section 2.2.

EXAMPLE 3.15 A MULTILEVEL FUNCTION

Let's consider the following multilevel Boolean function:

$$F = ABC + (A + D)(\bar{A} + \bar{C})$$

A quick application of the Distributive law yields

$$F_1 = ABC + A\bar{A} + A\bar{C} + \bar{A}D + \bar{C}D$$

This is the transient output function in sum-of-products form. Since A and its complement are treated as independent variables, the term $A \bullet \bar{A}$ is kept in the transient output function.

Once the function is in two-level form, we follow the procedure described in the previous subsection. First we check for static 1-hazards in the function. Note that the term $A \bullet \bar{A}$ can never cause a 1-hazard (it does indicate that a 0-hazard exists), so it can be eliminated from consideration when analyzing for 1-hazards. The K-map for the remaining terms is shown in Figure 3.45. The function contains three static 1-hazards, at the input transitions between $ABCD = 1111$ and 0111; 1111 and 1101; and 1110 and 1100.

The remedy is to add the necessary redundant prime implicants. In the K-map of Figure 3.45, this is achieved by adding the terms AB and BD to the sum of products form of F:

$$F_2 = A\bar{C} + \bar{A}D + \bar{C}D + AB + BD$$

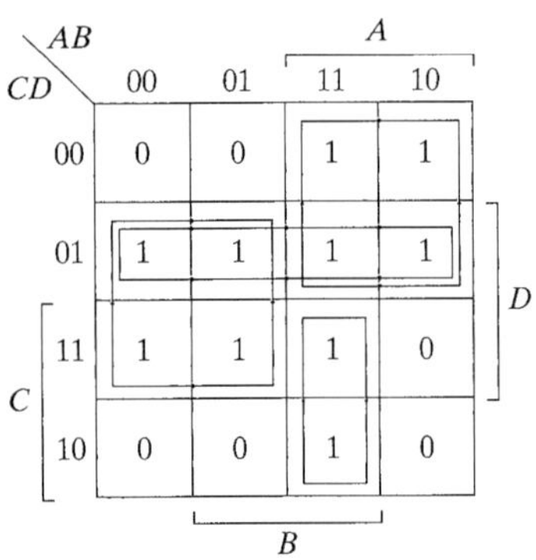

Figure 3.45 **K-map for circuit with 1-hazards.**

Because AB completely covers the term ABC, we have eliminated it from F_2.

Figure 3.46 compares the timing behavior of the original function, F, and its revised expression, F_2. Notice that F glitches on the input transitions 1111 to 0111 and 1111 to 1101, while F_2 does not have the glitches.

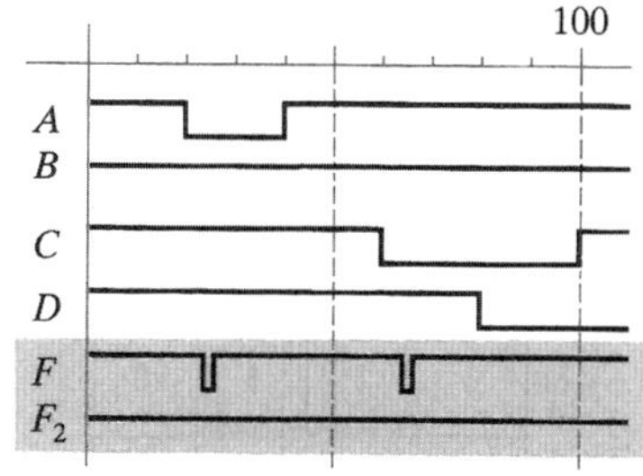

Figure 3.46 Waveforms with 1-hazards.

We use the shortcut method to verify that the new expression is free of static 0-hazards. For the original function F, $\overline{F}$ is

$$\overline{F} = \overline{ABC + (A + D)(\overline{A} + \overline{C})}$$
$$F = (\overline{A} + \overline{B} + \overline{C})(\overline{A}\overline{D} + AC)$$
$$F = \overline{A}\overline{D} + \overline{A}\overline{B}\overline{D} + A\overline{B}C + \overline{A}\overline{C}\overline{D}$$
$$F = \overline{A}\overline{D} + A\overline{B}C$$

This expression corresponds to the circled 0s in the K-map of Figure 3.47. The function has a 0-hazard on the transition from 1010 to 0010, as shown in the timing waveform of Figure 3.48. This problem can be fixed by adding the implicant $\overline{B}C\overline{D}$ to $\overline{F}$. The following is a two-level expression for F that is free of static 0-hazards:

$$F_3 = (A + D)(\overline{A} + B + \overline{C})(B + \overline{C} + D)$$

Expanding F_3 to place it into sum-of-products form yields F_2. Both expressions are simultaneously free of static 0- and 1-hazards.

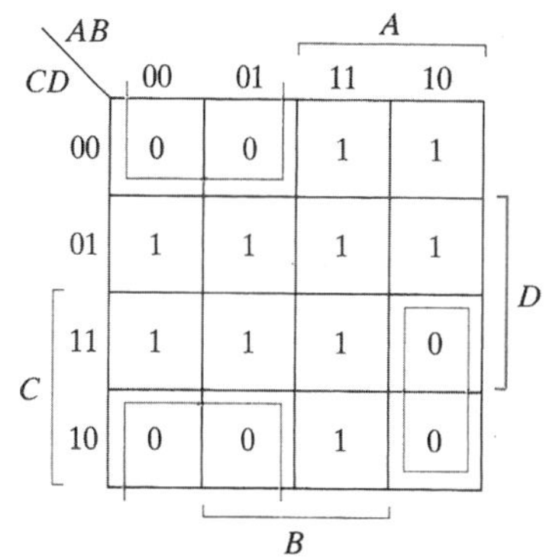

Figure 3.47 K-map for circuit with 0-hazards.

3.5.7 Designing Static Hazard-Free Multilevel Circuits

The procedure for *designing* a static hazard-free multilevel network is a straightforward application of the concepts we have just described. The key is to place the function in such a form that the transient output function guarantees that every set of adjacent 1s in the K-map are covered by a term, and that no terms contain both a variable and its complement. The former condition eliminates 1-hazards, and the latter eliminates 0-hazards.

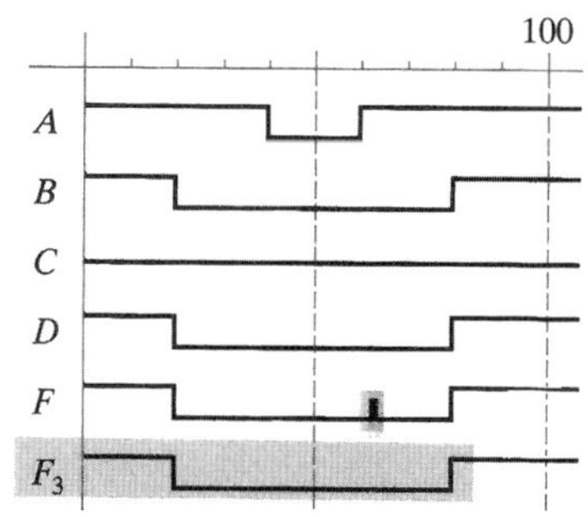

Figure 3.48 Waveform with 0-hazards.

Following this procedure will eliminate static hazards, at least for two-level implementations. We start with the truth table or the expression of the function in minterm shorthand form. Then we express the function in terms of prime implicants, which ensure that adjacent 1s are covered by a single term. To obtain a multilevel form, we factor the resulting expression using the laws and theorems of Boolean algebra, but treating a variable and its complement as independent variables. For example, the Distributive law can never introduce a hazard, so it can be used freely to simplify the function. The complementarity laws, on the other hand, cannot be used to simplify the function. As long as no terms in the resulting expression contain a variable and its complement, the function will be hazard-free.

Returning to the example function of this section, its shorthand minterm form is

$$F(A,B,C,D) = \Sigma\, m(1,3,5,7,8,9,12,13,14,15)$$

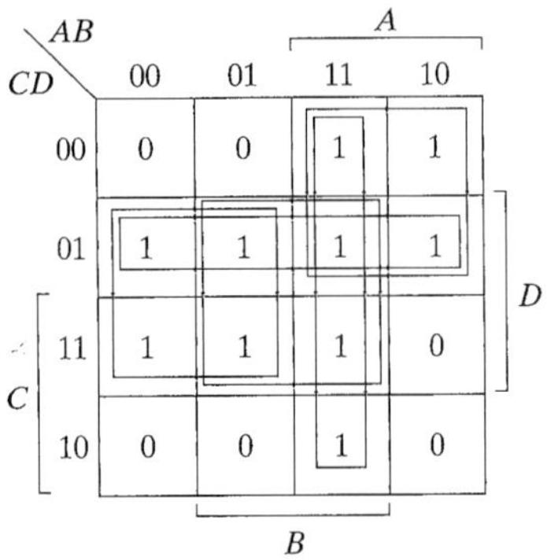

Figure 3.49 K-map for hazard-free function.

The K-map with circled terms is shown in Figure 3.49. This yields the expression

$$F = AB + \bar{A}D + BD + A\bar{C} + \bar{C}D$$

This result is the same as the expression for F_2 in the previous subsection. Factoring via the distributive law yields the following multilevel static hazard-free expression:

$$F = (\bar{A} + B + \bar{C})D + A(B + \bar{C})$$

This expression requires five gates, as was the case for the original expression for F given at the beginning of this section.

3.5.8 Dynamic Hazards

Dynamic hazards are defined as output transitions from 0 to 1 or 1 to 0, undergoing more than one change along the way. Dynamic hazards happen because of multiple paths in the underlying multilevel network, each with its own asymmetric delay. If there are three or more paths from an input or its complement to the output, the circuit has the potential for a dynamic hazard.

Figure 3.50 gives an example of a circuit with a dynamic hazard. Note that there are three different paths from B or $\bar{B}$ to the output. The following sequence of events can lead to a dynamic hazard at F. Suppose that the initial inputs are $ABC = 000$, $F = 1$ and the final configuration is 010 with $F = 0$. In the starting configuration, G1 = 0, G2 = 1, G3 = 1, G4 = 1, and G5 = 1. The initial gate inputs and outputs are shown in the figure.

Now suppose that B changes from 0 to 1. We assume that G1 is a slow gate, G4 is a very slow gate, and G2, G3, and G5 are fast. G2 changes from 1 to 0, followed by G3 going to 0 and G5 following it to 0 a short time later. So far, the output has changed from 1 to 0. This situation is shown by the bold values in the figure.

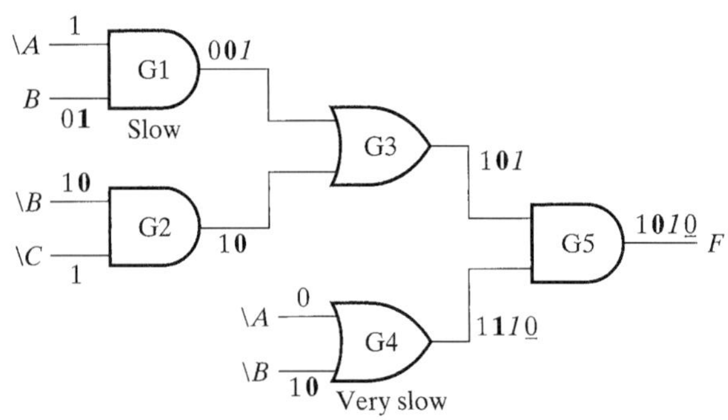

Figure 3.50 Circuit with a dynamic hazard.

Now G1 catches up. Its output goes to 1, causing G3 to go high, followed by G5 going high. At this point, the output has changed from 1 to 0 to 1. This is shown by italicized values in the figure.

Finally, G4's output changes from 1 to 0. This causes G5 to go low, reaching its final output. The output has changed from 1 to 0 to 1 to 0. This is shown by underlined values in the figure.

Although it is possible to extend the techniques for static hazard elimination to dynamic hazards, the process is rather complicated and goes beyond the scope of this text. It is not enough to eliminate the static hazards—a multilevel network free of static hazards may still have dynamic hazards. If you need a hazard-free network, it is best to design it as a two-level network using the techniques shown earlier in this section.

3.6 Hardware Description Languages

Up to this point, we have used Boolean expressions and schematic diagrams to describe our logic circuits. As you might imagine, it can become quite difficult to comprehend a large diagram or long Boolean expression, or for that matter, even to draw or write one. Hierarchy helps with this problem in that it lets us use smaller pieces to describe larger entities. We have seen this when we decompose a large Boolean expression into smaller subexpressions and we have seen it in schematic diagrams such as using two half adders to form the full adder of Figure 2.19.

But hierarchy is not enough. Diagrams take a lot of time to draw legibly. Computer-aided schematic editors make the job easier but a small change in a logic expression can cause a large change in its corresponding schematic drawing. Boolean expressions can be difficult to read and do not provide an easy way to comprehend the function being described. For example, consider the ease of reading the C program versus the sum-of-products expression for the leap month calculation of Section 1.4.1.

Hardware description languages were developed to deal with these issues. They provide a way for designers to textually describe logic circuits while exploiting some of the advantages of software languages, namely, variables, subprocedures, and conditional and iterative statements.

However, the most valuable use of hardware description languages is that they allow a design to be exercised without physically building it. Descriptions in these languages can be *executed*. That is, they run like software: a program, called a simulator, emulates the actual behavior of the circuit as faithfully as possible. Of course, it does not precisely match reality. Electrical effects require large amounts of computation to model at a high level of accuracy. A simulator strikes a balance between fidelity and performance. It focuses on how a circuit evolves over time. This is very different than a programming language that has no notion of time. Software is executed statement after statement and the computer does not keep track of how long the execution takes. A hardware simulator, on the other hand, has to keep track of how long the "execution" of a logic gate takes so that it can properly model propagation delay. The output of simulators is commonly shown as a waveform that explicitly models time.

Hardware description languages have been around for a long time. The first were designed to help processor designers describe their new designs and actually execute programs on them. A language called ISP, for instruction set processor, was developed at Carnegie-Mellon University in the mid-1970s. It radically shortened the amount of time needed to get a new design to work correctly by allowing designers to find most of the bugs in simulation rather than after building expensive hardware.

The success of these efforts was followed in the early 1980s by HDLs to describe arbitrary logic rather than just processor elements. These were mostly developed by industry (such as ABEL by Data-I/O and Verilog by Gateway). VHDL, developed under the leadership of Department of Defense, was developed to help in modeling even more aspects of complex systems composed of software as well as hardware. Both Verilog and VHDL are now IEEE standards.

By the early 1990s, with most designers using hardware description languages, attention turned to *compilers*, known as *synthesis tools*, that automatically could turn HDL descriptions into circuit implementations. More recently, the focus of research is on system-level description languages used to describe complex distributed systems and synthesis tools to generate the code and hardware that will make all the elements work together to perform the desired functions.

In this text, we will use Verilog simply because it makes it much easier for beginners to get started and it has a syntax similar to the C programming language. Users of HDLs are almost evenly split between those that use Verilog and those that use VHDL.

Hardware description languages appear quite similar to programming languages at first glance, but we'll see, in this and later chapters, that there are some fundamental differences. One of things they do share, however, is the notion of hierarchy. With HDLs, we define modules and then compose these into larger designs. A very basic module description in Verilog might look something like this:

```
module xor_gate (a, b, z);
   input   a, b;
   output  z;

   <module internals>

endmodule
```

The module has a name, in this case, `xor_gate`, and three wires: *a*, *b*, and *z*. Here *a* and *b* are inputs, while *z* is an output. We can think of *a* and *b* as the input parameters of the module and *z* as its output parameter or return value. There are two principal ways of describing the internals of the module: structure or behavior.

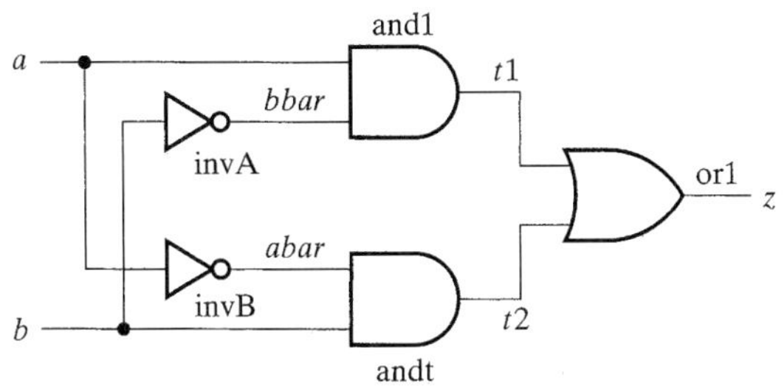

Figure 3.51 Schematic for an exclusive-or gate implementation.

3.6.1 Describing Structure

A structural description is simply a textual version of a schematic diagram. For an XOR gate constructed using the schematic of Figure 3.51, the module description would be completed as follows:

```
module xor_gate (z, a, b);
   input   a, b;
   output  z;
   wire    abar, bbar, t1, t2;

   inverter invA(abar, a);
   inverter invB( bbar, b);
   and_gate and1(t1, a, bbar);
   and_gate and2(t2, abar, b);
   or_gate or1(out, t1, t2);

endmodule
```

The module description lists all of the five gates of Figure 3.51 and uses variable names for the connecting wires. Each gate is an *instance* of another module. In this case, we use three different types of modules: `inverter`, `and_gate`, and `or_gate`. Descriptions of these modules would also be in the complete circuit description. However, just as in software, there will most likely be libraries of primitive elements so that descriptions can be made more concise. The five gates are given unique names (*invA*, *invB*, *and1*, *and2*, and *or1*) so that they can be referred to in later simulation and debugging. Finally, note that *abar*, *bbar*, *t1*, and *t2* had to be declared as internal variables to the module as they are wires that are neither inputs nor outputs but are used to connect the internal components of the module.

3.6.2 Describing Behavior

Rather than describing in detail how the function will be realized (by describing the gates to be used and how they will be interconnected), a behavioral description simply describes the function of the module without spelling out a specific implementation. A behavioral description of our `xor_gate` module may look as follows:

```
module xor_gate (a, b, z);
   input   a, b;
   output  z;
   reg     z;

   always @(a, b) begin
       z = a ^ b;
   end

endmodule
```

The module description contains an `always` block. This specifies *when* the values of the outputs of a module need to be updated and *how*. In this case, whenever the value of a or the value of b changes, then z should be assigned the value of the exclusive-or of a and b. The *sensitivity list* is specified within the parentheses after the @ symbol. The statement to be executed to determine how to update the value of z is within the `begin-end` block. The ^ symbol is used to signify the `exclusive-or` operation in Verilog. The sensitivity list of the *always* block indicates that if either a or b change in value then the statement within the *always* block should be executed. This is quite different than in programming languages where statements are executed sequentially. In this case, a statement is executed because of a change in the value of a wire. HDLs do this because this is how real circuits operate. The additional declaration for z makes this clear to the simulator (by indicating that z will have new values assigned to it continuously and a *register* should be created to keep track of when the statements that assign a new value to z need to be reevaluated).

We can also include more complex statements inside of `always` blocks that include iteration and conditionals. For example, we can rewrite the `always` block for our `xor_gate` as follows:

```
always @(a, b) begin
   if (a) then z = ~b else z = b;
end
```

The `if` statement also realizes the exclusive-or of a and b and assigns a value to z. The ~ symbol is the unary operator used to complement values in Verilog.

These types of `always` blocks are so common in Verilog that there is a special shorthand notation that is exactly equivalent:

```
module xor_gate (a, b, z);
   input   a, b;
   output  z;
   reg     z;

   assign z = a ^ b;

endmodule
```

The assign statement specifies that *z* should *continuously* be assigned a value that is the exclusive-or of *a* and *b*. This *continuous assignment* statement is quite different than an assignment statement in a programming language. It is not executed once, but many times. The statement is reevaluated whenever the value of *a* or *b* changes so that we can ensure that *z* always has the correct value. In effect, the assign statement takes the burden from the designer of having to spell out the sensitivity list. The statement is reevaluated whenever any variable used on the right side of the assignment changes value.

Of course, for a design to be complete, all its modules will require complete structural descriptions. However, there are several reasons for wanting behavioral descriptions as well. Early in the design process, designers may not want to put a lot of effort into a schematic that may have to be changed later. It is easier to simply state the function that will be required and move on to simulating and debugging the overall design. Moreover, automatic synthesis tools can be used to transform a behavioral description into a structural description. As you can imagine, these tools have radically improved the productivity of designers. Finally, some modules may never be realized in circuitry but are simply used to provide a simulation context for the design. We'll see an example of this shortly.

3.6.3 Delay

Sensitivity lists emulate how real logic circuitry works. Another aspect of real hardware that needs to be emulated is its delay characteristics. Verilog provides for this with a delay statement. For example, if we want our `xor_gate` to have a delay of six time units, then we can add a delay to its behavioral description as follows:

```
module xor_gate (a, b, z);
   input   a, b,
   output  z;
   reg     z;

   assign #6 z = a ^ b;

endmodule
```

```
module xor_gate (a, b, z)
   input   a, b;
   output  z;
   reg     z;

   always @(a, b) begin
      #6 z = a ^ b;
   end

endmodule
```

The effect of the *delay* statement is to postpone the assignment of a new value to *z* for six time units so that a change in *z* will occur six time units later than the change in *a* or *b* that caused it to happen. Note that, delay statements only make sense within a behavioral description. In a structural description it is the responsibility of the most primitive submodules (those not composed of other submodules) to have the appropriate delay specified between their inputs and outputs.

3.6.4 Event-Driven Simulation

We have mentioned simulation several times in this discussion of HDLs. A simulator is a tool that can read hardware descriptions written in languages such as Verilog and execute them for us. Most HDL simulators are *event-driven,* that is, they are based on the concept of an event occurrence. An event is simply the change in the logic value carried on a wire. It has an associated value and a time of occurrence.

Simulators are constructed to propagate events through the modules that constitute a design. In our example, any changes in the values of wires *a* and *b* constitute events, these are propagated through the `xor_gate` module and a new event on *z* may be generated. That, in turn, may affect other modules and cause them to generated other events. This continues until either there are no more events or the designer stops the simulation. *Delay* statements are used to advance time in event-driven simulators. Without delay, all events would occur in the same virtual instant. By modeling delay, we separate events in time in the same way they would be in a real circuit. Event-driven simulators take the burden from designers of having to manually check event propagation. This makes them invaluable and very popular tools for verifying a logic design.

Event-driven simulators need events to kick-off the process. These may be generated by a human designer who "drives" the simulation by changing values on wires through the simulator's user interface. Alternatively, we can include modules in our simulated design whose function is simply to generate events or react to events generated by the circuit. This commonly is called a test bench: an accompanying circuit to our design whose function it is to test that the circuit functions as intended by providing "stimuli." We can then check that our circuit responds appropriately. This is a common practice in engineering and scientific observation. Of course, these test-bench modules will not be part of a realization of the circuit.

Suppose we would like to exercise our `xor_gate` so that it is "tested" with the four combinations of possible input values. We could create a module that cycled through these four input combinations and connect it our `xor_gate` as follows:

```
module test-bench (x, y);
  output  x, y;
  reg     x, y;

  initial begin
    x = 0; y = 0;
    #10
    x = 0; y = 1;
    #10
    x = 1; y = 0;
    #10
    x = 1; y = 1;
    #10
    $finish
  end

endmodule
```

```
module both_together(z);
    output  z;
    wire    w1, w2;

    test-bench tb1(w1, w2);
    xor_gate xor1(w1, w2, z);

    always @(z) begin
        $display("At time: %d with
        inputs: %b and %b, the output
        is: %b", $time, w1, w2, z);
    end

endmodule
```

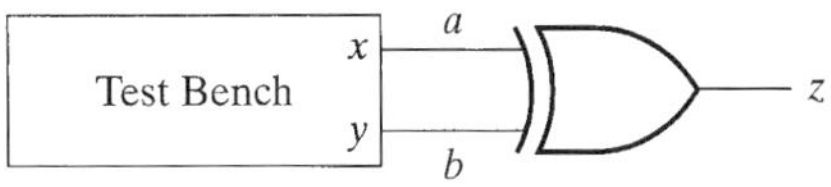

Figure 3.52 **Schematic of an exclusive-or gate connected to a stimulus generator.**

The schematic corresponding to the Verilog above is shown in Figure 3.52. The module on the left uses delay statements to change the values of its *x* and *y* outputs every 10 time units. The `initial` block is only executed once at the start of a simulation. The `$finish` statement halts the simulator. Note that the module has no inputs. The module on the right connects this stimulus generator (or test bench) to an instance of our `xor_gate`. It has a single output, coincidentally also labeled *z*, that will change at time 6, 16, 26, and 36 (due to the XOR gate's six time units delay). The simulator will halt at time 40. The `$display` statement is used as part of the user interface of the simulator. It is similar to a `printf` statement in *C*. By enclosing it in an `always` block, we will cause it to print its string every time *z* changes value and report to us the current time in the simulator as well as the values of the inputs and outputs of the `xor_gate` instance.

Of course, there are many more features to Verilog (and we'll see many more of them when we discuss sequential logic in later chapters). Unfortunately, there is not enough space here to provide all the details. One last example will serve to highlight some very useful elements of the language.

```
module test-bench (x, y);
   output  x, y;
   reg[1:0]  count;

   initial begin
        count = 0;
        repeat (3) begin
            #10 count = count +1;
        end
        #10
        $finish;
   end

   assign x = count[1];
   assign y = count[0];

endmodule
```

This module does precisely the same thing as the previous version of our test bench. It accomplishes it using a repeat loop and a two-bit variable called *count* (think of an *n* bit variable as an *n*-element array of 1-bit values). The outputs *x* and *y* are continuously assigned the respective bits of count. The high-order bit is connected to *x* and the low-order bit to *y* using the two continuos assignment statements at the bottom. The outputs *x* and *y* change whenever *count* is incremented.

CHAPTER REVIEW

In this chapter, we have discussed several ways to manipulate, analyze, and describe combinational logic. We expanded on two-level simplification and looked at how K-maps are difficult to scale to functions of large numbers of inputs. To address this problem we looked at algorithms that serve as the foundation of automated optimization tools. We also introduced the concepts of minimization in multilevel logic. Unfortunately, given the richness of criteria for multilevel logic implementations there are no algorithms that can guarantee optimality. Instead, we have methods that try to capture the knowledge of experienced designers and provide a means of applying it in the form of scripts that guide the multilevel simplification process.

In addition to simplification, we also began the process of mapping logic to available components. In this chapter, we focused on NAND and NOR gates and how we can map any expression to these most common of gates. In the next chapter, we'll learn about many more implementation technologies and will revisit the mapping problem for each of them.

The time behavior of our electronic components leads to both wanted and unwanted functionality. We analyzed pulse shaper circuits that react to extend the duration of a change in input values. We also looked at how glitches arise and developed an approach to designing circuits that are hazard-free. In later chapters, we'll develop methodologies that permit us to ignore hazards except at the periphery of our circuits where they communicate with each other and the outside world.

Finally, we introduced hardware description languages and their use in not only describing the structure and behavior of combinational logic but also in the simulation of designs before they are constructed. HDLs have allowed designers to experiment with their designs in a software environment and be much more effective when they actually get to the task of physically constructing their designs because they know that they are logically correct.

FURTHER READING

Boolean simplification is an important topic in all digital design textbooks. Virtually any other text can provide an alternative explanation to these topics. Of particular note are: C. H. Roth, Jr., *Fundamentals of Logic Design, Fourth Edition,* West Publishing Co., St. Paul, MN, 1992 with an indepth treatment of K-maps and the Quine-McCluskey method; R. F. Tinder, *Engineering Digital Design, Second Edition,* Academic Press, San Diego, CA, 2000 has an interesting discussion on

Boolean simplification targeting XOR gates; S. H. Unger, *The Essence of Logic Circuits,* Prentice Hall, Englewood Cliffs, NJ, 1989 describes symmetric and iterative combinational circuits; and E. J. McCluskey, *Logic Design Principles,* Prentice Hall, Englewood Cliffs, NJ, 1986 emphasizes hazard-free design.

For a detailed presentation of multilevel logic optimization techniques, the following papers are highly recommended: K. Bartlett, W. Cohen, A. DeGeus, G. Hachtel, "Synthesis and Optimization of Multi-Level Logic under Timing Constraints," *IEEE Transactions on Computer-Aided Design*, CAD-5, 4, pp. 582–596 (October 1986), and R. K. Brayton, R. Rudell, A. Sangiovanni-Vincentelli, A. R. Wang, "MIS: A Multiple-Level Logic Optimization System," *IEEE Transactions on Computer-Aided Design*, CAD-6, 6, pp. 1062–1081 (November 1987). A comprehensive text on the subject of computer-aided simplification methods is R. Brayton, G. Hachtel, C. McMullen, and A. Sangiovanni-Vincentelli, *Logic Minimization Algorithms for VLSI Synthesis,* Kluwer Academic Publishers, Dordrecht, The Netherlands, 1984.

There are many texts on Hardware Description Languages. For a more complete treatment of Verilog, see D. Thomas, P. Moorby, *The Verilog Hardware Description Language, Fourth Edition,* Kluwer Academic Publishers, Dordrecht, The Netherlands, 1998 and M. D. Ciletti, *Advanced Digital Design with the Verilog HDL,* Prentice Hall, Englewood Cliffs, NJ, 2002. For VHDL, see P. Ashenden, *The Student's Guide to VHDL,* Morgan-Kaufmann, San Francisco, CA, 1998. Several also treat the subject or writing HDL descriptions targeting the use of automated synthesis tools to realize the design: E. Sternheim, R. Singh, R. Madhavan, Y. Trivedi, *Digital Design and Synthesis with Verilog HDL,* Automata Publishing, San Jose, CA, 1993 or D. Smith, P. Franzon, *Verilog Styles for Synthesis of Digital Systems,* Prentice Hall, Upper Saddle River, NJ, 2000. Finally, a text with good side-to-side comparisons of Verilog and VHDL is D. Smith, *HDL Chip Design,* Doone Publications, Madison, AL, 1996.

EXERCISES

3.1 *(Boolean Simplification)* Use Karnaugh maps (K-maps) to simplify the following functions in sum-of-products form. How many literals appear in your minimized solutions?

(a) $f(V,W,X,Y,Z) = \Pi M(0,4,18,19,22,23,25,29)$
(b) $f(A,B,C,D) = \Sigma m(0,1,4,5,12,13)$
(c) $f(A,B,C,D,E) = \Sigma m(0,4,18,19,22,23,25,29)$
(d) $f(A,B,C,D,E,F) = \Sigma m(3,7,12,14,15,19,23,27,28,29,31,35,39,$ $44,45,46,48,49,50,52,53,55,56,57,59)$.

3.2 *(Boolean Simplification)* What are the prime implicants for each of the expressions in Exercise 3.1? Which are essential? Are any redundant?

3.3 *(Boolean Simplification)* Use Karnaugh maps (K-maps) to simplify the following functions in sum-of-products form taking advantage of the don't cares provided.

(a) $f(W,X,Y,Z) = \Pi M(4,7,8,11) \bullet \Pi D(1,2,13,14)$
(b) $f(A,B,C,D) = \Sigma m(0,1,4,10,11,14) + \Sigma d(5,15)$
(c) $f(A,B,C,D) = \Sigma m(1,2,3,5,8,13) + \Sigma d(0)$
(d) $f(A,B,C,D,E) = \Sigma m(3,7,12,14,15,19,23,27,28,29,31) + \Sigma d(4,5,6,13,30)$.

3.4 *(Boolean Simplification)* What are the prime implicants for each of the expressions in Exercise 3.3? Which are essential? Are any redundant? How many don't cares are set to 1 in each case?

3.5 *(Boolean Simplification)* Use the K-map method to minimize the following Boolean expressions in S-o-P form:

(a) $F(A,B,C) = \Sigma m(1,2,6,7)$
(b) $F(A,B,C,D) = \Sigma m(0,1,3,9,11,12,14,15)$
(c) $F'(A,B,C,D) = \Sigma m(2,4,5,6,7,8,10,13)$
(d) $F(A,B,C,D) = (ABC + \bar{A}\bar{B})(C + D)$
(e) $F(A,B,C,D) = (\bar{A} + B + C) \bullet (A + B + \bar{C} + D) \bullet (A + \bar{B} + C + \bar{D}) \bullet (\bar{A} + \bar{B})$

3.6 *(Quine-McCluskey Method)* Use the Quine-McCluskey method to find the minimum sum-of-products form for the following Boolean expressions. Show your process of deriving the prime implicants. Include the implication chart from which your minimum sum-of-products form is derived.

(a) $f(X,Y,Z) = \Sigma m(2,3,4,5)$
(b) $f(A,B,C,D) = \Sigma m(1,5,7,8,9,13,15) + \Sigma d(4,12,14)$
(c) $f(A,B,C,D) = \Sigma m(1,2,3,4,5,6,7,8,9,10,11,12)$
(d) $f(A,B,C,D) = \Sigma m(1,2,3,4,9,10,11,12) + \Sigma d(0,13,14,15)$

3.7 *(Espresso Method)* Use the Espresso method to simplify the expressions of Exercise 3.5. Which prime implicants are not found by Espresso's initial pass? Do you find that you need to do a REDUCE, EXPAND, IRREDUNDANT COVER iteration to get a more simplified expression? Show the K-map at each stage of the method.

3.8 *(Mapping to NANDs/NORs)* Map the following multilevel AND/OR Boolean functions into (i) NAND-only and (ii) NOR-only implementations. Do not simplify these expressions-there is no need to minimize them in this problem. You may insert inverters if you need them. Draw a schematic diagram for each realization.

(a) $F(A,B,C,D) = (ABC + \bar{A}\bar{B})(C + D)$
(b) $G(B,C) = \overline{BC}$
$F(A,B,C,G) = [(A + B)(\bar{A} + C) + G](\bar{A}\bar{C} + G)$
(c) $F(A,B,C,D) = [(\bar{A} + \bar{B} + C)(B + \bar{C})(A + \bar{B} + \bar{D})] + AC\bar{D} + B\bar{C}D$
(d) $H(A,B) = A + \bar{B}$
$G(B,C,D) = B\bar{C} + D$
$F(A,B,G,H) = (A + G)\overline{(B + H)}$

3.9 *(Mapping to NANDs/NORs)* Draw schematics for the following expressions, mapped into NAND-only networks. You may assume that literals and their complements are available:

(a) $A\bar{B}\bar{C} + \bar{A}C + \bar{A}B$
(b) $(\bar{A} + \bar{B} + \bar{C})(\bar{A} + \bar{B})(\bar{A} + \bar{C})$
(c) $\bar{A}B + A + \bar{C} + \bar{D}$
(d) $(\overline{AB})(\overline{\bar{A}\bar{C}})$
(e) $\overline{AB + \bar{A}\bar{C}}$

3.10 *(Mapping to NANDs/NORs)* Draw schematics for the following expressions, mapped into NOR-only networks. You may assume that literals and their complements are available:

(a) $(A + B)(\bar{A} + C)$
(b) $\overline{(A + B)\bullet(\bar{A} + C)}$
(c) $\overline{(A + B)}\bullet\overline{(\bar{A} + C)}$
(d) $(A + B)\bullet(\bar{A} + C + D)\bullet(\bar{A} + \bar{C})$
(e) $(A + B)\bullet(\bar{B}\bullet C)\bullet(\bar{A} + \bar{C})$

3.11 *(Mapping to NANDs/NORs)* Show how to implement the following Boolean functions using first NAND and Inverter gates only and then NOR and Inverter gates only:

(a) $F(A,B,C_{in}) = AB + BC_{in} + AC_{in}$ (e.g., the full-adder carry-out)
(b) $F(A,B,C_{in}) = A \text{ xor } B \text{ xor } C_{in}$ (e.g., the full-adder sum)
(c) $F(A,B,C,D) = 1$ (if the 2-bit binary quantity AB is strictly less than the 2-bit binary quantity CD in magnitude)

3.12 *(Multilevel Network Mappings)* Draw schematics for the following expressions, using mixed NAND and NOR gates only:

(a) $(AB + CD)E + F$
(b) $(A\bar{B} + C)E + DG$
(c) $\{A + [(B + C)(D + E)]\}\{[(F + G)(\bar{B} + \bar{E})] + \bar{A}\}$
(d) $(A + B)(C + D) + EF$
(e) $\overline{AB}(\bar{B} + C)\bar{D} + \bar{A}$

3.13 *(Multilevel Logic)* Using Boolean algebra, K-maps, or truth tables, verify that the multilevel forms for the full adder *Sum* and C_{out} (carry-out) obtained in Section 3.4.1 are logically equivalent to the two-level forms found in Section 2.2.

3.14 *(Multilevel Logic)* Using Boolean algebra, K-maps, or truth tables, verify that the multilevel forms for the 2-bit binary adder outputs, X, Y, and Z, of Section 3.4.1 are logically equivalent to the two-level forms found at the beginning of this chapter.

3.15 *(Multilevel Logic)* Reverse engineer the circuit shown in the schematic of Figure Ex. 3.15 in order to derive a two-level realization.

(a) Find the Boolean expression that describes the circuit.
(b) Construct the truth table for the function.
(c) Write the function in canonical sum-of-products form (little m notation).
(d) Simplify the function using K-maps.

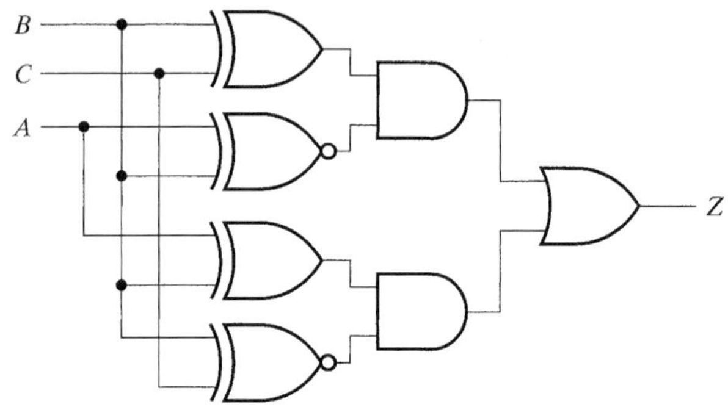

Figure Ex. 3.15

3.16 *(Multilevel Logic)* Consider the following multilevel Boolean expressions:

$$F(A,B,C,D) = (A + (BC))(\overline{C} + D)$$
$$G(A,B,C,D) = ((A + \overline{B})D) + (A + (BC))$$

Perform the following:

(a) Show how to implement each function as a multilevel NAND-only gate-level implementation.
(b) Repeat (a), but using NOR gates only.
(c) Find the two-level minimized sum-of-products forms.
(d) Find the two-level minimized product-of-sums forms.
(e) Briefly compare the implementation complexities in terms of gates and literals. For each function, which achieves the "simplest" implementation.

3.17 *(Multilevel Logic)* Consider the following multilevel Boolean expressions:

$$F(A,B,C,D,E) = (((AB) + C)(D + E)) + (\overline{A}\overline{D}))$$
$$G(A,B,C) = (AB\overline{C});\ H(A,B,C,D) = (D + G)\overline{G}$$

Perform the following:

(a) Show how to implement each function as a multilevel NAND-only gate-level implementation.
(b) Repeat (a), but using NOR gates only.
(c) Find the two-level minimized sum-of-products forms.
(d) Find the two-level minimized product-of-sums forms.
(e) Briefly compare the implementation complexities in terms of gates and literals. For each function, which achieves the "simplest" implementation.

3.18 *(Hazard-Free Design)* Given the following specifications of Boolean functions, implement them as hazard-free circuits:

(a) $F(A,B,C) = B\overline{C} + \overline{A}C$
(b) $F(A,B,C,D) = \Sigma m(0,4,5,6,7,9,11,13,14)$

(c) $F(A,B,C) = (A + B)(\bar{B} + C)$
(d) $F(A,B,C,D) = \Pi M(0,1,3,5,7,8,9,13,15)$
(e) $F(A,B,C,D,E) = \Sigma m(0,1,3,4,7,11,12,15,16,17,20,28)$

3.19 *(Time Response)* Consider the circuit in Figure Ex. 3.19(a). Write down its functions in minimized form. Given that XOR/XNOR gates have twice the delay of the NAND gates, what is the circuit's output response to the input waveforms in Figure Ex. 3.19(b)?

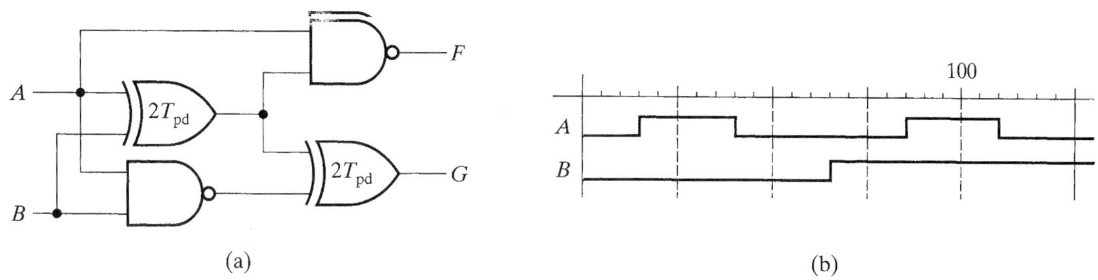

Figure Ex. 3.19

(Each 5-time unit division represents one NAND gate delay.)

3.20 *(Time Response)* Construct a timing diagram for the behavior of the circuit schematic in Figure Ex. 3.20.

(a) Start by finding a non-oscillating starting condition for the circuit with switch S in position 1 (up) as shown. Fill in the timing waveform with an initial steady–state condition for the circuit nodes labeled A, B, C, and D. (*Warning:* It is very easy to choose an initial configuration that oscillates. A unique non-oscillating configuration does exist. Start your reasoning with the tightest loop, or make an educated guess and verify that the assumed state is indeed non-oscillating.)

(b) At time T, the switch is moved from position 1 to position 2 (down). Fill in the rest of the timing diagram with the logic values of the signals at points A, B, C, and D in the given circuit.

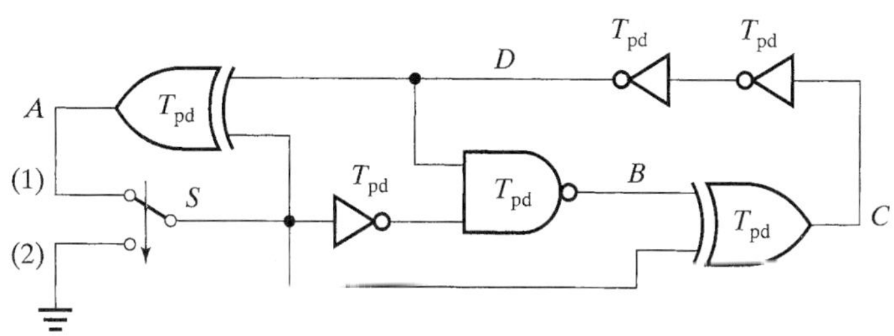

Figure Ex. 3.20

3.21 *(Time Response)* Consider the following circuits with feedback, that is, the output also serves as an input to the function.

(a) Determine the output of NOR gate U3 in Figure Ex. 3.21(a) when the switch to ground alternates between being closed and being open. Draw a timing diagram to illustrate your answer.

(b) Repeat your analysis for the circuit in Figure Ex. 3.21(b) that looks similar. Draw a timing diagram to explain its time dependent behavior. Does this circuit generate an oscillating output? If so, why? If not, why not?

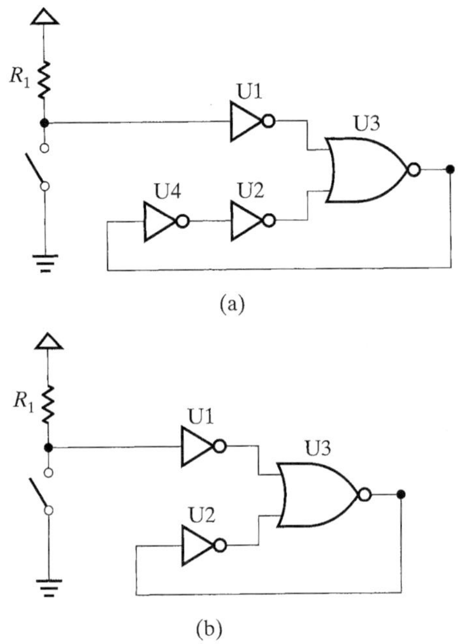

Figure Ex. 3.21

3.22 *(Design Problem)* Design a combinational circuit with three data inputs: D_2, D_1, D_0; two control inputs: C_1, C_0, and two outputs: R_1, R_0. R_1 and R_0 should be the remainder after dividing the binary number formed from D_2, D_1, D_0 by the number formed by C_1, C_0. For example, if $D_2, D_1, D_0 = 111$ and $C_1, C_0 = 10$, then $R_1, R_0 = 01$ (that is, the remainder of 7 divided by 2 is 1). Note that division by zero will never be requested.

(a) Fill in truth tables for the combinational logic functions R_1 and R_0.

(b) Derive minimized sum-of-product realizations of these functions using the Karnaugh-map method.

(c) Draw a circuit schematic that implements R_1 and R_0 using NAND gates only. You may assume any fan-in gates that you need.

(d) Derive multilevel functions for the two outputs. Try to share subexpressions as much as possible. Do you see any advantages to a multilevel realization for this example?

3.23 *(Design Problem)* Develop a minimized Boolean implementation of a 2-bit combinational divider. The subsystem has two 2-bit inputs *A*, *B* and *C*, *D*, and generates two 2-bit outputs, the quotient *W*, *X*, and the remainder *Y*, *Z*.

(a) Draw the truth tables for *W*, *X*, *Y*, and *Z*.
(b) Minimize the functions *W*, *X*, *Y*, *Z* using 4-variable K-maps. Write down the Boolean expressions for the minimized sum-of-products form of each function.
(c) Repeat the minimization process, this time deriving product-of-sums form.

3.24 *(Design Problem)* Consider a combinational logic subsystem that determines if a 4-bit binary quantity *A*,*B*,*C*,*D* in the range of 0000 (0) through 1011 (11 in base 10) is divisible by the decimal numbers two, three, or six. That is, the function is true if the input can be divided by the indicated amount with no remainder (e.g., By2(0110), By3(0110), and By6(0110) are all true). Assume that the binary patterns 1100 (12) through 1111 (15) are "don't cares."

(a) Draw the truth tables for By2(*A*,*B*,*C*,*D*), By3(*A*,*B*,*C*,*D*), By6(*A*,*B*,*C*,*D*).
(b) Minimize the functions using 4-variable K-maps to derive minimized sum-of-products forms.
(c) Can you further simplify the result by using a multilevel-logic implementation? If so, how?

3.25 *(Design Problem)* Develop a minimized Boolean implementation of a "ones count" circuit that works as follows. The subsystem has four binary inputs: *A*, *B*, *C*, *D*; and generates a 3-bit output: *XYZ*. *XYZ* is 000 if none of the inputs are 1, 001 if one input is 1, 010 if two are one, 011 if three inputs are 1, and 100 if all four inputs are 1.

(a) Draw the truth tables for *X*, *Y*, and *Z*.
(b) Minimize the functions *X*, *Y*, *Z* using 4-variable K-maps. Write down the Boolean expressions for the minimized sum-of-products form of each function.
(c) Repeat the minimization process, this time deriving product-of-sums form.

3.26 *(Design Problem)* Consider a combinational logic subsystem that performs a 2-bit addition function. It has two 2-bit inputs, *AB* and *CD*, and forms the 3-bit sum, *XYZ*.

(a) Draw the truth tables for *X*, *Y*, and *Z*.
(b) Minimize the functions using 4-variable K-maps to derive minimized sum-of-products forms.
(c) In this chapter, we have introduced the full-adder circuit. What is the relative performance to compute the resulting sum bits of the 2-bit adder compared to two full adders connected together? (*Hint:* Which has the worst delay in terms

of gates to pass through between the inputs and the final outputs, and how many gates is this?)

3.27 *(Design Problem)* Generalize Exercise 3.26 by adding a fifth input, C_{in} (carry in), and a fourth output, C_{out} (carry out). Minimize the three-sum bit functions and the C_{out} function using 5-variable K-maps.

3.28 *(Hardware Description Languages)* Write a Verilog module that describes the Boolean functions you derived for Exercise 3.22. Can you write a more abstract description that more closely corresponds to the problems statement rather than using specific Boolean expressions? (*Hint:* Use if-statements.)

3.29 *(Hardware Description Languages)* Write a Verilog module that describes the circuit of Exercise 3.23.

3.30 *(Hardware Description Languages)* Write a Verilog module that describes the circuit of Exercise 3.24.

3.31 *(Hardware Description Languages)* Write a Verilog module that describes the circuit of Exercise 3.25.

3.32 *(Hardware Description Languages)* Write a Verilog module that describes the circuit of Exercise 3.26.

CHAPTER

4

Combinational Logic Technologies

Introduction

In Chapters 2 and 3, you learned that Boolean functions can be represented as Boolean expressions and gate-level implementations in two-level and multilevel forms. A multilevel network potentially reduces the wiring complexity of a Boolean function's implementation. However, it was difficult to know transformations of Boolean functions will lead to a better or worse implementations. There are a wide variety of implementation technologies and each leads us toward very different choices for how to transform the functions we want to implement. In some technologies, two-level realizations are highly efficient. In others, we will be more concerned with minimizing delay and utilizing smaller gates.

The goal of this chapter is to provide an overview of combinational logic technologies. In addition we will look at several examples that will highlight the particular properties of each of these technologies. You will learn about fixed and customizable building blocks with a wide range of capabilities. In particular, you will:

- *Design with logic building blocks that are different from traditional logic gates.* In this chapter, we introduce several new components: multiplexers/selectors and decoders/demultiplexers. These are useful Boolean functions that are often easier to visualize as switching networks than as truth tables or logic gates. They can lead to designs that are highly flexible and easy to modify. You will learn the design methods for using them in digital systems.
- *Design with structured circuit implementation styles based on programmable array logic and memories.* We introduce PALs/PLAs and ROMs, which are particularly useful general-purpose digital building blocks that can be customized to implement specific functions. They are used to implement complex functions

in a regular structure that can be mass produced and easily reconfigured for different functions.

- *Design with field-programmable gate arrays.* Finally, we introduce a kind of system building block that allows you to program not only the on-chip logic itself, but also its internal interconnections and achieve very high levels of integration. These components have become ubiquitous in contemporary logic designs, largely replacing other forms of logic components in all but a few specialized applications.

4.1 History

4.1.1 From Switches to Integrated Circuits

There are three broad classes of building blocks for digital systems: combinational components such as logic gates, sequential components such as storage elements (which we will introduce later after dealing with the basics of sequential logic), and interconnection components such as wires. The underlying implementation technologies—be they mechanical switches, vacuum tubes, discrete transistors, or integrated circuits—dramatically influence the digital design process.

The very earliest programmable computers, such as Babbage's Analytical Engine in the mid-19th century, were entirely mechanical devices. They used complex gearing mechanisms to perform calculations and retain state (so complicated, in fact, that Babbage was never able to complete a prototype in his lifetime). Early 20th century systems, like the telephone switching network, were constructed from electrical and mechanical building blocks: *switches* and *relays* (briefly introduced in Chapter 1). Under a mechanical actuation—something that physically moves the device into one of its possible positions—a switch makes or breaks an electrical path between two contact points. Relays are switches that retain their last setting, so they can be used as primitive memory elements.

In the mid-20th century, *vacuum tubes* were used to implement simple gate logic. A vacuum tube is a kind of electrically controlled switch, which has the desirable ability to amplify weak electrical signals as they travel between connection points (this latter property was crucial for making radio and television affordable). A gate might require a half-dozen vacuum tubes and assorted resistors and capacitors to implement its logic function. Compared with their mechanical predecessors, vacuum tubes were faster, used less energy, and were much cheaper to manufacture.

There is little question that one of the greatest inventions of the 20th century is the *transistor*. Invented in 1947, it is a semiconductor analog of the vacuum tube. *Semiconductors* are materials with conductive properties that lie between electrical conductors and insulators. Silicon is an example. Silicon crystals with certain types of other atoms mixed in to their lattice structure, are most interesting in that their conductive properties can be adjusted by changing operating voltages and/or currents. Under the right electrical conditions, semiconductors can conduct like a closed switch. Or, they can break the flow of current, like an open switch. Even early, crude transistors were much smaller than vacuum tubes, and had speed, power, and reliability advantages. By the middle 1950s, the first digital computers were being built using transistors as their principal components.

The semiconductor industry is a remarkable success story of delivering more, for less year after year. In the same size integrated circuit, the industry has doubled the number of transistors packed together on a silicon substrate every 18–24 months. The actual size of the transistor structure has now shrunk down to fractions of a micron (10^{-6} meters). We have progressed from one transistor per package to 100 million (eight orders of magnitude) in fifty years. As an example, the history of the Intel microprocessor family is illustrated in Table 4.1.

But with this comes the challenge of what to do with those transistors. To be viable economically, packaged components have to be useful in a large number of designs. This leads designers to create highly flexible components such as microprocessors that can be used for many functions. Random access memories (RAMs) are another example as they are useful building blocks to store large amount of data for many different applications. However, these can often be much larger, much slower and/or more costly than required for any one particular function. When products are produced in high volume, profits and/or performance can often be increased by designing dedicated circuitry.

TABLE 4.1 Technology Advancements of the Intel Microprocessor Family, 1971–2003

Microprocessor	*Year*	*Application*	*Number of Transistors*	*MIPS (Millions of Instr. Per Sec.)*	*Clock Speed Range*
4004	1971	Calculator	2,300	0.1	108 KHz
8008	1972	Character terminal	3,500	0.1	200 KHz
8080	1974	Altair 8080 Homebrew PC	6,000	0.6	2 MHz
8086 (8088)	1978	Business PC (IBM PC)	29,000	0.8	5–10 MHz
80286	1982		134,000	3	6–12.5 MHz
80386	1985		275,000	10	16–33 MHz
80486	1989	Home PC	1.2 million	40	25–50 MHz
Pentium	1993		3.1 million	100	60–66 MHz
Pentium Pro	1995		5.5 million	200	150–200 MHz
Pentium II	1997		7.5 million	300	200–300 MHz
Pentium III	1999	High-performance workstations	9.5 million	500	333–900 MHz
Pentium IV	2001	Web servers, Graphics apps	42.0 million	1500	1–1.5 GHz
Pentium IV	2003	Multi-media PC, Hyperthreading	55.0 million	3000	2–3.5 GHz

As a middle ground to take advantage of the higher levels of integration, or where the requirements for performance or cost rule out using a large general-purpose microprocessor, the industry has developed a powerful solution: *programmable logic*. Basically, it trades some of those many transistors for increased flexibility and shortened design times. We introduce this concept later in this chapter as we outline the recent evolution of digital logic.

4.1.2 Packaged Logic, Configurability, and Programmable Logic

Standard Parts By the 1960s, the complexity that could be achieved in a single, *integrated circuit* (IC) chip was at the level of small numbers of simple logic gates. These simple building blocks, very similar to the gates that we have used so far, were very successful: the designer chose the right parts from a catalog, placed them on a breadboard or printed circuit board, and implemented the desired design by interconnecting them with external wires on the printed circuit board. Since the parts were primitive enough to be used in many designs, they could be produced in large quantities, stockpiled until needed, and were relatively inexpensive. Over time, more complicated components joined the standard catalogs, such as multi-bit adders and decoder functions. The term *small-scale integration* (SSI) refers to components with a complexity of 10 gates or less, while *medium-scale integration* (MSI) covers components with 10–100 gates. The most common example of these types of components is the Texas Instruments 7400 series, whose elements were the mainstay of logic designers for over 20 years.

ROMs and PLAs/PALs With increasing levels of integration, designers looked for a higher-density method for implementing logic functions. One solution was *read-only memories* (ROMs). A ROM is a fixed array of ones and zeros. Its address, representing a truth table input bit pattern, forms the index of the array. The contents at that index or ROM location, the ROM's output for that input combination, corresponds to a truth table entry. Thus a ROM is a large *look-up table* we can configure to hold a specific truth table. The inputs *look up* the row whose columns become the output values. Originally, ROMs were programmed once at the factory, that is, loaded with their contents when they were manufactured. This is suitable for a design that is produced in large numbers, but is too expensive for prototyping purposes or during development when the design may change often. The ability to set the contents in each individual ROM once it is ready to be used, in the *field*, (*programmable ROMs* or PROMs), or erase and reprogram the contents as needed (*erasable PROMs* or EPROMs) have been developed for designs that are produced in smaller numbers or require the extra flexibility that reprogrammability provides. ROMs have been an important component in many digital designs and still enjoy popularity today.

However, ROMs are not always the best way to implement a function. They may be too slow to meet the performance constraints of the design. Less expensive approaches such as *programmable logic arrays* (PLAs) and *programmable array logic* (PALs) make use of the same mechanisms as ROMs, but realize many designs in a less expensive and higher performance way. PLAs and PALs make use of an array of fixed logic gates, arranged in a standard two-level form such as AND/OR (with a possible final inverter stage). The designer adds or removes wires inside the chip by "programming" the part. For early generations, this involved using a high voltage to electrically fuse or destroy selected wire segments, so this could be performed only once for a given chip. Later generations, based on EPROM technology, can not only be programmed in the field, but can be reprogrammed many times over. Electrically erasable PROM (EEPROM) was developed to allow even faster reprogramming by no longer requiring the chip to be physically removed from the circuit.

PLAs allow the designer to program both the AND and OR gates. PALs are limited to AND gate programming only, and thus provide faster circuits with some limitation on the functions that can be implemented. Signetics introduced the PLA in 1975 followed in 1978 by Monolithic Memories introduction of the PAL.

Today, every more complex programmable logic falls into the category of *programmable logic devices* (PLDs). These can range to packages capable of implementing functions that require 100,000s of equivalent gates.

Application-Specific Integrated Circuits (ASICs) One widely used kind of ASIC is the *gate array* (see Figure 4.1). These components extend the PLA/PAL concept by allowing a more general form of programmable wires on chip, thus lifting the restriction that the gates be arranged in a two-level form. The approach is more expensive and time consuming than a PAL/PLA, but it achieves much higher densities and speeds. Gate arrays consisting of millions of logic gates can be built today. Unfortunately, the designer cannot program the wires rapidly, as represented by the dark areas in Figure 4.1. She can specify the interconnections between the logic cells for the final stage of the semiconductor manufacturing process that fabricates the chip. Since this involves many engineering expenses, the approach is only suitable for designs that are produced in very large volumes. Moreover, it is not a good technology for fast prototyping because of the relatively long turnaround times of days to weeks in obtaining parts from manufacturers.

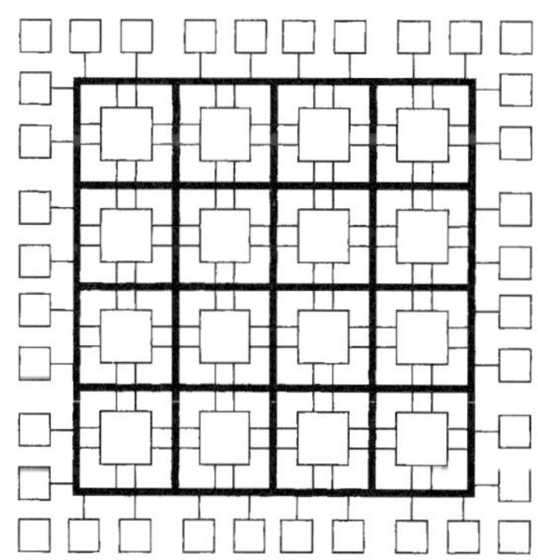

Figure 4.1 Conceptual drawing of a gate array showing logic cells surrounded by wiring resources and a periphery of specialized input/output cells.

Standard cells are an alternative ASIC technology to gate arrays. It provides even more flexible logic building blocks, corresponding to levels of complexity similar to the MSI components found in standard catalogs, with even less constrained wiring. The design approach is very similar to that used with standard components, but the resulting implementations fit within a single chip rather than on a circuit board. However, the approach incurs even higher costs because it customizes all manufacturing steps in order to gain the maximum flexibility. In gate arrays, the basic devices were pre-manufactured and only their interconnection was flexible. With standard cells, designers can specify even the mix of basic components. Thus, although they can achieve higher densities and better performance than gate arrays, it takes longer to implement a design using the standard cell approach, and demands even larger manufacturing batches to justify the front-end engineering costs. Again, this technology is not practical for prototyping, but rather for large-scale manufacturing of proven designs.

Field-Programmable Gate Arrays (FPGAs) FPGAs combine some of the density advantages of a gate-array structure with the rapid programmability of PLAs/PALs. FPGAs consist of basic logic blocks that can be configured to different functions. In addition, the wires that interconnect the logic blocks are also programmable. Therefore, unlike gate arrays or standard cells, whose wires must be manufactured on to the chip, FPGA wires as well as logic functions can be programmed in the field (hence the name) in a matter of seconds. This makes FPGAs the premier technology for rapid prototyping.

Some FPGA technologies are reprogrammable (e.g., Xilinx), while others are not (e.g., Actel). The Xilinx Logic Cell Arrays (LCA) have logic cells arranged in two dimensions with interconnection resources running both vertically and horizontally (see Figure 4.2). The Actel FPGAs have cells arranged in rows with the majority of the interconnect in horizontal tracks that alternate with the rows of logic cells (see Figure 4.3).

Xilinx FPGA configuration is table driven, and the tables' contents can be changed from outside the integrated circuit. Think of it as a collection of little memories that hold the bits that configure each of the logic blocks to realize specific functions and the wires between them to make specific connections. FPGAs trade gate density and reduced

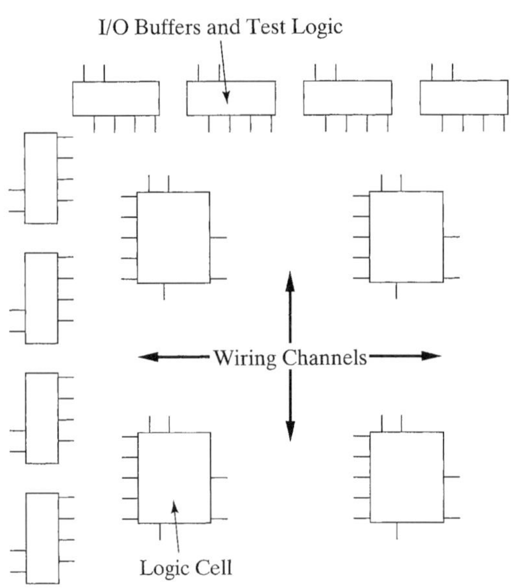

Figure 4.2 Array-based FPGA.

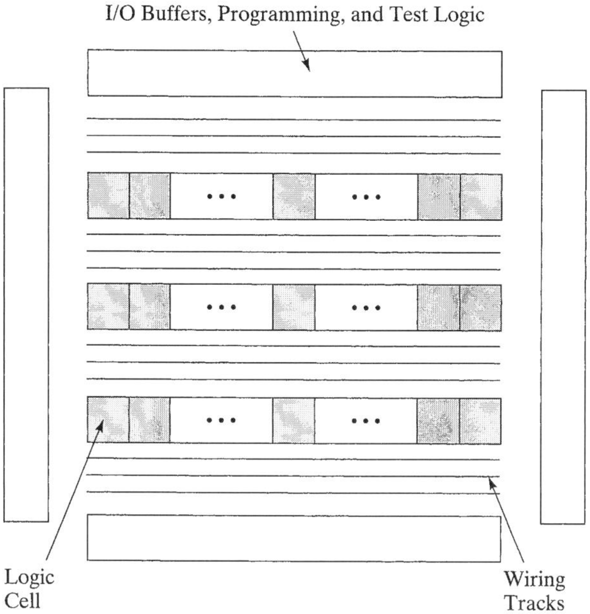

Figure 4.3 **Row-based FPGA.**

TABLE 4.2 Historical Development of Components (Adapted from Oldfield and Dorf, *Field Programmable Gate Arrays*, John Wiley, New York, 1995)

	1960s SSI/MSI	*1970s LSI*	*1980s VLSI*	*1990s Programmable Logic*
Components	Logic, Resistor/ Transistor elements	8-bit μprocessor, Memory, ROM	32-bit μprocessor, Gate arrays	64-bit μprocessor, PALs, FPGAs
Complexity Level (# of gates)	100s	10,000s	1 million	100,000s to millions
Pervasive Components	TTL 7400 series	Intel 8008, ROM	Intel 8086, Motorola 68000, Gate arrays, PALs	Pentium I, II, III, FPGAs, Complex PLDs
Dominant Trend	Standard catalog of components	Larger, General-purpose components, e.g., Microprocessors and Random-access memories	Application-specific integrated circuits	Field-programmable components

performance (because of the extra transistors to implement programmability) for the ability to easily change the detailed function of the component. Actel FPGAs are fuse programmable, much in the same way as PROMs, and can often be quite dense and fast.

Thanks to shrinking feature sizes and increased integration, FPGAs have reached very high densities. While originally they might implement designs with thousands of gates, some FPGA families can now implement designs with 100,000s of gates and even a million or more.

The history of component development is summarized in Table 4.2.

4.1.3 Technology Metrics

Metric	Bipolar	MOS
Gate delay	Low	Low
Integration	Low	Very high
Power	High	Low
Noise	Good	Good
Cost	Low	Low
Fan-out	Fair	Good
Drive	Good	Low

Figure 4.4 Comparison between bipolar and MOS technologies.

There are many aspects to consider in choosing an implementation technology. Two major technology families have been very popular: bipolar and MOS; named after the type of transistor types they use. How they stack up to each other along the principle technology metrics is summarized in Figure 4.4 and includes gate delay, degree of integration, power dissipation, noise margin, component cost, fan-out, and driving capability. In general, faster gates consume more power, generate more heat, cannot be packaged as densely, and are more sensitive to noise problems. Bipolar circuits come in a wide range of TTL (transistor–transistor logic) families, with different trade-offs in circuit speed and power. A super-fast (for its time) bipolar technology was emitter-coupled logic (ECL). The most popular MOS family is CMOS (complementary MOS), consisting of both *n*-channel and *p*-channel devices (see Appendix B). While bipolar technologies used to yield the fastest gates, this has no longer been the case for several years. Nevertheless, bipolar circuits do have the ability to drive large loads (i.e., many inputs of other chips and/or wires that go on to the circuit board) at high speed (and use a lot of power to do it). CMOS does not drive large loads as well but is a much lower power technology and leads to much longer battery life for portable devices. BiCMOS technology combines the best of both worlds: low power consumption and high speed for logical operations, with the ability to drive heavy external loads, such as long wires off of the integrated-circuit chip.

Today, CMOS is, by far, the dominant implementation technology. Let's discuss each of the metrics for this ubiquitous technology.

Gate Delay: If an input change causes an output change, the gate delay is the time delay between the changes. It usually is measured from the input reaching 90% of its final voltage value to the output reaching 90% of its final value. Today, MOS is the fastest technology, as evidenced by modern high speed microprocessors such as the 3.5 GHz Pentium IV (which performs its most basic operation in less than a third of a nanosecond—this corresponds to light travelling approximately 4 inches!).

Degree of Integration: The chip area and number of chip packages required to implement a given function in a technology is a measure of its degree of integration. While in older generation small-scale integrated (SSI) circuits, a package might contain up to 10 logic gates, and even medium-scale integrated (MSI) circuits might only get up to 100 gates, today's CMOS circuits pack millions of gates on to a single chip. Many custom designed large-scale gate arrays and very large scale integrated circuits (VLSI), contain hundreds of thousands of gates. Advanced programmable logic components are all implemented in CMOS technology because of this high degree of integration.

Power Dissipation: Gates consume power as they perform their logic functions, generating heat that must be dissipated. CMOS circuits can be designed to consume very little power, as evidenced by the lifetime of the battery powering a digital watch's CMOS circuitry. However, in CMOS circuits, power is in part a function of the frequency with which the circuit changes outputs. Very fast CMOS circuits can use as much power as the older bipolar technologies. Today's high-speed CMOS microprocessors can dissipate up to 20–25 Watts of peak power. That is enough heat in a small area to require special packaging techniques, such as heat sinks or "fins," fans, or even water channels, to help keep the chips cool. This quickly has become a limiting factor to further integration, as semiconductor transistors do not work as well at high temperatures.

Noise Margin: The maximum voltage that can be added to or subtracted from the logic voltages carried on wires and still have the circuit interpret the voltage as the correct logic value is called the "noise margin." Modern technologies have excellent noise margins. Over time, the voltage to represent a logic 1 has been scaling down, from 5 V in the 1970s to 3.3 V today to 1.8 V in the next few years, thus requiring designers to meet tighter noise margins but also using much less power.

Component Cost: CMOS is by far the cheapest technology we have available today on a per-transistor basis. It enjoys the highest levels of integration and economies of scale that come from being the most commonly used technology. TTL MSI components are extremely inexpensive. ROMs, PALs, and PLAs have become relatively inexpensive for implementing small amounts of logic and are often more cost-effective than MSI parts because of their higher levels of integration that lower manufacturing and assembly costs. FPGAs, with their wide range of capabilities, are the most expensive prototyping components but can pack what would have been many separate components in a single package.

Fan-Out: Fan-out is the number of gate inputs to which a given gate output can be connected because of electrical limitations. Fan-out is a metric related to the ease with which gates can be composed into more complex functions. CMOS circuits can be cascaded with a large number of other CMOS circuits without suffering degradation in the transmitted signals. However, signal delays increase with fan-out. The speed of bipolar circuits does not vary with fan-out, but signal quality is dramatically reduced as the fan-out increases.

Driving Capability: Discrete gates, as presented in Chapter 2, are usually placed with other gates in ready-to-use packages. Driving capability measures the speed of communications *between* packaged components. In general, CMOS circuits have more limited ability to drive other circuits than bipolar circuits, increasing their package-to-package delays. Increasing levels of integration make drive capability less important as the circuits are in the same package and thus are closer together. The only issues remain when connections must be made to the rest of the world using input/output devices such as switches, radios, networks, etc. This is why BiCMOS was developed. It uses CMOS for its digital core functions and bipolar circuits at the periphery for input and output.

Configurability: A major trend in digital logic has been towards components that have their on-chip logic and interconnection configured dynamically. CMOS is an excellent match for this capability, because of the ease with which high density switches and memory cells can be built. These circuit structures are essential for supporting configurability. We need to fit a lot of transistors in as small an area as possible to make this approach viable.

4.2 Basic Logic Components

A designer's task, after coming up with a particular logic function, is to figure out the right combination of available building blocks that will lead to the smallest, cheapest, fastest, and/or lowest power implementation. The building blocks available are more or less flexible. In this section, we will survey three major categories of basic logic components and then look at how they can be used to implement two-level and multilevel logic in the following section.

4.2.1 Fixed Logic

In Chapter 2, we described various kinds of logic gates with essentially no limits on their inputs. If you needed a 20-input AND gate, we assumed that you would have one. Unfortunately, this is not really the case. The reality is that there is a catalog, a library of readily available gates and frequently encountered combinations of gates. Designers pick and choose the best ones for their needs. This style of implementation is often called *cell-based design* and the entries in the catalog are called *standard cells.* Not all of the possible logic gates will be available in a given catalog.

Small- and Medium-Scale Integration (SSI and MSI) TTL is a family of packaged logic components that dominated logic designs between its introduction in the 1960s up to the 1980s. It is still used to upgrade the performance of legacy designs, as well as for certain niche applications. The TTL *standard catalog* is still so well known by designers that even libraries for the most modern of ASIC components, implemented in the radically different CMOS technology, provide the same kinds of logic functions from NAND gates to binary adders and beyond.

A TTL *integrated circuit* package typically contains from 10 to 100 gates. The Texas Instruments 7400-series components provide the standard numbering scheme used by industry. For example, the name for a package containing four 2-input NAND gates is "7400," while a "7404" contains six inverters.

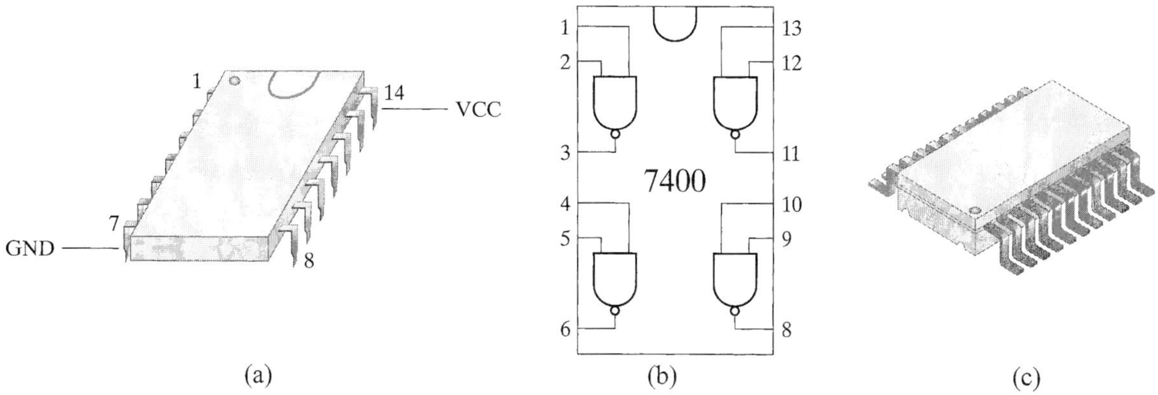

Figure 4.5 A 7400 part schematic (b) with a photo of a dual in-line package version (c) and a surface-mount package version (a).

In TTL technology, logic gates have been traditionally available in rectangular dual in-line packages (DIPs). Pins that connect internal logic to the outside are placed along the two long edges of the package, numbered counterclockwise starting with pin #1 at the upper left-hand corner (when the pins are pointing down). A photo of a 14-pin DIP package is shown in Figure 4.5(a), along with a diagram of its internal logic and pin connectivity (b). Today, most packages are surface-mounted on to printed circuit board and do not have pins that go through the board (requiring expensive drilling). A photo of a surface-mount version of the same package is shown in Figure 4.5(c). DIPs are still common for prototyping small circuits because they are so much easier to manipulate by humans and pin spacings are large enough for easy soldering by hand. Surface-mount parts require expensive equipment to connect them to a circuit board, as they have much tighter spacings.

Some of the kinds of combinational logic SSI building blocks found in the TTL catalog are: inverters; 2-input NAND, NOR, AND, OR, XOR, XNOR gates; 3- and 4-input NAND, NOR, AND, OR gates; and 5-, 8-, 12-, and, for some reason, 13-input NAND gates. The catalog also has a variety of more "exotic" gate-level building blocks. An example is the AND-OR-Invert "gate" constructed from two 2-input AND gates at the first level connected to a 2-input NOR (OR-Invert) gate at the second level. Variations include more inputs at the first level (e.g., 3- or 4-inputs to the AND stage) and more inputs at the second level (e.g., four AND gates feeding the second stage NOR gate). One of these was used in an implementation of the full adder in Figure 3.32.

The TTL catalog also includes even larger aggregations of logic gates that provide higher-level functions. For example, it includes a 4-bit adder function and even a BCD-to-seven segment display decoder in a single package. Catalog circuits of this complexity can easily make use of many tens of more primitive gates. They were added to the catalog as more and more designers found themselves implementing the same common functions.

There are many variations on TTL motivated primarily by the need to have faster parts or lower-power parts or parts that operate at a different voltage (ranging from 5 to 1.8 volts). Today there are more than a dozen different families in production with quite a variation in relative cost. One of the most common in the 70s was the 74LS series for low-power Schottky logic, incorporating a faster kind of transistor structure. Today, 74AS (advanced Schottky) and ALS (advanced low power Schottky) are the families most likely to be used. The *speed-power product* of these families, a measure of their efficiency varies from 0.6 pJ, $0.6*10^{-12}$ Joules, for today's 74AC (Advanced CMOS, 9 ns typical delay at 0.066 mW in power) to 132 pJ for the older 74H (High-speed, 6 ns typical delay—fast in the old days—at 22 mW in power). You can see why the newer CMOS technologies are so attractive when power consumption is a consideration.

Package/Cell Counts So far, we have considered literal count as the primary way to determine the simplicity of a design. Integrated circuit package count is another critical design metric. In older technologies such as SSI/MSI, a single package typically contains six inverters, four 2-input gates, three 3-input gates, and two complex gates per integrated package. The number of packages makes a big difference in manufacturing cost of a circuit. Designers would spend a lot of time making sure that their realization used the minimum cost combination of parts that still realized the function at the speed required.

Look back at the implementations of the full adder function *Sum* in Figure 4.6. The two-level implementation at the top of the figure uses

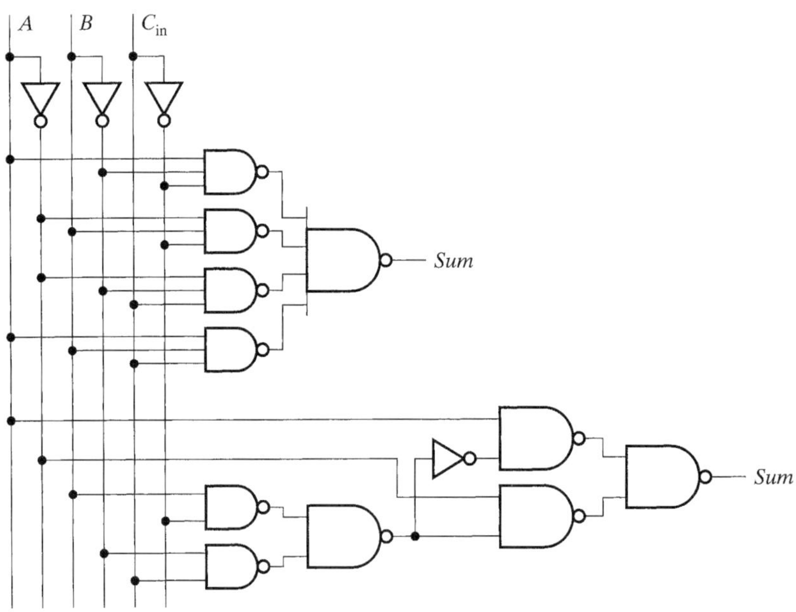

Figure 4.6 Two-level and multilevel implementations of the full-adder *Sum* output.

three inverters, four 3-input NAND gates, and one 4-input NAND gate. This can be realized in four packages: one for the inverters (with three left unused), two that hold three 3-input NAND gates each (with two left unused), and one that contains two 4-input NAND gates (with one left unused). Of course, we can use the leftover 4-input NAND gate to realize one of the 3-input NAND gates (by connecting its fourth input to 1) and this can save us one package of 3-input NAND gates. Therefore our total number of packages could be as low as three.

The multilevel implementation at the bottom of Figure 4.6 uses four inverters and six 2-input NAND gates and also requires three packages even though there are more gates. This is because it uses smaller gates of which there are more in one package (one package of six inverters with two left unused and two packages of four 2-input NAND with two left unused). Although this is the same number of packages as the two-level implementation, we have two leftover NAND gates that we hopefully will be able to use to implement other functions on our board.

If XOR gates are used, *Sum* can be implemented in just one package. In TTL, four XOR gates are integrated into a single package and *Sum* only needs two of them.

Let's now turn to the C_{out} function in Figure 4.7. The two-level implementation at the top of the figure uses no inverters, three 2-input NAND gates, and one 3-input NAND gate. This can be realized in two packages, one for each type of gate with some left unused in each package. The multilevel implementation at the bottom of Figure 4.7 uses two inverters and four 2-input NAND gates. This will also require two packages with four inverters left unused.

It is interesting to consider if we could do better if these two functions were implemented together as is quite likely to be the case.

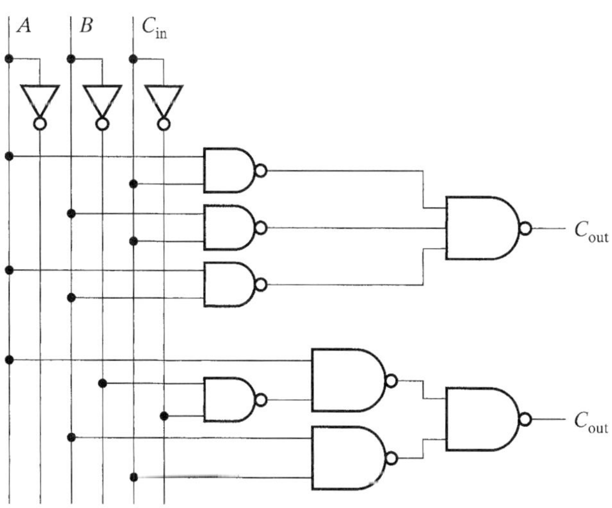

Figure 4.7 Two-level and multilevel implementations of the full-adder C_{out} output.

Immediately we can see that we won't need to duplicate inverters for use for *Sum* and C_{out} as they invert the same variables.

For the two-level implementation of *Sum* we used a single six-inverter package (three unused), one package of three 3-input NAND gates (all used), and one package of two 4-input NAND gates (all used). The 3-input NAND gate required for C_{out} now requires us to add an additional 3-input NAND gate package (with two left unused). We then only need to add the one package of 2-input NAND gates for the terms of C_{out} (with one left unused). For *Sum* and C_{out} together, we need a total of five packages (the same as if they had been realized independently).

For the multilevel implementation we only need one additional package of 2-input NAND gates because two were left unused for *Sum*. For *Sum* and C_{out} together, we need a total of four packages rather than five for the separate implementations. Thus, it turns out that the multilevel implementations leads to a 20% smaller design (four versus five for the two-level implementation).

It should be obvious that finding the best approach depends not just on minimizing literals or gate counts, but minimizing circuit area. In standard catalog-based technologies like TTL, the key attribute is the number of packages needed to realize the design. Often functions can be realized using different components than the ideal. For example, if we need an extra inverter, but have an unused 2-input NAND gate available, we can utilize the NAND gate as an inverter by connecting both inputs together. A 4-input NOR gate can be used to realize a 3-input NOR gate by tying one input to 0. Packing functions into the best mix of gates is quite an art and especially so for families with very large catalogs.

Packaging considerations are much easier when we are working with gate arrays or standard cells as we have available the individual parts we need on the same chip. The problem of finding the lowest package count is much simpler when each package has so much more logic within it.

4.2.2 Look-Up Tables

The second type of basic component is a *look-up table* (LUTs). The basic idea is that we store the output value of a function for each input combination in a table. We then use the current input value to *look-up* what the output should be by *indexing* the table. We already saw this in Section 4.1.2 as a ROM. This is a very flexible component and one that is less sensitive to the complexity of the function. An important feature of look-up tables is that we do not have to change any wiring to change the function, we simply change the values we store in the table. For fixed logic, we might have to change packages and/or wiring. Example of look-up table logic are read-only memories, multiplexers/selectors, and look-up table-based FPGAs.

Read-Only Memories (ROMs) A read-only memory is an array of values that are intended to be read many times but only written once.

Hence, the name. A ROM has a set of addresses, inputs that are used to index a row of the array, and a set of outputs that represent the values stored in that row. Recall that a programmable read-only memory (PROM) is one that can be *programmed,* that is, have its stored values set to a specific pattern of 1s and 0s, in the field rather than only by the manufacturer.

Any logic function can be expressed as a truth table, and we can program the output columns of the truth table directly into a ROM. When a new input is applied to the address lines of the ROM, the values of the corresponding row are read out as the values of the outputs. This is precisely what we do to determine the values of a set of functions that share inputs. Each additional input doubles the size of the memory array, so there are some practical and economic limits, but this is not a bad way to implement a complicated function that is difficult to simplify using the standard methods or is likely to change often.

Modern ROMs include the ability to reprogram the storage elements, at least occasionally. These are called erasable programmable ROMs (EPROMs). The process usually involves erasing the array and rewriting new values and it can take several minutes in some technologies. Therefore, you should definitely think of EPROMs as a ROM that you get to reprogram but still mostly just read.

Each row of the array is called a *word* and is selected by the control inputs, which are called the *address.* The number of columns in the array is called the *bit-width* or *word size.* The number of words and the word sizes of commercially available ROMs are usually powers of 2.

General Concept Figure 4.8 gives a block diagram view of a ROM. The ROM contains an internal *decoder* that maps the address lines into word select lines. Each line selects a unique ROM word, which in turn is gated onto the output lines. Each output of the decoder corresponds to each of the unique minterms of the input values at the address lines. There is a word line for every possible product term.

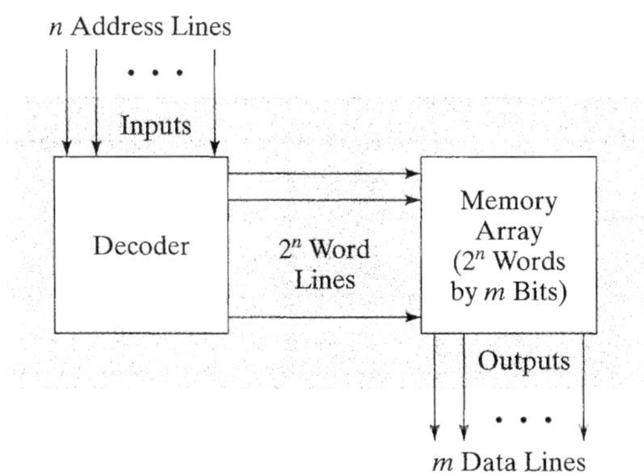

Figure 4.8 **Internal organization of a ROM.**

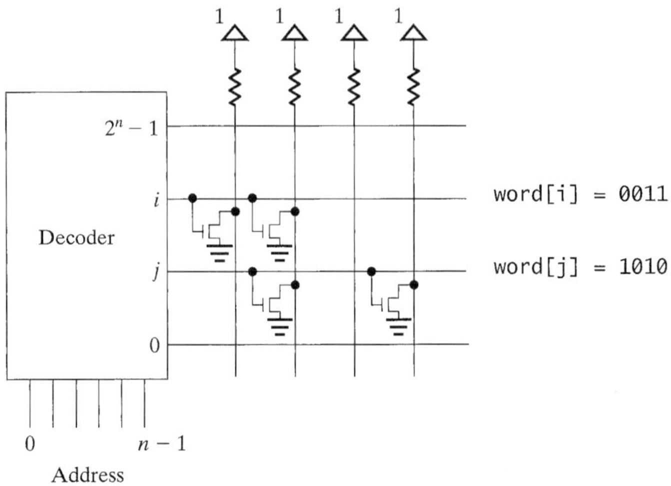

Figure 4.9 **ROM internals.**

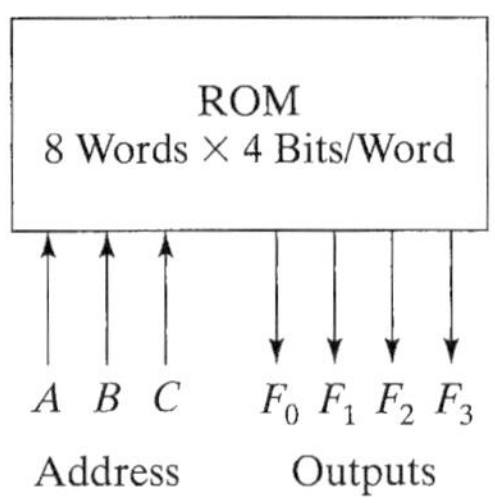

(a) Block diagram

A	B	C	F_0	F_1	F_2	F_3
0	0	0	0	0	1	0
0	0	1	1	1	1	0
0	1	0	0	1	0	0
0	1	1	0	0	0	1
1	0	0	1	0	1	1
1	0	1	1	0	0	0
1	1	0	0	0	0	1
1	1	1	0	1	0	0

(b) Truth table

Figure 4.10 **An 8-word by 4-bit ROM.**

Figure 4.9 shows the ROM in more detail. The n-input address decoder asserts one of the 2^n selection lines. These are also called *word lines*. Each vertical wire, representing an output from the ROM, is called a *bit line*. You can think of each bit line as being connected to a long distributed inverter. A transistor controlled by the word line *pulls down* the bit line (that is: connects it to ground) if a 0 is to be stored in that word. If there is no pull-down transistor in the selected word for that bit line, then this signifies a 1 is stored instead. Based on the placement of the pull-down transistors in Figure 4.9, word_i contains the value 0011 while word_j stores 1010.

Figure 4.10(a) shows a block diagram of an 8-word by 4-bit wide ROM. A ROM frequently is used to store the truth tables of a set of functions, as in the sample ROM contents of Figure 4.10(b). The variables over which the functions are defined (in this case A, B, and C) form the address. Depending on their values, the appropriate row of the truth table and the corresponding ROM word are selected. Each output function is associated with a column or bit position in the ROM. If the function is true for that combination of inputs, a 1 is stored in the selected word at the bit position for that function. Otherwise a 0 is stored. By examining Figure 4.10(b), we get the following set of Boolean equations for the outputs:

$$F_0 = \bar{A}\bar{B}C + A\bar{B}\bar{C} + A\bar{B}C$$
$$F_1 = \bar{A}\bar{B}C + \bar{A}B\bar{C} + ABC$$
$$F_2 = \bar{A}\bar{B}\bar{C} + \bar{A}\bar{B}C + A\bar{B}\bar{C}$$
$$F_3 = \bar{A}BC + A\bar{B}\bar{C} + AB\bar{C}$$

Multiplexers/Selectors A *multiplexer/selector* sets its single output to the same value as one of its many inputs under the direction of its control inputs. If the input to a circuit can come from several places, a multiplexer is one possible way to funnel the multiple sources selectively to a single wire.

More formally, a multiplexer, or *MUX,* is a combinational logic network with 2^n data inputs, n control inputs, and one data output. Depending on the settings of the control inputs, a single data input is selected and "steered" to the output. The 0s and 1s on the control inputs form the binary index of the input whose value is to be reflected on the output. Because a multiplexer does nothing more than select an input value to apply to the output, the terms *multiplexer* and *selector* are used interchangeably.

A *demultiplexer,* or *demux,* steers its one data input to exactly one of its many outputs. This is the reverse of the multiplexer. *Decoder* is another name for the component. We use the latter term when the data input is considered to be a special signal used to enable, or select, one of an array of similar components (as in a ROM).

Figure 4.11 shows how these components could be used in a simple adder subsystem. There are two sources for each of the adder's inputs A and B and two destinations for the resulting sum. The control input S_a chooses between the inputs A_0 and A_1, while S_b selects between B_0 and B_1. For example, to add A_0 to B_1, we would set $S_a = 0$ and $S_b = 1$. By placing the appropriate signal on S_s, the sum can be connected to one of the output nodes, S_0 or S_1, of the demultiplexer.

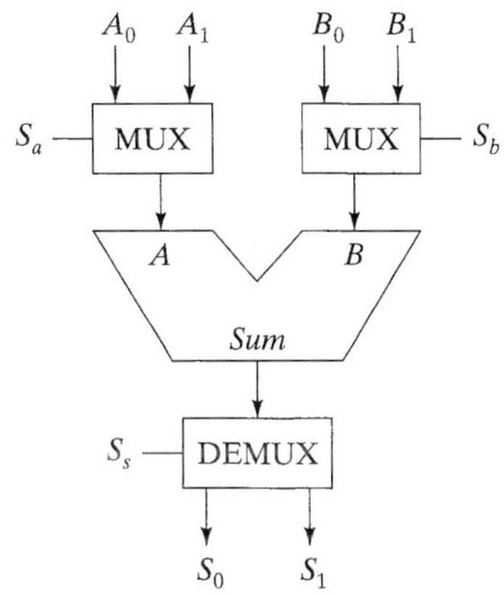

Figure 4.11 Illustration of the use of multiplexers and demultiplexers. Each arrow in the figure represents multiple bit lines for a n-bit binary number.

General Concept Figure 4.12 gives a *functional truth table* on the left and a more conventional truth table on the right for a multiplexer with two data inputs, I_0 and I_1, and one control input, A. Both communicate the same information, but the functional truth table is more clear. It really communicates the idea that we are setting the output to equal a selected input, independent of the value of that input. Restated as a Boolean equation, the two-input multiplexer can be described as

$$Z = \bar{A}I_0 + AI_1$$

If A is 0, the output is set to I_0. If A is 1, I_1 is the value used for the output.

A	Z
0	I_0
1	I_1

I_1	I_0	A	Z
0	0	0	0
0	0	1	0
0	1	0	1
0	1	1	0
1	0	0	0
1	0	1	1
1	1	0	1
1	1	1	1

Figure 4.12 A 2:1 multiplexer truth tables.

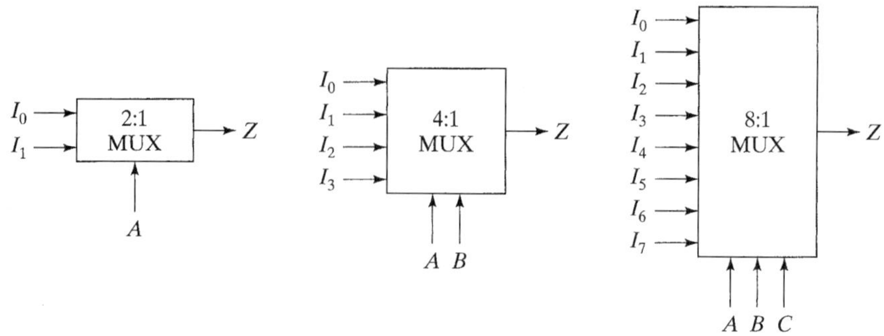

Figure 4.13 Multiplexer block diagrams.

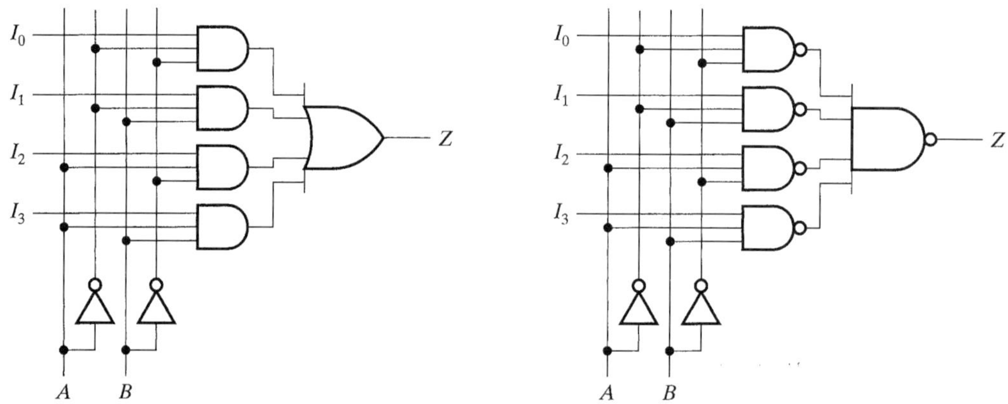

Figure 4.14 Gate-level implementations of a 4:1 multiplexer.

Multiplexers are described by the number of data inputs, since you can infer the number of control inputs from this. Thus, a 2:1 multiplexer has two data inputs, one data output, and one control input, while a 4:1 multiplexer has four data inputs, one output, and two control inputs. Figure 4.13 contains block diagrams for 2:1, 4:1, and 8:1 multiplexers. The Boolean equations for the 4:1 and 8:1 multiplexers generalize from the equation for the 2:1 multiplexer:

$$Z = \bar{A}\bar{B}I_0 + \bar{A}BI_1 + A\bar{B}I_2 + ABI_3$$

$$\begin{aligned} Z = {} & \bar{A}\bar{B}\bar{C}I_0 + \bar{A}\bar{B}CI_1 + \bar{A}B\bar{C}I_2 + \bar{A}BCI_3 \\ & + A\bar{B}\bar{C}I_4 + A\bar{B}CI_5 + AB\bar{C}I_6 + ABCI_7 \end{aligned}$$

The most general expression for the multiplexer, based on the minterm form of Chapter 2 (where m_k is the kth minterm of inputs of A, B, and C), is

$$Z = \sum_{k=0}^{2^n-1} m_k \cdot I_k$$

Figure 4.14 has the gate schematic for the 4:1 multiplexer. The circuit requires four 3-input gates (one for each data input), one 4-input

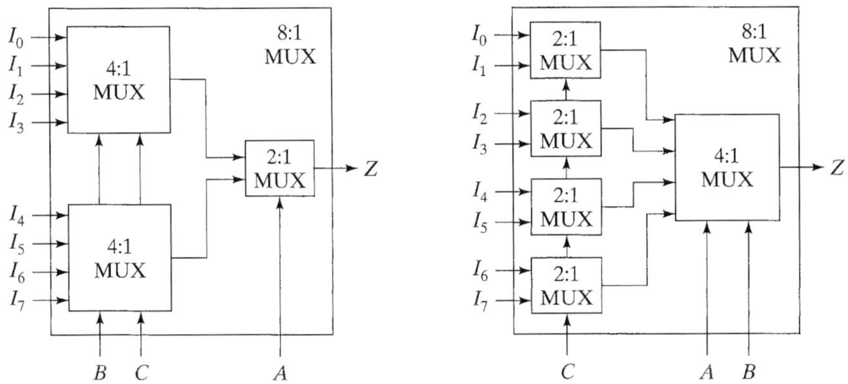

Figure 4.15 Alternative hierarchical implementations of an 8:1 multiplexer.

gate, and two inverters (one for each control input), a total of seven discrete gates. Assuming these are implemented in CMOS logic, the circuit requires 36 transistors (six transistors for each of four 3-input gates, eight transistors for the 4-input gate, and two transistors for each of the two inverters). An 8:1 multiplexer requires eight 4-input gates, one 8-input gate, and three inverters (86 transistors). In standard gate libraries, like the TTL catalog, the largest multiplexer is limited to 16 inputs. This is due to packaging limitations (we need 16 data pins, four control pins, one output pin, power, and a ground for a total of 23 pins on the package—most SSI parts were limited to 24 pins).

It is possible to construct an N:1 multiplexer from several multiplexers with fewer inputs. Figure 4.15(a) shows an implementation of an 8:1 multiplexer from two 4:1 and one 2:1 multiplexers. The first stage selects one input from the range I_0 through I_3 and one from I_4 through I_7 using the control inputs B and C. The second-stage multiplexer selects which first stage output should be gated to the final output using control input A.

This is not the only way to construct an 8:1 multiplexer, as Figure 4.15(b) shows. An alternative use of four 2:1 multiplexers and one 4:1 multiplexer. Control signal C chooses among the even or odd inputs of the 2:1 multiplexers. The remaining control signals, A and B, choose among the outputs of the first-level multiplexers.

Even if the component library limits you to multiplexers with a restricted number of inputs, these techniques can be used to build arbitrarily complex multiplexers from more primitive building blocks.

Multiplexer as a Logic Building Block Multiplexers certainly provide a useful function for implementing signal selection, but we have not yet shown how they implement a general-purpose logic building block. In fact, a multiplexer is another way to implement a truth table directly in hardware. The multiplexer's Boolean equation gives us a hint as to how we might use it to implement arbitrary logic.

EXAMPLE 4.1 USING MULTIPLEXERS TO IMPLEMENT A 3-VARIABLE FUNCTION

Consider the function $F(A,B,C) = m_0 + m_2 + m_6 + m_7$. We can implement it directly by an 8:1 multiplexer as shown in Figure 4.16. The input variables A, B, and C are connected to the multiplexer selection inputs. We simply wire I_i to 1 if the function includes minterm m_i. All other inputs are set to 0. For the example, I_0, I_2, I_6, and I_7 are set to 1, while I_1, I_3, I_4, and I_5 are set to 0.

To see how this works, let's consider the case of $A = B = C = 0$. This corresponds to the minterm m_0. Given these selection inputs, the multiplexer will select I_0, and F is set to 1. If $A = B = 0$ and $C = 1$, then I_1 is selected, and F is set to 0. This continues for all the other combinations of input values.

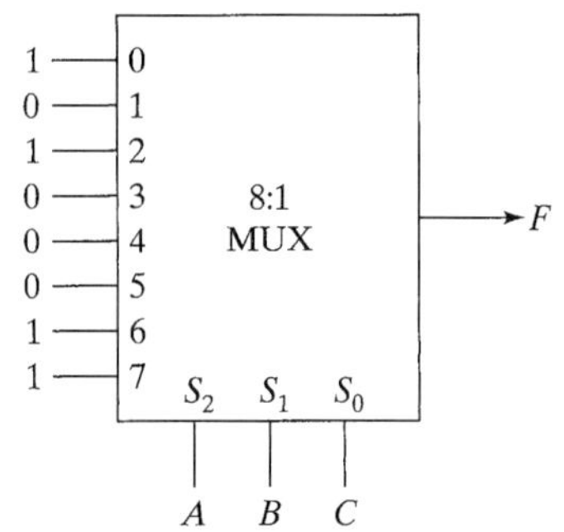

Figure 4.16 Implementation of a Boolean function with a look-up table.

We can get by with an even less-complex multiplexer to implement this function. We can use two of the variables, say A and B, for the selection inputs and connect the multiplexer data inputs to 0, 1, and the third variable C or its complement.

Consider the truth table for F given in Figure 4.17(a). It is arranged so that F is partitioned into groups of rows that share the same values of A and B. F can then be expressed as 0, 1, or a function of C for each of the four row groups defined by the values of A and B. This is equivalent to rewriting F as

$$F = \bar{A}\bar{B}(\bar{C}) + \bar{A}B(\bar{C}) + A\bar{B}(0) + AB(1)$$

We can now implement F with a 4:1 multiplexer, as shown in Figure 4.17(b). Figure 4.17(c) shows a similar implementation that uses A and C as the control signals for the multiplexer. In this case, the function can be written as:

$$F = \bar{A}\bar{C}(1) + \bar{A}C(0) + A\bar{C}(B) + AC(B)$$

This version does not require an inverter and demonstrates that a judicious choice of which inputs to use as the control signals of the multiplexer can make a difference in the cost of the implementation.

A	B	C	F	
0	0	0	1	C′
0	0	1	0	
0	1	0	1	C′
0	1	1	0	
1	0	0	0	0
1	0	1	0	
1	1	0	1	1
1	1	1	1	

(a)

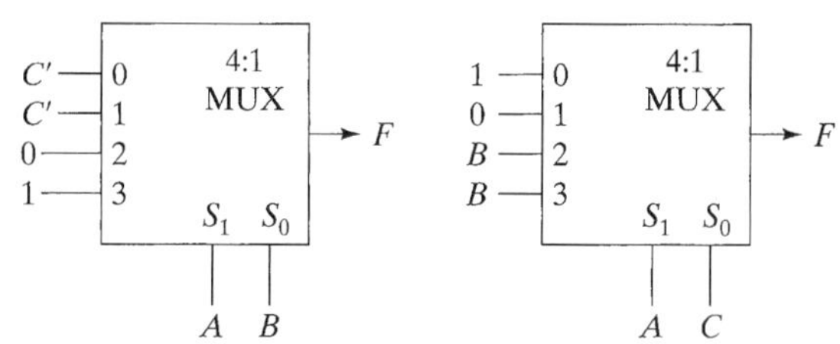

(b) (c)

Figure 4.17 Expressing F as a function of 0, 1, C, and $\bar{C}$ (a). A 4:1 multiplexer used to implement F using A and B as control signals (b), and A and C as control signals (c).

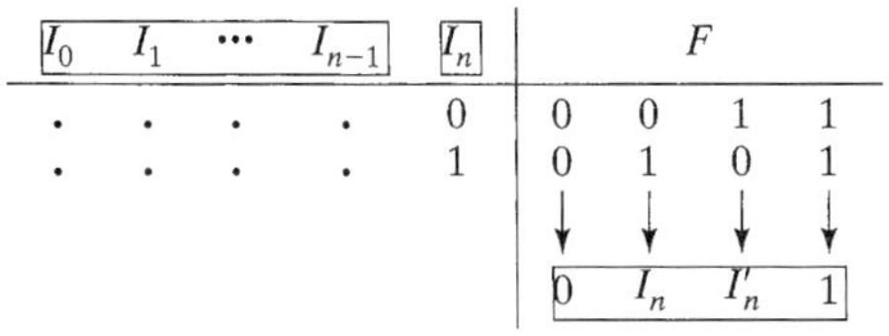

Figure 4.18 **Truth-table to multiplexer input mapping.**

A	B	C	D	G	
0	0	0	0	1	1
0	0	0	1	1	
0	0	1	0	0	D
0	0	1	1	1	
0	1	0	0	0	0
0	1	0	1	0	
0	1	1	0	1	1
0	1	1	1	1	
1	0	0	0	1	D'
1	0	0	1	0	
1	0	1	0	0	D
1	0	1	1	1	
1	1	0	0	1	D'
1	1	0	1	0	
1	1	1	0	1	D'
1	1	1	1	0	

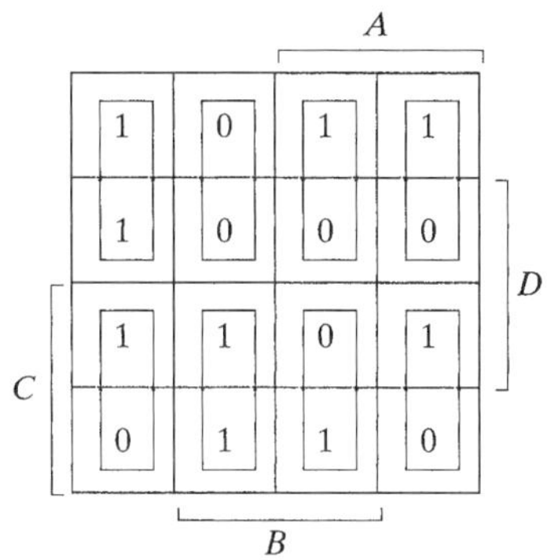

Figure 4.19 **Truth table and Karnaugh map for multiplexer implementation.**

The strategy presented here is easy to generalize. By selecting $n-1$ variables as control inputs to a 2^{n-1} input multiplexer, we can implement any Boolean function of n variables. Each choice of values for the $n-1$ variables selects exactly two rows of the truth table. The two truth-table rows have exactly one of four possible value pairs, as shown in Figure 4.18. If the truth-table rows are both 0 or both 1, the input is the constant 0 or 1, respectively. Otherwise the truth-table rows correspond to the variable or its complement, and the input should be wired to the variable or its complement, respectively.

EXAMPLE 4.2 USING AN 8:1 MULTIPLEXER TO IMPLEMENT A 4-VARIABLE FUNCTION

Consider the function $G(A,B,C,D)$ whose truth table is given in Figure 4.19. Since it is a function of four variables, it can be implemented by an 8-input multiplexer. Let's select A, B, and C as the control inputs. This immediately partitions the truth table into eight pairs of rows, each sharing common values for the three control inputs to be used as select inputs. Each pair can be replaced by the appropriate function of the remaining variable D, as shown in the truth table of Figure 4.19. In essence, we are rewriting the equation:

$$\begin{aligned} G = {} & \overline{A}\overline{B}\overline{C}(1) + \overline{A}\overline{B}C(D) + \overline{A}B\overline{C}(0) + \overline{A}BC(1) \\ & + A\overline{B}\overline{C}(\overline{D}) + A\overline{B}C(D) + AB\overline{C}(\overline{D}) + ABC(\overline{D}) \end{aligned}$$

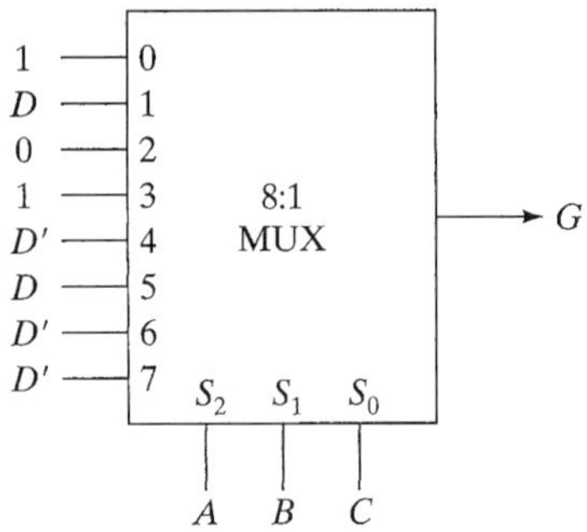

Figure 4.20 **Implementation of the K-map of Figure 4.19 by an 8:1 multiplexer.**

The multiplexer that realizes this function is shown in Figure 4.20.

An 8:1 multiplexer is typically available in a single TTL package. Now let's suppose an additional inverter is needed to form the complement of D, yielding at most two packages to implement the function. Using K-maps to find a minimized sum of products implementation of G, we get:

$$G = \overline{A}\overline{B}\overline{C} + \overline{A}CD + \overline{B}CD + AC\overline{D} + BC\overline{D}$$

This requires five 3-input gates and one 5-input gate plus four inverters, a total of 10 gates available in perhaps four TTL packages. Internally, the multiplexer may use many more gates, but in terms of TTL packages (or standard cell library elements) the use of a complex building block like the multiplexer can be more efficient. Of course, for functions with larger number of inputs, multiplexer-based implementations will be too inefficient—they may very well need lots of logic around the multiplexer. Breaking the function into smaller component pieces and mapping each piece to its own multiplexer may be a solution.

Another advantage of multiplexer logic, as it was for ROMs, is that it is easy to modify the logic function. This is true for look-up table-based components, in general. If we change our mind about an output value, we can simply rewire the inputs to the multiplexer. There is no need to come up with a new set of gates and possibly bring in new packages of devices, the change is small and highly localized. You can understand why multiplexer-based logic design has become quite popular.

More Examples The method of using multiplexers for the implementation of combinational functions may be clear at this point. To make sure, let's apply it to the implementation of three examples we have used in previous chapters: the full adder and 7-segment display decoder from Chapter 2 and the calendar system from Chapter 1.

EXAMPLE 4.3 IMPLEMENTING THE FULL ADDER USING MULTIPLEXERS

The full adder is a function of three inputs, A, B, and C_{in}, from which *Sum* and C_{out} are generated. Each of these functions can be implemented by a 4:1 multiplexer, with A and B on the control inputs and C_{in} and its complement at the inputs. The truth table and the multiplexer configurations are shown in Figure 4.21. In packaged logic, a single package typically contains dual 4:1 multiplexers controlled by the same signals so that they can be used easily to form two functions using the same control variables (as in this case) or to select corresponding inputs from two sets. Thus, the multiplexer approach is actually more efficient, requiring only two packages (one for the MUX and one for inverters), than the implementations based on more conventional gates, which we have seen require four or five packages.

A	B	C_{in}	Sum		C_{out}	
0	0	0	0	C_{in}	0	0
0	0	1	1		0	
0	1	0	1	$\overline{C_{in}}$	0	C_{in}
0	1	1	0		1	
1	0	0	1	$\overline{C_{in}}$	0	C_{in}
1	0	1	0		1	
1	1	0	0	C_{in}	1	1
1	1	1	1		1	

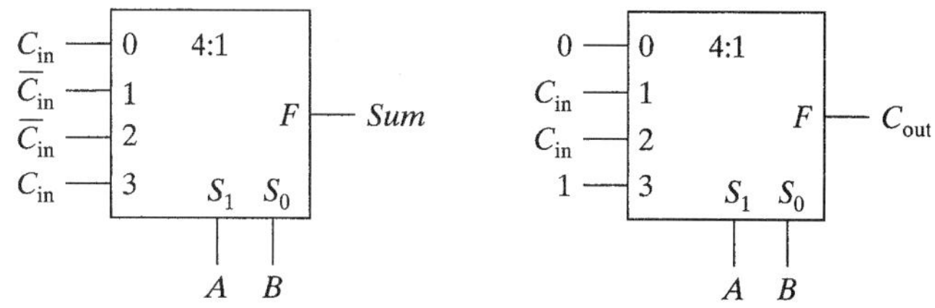

Figure 4.21 **Full-adder implementation with 4:1 multiplexers.**

A	B	C	D	C_2	
0	0	0	0	1	1
0	0	0	1	1	
0	0	1	0	0	D
0	0	1	1	1	
0	1	0	0	1	1
0	1	0	1	1	
0	1	1	0	1	1
0	1	1	1	1	
1	0	0	0	1	1
1	0	0	1	1	
1	0	1	0	X	X
1	0	1	1	X	
1	1	0	0	X	X
1	1	0	1	X	
1	1	1	0	X	X
1	1	1	1	X	

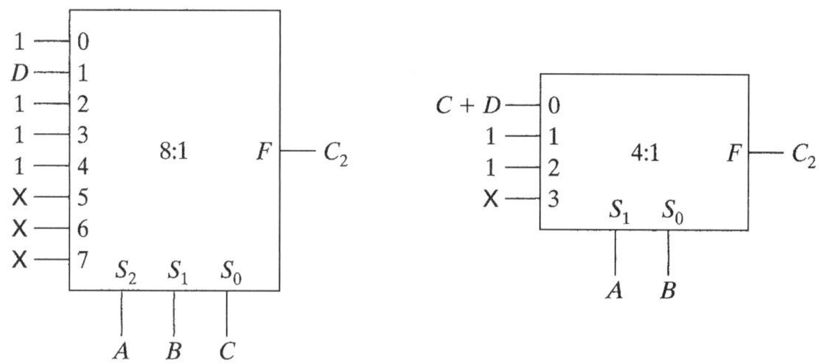

Figure 4.22 **A C_2 decoder output implemented with 8:1 and 4:1 multiplexers.**

EXAMPLE 4.4 IMPLEMENTING THE 7-SEGMENT DISPLAY USING MULTIPLEXERS

Figure 4.22 shows the implementation of the 7-segment display decoder's C_2 output using an 8:1 multiplexer. Note that we left don't cares at the inputs 5, 6, and 7. We arbitrarily can set these to any value.

The function is simple enough that we can get away with a 4:1 multiplexer by placing a little more complex logic outside the MUX. C_2 is zero only in the single case where $ABCD = 0010$. This can be expressed as $\bar{C} + D$ (the complement of $C\bar{D}$) in the selection case when $A = B = 0$. Given that the complete 7-segment display decoder computes seven functions (in terms of package count), using multiplexers for implementing all of these could turn out to be a rather expensive approach.

EXAMPLE 4.5 IMPLEMENTING THE CALENDAR SUBSYSTEM USING MULTIPLEXERS

The final example is the calendar system. In Figure 4.23, we show only the implementation for the d30 and d31 functions using 8:1 and 4:1 multiplexers (we can ignore the leap year input *leap* for these). Note the assignment of values to the don't cares. We chose values to make

m8	m4	m2	m1	d30		d31	
0	0	0	0	X	0	X	m1
0	0	0	1	0		1	
0	0	1	0	0	0	0	m1
0	0	1	1	0		1	
0	1	0	0	1	$\overline{\text{m1}}$	0	m1
0	1	0	1	0		1	
0	1	1	0	1	$\overline{\text{m1}}$	0	m1
0	1	1	1	0		1	
1	0	0	0	0	m1	1	$\overline{\text{m1}}$
1	0	0	1	1		0	
1	0	1	0	0	m1	1	$\overline{\text{m1}}$
1	0	1	1	1		0	
1	1	0	0	0	0	1	1
1	1	0	1	X		X	
1	1	1	0	X	0	X	1
1	1	1	1	X		X	

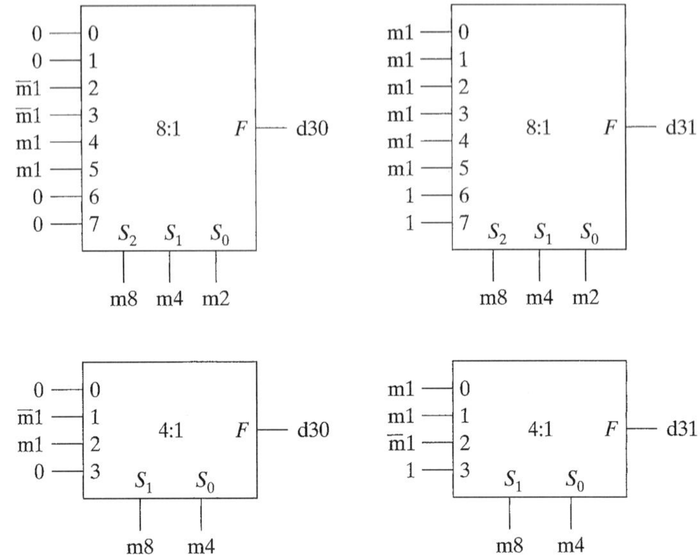

Figure 4.23 **d30 and d31 functions implemented with 8:1 and 4:1 multiplexers.**

the inputs to the multiplexer as regular as possible. Note that, for the 8:1 multiplexers, all pairs of inputs have the same value. For example, for the row for 1101, we chose a value of 0 for the output don't care for d30 and 1 for the don't care for d31, each was chosen so that the input to the multiplexer would be a constant rather than a function of m1 (it matches the output for the 1000 row). The last two rows' don't cares were then chosen to match these. The choices for the don't cares are what enabled us to easily move to 4:1 multiplexers—input m2 is not needed.

An interesting observation for look-up table-based design is that the package count depends only on the number of inputs; it is independent of the actual complexity of the function. Because d30 and d31 are functions of four variables, they can each be implemented in a single 8:1 multiplexer (and luckily, even a 4:1 multiplexer). The functions for d28 and d29 are a single-product term and may be best implemented with fixed components rather than a look-up table.

Look-up Table-Based FPGAs FPGAs offer another approach for implementing truth tables directly. The basic logic cell is a multiplexer with a few control signals. The FPGA chip has many of these arranged in an array. It provides the ability to program the value of the inputs of the multiplexer so that any function of its inputs can be implemented as a look-up table.

Xilinx is the company that popularized look-up table-based FPGAs. Their basic building block is a *Combinational Function Block* (CFB) that can be programmed to implement any combinational function of up to five inputs (there are 2^{32} possible functions of five inputs and 32 bits are required to program the cell's function—the inputs to the multiplexer). The CFB can also be configured to implement any two functions of four inputs (splitting the 32 programming bits into two independent sets of 16—two independent MUXes). This is accomplished by programming the appropriate truth tables and configuration bits into the CFB during the power-up sequence when the component is first turned on.

For example, a single CFB can implement a full-adder bit slice, using the three inputs A, B, and C_{in} to generate the two outputs *Sum* and C_{out}. Five input AND, NAND, OR, NOR, XOR, XNOR, etc. gates are easily implemented in a single CFB. Note that there might be advantages to simplifying a design to require gates of 4-inputs or less, since each CFB can implement two such gates. On the other hand, making the gates even simpler, such as 3- or 2-inputs, really buys the designer little but wiring simplicity, since the number of CFBs to implement a given function is the same whether the gates are 4-, 3-, or 2-inputs.

This illustrates just some of the practical issues related to how far to push logic simplification. Fortunately the more advanced logic families, like FPGAs, are supported with computer-aided design software to automatically map the specification of a collection of Boolean functions into an optimized design for the target family. The specification can take the form of Boolean algebra, gate-level drawings, or a description written in a high-level hardware description language. Such CAD tools are designed to make good use of the catalog of available building blocks for the particular FPGA family. FPGAs such as Xilinx are programmed by changing the values at memory locations and can be easily reconfigured to implement different functions without having to remove the chip from the board.

4.2.3 Template-Based Logic

The third and last type of basic component is one based on a template that we can customize to implement one of a large set of possible functions. This is a more flexible component in that the function is not fixed *a priori*. We can make the component be one of many possible functions depending on what we need. These parts typically fall into the range of *medium-scale integration* to *large-scale integration*, in that they typically involve a dozen to even thousands of gates in a single package or logic cell. If you can use these components as general-purpose logic building blocks, you usually can achieve an implementation with a smaller package count. As examples of these kinds of components, we will examine *decoder/demultiplexer functions, programmable array logic (PALs)*, and *multiplexer-based FPGAs*.

Decoder/Demultiplexer A demultiplexer is the opposite of a multiplexer: a single input is gated to exactly one of several outputs. It is an excellent device for generating mutually exclusive signals, that is, a set of signals where only one is asserted at any one time.

General Concept A decoder/demultiplexer takes as input a single data input (an enable signal) and n control signals and uses the latter to assert one of 2^n output lines. For example, a 1:2 decoder/demultiplexer has two inputs, G (enable) and S (select), and two outputs, O_0 and O_1.

The Boolean equations for the outputs are as follows:

$$O_0 = G \bullet \overline{S}$$
$$O_1 = G \bullet S$$

If G is unasserted (set to 0), then both outputs will be 0. Otherwise, the value of S determines which of the two outputs will be set to 1 (the value of the input G). The equations for the 2:4 and 3:8 decoders are obvious generalizations:

$$\begin{aligned} O_0 &= G\overline{S}_1\overline{S}_0 \\ O_1 &= G\overline{S}_1 S_0 \\ O_2 &= GS_1\overline{S}_0 \\ O_3 &= GS_1 S_0 \end{aligned} \qquad \begin{aligned} O_0 &= G\overline{S}_2 S_1\overline{S}_0 \\ O_1 &= G\overline{S}_2\overline{S}_1 S_0 \\ O_2 &= G\overline{S}_2 S_1\overline{S}_0 \\ O_3 &= G\overline{S}_2 S_1 S_0 \\ O_4 &= GS_2\overline{S}_1\overline{S}_0 \\ O_5 &= GS_2\overline{S}_1 S_0 \\ O_6 &= GS_2 S_1\overline{S}_0 \\ O_7 &= GS_2 S_1 S_0 \end{aligned}$$

Consider the following design situation. We have several digital subsystems, each of which implements its own particular function. But the overall system is designed so that only one of these is to be active at any given time. Now let's suppose that each subsystem has a special "enable" signal which indicates that it has been selected to perform its function. In addition, we have some control signals whose binary value indicates which subsystem to enable. We connect these signals to the demultiplexer control lines.

A demultiplexer is exactly the correct device to use in this situation. First we connect a global enable signal to the demultiplexer input. The corresponding demultiplexer output will now be asserted (that is, set to 1), while all of their other outputs are left unasserted (that is, set to 0). Thus, one subsystem will be enabled while all the others are not. When used in this manner, the demultiplexer is often called a *decoder.*

A decoder/demultiplexer is typically named by the number of control signals and the number of output signals (for example, 1:2, 2:4, 3:8). Contrast this with the multiplexer naming: the number of data inputs and the number of data outputs (for example, 2:1, 4:1, 8:1). This indicates that the demultiplexer is traditionally thought of as a decoder.

The gate-level implementation of a 1:2 decoder is straightforward. The two inputs are G (the enable signal) and a control signal S. When S is 0, G is gated to O_0 and O_1 is driven to 0. When S is 1, the opposite happens: G is steered to O_1 and O_0 is set to 0. If G is unasserted, both outputs are left unasserted (set to 0) no matter what the value of S. We can say that the decoder has an active-high enable (there is a single output at a logic 1 when G is asserted, all outputs are 0 when G is unasserted) and uses AND gates for its implementation. This is not the only way to construct such a decoder. We could have also used NOR gates. Figure 4.24 gives two implementations (one with AND gates and one with NOR

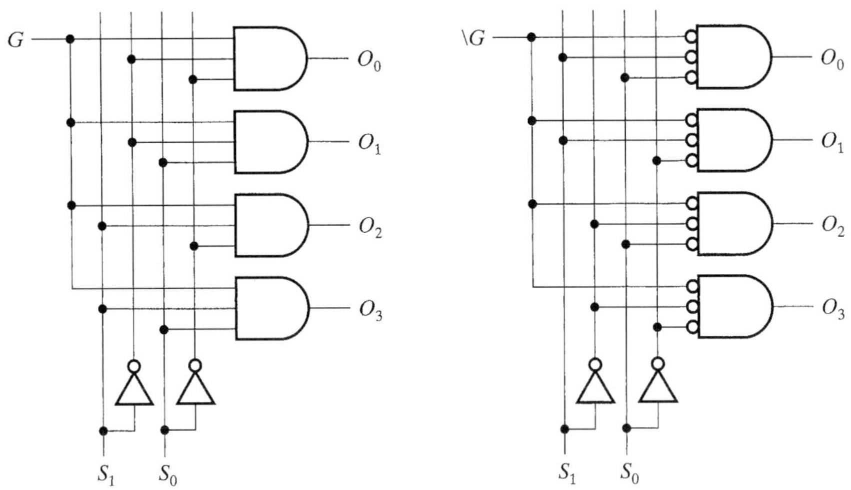

Figure 4.24 **Gate-level implementations of a 2:4 decoder.**

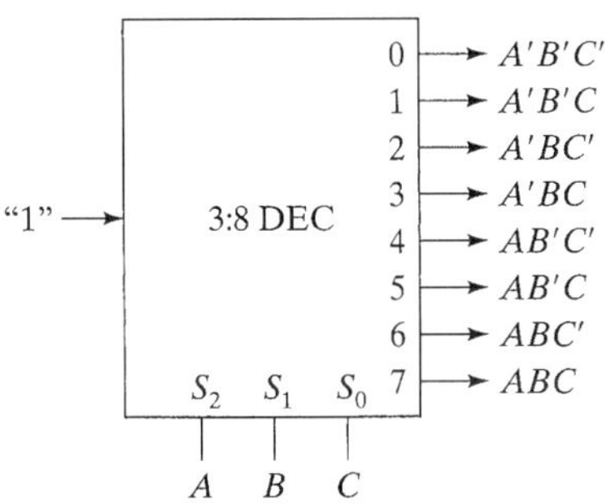

Figure 4.25 **A 3:8 decoder.**

gates) for a 2:4 decoder. For a given value of the select lines, only one of the AND gates will be able to have its output at 1 (if the enable input, G, is also asserted). When G is asserted, the output associated with the select inputs will be driven to 1. All other outputs will be driven to 0. Note that the NOR implementation requires the complements of the select inputs (accomplished via re-wiring), as well as an active-low enable signal, $\overline{G}$. For example, in the AND implementation, the inputs to the O_0 gate are $\overline{S}_0, \overline{S}_1$, and G. In the NOR case, these become S_0, S_1, and $\overline{G}$.

Decoder/Demultiplexer as a Logic Building Block A decoder can also be viewed as a "minterm generator." Figure 4.25 shows a block diagram of a 3:8 decoder where the select lines are labeled with the signals A, B, C. Each output can be labeled with the select-line combination that causes that output to be asserted. For example, suppose the control signals A, B, and C are set to 0, 1, and 0, respectively. This corresponds to the minterm $\overline{A}B\overline{C}$, and output 2 is enabled. In other words, if the inputs correspond to minterm m_i then decoder output$_i$ will be the single asserted output (if the decoder is enabled).

Although the decoder is not intended as a general-purpose combinational logic building block, it can be used as such. Any function expressed in sum-of-products form over n variables can be implemented by an $n{:}2^n$ decoder in conjunction with an OR gate to combine the product terms.

EXAMPLE 4.6 IMPLEMENTING THREE FUNCTIONS WITH A SINGLE DECODER

To see how this can be accomplished, let's consider the following three functions of the Boolean variables: A, B, C, D:

$$F_1 = \bar{A}B\bar{C}D + \bar{A}\bar{B}CD + ABCD$$
$$F_2 = ABC\bar{D} + ABC$$
$$F_3 = (\bar{A} + \bar{B} + \bar{C} + \bar{D})$$

It is more convenient to reexpress the functions as:

$$F_1 = \bar{A}B\bar{C}D + \bar{A}\bar{B}CD + ABCD$$
$$F_2 = AB\bar{C}\bar{D} + ABC\bar{D} + ABCD$$
$$F_3 = \overline{(ABCD)}$$

F_1 is already in canonical sum of products form. To get F_2 into canonical sum of products form, we expand the term ABC to two 4-variable terms: $ABC\bar{D} + ABCD$. Finally, we represent F_3 as the complement of the sum-of-products function $ABCD$.

Figure 4.26 shows how to implement these functions using a 4:16 decoder. F_1 is asserted whenever any of its three minterms are asserted. By connecting A, B, C, and D to the decoder select lines, the output O_3 $(\bar{A}\bar{B}CD)$, O_5 $(\bar{A}B\bar{C}D)$, or O_{15} $(ABCD)$ will be asserted if the inputs correspond to those minterms. F_1 is realized by an OR gate that combines these decoder outputs. For example, if $A = B = C = D = 0$, the decoder output O_0 is asserted and F_1 is not asserted. But if $A = B = C = D = 1$, then the decoder output O_{15} is asserted and F_1 is asserted.

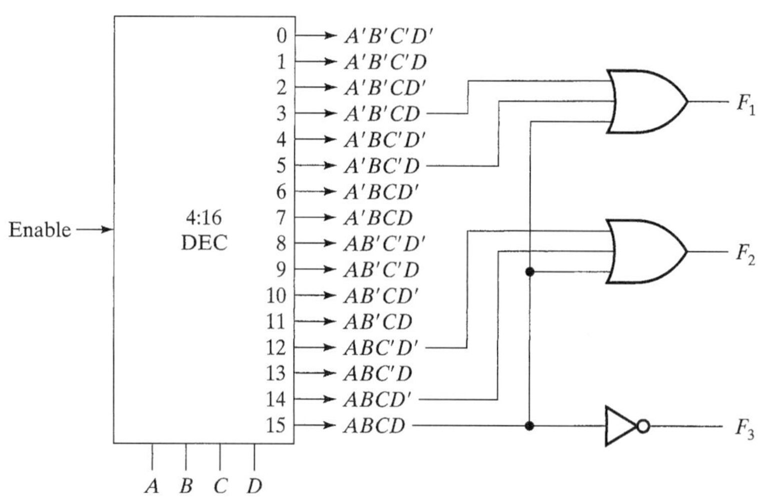

Figure 4.26 **Combinational logic implemented by a decoder.**

Similarly, F_2 can be implemented by a three-input OR gate connected to the decoder outputs O_{12} ($AB\overline{C}\overline{D}$), O_{14} ($ABC\overline{D}$), and O_{15} ($ABCD$). F_3 is implemented by an inverter driven by the O_{15} decoder output.

This approach to implementing logic is most advantageous for functions of a relatively small number of variables (decoders of more than four select inputs are not available in a single package) and a small number of minterms per function (so that an OR gate with a small fan-in can be used).

Programmable Logic Arrays A generalization of the decoders (and a variation of ROMs, which use an internal decoder to select a word) are devices called *programmable logic arrays* (PLAs). These consist of predesigned AND and OR arrays where the designer needs to specify the connections between the inputs and the AND gates to form product terms, and the product terms formed by the AND gates and the OR gates. Unlike decoders and ROMs, not all product terms are formed automatically via a general-purpose decoder for the inputs. Instead, only the product terms that are actually needed are generated and then combined with OR gates. This is especially advantageous when we have to implement functions with so many inputs as to make a full decoder prohibitively large.

There are practical limits on the number of inputs to the AND and OR gates, but, in general, they are much larger than are found in the basic fixed logic building blocks or in ROM chips. On the order of sixteen inputs to the AND gates and eight inputs to the OR gates, this is very commonly supported in PLAs. Given the support for relatively large numbers of inputs, it does not make sense to use a PLA to implement a simple function of AND and OR gates. But it usually is cheaper in cost, design time, and circuit area to implement a complex set of logic equations using a PLA than discrete gates. Moreover, it is relatively much easier to change logic within a reprogrammable PLA than adding or replacing MSI circuit packages that may require alterations to the circuit board.

Integrated circuit (IC) manufacturers have been phenomenally successful in packing more and more transistors into a single chip. One of the ways this richness of resources has been put to use is in increasing levels of integration that allow designers to pack more functions into a single package. To accomplish this for the many functions designers may want to realize, we need a regular structure to which we can map many functions easily.

The key is to find logic building blocks that are sufficiently general purpose to be used in many designs. It is easy to identify generally useful building blocks with primitive structures, such as a handful of gates implementing the commonly encountered AND-OR-Invert structure. But such building blocks reduce the component count by only a modest amount, and we have to have a large number of them in our catalog. We could get a significant reduction if we could have a single building block that contained the equivalent of hundreds of gates and could be used to realize billions of possible functions. But what should this building block look like?

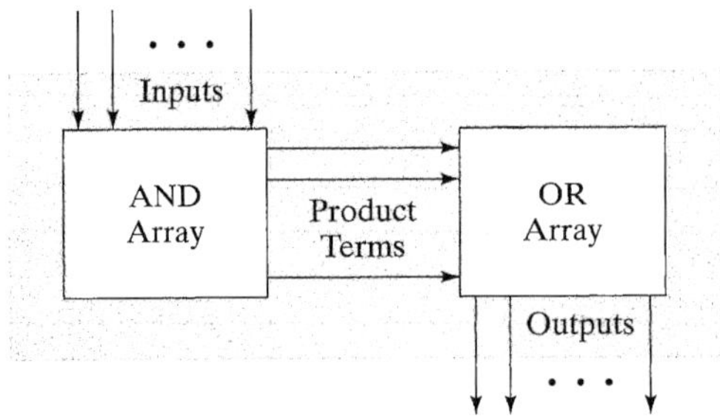

Figure 4.27 **Programmable logic array (PLA) organization.**

The ingenious solution to this dilemma is to exploit the regular structure of two-level logic and arrange AND and OR gates (or NOR gates, or NAND gates) into a generalized array structure whose connections can be *personalized* or *programmed* to implement a specific function. Such general-purpose logic building blocks are called PLAs or PALs (*programmable array logic*).

PALs and PLAs Figure 4.27 shows a general block diagram for an array logic component. Such components are multi-input/multi-output devices, typically organized into an AND subarray and an OR subarray. The AND subarray is used to compute particular product terms of the inputs, depending on the programmed connections. The OR subarray takes these terms and ORs them together to produce the final sum-of-products expression.

A PLA device can implement a modest collection of functions of considerable complexity. This complexity is determined by the number of inputs, the number of product terms (number of AND gates), and the number of outputs (number of OR gates) that the PLA can support.

For example, a typical TTL field-programmable logic array (FPLA) might have 16 inputs, 48 product terms, and eight outputs. In a package with 24 data pins, this PLA contains the equivalent of forty-eight 16-input AND gates and eight 48-input OR gates. When you consider that an 18-data-pin SSI package gives you only four 4-input gates, you can see the enormous packaging efficiency of array logic.

EXAMPLE 4.7 DETERMINING THE MINIMUM SIZE FOR A PLA

Suppose you want to implement the following Boolean equations, defined for A, B, and C:

$$F_0 = A + \overline{B}\overline{C} \qquad F_2 = \overline{B}\overline{C} + AB$$
$$F_1 = A\overline{C} + AB \qquad F_3 = \overline{B}C + A$$

We can characterize them by the number of variables (A, B, C), the unique product terms $(A, \overline{B}\overline{C}, A\overline{C}, AB, \overline{B}C)$, and the functions ($F_0$, F_1, F_2, F_3). These correspond to the number of inputs to the AND array, the number of outputs from the AND array (which are the inputs to the OR array), and the number of outputs from the OR array. To implement these functions, we would need (at least) a 3-input, 5-product term, and 4-output PLA device.

Product Term	Inputs			Outputs			
	A	B	C	F_0	F_1	F_2	F_3
AB	1	1	–	0	1	1	0
$B'C$	–	0	1	0	0	0	1
AC'	1	–	0	0	1	0	0
$B'C'$	–	0	0	1	0	1	0
A	1	–	–	1	0	0	1

Figure 4.28 PLA personality matrix.

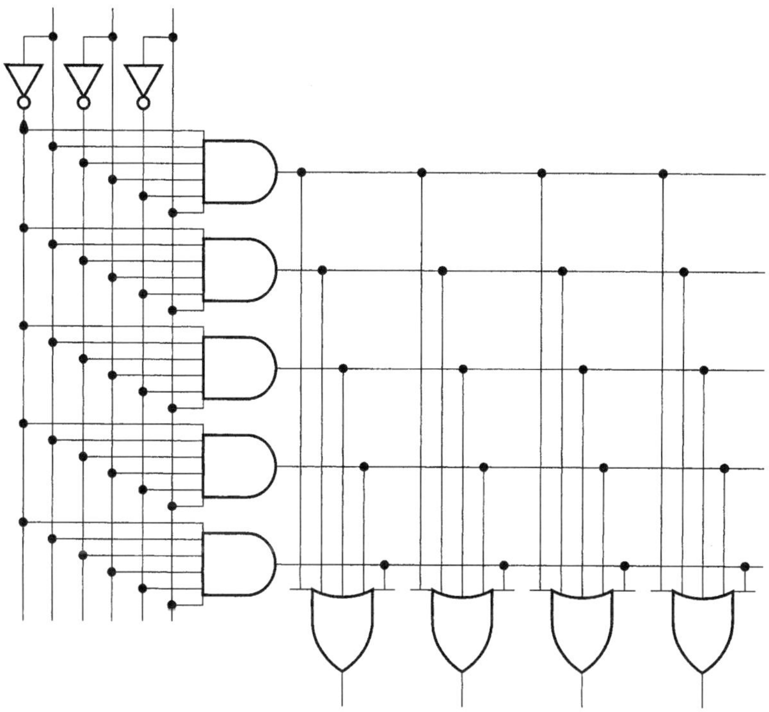

Figure 4.29 PLA before programming.

A convenient way to describe the functions is by a *personality matrix:* a minor reformulation of the truth table shown in Figure 4.28. It describes which input literals should be connected to each AND gate to form the desired product term (1 = asserted variable, 0 = complemented variable, – = no connection), and which of these should be ORed together to form the final outputs (1 = connect product term to OR, 0 = no connection). Rows determine product terms; columns represent inputs and outputs. A product term participates in more than one function if there are multiple 1s in its row in the personality matrix's output columns. In Figure 4.28, it is easy to see that $\overline{B}\,\overline{C}$, AB, and A are used by more than one function.

Gate-equivalent Representation

Figure 4.29 and Figure 4.30 show gate-equivalent diagrams of the implementation of our example functions in programmable array

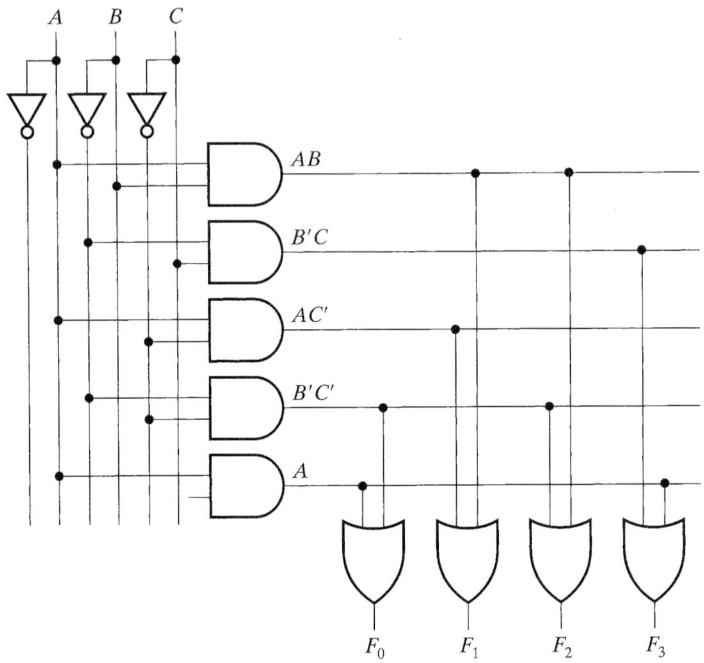

Figure 4.30 **PLA after programming.**

logic. Figure 4.29 shows the array before programming, with all possible connections between inputs and gates prewired "at the factory." You personalize the array by using a hardware device called a *programmer.*

The details of the programming process depend on the particular integrated-circuit technology. A frequently used technique places fuses between all possible inputs to a gate and the gate itself. A fuse is an electrical connection specially designed to break down or "blow" under high current. By placing a high current across selected fuses, the programmer hardware breaks those connections. The programmer software analyzes your Boolean equations to determine which fuses should be blown and which should be left alone. Figure 4.30 shows the same array after it has been personalized.

Figure 4.31 gives a commonly used notation for representing the topology of array logic. You should interpret the single wires entering the AND and OR gates as representing multiple inputs. This is just a shorthand notation that makes the drawing less cluttered. The Xs represent the locations of the programmable elements. These may be fuses that are blown if a connection is to be severed or may be memory cells of some kind to which a value of 0 or 1 can be stored indicating whether there is to be a connection or not. The memory-cell value controls a switch that makes that connection. PLA technologies may differ on how the memory-cell contents is written and rewritten. As with ROMs, PLAs can be write-once or erasable.

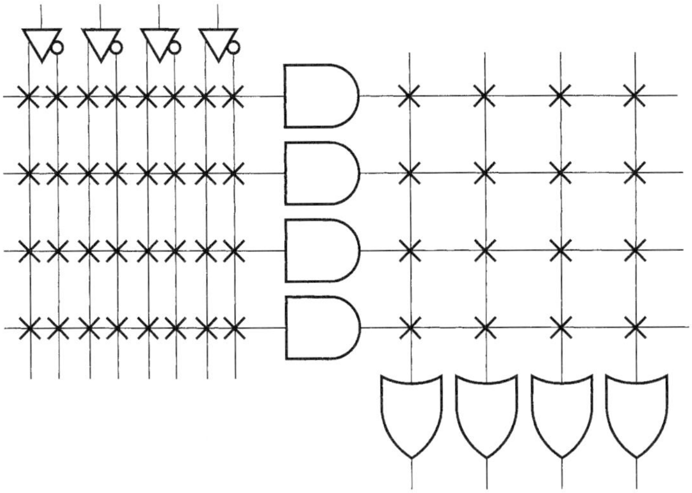

Figure 4.31 **Notation for four-input, four-output, four-product term array.**

EXAMPLE 4.8 IMPLEMENTING A FUNCTION GENERATOR WITH A PLA

Now let's consider the design of a function generator of three inputs. The circuit should implement the logic functions AND, OR, NAND, NOR, XOR, and XNOR:

$$F_1 = ABC$$

$$F_2 = A + B + C$$

$$F_3 = \overline{ABC} = \bar{A} + \bar{B} + \bar{C}$$

$$F_4 = \overline{A + B + C} = \bar{A}\bar{B}\bar{C}$$

$$F_5 = A \oplus B \oplus C = \bar{A}\bar{B}C + \bar{A}B\bar{C} + A\bar{B}\bar{C} + ABC$$

$$F_6 = \overline{A \oplus B \oplus C} = AB\bar{C} + \bar{A}BC + A\bar{B}C + \bar{A}\bar{B}\bar{C}$$

Figure 4.32 shows how this collection of functions is implemented in a PLA. The terms ABC and $\bar{A}\bar{B}\bar{C}$ are shared among more than one output function. ABC is used to implement F_1 (AND) and F_5 (XOR), while $\bar{A}\bar{B}\bar{C}$ is used by F_4 (NOR) and F_6 (XNOR).

The Difference Between PLAs and PALs Figure 4.31 shows that both the AND and OR subarrays can be personalized in any way the designer wants (there are programmable connections shown in both arrays). Devices with this generality are called *programmable logic arrays* (PLAs).

However, not all programmable logic supports full programmability. For instance, Monolithic Memories' *programmable array logic* (PAL) devices have a programmable AND array, but the connections between product terms and specific OR gates are hardwired. The number of product term inputs to an OR gate usually is limited to 2, 4, 8, or 16.

There is a fundamental trade-off in PAL devices between the complexity of the functions in terms of the product terms per OR gate and the

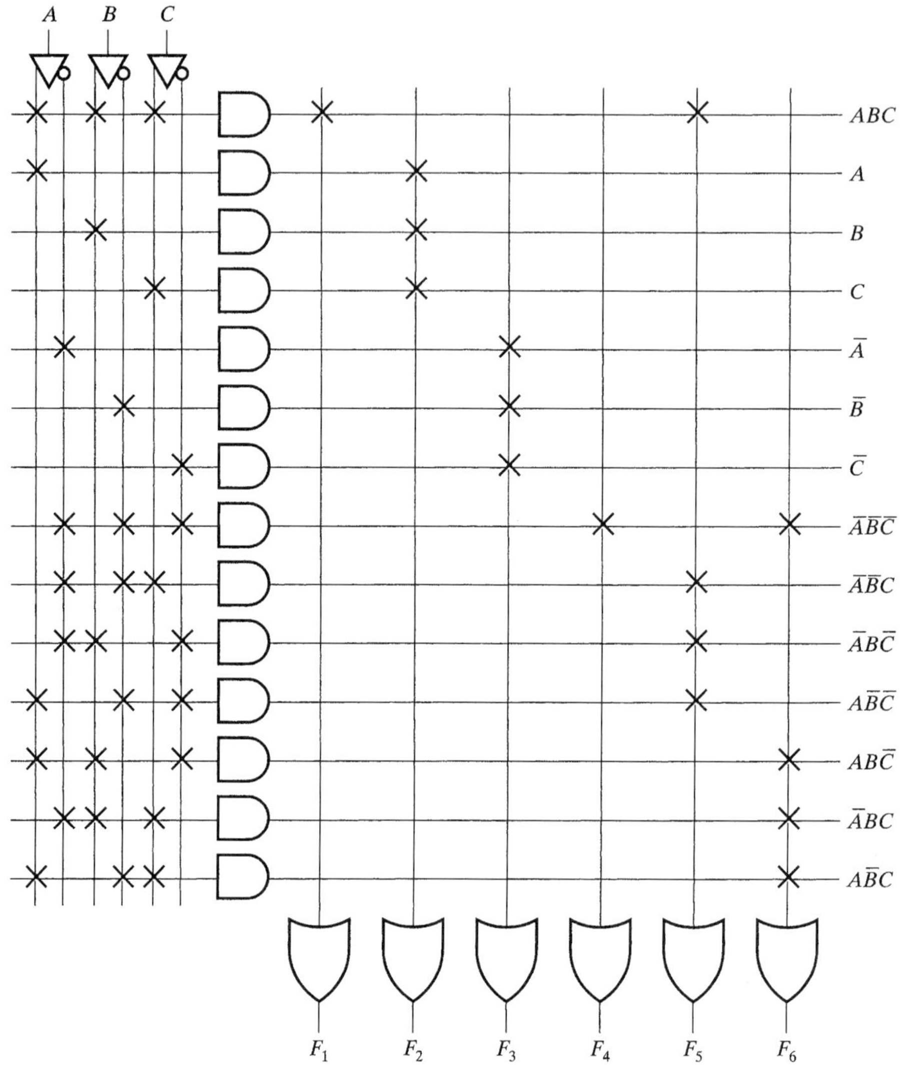

Figure 4.32 **Example implementation.**

number of independent functions the device can implement. The higher the OR gate fan-ins, the fewer the functional outputs from the PAL.

For example, a PAL family might include three alternative devices, each with 16 inputs and 16 product terms, but differing in their OR array organization (and, thus, their number of outputs): four OR gates with 4 inputs each, two OR gates with 8 inputs each, and one OR gate with 16 inputs. The AND subarrays remain completely programmable. Figure 4.33 shows a 4-input/4-product term/2-output PAL organization with a particular fixed choice for the OR array. In this case, the OR gates are limited to two product terms each.

A key difference between PLAs and PALs is that the former can take advantage of shared product terms and the latter cannot. Let's revisit the function generator design we just completed. Since no product terms can be shared among the PAL outputs, the functions need a

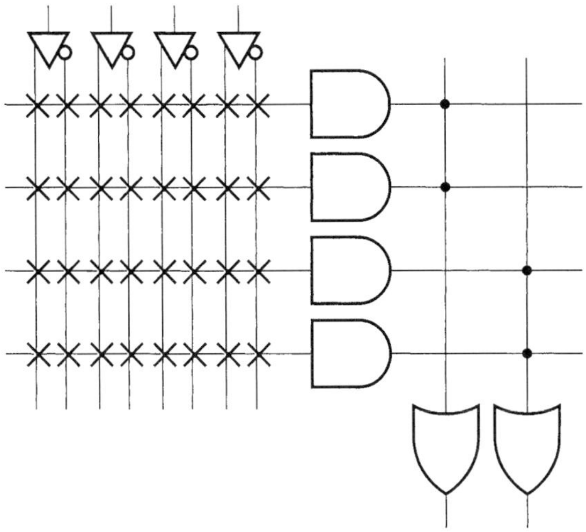

Figure 4.33 **Example of a PAL organization with a constrained OR array and programmable AND array.**

PAL that can compute 16 product terms (ABC and $\bar{A}\bar{B}\bar{C}$ are duplicated, each has to be implemented twice because terms cannot be shared by the OR gates). A PLA needs to compute only 14 product terms, the number of unique terms. For devices with comparable internal resources, a PLA should be able to implement a more complex collection of functions than a PAL if many product terms are shared.

On the other hand, the PLA will be slower because of the electrical characteristics of programmable versus hardwired connections. Programmable connections have higher resistance than standard wired connections. Thus, signals pass through two programmable connections in the PLA, incurring worse delays than the single programmable connection in the PAL. This speed performance difference is actually what led to the original development of the PAL concept.

Let's turn to some examples to help clarify the use of PALs/PLAs.

EXAMPLE 4.9 BCD-TO-GRAY-CODE CONVERTER

In this example, we will design a code converter that maps a 4-bit binary-coded decimal (BCD) number into a 4-bit Gray-code number. Each number in a Gray-code sequence differs from its predecessor by exactly 1 bit. The circuit has four inputs, $ABCD$, representing the BCD number, and four outputs, $WXYZ$, the 4-bit Gray-code word.

The truth table for the translation logic is shown in Figure 4.34. Minimization can be applied (see the K-maps of Figure 4.35), resulting in the following reduced equations:

$$W = A + BD + BC$$

$$X = B\bar{C}$$

$$Y = B + C$$

$$Z = \bar{A}\bar{B}\bar{C}D + BCD + A\bar{D} + \bar{B}C\bar{D}$$

A	B	C	D	W	X	Y	Z
0	0	0	0	0	0	0	0
0	0	0	1	0	0	0	1
0	0	1	0	0	0	1	1
0	0	1	1	0	0	1	0
0	1	0	0	0	1	1	0
0	1	0	1	1	1	1	0
0	1	1	0	1	0	1	0
0	1	1	1	1	0	1	1
1	0	0	0	1	0	0	1
1	0	0	1	1	0	0	0
1	0	1	0	X	X	X	X
1	0	1	1	X	X	X	X
1	1	0	0	X	X	X	X
1	1	0	1	X	X	X	X
1	1	1	0	X	X	X	X
1	1	1	1	X	X	X	X

Figure 4.34 Truth table for the BCD-to-Gray-code converter.

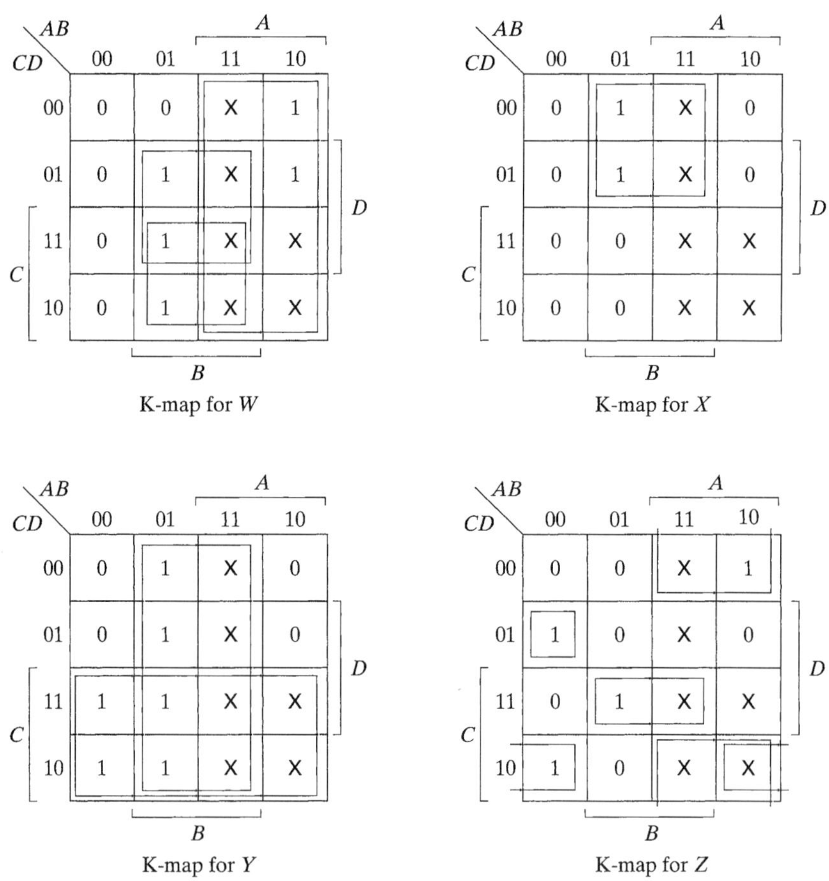

Figure 4.35 Karnaugh maps for code converter.

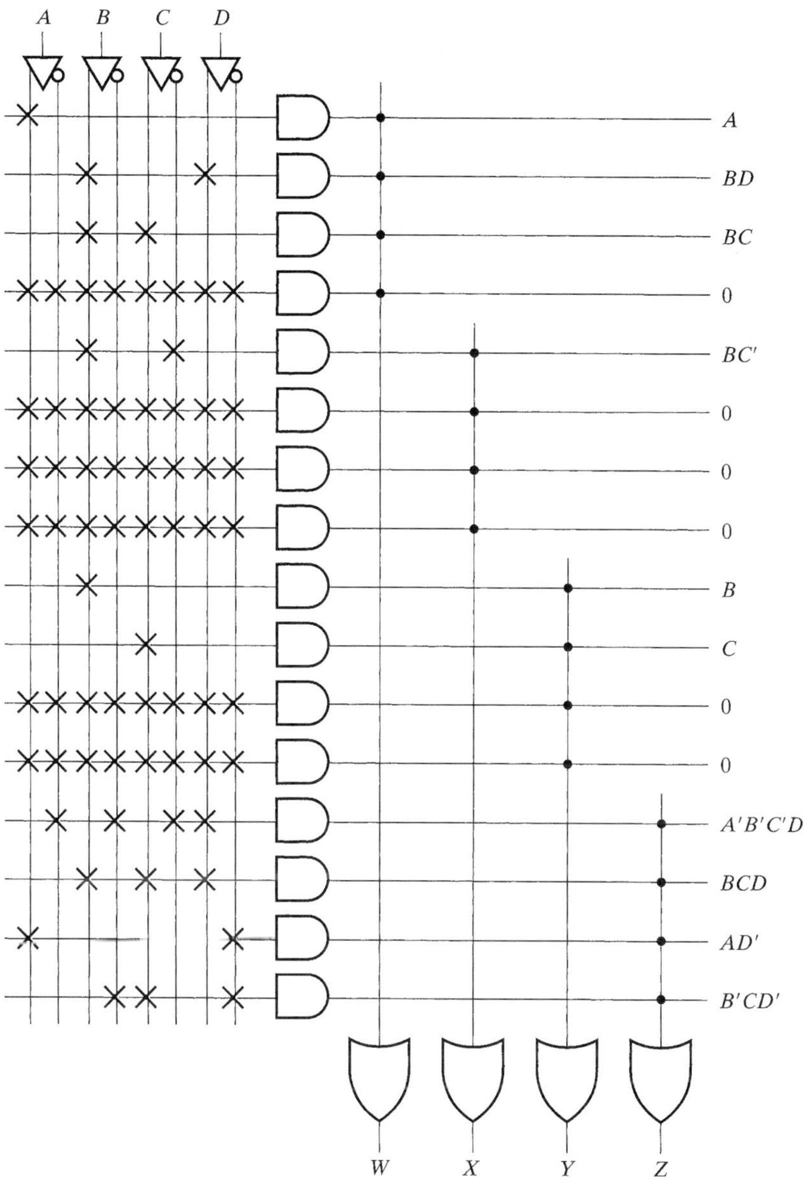

Figure 4.36 **PAL Implementation of BCD-to-Gray-code converter.**

Since there are no shared product terms, these functions are best suited for a PAL implementation. The implementation is shown in Figure 4.36. The PAL contains four 4-input OR gates. You can see that many of the AND gates are being wasted implementing functions that are always 0. The same function could be implemented with a more compact PLA, but it would likely be slower and more expensive.

The programmable logic approach implements the functions in a single integrated-circuit package. Let's consider an equivalent implementation with discrete TTL gates. We will restrict ourselves to NAND gates and inverters only.

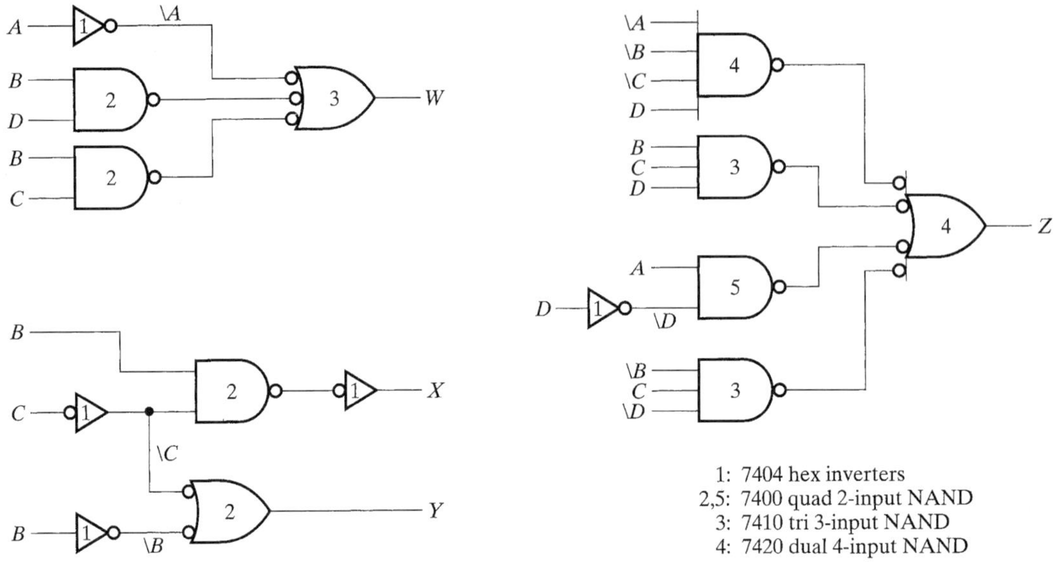

Figure 4.37 **Discrete gate implementation of code converter.**

The resulting discrete gate circuit is shown in Figure 4.37. The figure includes a parts list at the lower right. We have labeled gates from the same package with the same number. The circuit uses 15 gates in five packages. Clearly, the programmable logic approach has a big advantage. Of course, only one of the four gates in package 5 is being used, leaving three gates that could be used by other circuits.

EXAMPLE 4.10 TWO-BIT MAGNITUDE COMPARATOR

Our next task is to design a comparator circuit. The circuit takes two 2-bit binary numbers as inputs, denoted by AB and CD, and computes the four functions of $AB = CD$ (EQ), $AB \neq CD$ (NE), $AB < CD$ (LT), and $AB > CD$ (GT).

Figure 4.38 shows the truth table for the four functions. We can obtain the following reduced equations for the output functions (using the K-maps of Figure 4.39):

$$EQ = \bar{A}\bar{B}\bar{C}\bar{D} + \bar{A}B\bar{C}D + ABCD + A\bar{B}C\bar{D}$$

$$NE = A\bar{C} + \bar{A}C + \bar{B}D + B\bar{D}$$

$$LT = \bar{A}C + \bar{A}\bar{B}D + \bar{B}CD$$

$$GT = A\bar{C} + AB\bar{D} + B\bar{C}\bar{D}$$

The functions use 14 product terms, of which two terms ($A\bar{C}$ and $\bar{A}C$) are used twice. Because product terms are shared, a PLA-based implementation may be more attractive than the PAL-based method. The programmed PLA is shown in Figure 4.40. This single-chip implementation compares very favorably with a multiple-package implementation formed using TTL SSI gates.

A	B	C	D	EQ	NE	LT	GT
0	0	0	0	1	0	0	0
0	0	0	1	0	1	1	0
0	0	1	0	0	1	1	0
0	0	1	1	0	1	1	0
0	1	0	0	0	1	0	1
0	1	0	1	1	0	0	0
0	1	1	0	0	1	1	0
0	1	1	1	0	1	1	0
1	0	0	0	0	1	0	1
1	0	0	1	0	1	0	1
1	0	1	0	1	0	0	0
1	0	1	1	0	1	1	0
1	1	0	0	0	1	0	1
1	1	0	1	0	1	0	1
1	1	1	0	0	1	0	1
1	1	1	1	1	0	0	0

Figure 4.38 **Truth table for magnitude comparator.**

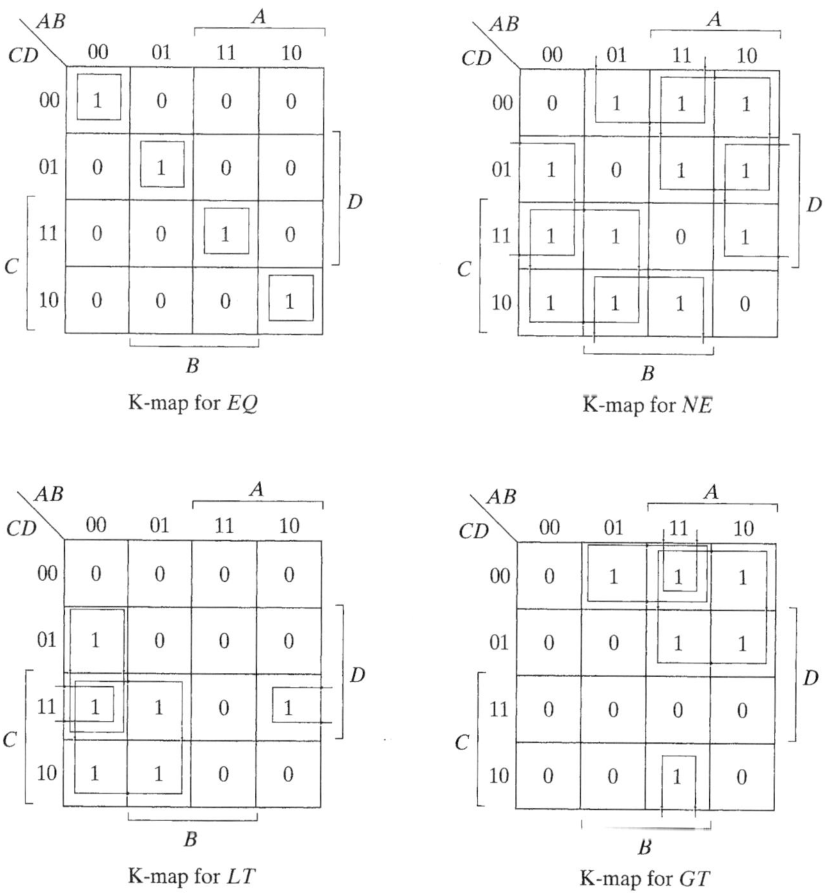

Figure 4.39 **Karnaugh maps for magnitude comparator.**

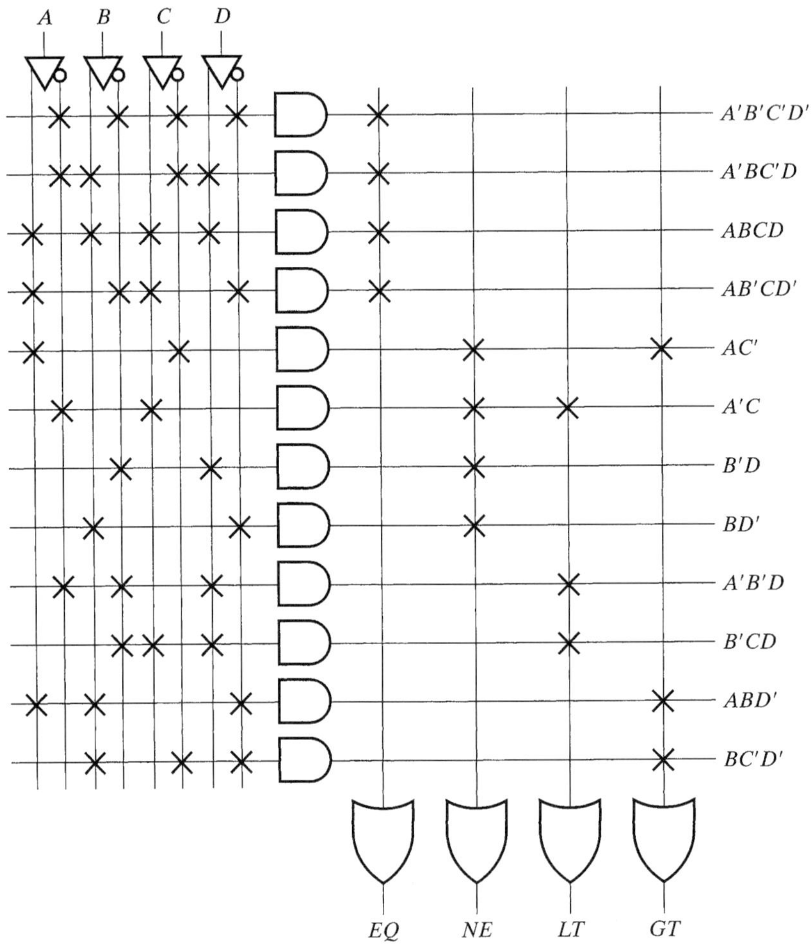

Figure 4.40 PLA implementation of magnitude comparator.

It is not unusual for a PAL or PLA implementation to be equivalent to up to 20 TTL packages containing 100 gates. This is why many real-world designs make extensive use of two-level programmable logic. It is an effective way to reduce an implementation's component count and the corresponding manufacturing costs.

EXAMPLE 4.11 CALENDAR SUBSYSTEM

The calendar subsystem was introduced in Chapter 1, and the equations for its four outputs are provided below. Again, there are no opportunities for shared terms, so a 5-input PAL is an acceptable implementation choice. The output functions d28 and d29 are formed from a single product term, d30 and d31 from two product terms.

$$d28 = m8'm4'm2m1'\,leap'$$

$$d29 = m8'm4'm2m1'\,leap$$

$$d30 = m8'm4m1' + m8m1 = d28'd29'd30'$$

$$d31 = m8'm1' + m8m1'$$

Note that although d30 could be expressed in terms of the three other functions, this would require feedback that would slow down our circuit as d28, d29, and d31 would be computed first, connected to an array input, and then used to compute d30.

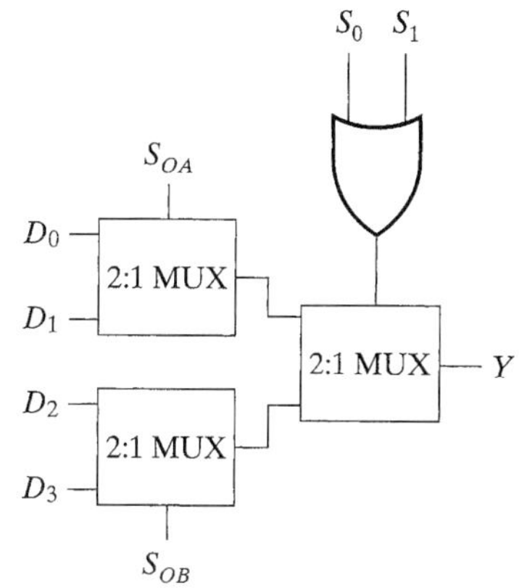

Figure 4.41 Actel logic module.

Multiplexer-Based FPGAs Actel offers an alternative logic template based on a modified four-to-one multiplexer. Its block structure is shown in Figure 4.41. D_0, D_1, D_2, D_3, S_{OA}, S_{OB}, S_0, S_1 are inputs selected through programmable connections from wiring tracks either above or below the logic block. Inputs can also be connected to 0 or 1, if needed. Y is the single output. It can be connected to wiring resources through programmable connections. Note that the cell is a bit more than a 4:1 multiplexer. A 4:1 multiplexer can be derived easily from the cell by connecting together S_{OA} and S_{OB} as well as S_0 and S_1 to form the two control inputs.

Interestingly, this simple circuit can implement all functions of one and two variables, 223 of the 256 functions of three variables, most functions of four variables, some of five variables, few of six and seven variables, and one of eight variables. The function of eight variables is simply the logic function of Figure 4.41, namely,

$$Y = (S_0 + S_1)'(S'_{OA}D_0 + S_{OA}D_1) + (S_0 + S_1)(S'_{OB}D_2 + S_{OB}D_3)$$

EXAMPLE 4.12 USING A LOGIC MODULE

Let's see how the module can be used to implement a two-input AND gate with inputs A and B. We simply wire A to D_1, 0 to D_0, and B to S_{OA}. Then wire S_0 and S_1 to 0. If B is 1 then Y is equal to A; otherwise it is equal to 0. This is essentially an AND function. Using the function of the cell as a template we can substitute our connections:

$$Y = (S_0 + S_1)'(S'_{OA}D_0 + S_{OA}D_1) + (S_0 + S_1)(S'_{OB}D_2 + S_{OB}D_3)$$

$$Y = (0 + 0)'(B'0 + BA) + (0 + 0)(S'_{OB}D_2 + S_{OB}D_3)$$

$$Y = AB$$

As a second example, let's see how to implement a 2-input XOR function. Set S_{OA} and S_{OB} to input A. Wire up D_0 to 0, D_1 to 1, D_2 to 1, and D_3 to 0. The S_{OA} multiplexer then generates A (if A is 0, it passes a 0; if A is 1, it passes a 1), while the S_{OB} multiplexer, in a similar way, generates $\bar{A}$. Wire up S_0 to B and S_1 to 0. If B is 0 it selects A from the S_{OA} MUX. If it is 1, it selects $\bar{A}$ from the S_{OB} MUX. This is exactly the behavior of the XOR operation. The logic function is used as a template as follows:

$$Y = (S_0 + S_1)'(S'_{OA}D_0 + S_{OA}D_1) + (S_0 + S_1)(S'_{OB}D_2 + S_{OB}D_3)$$

$$Y = (B + 0)'(A'_0 + A_1) + (B + 0)(A'_1 + A_0)$$

$$Y = AB' + A'B$$

EXAMPLE 4.13 A FULL ADDER

The *Sum* function of the full adder requires two XOR gates. A single logic module can be easily used for both. Using A, B, and C as inputs, we can substitute into our template as follows:

$$Y = (S_0 + S_1)'(S'_{OA}D_0 + S_{OA}D_1) + (S_0 + S_1)(S'_{OB}D_2 + S_{OB}D_3)$$

$$Y = (A + A)'(B'C + BC') + (A + A)(B'C' + BC)$$

$$Y = A \oplus B \oplus C$$

Note that in this case, we will require the complement of C as an input to the module (for D_1 and D_2). Some logic modules provide extra flexibility with programmable inverters at the inputs.

Clearly, it is not a simple task to find the particular arrangement of inputs to obtain the desired function from this template. The template was defined by a combination of what efficiently could be implemented and packed in large numbers onto a single chip and by the looking at the functions designers find themselves implementing the most often. Fortunately, computer-aided design tools are also available to perform this type of technology mapping for us.

4.3 Two-Level and Multilevel Logic

Of course, two-level logic can be implemented with any and all of the logic components of the previous section. However, some can do so more efficiently or more flexibly than others. Multiplexer and look-up table techniques are very well suited as are approaches built directly using two-level logic principles, such as PALs/PLAs (you can think of a ROM as a PAL/PLA with a complete AND array that generates every possible product term and a fully programmable OR array). Again, let's turn to some examples.

EXAMPLE 4.14 7-SEGMENT DISPLAY DECODER

Let's look at some alternative ways we can realize the complete 7-segment display decoder we first introduced in Chapter 2.

The full subsystem's truth table was given in Figure 2.33. In this subsection, our goal is to examine how to implement this set of Boolean functions using different approaches, such as via discrete gates and programmable logic.

Using Boolean algebra, we have already obtained minimized equations for the LED segment control outputs. They are:

$$C_0 = A + BD + C + \overline{B}\overline{D}$$

$$C_1 = A + \overline{C}\overline{D} + CD + \overline{B}$$

$$C_2 = A + B + \overline{C} + D$$

$$C_3 = \overline{B}\overline{D} + C\overline{D} + B\overline{C}D + \overline{B}C$$

$$C_4 = \bar{B}\bar{D} + C\bar{D}$$
$$C_5 = A + \bar{C}\bar{D} + B\bar{D} + B\bar{C}$$
$$C_6 = A + C\bar{D} + B\bar{C} + \bar{B}C$$

We would need a 4-input, 7-output PAL with at least four product terms per output to implement this function. A P16H8 PAL, for example, has sufficient resources to implement it: ten external inputs plus six feedback inputs (which we don't need to use in this case), eight outputs, and seven product terms per output. The simpler P14H8 PAL cannot do the job. It has 14 inputs and eight outputs, but only two of the outputs are computed as the OR of four product terms while the remaining six outputs have only two product terms each. See Figure 4.42 and Figure 4.43 for the PAL programming maps.

Suppose the target implementation is for a PLA structure rather than a PAL. The limiting factor in a PLA is the number of unique product terms to implement the outputs. These correspond to the number of horizontal "wires" in the PLA AND array. We have fifteen unique product terms in the above set of equations. A typical PLA component can handle sixteen inputs, eight outputs, and forty-eight product terms (for example, see Figure 4.44 for the F100 PLA programming map).

Using a formal two-level minimization computer-aided design (CAD) tool (that uses the don't cares quite differently) we obtain the following set of Boolean equations:

$$C_0 = B\bar{C}D + CD + \bar{B}\bar{D} + A + BC\bar{D}$$
$$C_1 = \bar{B}D + \bar{C}\bar{D} + CD + \bar{B}\bar{D}$$
$$C_2 = \bar{B}D + B\bar{C}D + \bar{C}\bar{D} + CD + BC\bar{D}$$
$$C_3 = \bar{B}C + B\bar{C}D + \bar{B}\bar{D} + BC\bar{D}$$
$$C_4 = \bar{B}\bar{D} + BC\bar{D}$$
$$C_5 = B\bar{C}D + \bar{C}\bar{D} + A + BC\bar{D}$$
$$C_6 = B\bar{C} + \bar{B}C + A + BC\bar{D}$$

Remarkably, this tool seems to have come up with even *more* complex expressions for the outputs than our manual method! A closer inspection reveals some interesting properties of this solution. For one, the number of *unique* product terms has been reduced from 15 to only 9! Second, although the individual expressions now have more product terms (up to five terms for two of the outputs), there is a high degree of sharing of terms among the outputs. Remember that the size of the PLA is determined primarily by the number of unique product terms, so the better design is the one with fewer unique terms. The minimization of these functions was performed so as to maximize sharing. This is an aspect of minimization that we have not yet discussed. When terms can be shared, it may make particular sense to use less than the minimum set of prime implicants to cover a function. The tool we discussed in Chapter 3 (*espresso*) supports this multiple-output minimization mode. The details of its algorithms are outside the scope of this text.

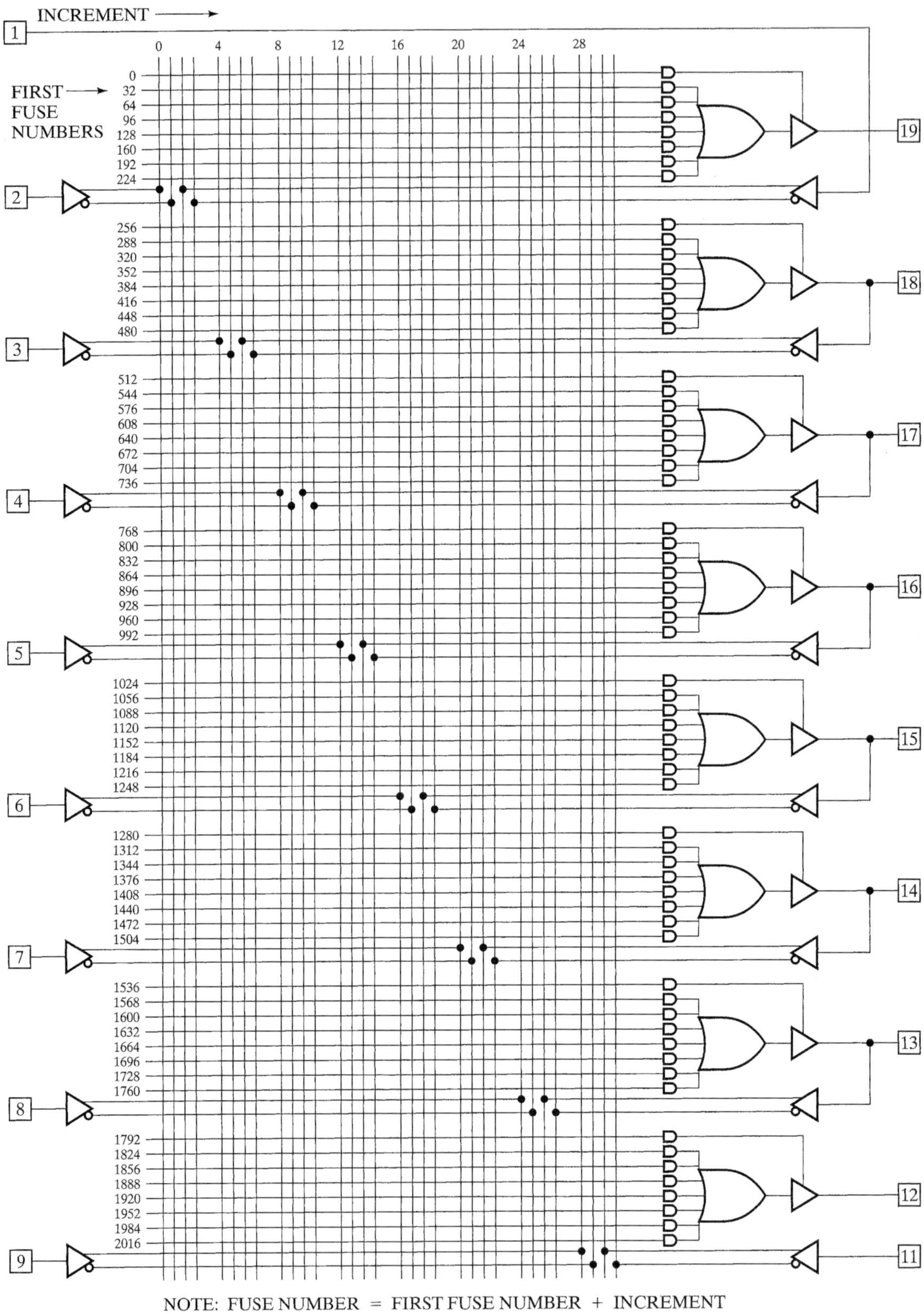

Figure 4.42 **P16H8 PAL programming map.**

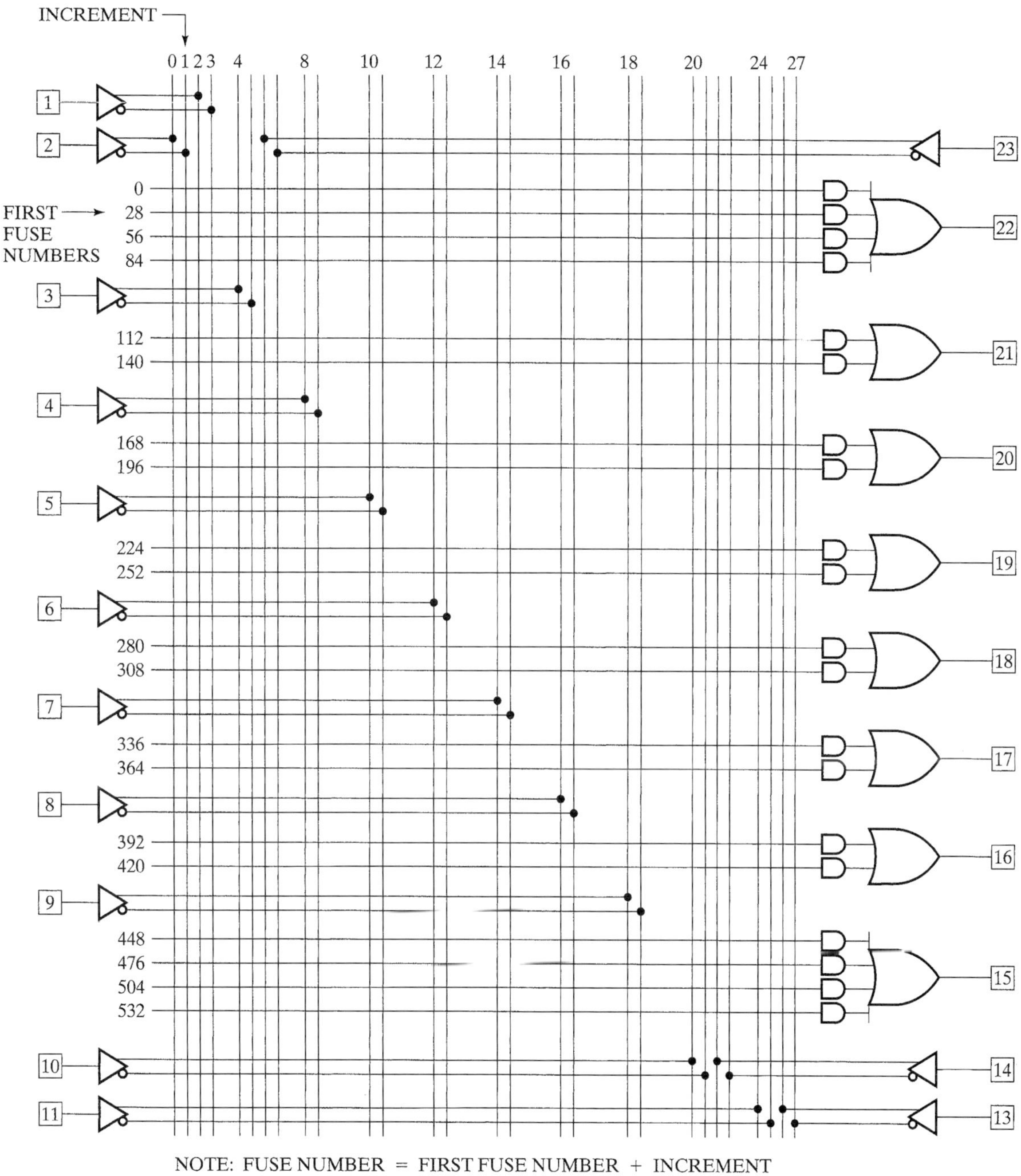

Figure 4.43 **P14H8 PAL programming map.**

Term sharing provides no assistance for a PAL based implementation, since the topology of a PAL makes it impossible to share terms among the different output functions. In fact, by increasing the number of terms per output, this CAD tool would have us use a more

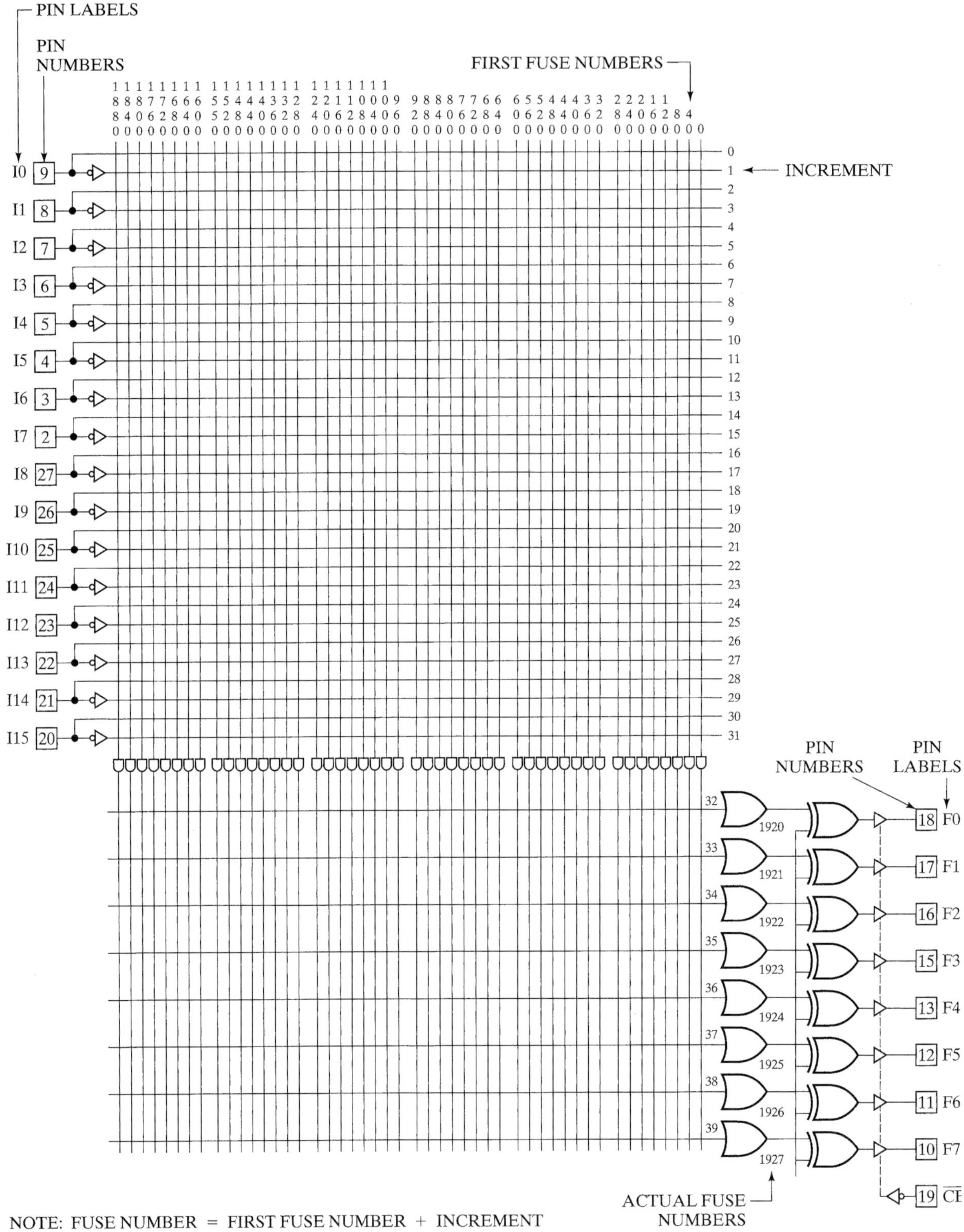

NOTE: FUSE NUMBER = FIRST FUSE NUMBER + INCREMENT

Figure 4.44 F100 PAL programming map.

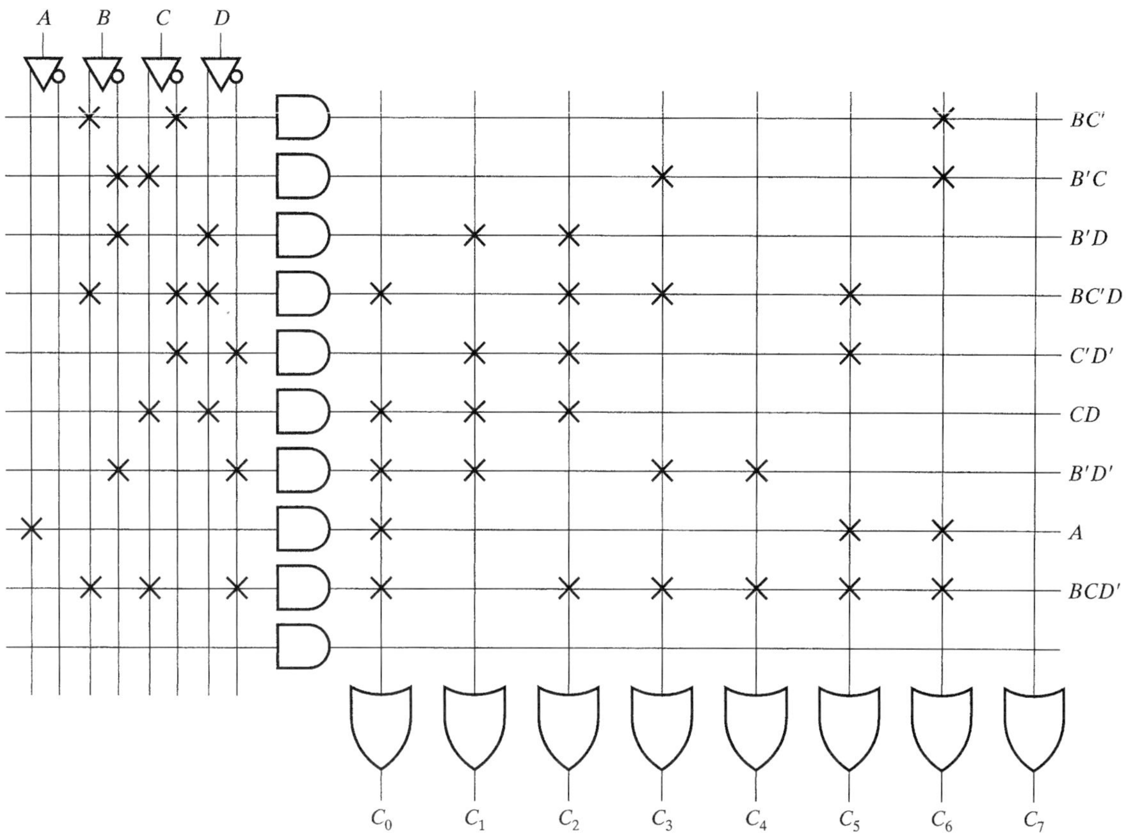

Figure 4.45 PLA implementation of BCD-to-seven-segment display decoder.

complex PAL than is really needed (it is possible to ask these tools to perform the optimization on the basis of individual outputs, and this approach should be followed if the implementation technology is going to be a PAL). Figure 4.45 shows the programming map for a PLA implementation.

Minimized Multilevel Implementation How does this compare with a multilevel approach? Multilevel logic generally is not well suited to PAL or PLA implementation, despite the ability to feedback outputs as inputs to build up multilevel structures. So we will focus on discrete gate implementations to compare optimized two- and multilevel implementations.

EXAMPLE 4.15 7-SEGMENT DISPLAY DECODER REVISITED

First, let's examine the complexity of the solution derived by our two-level implementation optimizer in a little more detail. The number of literals is 63. The number of gates is 7 (an OR gate for each output) plus 8 (an AND gate for each unique product term except the simple term A) plus 3 (an inverter for the inputs B, C, D), for a total of 18 gates.

Each output is at the second level of the network, experiencing two to three gate delays.

A standard multilevel optimization tool might yield the following set of multilevel equations (note that hand optimization may do better):

$$X = \bar{C} + \bar{D} \qquad C_3 = C_4 + BDC_5 + \bar{A}\bar{B}\bar{X}$$

$$Y = \bar{B}\bar{C} \qquad C_4 = \bar{D}Y + \bar{A}C\bar{D}$$

$$C_0 = C_3 + \bar{A}B\bar{X} + ADY \qquad C_5 = \bar{C}C_4 + AY + \bar{A}BX$$

$$C_1 = Y + \bar{A}\bar{C}_5 + \bar{C}\bar{D}C_6 \qquad C_6 = AC_4 + CC_5 + \bar{C}_4C_5 + \bar{A}\bar{B}C$$

$$C_2 = C_5 + \bar{A}\bar{B}D + \bar{A}CD$$

X and Y are intermediate terms "factored out" and inserted into the logic network. This yields a total of 52 literals and 33 gates (one for each of outputs C_0 through C_6, X and Y, plus seven inverters including $\bar{X}$, $\bar{C}_4$, and $\bar{C}_5$, plus 17 other product terms with more than one literal). In terms of delay, the slowest output is C_1, since it is a function of C_6, which is a function of C_5, which in turn is a function of C_4. C_4 already experiences three gate delays since it is a function of Y. Thus C_1 experiences nine or ten gate delays! Although the CAD tool has done an admirable job of reducing the area to implement the functions (as measured via the number of literals and their corresponding effect on wiring and switch area), it has not done so well on reducing the gate count, and its implementation should be quite a bit slower than an optimal two-level realization.

Mapping to Look-up Table Logic Each of the 7-segment display decoder outputs is a function of four inputs. This means that each can be implemented by one 8:1 multiplexer. Compared with a single PAL or PLA component, such an implementation requires considerably more space and resources.

Pairs of output functions can be implemented easily in a single FPGA building block such as a Xilinx CLB. They are each functions of the same four variables. The complete decoder function of seven outputs would require three and one-half Xilinx CLBs. The multilevel implementation with nine functions would needlessly waste another CLB. However, if we looked at logic with more inputs per function, a multilevel implementation would be a necessity to simplify each of the functions so it would have a number of inputs equal to or less than the number of inputs of the look-up table cell.

Mapping to Template Logic Determining the number of Actel-like multiplexer modules is a somewhat more complex exercise. To a first approximation, a single such building block is comparable to a 4:1 multiplexer. In a worse-case scenario, it should require no more than three such modules to implement an 8:1 multiplexer. With this reasoning, it should be possible to implement the seven output functions in not more than 21 logic modules. Of course, this is a pessimistic *upper-bound*. Optimizations probably are possible to reduce this number.

EXAMPLE 4.16 CLB APPLICATION

Consider how we might implement a 4-bit binary adder. A full adder can be implemented in a single CLB. The C_{out} and S_i outputs are functions of the three inputs: A_i, B_i, and C_{in}. To get a 4-bit adder, we simply cascade four CLBs. This is shown in Figure 4.46(a). In many FPGA chips, these can be paired so that each CLB implements two functions.

An alternative approach is to use a 2-bit binary adder as a building block. This circuit has five inputs, A_1, A_0, B_1, B_0, and a C_{in}. The outputs, S_1, S_0, and C_{out}, are each functions of these five variables. Thus, the 2-bit adder can be implemented with three CLBs. To construct a 4-bit adder, we cascade two 2-bit units for a total of six CLBs. This is shown in Figure 4.46(b).

The second implementation may not look attractive, but it has some advantages. It incurs two CLB delays in computing the 4-bit sum, one to compute the carry between the low-order 2 bits and the high-order sums and one to compute the final sums of the high-order bits. This compares with four CLB delays in the implementation based on one complete full adder per CLB.

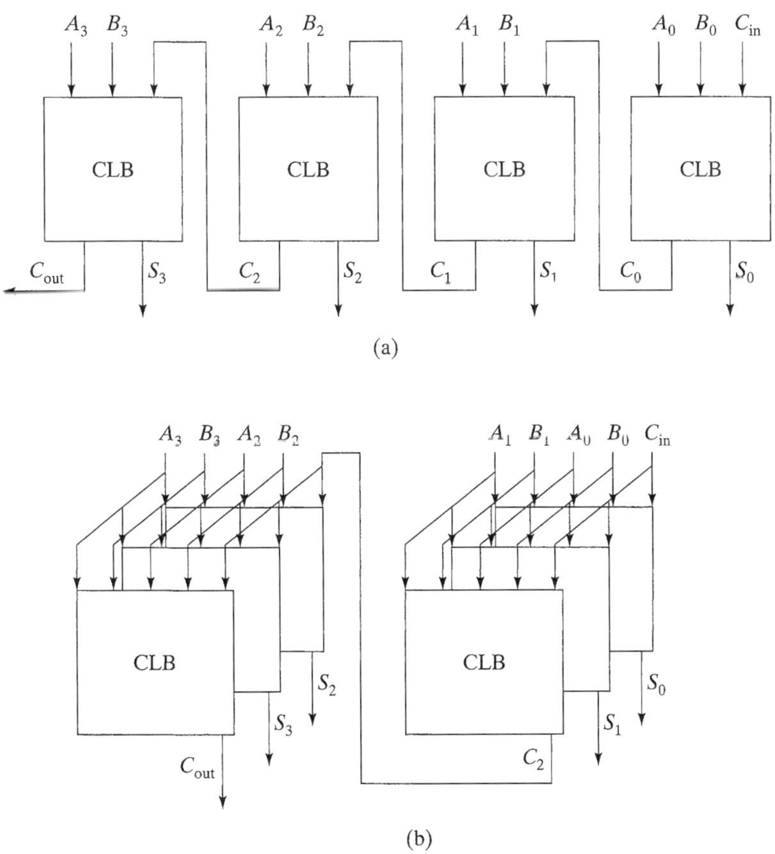

Figure 4.46 **Alternative implementations of a 4-bit adder.**

This example illustrates some of the trade-offs between CLB resources and delay. The delay through the logic block is fixed, independent of the function it is implementing. The first approach uses less CLB resources than the second, but it substantially is slower due to more CLBs being used in series and the programmable wiring resources between them which cause further delay.

ROMs versus PALs/PLAs Recall Figure 4.34, which shows a truth table for a conventional circuit mapping from BCD-to-Gray code. To implement this as a read-only memory, we need a four address line (16 words) by 4-bit ROM. Each row of the truth table maps to a ROM location holding a 4-bit Gray-code word, addressed by the input word in 4-bit BCD code. Don't-care entries, such as the input configurations 1010 through 1111, are of no particular advantage in reducing the hardware needed for the implementation, since the ROM comes with locations for those words whether we store 1s and 0s there or not.

Code converters are used to translate from one data representation to another. For example, a calculator uses standard binary codes for its internal calculations. But it may convert these codes into BCD for input/output and display.

Code conversions are a particularly ideal application for ROMs, since most, if not all, combinations of the inputs need to be generated anyway and the decoder can do this quite efficiently. So how do you choose between a ROM and a PAL/PLA-based implementation?

The answer depends on the structure of the functions being implemented. This can be quantified in terms of (1) the number of unique product terms that must be generated to implement all of the output functions, (2) the degree to which these terms can be shared among multiple output functions, and (3) the number of terms that must be ORed together to implement a given output function.

PLAs are effective when the number of unique terms is small and a given term generated in the AND array can be shared among multiple outputs (this contributes to keeping down the number of unique terms). You should remember that PAL/PLA structures do not always provide enough outputs from the AND array if you need to generate many terms. And PALs are limited in the number of terms that can contribute to realizing any single output function. It will take longer to design a circuit with a PAL or PLA, since minimization methods are needed to reduce the number of terms and to maximize the number of terms that can be shared.

ROMs are effective when the number of unique terms is large. You can reduce your design time because there is no need to perform minimization: the size of the ROM is determined solely by the number of inputs and outputs, not by the number of product terms. However, the size of the ROM doubles with each additional input, so the output functions had better be of a form that requires a large fraction of all possible product terms if this approach is to make sense.

In the BCD-to-Gray-code example, 10 of 16 minterms were needed, with the rest being don't-cares. Boolean minimization reduced the number of literals in the terms, but not the number of unique terms, which was still 10. This is a good hint that a ROM-based implementation is probably the best.

Using PALs for Multilevel Logic At this point, we should be very familiar with the Boolean equations for the full adder:

$$Sum = \bar{A}\bar{B}C_{in} + \bar{A}B\bar{C}_{in} + A\bar{B}\bar{C}_{in} + ABC_{in}$$

$$C_{out} = AB + BC_{in} + AC_{in}$$

There are no shared product terms, so a PAL is an acceptable design approach. The structure needs to support three inputs, seven product terms, a given gate in the OR plane should accept at least four product terms, and it should be possible to generate two outputs.

These are modest constraints easily met by commercially available PALs. But for the purposes of an example, suppose that the PAL at hand is limited to two product terms per gate in the OR plane. The cascaded half adder offers a solution for implementing the full adder for such a highly constrained PAL. The equations are:

$$X = A\bar{B} + \bar{A}B \qquad Y = AB$$

$$Sum = X\bar{C}_{in} + \bar{X}C_{in} \qquad C_{out} = XC_{in} + Y$$

While this implementation still requires seven distinct product terms, no output function requires more than two such terms to be ORed together. However, the intermediate outputs X and Y do need to be fed back as inputs to the PAL, introducing additional delay in computing the outputs. Functions constructed to use X and Y will take more time to compute their results because they need to wait for X and Y to be computed and fed back into the PAL. This is shown in block diagram form in Figure 4.47. Some PAL architectures even include such feedback paths on-chip, to facilitate multilevel designs without using up a precious output pin on the IC package and speeding up the feedback path by not requiring it to go outside the chip and then back in, which is considerably slower.

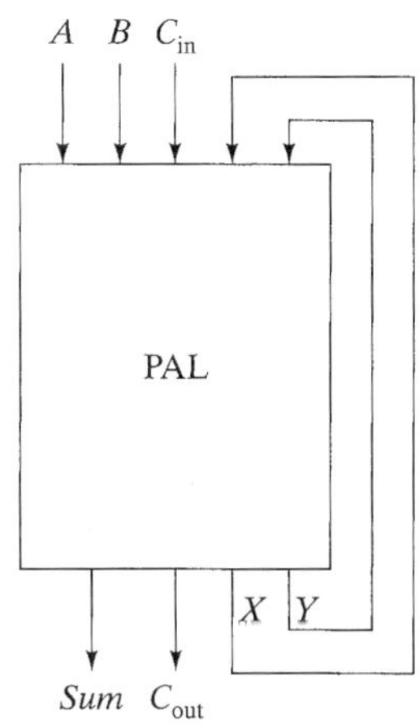

Figure 4.47 Feedback PAL implementation of a full adder.

4.4 Non-Gate Logic

Multiplexers allow us to build circuits in which more than one signal can be gated to a particular output wire under the influence of control signals. This is important, because in most technologies it is not possible to tie more than one gate output to the same wire without dire results. If one gate attempts to draw the line to 0 while another actively pulls it to 1, a battle ensues in which one or both of the gates are likely to "melt down." This circuit configuration makes a relatively low-resistance path between power and ground, leading to dangerously high currents within the gates. We now examine alternatives to the multiplexer for selectively connecting one of many signals to the same wire.

4.4.1 Tri-State Outputs

A	OE	F
X	0	Z
0	1	0
1	1	1

Figure 4.48 Truth table for tri-state buffers.

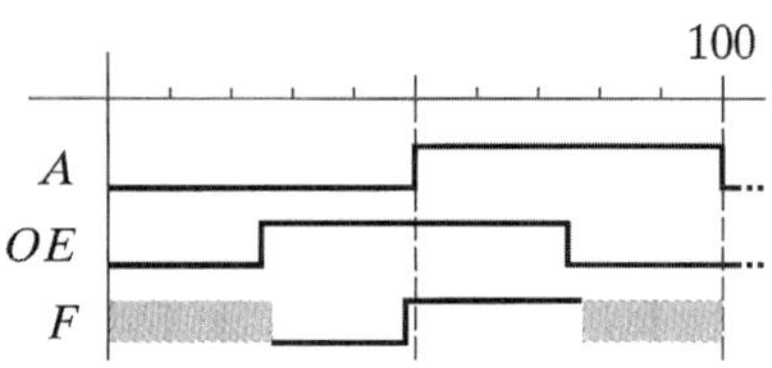

Figure 4.49 Waveform with a high-impedance output.

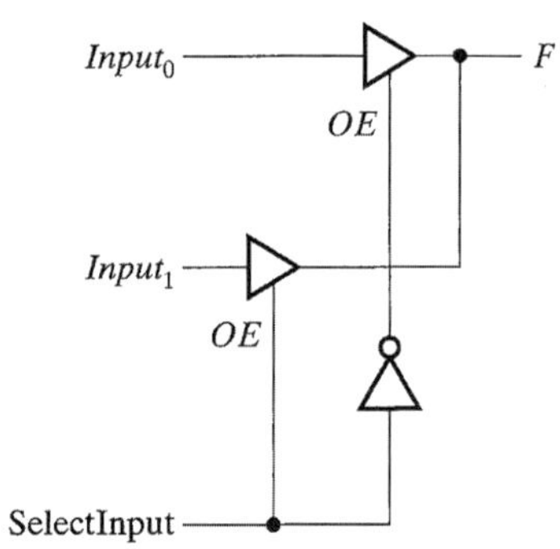

Figure 4.50 Tri-state circuit fragment: tri-state buffers with active-high enables.

So far, we have seen only two possible logic values: 0 and 1. A third value we often encounter during design is the *don't-care* value, denoted by X. Of course, any realization of a circuit involving don't-care values in the output maps each to a 0 or 1. We never encounter a don't-care value in a real circuit.

However, there is a third signal value in digital systems: the *high-impedance state,* usually denoted by the symbol *Z*. When a gate's output is in a high-impedance state (that is, a very high, essentially infinite, resistance exists between the power supply and ground lines and the output), it is as though the gate's internal circuitry were disconnected from its output. Gates that can be placed in such a state are called *tri-state gates:* they can produce as outputs the three values of 0, 1, and *Z*. In addition to its conventional inputs, a tri-state has one more input called *output enable*. When this input is unasserted, the output is in its high-impedance state and the gate effectively is disconnected from the output wire. When the output enable is asserted, the gate's output is determined by its data inputs.

Z is not a value between 0 and 1. It isn't a voltage level at all. Rather, it is a disconnected state of the gate's output. However, it does have some interesting uses. Let's start with the truth table of a tri-state buffer gate is shown in Figure 4.48. When the output enable (OE) signal is unasserted, no matter what the input is, the output will be *Z*. When the OE input is asserted, the buffer simply passes its input to the output.

Figure 4.49 gives a timing waveform for a buffer with a tri-state output. The input signal *A* changes from 0 to 1 and back between time unit 0 and 100. The *OE* signal is asserted between time units 25 and 75. A high-impedance output is represented by a patterned waveform, as shown in the figure. When *OE* is low, *F* is in its high-impedance state, regardless of the value of *A*. When *OE* is high, *F* outputs the same value as that presented at *A*.

Application of Tri-State Gates Let's consider the circuit fragment shown in Figure 4.50, constructed from tri-state buffers (with active-high enables as their second input—on the bottom of the triangle symbol) and a conventional inverter. The control signal *SelectInput* determines which of $Input_0$ or $Input_1$ is steered to the output node *F*: if *SelectInput* is 0 then $Input_0$ is steered to *F*; if it is 1 then $Input_1$ is steered to *F* instead.

The circuit works in the following way. A tri-state buffer passes its input to its output when the output enable signal is driven high. Otherwise, the output is *Z* (disconnected). $Input_1$'s tri-state buffer is controlled by *SelectInput,* while $\overline{SelectInput}$ controls $Input_0$'s tri-state buffer. When *SelectInput* goes high, the $Input_1$ tri-state buffer is enabled and the $Input_0$ tri-state buffer is disabled, effectively disconnecting it from the output node *F*. When *SelectInput* goes low, the roles are reversed, and now the $Input_1$ tri-state buffer is disabled/disconnected and the $Input_0$ tri-state buffer drives the output node.

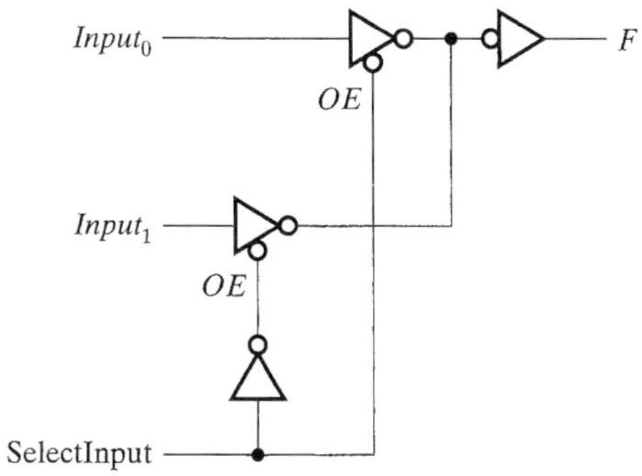

Figure 4.51 **Tri-state circuit fragment: inverting tri-state buffers with active-low.**

To complicate matters a bit, suppose we constructed a functionally identical circuit from inverting tri-state buffers with active-low enables. Tri-state gates often come in this more complicated form (due to the arrangement of their internal transistors). Figure 4.51 gives the revised circuit. This provides us with a good opportunity to test our understanding of positive and negative logic, for both data and control signals.

The tri-stated inverters now have active-low output enable signals, $\overline{\text{OE}}$. When the active-low enable signal is driven to 1, their output is *Z*, and when it is driven to 0, the output becomes the complement of the input. *SelectInput* is now the wrong logic polarity for enabling $Input_1$'s tri-state inverter: it is a positive logic control signal, but the gate expects negative logic. This is corrected easily by placing an inverter between *SelectInput* and the $\overline{\text{OE}}$ signal for the $Input_1$ tri-state inverter. Now, when *SelectInput* is asserted in positive logic, the inverter maps it into an asserted signal in negative logic, thus enabling the tri-state buffer. The relationship between *SelectInput* and $Input_0$'s tri-state enable signal is complementary. If *SelectInput* is not asserted in positive logic, it is asserted in negative logic, and it is already in the desired form for controlling Input_0's tri-state enable.

The second complexity of Figure 4.51 is that the input data are complemented by the tri-state inverters. To revert to positive logic, we can place a second-stage inverter between the tri-states and the output node *F*.

Figure 4.52 shows a switching network implementation of an inverting tri-state gate with active low enable. It looks very much like the CMOS implementation of an inverter, but with one critical difference. Switches controlled by the output enable signal and its complement isolate the output node. Let's suppose the $\overline{\text{OE}}$ signal is at logic 1, unasserted in negative logic. Then the PMOS transistor controlled by $\overline{\text{OE}}$ is nonconducting and the NMOS transistor controlled by its complement is also nonconducting. This leaves the output node *F* floating. Now let's suppose that the $\overline{\text{OE}}$ signal is driven low, becoming asserted in negative logic. The switches controlled by $\overline{\text{OE}}$ and its complement are now both closed and conducting. If *I* is a logic 1, *F* is connected to 0; if *I* is 0, then *F* is connected to 1.

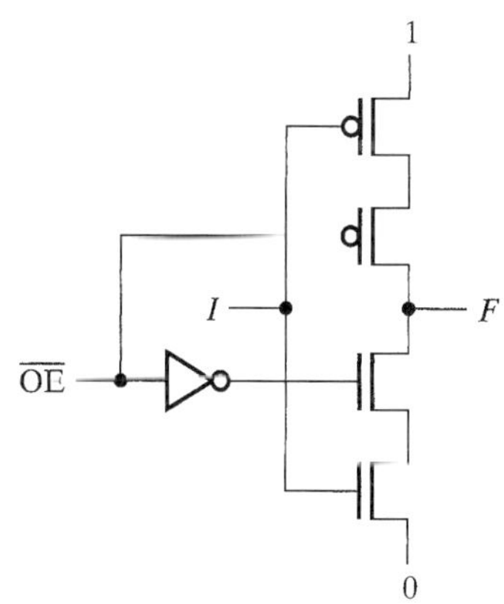

Figure 4.52 **Switch implementation of a tri-state gate.**

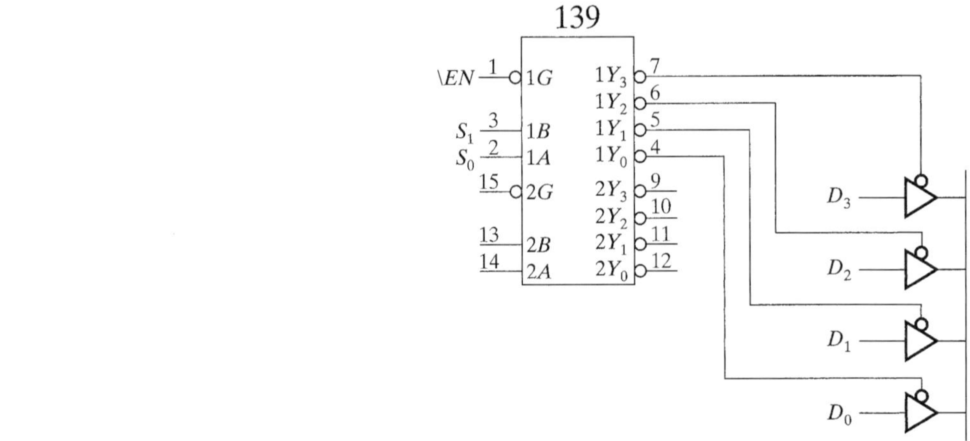

Figure 4.53 **Implementation of a 4:1 multiplexer with tri-state buffers and a decoder.**

The principle of using tri-state gates to connect selectively an input to an output wire generalizes for connecting n input signals to the same output wire. For large n, this represents a considerable savings in gates. Figure 4.53 shows one way to implement a multiplexer with a decoder and tri-state gates. Here, we use a common TTL component, the 139 dual 2:4 decoders, that has active-low outputs (Y_3, Y_2, Y_1, and Y_0) and active-low enables (G). Based on the values of S_1 and S_0, only one of D_3, D_2, D_1, or D_0 will be gated through to the shared output wire. It is very common to use tri-state logic to connect the outputs of many logic cells in an FPGA without having to get all of the wires to a multiplexer for selection.

ROM Components ROMs come in a variety of sizes and word widths. A relatively large ROM is the 27512 EPROM, which provides 2^{16} 8-bit words, or one-half million ROM bits in a single 28-pin package! These can be further combined to form larger ROMs. This is accomplished using tri-state logic on the data outputs of the ROM.

PAL Components Many PALs also have tri-state capability. In this case, the utility comes from being able to split the function for one logic output across two chips. For example, if the logic in one chip computes part of a function and the logic in another computes the rest, then these can be combined by enabling the tri-state buffer on each of the two separate outputs selectively and thus have either drive the output. A simple example is a function of the form $S(\ldots) + S'(\ldots)$. The signal S can be used to select one PAL's output when true and another when false. Both outputs can be wired together because only one will be enabled at one time. This lets the two chips implement the logic corresponding to each case independently. Both the P16H8 and the F100 PALs of Figure 4.42 and Figure 4.43 have tri-state drivers built in. The P16H8 allows for one product term to be used to enable the tri-state driver of each output separately for maximum flexibility. In the case of the F100, all the output tri-state drivers are controlled by a single input signal, $\overline{CE}$ —active-low chip enable, much like a ROM's output data pins.

EXAMPLE 4.17 USING OPEN-COLLECTOR LOGIC TO INTERCONNECT COMPONENTS

As an example, let's concentrate on the smaller (and somewhat ancient) 2764 8192 (2^{13}) word by 8-bit ROM (just to save space in the figures), whose schematic symbol is shown in Figure 4.54. The 2764 component has 13 address lines, to identify each of the possible 8192 words, and eight tri-state output lines, one for each bit in the ROM word. These are special outputs that permit the interconnection of several outputs to the same wire. The *output enable* ($\overline{OE}$) controls whether the outputs really connect to the wire or not.

There are four additional inputs for control. The component also comes with an active-low chip select input ($\overline{CS}$). This can be used to cascade smaller/narrower ROMs to form larger/wider memories. We will see how this can be used in a moment. Since EPROMs are programmed by electrical pulses, an additional input places the ROM in programming mode rather than reading mode ($\overline{PGM}$), while the final input (VPP) provides the necessary high-voltage source used during the programming process. Outside the actual PROM programming process (which usually takes place on a specialized unit), these inputs are hardwired to a logic 1 so that they are unasserted.

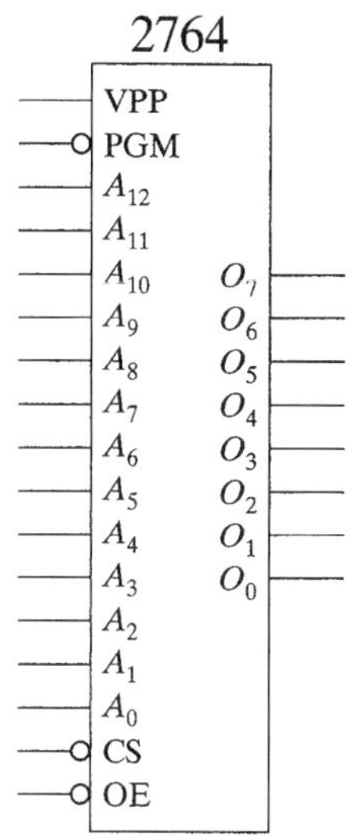

Figure 4.54 2764 EPROM schematic shape.

Figure 4.55 shows how the chip select lines can be used to construct a larger memory. Suppose that we are to build a memory that is 2^{14} by 16 bits wide using the 2764 EPROM. Such a subsystem has 14 address lines (A_{13}:A_0) and 16 data lines (D_{15}:D_0). An additional $\overline{OE}$ signal enables the entire subsystem for output.

Because this has twice the width and twice the number of words of a single 2764, we will need four of them to implement the subsystem. These components are labeled U0 through U3. The thick lines in the figure denote *buses,* several logically related wires that share a common function. For example, buses are used to transmit related data and address bits. This is a shorthand notation used to avoid drawing individual signal lines.

The high-order address bit, A_{13}, selects between the higher 8192 ROM words (addresses 8192 through 16383) and the lower 8192 words (addresses 0 through 8191). Address line A_{13}, if low, enables U1 and U0. If high, it enables U3 and U2. The remaining address lines, A_{12}:A_0, are wired directly to the address line inputs of the ROM chips, as is the output enable signal. The output of the ROM will be driven only if the chip is selected and the output is enabled. Because the outputs are tri-stated and the upper and lower ROM chips are never asserted at the same time, it is perfectly acceptable to wire their respective outputs together, forming very efficient multiplexers. U3 and U1 are connected to D_{15}:D_8 providing the high-order 8 bits of the word, while U2 and U0 are connected to D_7:D_0 to provide the low-order 8 bits.

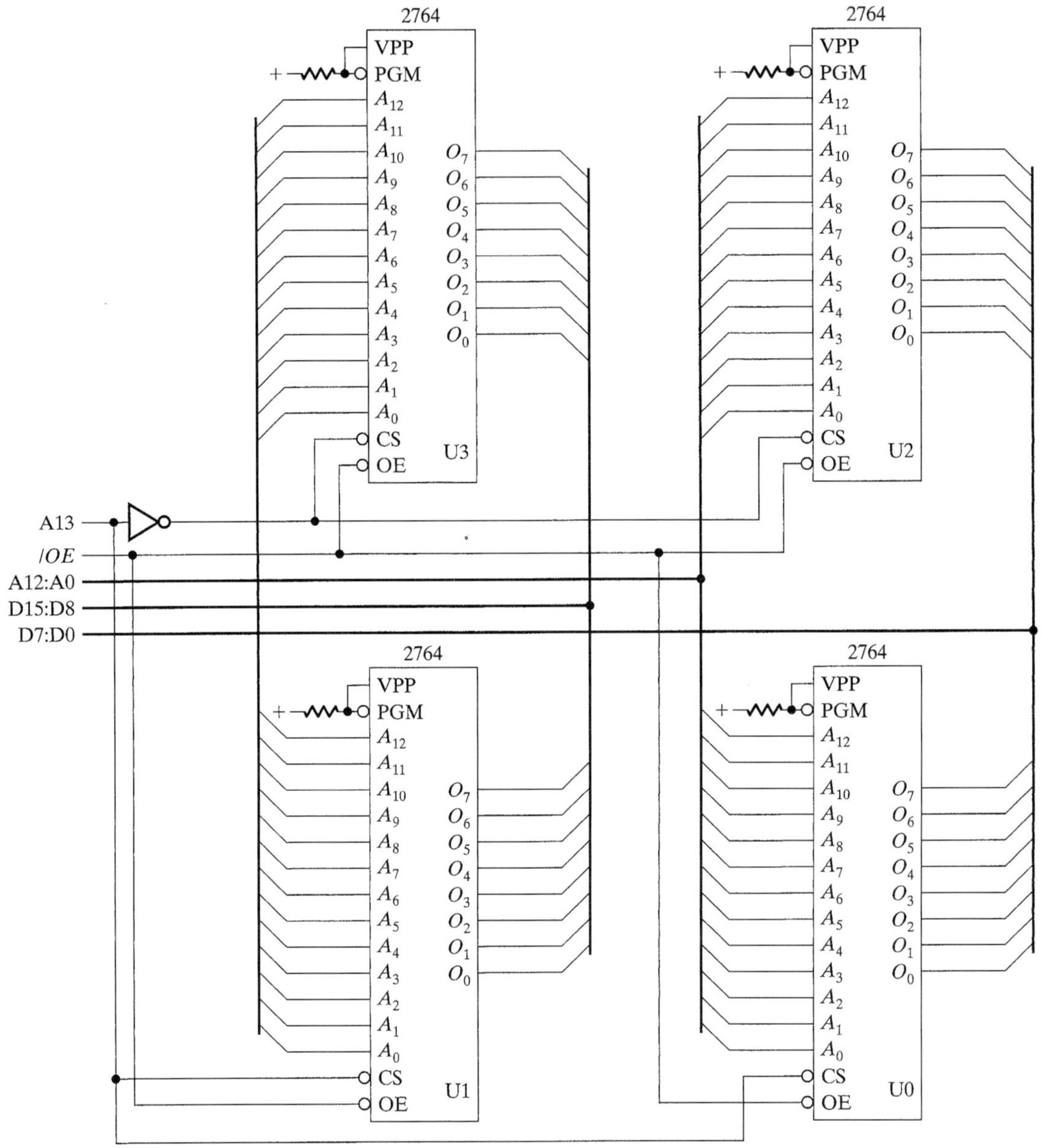

Figure 4.55 **16384 × 16 ROM memory subsystem.**

4.4.2 Open-Collector Outputs and Wired Logic

There is one more way to implement multiple simultaneous connections to a single wire. The idea is to provide a gate that can only pull its output down to 0. If the output is to be 1, the voltage on the wire is left floating. To make this behave properly, the output wire must be attached to a resistor that pulls it to a logic 1 voltage if none of the connected gates are attempting to pull the wire low. The resistor is chosen to be large enough that it can be overcome easily by the pull-down resistances of any attached gates. These gates are called *open-collector* gates.

Figure 4.56 shows the switch-level circuitry for an open-collector NAND gate. The output is 0 only when A and B are both asserted. Otherwise the node, F, is disconnected from the gate and the resistor will pull it up to a logic 1 voltage.

Figure 4.57 shows two open-collector NAND gates in what is called a *wired-AND* configuration. When both inputs to the NAND gate are 1, the output is low, and node F is pulled low. In the other input configurations, the gate's output *floats* (another term often used synonymously for disconnected) and is pulled up to 1 by the resistor.

The open-collector output is active only in cases in which the NAND gate is unasserted! The configuration is called a wired-AND because the effect on F is just as though both NAND gates had been ANDed together: if one or both gates have an output of 0 (that is, both inputs to a gate are 1), then F will be at 0; only when both gates have their outputs asserted is F at a logic 1.

We show a possible circuit to implement a mux in Figure 4.58. The decoder maps its two selection inputs, S_1 and S_0, into one of its four outputs to be asserted active low. Note that the data inputs, $\overline{I}_0$, $\overline{I}_1$, $\overline{I}_2$, and $\overline{I}_3$, are also active low. Suppose the decoder selects the topmost OR gate ($\overline{Y}_3$ is driven to 0) and $\overline{I}_3$ is also asserted (active low). The OR gate pulls the shared wire low, and the inverter asserts the output F as active-high. If $\overline{I}_3$ is not asserted, the shared wire stays high and the output is driven low.

We can think of the wire shared by the OR gates as active-low, that is, a signal is asserted by pulling the wire low. The pull-up provides an unasserted default value. It pulls the wire high if none of the OR gates pull it low. The inverter hanging off the wire restores the signal to positive logic. This is a common usage with open-collector circuit nodes: they are designed to be active-low, with inverters converting the signals back to positive logic where needed.

In general, open-collector gates and tri-state gates are used for the same kinds of functions. But tri-states are more economical because

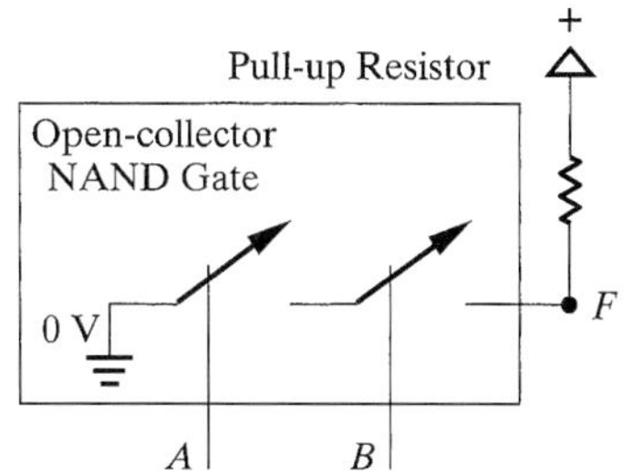

Figure 4.56 Switch representation of an open-collector NAND gate.

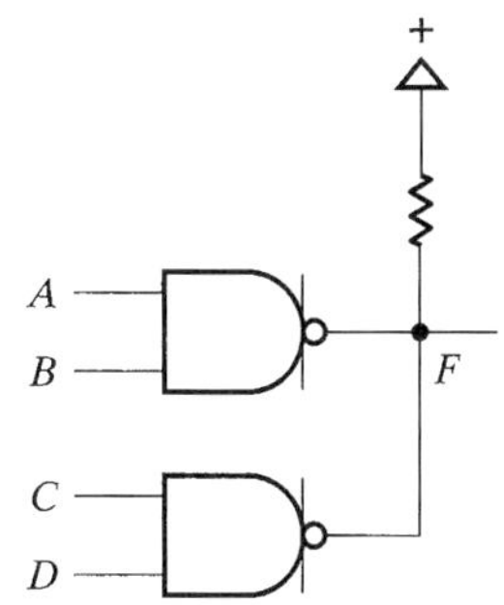

Figure 4.57 Open-collector NANDs in wired-AND configuration.

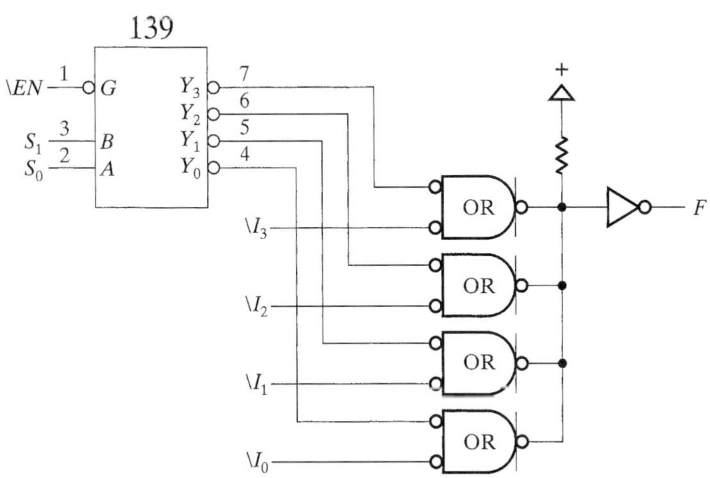

Figure 4.58 A 4:1 multiplexer constructed from open-collector OR gates.

they do not require the external resistor. They have largely replaced open-collector gates in modern designs.

Open-collector gates are available in packages similar to those of conventional discrete gates. The TTL components 7406 and 7407 contain six open-collector inverting and noninverting buffers, respectively. The 7426 and 7433 contain four 2-input NAND and NOR gates, respectively. The 74240 octal buffer with three-state outputs is organized into two independent 4-bit sections, each with its own enable, data inputs, and data outputs.

CHAPTER REVIEW

The theme of this chapter has been to detail how to construct digital systems with more complex logic building blocks than discrete gates. We began with an overview of the different classes of basic components: fixed logic, look-up table-based logic, and template-based logic. Within each category we looked at several approaches, including: multiplexers, decoders, ROMs, PALs, PLAs (which are most appropriate for two-level logic), FPGA cells, and fixed logic for multilevel logic implementations.

Many of these components have the important capability of being *programmable*. Either the connections among gates and logic modules can be wired *on demand*, such as PALs, PLAs, and FPGAs; or the logic functions themselves can be personalized by a user-supplied truth table, such as ROMs, multiplexers, and FPGA look-up tables.

We looked at discrete gate implementations of multiplexers/selectors and demultiplexers/decoders and described methods for realizing arbitrary logic functions from such building blocks.

We spent considerable time on two-level logic. PLAs are attractive when there is a high degree of sharing among the product terms of the multiple output functions. However, a two-level minimization method is critical for using PLAs (or PALs) effectively. ROM-based designs require nothing more than a truth table, but a ROM does not exploit don't-care conditions or shared product terms among output functions. It can also end up being quite large and slow. On the other hand PLAs can also be slow because of their high level of programmability. PALs provide an often used middle ground.

FPGAs require a careful consideration of multilevel implementations, since it is desirable to use as few on-chip modules as possible. The goal is to share common subfunctions among as many output functions as possible. On the other hand, there are natural ways to aggregate logic functions to make best use of the logic modules at hand. For example, a Xilinx CLB can implement one 5-input function or two related 4-input functions, so it is desirable to think of these as the right primitives for the multilevel implementation. In other words, decomposing a design into a multilevel implementation of 2-input gates would not make the most efficient use of CLB building blocks.

We concluded with a discussion of non-gate, tri-state logic. It enables efficient multiplexing of large numbers of signals and is quite useful in constructing larger memory structures from smaller components.

FURTHER READING

Charles Babbage's *Analytical Engine* was the first mechanical computer and was programmed by Ada Lovelace, whose first name was used for the U.S. Department of Defense's Ada programming language, which strongly influenced the VHDL hardware description language. The *Analytical Engine* is prominent in many science and computer museums. The definitive reference is C. Babbage, *The Works of Charles Babbage*. Eleven volumes, Campbell-Kelly, Martin, Editors, William Pickering, London, 1989. Several other pointers to works of special historical note are at the end of Chapter 1.

Texas Instruments' *TTL Data Book* was known as the "yellow bible" of digital designers in its heyday. It is still published and updated yearly.

Books on digital design with programmable devices have appeared only in the past several years. R. Alford, *Programmable Logic Designer's Guide,* Sams and Co., Indianapolis, IN, 1989, provides a good discussion of programmable technology and design methods. Chapter 3 describes the wide range of programmable devices currently available, including more general logic structures than PALs and PLAs. Another good book on programmable logic is G. Bostock, *Programmable Logic Devices: Technology and Applications,* McGraw-Hill, New York, 1988. Chapter 2 describes the processing and electronic circuit technology of programmable devices. This chapter covered logic design with PROMs, PLAs, and PALs.

Another interesting text specifically focusing on FPGAs is J. Oldfield and R. Dorf's *Field-Programmable Gate Arrays: Reconfigurable Logic for Rapid Prototyping and Implementation of Digital Systems,* John Wiley & Sons, New York, 1995.

The standard handbook on PLD components is Monolithic Memories' *Programmable Logic Handbook, Fourth Edition,* Monolithic Memories, Inc., Santa Clara, CA, 1985. The book contains a wealth of information on PLD parts from the company that invented the concept.

EXERCISES

4.1 *(Fixed Logic)* How many packages of 2-input NAND gates (assume there are four to a package) does it take to implement a 4-bit even-parity function (defined as 1 if an even number of inputs are 1)? How many packages of 2-input XOR gates would be required (again, assume there are four to a package).

4.2 *(Fixed Logic)* You got a special deal on some cheap logic packages. Unfortunately, they are all identical and contain two complex logic gates each. The gate's function is $Z = \overline{(AB + CD)}$.

Implement the following functions using only these parts and state how many packages of this gate you will need for each. Feel free to set any inputs to 0 or 1 as needed.

(a) $f(P,Q,R) = \overline{(PQ + R)}$
(b) $f(P,Q,R,S,T) = (P + Q)S + (R + T)\overline{S}$
(c) $f(P,Q,R,S,T) = PQR + ST$
(d) $f(P,Q,R,S,T) = \overline{P(Q + RST)}$

4.3 *(ROM Logic)* Show how to program an appropriately sized ROM (state exactly the minimum number of address bits and minimum number of bits per word that the ROM will require) to implement the functions for the 7-segment display decoder of Section 4.3. Provide a personality matrix for the ROM.

4.4 *(PAL Logic)* Show how to program the P16H8 PAL of Figure 4.42 to implement the functions for the 7-segment display decoder of Section 4.3 (use the unoptimized equations at the start of the section). Use the shorthand notation developed in Section 4.2.

4.5 *(PAL Logic)* Show how to program the P14H8 PAL to implement the functions for the 7-segment display decoder of Section 4.3 (use the unoptimized equations at the start of the section). Note that you may have to use multilevel logic and feedback some of the PAL's outputs back in as inputs. Can you get the logic to fit in a single P14H8? Use the shorthand notation developed in Section 4.2.

4.6 *(PAL Logic)* Compare your implementations for Exercises 4.3 and 4.4 as well as the PLA implementation of Figure 4.45 and discuss the pros and cons of each. Which is likely to be fastest and why?

4.7 Implement the following functions using multiplexers:

(a) $f(P,Q,R) = \overline{(PQ + R)}$ using an 8:1 multiplexer
(b) $f(P,Q,R,S,T) = (P + Q)S + (R + T)\overline{S}$ using a 2:1 multiplexer and two OR gates
(c) $f(P,Q,R,S,T) = PQR + ST$ using a 32:1 multiplexer
(d) $f(P,Q,R,S,T) = \overline{P(Q + RST)}$ using a 32:1 multiplexer

4.8 *(Multiplexers versus Demultiplexers)* Multiplexers and demultiplexers are related closely, but there are important differences.

(a) Briefly, define and differentiate the following terms: *decoder, demultiplexer,* and *multiplexer.* Mention the number of inputs, outputs, enable, and select bits, if any.
(b) Using AND and OR gates, design a circuit to gate a single data input to one of four output lines, determined by the binary-encoded index on the two control lines. Which of the above parts did you design?

4.9 *(Multiplexer Logic)* Implement the function:

$$f(A,B,C,D,E) = A + \overline{C}D + B\overline{D} + \overline{B}D + \overline{B}CE$$

using a multiplexer and no other logic. The constants logic 1, logic 0, and the variables (but not their complements) are available. Try to use the smallest possible multiplexer.

4.10 *(Multiplexer Logic)* You are asked to implement the function:

$$f(A,B,C,D,E) = A + \bar{C}D + B\bar{D} + \bar{B}D + \bar{B}CE$$

using a 4:1 multiplexer and as much other logic as you need, but you want to minimize that extra logic. Which two signals should you use to control the multiplexer? Implement the function using a multiplexer controlled by the following combinations of input signals. Which ends up requiring the least logic? Why?

(a) Use A and B as control inputs to the 4:1 multiplexer
(b) Use B and C as control inputs to the 4:1 multiplexer
(c) Use B and D as control inputs to the 4:1 multiplexer
(d) Use C and D as control inputs to the 4:1 multiplexer

4.11 *(Multiplexer Logic)* Implement the 2-bit adder function (i.e., 2-bit binary number *AB* plus 2-bit binary number *CD* yields 3-bit result *XYZ*) using three 8:1 multiplexers. Show your truth table and how you derived the inputs to the multiplexers.

4.12 *(Multiplexer Logic)* Because 32:1 multiplexers do not exist in standard component catalogs, design a two-stage multiplexer network that realizes the 6-variable function

$$f(A,B,C,D,E,F) = \Sigma\, m(3,7,12,14,15,19,23,27,28,29,31,35,39,$$
$$44,45,46,48,49,50,52,53,55,56,57,59)$$

(a) Assuming that there is one 8:1 multiplexer or two 4:1 multiplexers per logic package, how many packages are used?
(b) How many packages are required to implement the function using conventional inverters and NAND gates in a two-level network? Assume there are four 2-input NAND gates, three 3-input NAND gates, or two 4-input gates per package and six inverters to a package.

4.13 *(Multiplexer Implementation)* Show how to implement the full adder $Sum(A,B,C_{in})$ and $C_{out}(A,B,C_{in})$ in terms of:

(a) Two 8:1 multiplexers
(b) Two 4:1 multiplexers
(c) If you are limited to 2:1 multiplexers (and inverters) only, how would you use them to implement the full adder and how many 2:1 multiplexers would you need?
(d) It is possible to implement the full-adder function with only 5 2:1 multiplexers. Do you see how?

4.14 *(Decoder Implementation)* Implement the following functions using decoders and any size OR gates you may need:

(a) $f(P,Q,R) = \overline{(PQ + R)}$ using a 3:8 decoder
(b) $f(P,Q,R,S,T) = (P + Q)S + (R + T)\overline{S}$ using two 2:4 decoders (*Hint:* Use the enable signals of the decoders.)
(c) $f(P,Q,R,S,T) = P'Q'R'(S + T)$ using a 2:4 decoder
(d) $f(P,Q,R,S,T) = \overline{P(Q + RST)}$ using a 5:32 decoder

4.15 *(Decoder Implementation)* Demonstrate how to implement a 6:64 decoder using smaller 2:4 and 4:16 decoders.

4.16 *(Decoder Implementation)* We have seen how to implement decoders using AND gates and OR gates. Show how to implement the truth table of a 4:16 decoder, including an enable input using only NAND gates and inverters.

4.17 *(Decoder Logic)* A decoder together with an OR gate connected to its output terminals can be used in the synthesis of combinational networks.

(a) Implement the function

$$f(A,B,C,D) = \overline{A}\overline{B}D + \overline{A}BD + A\overline{C}\overline{D} + AC\overline{D}$$

(not necessarily in minimized form) using one 4:16 decoder and a very large fan-in OR gate.

(b) Compare the resulting number of IC packages with a solution using discrete gates only.

4.18 *(Regular Logic Implementation Methods)* Given the following function in sum-of-products form (not necessarily minimized):

$$F(A,B,C,D) = A\overline{B}C + AD + AC$$

Implement the function F using

(a) An 8:1 multiplexer
(b) A 4:16 decoder with a 16-input OR gate
(c) A 16-word ROM
(d) A PLA-like structure using the notation of Section 4.2
(e) A 4-input look-up table
(f) An multiplexer-based logic module, such as that of Figure 4.41

4.19 *(Regular Logic Implementation Methods)* Given a four-input Boolean function, $f(A,B,C,D) = \Sigma\, m(0,3,5,7,11,12,13,15)$

(a) Implement the function using a 16:1 multiplexer
(b) Implement using an 8:1 multiplexer (use D, $\overline{D}$ as MUX data inputs and A, B, C as MUX control inputs)
(c) Implement the function using a 4:1 multiplexer. (*Hint:* Place A and B on the select inputs. Assume $\overline{C}, \overline{D}$ are available and use an OR gate to form one of the inputs to the multiplexer.)
(d) Implement the function using a 4:16 decoder and an OR gate.

4.20 *(Regular Logic Implementation Methods)* Given the function

$$f(A,B,C,D) = (\bar{A} + B)(\bar{A} + C + D)(A + \bar{C} + D)$$

in minimized product-of-sums form and the don't-care set, $D = \{M_0, M_2, M_9, M_{10}\}$, do the following:

(a) Write f in canonical product-of-sums form
(b) Write f in canonical sum-of-products form
(c) Write f in minimized sum-of-products form
(d) Show how to implement f with a single three-stack by three-input AND-OR-Invert gate
(e) Show how to implement f with an 8:1 multiplexer
(f) Show how you might use the don't cares when implementing f using the multiplexer-based logic module of Figure 4.41

4.21 *(Regular Logic Implementation Methods)* Given the three functions X, Y, and Z, defined by

$$X(A,B,C,D) = \Sigma\, m(1,2,3,5,7,9,11,13,15)$$
$$Y(A,B,C,D) = \Pi\, M(2,3,4,5,6,7,8,10,12,14)$$
$$Z(A,B,C,D) = \Sigma\, m(0,1,2,3,5,7)$$

(a) Find the minimum sum-of-products form for each of these functions. How many unique product terms are there in your answer?
(b) Find an alternative sum-of-products form for X, Y, and Z that minimizes the number of unique product terms to implement all three functions simultaneously. How many unique product terms do you find in this implementation?
(c) Show how to implement your solution to part (b) in a PLA structure.

4.22 *(ROM-based Implementation)* Design a schematic for a read-only memory subsystem with the size of 65536 words by 8 bits wide, using 2764 8 K-by-8-bit ROMs.

(a) Use a single 3:8 decoder and inverters.
(b) Use a single 2:4 decoder and inverters. Is there a clever way to make use of the output enable inputs to the ROM as well as the chip select lines?

4.23 *(Non-Gate Logic)* The 2:1 multiplexer function has two data inputs A and B, a select control input S, and a single positive-logic output Z that operates as follows. When S is unasserted, input A is gated to the output. When S is asserted, input B is gated to the output Z. Draw schematics that implement the multiplexer function using *only* the following components:

(a) Inverting tri-state buffers and conventional inverters.
(b) Open-collector NAND gates, conventional inverters, and pull-up resistors.

(c) Repeat parts (a) and (b) for a four-input multiplexer. S_1, $S_0 = 00$ gates A to Z; S_1, $S_0 = 01$ gates B to Z; S_1, $S_0 = 10$ gates C to Z; S_1, $S_0 = 11$ gates D to Z.

4.24 *(Regular Logic Implementation Methods)* A logic network has four inputs ($Input_0$, $Input_1$, $Input_2$, $Input_3$) and two outputs ($Output_0$, $Output_1$). At least one of the inputs always is asserted high. If a given input line has a logic 1 applied to it, the output signals will encode its index in binary. For example, if $Input_2$ is asserted, the output reads $Output_1 = 1$, $Output_0 = 0$. If two or more inputs are at logic 1, the output will be set according to which input has the highest index ($Input_3 > Input_2 > Input_1 > Input_0$).

(a) Fill in the truth table for this function.
(b) Fill in K-maps for $Output_1$ and $Output_0$, and find the Boolean expression for the minimum sum of products implementation.
(c) Show how you would implement this function using a PLA that is as small as possible.

4.25 *(Regular Logic Implementation Methods)* You are to implement a combinational multiplier. It has two 2-bit inputs and a 4-bit output. The first 2-bit input is represented by the variables A, B; the second 2-bit input is represented by C, D. The outputs are W, X, Y, Z, from the most-significant bit to the least.

(a) Complete a truth table that describes the functional behavior of the multiplier.
(b) Find the minimum sum of products forms for the outputs using the K-map method.
(c) Implement the multiplier using the P16H8 PAL of Figure 4.42. How many of these PALs do you need?

4.26 *(Regular Logic Implementation Methods)* An n-input majority function asserts its output whenever more than half of its inputs are asserted. You are to implement a seven-input majority function, which will assert its output whenever four or more of its inputs are asserted.

Don't panic just because this is a seven-variable function. Build it up as a multilevel function whose subfunctions each have less than six variables. As a block diagram, it looks like Figure Ex. 4.26. Circuits #1 and #2 tally the number of their inputs that are asserted, providing the count in binary on the outputs (V, Y are the most significant bits; W, Z are the least significant bits). Based on these second-level inputs, Q determines if more than four or more of the original inputs are 1.

(a) Find the minimized sum-of-products form for Circuit #1 (Circuit #2 is identical). The functions V and W should look familiar. What do they implement?

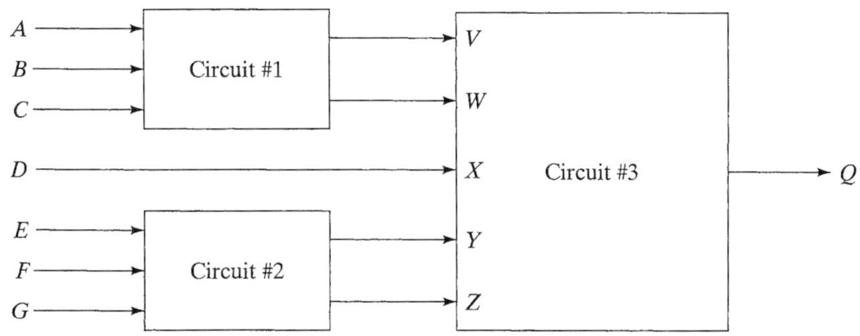

Figure Ex. 4.26 **Block diagram for the tally circuit.**

(b) Complete a five-variable truth table for Circuit #3.

(c) Find the minimum sum-of-products form for Q using the K-map method.

(d) Find the minimum product-of-sums form for Q using the K-map method.

(e) How many 5-input look-up table CLBs would be required to implement this circuit? What if you only had 4-input CLBs?

4.27 *(Regular Logic Implementation Methods)* Verify that the *multi-level* equations for the BCD-to-seven-segment LED decoder in Section 4.3 (those including two intermediate functions X and Y) really do map onto the same on-set as the original *two-level* equations at the start of the section. This can be accomplished by expanding the equations into two-level sum of products form and filling in four-variable K-maps from the equations thus derived. Show how to implement the functions in a minimum sized PLA?

4.28 *(Regular Logic Implementation Methods)* We wish to extend the BCD-to-seven-segment LED display decoder to become a hexadecimal LED display decoder. The LED representations of the hex digits 0, 1, 2, 3, 4, 5, 6, 7, 8, and 9 are exactly the same as for the equivalent BCD digits. Figure Ex. 4.28 shows how the segment displays should be illuminated to denote the hex digits A (1010), B (1011), C (1100), D (1101), E (1110), and F (1111).

(a) Obtain the minimized sum-of-products implementations for the display inputs.

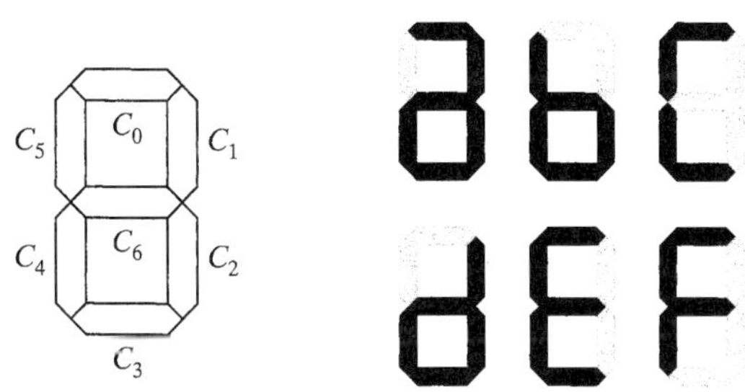

Figure Ex. 4.28 **Hexadecimal displays.**

(b) Show how to implement the logic for the extended design as a PLA with four inputs and seven outputs. Draw the AND array and OR array, and indicate which connections must be made to implement the function. For each output from the AND array, indicate along the wire the product term it is implementing.

4.29 *(Regular Logic Implementation Methods)* Your task is to design a combinational logic subsystem to decode a BCD digit in the range of 0 through 9 in order to drive a seven-segment display for the Klingon number system (*Note:* Klingons, fortunately, also have ten fingers and use base 10). The Klingon numerals are shown in Figure Ex. 4.29. Design a minimized implementation in PLA form. That is, look for common terms among the seven output functions.

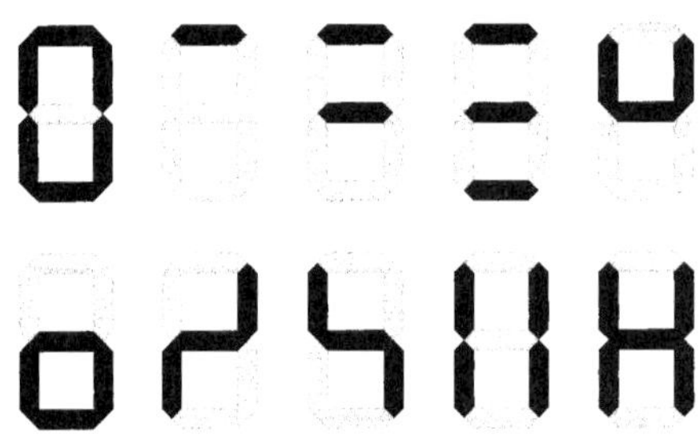

Figure Ex. 4.29 **The Klingon number system.**

4.30 *(Regular Logic Implementation Methods)* Your task is to design a combinational logic subsystem to decode a hexadecimal digit in the range of 0 through 9 and *A* through *F* to drive a seven-segment display. The display of hexadecimal numerals is specified in Figure Ex. 4.30. Design a minimized implementation in PLA form. That is, look for common terms among the seven output functions.

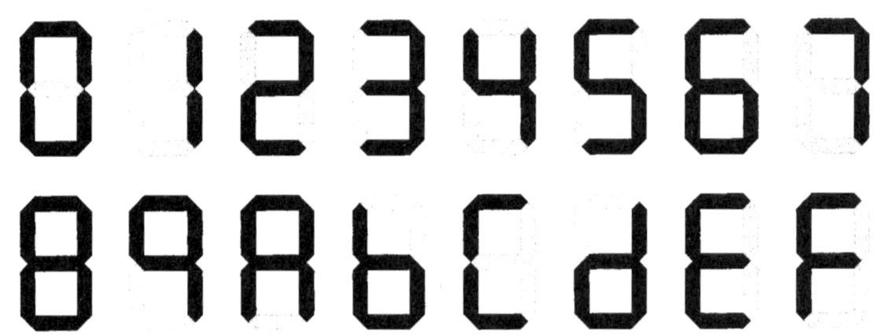

Figure Ex. 4.30 **Specification for the 7-segment display of hexadecimal digits.**

4.31 *(Regular Logic Implementation Methods)* You are to design a converter that maps a 4-bit binary code into a 4-bit Gray code. The 4-bit Gray-code sequence is defined as follows: 0000, 0001, 0011, 0010, 0110, 0111, 0100, 1100, 1101, 1111, 1110, 1010, 1011, 1001, 1000. Give the truth table, and show how to implement this code converter as a ROM circuit and as a PLA circuit.

CHAPTER 5

Case Studies in Combinational Logic Design

Introduction

In this chapter, we will examine several examples of combinational circuit design in great detail. We'll begin with a basic design procedure and then use it on a variety of problems ranging from the small to the large, several will be drawn from examples of earlier chapters. Most of the case studies will start with a written description of the problem. For each, we'll look at the appropriateness of the various implementation technologies we saw in the previous chapter and discuss the reasons for selecting one over another.

Arithmetic circuits provide excellent examples of the trade-offs between circuit speed and complexity. Up to this point, we have concentrated on circuits that manipulate binary numbers as unsigned magnitudes only. We will examine two approaches for high-speed addition, *carry-lookahead logic* and *carry-select addition*. These allow us to sum numbers very quickly, but at the cost of a more complex circuit with many more gates. In this chapter, we'll be looking at ways to represent both positive and negative numbers in the binary system, and at the various schemes for adding and subtracting signed numbers. For an overview of these representations see Appendix A. Then we will revisit the circuits for the half and full adder, as these form the basis for just about every arithmetic circuit we will meet.

We will also learn how to design one of the most important subsystems in a digital computer: the *arithmetic logic unit* or *ALU*. The ALU comprises the combinational logic that implements logic operations, such as AND and OR, and arithmetic operations, such as ADD and SUBTRACT. It is at the heart of the instruction execution engine of every computer that has ever been built.

5.1 Design Procedure

Mapping circuit specifications into gate-level realizations is truly an "art". Sometimes the problems may seem more like puzzles than anything having to do with digital design. The art is in understanding the problem and formulating the solution; the implementation might well be very simple or at the very least straightforward.

In tackling word problems, we recommend following a *standard procedure* for extracting the key points and constraints of the problem and formulating its solution. Our method consists of the following steps. First, try to understand the problem. Second, reformulate it in terms of a standard digital design representation, the encoding that we will need to describe the values of our inputs and outputs. Third, pick a technology that you believe is a good candidate to implement your design. And last, follow an implementation algorithm suitable for your chosen implementation approach. You may do this for several technologies to determine which may be best suited. It is not possible to give a precise algorithm for some of these steps; you will need to rely on practice and experience. Let's look at each of them in more detail.

Step 1: Understand the Problem The very first thing you must do is to understand the problem. There is no single approach for understanding a complex problem statement. Word problems are inherently difficult because they are presented in imprecise, ambiguous, and sometimes confusing statements. In the real world of design, what needs to be built may not be particularly well understood or clearly defined at the start of the design process. A good place to start is with the input/output behavior of the object being designed. Can you identify its inputs, outputs, and control signals? How do the control signals operate on the inputs to generate the outputs? If the problem gives example output behaviors based on certain input streams, make sure you understand how those outputs are derived from the inputs. Sometimes it is helpful to draw a diagram, relating inputs, outputs, and control, to obtain a better understanding of the problem. Or perhaps the diagram can help you understand the different configurations of the inputs you are supposed to recognize.

For example, suppose you are asked to design a logic circuit to control a hall light, with light switches at either end of the hall. It should be possible to turn the light on from either end of the hall, then walk through the hall and turn it off at the other end.

You start by identifying the inputs and outputs. The inputs are the two light switches and the output is the light. Let's call the switches A and B. Sometimes you just need to make reasonable starting assumptions. We will assume that when the light switch is in the down position the input is 0, and when the switch is up the input is 1. The light is on when the logic function is 1. Otherwise it is off.

Let's assume that the system begins with both light switches in the down position. Clearly, the light should be off when both switches are down. When one switch goes up, the light should go on. At the other end of the hall, placing the second switch in the up position should turn the light off.

So if both switches are up, the light is off. What happens when you walk back through the hall? Putting one switch down, the light should go on. Putting the far switch down should turn the light off.

Step 2: Formulate in a Standard Representation Once we have a feeling for what the combinational logic circuit is supposed to do, we must describe it more formally. Digital-design representations are the precise ways of representing a digital system. For combinational-logic problems, the appropriate representations are almost always Boolean equations or truth tables. These representations capture the relationships among inputs, control, and output as algebraic statements or as tabulations of input/output behavior. The key challenge is to extract these relationships from the word statement of the problem.

For the hall light example, simply tabulate the input conditions. When both switches are in the same position, the light is off. When they are in different positions, the light is on.

Step 3: Choose an Implementation A Boolean equation or a truth table is an abstract representation of the digital system. The next step is to map this into something more concrete, like logic gates. Before you can implement the system, however, you must make a critical design decision: you must choose a technology for implementation. The kinds of choices available are two-level combinational networks of discrete gates, PALs or PLAs, memories such as ROMs, or various types of FPGAs. The detailed choice is usually based on economics or performance, but it may be constrained by the kind of technology available to you. For example, if you can only use discrete gates then you'll need a more complex optimization procedure than a ROM.

The hall light circuit is simple enough to implement with a small number of discrete gates.

Step 4: Apply the Design Procedure The last step is perhaps the most mechanical. You have formulated the solution in terms of Boolean equations or truth tables, you have chosen an implementation approach, and now you must map your digital representation into an actual implementation. For two-level networks, you will apply the techniques of Chapter 2 to derive a circuit with the fewest number of product terms. The approach is the same whether the implementation target is discrete gates or programmable logic. For a ROM-based design, only the truth table is needed; there is no need to minimize your logic description first. If you choose a multilevel implementation approach, your best bet is to use computer-aided design tools to factor the Boolean equations into their best form to minimize gate and literal counts. PALs, PLA, and FPGAs are almost exclusively utilized with

CAD tools as these technologies are often used for larger circuits where humans find it hard to handle all the detail.

We are fortunate with the hall-light circuit; it can be implemented with a single XOR gate.

Introduction to the Case Studies In the following sections, we will tackle seven different word problems/case studies, applying the four-step approach we have just outlined. These will vary dramatically in size and complexity. The first problem is a simple process-control application. You must determine the particular Boolean conditions, as indicated by the condition of input light sensors, which are needed to characterize the length of rods moving down an assembly line. When a particular rod length is detected, the logic must generate a signal to move a mechanical arm. The second case study is a decoder application: the combinational logic circuit must map encoded input signals to specified output signals. We'll be looking at a telephone keypad with buttons arranged in two dimensions. We will need to translate the button press into a code for the number or symbol. The third case study is a more complex decoder application. In this case, it is an extension of our leap year calendar system with which we began in Chapter 1. We will be computing the value of the leap-year flag, given the year. The fourth case study describes a logical-function unit that implements several different logic functions of its inputs. This kind of system could be of use in a simple microprocessor. The outputs are defined as combinational-logic functions of the inputs and control signals. The fifth case study will introduce us to the speed/area trade-offs that are possible in combinational-logic design. We'll use circuits that can do addition and subtraction as a vehicle for this lesson. The sixth case study will combine what we learn from the logic unit and arithmetic circuits to the design of a general-purpose arithmetic-logic unit that is a fundamental building block of all computer systems. Finally, the seventh case study will give us a glimpse into the design of larger-scale combinational circuits. An 8-bit by 8-bit multiplier will be our example. Because of the limited space in this text, we won't be able to show all of the variations on these examples nor the mapping to all the possible implementation technologies. You may want to try your hand at these as we go along.

5.2 A Simple Process Line-Control Problem

Rods of varying length (±10%) travel one at a time in the direction of their longest axis on a conveyor belt. A mechanical arm pushes rods that are within specification (±5%) off the belt to one side. A second arm pushes rods that are too long to the other side. Rods that are too short remain on the belt. Three light barriers (light source + photocell) are available as sensors. Our task is to design the combinational circuits that activate the two mechanical arms.

STEP 1 Understand the problem.

In a problem like this, it is important to make sure that you understand the problem specification. Suppose the perfect

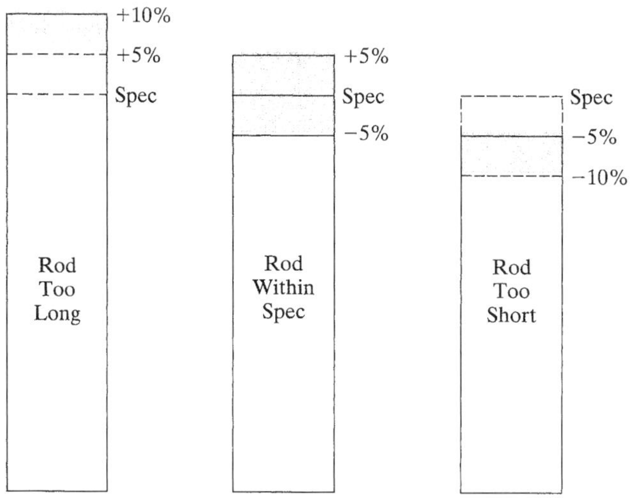

Figure 5.1 **Problem specification.**

length is 100 inches. All rods will be within 10% of this specification, or from 90 to 110 inches in length. "Too long" rods are longer than the specified length by more than 5%—more than 105 inches but less than or equal to 110 inches. These rods are supposed to be pushed to one side of the conveyor belt. "Too short" rods are more than 5% shorter than the desired length—greater than or equal to 90 inches but less than 95 inches. These are destined to remain on the belt. Rods in the range of 95 to 105 inches are "within spec" and will be pushed to the other side by a second arm. A picture should make this more obvious. (See Figure 5.1.)

Next, let's identify the inputs and outputs. The only inputs we have are the readings from the three light sensors. The outputs are the signals that actuate the two positioning arms. Since it is not stated explicitly in the problem, let's assume that the light barriers operate by reading out a 0 when the light beam is uninterrupted and a 1 when it is tripped by the passing rod. The inputs and outputs are fairly straightforward. The key design question becomes how to arrange the placement of the sensors so we can distinguish between the three different classes of rods.

We will call the light barriers A, B, and C and arrange them so that the sensor C is the first to be tripped by the passing rod, then B, and finally, A. Let's make one critical assumption: we will fix the placement of barrier A and arrange barriers B and C so as to be able to decide on the rod length when the rod breaks the light barrier at A. A second assumption is that the rods are spaced adequately on the conveyor belt, at least a maximum rod length, so that it is never possible to interpret two rods accidentally as a single rod.

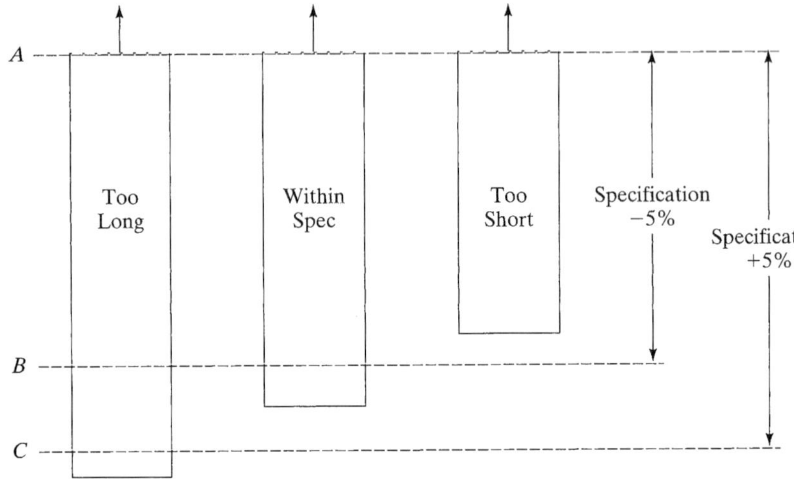

Figure 5.2 **Placement of the light-sensor barriers for the three cases of rod length.**

By redrawing Figure 5.1 slightly, this time incorporating the light barriers, it becomes a little clearer where to place *B* and *C* to distinguish the rod types. See Figure 5.2. We will use barrier *B* to identify rods that are too short and barrier *C* to detect those that are too long. *B* should be placed just far enough from *A* so that a too-short rod will trip *A* but not *B*. Thus, *A* and *B* should be separated by the specified length minus 5%. For our hypothetical specification of 100 inches, this means that *A* and *B* should be placed 95 inches apart.

Now we must place barrier *C* to detect the rods that are too long. Let's place it a distance that is the specification plus 5% from *A* (105 inches in our example). The barriers now distinguish between the three cases. Whenever *A* is tripped, but not *B* or *C*, then the rod is clearly too short ($ABC = 100$). If *A* is tripped, and both *B* and *C* are tripped as well, then the rod is too long ($ABC = 111$). This condition activates the arm to move the rod to one side of the belt. The remaining case is handled by *A* being tripped and *B*, but not *C*, being tripped at that point in time ($ABC = 110$). Then the rod is within the desired specification. This is the condition to move the rod to the other side of the belt.

This represents only three of the eight possible barrier input configurations. Because of our first assumption, that we will make our decision as soon as barrier *A* is triggered, the rod must still be in transit if the barrier configuration is $ABC = 0XX$. Since we also assume that the rods are separated by a maximum rod length, the remaining input condition, $ABC = 101$, can never occur. After we make a decision on a rod, we have to wait until the rod clears the *A* sensor (in other words, letting that rod pass by) so that we can start looking for the next rod to trip A.

STEP 2 Formulate in a standard representation.

At this point, most of the hard work has been done. A summary truth table, derived from the discussion above, appears in Figure 5.3. This can be written in Verilog very directly:

```
module rodcontrol (A, B, C, TooShort, InSpec, TooLong);
    input  A, B, C;
    output TooShort, InSpec, TooLong;

    assign TooShort = A && !B && !C;
    assign InSpec = A && B && !C;
    assign TooLong = A && B && C;

endmodule
```

A	B	C	Function
0	0	0	X
0	0	1	X
0	1	0	X
0	1	1	X
1	0	0	Too short
1	0	1	X
1	1	0	In spec
1	1	1	Too long

Figure 5.3 **Rod-control function.**

STEP 3 Choose implementation technology.

The actual logic really is quite simple, so we can implement it as straightforward gate logic.

STEP 4 Follow implementation procedure.

The "too long" condition is represented by the function $F = ABC$. The "within spec" condition is represented by the function $G = AB\bar{C}$. These can be implemented via two 3-input AND gates and one inverter.

5.3 Telephone Keypad Decoder

You are to design a combinational circuit that decodes a button press on a telephone keypad that has four rows and three columns of buttons. The buttons are labeled 0 through 9, *, and #. Figure 5.4 shows the keypad. Its buttons need to be decoded into a 4-bit binary number corresponding to the number on the button that was pressed; 10 and 15 will correspond to * and #, respectively.

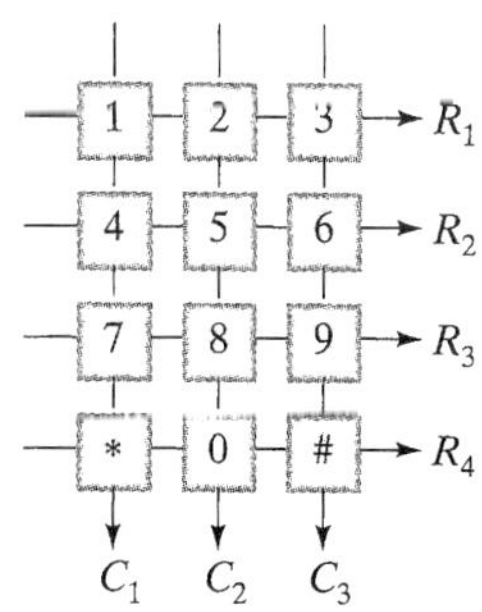

Figure 5.4 **Telephone keypad.**

STEP 1 Understand the problem.

Begin by asking yourself what the circuit is supposed to do. What are the inputs and outputs? There are input signals for the four rows and four columns. There are four outputs to represent the binary number corresponding to the button pressed. We'll need another output to indicate that a button has been pressed so that a 0 output is not confused between nothing being pressed and the number 0 being pressed. We'll assume that when a button is pressed its corresponding row and column signals go to a logic 1 level while all the others are at 0. We also need to consider what will happen if two buttons or more are pressed simultaneously. We can choose to assume that this can't happen, but given that this keypad may be used on a very small cell phone, we'll be cautious and make sure we ignore multiple button presses.

STEP 2 Formulate in a standard representation.

We could use a truth-table to tabulate all of the possible values of the seven inputs, but this would be a very large truth table (128 rows!). This is a case where we see the real benefit of a hardware description language. In Verilog:

```
module keypaddecoder (R1, R2, R3, R4, C1, C2, C3, K8,
   K4, K2, K1, KP);
   input    R1, R2, R3, R4, C1, C2, C3;
   output   K8, K4, K2, K1, KP;
   reg[3:0] key;

   always @(R1, R2, R3, R4, C1, C2, C3) begin
      if R1 & C1 key = 1;
      if R1 & C2 key = 2;
      if R1 & C3 key = 3;
      if R2 & C1 key = 4;
      if R2 & C2 key = 5;
      if R2 & C3 key = 6;
      if R3 & C1 key = 7;
      if R3 & C2 key = 8;
      if R3 & C3 key = 9;
      if R4 & C1 key = 10;
      if R4 & C2 key = 0;
      if R4 & C3 key = 15;
      KP = ((R1 + R2 + R3 + R4) == 3b'001)
              && ((C1 + C2 + C3) == 3b'001);
   end

   assign K8 = key[3];
   assign K4 = key[2];
   assign K2 = key[1];
   assign K1 = key[0];

endmodule
```

The Verilog module consists of an "always" block and four "assign statements." Recall that these five statements execute continuously and in parallel. The four assign statements are simply used to connect the four bits of a 4-bit number, *key,* to the four outputs, K_8, K_4, K_2, and K_1, so named for their positional value. The always block does the bulk of the work. Note that its sensitivity list includes all seven inputs. The always block will execute whenever any of the seven inputs change value. It consists of 12 very simple conditional statements that check for each button being pressed and assign the appropriate value to *key.* Note that more than one of the if-statement conditions may be true if more than one button is pressed simultaneously. The value of *key* will be the last one to be checked, as the statements within an always block are executed sequentially. The last statement of the always block is the most complex. It checks that, in fact, one and only one button was pressed. It does this by separately "adding" up the values of the row inputs and the values of the column inputs. It then compares each of these two results to the number 1 as a 3-bit binary number (3b'001). If both of these conditions are true, namely, one row and one column, then *KP* is set to true.

STEP 3 Implementation target.

Let's consider three possible implementation technologies: a ROM, a look-up table-based FPGA with 4-inputs per look-up table; and a PAL with six product terms per output.

It's a bit difficult to tell at this point which of these is likely to be best. Its difficult to visualize the logic functions we'll have to implement given the behavioral description in Verilog. Before we can proceed to an evaluation of these technologies we'll need to transform the Verilog into logic equations. Of course, we typically would use CAD tools to do this task for us. But, for now, lets transform the Verilog by hand.

We'll start with K_1. Note that this output will be true if *key* is set to 1, 3, 5, 7, 9, or 15. Therefore, it will require six product terms with two literals in each. K_2, K_4, and K_8 are similar (K_2 is true when the key is 2, 3, 6, 7, 10, and 15; K_4 is true when the key is 4, 5, 6, 7, and 15; K_8 is true when the key is 8, 9, 10, and 15) and together lead to the following equations:

$$K_1 = R_1C_1 + R_1C_3 + R_2C_2 + R_3C_1 + R_3C_3 + R_4C_3$$
$$K_2 = R_1C_2 + R_1C_3 + R_2C_3 + R_3C_1 + R_4C_1 + R_4C_3$$
$$K_4 = R_2C_1 + R_2C_2 + R_2C_3 + R_3C_1 + R_4C_3$$
$$K_8 = R_3C_2 + R_3C_3 + R_4C_1 + R_4C_3$$

The equation for *KP* is a bit less intuitive. There will be 12 terms for *KP*, one for each button. Each one will have the form $R_1R_2'R_3'R_4'C_1C_2'C_3'$, which corresponds to the "1" button being pressed. We'll need one of these terms for each of the 12 buttons leading to an equation of 12 product terms of seven literals each. Another approach is to notice that we can just as easily implement the complement of *KP* which is true if there are ever two rows or two columns that are true; simultaneously indicating multiple buttons being pressed. This leads to the following equation for *KP'*:

$$KP' = R_1R_2 + R_1R_3 + R_1R_4 + R_2R_3 + R_2R_4 + R_3R_4 + C_1C_2 + C_1C_3 + C_2C_3$$

This equation is considerably simpler and requires only nine product terms of two literals each. You may be wondering how such an expression could ever arise from the description of *KP* in Verilog. In this case, it is highly unlikely that CAD tools could have come up with such a reduced expression. The "+" signs in the Verilog will cause full adders to be generated to implement the circuit. Although the logic for these adders will be optimized, it is difficult to imagine today's tools achieving this result. In fact, the immense expressive power of HDLs is a serious issue in today's design environment. HDLs make it easy to express behavior, but the resulting circuit may end up much larger than intended or expected. The experienced designer knows the limitations of the tools and will watch for those cases where a reformulation or decomposition of the problem will lead to a more efficient implementation.

STEP 4 Implementation procedure.

Let's start with the simplest: the ROM. For a ROM implementation, we'll require a ROM with 128 words for each of the seven input combinations that are possible. Each of these words requires five bits, four for the number and one for *KP*. Filling in the ROM should be straightforward.

For a PAL with six product terms per output, we'll require six different outputs to be available (along with the seven inputs): four for the four number outputs and two for *KP'*. *KP'* has nine product terms and must be implemented using two outputs, the first combining the first six product terms (for the rows) and then fed back into the logic array so that we can combine it with the last three product terms (for a four product-term function). If the PAL does not have programmable output polarity, then we'll need an external inverter for generating *KP* from *KP'* or feed *KP'* back into the PAL to be inverted and, thereby, using up a seventh output block and adding another delay through the PAL in addition to the two already there.

In an FPGA, we can see that the five equations each require up to seven inputs (K_1 requires all seven, so does K_2, K_4 requires six, K_8 only requires five, and *KP* requires all seven as well). Unfortunately, we only have look-up tables with four inputs. We must now look at each of the equations and see if it possible to decompose them into parts that each have no more than four inputs. For K_1 we can implement the following equations:

$$X_1 = R_1C_1 + R_1C_3 + R_3C_1 + R_3C_3$$

$$Y_1 = X_1 + R_2C_2$$

$$K_1 = Y_1 + R_4C_3$$

X_1 and Y_1 are intermediate terms that, together with K_1, are the outputs of the three 4-input logic blocks we'll need to implement K_1 in our FPGA. Similar decompositions can be done for the other equations yielding: three blocks for K_2, two for K_4, two for K_8, and two for KP; for a total of 12 logic blocks. Note that for KP, the inversion does not cost us an extra logic block because any logic block can invert its output for free—it still is a function of the same inputs. Thus, for KP, the equations are:

$$X_P = R_1R_2 + R_1R_3 + R_1R_4 + R_2R_3 + R_2R_4 + R_3R_4$$
$$KP = (XP + C_1C_2 + C_1C_3 + C_2C_3)'$$

Doesn't it become clear quite quickly that we need CAD tools to help us optimize these decompositions?

5.4 Leap Year Calculation

In Chapter 1, and then again in Chapter 4, we derived implementations for a circuit that can tell us the number of days in a month given the number of the month and the leap-year flag. Now, we are tasked to construct the circuit that determines the value of the leap-year flag, given the year.

STEP 1 Understand the problem.

First we need to do some research into how leap years work and when they began being used. The current leap year system was established in 1582 by Pope Gregory I (hence, the name "Gregorian calendar" for our current calendar). To compensate for the inaccuracies of the previous calendar (the "Julian calendar" established by Julius Caesar), it was decided to add a day to every fourth year. However, this was a bit too large of a correction. To make up for the difference, it was decreed that every year divisible by 100 would not be a leap year. As you can guess, this also turned out to be too big of a correction. Therefore, every year that is divisible by 400 is a leap year nonetheless. In summary, the rule is: a year is leap if it is divisible by 4 and greater than 1582, unless it is divisible by 100 but not by 400.

For our purposes, let's assume that we won't be given a year before 1582 as input.

STEP 2 Formulate in a standard representation.

Before we can come up with a representation for our circuit, we need to decide on how we will encode the year. We could choose a straightforward binary form (11 bits will allow us to represent years up to 2047). Alternatively, we could represent it using binary coded decimal (BCD) notation and use 16 bits, four decimal digits of 4 bits each, to represent the year. Given that we are likely to also want to display the year and that people will want to see it as four decimal digits, we'll choose the BCD encoding of the year. This will allow our circuit to work correctly for the range of years between 1582 and 9999.

We can directly derive the equations for our circuit if we first break it up into four pieces: (a) a circuit that determines if the year is divisible by 4, (b) a circuit that determines if the year is divisible by 100, (c) a circuit that determines if the year is divisible by 400, and (d) a circuit that combines these three intermediate results to generate the value for the leap-year flag.

If we had used binary notation, we would have been able to tell if a year was divisible by 4 if its two low-order bits were both 0. For a BCD encoded year, this is not so simple. It seems as if our choice of BCD encoding may have been a mistake. But let's pursue this further. We only need to look at the low-order two digits of the BCD encoded year to determine if it is divisible by 4. All years ending in 00, 04, 08, 12, 16, 20, etc. are divisible by 4. One way to determine this is as follows: if the tens digit is even, then the number is divisible by 4 if the ones digit is 0, 4, or 8; if the tens digit is odd, then the number is divisible by 4 if the ones digit is 2 or 6. This easily translates into the following Boolean expression (where YT1 is the year's tens digit low-order bit, YO8 is the high-order bit of the year's ones digit, etc.):

$$\begin{aligned} D4 = {} & YT1' \, (YO8' \, YO4' \, YO2' \, YO1' + YO8' \, YO4 \, YO2' \, YO1' \\ & + YO8 \, YO4' \, YO2' \, YO1') + YT1 \, (YO8' \, YO4' \, YO2 \, YO1' \\ & + YO8' \, YO4 \, YO2 \, YO1') \end{aligned}$$

If we then consider that digits with values of 10 to 15 will never occur in BCD encodings, we can simplify further to yield:

$$D4 = YT1' \, YO2' \, YO1' + YT1 \, YO2 \, YO1'$$

The next step demonstrates why BCD may have been a good idea after all. Determining if a binary number is divisible by 100 is not at all straightforward (think about this for a minute or two). On the other hand, it is trivial for a BCD-encoded number. We can simply check that the two low-order decimal digits are both 0, that is, all their 8 bits are 0. This yields:

$$D100 = YT8' \, YT4' \, YT2' \, YT1' \, YO8' \, YO4' \, YO2' \, YO1'$$

The third part of our circuit must determine if the year is divisible by 400. This is a combination of the previous two circuits. A 4 digit decimal number is divisible by 400 if it is divisible by 100 and its two high-order digits are divisible by 4. Thus we can compute D400 with the equation for D100 and the equation for D4 applied to the two high-order digits. This yields:

$$D400 = (YM1' \, YH2' \, YH1' + YM1 \, YH2 \, YH1')D100$$

where YM and YH are the thousands and hundreds digits of the year, respectively.

Finally, we can combine these three results with an equation for *leap* that directly translates the leap-year rule into a logic expression:

$$\begin{aligned} \text{leap} = \text{D4}\,(\text{D100}\,\text{D400}')' &= \text{D4}\,\text{D100}' + \text{D4}\,\text{D400} \\ &= \text{D4}\,\text{D100}' + \text{D400} \end{aligned}$$

Note that we can simplify D4 D400 to just D400 because if a year is divisible by 400 then it is also divisible by 4, therefore, D4 is redundant.

STEP 3 Implementation target.

Let's consider an FPGA with 5-input look-up tables, a PAL with only two product terms per output, and discrete gates as our implementation targets.

STEP 4 Implementation procedure.

Fortunately, the equations we have derived for this case study are simple enough that it should be straightforward to see how we can do our mappings manually.

For the FPGA, we can see that D4 will require a single logic block, as it only has three inputs. D100 will require two logic blocks: one to combine the inputs of YO and another to combine that result with the inputs of YT. D400 has four inputs requiring a single logic block. Finally, leap will require another logic block based on its three inputs. However, if we look a bit more carefully, we'll see that we do not need a separate logic block for D400. Since it is only used by leap, its expression can be substituted into the equation for leap to yield an expression with four inputs instead of three, but one that will still fit into a single logic block. Trying to substitute D100 or D4 yields equations with more than five inputs so we can't improve any further. Our realization will require only three logic blocks.

Although a two product term per output PAL seems very constraining, our equations are each two terms or less and can easily fit. We will require 11 inputs into the PAL and 4 outputs. The more important characteristic, however, is that the PAL implementation will require a high delay as leap depends on D400 which depends on D100, giving it a delay more than three times that of a single output block in our PAL. Moreover, D100 and D400 along with D4 will have to be fed back into the logic array and will increase the number of inputs needed for each term to 14.

A discrete gate implementation for *leap* is shown in Figure 5.5. This realization uses six 2-input gates, four 3-input gates, and two 4-input gates. Several inverters will be needed as well. The delay of the implementation is four gate delays.

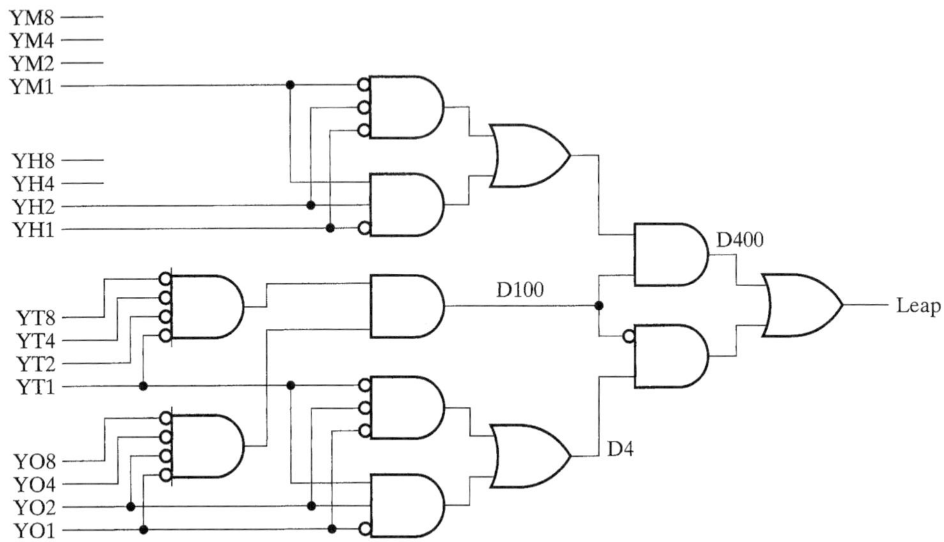

Figure 5.5 **Discrete gate implementation for leap-year flag.**

C_0	C_1	C_2	F	Comments
0	0	0	1	Always 1
0	0	1	$A + B$	Logical OR
0	1	0	$(AB)'$	Logical NAND
0	1	1	A XOR B	Logical XOR
1	0	0	A XNOR B	Logical XNOR
1	0	1	AB	Logical AND
1	1	0	$(A + B)'$	Logical NOR
1	1	1	0	Always 0

Figure 5.6 **Specification of logic function unit.**

5.5 Logic Function Unit

You are to design a logic network that has two data inputs, A and B, and three control inputs, C_0, C_1, and C_2. The network should implement the logical function, F, specified in Figure 5.6. Components of this type find wide application in microprocessor data paths for computing a variety of bit-wise logic functions of two operands.

STEP 1 Understand the problem.

The problem specification mentions three control inputs (C_0, C_1, C_2), two data inputs (A, B), and a single output function (F). Thus, F is a combinational logic function of these five variables.

STEP 2 Formulate in a standard representation.

The natural thing to do next is to create a five-variable truth table, in preparation for solving a five-variable K-map. In terms of the truth-table representation, there really is no distinction between control and data inputs. We begin by listing the

32 different combinations of five variables. Next, let's partition the truth table into eight groups of four rows each. Each group represents a unique setting of the three control inputs (000 through 111), and each row within the group represents a unique configuration of the data inputs A and B (00 through 11). Since the first truth-table group ($C_0C_1C_2 = 000$) represents the constant function of 1, we fill in all four rows with a 1. The second group ($C_0C_1C_2 = 001$) is the function OR. We fill in the first row ($AB = 00$) with a 0 and the remaining three rows ($AB = 01, 10, 11$) with a 1. Continuing in this fashion, we can complete the rest of the table. This is shown in Figure 5.7.

C_0	C_1	C_2	A	B	F
0	0	0	0	0	1
0	0	0	0	1	1
0	0	0	1	0	1
0	0	0	1	1	1
0	0	1	0	0	0
0	0	1	0	1	1
0	0	1	1	0	1
0	0	1	1	1	1
0	1	0	0	0	1
0	1	0	0	1	1
0	1	0	1	0	1
0	1	0	1	1	0
0	1	1	0	0	0
0	1	1	0	1	1
0	1	1	1	0	1
0	1	1	1	1	0
1	0	0	0	0	1
1	0	0	0	1	0
1	0	0	1	0	0
1	0	0	1	1	1
1	0	1	0	0	0
1	0	1	0	1	0
1	0	1	1	0	0
1	0	1	1	1	1
1	1	0	0	0	1
1	1	0	0	1	0
1	1	0	1	0	0
1	1	0	1	1	0
1	1	1	0	0	0
1	1	1	0	1	0
1	1	1	1	0	0
1	1	1	1	1	0

Figure 5.7 Truth table for logic function unit.

We can write this in Verilog to demonstrate some more features of the language (in this example, the *case* statement and some logical operators):

```
module logicfunctionunit (A, B, C0, C1, C2, F);
    input   A, B, C0, C1, C2;
    output  F;
    reg[2:0] function;


    assign function = {C0, C1, C2};

    always @(A, B, function) begin
       case (function):
           3b'000: F = 1;
           3b'001: F = A || B;
           3b'010: F = !(A && B);
           3b'011: F = A ^ B;
           3b'100: F = !(A ^ B);
           3b'101: F = A && B;
           3b'110: F = !(A || B);
           3b'111: F = 0;
       endcase

endmodule
```

STEP 3 Choose an implementation technology.

The basic choice is whether to implement the function with a gate-oriented approach or with memories. Note that using discrete (or even programmable) gates is not the only "gate-oriented" approach. An implementation based on a multiplexer or a multiplexer-based FPGA might be worth considering. See the block diagram in Figure 5.8. We could have come up with this implementation without even deriving the truth table of Step 2.

To a large extent, the implementation decision depends on two factors: design time versus component cost. ROM-based designs usually require less time for design and implementation. For one thing, a ROM-based design starts with the same truth table as a discrete-gate design, plus it has the advantage that we need not minimize the function. However, a ROM

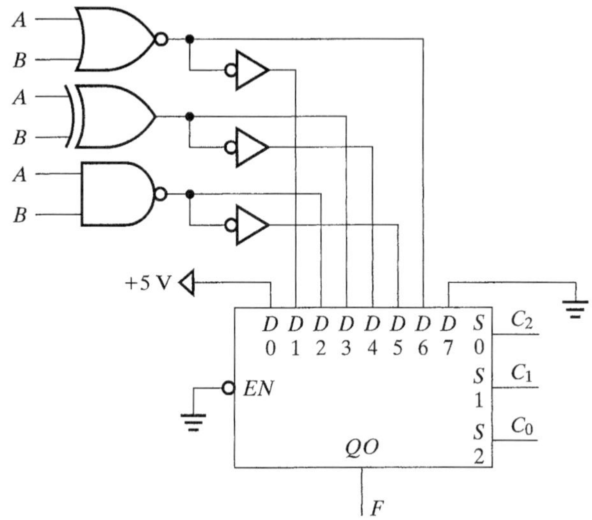

Figure 5.8 **Multiplexer-based implementation of logic unit.**

package is about five times as expensive as a discrete-gate package (of course, this depends on the number of bits in the ROM). On the positive side, the function can be implemented in a single ROM, but it will take several packages to implement it using discrete gates. For example, the multiplexer-based design of Figure 5.8 requires four TTL packages: four 2-input NAND, four 2-input NOR, two 2-input XOR, and an 8:1 multiplexer (inverters can be formed from leftover NAND, NOR, or XOR gates). If the cost differential is a factor of 5, the multiplexer-based design is still less expensive than ROM. Yet the multiplexer-based design takes up more circuit area than the ROM-based one, with more opportunities for wiring errors and component failures.

If cost is the overriding criterion, we will choose packaged logic. But which requires fewer packages, the discrete gate design or the multiplexer design? The minimized gate-level design cannot be evaluated without performing the K-map minimization, which we do next.

STEP 4 Follow the implementation procedure.

To continue the analysis, we will minimize the function's K-map. This is shown in Figure 5.9, and the function for F turns out to be a rather simple equation:

$$F = \bar{C}_2\bar{A}\bar{B} + \bar{C}_0 A\bar{B} + \bar{C}_0\bar{A}B + \bar{C}_1 AB$$

In packaged logic like TTL, this requires five gates (four 3-input, one 4-input), plus five inverters. This can be implemented in three TTL packages (one package of three 3-input gates, one package of two 4-input gates, and one package of six inverters), so there is a slight advantage for the gate approach.

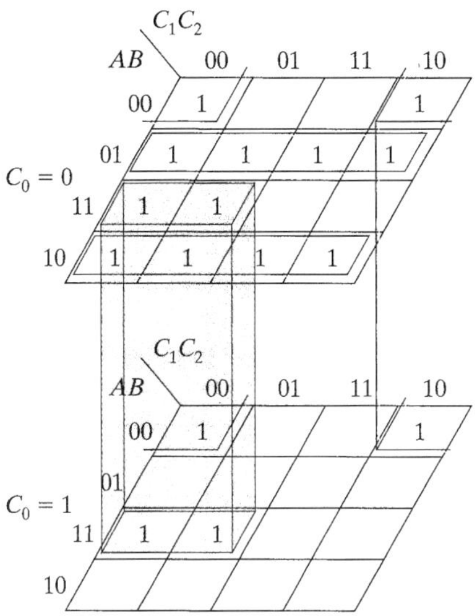

Figure 5.9 **K-map for logic unit.**

Note that the function can be simplified further, assuming that XOR gates are available:

$$F = \bar{C}_2\bar{A}\bar{B} + \bar{C}_0(A \oplus B) + \bar{C}_1AB$$

We still need five inverters and five more gates: three 3-input, one 2-input, and one 2-input XOR. This time we need four packages: one package of three 3-input gates, one package of four 2-input gates, one package of two 2-input XOR gates, and one package of six inverters. This seeming simplification actually increases the implementation costs unless the leftover gates can be used for other purposes.

If the criterion is circuit area rather than cost, the clear winner is the ROM-based implementation, since it requires a single package. In a gate-array design style, the 10-gate version probably has the area advantage. This is because it would be difficult to implement a 32-word ROM in an equivalent area using the primitive gate structures available in a typical gate array.

In a multiplexer-based FPGA with 4-input multiplexer blocks, we are likely to use approximately six logic blocks. Three will be used to implement the 8-input multiplexer and one each for the control gates (assuming that the output is available in complemented form as well—a common feature of FPGA blocks). If we implement the optimized equation rather than the multiplexer-based solution, we are likely to use fewer cells but will have a much more difficult time completing the mapping by hand. Suffice it to say, that no more than four logic blocks will be needed, as each of the terms can be implemented in one block with a fourth devoted to ORing them together.

5.6 Adder Design

In this section, we review the half-adder and full-adder circuits and show how these can be cascaded to form adder circuits over multiple bits. These circuits have no difficulty working with the twos complement number scheme described in Appendix A. We'll stick to our design procedure a bit less formally from this point on.

5.6.1 Half Adder/Full Adder

In this subsection, we reexamine the adder structures first introduced in Chapter 2.

When we do addition with pencil and paper, the carry from one column of digits is added to the sum of the column to its left. The same works for binary addition. We form the ith sum from the addition of the ith bits and the carry-out to the $(i+1)^{st}$ column. The idea is shown in Figure 5.10 for a 4-bit adder. The right-most "adder slice" can be a half adder, but each of the adders to the left has three inputs: A_i, B_i, and the carry-out of the preceding stage, C_i. The best way to construct this multi-bit adder is to use a single building block that can be cascaded to form an adder of any number of bits: the *full adder*.

The full adder has three inputs—A, B, and CI (carry-in)—and two outputs—S (sum) and CO (carry-out). S is written as

$$S = CI \oplus A \oplus B$$

while CO can be expressed in two-level and multilevel forms as

$$CO = BCI + ACI + AB = CI(A + B) + AB$$

The implementation of the full adder suggested by the multilevel expressions is shown in Figure 5.11. This implementation requires six gates.

In Chapter 2, we saw how to implement a full adder in terms of cascaded half adders. In this scheme, we need only five gates: two for each of the two half adders and one OR gate for the carry-out. This is one fewer than in Figure 5.11. However, with cascaded half adders, the

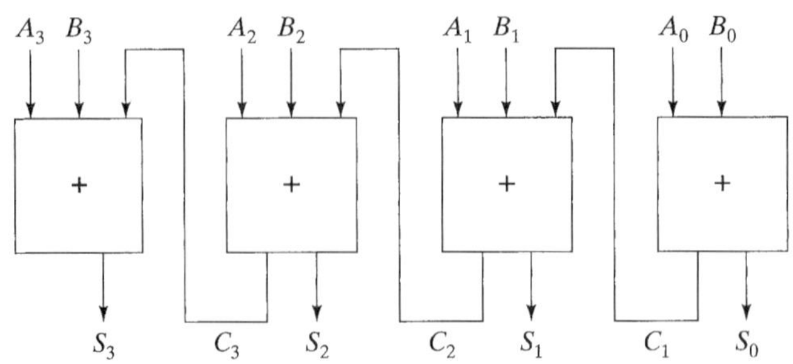

Figure 5.10 **Multi-bit-adder block diagram.**

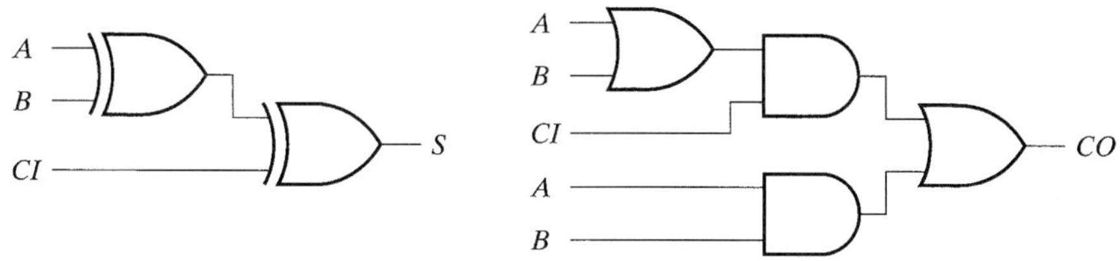

Figure 5.11 **Schematic for multilevel full adder.**

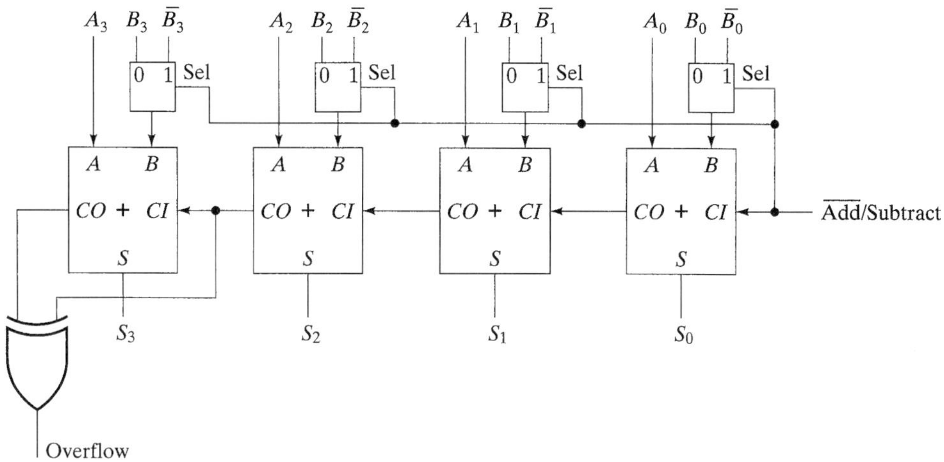

Figure 5.12 **A 4-bit adder/subtractor.**

carry output passes through three gate levels: an XOR (first-stage sum), an AND gate (second-stage carry), and a final stage OR gate (final carry). This compares with an OR, AND, OR path in Figure 5.11. Since an OR gate is considerably faster than an XOR gate in most technologies, the multiple half adder implementation is probably slower.

Adder/Subtractor Figure 5.12 shows the circuit for a 4-bit adder/subtractor constructed from full-adder building blocks. Besides the A_i and B_i inputs, we have introduced a control input $\overline{Add}/Subtract$. When this signal is 0, the circuit performs addition. When it is 1, the circuit becomes a subtractor.

The $\overline{Add}/Subtract$ input feeds the low-order carry-in and the selection input of four 2:1 multiplexers. When it is asserted, the multiplexers deliver the complements of the B_i inputs to the full adders and set the low-order carry-in to 1. This is exactly the way to form $A + \overline{B} + 1$, the sum of A and the twos complement negation of B. When $\overline{Add}/Subtract$ is unasserted, the result is simply $A + B + 0$.

The circuit includes an XOR gate whose inputs are the carry-in and carry-out of the highest-order (left-most) adder stage. When these bits differ, an overflow has occurred which corresponds exactly to an XOR function. See Appendix A for a discussion of overflow when adding two's complement numbers.

5.6.2 Carry-Lookahead Circuits

Carry-lookahead circuits are special logic circuits that can dramatically reduce the time to perform addition. We study them next.

Critical Delay Paths in Adder Circuits The adder of Figure 5.12 sums the A_i and B_i bits in parallel. But for the S_i outputs to be correct, the adder requires serial propagation of the carry outputs from the right-most, lowest-order stage to the left-most, highest-order stage. The *rippling* of the carry from one stage to the next determines the adder's ultimate delay.

Let's analyze the delays in the ripple adder by counting gate delays. We will assume that the adder stages are implemented as in Figure 5.11 and, for simplicity, that all gates have the same delay. At time 0 the inputs A_i, B_i, and C_0 are presented to the adder. Within two gate delays, one delay for each XOR gate, S_0 will be valid.

The carry signal is more complex, and we must examine it on a case-by-case basis. When $A_i = B_i = 1$, the carry is computed in two gate delays and is independent of the carry-in (carry-out will always be 1 in this case). When $A_i = B_i = 0$, the carry is also valid after two gate delays and is still independent of the carry-in (carry-out will always be 0). When $A_i \neq B_i$, the calculation of the carry takes three gate delays (carry-out depends on the carry-in).

This is the base case for the zeroth bit. When adders are cascaded, the critical delay is the time to compute the carry-out after the arrival of a valid carry-in. If the carry-in arrives after N gate delays, the carry-out will be computed by time $N + 2$. This is shown in Figure 5.13(a). The @ notation indicates the number of gate delays before a given signal is valid.

Figure 5.13(b) shows the delays in the cascaded logic for the worst-case addition, $1111_2 + 0001_2$, since each bit sum generates a carry into the next position. With the inputs arriving at time 0, the zeroth-stage sum and carry are generated at time 2. In the first stage, the sum is computed after one delay (only one XOR gate) and the carry-out takes

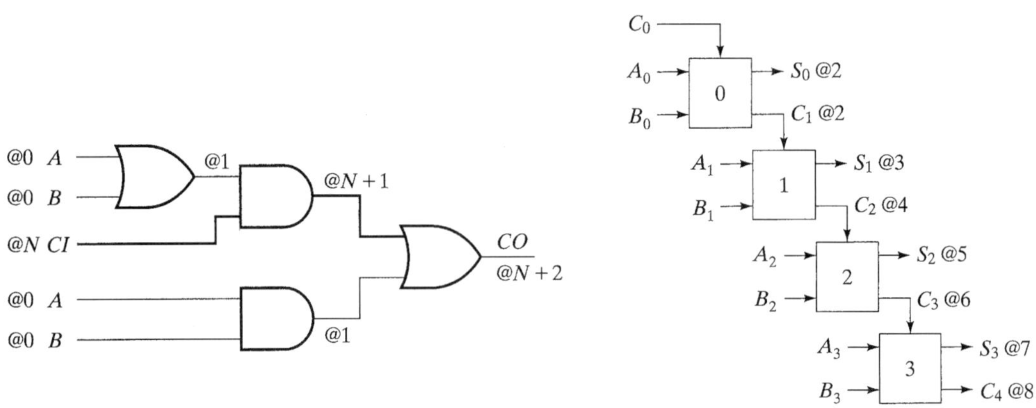

(a) Critical delay path in the carry function

(b) Critical delay in four-stage adder

Figure 5.13 **Critical delay paths.**

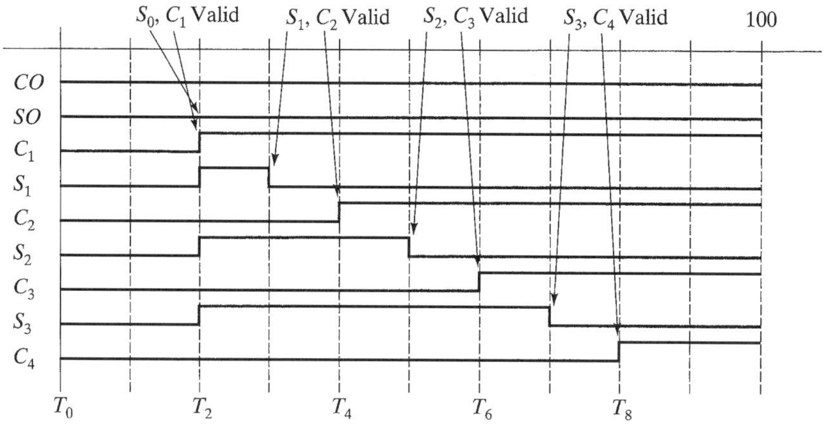

Figure 5.14 Waveforms for 1111 + 0001.

two. The valid carry-out of stage 1 at time 4 generates a valid sum and carry-out at times 5 and 6, respectively, for the third stage. This leads to a final stage sum and carry at times 7 and 8, respectively.

In general, the sum output from stage i, S_i, will be valid after two gate delays if $i = 0$ and $(2*i + 1)$ if $i > 0$. For $i \geq 0$, the carry-out C_{i+1} will be valid one delay after the sum is valid.

Because this analysis assumes that each stage experiences the worst-case delay, it is often called the *upper bound*. The timing waveforms for the worst case are shown in Figure 5.14. The actual delay depends on the particular pattern of the inputs. For example, for the 4-bit sum $0101_2 + 1010_2$, the output will be valid after only two gate delays, since there are no carries between stages.

Although an eight-gate delay may not seem so bad for a 4-bit adder, the cascaded delays become intolerable for adders of greater widths, such as 32 or 64 bits. An 8-stage adder takes 16 gate delays, a 16-stage adder 32 gate delays, and a 32-stage adder 64 gate delays (to the final stage carry-out) in the worst case. This observation led hardware designers to develop *carry-lookahead schemes*. These are ways to calculate the carry inputs in parallel, rather than in series.

Carry-Lookahead Logic In the 4-bit *ripple adder*, the carry-out of each stage, C_{i+1}, is expressed as a function of A_i, B_i, and C_i. The basic idea of carry-lookahead logic is to express each C_i in terms of $A_i, A_{i-1}, \ldots, A_0$, $B_i, B_{i-1}, \ldots, B_0$, and C_0 directly. This is a much more complex Boolean function, but it can always be expressed in two-level logic form. Thus, it should never take more than two gate delays to compute any of the carry outputs. Of course, the gates may get large and slow if the two-level form has lots of terms and each has many literals.

We begin by introducing two new functions from which we will construct the lookahead carry. These are called *carry generate*, written as G_i, and *carry propagate*, written as P_i. They are defined as

$$G_i = A_i \bullet B_i \qquad P_i = A_i \oplus B_i$$

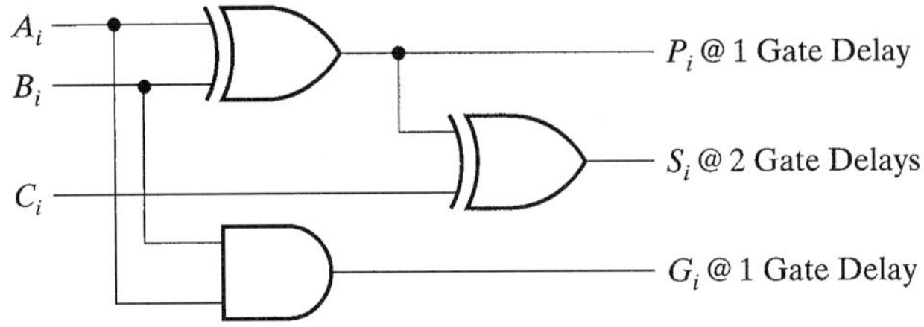

Figure 5.15 **Add with propagate and generate.**

In our previous analysis of the carry function, when A_i and B_i are both 1, a carry-out must be asserted, independently of the carry-in. Hence, we call the function a *carry generate*. If one of A_i and B_i is 1 while the other is 0, then the carry-out will be identical to the carry-in. In other words, when the XOR is true, we pass or *propagate* the carry across that stage.

Interestingly, the sum and carry-out can be expressed in terms of the carry-generate and carry-propagate functions:

$$
\begin{aligned}
S_i &= A_i \oplus B_i \oplus C_i = P_i \oplus C_i \\
C_{i+1} &= A_iB_i + A_iC_i + B_iC_i \\
&= A_iB_i + C_i(A_i + B_i) \\
&= A_iB_i + C_i(A_i \oplus B_i) \\
&= G_i + C_iP_i
\end{aligned}
$$

When the carry-out is 1, either the carry is generated internally within the stage (G_i) or the carry-in is 1 (C_i) and it is propagated (P_i) through the stage.

Expressed in terms of carry propagate and generate, we can rewrite the carry-out logic as follows:

$$
\begin{aligned}
C_1 &= G_0 + P_0C_0 \\
C_2 &= G_1 + P_1C_1 = G_1 + P_1G_0 + P_1P_0C_0 \\
C_3 &= G_2 + P_2C_2 = G_2 + P_2G_1 + P_2P_1G_0 + P_2P_1P_0C_0 \\
C_4 &= G_3 + P_3C_3 = G_3 + P_3G_2 + P_3P_2G_1 + P_3P_2P_1G_0 + P_3P_2P_1P_0C_0
\end{aligned}
$$

The *i*th carry signal is the OR of $i + 1$ product terms, the most complex of which has $i + 1$ literals. This places a practical limit on the number of stages across which the carry lookahead logic can be computed. Four-stage lookahead circuits commonly are available in parts catalogs and cell libraries. Eight-stage lookahead circuits are difficult to find because of the scarcity (and slowness) of 9-input gates.

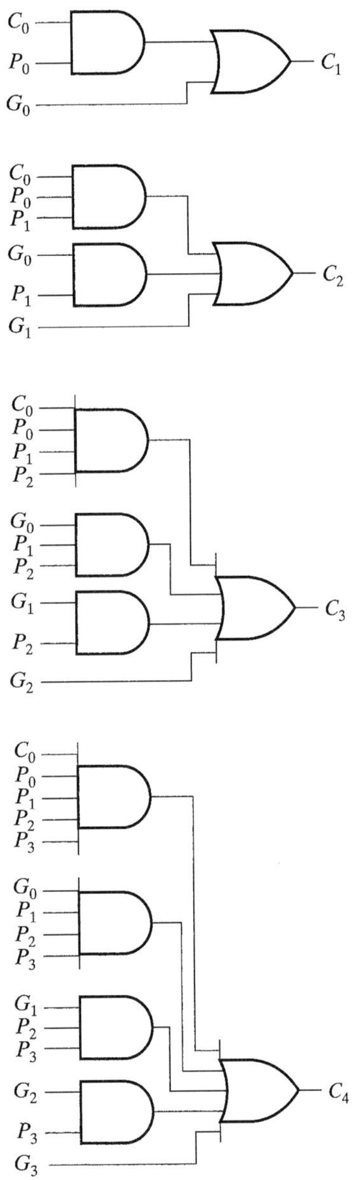

Figure 5.16 **Four-bit carry-lookahead logic.**

Implementing Carry-Lookahead Logic Figure 5.15 shows the schematic for an adder stage with propagate and generate outputs. The carry-lookahead circuits for a 4-bit adder are given in Figure 5.16. If the inputs to the zeroth adder stage are available at time 0, it takes one gate delay to compute the propagate and generate signals and two gate delays to compute the sum. When the P_i and G_i are available, the subsequent carries, C_1, C_2, C_3, and C_4, are computed after two more gate delays (three gate delays total). The sum bits can be computed in just one more gate delay.

The cascaded delay for the 4-bit adder with carry lookahead is shown in Figure 5.17. The final sum bit is available after four gate delays, compared with seven gate delays in the adder without carry lookahead. This analysis assumes that 5-input gates have the same delay as 2-input gates, which is not usually the case. Based on this simplifying assumption, we have been able to cut the time to do a 4-bit addition almost in half.

Direct calculation of the carry-lookahead logic beyond four bits becomes impractical because of the very high fan-ins that would be required. So how would we apply carry-lookahead techniques to a 16-bit adder?

We take a hierarchical approach. We can implement 16-bit sums with four 4-bit adders, each employing its own internal 4-bit carry-lookahead logic. Each 4-bit adder computes its own "group" carry-propagate and carry-generate functions: the group propagate is the AND of P_3, P_2, P_1, P_0; while the group generate is the expression $G_3 + G_2P_3 + G_1P_3P_2 + G_0P_3P_2P_1$.

A second-stage circuit computes the lookaheads and the output carries between first-stage 4-bit adders, its logic is identical to the carry logic within each 4-bit adder (Figure 5.16). Figure 5.18 shows the block diagram for the 16-bit address. The single-bit propagate and generate functions are computed from the inputs in just one gate delay. For each 4-bit adder, it takes one more gate delay to computer the group propagate while the group generate requires two delays. These

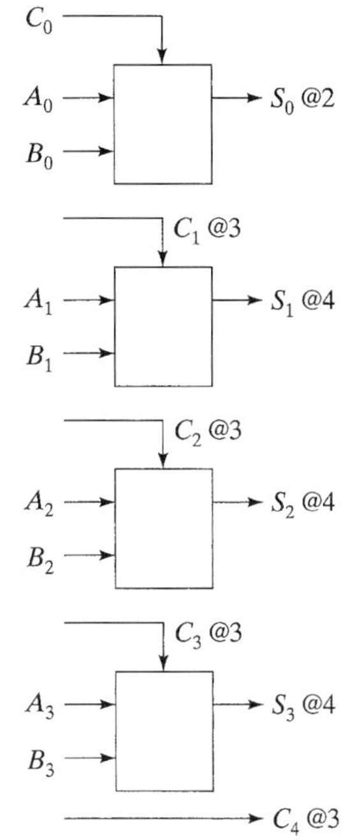

Figure 5.17 Critical delay in four-stage adder with carry lookahead.

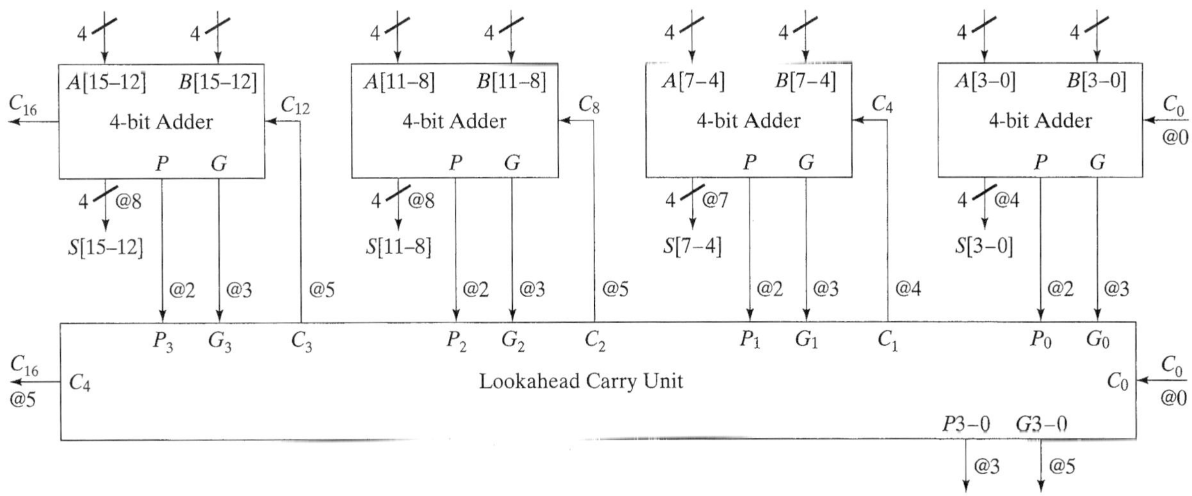

Figure 5.18 16-bit adder using hierarchical carry lookahead.

signals are presented to the second-level carry-lookahead logic at times 2 and 3, respectively. The group carries are computed with one additional gate delay for the first group and two gate delays for the others. Thus, the group carries that go back to the 4-bit adders are ready at times 4 and 5, respectively.

Once the carry-in to a 4-bit adder is known, the internal carry-lookahead logic computes the sums in three more gate delays. The zeroth stage takes four gate delays because its sums must wait one gate delay for the propagates and generates to be computed in the first place. The higher-order stages overlap the propagate and generate computations with the carry calculations in the external carry-lookahead unit.

To the external second-level carry lookahead, C_0 arrives at time 0, P_i arrives at time 2, and G_i arrives at time 3. Using the schematics of Figure 5.16, C_4 becomes valid at time 4, while C_8, C_{12}, and C_{16} become valid at time 5. The sums of the second-, third-, and fourth-adder stages are computed in three more gate delays. Thus, sum bits 7 through 4 become valid at time 7, while bits 15 through 8 are available at time 8.

So in eight delays, we can calculate a 16-bit sum, compared with 32 delays in a simple 16-bit ripple adder. We can generalize the approach for adders spanning any number of inputs. For example, a 64-bit adder can be constructed from sixteen 4-bit adders, four 4-bit carry-lookahead units at the second level, and a single 4-bit carry-lookahead unit at the third level.

5.6.3 Carry-Select Adder

The circuits of the last subsection trade more gates and hardware complexity for a faster method to compute the inter-stage carries. In this section, we examine the *carry-select adder,* an adder organization that introduces redundant hardware to make the carry calculations go even faster.

Figure 5.19 illustrates the concept by showing the organization of an 8-bit carry-select adder. The 8-bit adder is split in half. The upper

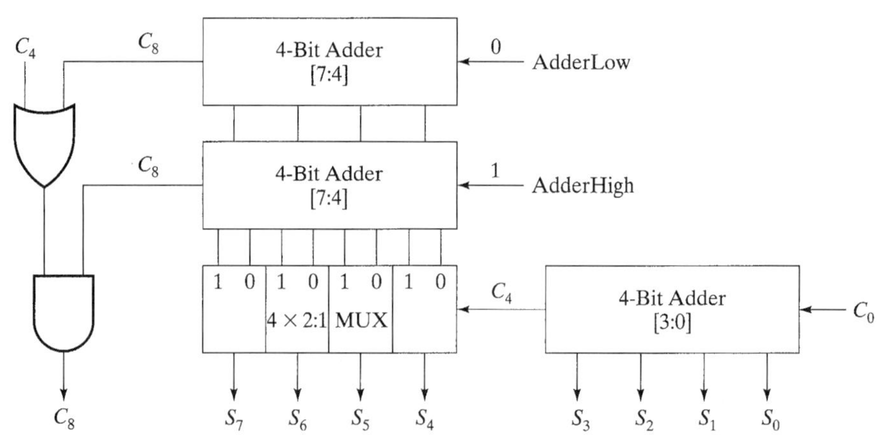

Figure 5.19 **8-bit carry-select adder.**

half is implemented by two independent 4-bit adders, one whose carry-in is hardwired to 0 (AdderLow), another whose carry-in is hardwired to 1 (AdderHigh). In parallel, these compute two alternative sums for the higher-order bits. The carry-out of the lower-order 4-bit adder controls multiplexers that select between the two alternative sums.

The circuit for computing the final C_8 from the AdderLow and AdderHigh could be a 2:1 multiplexer, but a simpler circuit that reduces the gate count also does the job. If C_4 is 1, the final C_8 is simply the carry out of AdderHigh. If C_4 is low, the final C_8 will only be 1 if the carry-out of AdderLow is 1 (and if that is 1, then the carry-out of AdderHigh must also be 1).

How long does the carry-select adder take? Assuming internal carry-lookahead logic is used, the 4-bit adders in Figure 5.19 take four gate delays to compute their sums and three gate delays to compute the stage carry-out. The 2:1 multiplexers add two further gate delays to the path of the high-order sum bits. Thus, the 8-bit sum is valid after only six gate delays. This saves one gate delay over the standard two-level carry-lookahead implementation for an 8-bit adder.

As you may imagine, the carry chain used for adders can be a big factor in determining the performance of logic designs. For this reason, most FPGA have special optimized circuitry that allows the logic blocks to implement fast adders. This is true for all of the major manufacturers of FPGAs; both look-up table and multiplexor based.

5.6.4 BCD Adder Design

It is possible to build digital hardware that manipulates BCD directly, and such hardware could be found in early computers and many handheld calculators. The BCD system was chosen for the internal number system in these machines, because it is easy to convert BCD numbers to alphanumeric representations for printouts and displays. The compelling advantages of BCD have waned over time, and these digits are supported by more modern hardware simply to provide backward compatibility with earlier generations of machines. In this section, we briefly examine the approaches for constructing BCD arithmetic elements.

Appendix A describes addition for BCD numbers. The principal concept we need to recall is that the carry generation from one BCD digit to the next is quite different than within a digit because a digit only goes up to 9 even though it is four bits and could go to 15.

Figure 5.20 gives a block diagram implementation for a BCD adder. The first row of full adders implements a conventional 4-bit binary adder. The second row provides the capability to add 0110_2 when the sum obtained by the first row exceeds 9 (1001_2). See Appendix A for details.

Here is how it works. The adders of the second row add the carry-out bit to the sum bits S_2 and S_1. Carry-out should be asserted in

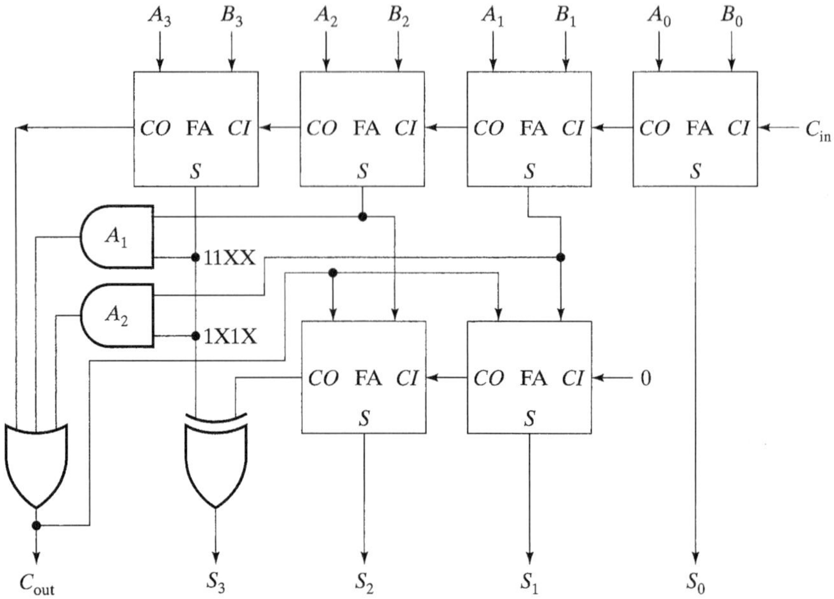

Figure 5.20 **BCD adder block diagram.**

cases in which we need to add the correction factor. What are these cases?

The AND gates labeled A_1 and A_2 detect the conditions under which the first-level sum matches the patterns $11XX_2$ and $1X1X_2$. These are exactly the cases in which this sum exceeds 9. When carry-out is asserted, the XOR gate and the adders in the second row effectively add 0110_2 to the first row's sum.

There is one further case to consider. The correction factor should also be applied whenever the first-row sum exceeds 15. This can happen with the sum of 9 and 7, for example. This case is easy to detect: the carry-out of the first-row adders will be asserted.

Thus, the sum exceeds 9 if either the first-row carry-out is asserted, or the sum matches the pattern $11XX_2$, or the sum matches the pattern $1X1X_2$. These are precisely the inputs to the OR gate that computes the BCD carry-out.

A BCD adder requires over 50% more hardware than a comparable binary adder. Since binary adders are faster, it is no surprise that they have replaced BCD adders in most cases.

5.7 Arithmetic Logic Unit Design

An arithmetic logic unit, or ALU (sometimes pronounced "Al Loo"), is a combinational network that implements a function of its inputs based on either logic or arithmetic operations. ALUs are at the heart of all computers as well as most digital hardware systems. In this section, we learn how to design these very important digital subsystems.

S_1	S_0	Function	Comment
$M = 0$, Logical Bitwise Operations			
0	0	$F_i = A_i$	Input A_i transferred to output
0	1	$F_i = \text{NOT } A_i$	Complement of A_i transferred to output
1	0	$F_i = A_i \text{ XOR } B_i$	Compute XOR of A_i, B_i
1	1	$F_i = A_i \text{ XNOR } B_i$	Compute XNOR of A_i, B_i
$M = 1$, $C_0 = 0$, Arithmetic Operations			
0	0	$F = A$	Input A passed to output
0	1	$F = \text{NOT } A$	Complement of A passed to output
1	0	$F = A$ plus B	Sum of A and B
1	1	$F = (\text{NOT } A)$ plus B	Sum of B and complement of A
$M = 1$, $C_0 = 1$, Arithmetic Operations			
0	0	$F = A$ plus 1	Increment A
0	1	$F = (\text{NOT } A)$ plus 1	Twos complement of A
1	0	$F = A$ plus B plus 1	Increment sum of A and B
1	1	$F = (\text{NOT } A)$ plus B plus 1	B minus A

Figure 5.21 Tabular specification of an ALU.

5.7.1 A Sample ALU

An n-bit ALU typically has two input words, denoted by $A = A_{n-1}, \ldots, A_0$ and $B = B_{n-1}, \ldots, B_0$. The output word is denoted by $F = F_n, F_{n-1}, \ldots, F_0$, where the high-order output bit, F_n, is actually the carry-out. In addition, there is a carry-in input C_0.

Besides data inputs and outputs, an ALU must have control inputs to specify the operations to be performed. Often, one input is a mode selector; M in our case. When $M = 0$, the operation is a logic function; when $M = 1$, an arithmetic operation is indicated. In addition, there are operation selection inputs, S_i, which determine the particular logic or arithmetic function to be performed.

To make the discussion more concrete, Figure 5.21 contains the specification of a simple ALU *bit slice,* that is, the behavior of a single bit of the ALU. The list of operations is partitioned into three sections: logic operations, arithmetic operations where the carry-in is 0, and arithmetic operations where the carry-in is 1. Some of the operations do not appear to be useful, such as the sum of B and the ones complement of A. However, if we set carry-in to 1, we obtain a very useful operation indeed: B minus A (B plus the twos complement of A).

Implementation of an ALU ALUs are relatively simple to implement: design a 1-bit slice and cascade as many of these as you need to build a multi-bit structure. Of course, the limiting performance factor will be the propagation of carries among the ALU stages.

Using the specification of Figure 5.21, a single bit slice has six inputs: A_i, B_i, C_i, M, S_1, and S_0, and two outputs: F_i and C_{i+1}. This may appear daunting, but the truth table shows a relatively simple structure with many don't cares (see Figure 5.22).

M	S_1	S_0	C_i	A_i	B_i	F_i	$C_i + 1$
0	0	0	X	0	X	0	X
			X	1	X	1	X
	0	1	X	0	X	1	X
			X	1	X	0	X
	1	0	X	0	0	0	X
			X	0	1	1	X
			X	1	0	1	X
			X	1	1	0	X
	1	1	X	0	0	1	X
			X	0	1	0	X
			X	1	0	0	X
			X	1	1	1	X
1	0	0	0	0	X	0	X
			0	1	X	1	X
	0	1	0	0	X	1	X
			0	1	X	0	X
	1	0	0	0	0	0	0
			0	0	1	1	0
			0	1	0	1	0
			0	1	1	0	1
	1	1	0	0	0	1	0
			0	0	1	0	1
			0	1	0	0	0
			0	1	1	1	0
1	0	0	1	0	X	1	0
			1	1	X	0	1
	0	1	1	0	X	0	1
			1	1	X	1	0
	1	0	1	0	0	1	0
			1	0	1	0	1
			1	1	0	0	1
			1	1	1	1	1
	1	1	1	0	0	0	1
			1	0	1	1	1
			1	1	0	1	0
			1	1	1	0	1

Figure 5.22 **ALU truth table.**

We can use the *espresso* algorithm to find an optimized two-level implementation. It turns out that the F_i output still requires 18 product terms. A PLA-based design is feasible, but a design based on discrete gates is probably too complex even to consider.

Multilevel logic optimization does a bit better. The circuit derived by the MisII tool appears in Figure 5.23. It requires 12 gates plus five inverters, which are not shown in the schematic.

The schematic in Figure 5.24 offers an alternative multilevel implementation for the ALU. We obtained it by hand after a careful evaluation of how operations are encoded by the M and selection inputs. This implementation is based on the observation that when S_1 is 0, B_i is blocked from affecting the outputs by gate A1. This happens whenever the operation deals only with A_i. The same is true for C_i when M is 0 (gate A2). These are exactly the nonarithmetic operations that do not concern themselves with carries.

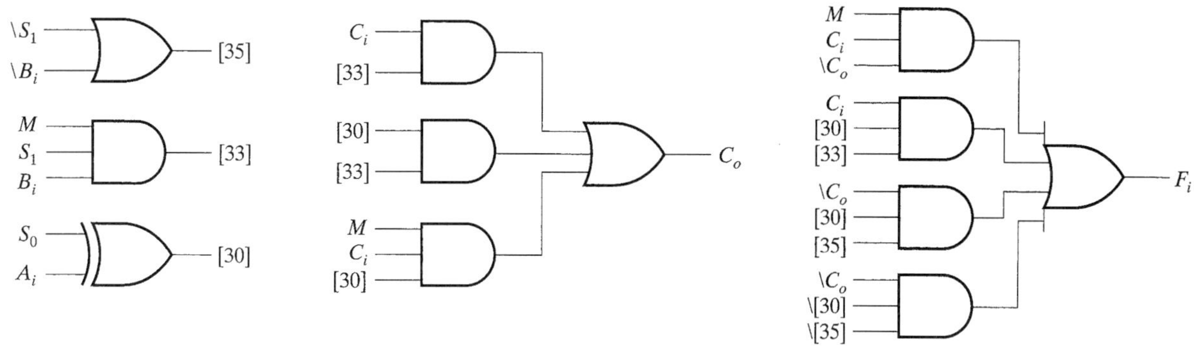

Figure 5.23 **Multilevel schematic drived by *MIS*.**

Addition is indicated whenever $M = 1$. In these cases, B_i (assuming $S_1 = 1$) and C_i are passed to the inputs of the XOR gates X3 and X2. When S_0 is 0, the topmost XOR gate, X1, simply passes A_i. When S_0 is 1, it passes A_i's complement. Thus, the three cascaded XOR gates form a proper sum of A_i (or its complement if $S_0 = 1$), B_i (or 0 if $S_1 = 0$), and C_i whenever the ALU is in its arithmetic mode.

How about the carry output? When the ALU is in the arithmetic mode, the output of gate O1 is the function $A_i C_i + B_i(A_i \oplus C_i)$. The first product term is formed from gate A3: the second from A4. This is a valid form of the carry.

When the ALU is in the logic mode, $M = 0$, we can concentrate on the cascaded XOR gates. When S_0 and S_1 are both 0, A_i is passed through to the output. If S_0 is 1 while S_1 is 0, the complement of A_i is passed. When S_0 is 0 while S_1 is 1, the inputs to X3 are A_i and B_i, and their XOR is computed. In the last case, $S_0 = 1$, $S_1 = 1$, the inputs to X3 are $\overline{A}_i$ and B_i. This function is equivalent to the XNOR function. The circuit does the right thing!

A careful hand design can sometimes do better than a sophisticated CAD tool. Typical CAD tools are unlikely to come up with this schematic, in part because of their inability to exploit don't-care conditions or to use XOR functions effectively.

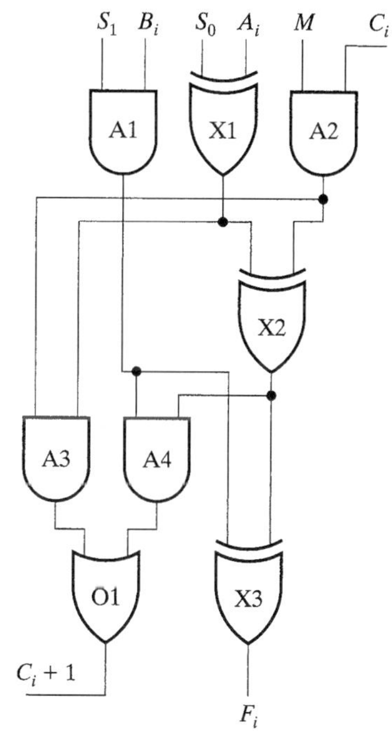

Figure 5.24 **Multilevel ALU bit.**

5.8 Combinational Multiplier

In this section, we look at the design of multiplier circuitry. The methods we introduce are combinational, although alternative methods based on circuits with state are also possible. However, the fastest circuits for multiplication use just the techniques we will be discussing here.

Basic Concept Throughout this section, we will look only at multiplication techniques for unsigned numbers. Alternatively, the hardware we present is suitable for sign and magnitude multiplication, but we concentrate on the manipulation of the magnitude part. Recall that the two numbers involved in a multiplication are called the *multiplicand* and the *multiplier.*

The process of binary multiplication is best illustrated with an example. In this case, the multiplicand is 1101_2 (13) and the multiplier is 1011_2 (11):

Multiplicand	1101	(13)
Multiplier	* 1011	(11)
	1101	
	1101	
	0000	
	1101	
	10001111	(143)
	128 + 8 + 4 + 2 + 1 = 143	

Each bit of the multiplier is multiplied against the multiplicand, the product is aligned according to the position of the bit within the multiplier, and the resulting products are then summed to form the final result. One attraction of binary multiplication is how easy it is to form these intermediate products: if the multiplier bit is a 1, the product is an appropriately shifted copy of the multiplicand; if the multiplier bit is a 0, the product is simply 0.

For an n-bit multiplicand and multiplier, the resulting product will be $2n$ bits. Stated differently, the product of 2^n and 2^n is $2^{n+n} = 2^{2n}$: a $2n$-bit number. Thus, the product of two 4-bit numbers is represented in 8 bits, while multiplying two 8-bit numbers generates a 16-bit output, and so on.

Partial Product Accumulation We can construct a combinational circuit that directly implements the process described by the preceding example. The method is called *partial product accumulation.*

First, we rewrite the multiplicand bits as A_3, A_2, A_1, A_0 and the multiplier bits as B_3, B_2, B_1, B_0. The multiplication of A and B becomes

			A_3	A_2	A_1	A_0
			B_3	B_2	B_1	B_0
			$A_3 \cdot B_0$	$A_2 \cdot B_0$	$A_1 \cdot B_0$	$A_0 \cdot B_0$
		$A_3 \cdot B_1$	$A_2 \cdot B_1$	$A_1 \cdot B_1$	$A_0 \cdot B_1$	
	$A_3 \cdot B_2$	$A_2 \cdot B_2$	$A_1 \cdot B_2$	$A_0 \cdot B_2$		
$A_3 \cdot B_3$	$A_2 \cdot B_3$	$A_1 \cdot B_3$	$A_0 \cdot B_3$			
S_6	S_5	S_4	S_3	S_2	S_1	S_0

Each of the ANDed terms is called a *partial product.* The resulting product is formed by accumulating down the columns of partial products, propagating the carries from the rightmost columns to the left.

A combinational circuit for implementing the 4-bit multiplier is shown in Figure 5.25. The first level of 16 AND gates computes the individual partial products. The column sums are formed by a mixture of cascaded half adders and full adders. For the adders in the figure, inputs from the top are the bits to be added and the input from the

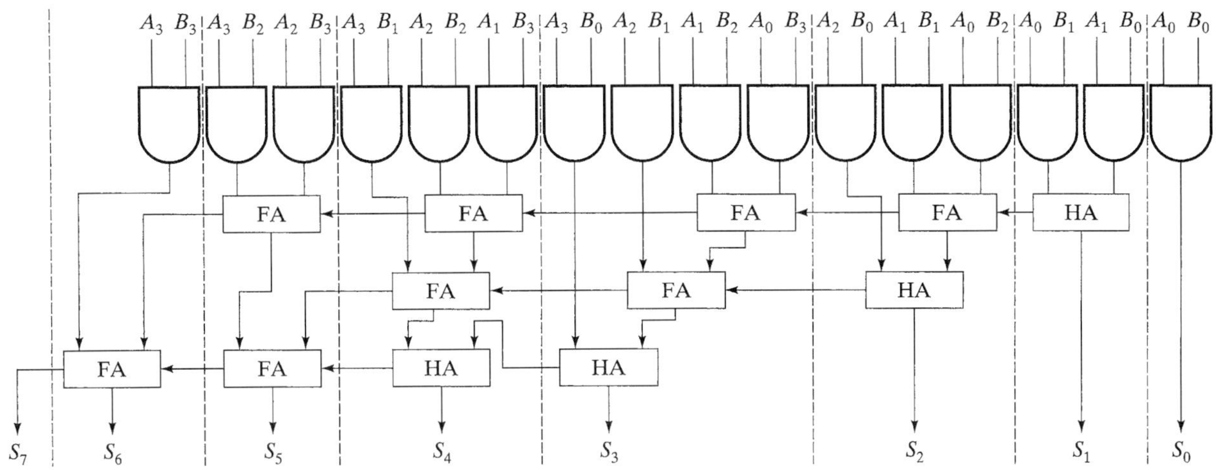

Figure 5.25 **4 × 4 combinational adder.**

right is the carry-in. The output from the bottom is the sum and to the left is the carry-out.

To see how the partial products are accumulated, let's look at the circuit of Figure 5.25 in a little more detail. S_1 is the straightforward sum of just two partial products, $A_1 \bullet B_0$ and $A_0 \bullet B_1$. S_2 is the sum of three products. We implement this with two cascaded adders, one of which takes the carry-out from S_1's column.

S_3 is a little more complicated, because there are two different carry-outs from the previous column. After all, the previous column was adding four 1-bit numbers (three partial products and one carry-in from the previous column and this could add up to as much as four or two carries into the S_3 column). We use three cascaded adders, two full adders and one half adder to implement the sum. The two carry-outs from S_2 are accumulated through the carry-in inputs of the two full adders. The sum in this column can be as high as six from the four partial products and two carry-ins and, thus, has three carry-outs.

S_4 is the sum of three products and three possible carry-outs from S_3. The solution for this column is similar to that for S_3—two full adders and one half adder. The two full adders sum the three products and two of the carry-ins. The half adder adds to this result the third possible carry-in.

The logic for S_5 continues similarly. Here we must sum two products and three carry-ins. Two full adders do the job. Note that the second full adder sums two of the three carries from the previous column with the result of the first full adder. A similar analysis applies to S_6 and S_7.

The delay through the multiplier is determined by the ripple carries between the adders. We can use a carry-lookahead scheme to reduce these delays.

Clearly, the full combinational multiplier uses a lot of hardware. The dominating costs are the adders—four half adders and eight full

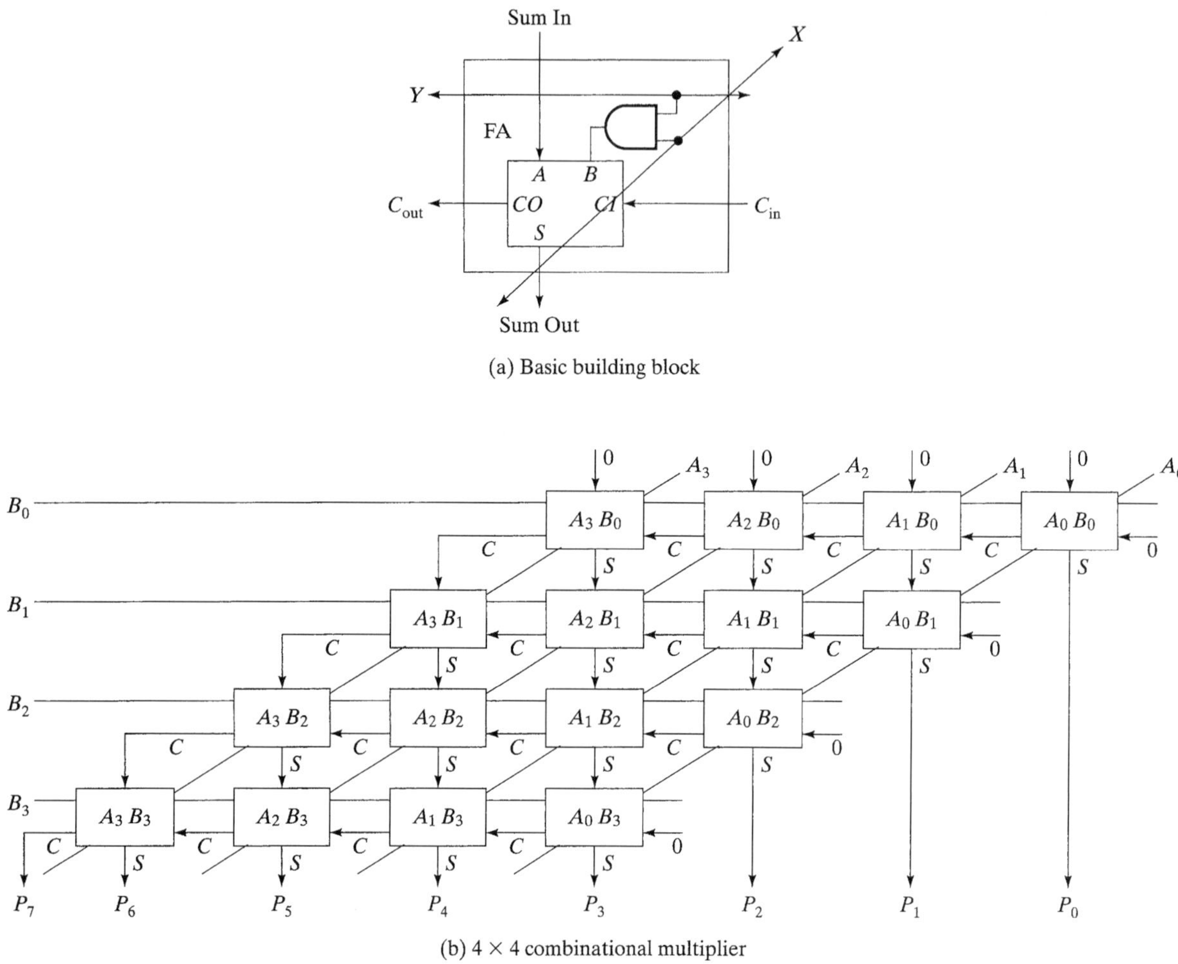

Figure 5.26 **4 × 4 combinational multiplier.**

adders. To simplify the implementation slightly, a designer may choose to use full adders for all of the adder blocks, setting the carry input to 0 where the half-adder function is required. Given the full-adder schematic of Figure 5.11, this is 12 adders of six gates each, for a total of 72 gates. When we add to this the 16 gates forming the partial products, the total for the whole circuit is 88 gates. It is easy to see that combinational multipliers can be justified only for the most high performance of applications.

A slightly different implementation of the 4-by-4 combinational multiplier is shown in Figure 5.26. Figure 5.26(a) gives the basic building block, a full-adder circuit that sums a locally computed partial product ($X \cdot Y$), an input passed into the block from above (*Sum In*), and a carry passed from a block diagonally above. It generates a carry-out (C_{out}) and a new sum (*Sum Out*). Figure 5.26(b) shows the interconnection of 16 of these blocks to implement the full multiplier function. The A_i values are distributed along block diagonals and the B_i values are passed along rows. This implementation has a larger gate count as the previous one for greater regularity: 16 AND gates and 16 full adders (we did not bother with half adders so that every block is identical).

CHAPTER REVIEW

This chapter serves to cap the discussion of combinational logic with a wide range of design case studies. We examined seven different combinational logic design problems. We began with a discussion of a general design procedure and followed it through each of the examples. The discussions centered on issues of encoding, representation, and implementation.

The latter case studies centered around the representation of numbers in digital systems and the primitive circuits for their arithmetic manipulation. A half or full adder will be found at the center of all arithmetic circuits. The full adder can be constructed from cascaded half adders. Multi-bit adders are constructed from cascaded full adders. Since the performance of such adders is limited by the serial rippling of the carry from one adder stage to the next, designers have developed parallel carry-lookahead circuits for fast adders. This is a very good example of the trade-off between hardware and speed. The ripple adder is much slower than the lookahead adder, but the latter achieves its high speed at the expense of many more gates. Adders for binary-coded decimal numbers are built from cascaded full adders. They look much like conventional binary adders, with extra circuitry to correct the sum of two BCD digits when it exceeds nine. A 4-bit BCD adder needs six full adders.

The arithmetic logic unit, or ALU, is an ubiquitous circuit component that implements both logic and arithmetic operations over data inputs. ALUs can be found embedded in just about every digital system that manipulates numbers. Any of the design strategies we have covered can be used to implement ALUs, including two-level, multilevel, and even ROM-based approaches.

Multipliers are also formed by combining cascaded full and half adders. We saw how to construct a 4-by-4 bit multiplier using 12 full adders. The circuits we have shown for multiplication are based on number magnitudes. This means that a system that uses twos complement for addition/subtraction and sign and magnitude for multiplication and division must provide additional circuitry to convert between the forms and handle the sign correctly.

FURTHER READING

Elements of this chapter were drawn from several previous logic design textbooks, primarily C. H. Roth's *Fundamentals of Logic Design,* West, St. Paul, 1985, and Johnson and Karim's *Digital Design: A Pragmatic Approach,* PWS, Boston, 1987. The 8-by-8 bit multiplier case study is based on the application note associated with the 74284/285 data sheet in the Texas Instruments' *TTL Data Book,* Volume 2, 1985.

Computer arithmetic is a complex topic worthy of its own advanced course. There are several comprehensive textbooks on the underlying mathematics of number systems and the hardware to implement arithmetic operations over them. K. Hwang's *Computer Arithmetic,* Wiley, New York, 1979, is one of the best known.

Appendix A provides a brief overview of the background necessary in binary number systems to comprehend the examples in this book.

EXERCISES

5.1 *(Word Problem)* Your task is to design a combinational-logic subsystem as part of a larger system that makes change from quarters. There is a large reservoir of dimes and another one of nickels. There are also two binary counters that keep track of the number of dimes and nickels in each (these are outside your subsystem, and you won't need to design them—but you will need them as inputs). Your subsystem needs to work as follows. It takes the low-order bits of the dimes counter and the nickels counter, and generates the number of dimes to give as change and the number of nickels. In general, dimes should be given before nickels, e.g., if there are at least two dimes left, these should be given with a single nickel, rather than five nickels. There is definitely a possibility that no change can be given before the reservoirs are (almost) exhausted of coins, and a "no change available" sign should be illuminated.

- **(a)** Identify your inputs and outputs.
- **(b)** Specify the encoding of your outputs: what do they mean?
- **(c)** Develop the minimized gate-level implementation using the K-map method.

5.2 *(Word Problem)* Consider the following logic unit. It has three operation inputs (A, B, C), two data inputs (D_1, D_0), and a single output (Z). The logic unit is defined as follows: when $ABC = 000$, $Z = 0$; $ABC = 001$, $Z = D_0$; $ABC = 010$, $Z = D_1$ AND D_0; $ABC = 011$, $Z = D_1$ NAND D_0; $ABC = 100$, $Z = D_1$ NOR D_0; $ABC = 101$, $Z = D_1$ OR D_0; $ABC = 110$, $Z = D_1$; $ABC = 111$, $Z = 1$.

- **(a)** Implement to obtain a minimized implementation using the K-map method (*Note:* This uses 5 variables!).
- **(b)** Implement using a 16:1 MUX (use A,B,C,D_1 as your MUX control inputs).
- **(c)** Implement using an 8:1 MUX plus whatever additional gate logic you need (use A,B,C as your MUX control inputs).
- **(d)** Which implementation is "best"? Why?

5.3 *(Word Problem)* Design a combinational logic subsystem with three inputs, I3, I2, I1, and two outputs, O1, O0, that behaves as follows. The outputs indicate the highest index of the inputs that is driven high. For example, if I3 is 0, I2 is 1, I1 is 1, then O1, O0 would be 10 (i.e., I2 is the highest input set to 1).

(a) Specify the function by filling out a complete truth table.
(b) Develop the minimized gate-level implementation using the K-map method.
(c) Develop an implementation using two 4:1 multiplexers.
(d) Compare your implementation for 1(b) and 1(c). Which is better and under what criterion?

5.4 *(Word Problem)* Consider a variation on the calendar combinational subsystem that works as follows. Given the inputs MONTH (1–12), DAY (1–31), and LEAP_YEAR flag, the subsystem generates the output DAY_OF_YEAR (1–365 or 366). In this problem, you will design the subsystem to the block-diagram level only.

(a) One block maps the month into a day offset into the year. Identify the inputs and outputs and their bit widths. Use any formal specification method you wish (e.g., truth tables, ROM contents, equations, hardware description language, etc.) to describe the function of this block.
(b) You may assume any width binary adder you may require. Indicate how the adder is composed with the block of part (a) and any other blocks or inputs to compute the correct output. Be sure to describe how you deal with the LEAP_YEAR input
(c) How many 4-input lookup tables will be needed to implement your design?

5.5 *(Subtraction Logic)* The truth table for a 1-bit combinational binary subtractor, analogous to the half adder, computing D(ifference) = A minus B, with BL (borrow-from-left), is

A	*B*	*D*	*BL*
0	0	0	0
0	1	1	1
1	0	1	0
1	1	0	0

(a) Design a 1-bit combinational binary subtractor, comparable to the full adder, with two data inputs (A, B), a borrow from the right input (BI), a borrow request to the left output (BL), and a difference output (D).
(b) Show how your design can be cascaded to form multi-bit subtractors.
(c) Does the subtractor work correctly for negative two's complement numbers?
(d) How is a subtraction *underflow* condition indicated?

5.6 *(Adder/Subtractor Logic)* Design a fully combinational adder/subtractor that can be cascaded to form a multi-bit circuit. The inputs are data inputs, A and B, carry-in, CI, and borrow-in, BI. The outputs are data output, F, carry-out, CO, and borrow from left, BL. A mode input $M = 0$ indicates addition and $M = 1$ indicates subtraction. Can the carry and borrow inputs and outputs be combined?

5.7 *(Adder Design)* Using comparators, multiplexers, and binary adder/subtractor logic blocks, design a 4-bit sign and magnitude adder (that is, one sign bit and three data bits). Include an overflow indicator in your design.

5.8 *(Remainder Function)* Design a combinational circuit with three data inputs D_2, D_1, D_0, two control inputs C_1, C_0, and two outputs R_1, R_0. The R_1, R_0 should be the remainder after dividing the unsigned binary number formed from D_2, D_1, D_0 by the unsigned binary number formed by C_1, C_0. For example, D_2, D_1, $D_0 = 111$, C_1, $C_0 = 10$, then R_1, $R_0 = 01$ (7 divided by 2 yields a remainder of 1). *Note:* Division by zero will never occur—take advantage of don't-care conditions in this case.

(a) Write the truth table for R_1 and R_0.
(b) Fill in K-maps and write Boolean expressions for R_1 and R_0.
(c) Draw a schematic that implements the functions for R_1 and R_0 using a minimum number of NAND gates. Assume complemented literals are available and that there are no limits on NAND gate fan-ins.

5.9 *(Multi-Bit Adders)* Figure 5.18 shows how to use 4-bit adders and a 4-bit carry-lookahead unit to implement a fast 16-bit adder. Using these as primitive building blocks, show how to construct 32- and 64-bit adders with carry lookahead.

(a) Draw block diagrams for the 32- and 64-bit adders, showing all interconnections.
(b) Analyze the worst-case gate delays encountered in 32- and 64-bit addition. Use the simple delay models as in Section 5.6.

5.10 *(Carry-Select Adder)* Consider a 16-bit adder implemented with the carry-select technique described in Section 5.6. The adder is implemented with three 8-bit carry-lookahead adders and eight 2:1 multiplexers. Estimate the gate delay and compare it against a conventional 16-bit ripple adder and a 16-bit carry lookahead adder.

5.11 *(Carry-Select Adder)* Argue why, using explicit 8-bit test cases, the implementation of C_8 in Figure 5.19 is correct.

5.12 *(ALU Design)* Implement to the gate level an ALU bit slice with three operation selection inputs, S_2, S_1, S_0, that implements the following eight functions of the two data inputs, A and B (and carry-in Cn):

S_2	S_1	S_0	*ALU Operation*
0	0	0	$F_i = 0$
0	0	1	$F_i = B$ minus A
0	1	0	$F_i = A$ minus B
0	1	1	$F_i = A$ plus B
1	0	0	$F_i = A$ XOR B
1	0	1	$F_i = A$ OR B
1	1	0	$F_i = A$ AND B
1	1	1	$F_i = 1$

Assume a simple ripple–carry scheme between bit slices.

5.13 *(ALU Design)* Implement to the gate level an ALU bit slice with three selection inputs, S_2, S_1, S_0, and a logic/arithmetic mode input, M, that implements the following 16 functions of the two data inputs A and B (and carry-in C_0):

M = 0, *Logic Mode*				*M* = 1, *Arithmetic Mode*			
S_2	S_1	S_0	*ALU Operation*	S_2	S_1	S_0	*ALU Operation*
0	0	0	$F_i = 0$	0	0	0	$F_i =$ not A
0	0	1	$F_i = A$ XOR B	0	0	1	$F_i =$ not B
0	1	0	$F_i = A$ XNOR B	0	1	0	$F_i = A$ minus B
0	1	1	$F_i = A$ OR B	0	1	1	$F_i = B$ minus A
1	0	0	$F_i = A$ AND B	1	0	0	$F_i = A$ plus B
1	0	1	$F_i = A$ NOR B	1	0	1	$F_i = A$ plus 1
1	1	0	$F_i = A$ NAND B	1	1	0	$F_i = B$
1	1	1	$F_i = 1$	1	1	1	$F_i = A$

Assume a simple ripple-carry scheme between bit slices.

5.14 *(ALU Design)* Revise your design for the ALU of Exercise 5.12 to include carry-lookahead logic that can operate across four bits.

5.15 *(ALU Design)* Revise your design for the ALU of Exercise 5.13 to include 4-bit carry-lookahead logic.

5.16 *(ROM-Based ALU Design)* Describe how a two-output ROM can be used to implement the ALU bit slice of Figure 5.22. Give a block diagram showing how a 4-bit ALU can be implemented from four cascaded ROMs. In terms of the total count of ROM bits, how does the bit-slice approach compare with a solution that uses a single ROM with full 4-bit data inputs and outputs?

5.17 *(ALU Design)* Verify that the multilevel circuit of Figure 5.24 implements the ALU specification of Figure 5.21. Do this by expanding the circuit into its equivalent two-level sum-of-products expressions for C_0 and F_i. Then show that the truth table/K-map for these expressions is compatible with the truth table of Figure 5.22.

5.18 *(BCD Addition)* Perform the following binary additions assuming the numbers are in BCD form:

(a) 0001 + 0100
(b) 1000 + 1001
(c) 0111 + 0011
(d) 1001 1001 + 0001 0001

5.19 *(BCD/Binary Adder Design)* Design a 4-bit circuit that can perform either BCD or binary addition under the control of a mode setting, M. When $M = 0$, the circuit's outputs implement binary addition. When $M = 1$, the outputs are BCD addition. Your solution should require minimal changes to the circuit of Figure 5.20.

5.20 *(Combinational Multiplier)* Verify that the operation of the combinational multiplier of Figure 5.25 is correct by tracing all intermediate signals for the multiplication of 11 (1011_2) by 13 (1101_2).

5.21 *(Combinational Multiplier)* Verify that the operation of the combinational multiplier of Figure 5.26 is correct by tracing all intermediate signals for the multiplication of 11 (1011_2) by 13 (1101_2).

5.22 *(Combinational Multiplier)* What is the worst-case propagation delay through the combinational multiplier of Figure 5.25, assuming the hardware is implemented with 12 full adders as in Figure 5.11?

5.23 *(Combinational Multiplier)* Assume that you have a 4-by-4 magnitude multiplier available as a primitive building block. Discuss the design issues in building a 4-by-4 twos complement multiplier. Consider such issues as (a) determining the signs of the inputs and outputs, (b) putting the inputs into the appropriate form for the magnitude multiplier, and (c) putting the magnitude multiplier's output into a form suitable for twos complement representation.

5.24 *(Combinational Divider)* Design a combinational divider, following the same general approach as used for the combinational multiplier of Section 5.8. Of course, this time you will use binary subtractors rather than adders.

5.25 *(Combinational Multiplier)* Consider the 4-by-4 combinational multiplier of Figure 5.26. Is there a way to make it faster by rearranging how the column additions are performed? *Hint:* Consider forwarding carries to a different adder in the next column so that the effects of ripple carries are minimized. You may need to add other full adders.

CHAPTER

6

Sequential Logic Design

Introduction

To this point, we have focused on those circuits whose outputs are solely a function of their inputs—*combinational* logic circuits. We are now ready to expand the discussion to circuits, like the combination lock we studied in Chapter 1, whose outputs are a function of the current as well as the past sequence of inputs—*sequential* logic circuits. The particular parts of these circuits that store the information about the previous history of inputs are called *storage, memory,* or *state elements*. The structure and behavior of these most primitive sequential logic elements will be our primary topic in this chapter. These are building blocks for more complex sequential circuits to be introduced in the next chapters. In particular, we shall cover:

- *Simple circuits with memory.* A primitive storage element can be constructed from a small number of gates whose outputs loop back around as inputs to the same gates. We will examine how to build such memory elements, including the two most popular varieties: *latches* and *flip-flops*.
- *Use of clocks.* Memory elements must replace their current value (or state) with a new value in a controlled and predictable way. A *clock* is a periodic signal distributed throughout a circuit that can ensure that all memory elements change state at approximately the same instant of time and only once for each clock tick. Systems that use a clock are called *synchronous*. In contrast, circuits that do not utilize a clock are called *asynchronous*.
- *Building more complex circuits from flip-flops.* We will develop design methods for combining the primitive building blocks into more complex clocked circuits. We will build registers, combinations of flip-flops that work in concert on a number of bits simultaneously.

- *The metastability problem.* A problem often arises when two clocked circuits (synchronized by different clocks) or a synchronous and an asynchronous circuit need to communicate with one another. The output from one may be in transition when the other samples it as an input and may be wrongly interpreted. The problem is intrinsic in communicating systems, but we will present design techniques that can reduce its effect. We also will highlight briefly techniques for creating locally synchronous but globally asynchronous systems that are increasing in importance as chips get larger.
- *Hardware description languages.* We show how we can describe the building blocks of sequential logic using HDLs such as Verilog and introduce a few new Verilog constructs specifically designed to model sequential logic.

6.1 Sequential Logic Elements

In Chapter 1 we defined sequential circuits as those in which some outputs feed back as inputs. Feedback is a necessary condition for sequential circuits, it can make the outputs depend on the entire history of input sequences.

A digital alarm clock is a sequential circuit. The output of the clock is the current time of day, updated every second. The current setting of the clock determines the next output. The circuitry uses an independent, periodic reference signal to determine exactly when to change to a new output. This signal could be provided by an internal crystal that oscillates at a known frequency, or it could be the 60-cycles-per-second (60 Hz) alternating current that is delivered to your house by the power company. Appropriately, this independent reference signal oscillating at a known frequency is often called a *clock*. The alarm clock also includes memory to remember its alarm setting and compares this value to the current time. If it matches, it sounds the alarm.

Now let's recall the combination lock of Chapter 1. This circuit must *remember* which digit of the combination the user is going to enter next. It could be the first, second, or third. The current state of the lock, which digit it is waiting for, determines which configurations the lock could advance to next depending on whether the digit is entered correctly or wrongly. The unique configurations of the circuit are called its *states*. The combination lock has exactly five states (S_1, S_2, S_3, *ERR*, and *OPEN*), stored in special circuit structures called *memory* or *state elements*. Although the combination lock is a sequential circuit in its own right, these memory elements are the simplest, most *primitive* sequential circuits.

6.1.1 Simple Circuits with Feedback

We can implement a simple memory element from cascaded inverters. This is the basic circuit structure used in all static RAM (random-access memory) designs. Alternatively, we can build a simple memory structure from cross-coupled NOR and NAND gates. These two types of memory elements form the basic building blocks of the latch and flip-flop memory elements we will discuss in this chapter. We begin our examination of sequential logic with these primitive structures.

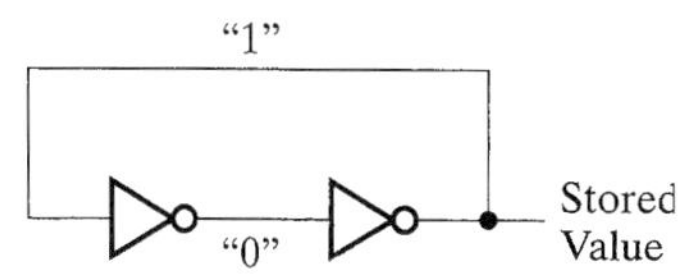

Figure 6.1 Cascaded inverters as a storage element.

Inverter Chains Consider the almost trivial circuit of Figure 6.1. It contains nothing more than two inverters in series, with the output of the second-stage inverter fed back as input to the first-stage inverter. A logic 1 at the input of the first inverter becomes a 0 at its output. The 0 is mapped to a logic 1 at the output of the second stage, which reinforces the value at the first inverter's input. A similar argument holds for a 0 at the input. Once a value is inserted at the input, it can be held indefinitely by the circuit.

Of course, the problem with this circuit is that there doesn't seem to be a way to get a value into the memory element in the first place. Some value will be there when the circuit first gets powered up, but for this circuit to be an effective building block, we must be able to choose the value to be stored. We need extra logic to set the memory element to a specific value. The feedback path must be broken while a new value is connected to the input. One way to build such a memory element is to use switches to create two alternative input paths to the first inverter. Figure 6.2 shows two switches that are used to implement a 2:1 multiplexer on the inputs to the memory element. We close only one of the two switches at a time. We assert *load* and close one switch when we want to load a new data value into our memory element, while the *remember* path is broken and has no effect. We assert *remember,* and close the other switch, when we want the element to maintain its current value, while the *load* path is broken.

Cascaded inverters can serve a purpose besides storage. They can be used to build circuits whose outputs oscillate between low and high voltages. Such circuits are called *ring oscillators.* Figure 6.3(a) shows an inverter chain, and Figure 6.3(b) gives the associated timing waveform. The timing waveform begins with node A (also labeled X) about to switch from 0 to 1.

The odd number of inverters leads to oscillatory behavior that repeats every t_p time units. This is called the *period* of the signal. *Duty cycle* is defined as the percentage of time a signal is high during its period. In the figure, the periodic signal in this circuit has a 50% duty cycle, it is low for the same amount of time it is high.

In the ring oscillator, the duration of the period depends on the number of inverters in the chain. In this case, we have five inverters

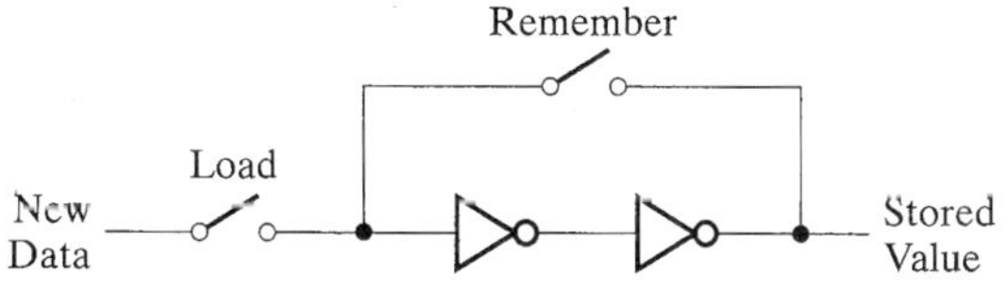

Figure 6.2 Storage element with load and refresh paths.

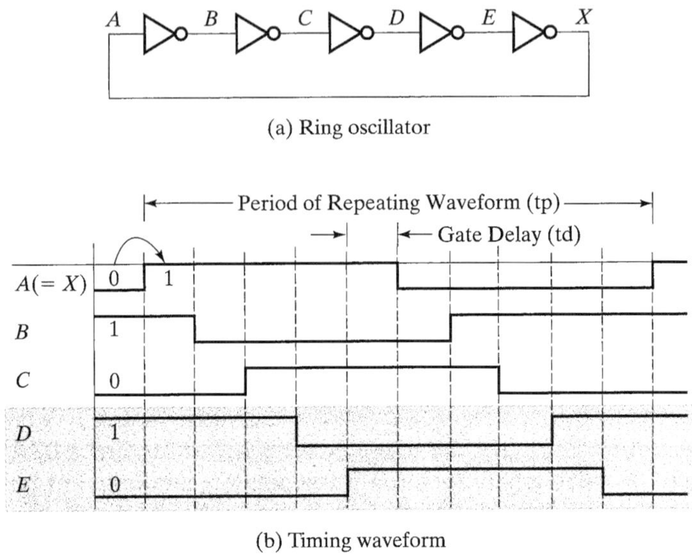

Figure 6.3 Inverter chain and timing waveform.

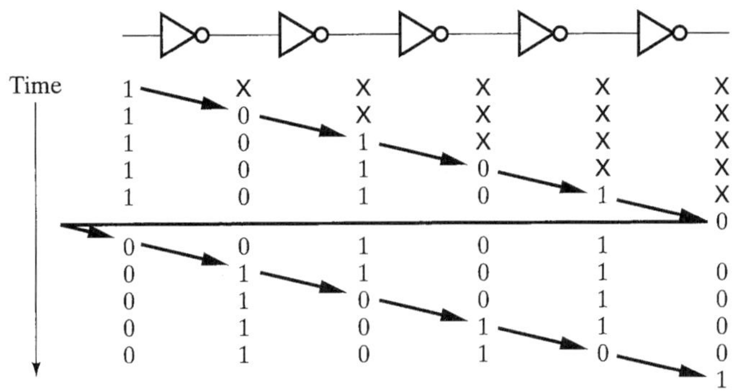

Figure 6.4 Propagation of signals through the inverter chain.

with unit delay (that is, a gate delay of 1 time unit). The high time of the waveform is five time units: five times the gate delay of a single stage. If we had a seven-inverter chain, the period would be 14 time units: seven units high, seven units low.

To see why this is the case, let's examine Figure 6.4. Each row gives the logic state associated with the nodes of the circuit. The rows differ in time by a single gate delay. Starting with a 1 at the input to the first stage (an X signifies an unknown value for all the other wires in the circuit), the signal propagates through the inverters, alternating its logic value between the stages. Once the signal emerges from the last stage, it is fed back to the first stage as the complemented value. The propagation repeats. In examining any node in the circuit, we discover that it alternates between high and low for exactly five gate delays at each value. Circuits like this are sometimes used to generate periodic clock signals.

Cross-Coupled NOR Gates as a Basic Memory Element An alternative method for building circuits with state is to use cross-coupled NOR gates (NAND gates can be used in a similar fashion). Figure 6.5 shows two alternative ways to represent cross-coupled NORs. Recall that a NOR gate with a 0 input acts like an inverter with respect to the other input: if the nonzero input is 1, the NOR's output is 0, and if it is 0, the NOR's output is 1. Similarly, if one of the NOR's inputs is 1, the output is always 0. One of the NOR gates of Figure 6.5 acts like an inverter while the other injects a 0 into the feedback loop, depending on the settings of the R and S inputs.

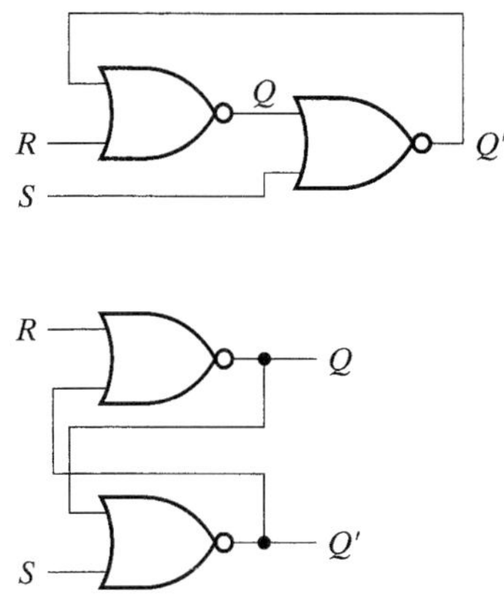

Figure 6.5 ***R-S* latches.**

Suppose $R = 1$ and $S = 0$. Since the R input is 1, the Q output is *reset* to 0 independent of the $\bar{Q}$ input to the first NOR gate. With S at 0, the Q input is inverted to form the $\bar{Q}$ output. This is why R is called the reset input, the new value of Q is 0.

Now suppose that $R = 0$ and $S = 1$. The same arguments apply as in the previous case. $\bar{Q}$ is reset to 0. When this output is fed back to the first NOR gate, it is inverted, and Q is *set* high. Hence, S is called the set input. When R and S are both 0, the NOR gates behave like chained inverters and will hold their current output values indefinitely. This configuration of NOR gates is called an *R-S latch*.

Timing Behavior of the Cross-Coupled NOR Gates The timing behavior of these gates is shown in the timing waveform of Figure 6.6. Q is set high when S is asserted and is reset low when R is asserted. Whenever R and S are both zero, the outputs remain unchanged.

What happens when both R and S are asserted? Q cannot be both 1 and 0 simultaneously! This input condition is *forbidden* in normal operating conditions. Both Q and $\bar{Q}$ are driven to 0, violating the assumption that the two outputs are always complements of each other.

When one of R or S is returned to the unasserted state, the remaining asserted signal determines the steady–state output values of Q and $\bar{Q}$. In the first case of forbidden inputs in Figure 6.6, R remains asserted while S is unasserted. Q stays at 0, while $\bar{Q}$ goes to 1. Comparable behavior will be observed if R is unasserted while S stays asserted.

Now suppose that R and S return to 0 simultaneously. This is the second case of forbidden inputs in Figure 6.6. The outputs actually

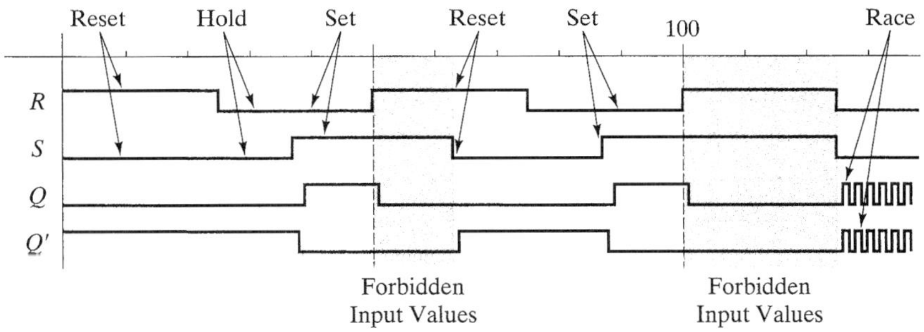

Figure 6.6 **Cross-coupled NOR gate timing waveform.**

oscillate. Q and $\bar{Q}$ are initially 0. When R and S go to 0, the NOR gate outputs go to 1. But when these are fed back to the inputs, the NOR gates behave like inverters, switching the outputs back to 0. This oscillatory behavior is called a *race condition*.

Theoretically, the race condition can continue as long as R and S are 0. However, the delay through the two NOR gates is not perfectly matched, and one of Q and $\bar{Q}$ will be driven to a new value before the other, stopping the oscillations (hence, the term "race condition"—which will win?). To avoid the race condition, we restrict the R and S inputs never to be 1 at the same time.

S	R	Q
0	0	Hold
0	1	0
1	0	1
1	1	Unstable

Figure 6.7 Functional truth table of cross-coupled NOR gates.

We summarize the behavior of the cross-coupled NOR gates in the functional truth table of Figure 6.7. When $R = S = 0$, the circuit holds its current output—in other words, the output is the same as it was for the last setting of the inputs. When one of R or S is set to 1, the output Q is forced to 0 or 1, respectively. When both inputs are 1, the outputs are both 0 and may oscillate between 0 and 1.

Another representation of the behavior of cross-coupled gates is called the *state diagram*. The diagram consists of nodes and arcs. The nodes represent unique configurations or states of the circuit. The arcs are labeled with the input combinations that cause a transition from one state to another.

Figure 6.8 shows the state diagram for the cross-coupled NOR gates. The circuit's state depends on the values of the outputs Q and $\bar{Q}$, so there are four possible states, one for each combination of possible values for the two outputs. Since there are two inputs, S and R, there are exactly four transitions from each state: one for each of the possible input combinations.

The states labeled 01 and 10 are the normal configurations of the circuit. When set (S) is asserted, we enter state 10 ($Q = 1$, $\bar{Q} = 0$). When reset (R) is asserted, we change to state 01 ($Q = 0$, $\bar{Q} = 1$). When S and R are both 0, we hold in the current state.

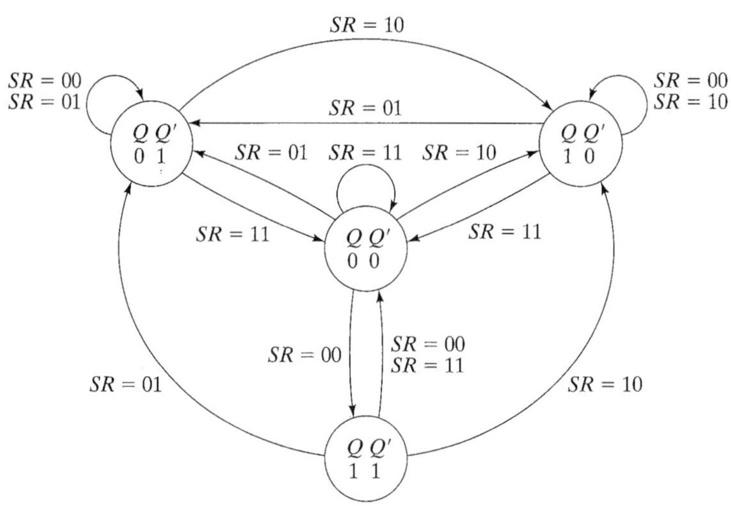

Figure 6.8 Theoretical state diagram of cross-coupled NOR gates.

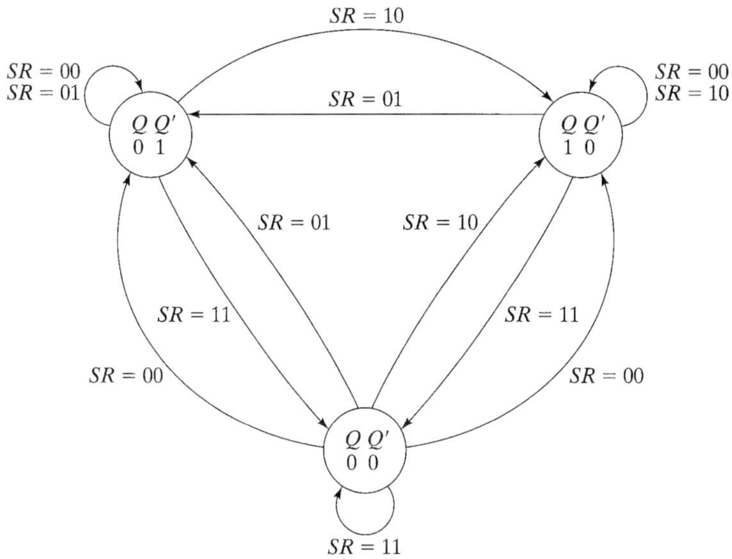

Figure 6.9 Observed behavior of cross-coupled NOR gates.

When we encounter the forbidden input configuration $S = R = 1$, the circuit enters state 00. It stays in this state as long as these inputs are asserted. As soon as one input returns to 0, the circuit returns to state 01 or 10.

If the current state is 00 and $S = R = 0$, the circuit enters state 11. It does not stay there very long, immediately returning to state 00 if S and R remain 0. If the delays through the gates are perfectly matched, the circuit can oscillate between these two states forever. Of course, this does not make a very useful memory element.

Figure 6.9 gives a state diagram we constructed by observing the behavior of a real cross-coupled NOR gate circuit. The actual circuit does not oscillate between 00 and 11 indefinitely, but rather ends up sometimes in state 10 and sometimes in state 01. This is the true meaning of a race condition in sequential logic: the resulting state depends on the circuit's time-dependent behavior and cannot be predicted in advance. The race condition is most easily avoided by never putting the circuit into state 00 in the first place and avoiding this *nondeterministic* situation. This condition is represented in the state diagram by having two arcs leaving the 00 state with the same input condition, namely, S and R both low.

6.1.2 Basic Latches

The R-S latch can hold its current state, reset the state to 0, or set the state to 1. We can derive a truth table for the R-S latch by cutting the feedback loop. We'll label the input with the time at which it changes (t) and the output with the time at which it changes ($t + \Delta$), where Δ is the propagation time of the circuit. Of course, since this is a feedback loop, the output will become the input and time will advance. The cross-coupled NOR

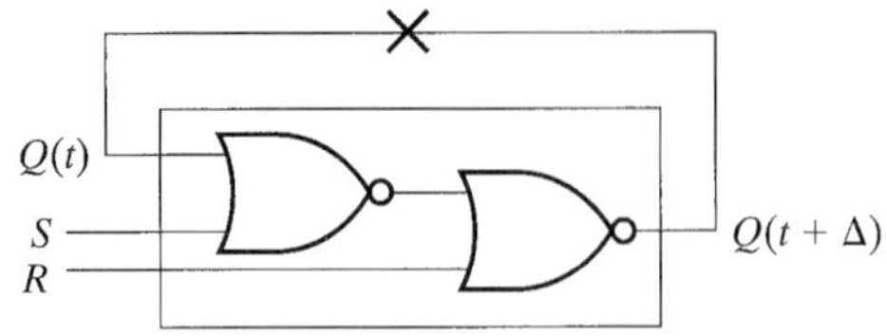

Figure 6.10 ***R-S* latch with cut feedback loop.**

$S(t)$	$R(t)$	$Q(t)$	$Q(t+\Delta)$	
0	0	0	0	Hold
0	0	1	1	
0	1	0	0	Reset
0	1	1	0	
1	0	0	1	Set
1	0	1	1	
1	1	0	X	Not Allowed
1	1	1	X	

Figure 6.11 ***R-S* latch truth table.**

			S	S
	0	0	X	1
$Q(t)$	1	0	X	1
		R	R	

Figure 6.12 **K-map for $Q\ (t + \Delta)$.**

gates with their feedback loop cut are shown in Figure 6.10. The detailed truth table for the *R-S* latch is shown in Figure 6.11. Note that the inputs are S, R, and Q at a given time t, and the output is Q at time $t+\Delta$. $Q(t)$ is the *current state* of the latch, and $Q(t+\Delta)$ is its *next state*. For example, if $S = R = 0$ (hold) then $Q(t+\Delta) = Q(t)$, the latch's state does not change. Note that $S = R = 1$ is not allowed, so we use those two rows as don't cares.

A *K*-map for the truth table is given in Figure 6.12. We can derive the following *characteristic equation* to describe the next state in terms of the inputs and current state:

$$Q(t+\Delta) = S(t) + \overline{R(t)} \bullet Q(t)$$

$$Q^{+} = S + \bar{R}Q$$

This equation is a convenient shorthand for describing the memory element's behavior. For example, if $S = 1$ and $R = 0$, the next state Q^{+} becomes 1 independent of the current state. When $S = 0$ and $R = 1$, the next state is forced to be 0, independent of the current state. When $S = R = 0$, $Q^{+} = Q$ and the state does not change.

Gated* R-S *Latch An interesting extension of the basic *R-S* latch is the *gated R-S* latch in Figure 6.13. It adds two NOR gates to the R and S inputs of Figure 6.5. With this device it is now possible to force the R and S inputs to both be 0. This is accomplished through another input signal, called $\overline{enable}$. When the *enable* signal is asserted ($\overline{enable}$ is 0), the latch is allowed to change state as the new $\bar{R}$ and $\bar{S}$ inputs are inverted and propagated to the original R and S inputs of the basic latch. When enable is not asserted ($\overline{enable}$ is 1), the latch holds its current state because the R and S inputs are both forced to 0. Thus, regardless of the values of $\bar{R}$ and $\bar{S}$, the latch does not change its value while enable is deasserted.

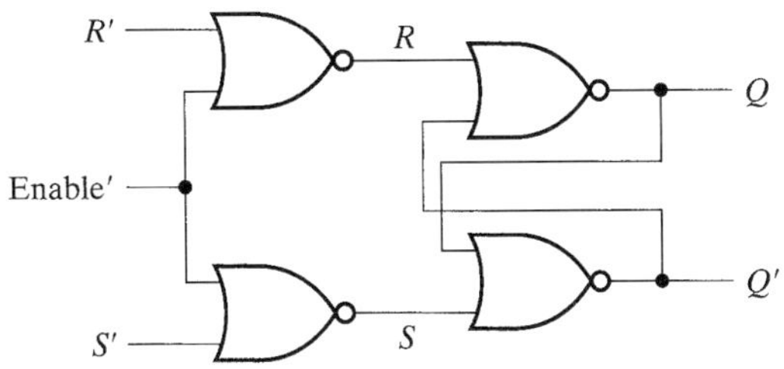

Figure 6.13 **Gated or level-sensitive *R-S* latch with enable input.**

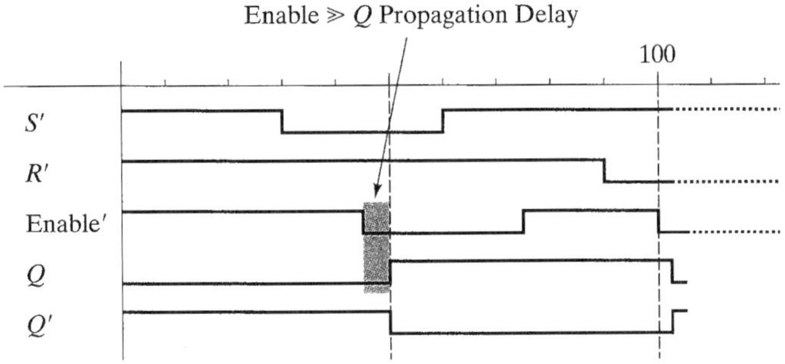

Figure 6.14 **Timing diagram for the level-sensitive *R-S* latch.**

This is a useful circuit in that it allows us to control when the latch can change its stored value. The timing behavior of the gated, or *level-sensitive, R-S* latch is shown in Figure 6.14. The propagation delay from when enable is asserted to when the output changes (*enable* ≫ *Q*) is shown in the diagram.

6.1.3 Clocks

With the control we got from gated latches, we can now set up a collection of latches to all change their value under the control of a single enable signal. We call these systems, whose sequential elements change their values together, *synchronous* and rename the enable signal as a *clock*. A clock signal is a periodic signal that is used to keep time in synchronous sequential systems. Its *ticks* determine when the stored values can change. These changes can only happen at clock ticks so that we can be assured the circuit's state is unchanging between clock ticks.

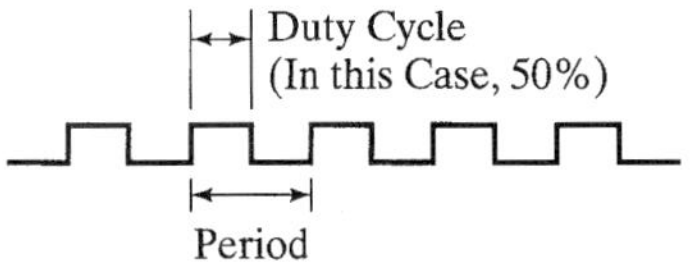

Figure 6.15 **Periodic clock signal with 50% duty-cycle.**

A clock signal is shown in Figure 6.15. It has two parameters: its *period* and its *duty-cycle*. The clock's period is the time between iterations, for example, between one rising edge (when the clock goes from low to high) and the next. The duty-cycle is the percentage of the time during that period that the clock signal is asserted. In the case of Figure 6.15, the duty-cycle is 50%.

How do we decide on the period and duty-cycle of a clock signal? We can begin to answer this question if we consider the clock signal as

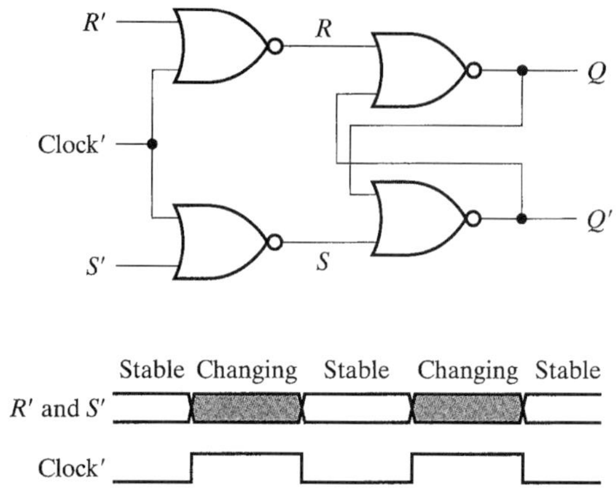

Figure 6.16 Level-sensitive latch controlled by a clock. The timing waveforms show when the $\overline{R}$ and $\overline{S}$ inputs can change and when they should be stable.

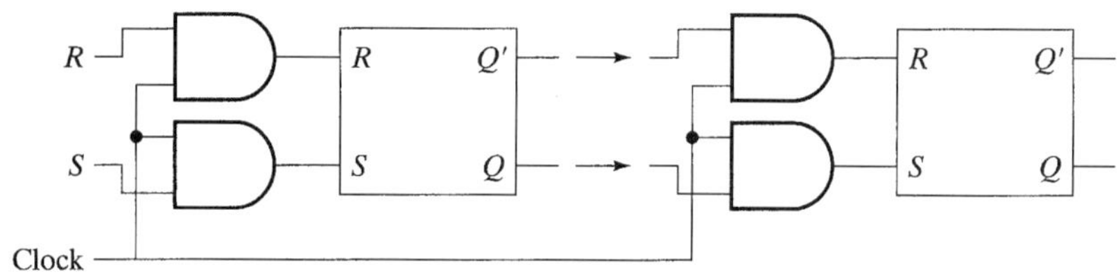

Figure 6.17 Cascading level-sensitive latches controlled by the same clock signal.

the enable input of our level-sensitive latch as shown in Figure 6.16. When the latch is enabled (*clock* is 1, $\overline{clock}$ is 0), the inputs will be able to affect the state of the latch. During this time interval, the inputs should be stable and unchanging so that the proper value is set in the latch. When the clock is 0 ($\overline{clock}$ is 1), the inputs can do as they please and can use this time to change to the values they will need to be set to before the clock will tick again (*clock* is 1). Therefore, the period of the clock determines the amount of time we have to compute the next values of the inputs. The inputs must be stable while they are affecting the latch, this percentage of the period is the duty-cycle. Clearly, we want to make the duty-cycle small so that we can have more time to compute the next values of the inputs and less time during which we have to worry about the stability of the inputs. This leads to what is referred to as *narrow-width clocking:* namely, keeping the clock high for only a narrow pulse.

6.1.4 Combining Latches

Our next concern is combining latches so that we can transfer a stored value from one latch to another. Consider the two latch combination shown in Figure 6.17. Note that when the clock is high, both level-sensitive latches (slightly redrawn for clarity to show the basic latch core and two AND gates at the inputs) will be enabled. That means that the inputs will affect the value stored in the latch, the value at the Q and $\bar{Q}$ outputs. What will stop the change in one latch's outputs affecting the next latch? How do we control the propagation of these changes?

One approach is to carefully make the delay of the intervening wires and logic (indicated by the arrows between the latches) be greater than the duty-cycle of the clock. By doing so we delay the change in the first latch's output so that it doesn't reach the second latch until its too late, that is, the clock has been deasserted. As you can imagine, this can lead to very delicate logic design. All paths between latches throughout our system must be carefully measured to be greater than the duty-cycle of the clock. This also puts pressure to make the duty-cycle as small as possible. However, the duty-cycle can't be made infinitesimally small because latches have to be enabled for enough time for the inputs to have the desired affect on the latch's state.

A second approach is to use different polarities of the clock on the two latches so that while one is enabled the other is not. This can be accomplished by inverting the clock to one of the latches as show in Figure 6.18. This is still problematic, however. We now have two flavors of latches: those enabled when the clock is high and those enabled when the clock is low. Again, this can become rapidly unmanageable as we build larger circuits because we now have to make sure to only connect latches with opposite polarities of the clock.

A more robust and regular way to deal with this problem that leads to a cleaner abstraction for a basic sequential logic element is to double up on the number of latches. Instead of a single latch for every value we store, we will use two as shown in Figure 6.19. They will be enabled on different polarities of the clock. The first latch, the *master,* will be set to the

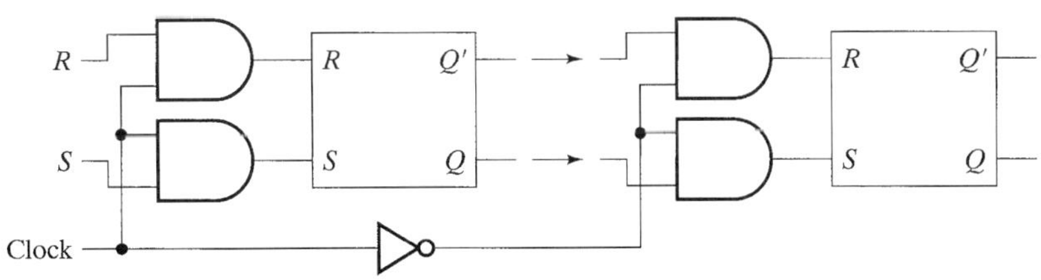

Figure 6.18 Cascading level-sensitive latches controlled by different polarities of the same clock signal.

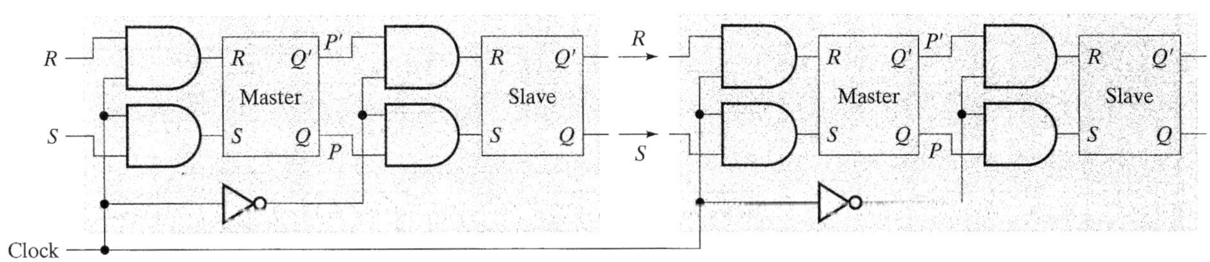

Figure 6.19 Master–slave latches controlled by the same clock signal.

value desired and this will be transferred to the second latch, the *slave*, when the clock changes polarity. This is referred to as an *edge-triggered* arrangement because the change in output occurs soon after the clock goes from high to low, the falling edge of the clock, making this a *negative edge-triggered master–slave latch*. We can also think of the clock edge as *sampling* the inputs and *applying* them to update the latch value.

6.1.5 Master–Slave Latches and Edge-Triggered Flip-Flops

The master–slave arrangement has some big advantages for its extra cost. Value changes are confined to a single master–slave latch and can't propagate to a second master–slave latch until the next clock period. We are no longer concerned with the delay of logic circuits having to be greater than the clock's duty-cycle. However, we now have some new trade-offs for the shape of the clock.

As outputs change on the falling edge of the clock, intervening logic between latches must compute its new values, the inputs to other latches, before the clock goes low again. The period of the clock must be greater than the slowest path through this logic. If we can improve the worst-case path delay, we can speed up our clock while being assured that our entire circuit will still function correctly.

However, we have one problem remaining. We still have to ensure that the inputs to each latch are stable during the time when the master stage is enabled. One would think that as long as the right values are there before the clock goes low, we should be ok, but it is not that simple. Consider the case when we want a latch to hold its current value. The inputs would both be 0. However, if there is a static-0 hazard (a momentary glitch to 1) in the logic for one of the signals, then we may erroneously set or reset the latch. This is referred to as the *ones catching problem*. Figure 6.20 illustrates this problem. Note that in the third clock cycle there is a glitch on S that ends up setting the latch output even though S went back to 0 before the falling edge of the clock. If the latches were constructed from NAND gates instead of NOR gates (that is, $\bar{R}$-$\bar{S}$ latches), there is the dual problem called *zeros catching*. If the clock is asserted and the $\bar{R}$ or $\bar{S}$ input temporarily changes through zero, the associated set or reset operation will take place.

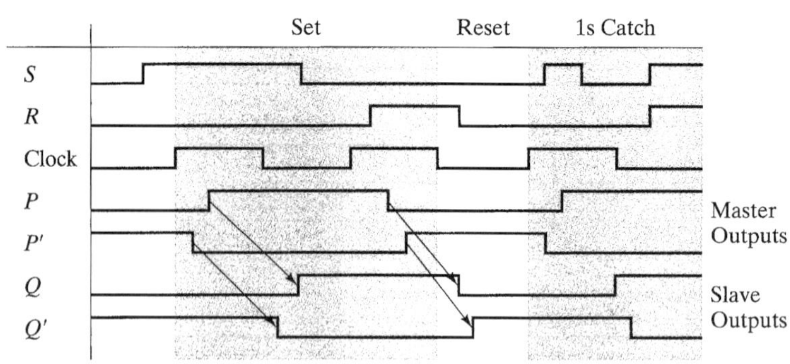

Figure 6.20 **Illustration of 1s catching problem in *R-S* master–slave latch.**

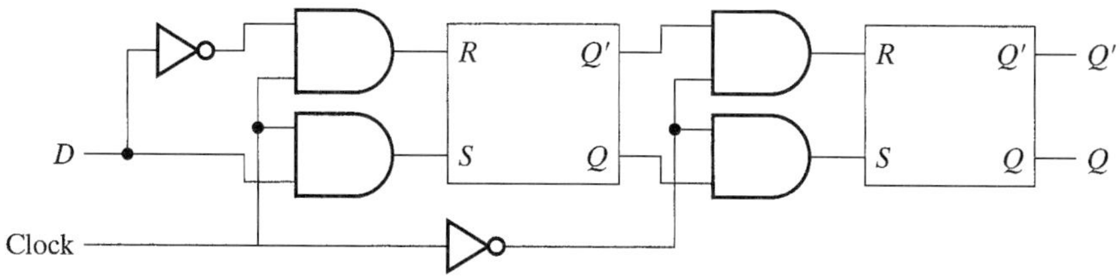

Figure 6.21 **Master–slave negative edge-triggered *D*-type flip-flop.**

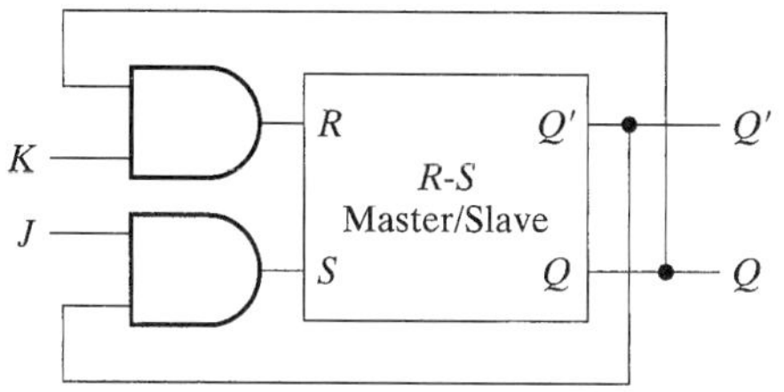

Figure 6.22 **Master–slave negative edge-triggered *J-K*-type flip-flop.**

To eliminate this problem, designers have taken to using an inverter to always make the *R* and *S* inputs be opposites as shown in Figure 6.21. We now have only a single input *D* and the latch has lost the ability to hold a previous value. Every clock period the latch will now be either set or reset. This means that if we want a value to stay the same, we must include logic that brings it to the *D* input. The letter *D* is an abbreviation for *data*. Whatever data value is at the *D* input, that is the value that will be stored in the latch at the next falling clock edge. The only constraint on the logic is that it get the value there in time. There is no longer any need to worry about hazards as a momentary glitch will be cancelled out as the value of *D* settles to what it should be. The characteristic equation for the D flip-flop is very simple (we call it a flip-flop because it is no longer sensitive to glitches on its inputs as in the case of a latch):

$$Q^{+} = D$$

After this long evolution from our simple *R-S latch,* we have now derived the *D negative edge-triggered flip-flop*. There are many other kinds of flip-flops. One that has been traditionally popular is the *J-K* flip-flop which uses a bit more logic than the D flip-flop. The circuit is shown in Figure 6.22. The *S* input is $J \cdot \bar{Q}$ and the *R* input is $K \cdot Q$. By feeding the *Q* and $\bar{Q}$ outputs back to the inputs and gating these with external set and reset control inputs (now called *J* and *K*, respectively), we guarantee that the internal *R* and *S* are never simultaneously 1 (this assumes that *Q* and $\bar{Q}$ are never both 1). The truth table for the *J-K* flip-flop is shown in Figure 6.23. Note that the previously prohibited input combination 1-1 is now used to *toggle* the stored value. For example, if *Q* is 1, $\bar{Q}$ is 0, and *J* and *K* are both 1,

$J(t)$	$K(t)$	$Q(t)$	$Q(t+\Delta)$	
0	0	0	0	Hold
0	0	1	1	
0	1	0	0	Reset
0	1	1	0	
1	0	0	1	Set
1	0	1	1	
1	1	0	1	Toggle
1	1	1	0	

Figure 6.23 ***J-K* truth-table.**

then the inputs presented to the internal latch are $R = 1$, $S = 0$. This flips Q to 0 and $\bar{Q}$ to 1. If Q starts out as 0, then $\bar{Q}$ is asserted, as is the internal S signal. Once again, this will cause Q to toggle, this time to 1. All input combinations lead to useful functions for the *J-K* flip-flop: hold, reset, set, and toggle. The characteristic equation for the *J-K* flip-flop is:

$$Q^+ = S + \bar{R}Q = J\bar{Q} + \overline{(KQ)}Q = J\bar{Q} + (\bar{K} + \bar{Q}) = J\bar{Q} + \bar{K}Q$$

Notice how nicely the characteristic equation summarizes the behavior of the *J-K* flip-flop. When $J = 1$, $K = 0$, $Q^+ = Q + \bar{Q}$, which is always 1 (set). Q^+ is always 0 when $J = 0$, $K = 1$ (reset). Finally, when $J = 1$, $K = 1$, $Q^+ = \bar{Q}$ (toggle). See Appendix C for more detail of *J-K* and other flip-flop types.

One last optimization will complete our evolutionary story. Master–slave *D* flip-flops use eight 2-input gates and two inverters. A more efficient form that uses only six gates (all 2-input except for a single 3-input gate) has become the norm. Let's look at the circuit schematic for the optimized negative edge-triggered *D* flip-flop in Figure 6.24 (this can be changed to a positive edge-triggered device by using the clock's complement). This flip-flop stores whatever value is on the *D* input when the clock signal performs the appropriate (in this case, negative) transition. Let's see how it works.

Operation of the D Flip-Flop The operation of this optimized edge-triggered device is considerably more complex than that of the master–slave flip-flop. The circuit of Figure 6.24 contains three communicating latches. The bottom latch samples the *D* input while the top stage

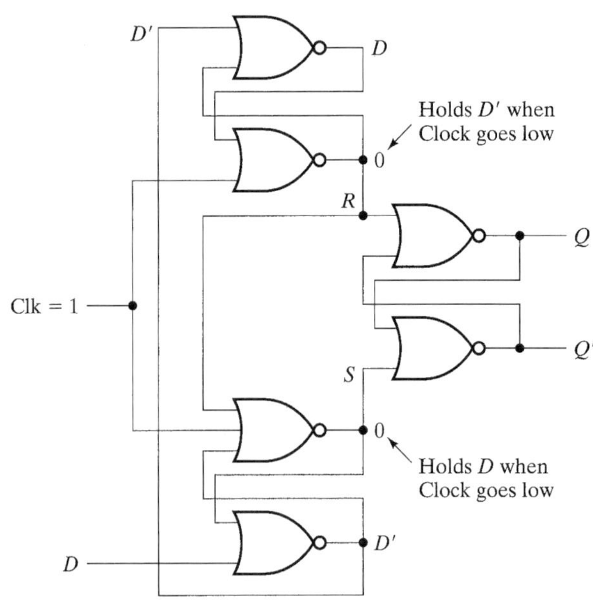

Figure 6.24 **Negative edge-triggered *D* flip-flop when clock is high.**

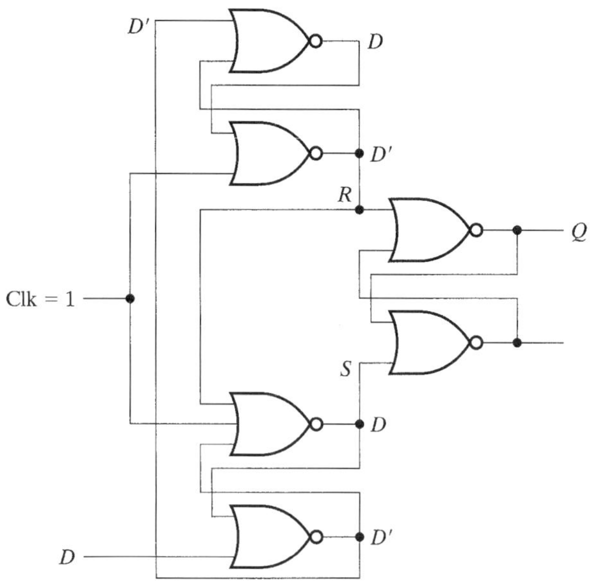

Figure 6.25 **Negative edge-triggered *D* flip-flop when clock goes from high to low.**

holds $\bar{D}$. The output from the bottom latch drives the set input of the final stage latch, while the top latch provides the reset input.

The figure shows the state of the circuit when the clock is high. The clock forces the outputs of the top and bottom latches to zero, thus keeping the final stage *R-S* latch in its holding state. Any change in the *D* input will be sampled by the top and bottom latches, but these changes are inhibited from affecting the final stage latch.

Figure 6.25 shows what happens when the clock initially changes from high to low. The output NOR gates in the top and bottom latches now act like inverters. The previously sampled value of *D* is presented to the *S* input, while $\bar{D}$ drives the *R* input. If $D = 1$, the *R-S* latch now stores a 1. Otherwise it stores a 0.

What happens when the clock is low, but *D* changes? Figure 6.26 shows the state when the input changes to D_{new}, which is different from the previously sampled value *D*. This new input forces the output of the NOR gate to change from 1 to 0, since one input must be a 1 and the other a 0. By driving this circuit node to 0, gates 2, 4, and 5 are guaranteed to hold their previous values. D_{new} can affect the circuit only when the outputs of gates 2 and 4 are forced to 0. This will happen only when the clock next goes high. Therefore, a change in *D* to D_{new} while the clock is low has no effect.

6.1.6 Timing Definitions

We are now ready to introduce some definitions about the timing waveforms of sequential circuits. The output of a sequential circuit is a function of the current inputs and any signals that are fed back to the

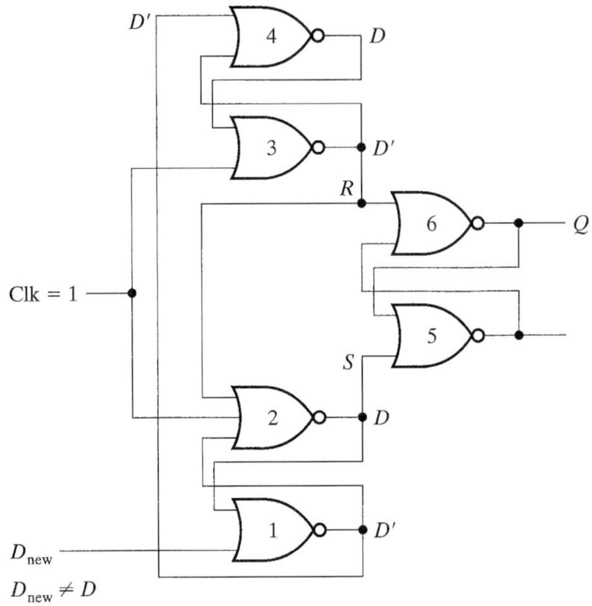

Figure 6.26 **Negative edge-triggered *D* flip-flop when clock is low.**

inputs. We call these feedback signals the *current state* of the circuit. A periodic external event, called a *clock,* determines when the circuit can change the current state to a new state. When the *clocking event* occurs, the circuit *samples* this new value and stores it in its flip-flops. The combinational logic can then compute the next state value from the circuit's current inputs and current state.

As an example, consider the digital alarm clock again. Part of its state is the current time of day. Other parts of the state include the alarm's configuration as to whether or not it is set and the time the alarm should go off. Suppose that our oscillating reference signal, the clock signal, becomes asserted once every second. When the clock is asserted, the digital alarm advances by 1 second and checks whether it matches the alarm time. If it does and the alarm is set, then the sequential circuit sounds the alarm. The digital alarm's new state now indicates that the alarm is on.

We can designate the clocking event as either the low-to-high or high-to-low transitions of the clock. It is important that the inputs determining the new state remain unchanged around the clocking event. We define a special window of time by two constraints: the *setup time* before the clocking event and the *hold time* after the event. If the inputs do not change within this window, the state will be updated in a correct and unambiguous way. If they do change, the effect on the state is undefined.

Let's be more precise. The setup time, T_{su}, is the minimum time interval *before* the clocking event during which the input must be stable to be validly recognized. The hold time, T_h, is the minimum time interval *after* the edge of the clocking event during which the input signal must be stable to be validly recognized.

These concepts are shown in Figure 6.27 for a rising clock edge event. The input must remain stable at least a setup time before the reference clock edge and at least a hold time after the edge for the input signal to be recognized as a logic 1. If these constraints are not satisfied, the input may be interpreted as a 1 or a 0 or some unrecognizable value between 0 and 1. It is extremely dangerous to allow input signals to change very close to the sampling event, as we shall see in Section 6.4.

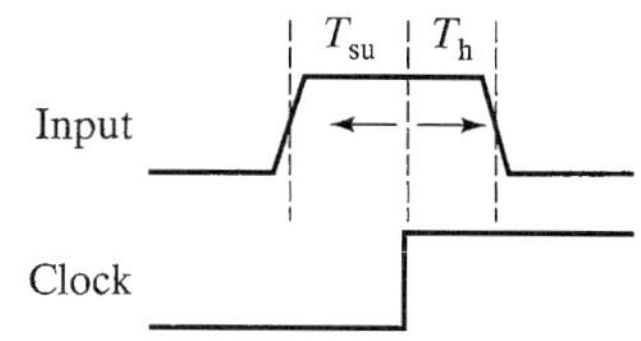

Figure 6.27 Setup and hold times.

Latches and Flip-Flops Think of the sampling event as an enable signal that instructs the memory element to examine its input and change its internal state to the same value. We can use a single clock to synchronize the update of many memory elements in our system.

Primitive memory elements actually fall into two broad classes: *latches* and *flip-flops*. When the memory element's outputs immediately change in response to input changes, they are called *transparent* outputs. An example of a memory elment with transparenet outputs is the basic *R-S* latch of Figure 6.5.

If a latch has only data inputs, like *R* and *S*, it is called an *unclocked* latch. We have seen *level-sensitive* latches that have an additional enable input, sometimes called the *clock* as in Figure 6.16.

Level-sensitive latches continuously sample their inputs while they are enabled. Any change in the level of the input is propagated through to the output. When the enable signal is unasserted, the last value of the inputs determines the state value to be held by the latch. The *latched value* is determined by the window formed from the setup and hold times around the falling (high-to-low) edge of the enable signal.

Flip-flops differ from latches in that their outputs change only with respect to the clock, whereas latches may change their output when their inputs change. We can characterize flip-flops on the basis of the clock transition that causes the output change: there are *positive edge-triggered* and *negative edge-triggered* flip-flops.

A positive edge-triggered flip-flop samples its inputs on the low-to-high clock transition. To be properly recognized, the input must be stable within the setup and hold time window around the clock edge as in Figure 6.27. The outputs change within a propagation delay after the triggering clock transition.

A negative edge-triggered device works in a similar fashion, with the input sampled on the high-to-low clock transition. The outputs change within a propagation delay after the falling edge of the clock. Under this classification, a master–slave flip-flop is indistinguishable from the negative edge-triggered device, except that it can exhibit the strange behavior we called *ones catching*. Table 6.1 summarizes the different attributes of latches and flip-flops.

TABLE 6.1 Input/Output Behavior of Latches and Flip-Flops

Type	*When Inputs are Sampled*	*When Outputs are Valid*
Unclocked latch	Always	Propagation delay from input change
Level-sensitive latch	Clock high (T_{su}, T_h around falling clock edge)	Propagation delay from input change
Positive-edge flip-flop	Clock low-to-high transition (T_{su}, T_h around rising clock edge)	Propagation delay from rising edge of clock
Negative-edge flip-flop	Clock high-to-low transition (T_{su}, T_h around falling clock edge)	Propagation delay from falling edge of clock
Master–slave flip-flop	Clock high-to-low transition (T_{su}, T_h around falling clock edge)	Propagation delay from falling edge of clock

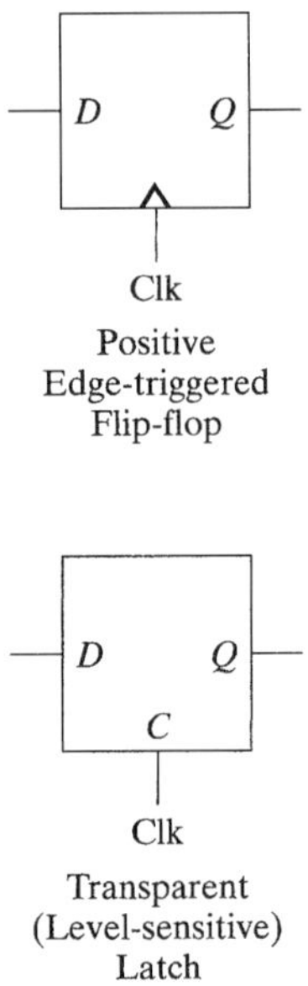

Figure 6.28 Flip-flop and latch symbols.

Timing Examples To better understand the terms just introduced, let's consider two memory elements: a positive edge-triggered flip-flop (such as the Texas Instruments low-voltage CMOS SN74ALVCH374) and a level-sensitive latch (such as the Texas Instruments low-voltage CMOS SN74ALVCH373). Each of these parts has eight elements where each has a single data input, *D*, a shared clock input, *C*, and an output, *Q*. The block diagrams of one of each of the corresponding elements are shown in Figure 6.28. We normally represent edge-triggered devices with the clock input as a triangle. A negative edge-triggered device has an additional negative logic bubble at the clock input.

We compare the timing behavior of the two types of devices in Figure 6.29. The outputs differ only when the input changes while the clock is asserted. This is because the latch immediately responds to changes in the input while the clock is high but the flip-flop does not.

The definitions of the setup and hold times are quite different for these two devices. Figure 6.30 gives the timing waveforms for the 374 positive edge-triggered flip-flop. T_{su} is 1.8 ns and T_h is 0.5 ns. In addition, the clock signal has a minimum duration, $T_w = 3.3$ ns. The figure also defines T_{pd}, the propagation delay between the rising edge of the clock and the change in the output. In this particular technology, T_{pd} is between 1.1 and 3.6 ns (it varies depenting on the particular chip, operating temperature, humidity, etc.) and is the same whether the output is rising or falling. This is typical for a CMOS technology. For

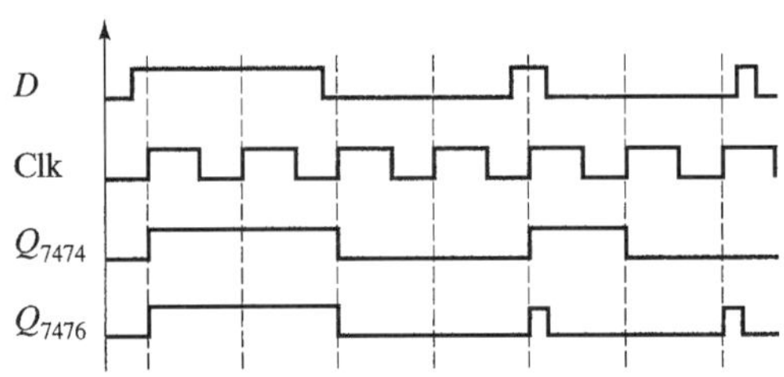

Figure 6.29 Waveform behavior of a positive edge-triggered flip-flop versus a level-sensitive latch.

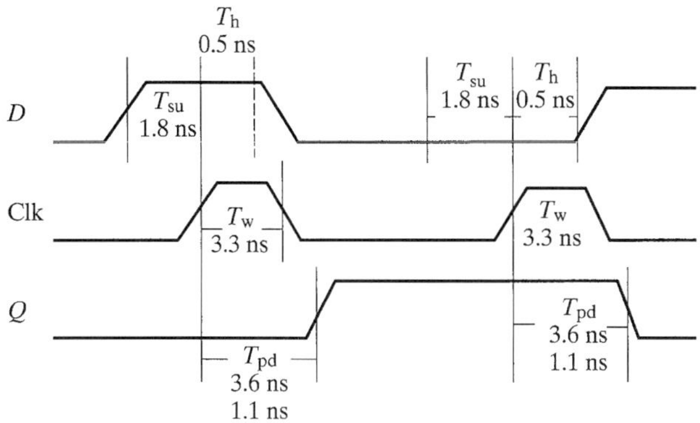

Figure 6.30 Timing constraints and specifications for the 374 positive edge-triggered flip-flop.

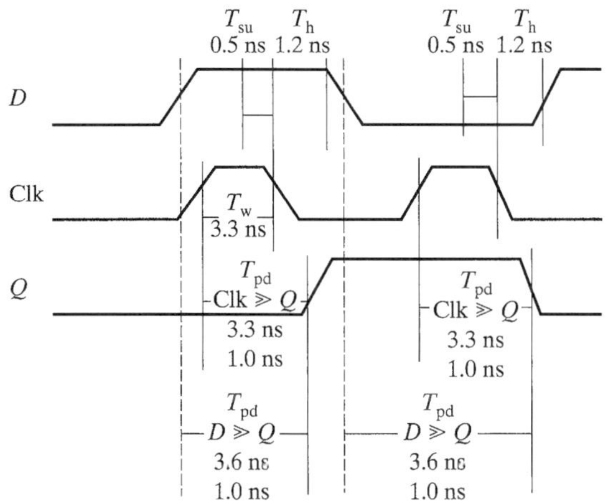

Figure 6.31 Timing constraints and specifications for the 373 clocked transparent latch.

bipolar technologies we would have two value for T_{pd}, T_{phl} and T_{plh}, the propagation delay from the rising clock edge to a high-to-low and a low-to-high change in the output, respectively, and these could differ by as much as a factor of 2.

Figure 6.31 displays the timing specifications for the 373 clocked transparent latch. The quantities T_{su}, T_h, and T_w are slightly different and are defined to be 0.5 ns, 1.2 ns, and 3.3 ns, respectively. However, the setup and hold times are defined relative to the falling edge. The specification for T_{pd} is more complex, because it must be defined for two different events: from when the input changes to the corresponding change in the output ($D \gg Q$) and from when the clock becomes asserted to when the output changes ($Clk \gg Q$). Of course, the clock must be asserted before the input can influence the output. T_{pd} ($D \gg Q$) is 1.0 ns minimum to 3.6 ns maximum, T_{pd} ($C \gg Q$) is similar with 1.0 to 3.3 ns delay.

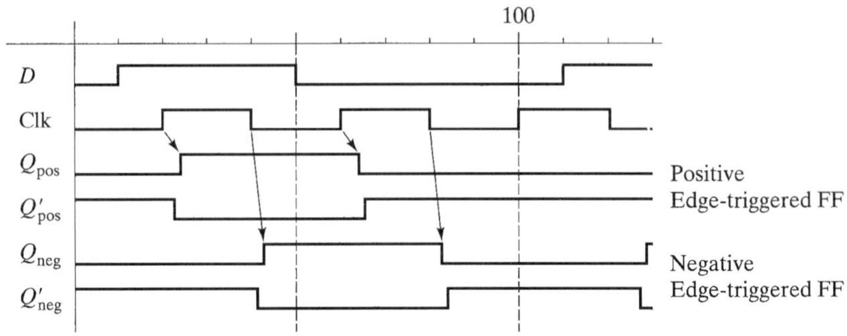

Figure 6.32 **Timing diagram for positive and negative edge-triggered *D* flip-flops.**

An examination of the edge-triggered *D* flip-flop of Figure 6.24 should make you appreciate the need for setup and hold time specifications for flip-flops. If the *D* input changes too close to the triggering clock edge, its value may not be held correctly by the top and bottom latches. Also, you can see why the propagation delays may be rather substantial for these kinds of devices as signals need to propagate through several gates before stabilizing at the final values.

We give a timing diagram for both positive and negative edge-triggered devices in Figure 6.32. The outputs change after a small propagation delay from the rising or falling edge. The last transition (time step 110) shows how edge-triggered devices sample their inputs. In this case, the input changes from a 0 to a 1 midway through the clock period. The change goes undetected by the positive edge-triggered device—no ones catching here—but is recognized by the negative edge-triggered flip-flop. Of course, if *D* stays high it will affect the positive edge-triggered flip-flop at its next clocking event.

As we'll see in Chapter 9, logic design technologies include a wide assortment of flip-flop types. The most universal are the positive edge-triggered *D* flip-flop and the transparent latch. Traditionally, *J-K* flip-flops and *T* (or toggle) flip-flops were also popular (see Appendix C for details). Negative edge-triggered versions of flip-flops are sometimes also available.

Flip-flops include a variety of special features including additional *preset* and/or *clear* inputs. These are analogous to *set* and *reset* but we use differnet names so that they are not confused with the inputs of the *R-S* latch. *Preset* and *clear* can be synchronous, that is, their effect is not seen until the clocking event, or they can be asynchronous and have immediate effect. Of course, with three inputs: *D, preset,* and *clear*, there has to be a precedence established. Typically, preset and clear take precedence over *D*. Flip-flops may be *set-dominant* (if preset takes precedence over clear) or *reset-dominant.*

6.2 Timing Methodologies

Timing in combinational circuits was relatively straightforward. Propagation delays through logic gates are added to provide an overall delay for the logic. In sequential logic, on the other hand, we must examine not only the inputs but also their changes relative to clocking events of the flip-flops. Transparent latches are more difficult to analyze than edge-triggered flip-flops because input changes may propagate to the output while the clock is asserted rather than just at a clocking event.

In the synchronous systems we will construct in the remainder of this text, we will use edge-triggered flip-flops exclusively because they are easier to deal with. For a synchronous system comprising edge-triggered flip-flops to function properly, the inputs to the flip-flops must not change around the clocking event. This window is defined by the setup and hold times, during which the inputs must be stable.

In this section, we will describe the timing methodologies associated with proper synchronous system design. A *timing methodology* is nothing more than a set of rules for interconnecting components and clock signals that, when followed, guarantee proper operation of the resulting system.

6.2.1 Cascaded Flip-Flops and Setup/Hold/Propagation

Timing methodologies guarantee "proper operation," but just what does this mean? For synchronous systems, we define *proper operation* as follows. For each clocking event, all flip-flops controlled by the same clock signal simultaneously examine their inputs and determine their new states. This means that: (1) the correct input values are provided to the flip-flops and are stable around the clocking event, and (2) no flip-flop should change its state (and thus its output) more than once for each clocking event.

What rules should we follow for composing synchronous systems to guarantee these two properties? Figure 6.33 illustrates the problem. Here, we cascade two D flip-flops so that the output from the first stage feeds the input to the second stage. Both flip-flops are controlled by the same clock signal. The purpose of this circuit is to transfer the current state of the first stage to the second stage while the first stage receives a new value. In other words, the second stage contains the value stored in the first stage during the previous clock period. This is an example of a *shift register*, a multi-bit memory with a capability of transferring a single memory element's contents to its neighbor. We will see much more of shift registers later in this chapter.

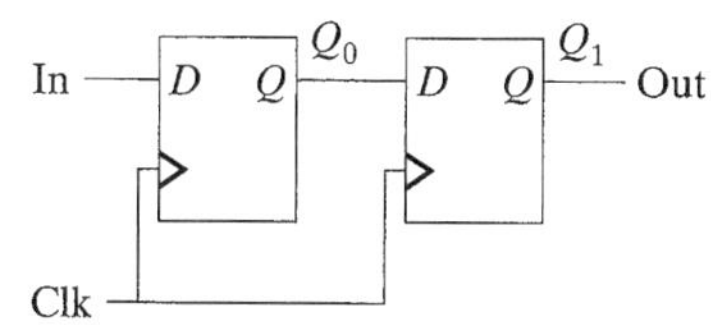

Figure 6.33 Cascaded D flip-flops to implement a two-stage shift register.

The proper logic operation of the circuit is shown in Figure 6.34, assuming that the flip-flops are positive edge-triggered and that the input stream is 0101011 where each value has a duration of one clock period. Initially, both flip-flops contain unknown values (the shaded regions early in the timing diagram). On the first rising edge of the clock,

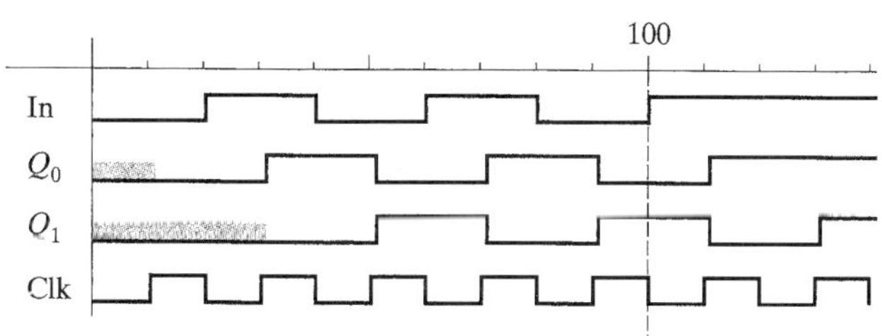

Figure 6.34 Timing diagram for the two-stage shift register.

the input is 0 and appears at output Q_0 a short propagation delay after the clocking event. The state of the second flip-flop is still unknown.

On the second rising clock edge, the input is 1, and Q_0 takes on this value some propagation delay after the clock edge. However, at the clock edge, the second stage sees the old value of Q_0 as its input. Thus, Q_1 becomes zero shortly after the rising edge. The pattern continues through the diagram. Q_1 always displays the value that Q_0 had just before the rising clock edge.

We would not observe proper transfer of data between the stages if the first was positive edge-triggered while the second stage was negative edge-triggered. After the first clock cycle, we would have 0 in both flip-flops. After the second cycle, both flip-flops would hold a 1, and so on. It is not good design practice to mix flip-flops that are sensitive to different timing events within the same circuit!

Proper Cascading of Flip-Flops In general, we assume that the propagation of the clock signal is infinitely fast and that all flip-flops have identical timing. In real circuits, this isn't true. Some components may be faster than others, and the wire delay for distributing a signal to all points where it is needed may vary substantially. Let's suppose that the first-stage flip-flop has a very fast propagation delay, so fast in fact that the new value of Q_0 appears at the input to the second stage before it had a chance to observe the previous value of Q_0. We would not be properly passing the value from stage to stage in this case.

The same problem arises if the connection between the clocks of the two flip-flops is a long meandering wire, while the output of stage one and the input to stage two are connected by a very short wire. It is only after the first stage has changed its value that the second stage receives the clock transition. Thus, the stages will have latched the same value incorrectly. Such a circuit violates our basic assumption that all flip-flop inputs are examined simultaneously. This is a particularly serious problem in modern intergrated circuits where wire delays are often greater than delays through logic. Designers increasingly are concerned with proper clock distribution.

Fortunately, the designers of flip-flop components have built them so they can be cascaded without timing problems (this is true as long as the same families of components and consistent clocking events are used). It is important to remember that the inputs must be held stable for a setup time before and a hold time after the clock edge. Let's use the low-voltage CMOS parts of Figure 6.30 and Figure 6.31 as an example. For the 374 positive edge-triggered flip-flop, the setup and hold times are 1.8 ns and 0.5 ns, respectively. Fortuitously, the propagation delay exceeds the hold time. In the case of the 374, the delay for an output transition is 1.1 to 3.6 ns. Unless the clock signal to the second stage is delayed by more than 0.6 ns, the first stage cannot change its value and propagate it to the second stage before the hold time has expired. By then, the second stage has successfully latched the original Q_0 value. This timing behavior is shown in Figure 6.35. Thus, all logic families have propagation delay greater than their corresponding hold time.

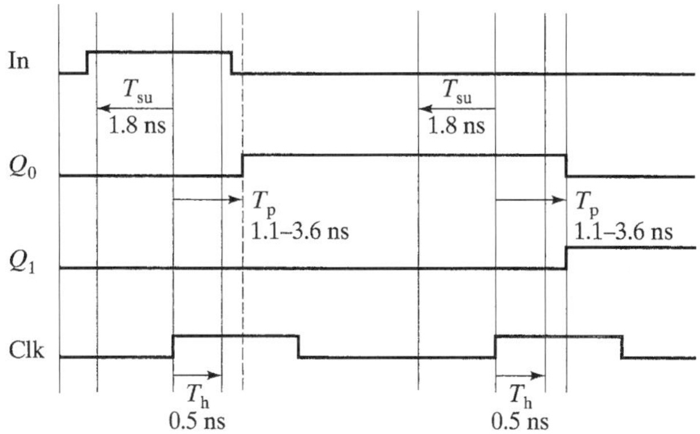

Figure 6.35 **Timing behavior of cascaded positive edge-triggered flip-flops.**

The hold time affects the amount of clock skew that can be tolerated. More tolerance is good as it gives our clock distribution more flexibility, but it makes the circuit slower (a longer clock period). This is a typical tradeoff in synchronous sequential logic design. In summary, the following relations must be true for proper operation:

$$T_p > T_h$$
$$T_{period} > T_p + T_{su}$$

6.2.2 Clock Skew

Proper operation of synchronous systems requires that the next state of all storage elements be determined at the same instant. In effect, the clock signal must appear at every storage device at the same time. Unfortunately, this condition cannot always be guaranteed. A single clock signal may fan out from more than one physical circuit, each with its own timing characteristics.

EXAMPLE 6.1 THE SKEW PROBLEM

Clock skew can introduce subtle bugs into a synchronous system. As an example, refer back to the two-stage shift register of Figure 6.33. This time, think of the two flip-flops being clocked by signals Clk0 and Clk1, respectively, whose arrival is skewed slightly in time.

Suppose that the original state of both flip-flops is 1 and that the input to the first flip-flop is 0. If the circuit behaves correctly, the state of the first flip-flop should change from 1 to 0 while the second flip-flop stays at 1 (its current value of 1 is replaced by the old value of the first flip-flop, which is also 1).

The timing diagram of Figure 6.36 shows what really happens. The first flip-flop is reset to 0, but so is the second flip-flop! This occurs because the second stage sees the new state of the first stage (0), rather than its current state (1), by the time Clk1 arrives.

One of our constraint equations is changed as follows:

$$T_p > T_{skew} + T_h$$

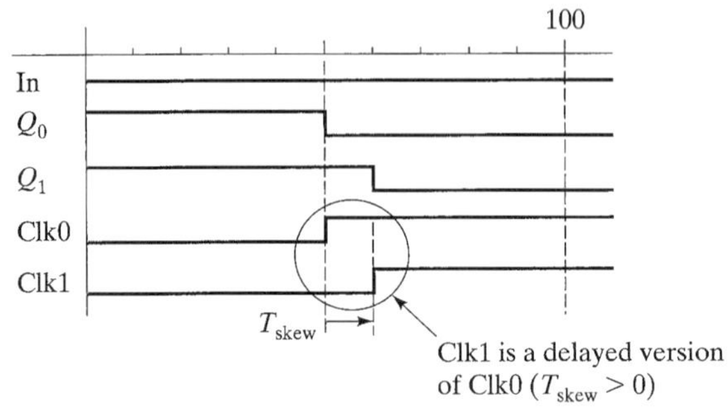

Figure 6.36 Timing diagram for cascaded flip-flops, illustrating the effects of clock skew.

Obviously, if the skew gets large enough this inequality will no longer be true.

If the clock skews in the other direction ($T_{skew} < 0$) then the other constraint equation becomes:

$$T_{period} + T_{skew} > T_p + T_{su}$$

Again, if the skew is negative enough then this constraint is violated.

Avoiding Clock Skew One way to avoid clock skew is to carefully arrange the wires that carry the clock signal so that it incurs an identical delay in reaching all the flip-flops it must trigger. As technologies get faster, that is, their propagation delays decrease, a smaller amount of clock skew can cause problems. Minimizing clock skew is now one of the dominant problems in the design of high-performance systems.

6.2.3 Asynchronous Inputs

So far, we have described only simple clocked synchronous circuits. These are circuits in which all components are driven from a common-reference signal. The state of the circuit changes only in relation to the clock. The clock event determines when inputs are sampled and when outputs can change. If the setup, hold, and propagation delays are designed appropriately, there are no problems in composing components by connecting the output of one to the input of another.

An *asynchronous* circuit is one whose inputs, state, and outputs can be sampled or changed independently of any clock reference. Asynchronous circuits lie at the heart of every synchronous circuit. The basic *R-S* latch is an asynchronous circuit, whereas the edge-triggered *D* flip-flop, constructed from several such latches, is synchronous.

Synchronous versus Asynchronous Inputs Even a supposedly synchronous circuit like the *D* flip-flop can have asynchronous inputs, such as preset and clear. These set the output (preset) or reset it (clear) whenever they are asserted, independent of the clock. *Synchronous inputs* are active only while the clock edge or level is active; at all other times, changes on these inputs are not noticed by the memory element. *Asynchronous inputs,* on the other hand, take effect immediately and are independent of the clock.

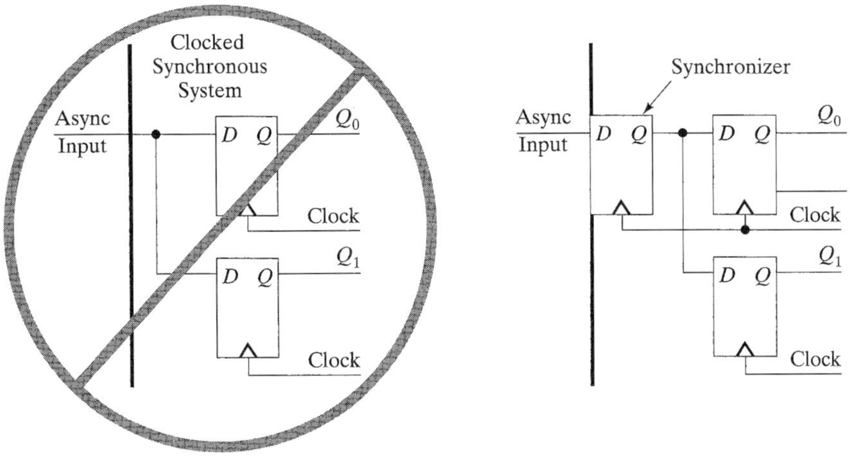

(a) Incorrect fan-out to multiple flip-flops (b) Correct synchronization

Figure 6.37 Incorrect and correct circuits for handling an asynchronous input.

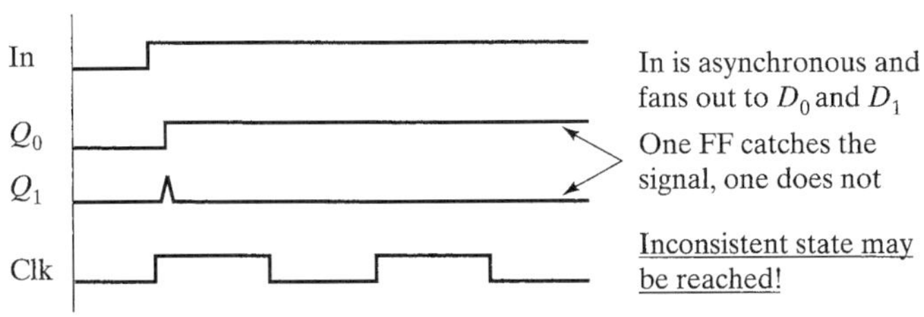

Figure 6.38 Outputs of two positive edge-triggered flip-flops when their input changes close to the clock edge.

Glitches make asynchronous inputs extremely dangerous and should be avoided whenever possible. A glitch on the logic that drives an asynchronous input can cause a flip-flop to be cleared or set when no state change was called for. It is good design practice to choose components that have only synchronous inputs.

Sometimes asynchronous inputs cannot be avoided—for example, when a signal must pass from the outside world into the synchronous system. An example might be a reset signal, triggered by a user pressing a push-button. It is particularly dangerous to fan out an asynchronous input to many points in the clocked system: if the input changes close to the clock event, it may be seen as one value at some flip-flops and a different value at others, leading to an "impossible" state that should not occur.

An incorrect circuit for handling an asynchronous input is shown in Figure 6.37(a). Two positive edge-triggered D flip-flops are driven by the same asynchronous input. You would expect both devices to hold the same state, yet because of different wiring and other internal delays, one flip-flop is set while the other remains reset. The assumption that both flip-flops hold the same state is now invalid. The timing waveform in Figure 6.38 tells the sad tale.

A better way to deal with an asynchronous signal is to *synchronize* it to the clocked system. This synchronization is accomplished by placing a single D flip-flop between the input source and the rest of the system. The proper circuit is shown in Figure 6.37(b). The flip-flop's output Q will change only in relation to the clock and can be properly fanned out and distributed to other points in the circuit in a synchronous manner.

6.2.4 Metastability and Synchronizer Failure

What if the setup and hold times of the synchronizer flip-flop are not met by the asynchronous signal? Under such conditions, the output of the synchronizer is undefined.

Normally, we minimize this problem by choosing the synchronizer flip-flop from the fastest available logic family, with the shortest possible setup and hold times. Unfortunately, the problem cannot be eliminated completely. The behavior of this flip-flop is worse than unpredictable: it can result in input values injected into the system that cannot be interpreted as either a 1 or a 0. Figure 6.38 gives a hint of this: Q_1 exhibits a partial transition that falters back to 0. This "inbetween" voltage is called the *metastable* state. Under the right (or wrong) conditions, the flip-flop can hang in this state indefinitely, a *synchronizer failure*.

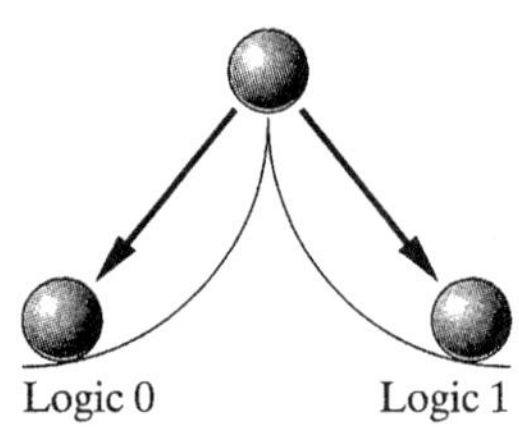

Figure 6.39 **Analogy for explaining synchronizer failure.**

An Analogy for Understanding Metastability Figure 6.39 provides a useful analogy for describing the nature of synchronizer failure. The states of the flip-flop are represented by two flat regions separated by a steep slope. The flat parts represent the stable states: logic 0 and logic 1. For the purpose of this analogy, we will represent the state of the flip-flop by a ball in one plateau or the other. To change the state, energy must be exerted to push the ball up and over the slope to the other side.

When setup and hold time constraints are met, there is sufficient energy to cause the state change. If these constraints are not met, three cases are possible, two that yield acceptable behavior and one that does not. In the first case, there isn't enough energy to get the ball over the summit, and it rolls back—the state is not changed. In the second case, the energy might be just enough to get the ball over the top—and the state changes from 0 to 1. Both of these are acceptable outcomes. However, there is a small probability that just enough energy is imparted that the ball can be pushed up the slope but remains tottering at the top, not able to return to one or the other of the stable states. This is the metastable state.

Theoretically, a flip-flop can remain in the metastable state indefinitely. However, thermal disturbances and asymmetries in signal delays within the transistor-level implementation of the flip-flop usually make it settle in one state or the other after some period of time.

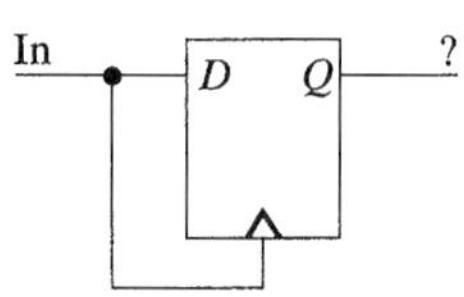

Figure 6.40 **Circuit under test for Figure 6.41.**

Figure 6.40 shows a circuit with its data input and clock input tied together, a perfect circuit in which to violate setup constraints. An oscilloscope trace is shown in Figure 6.41 that superimposes many clocking events on top of each other. The end result is always 0 or 1, but the output value stays in an indeterminate state for some time. Even though a stable state usually is obtained (after a potentially long

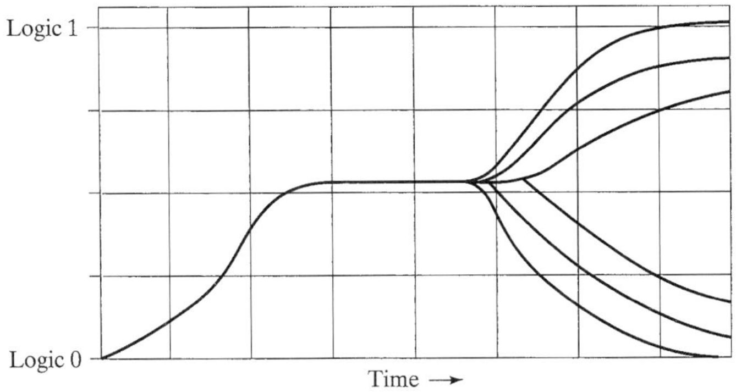

Figure 6.41 Oscilloscope sketch of metastable behavior.

delay), this can still cause a system failure if the flip-flop has not left the metastable state by the end of the system's clock period.

Reducing the Chance of Synchronizer Failure The only way to recover from synchronizer failure is to reset the entire circuit. While the probability of synchronizer failure can be made small, it can never be eliminated as long as there are asynchronous inputs.

One way to reduce the probability of synchronizer failure is to lengthen the system's clock period. This gives the synchronizer flip-flop more time to make its decision and enter a stable state. The longer the clock period is stretched, the lower the probability of failure. Unfortunately, this is not an adequate solution for high-performance systems in which a short clock period is critical.

A second strategy places two synchronizers in series between the asynchronous input and the rest of the synchronous system. Both flip-flops must be metastable before the synchronization fails, an event with low probability. However, this slows down any affect the input can have on a circuit by approximately two clock periods. This may be too costly when two systems with different clocks are trying to communicate at high speed.

A third strategy does away with the clock altogether and follows a timing strategy that is independent of the speed of the individual circuits. We examine this intriguing approach to digital design next.

6.2.5 Self-Timed and Speed-Independent Circuits

Although synchronous system design is our preferred approach, it is not always possible to build a complete system this way. As digital systems become larger, incorporating more components, or faster, running at higher clock rates, the limiting problem becomes how to distribute a single global clock without introducing intolerable clock skew.

When a global clocking approach is inappropriate, the alternative is to partition the digital system into locally clocked pieces that communicate with each other using *delay-insensitive* signaling conventions. A delay-insensitive protocol allows the sender and receiver to communicate with each other without the need for a global clock. Each proceeds at its own speed, synchronizing its communications when it needs to interact.

A good example of this locally clocked, globally delay-insensitive approach can be found in just about every digital system constructed from multiple printed circuit boards. Each board contains its own independent clock, and the logic on the board is designed to fully be synchronous. When one board needs to communicate with another, special signaling conventions are used that have been designed to be insensitive to delay. We examine these conventions next.

Delay-Insensitive Signaling The concept of delay-insensitive signaling is shown graphically in Figures 6.42 and 6.43. One subsystem is called the *requester* or *client*, while the other is called the *provider* or *server*. It is common for a given component to contain logic that allows it to be either a client or a server. For data to flow from the client to the server, the client must make an explicit request to the server. If the client requests a read, the server responds by providing the requested data and signals an acknowledgment when the data is ready for the client.

Synchronous Delay-Insensitive Signaling There is no reason why the communication between the client and the server cannot proceed in a completely synchronous manner. This is demonstrated by the timing diagram of Figure 6.44. The protocol is synchronous because the request and acknowledgment signals are asserted with the rising edge of the clock. They can be sampled on the falling edge.

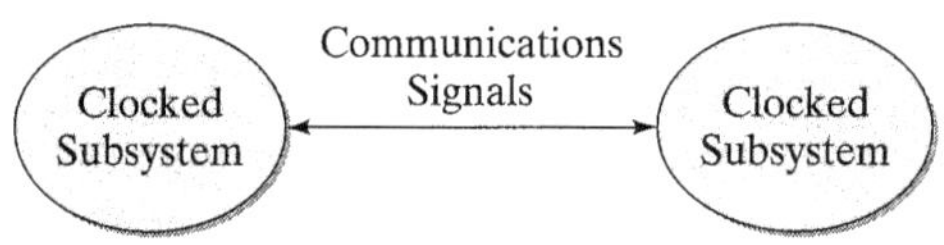

Figure 6.42 **Independently clocked subsystems.**

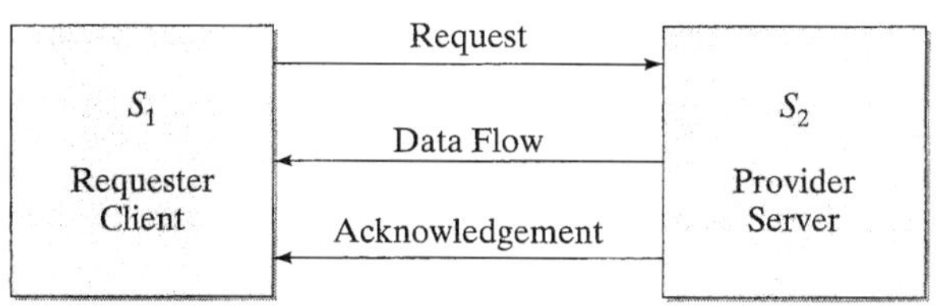

Figure 6.43 **Request/acknowledgment signaling.**

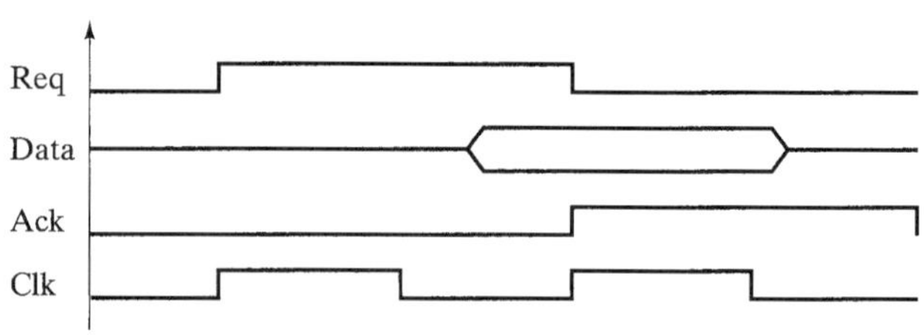

Figure 6.44 **Synchronous request/acknowledgment signaling.**

If it is known that the server will respond in a certain number of cycles, an acknowledgment signal is not strictly needed. However, this method can be inefficient if there is a wide variation among the speeds of potential servers. We would have to set the number of cycles to that of the slowest server.

To provide at least some degree of speed independence, an alternative signaling convention allows the server to delay the client by asserting a wait signal if it cannot satisfy the request in the allotted number of cycles. This is shown in Figure 6.45. When the server unasserts this signal, it implicitly acknowledges that the data is now available for the client. This signaling still assumes a single, global clock and that all interface signals change synchronously with clock edges.

Delay Insensitivity Without a Clock: Four-Cycle Handshaking True speed-independent signaling can no longer assume that signals change with respect to a global clock. A component that asserts a signal must ensure that the signal has been observed by the component for which it is intended. This requires that the signals be *interlocked*: a request cannot be unasserted until the target acknowledges that it has been seen and acted upon.

A four-cycle interlocked handshake, sometimes called *return-to-zero signaling*, is shown in Figure 6.46. Initially, both request and acknowledge are at 0. The request for data is initiated by asserting the request signal. Because the clocks of the client and server are unrelated, the client can make no assumptions about how long it takes for the server to notice the request. That the request has been seen and acted on is indicated only by the server's assertion of its acknowledgment signal. This is driven high only after the requested data has been placed on the data-signal lines.

Similarly, the server cannot make assumptions about whether the client has seen its acknowledgment signal. It must wait, continuing to

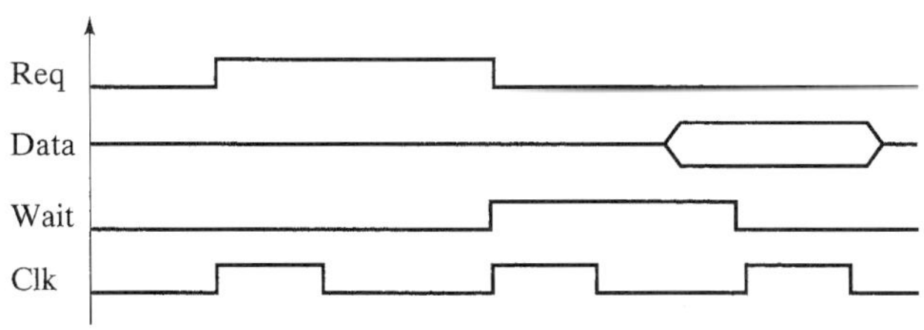

Figure 6.45 Synchronous request with wait signaling.

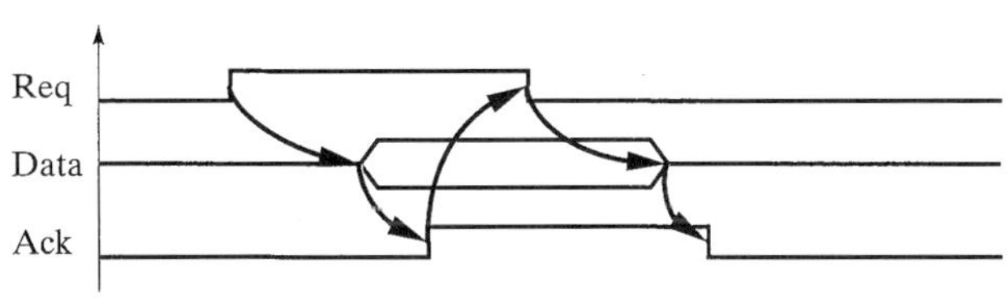

Figure 6.46 Four-cycle asynchronous signaling.

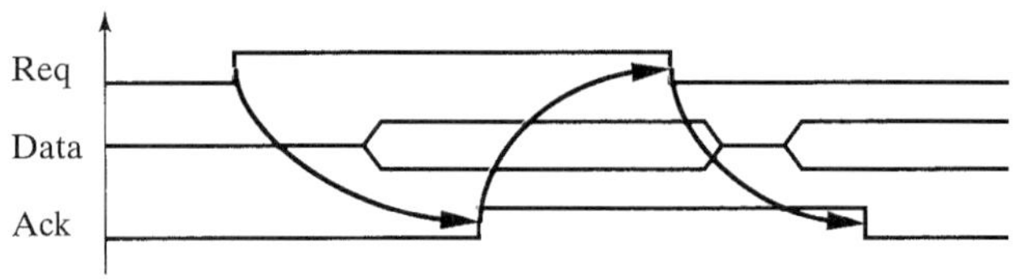

Figure 6.47 **Two-cycle asynchronous signaling.**

drive the data and the acknowledgment until the client communicates that the data has been latched by unasserting its request signal. Now the server can stop driving the data and can reset its acknowledgment to 0. The acknowledgment signal must become unasserted before a new request can be made.

Four-cycle signaling guarantees that the client and server will behave properly in all timing scenarios. The request signal is asynchronous to the server and the acknowledgment is asynchronous to the client: both must be synchronized and have the unavoidable potential for synchronizer failure.

Two-Cycle Handshaking An alternative to four-cycle signaling is called two-cycle or *non-return-to-zero signaling*. The waveforms are shown in Figure 6.47. Initially, both request and acknowledge are at 0. The client makes a request by complementing the request line, in this case driving it high. Eventually, the server notices the change in request state, services the request, and acknowledges that the data is available by complementing its acknowledgment line, driving it high.

The client makes its next request by complementing the request line once again, driving it low. The server stops driving the data lines with their old data, services the new request, places new data on the data lines, and signals that it has accomplished its task by complementing the acknowledge line, driving it low.

Although two-cycle signaling involves fewer transitions to accomplish the same function as four-cycle signaling, it requires that the client and server contain an additional state to recognize a change in the request/acknowledge signal lines. Four-cycle signaling usually requires less hardware for its implementation.

Self-Timed Circuits Using the request/acknowledgment signaling conventions just described, we can build internally clocked circuit components that communicate with each other in a speed-independent manner. It is also possible to build these components so that they do not contain any internal clocks.

These unclocked circuits are sometimes called *self-timed circuits*. A self-timed circuit can determine on its own when a given request has been serviced. This usually involves mimicking the worst-case propagation delay path by using special logic to delay the request signal.

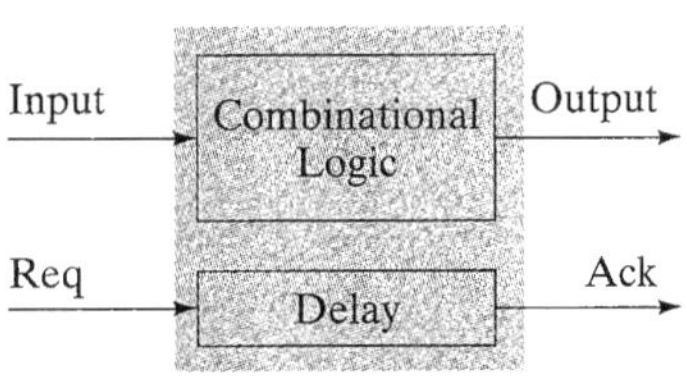

Figure 6.48 **Self-timed combinational logic block.**

This scheme for a self-timed combinational logic block is shown in Figure 6.48. The delay line merely slows down the acknowledgment signal, derived from the request signal, long enough to guarantee that the combinational logic has sufficient time to compute the correct output. If the worst-case delay path is n gate levels deep, the delay line should be a comparable number of gate levels.

Self-timed systems continue to interest designers as integrated circuits become denser and the problems of clock distribution become even more significant. Many complex VLSI circuits, such as dynamic memory chips, contain self-timed elements, and an entire microprocessor can be constructed in a self-timed manner. More information about self-timed circuits and systems can be found in the references at the end of this chapter.

6.3 Registers

In this section, we will examine more deeply a most basic kind of storage element found in digital logic: registers. We will also look at how to combine register elements to perform interesting functions involving the storage of multiple bits of data. Registers will form the basis of all the sequential logic circuits we'll see in the rest of the text.

6.3.1 Storage Registers

A *register* is a group of storage elements read or written as a unit. The simplest way to construct a register is by grouping together as many D flip-flops as you need to obtain your desired bit width. Figure 6.49 shows a 4-input register built by wiring together the clocks of four flip-flops.

The individual flip-flops have asynchronous reset (R) and set (S) inputs. Note that these are connected to a logic 0 so that they do not have any effect in this register. The D data inputs are synchronous. They are sampled only in conjunction with clocking events.

Circuits such as the one in Figure 6.49 are at the heart of almost all packaged registers. However, there are many variations on the basic theme, usually involving additional control signals for the inputs and outputs. One extension provides a load signal. The register's internal logic holds the current state as long as an input enable signal (it may be called EN for enable, LD for load, or WE for write enable) is not asserted. When enable is asserted, the new inputs are gated to the

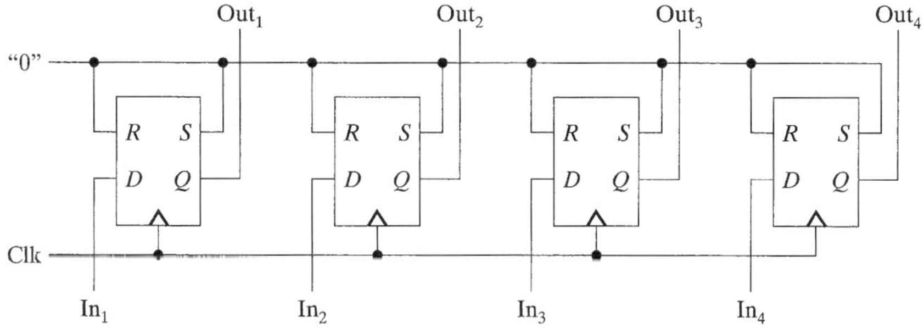

Figure 6.49 **Possible implementation of a four-input register.**

internal inputs of the flip-flops. These replace the current state when the clock edge arrives. This can often save some external logic because the outputs to not have to be recirculated to the inputs.

Another extension allows multiple registers to share output wires so that the values stored in different registers can be selectively connected to the same wires and used as inputs elsewhere. The register's outputs are tri-state or open collector logic. The register package includes a common control signal that enables the flip-flop state to be connected or disconnected from the output wires. This control signal is often referred to as an output enable signal (*OE* for output enable or *RE* for read enable). External logic selectively enables a single register to place its outputs on the shared wires and ensures that two registers aren't connected to the common wires simultaneously.

Register Files If registers group multiple storage elements into a unit, then register files extend the grouping to multiple registers. Each register in the register file is called a *word* and is identified by a unique index or *address.*

The component contains its own internal decoders. In conjunction with an externally specified address, the decoders select a specific register-file word to be written. On a read, the selected word is multiplexed to the outputs.

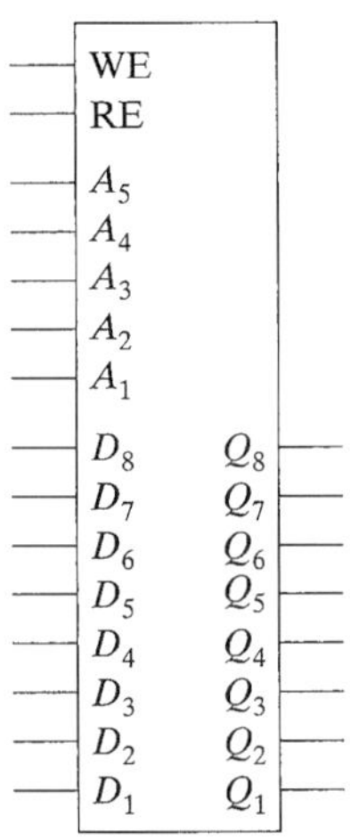

Figure 6.50 A 32-by-8 register file with write (WE) and read (RE) enable and five shared address lines.

An example could be a 32-by-8 register file with tri-state outputs. The device would contain 256 flip-flops organized into 32 words of 8 flip-flops each. Five address lines would be required to select a register to be written from the inputs and five address lines would be required to select a register to be read and connected to the output wires. A write enable is used to indicate that the particular register corresponding to the index on the address lines should take on the values at the inputs (a load enable). A read enable is used to indicate that the particular register corresponding to the index on the address lines should be connected to the output wires (an output enable). If the write and read address lines are different (10 wires in all) then a register can be written while a different one can be read. If they are the same (to save on package pins) then reads and writes must be done one after the other rather than in parallel. The diagram of Figure 6.50 shows the possible pinout of such a register-file package.

Sometimes, such a device may not have a separate distinguished clock signal. It uses the *WE* signal and the address to generate the clock inputs to the internal flip-flops. If the component lacks an explicit clock, it should be used with great care. The address lines must be stable before the *WE* signal arrives, otherwise the wrong word might be written. Any glitches on the *WE* signal could also lead to unwanted writes.

Random-Access Memories Registers are convenient for holding small amounts of information, usually in the range of 4 to 16 (or maybe even 32) bits wide. However, many digital systems require substantially more storage elements than this. *Random-access memory,* or RAM, offers a solution. By using very transistor-efficient methods for implementing

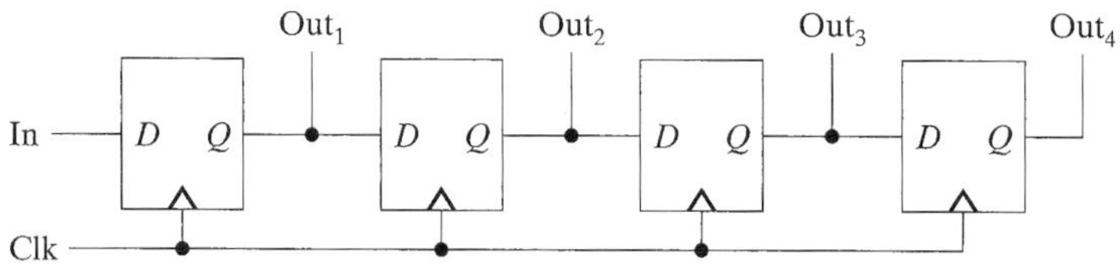

Figure 6.51 **Quad right-shifting shift register.**

storage elements, RAM generalizes the register file concept to make many more words available in a single integrated circuit package. A small RAM might hold 1024 four-bit words (4096 storage elements), a state-of-the-art device contains over 128 million storage elements (16 megabytes), and even larger devices are on the way. We'll cover RAMs in more detail in Chapter 9.

6.3.2 Shift Registers

Registers can be used for other applications besides simply storing bits. They are often used to circulate, or *shift*, values among the storage elements. In this subsection, we concentrate on register components that shift as well as store. These are called *shift registers.*

Figure 6.51 shows the logic of a simple right-shifting shift register constructed from D-type flip-flops. Data moves from left to right. On every shift pulse, the contents of a given flip-flop are replaced by the contents of the flip-flop to its left. The leftmost device receives its inputs from an input signal. Because flip-flop propagation times far exceed hold times, the values are passed correctly from one stage to the next (we discussed cascaded flip-flops in Section 6.2.1).

The step-by-step operation of the shift register is shown graphically in Figure 6.52, for an initial configuration of $Q_1 = 1$, $Q_2 = 0$, $Q_3 = 0$, $Q_4 = 0$ followed by three shifts with the input being 1 for the first and 0 for the following two, leaving a final state of $Q_1 = 0$, $Q_2 = 0$, $Q_3 = 1$, $Q_4 = 1$.

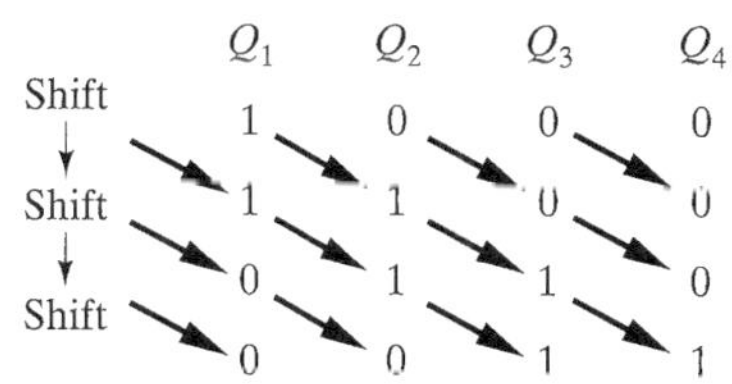

Figure 6.52 **Shift behavior.**

Variations: Parallel Versus Serial Inputs and Outputs The circuit of Figure 6.51 provides a primitive ability to shift. This can be used to set the initial state of the flip-flops. The leftmost input can be used in four successive clock periods to load four specific values to serve as the starting state of the shift register. In the simple example above, we would have used the previous four cycles to load the values of 0, 0, 0, and 1 to form the $Q_1 = 1$, $Q_2 = 0$, $Q_3 = 0$, and $Q_4 = 0$ initial condition.

This *serial input* approach is quite simple and requires only enough external circuitry to get those values at the input of the shift register. However, it does take time. In this case, four cycles are needed to set the register to a specific set of values. Often, we need to do this much more quickly.

A shift register with *parallel inputs* contains multiplexing logic at the input to each internal flip-flop. The flip-flop receives a new value from its left neighbor when the shift register is in shift mode or from an external input when the register is in load mode. This permits a new set of values to be loaded in a single cycle, in parallel.

A similar variation distinguishes between parallel and serial outputs. *Parallel outputs* mean that the outputs of the internal flip-flops are visible at the pins of the register's package. With *serial outputs,* only the value of the last element is visible outside the register, such as Q_4 in Figure 6.51. The primary motivation for serial outputs is to reduce the number of output connections, since the flip-flop outputs are always available inside the shift-register package.

EXAMPLE 6.2 UNIVERSAL SHIFT REGISTER

To illustrate these concepts, let's consider the design of a universal shift register that has serial inputs and outputs, parallel inputs and outputs, and, in addition, can shift either left or right. A block diagram of this building block is shown in Figure 6.53, including the encoding of control signals for choosing which function is to be performed.

The device has four parallel inputs that will be multiplexed into the inputs of its four flip-flops, the four outputs of the flip-flops are available as parallel outputs, there is a left input and a right input that are input to the leftmost and rightmost internal flip-flops, the left and right outputs are simply the outputs of the leftmost and rightmost flip-flops (there is some duplication here with the parallel outputs—but we list them separately in the interest of conceptual simplicity), and there are three control signals: *clear,* S_0, and S_1. Of course, there is also a clock input.

We can simplify our design task by looking at a single bit of the universal shift register in isolation. If we design it properly, we can then connect four of them together to complete the final circuit. Figure 6.54 shows a possible design for a single cell. It includes a single flip-flop at its core. Its output is one of the parallel outputs but is also sent to the adjoining cells to implement the shift left and shift right operations.

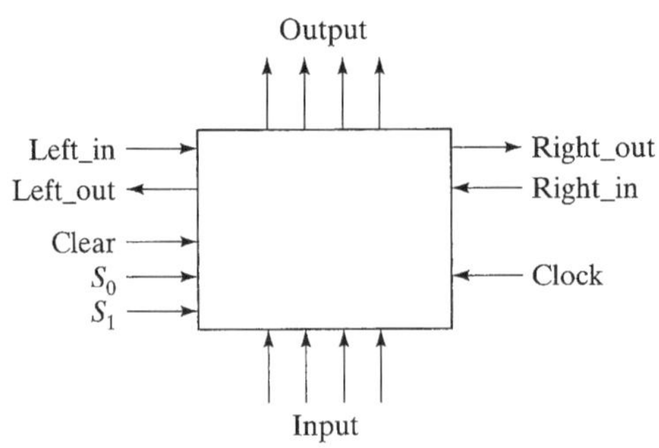

Clear Sets the Register Contents and Output to 0

S_1 and S_0 Determine the Shift Function

S_0	S_1	Function
0	0	Hold state
0	1	Shift right
1	0	Shift left
1	1	Load new input

Figure 6.53 Block diagram of a universal shift register showing all inputs and outputs and encoding of control signals for its four different functions.

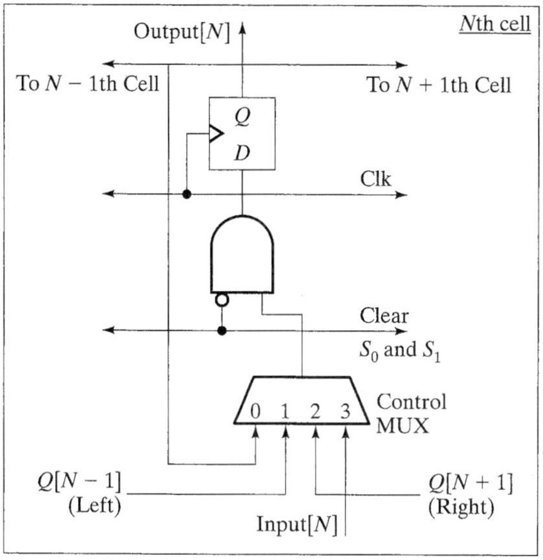

Figure 6.54 Design for a 1-bit cell of the universal shift register of Figure 6.53.

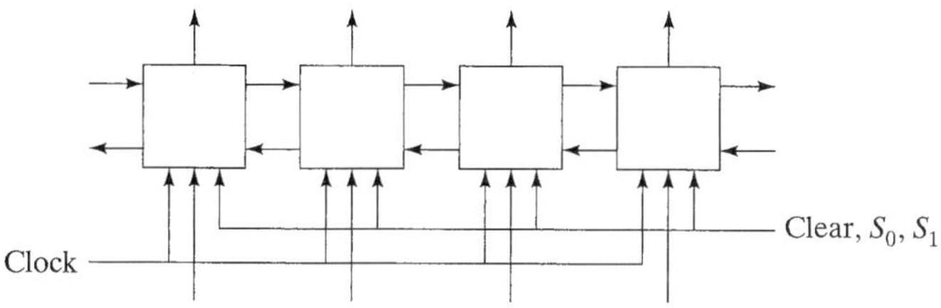

Figure 6.55 Design of a 4-bit universal shift register utilizing the basic cell of Figure 6.54.

The input to the flip-flop comes through an AND gate whose output is 0 when *CLEAR* is asserted. This implements a synchronous clear operation because the 0 value will not be loaded into the flip-flop until the next clocking event. When *CLEAR* is not asserted, the input to the flip-flop comes from a 4:1 multiplexer that is controlled by the two signals S_0 and S_1. When the 0-input is selected, we simply recirculate the current value of the flip-flop to implement the hold operation. When the 1-input is selected, we take as input the value of the cell to the left, thus implementing a shift right. When the 2-input is selected, we take as input the value of the cell to the right, thus implementing a shift left. When the 3-input is selected, we take as input the value of the corresponding parallel input to implement a load operation. Changes to the flip-flops only occur at the clocking event so that all inputs must respect setup and hold time requirements.

Four identical cells can be assembled as shown in Figure 6.55 to implement the four-bit version of this universal shift register. This device is completely synchronous.

EXAMPLE 6.3 SERIAL COMMUNICATION APPLICATION

Consider the problem of communicating between a terminal and a computer over phone lines. The terminal expects its data to appear in a byte-wide parallel form, but the data must be sent over the line in bit-serial form. Shift registers play a key role in such communication systems because they can convert between parallel and serial formats. We design the hardware to load the data from the computer in parallel and shift it out serially over the communications link. On the return trip, serial data from the terminal is captured by the shift register, bit by bit, and presented to the computer via the shift register's parallel outputs.

A partial implementation of this subsystem is shown in Figure 6.56. On the sender side, we load 8 bits of parallel data into the cascaded shift register by setting S_1, S_0 to 1, 1. The sender and receiver then enter their shift-left mode when S_1, S_0 are set to 0, 1. We do this for eight clock cycles, transmitting D_0 through D_7, 1 bit at a time. Note how

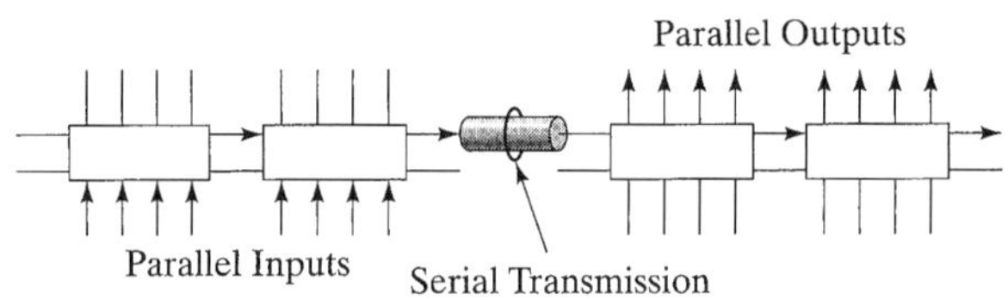

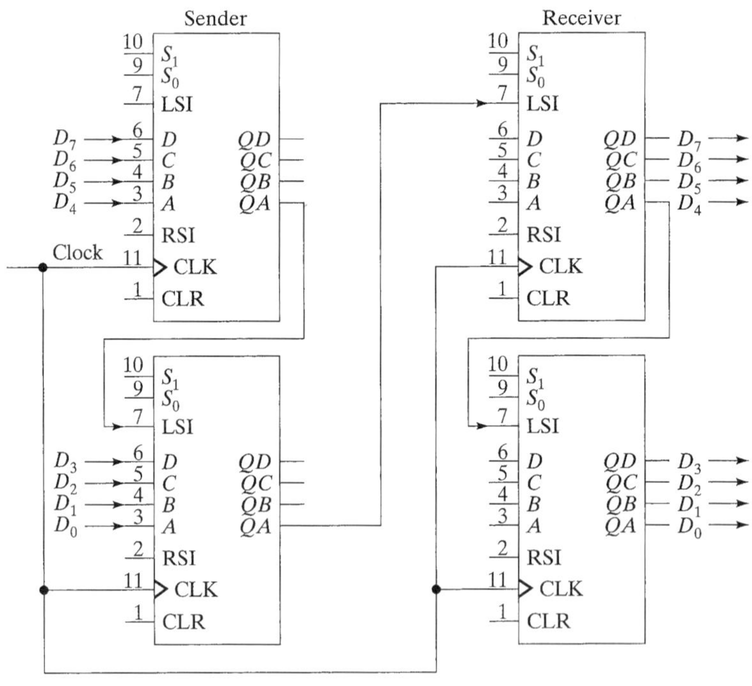

Figure 6.56 Partial implementation of 8-bit parallel/serial transmission subsystem.

the shift registers are chained together using the *LSI* (left-shift-in) inputs. Once all 8 bits have been sent, we can place the receiver in its hold mode by setting S_1, S_0 to 0, 0.

6.4 Hardware Description Languages

Of course, hardware description languages can also be used to describe sequential logic. Verilog has several constructs that enable it to efficiently describe the functions of latches, flip-flops, and registers.

The *R-S* latch is quite straightforward since we can easily represent it in terms of two NOR gates (see Figure 6.5):

```
module rs_latch (r, s, q);
    input   r, s;
    output  q;
    wire    qbar;

    nor_gate nor1(s, q, qbar);
    nor_gate nor2(r, qbar, q);

endmodule
```

We could describe an edge-triggered *D* flip-flop the same way. However, since it is used so commonly, Verilog provides a more direct construct:

```
module dff (clk, d, q);
    input   clk, d;
    output  q;
    reg     q;

    always @(posedge clk)
       q = d;

endmodule
```

The `always` statement is triggered whenever the signal *clk* has a positive edge (the triggering event). Thus, Verilog generalizes its sensitivity list to be selective to the direction of the change on a signal rather than any change at all, as we saw in Chapter 3. If we had written just `@(clk)` then we would make a new assignment to *q* twice per clock period (on both the rising and falling edges) and this clearly is incompatible with our flip-flop design principles.

When the triggering event occurs the value of the input *d* is assigned to the internal *register* *q* which is also an output. This is exactly how we want our flip-flop to behave. We should note that if there are multiple `always` blocks sensitive to the positive edge of the clock: they are all triggered in parallel. Thus, Verilog makes it easy to describe combinations of flip-flops in registers by following the same rules of composition (only one change-in-state per triggering event). We'll see shift registers soon. First, we'll look at a few variations on flip-flops.

A flip-flop with synchronous reset and set is described as follows:

```
module dff_sync_rs (clk, r, s, d, q);
   input  clk, r, s, d;
   output q;
   reg    q;

   always @(posedge clk)
      if (r)        q = 1'b0;
      else if (s)   q = 1'b1;
      else          q = d;

endmodule
```

This, of course, is a reset-dominant flip-flop as we look at the reset input first, then set, and then the default behavior that loads in a new data value from the input. By reversing the first two clauses in the `if` statement we can make this a set-dominant flip-flop.

A flip-flop with asynchronous reset and set is only a little bit more complex:

```
module dff_async_rs (clk, r, s, d, q);
   input  clk, r, s, d;
   output q;
   reg    q;

   always @(posedge r)
      q = 1'b0;
   always @(posedge s)
      q = 1'b1;
   always @(posedge clk)
      q = d;

endmodule
```

Note that in this description, only the rising edge of the set and reset inputs cause an immediate change in the state of the flip-flop. If the reset input stays high, the set input can still have an effect.

A shift register is easy to specify in Verilog. The shift register of Figure 6.51 is described as follows:

```
module shift4bit (clk, in, out1, out2, out3, out4)
   input  clk, in;
   output out1, out2, out3, out4;
   reg    out1, out2, out3, out4;

   always @(posedge clk) begin
      out4 = out3;
      out3 = out2;
      out2 = out1;
      out1 = in;
   end

endmodule
```

Note that we had to be careful about the order in which we wrote the statements in the always block. Our shift register would not function correctly if we had written the assignment statements in reverse order. This is because, in Verilog, statements within an always block are executed sequentially. Thus, if we had already assigned a new value to out_1 before using its value as the new value for out_2, we would lose the original value of out_1 forever and our circuit would not function as intended.

Rather than requiring designers to have to exercise so much care, Verilog has been designed to include a *delayed* or *non-blocking* assignment (in contrast to the *sequential* or *blocking* assignment above). The delayed assignment, indicated as <= instead of simply =, takes effect at the end of the `always` block rather than in the order it is encountered. Our shift-register description can now be rewritten as:

```
module shift4bit (clk, in, out1, out2, out3, out4)
   input   clk, in;
   output  out1, out2, out3, out4;
   reg     out1, out2, out3, out4;

   always @(posedge clk) begin
      out1 <= in;
      out2 <= out1;
      out3 <= out2;
      out4 <= out3;
   end

endmodule
```

Now the order of the statements does not matter. They will all execute in parallel at the end of the `always` block. This is very compatible with our synchronous timing methodology and the design of real flip-flops, which all change their states simultaneously at the common clocking event. A Verilog shorthand for this group of statements is:

```
{out1, out2, out3, out4} <= {in, out1, out2, out3};
```

Finally, lets describe the register file of Figure 6.50 in Verilog:

```
module register_file_32_by_8 (we, re, abus, dbus, obus)
   parameter mem_size = 32;
   parameter data_bus_width = 8;
   parameter addr_bus_width = 6;
   input we, re;
   input [addr_bus_width-1:0] abus;
   input [data_bus_width-1:0] dbus;
   output [data_bus_width-1:0] obus;
   reg [data_bus_width-1:0] obus;
   reg [data_bus_width-1:0] out;
   reg [data_bus_width-1:0] regfile [mem_size-1:0];
```

```
    always @(posedge we) begin
       regfile[abus] = dbus;

    always @(posdege re) begin
       out = regfile[abus];

    assign obus = (re) ? out : 8'bZZZZZZZZ;

endmodule
```

We make use of bit vectors to group signals into busses for convenience (for example, a_5 through a_1 into abus, a 5-bit bus) so that we can assign entire groups of signals in one assignment statement. We use a parameter for width of the busses. We also introduce some new Verilog constructs for conditional assignment at the end of the module description.

There is no clock for this module. Rising edges on the read- and write-enable inputs are what trigger all the activity. When the write-enable signal goes high, the value on the inputs (grouped into a single 8-bit vector called dbus is written into the word in the register indexed by the abus inputs. When the read-enable signal goes high, the value of the register indexed by the abus inputs is placed on the internal variable *out*. The value of *out* are assigned to the eight outputs of the register file if the read enable signal (*re*) is asserted, otherwise, they are tri-stated to the high-impedance value.

CHAPTER REVIEW

In this chapter, we have begun our study of circuits with state by looking at the fundamental building block of such circuits: the latch, its more complex derivative, and the flip-flop. We introduced the basic *R-S* latch and the level-sensitive *R-S* latch. We proceded through an evolution of this basic storage element into a master–slave flip-flop and then an edge-triggered *D* flip-flop that is the foundation of modern synchronous sequential circuit design. Along the way, we considered problems in composing latches that led us to develop robust timing methodologies that make it possible to easily build circuits with large numbers of flip-flops. A key part of this was the concept of master–slave devices that construct an "air-lock" between storage elements so that each clocking event causes all flip-flops to simultaneously update their state.

Next we discussed issues related to timing methodologies, starting with the meaning of setup and hold time constraints as they relate to latches and flip-flops. If an input changes within the window formed by the setup time before the clocking event and the hold time after the clocking event, we do not know what value will be stored in the memory element. There is even a chance that the flip-flop will be caught in an in-between state, called the *metastable state,* should the input change too near the clock edge. Theoretically, it could be stuck in this state forever. Fortunately, when a system is constructed from a compatible family of logic, such as TTL LS logic, flip-flops can be cascaded without fear of setup and hold time violations.

We briefly described asynchronous inputs and the dangers in using them. Unfortunately, they cannot be eliminated completely. We introduced the synchronizer concept to reduce their danger. We also showed the four-cycle signaling convention as a protocol for communication among independently clocked subsystems. We presented the concept of a self-timed circuit, which determines on it own when it has finished computing its function. These circuits will provide the basis for future systems that may have no clocks in them at all.

Finally, we introduced the basic building block of all the sequential logic circuits that we will learn much more about in the next chapter, the register. We considered the use of registers for storage and for shifting values. We designed a universal shift register and looked at how it could be applied to a very practical application in communication: taking a set of values provided in parallel and sending them over a serial communication link to be collected into a parallel set on the other side.

Finally, we described how hardware description languages can be used to specify basic latches, flip-flops, shift registers, and register files. Two new constructs were introduced: sensitivity to a particular change in the value of an input and assignments that take effect simultaneously rather than sequentially so as to better model what happens in real flip-flops.

FURTHER READING

Most logic-design textbooks provide an extensive discussion of flip-flops along the lines of the presentation we have given here. However, clocking issues and metastability are not usually covered as thoroughly. An unusual exception to this is T. R. Blakeslee's book, *Digital Design with Standard MSI & LSI, Second Edition,* Wiley, New York, 1979. Chapter 6, "Nasty Realities I: Race Conditions and Hang-up States," formed the basis for our discussion of asynchronous inputs, clock skew, and metastable states.

The concept of narrow-width clocking and other more complex timing methodologies are described in C. Mead and L. Conway's classic text, *Introduction to VLSI Design,* Addison-Wesley, Reading, MA, 1980. Chapter 7, "System Timing," contributed by C. Seitz of the California Institute of Technology, provided the motivation for the discussion of clocking strategies, metastable behavior, and self-timed circuits in this chapter. Although it is an advanced presentation, it should be read by every designer attempting to build a high-performance system.

Metastability has plagued digital designers for many years but was not well understood until the mid-1970s. The classic papers describing the phenomenon include Chaney, Ornstein, and Littlefield,

"Beware the Synchronizer," *Proceedings of the Spring COMPCON Meeting,* San Francisco, September 1972, and Chaney and Molnar, "Anomalous Behavior of Synchronizer and Arbiter Circuits," *IEEE Transactions on Computers,* C-22:4, pp. 421–422 (April 1973).

The concepts of self-timed circuits date back to the early days of computers, but the increased difficulty of building such systems has limited their application in digital design. Self-timed concepts, however, are used extensively in advanced dynamic memory components. The best place to start in finding out more about self-timed circuits is in Seitz's chapter in Mead and Conway's book referenced above. A research group, led by A. Martin of the California Institute of Technology, has succeeded in implementing a complete modern microprocessor using self-timed techniques. Their work is reported in the *10th CALTECH VLSI Conference Proceedings;* the conference was held in March 1989 in Pasadena, CA.

EXERCISES

6.1 *(Simple Circuits with Feedback)* Build a feedback circuit with cross-coupled NAND gates. What input conditions cause the state of this latch-like device to be reset? To be set? Does this circuit have forbidden inputs? If so, what are they?

6.2 *(Simple Circuits with Feedback)* An *R-S* latch can be used to determine which of two events has occurred first. Design a circuit with three inputs and three outputs that determines which of three single pole/single throw switches connected to the inputs has been opened first. The circuit will produce a logic 1 on the output that corresponds to that input. Discuss how you would expand this circuit to a larger number of inputs, say 12 inputs or 30 inputs.

6.3 *(Setup and Hold Times)* Imagine that it is possible to have storage devices with negative setup and hold times. What do you think such a concept would be? Draw timing diagrams to illustrate your answer.

6.4 *(Flip-Flops)* Add asynchronous preset and clear inputs to the edge-triggered *D* flip-flop of Figure 6.24. Draw the logic schematic of the revised circuit.

6.5 *(Flip-Flops)* Add synchronous preset and clear inputs to the edge-triggered *D* flip-flop of Figure 6.24. Make it preset-dominant. Draw the logic schematic of the revised circuit.

6.6 *(Flip-Flops)* How would you implement a negative edge-triggered *D* flip-flop using NAND gates only? What changes are necessary to make this a positive edge-triggered device?

6.7 *(Flip-Flops)* Starting with the basic circuit schematic for the master–slave *R-S* flip-flop like the two shown in Figure 6.19, show how to add asynchronous preset and clear inputs to force the flip-flop into a 1 (preset) or 0 (clear) state. Draw a timing waveform for the preset input, clear input, clock, master-stage outputs (P, $\bar{P}$), and slave-stage outputs (Q, $\bar{Q}$) showing the operation of preset and clear.

6.8 *(Flip-Flops)* *R-S* master–slave flip-flops exhibit the phenomenon of ones catching. Briefly explain why this takes place. Can a master–slave flip-flop catch 0s? Explain why or why not.

6.9 *(Flip-Flops)* The basic functionality of a *D* flip-flop can be implemented by a *J-K* flip-flop simply by connecting the input signal *D* to the *J-K* flip-flop's *J* input and $\overline{D}$ to the *K* input. Show that this is true by comparing the characteristic equations for a *D* flip-flop and a *J-K* flip-flop.

6.10 *(Flip-Flops)* Given the input and clock transitions in Figure Ex. 6.10, indicate the output of a *D* device assuming:

(a) It is a negative edge-triggered flip-flop.
(b) It is a master–slave flip-flop.
(c) It is a positive edge-triggered flip-flop.
(d) It is a clocked latch. You may assume 0 setup, hold, and propagation times.

Figure Ex. 6.10 **Timing diagram for Exercise 6.10.**

6.11 *(Flip-Flops)* Identify the following statements as either true or false:

(a) The inputs to a level-sensitive latch always affect its outputs.
(b) Flip-flop delays from the change in the clock edge to the change in the output typically are shorter than flip-flop hold times, so shift registers can be constructed from cascaded flip-flops.
(c) Assuming zero setup and hold times, clocked latches and flip-flops produce the same outputs as long as the inputs do not change while the clock is asserted.
(d) A master–slave flip-flop behaves similarly to a clocked latch, except that its output can change only near the rising edge of the clock.
(e) An edge-triggered *D* flip-flop requires more internal gates than a similar device constructed from a *J-K* master–slave flip-flop.

6.12 *(Flip-Flops)* Match each of the following five circuits (Figure Ex. 6.12) with the phrase that best describes it from the list: (1) clocked *R-S* latch, (2) clocked *D* latch, (3) master–slave *R-S*

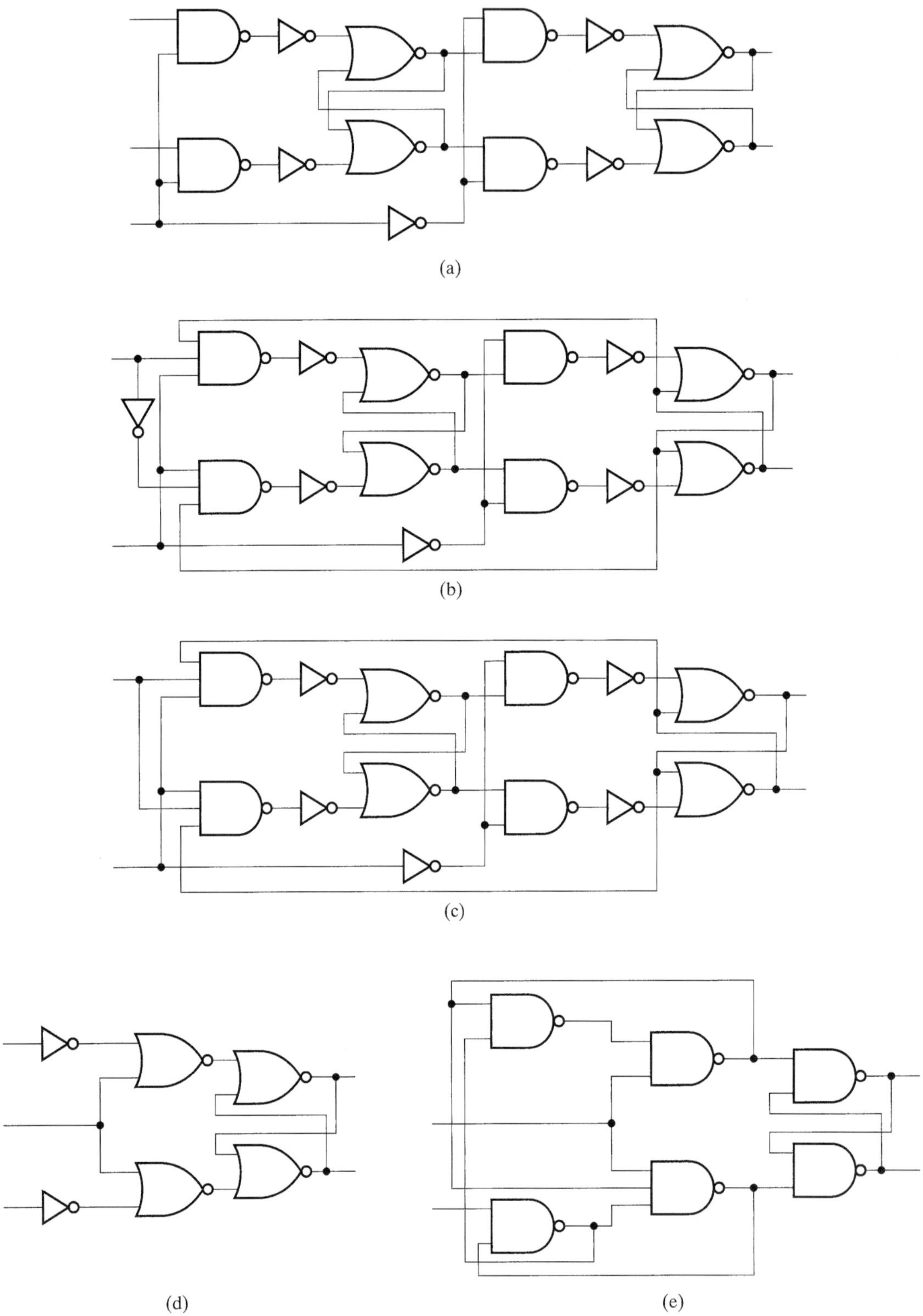

Figure Ex. 6.12

flip-flop, (4) positive edge-triggered *R-S* flip-flop, (5) negative edge-triggered *R-S* flip-flop, (6) master–slave *D* flip-flop, (7) positive edge-triggered *D* flip-flop, (8) negative edge-triggered *D* flip-flop, (9) master–slave *J-K* flip-flop, (10) positive edge-triggered *J-K* flip-flop, (11) negative edge-triggered *J-K* flip-flop.

6.13 *(Flip-Flops)* Any flip-flop type can be implemented from another type with suitable logic applied to the latter's inputs. Show how to implement a *J-K* flip-flop starting with a *D* flip-flop.

6.14 *(Flip-Flops)* Show how to implement a *D* flip-flop starting with a *J-K* flip-flop.

6.15 *(Flip-Flops)* A *T* flip-flop has a single input. When its input is 0, it holds its current value. When its input is 1, it toggles the stored value. Show how to implement a *D* flip-flop starting with a *T* flip-flop.

6.16 *(Flip-Flops)* Show how to implement a *T* flip-flop starting with a *J-K* flip-flop.

6.17 *(Flip-Flops)* Show how to implement a *T* flip-flop starting with a *D* flip-flop.

6.18 *(Clock Skew)* Given the timing specification of the 74LS74 flip-flop of Figure 6.30, what is the worst-case skew in the clock that could be tolerated when one 74LS74 needs to pass its value to another 74LS74, as in Figure 6.33?

6.19 *(Clocking Issues)* Given the sequential-logic circuit of Figure Ex. 6.19, where the flip-flops have worst-case setup times of 20 ns, propagation delays of 13 ns, and hold times of 5 ns, answer the following questions:

(a) Assuming 0 propagation delay through the combinational-logic block, what is the maximum allowable frequency of the clock that controls this subsystem?

(b) Assuming a typical combinational-logic delay of 75 ns and a worst-case delay of 100 ns, how does your answer to part (a) change?

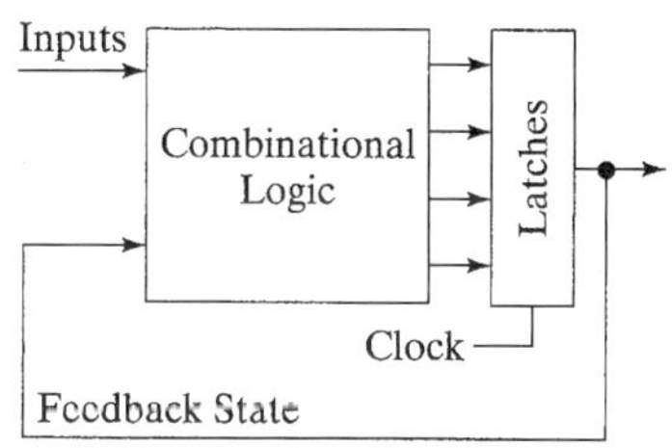

Figure Ex. 6.19 **Sequential circuit for Exercise 6.19.**

6.20 *(Metastability)* You have designed a high-performance disk drive interface. The interface has an internal clock rate of 25 MHz, and asynchronous commands from a computer with a different clock are presented every 200 ns. It works fine, but every few days or weeks it has random operational failures that causes a loss of data. There are no component failures, software bugs, or power glitches; and the errors occur mainly for customers who use the interface heavily. Suggest a possible cause of these failures and how you could change the design to reduce the failure rate.

6.21 *(Metastability)* One way to reduce the probability of synchronizer failure is to place two synchronizer flip-flops in series between the asynchronous input and the rest of the synchronous digital system. Why do you think this reduces the problem of metastability? Why not add more synchronizers in series?

6.22 *(Delay Insensitive Handshaking)* Draw a simple flowchart for the client-side and the server-side algorithms of the four-cycle handshake. Repeat for the two-cycle handshake. How does the complexity of the two approaches compare?

6.23 *(Shift-register Design)* Design the basic cell of a universal shift register to the following specifications. The internal storage elements will be positive edge-triggered D flip-flops. Besides the clock, the shifter stage has two external control inputs, S_0 and S_1, and three external data inputs, SR, SL, and DI. SR is input data being shifted into the cell from the right, SL is data being shifted from the left, and DI is parallel-load data. The current value of the flip-flop will be replaced according to the following settings of the control signals: $S_0 = S_1 = 0$: replace D with DI; $S_0 = 0$, $S_1 = 1$: replace D with SL; $S_0 = 1$, $S_1 = 0$: replace D with SR; $S_0 = S_1 = 1$: hold the current state. Draw a schematic for this basic shifter cell.

6.24 *(Shift-register Design)* Shifters normally are used to shift data in a *circular* pattern (the data that shifts out at one end of the shifter is shifted back into the other end), or as a *logic* shift (fill the shifted positions with 0s) or an *arithmetic* shift (propagate the high-order sign bit to the right or shift in 0s to the left). For example, if a 4-bit register contains the data 1110, the effects of the six kinds of shifts are the following:

Circular shift right: 1110 becomes 0111
Circular shift left: 1110 becomes 1101
Logical shift right: 1110 becomes 0111
Logical shift left: 1110 becomes 1100
Arithmetic shift right: 1110 becomes 1111
Arithmetic shift left: 1110 becomes 1100

Show how to wire up a 4-bit universal shift register (TTL component 74194) to perform the following kinds of shifts:

(a) Circular shift right
(b) Circular shift left
(c) Logic shift right
(d) Logic shift left
(e) Arithmetic shift right
(f) Arithmetic shift left

6.25 *(Shift-register Design)* Your task is to design a shift-register subsystem based on the TTL 74194 component that can implement the six kinds of shifts described in Exercise 6.24. The subsystem has three control inputs, S_2, S_1, and S_0, that are interpreted as follows: S_2, S_1, $S_0 = 000$ is hold; 001 is circular shift right; 010 is circular shift left; 011 is logic shift right; 100 is logic shift left; 101 is arithmetic shift right; 110 is arithmetic shift left; and 111 is parallel load.

(a) Show the data path for the shifter subsystem. You may use multiplexers at the shift inputs to the 74194.

(b) Show the combinational logic (equations or schematics) to decode the global S_2, S_1, and S_0 control inputs into the appropriate detailed control signals for the 74194 shifter and the external data-path logic for handling the serial-shift inputs.

6.26 *(Register Design)* A FIFO (first in, first out) queue is a special-purpose register file n words deep and m bits wide that operates as follows (see the block diagram in Figure Ex. 6.26(a)). When a PUSH_DATA control input is asserted, new data at the inputs at the right is read into the end of the queue. When a POP_DATA control input is asserted, existing data at the head of the queue becomes available at the outputs at the left. Since the FIFO has finite capacity, two status outputs indicate whether the FIFO is empty or full. PUSH_DATA is inhibited in a full FIFO, while POP_DATA is inhibited in an empty FIFO. On reset, the FIFO should be set to empty.

A "flow-through" FIFO is the simplest form of this kind of device. The FIFO must fill up with data before any data can be removed. Furthermore, the FIFO must be emptied completely before new data can be placed in it. Using only shift-register components and combinational logic, design a flow-through 4-word by 4-bit FIFO. Consider carefully how to represent the empty/full status of the FIFO. (*Hint:* Consider adding an $(m+1)$st bit to the FIFO to indicate whether the FIFO word is valid.)

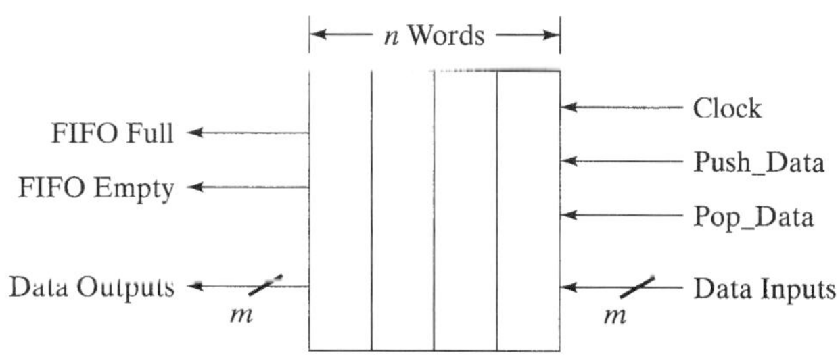

(a) FIFO block diagram

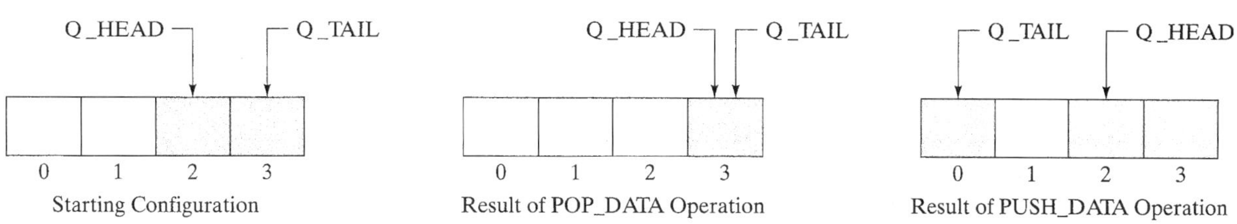

(b) PUSH_DATA and POP_DATA FIFO operations

Figure Ex. 6.26

6.27 *(Register Design)* A LIFO (last in, first out) stack is similar in concept to the FIFO queue, except that the most recently pushed data is the first to be popped. The block diagram is identical to the FIFO, except that the data inputs and outputs are the same lines. Design a 4-word by 4-bit LIFO stack using shift registers and combinational logic only. Draw your schematic, indicating the components used. How do you distinguish between a full stack and an empty stack in this implementation?

6.28 *(Register Design)* One way to compute the twos complement of a number is examine the number bit by bit from the lowest-order bit to the highest-order bit. In scanning from right to left, find the first bit that is 1. All bits to the left of this should now be complemented to form the twos complement number. For example, the twos complement of 0010 is formed as follows:

```
0010 ──────> 1101  (Ones complement)
  ↓           + 1
1110         1110  (Twos complement)
```

Your task is to draw a schematic for a 4-bit register with parallel inputs and outputs and the synchronous control signals HOLD, CLEAR, LOAD, and COMPLEMENT. When the complement signal is asserted, the register's contents will be replaced by its twos complement. You can assume that COMPLEMENT will be asserted for a number of cycles equal to the width in bits of the register.

6.29 *(Serial Communication Application)* Figure 6.56 shows the partial implementation of a bit-serial transmission subsystem. This subsystem is unidirectional: it can transmit only from the left shift registers to the right.

Extend this data-path implementation to make it bidirectional. Under external control, include logic (such as multiplexers) that allows the subsystem to transmit from right to left as well as left to right.

6.30 *(Hardware Description Languages)* Describe a master–slave flip-flop with its ones-catching problem as a Verilog module. Do not use gates but only behavioral constructs.

6.31 *(Hardware Description Languages)* Describe a circular shift register using Verilog blocking and non-blocking assignments. There should be 3 flip-flops in your register with the value of FF_1 going to FF_2, FF_2 to FF_3, and FF_3 to FF_1.

6.32 *(Hardware Description Languages)* Describe the universal shift register of Figure 6.53 in Verilog. Use the same hierarchy as in the example: create a module for a single bit and then instantiate four 1-bit modules to create the 4-bit module.

CHAPTER 7

Finite State Machines

Introduction

In this chapter, we will use the sequential logic elements of the previous chapter to build *finite state machines*. Finite state machines, so named because the sequential logic that implements them can be in one of a fixed number of possible states, are the most important abstraction for sequential logic. We'll begin with some of the simplest finite state machines, namely, *counters*, and use them to gently introduce the key concepts. Counters are registers with additional logic that cycle the register's contents through a predefined sequence of states. The pattern of register output values in each state can be interpreted as a binary or binary-coded decimal (BCD) number. If the sequence of values increases then we have an *up-counter;* if it decreases then we have a *down-counter*. For a counter, we think of its output value as being identical to its state.

We'll then move on to the more general concept of finite state machines. The outputs and next state of a finite state machine are combinational logic functions of their inputs and present state. The choice of next state can depend on the value of an input, leading to richer possible behaviors than that of simple counters which are only dependent on the previous state. These, more general, finite state machines are critical for realizing the control and decision-making logic in digital systems.

We will develop a general design procedure for finite state machines that will translate an abstract representation into gates and flip-flops. We will then motivate the possible optimizations that can be realized in the design of finite state machines. We will cover:

- *Construction of various kinds of counters.* Starting with the flip-flop building blocks introduced in the previous chapter, we show how to build a variety of different counters.
- *Counter design procedure.* We develop a procedure for mapping a behavioral description of a counter into an implementation based on flip-flops and combinational logic. This is a primitive form of the more general finite-state-machine design procedure we will develop later in the chapter.

- *Methods for describing the behavior of finite state machines.* These include state diagrams and unencoded and encoded state tables. In the next chapter, we will discuss how to describe finite state machines using hardware description languages.
- *Basic design procedure for finite state machines.* We will use several examples to illustrate the process of mapping a high-level abstract description into a collection of flip-flops and combinational logic.
- *Motivation for optimization of finite state machines.* We will show how there are several steps in the design procedure where different choices can lead to very different realizations and help designers meet design constraints. The optimization methods themselves are the subject of the next chapter.

7.1 Counters

Counters are sequential logic circuits that proceed through a well-defined sequence of states. We study them in more detail in this section.

Up-Counters and Down-Counters An *up-counter* proceeds from a value to the next larger value in the sequence in response to a count pulse. The counter outputs are identical to the state of the internal flip-flops. Once the largest possible value has been reached, the sequence restarts with the smallest value. For example, a 3-bit *binary up-counter* begins in state 000; goes through the sequence 001, 010, 011, 100, 101, 110, 111; and returns to 000, 001, and so on.

As you might expect, a *down-counter* operates similarly, except that it sequences from large values to smaller values. The counter wraps from the smallest possible value back to the largest value in the sequence.

Binary, Decade, and Gray-Code Counters Although we have been describing binary counters, a counter can be designed to advance through any periodic sequence. Popular alternatives to binary counters are *decade* counters. These sequence through the binary-coded decimal (BCD) digits: 0000, 0001, 0010, 0011, 0100, 0101, 0110, 0111, 1000, and 1001.

In some applications, to avoid circuit hazards, it is important that only a single bit of the counter changes at a time. The binary and decade counters do not exhibit this property. However, a Gray-code counter does. A 4-bit Gray-code up-counter would sequence through the states 0000, 0001, 0011, 0010, 0110, 0111, 0101, 0100, 1100, 1101, 1111, 1110, 1010, 1011, 1001, 1000 and repeat. Note that this pattern is similar to what we've seen in K-maps, where adjacent cells differ in only 1 bit.

Ring Counters The shift registers of the previous chapter can also be used as a kind of primitive counter, a *ring counter,* by connecting the output of the last stage to the input of the first thereby forming a ring as shown in Figure 7.1. Assuming we start the shift register in the state 1000, through appropriate reset logic, it sequences through the states 0100, 0010, 0001 and then repeats to 1000. The ring counter uses minimal hardware for its implementation, but it does not encode its states

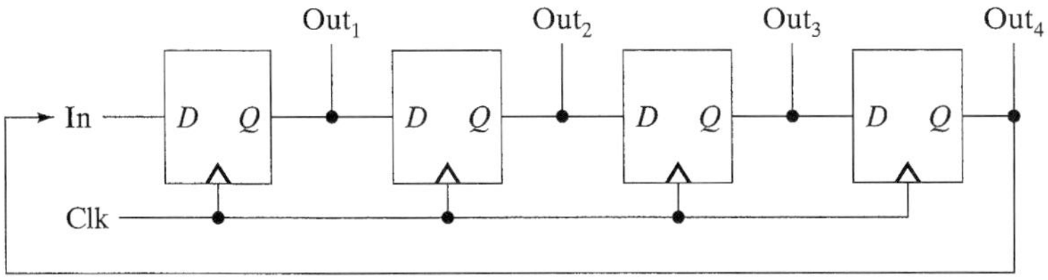

Figure 7.1 **Four-bit simple ring counter.**

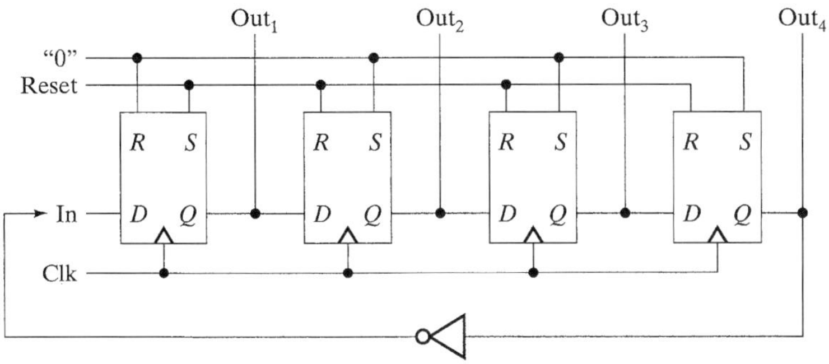

Figure 7.2 **Four-bit Johnson counter.**

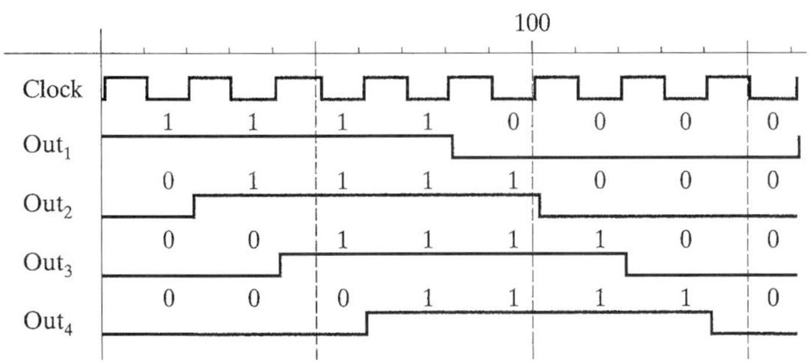

Figure 7.3 **Timing waveforms for 4-bit Johnson counter.**

as efficiently as the counters we have just described. The four-element ring counter sequences through only four states, compared with the 16 states of the 4-bit binary counter.

A *Johnson counter* (also known as a *Mobius counter*) requires only one inverter more than the basic ring counter but can sequence through twice as many states. The idea is to complement the feedback output of the rightmost flip-flop to yield the shifter/ring counter. This has the effect of complementing the bit shifted out to the right before it is reinserted into the shifter from the left.

We show an implementation of the 4-bit Johnson counter in Figure 7.2. We use the asynchronous set and reset inputs to force the flip-flops into the initial state 1000 when reset is asserted.

Figure 7.3 shows the timing diagram. The counter sequences through the states 1000, 1100, 1110, 1111, 0111, 0011, 0001, 0000 and repeats. It just so happens that this sequence satisfies the same property as the Gray code: only a single bit changes its value from one state to the next, but unlike a Gray-code counter, the Johnson counter has only eight states, rather than 16.

7.1.1 Counter Design Procedure

This section begins our study of designing an important class of clocked sequential logic circuits—*synchronous finite state machines.* Like all sequential circuits, a finite state machine determines its outputs and its next state from its current inputs and current state. A synchronous finite state machine changes state only on a clocking event.

Counter design is a good place to start understanding the design process for finite state machines. Counters are the simplest possible finite state machines. They typically have only a single input instructing them to count (often just the clock, that is, they count at each and every clock tick), and their outputs are nothing more than their current state.

EXAMPLE 7.1 THREE-BIT UP-COUNTER

Let's start with a simple counter, a 3-bit binary up-counter. We begin the design process by understanding how the counter is to operate. A convenient way to describe this is with a graphical specification called a *state transition diagram.*

The state transition diagram is a graph with nodes and directed arcs. Each node represents a unique state of the counter. A directed arc connects two nodes representing the present state and the next state. If the counter is in the state at the tail of the arc, it will advance to the state at the head of the arc at the next count request.

For the example design, we will simply advance the counter to its next state on each clock pulse. Figure 7.4 shows a state transition diagram for the example. The nodes are labeled with the bit pattern of the counter state they represent, and the arcs connect the nodes in the sequence we want the counter to implement. We can describe the behavior of any finite state machine with a state transition diagram, although the diagrams typically are more complex than those for counters.

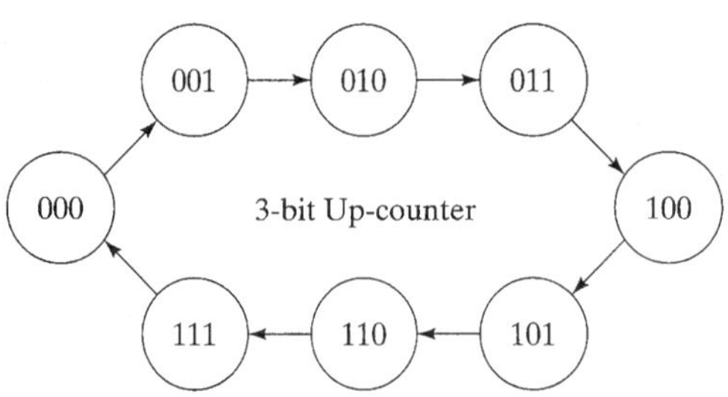

	Present State			Next State			
	C	B	A	$C+$	$B+$	$A+$	
0	0	0	0	0	0	1	1
1	0	0	1	0	1	0	2
2	0	1	0	0	1	1	3
3	0	1	1	1	0	0	4
4	1	0	0	1	0	1	5
5	1	0	1	1	1	0	6
6	1	1	0	1	1	1	7
7	1	1	1	0	0	0	0

Figure 7.4 **State transition diagram and table for a 3-bit binary up-counter.**

An alternative formulation of the state transition diagram is the *state transition table* or, simply, *state table,* which shows the present state and its corresponding next state. Each row corresponds to an arc in the state transition diagram. The state transition table for the up-counter is also shown in Figure 7.4.

Each bit of the state is held by a single storage element. In this example, the counter proceeds through eight states each corresponding to a binary code. Thus, we need exactly three storage elements. We have named the storage elements *C*, *B*, and *A*, from the highest- to the lowest-order bit (the outer columns of the state table are simply the decimal value of the state bit pattern).

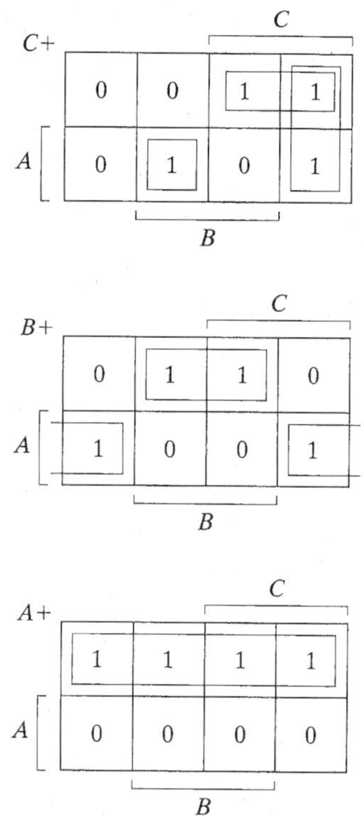

Figure 7.5 **K-maps for up-counter flip flops.**

Next-State Logic

Our task now is to design combinational logic whose input is the current state of the counter and whose output is the inputs to the state flip-flops (the "next state" to be loaded into the flip-flops at the next clocking event). For this simple example, we can determine the logic just by examining the transition table. Flip-flop *A* toggles on each state transition, *B* toggles whenever *A* is asserted (every other count), and *C* toggles whenever *A* and *B* are asserted (every fourth count).

For more complex examples, we can view the transition table as a truth table that specifies the flip-flops' inputs as a function of *C*, *B*, and *A*. We use standard K-map methods to obtain the reduced Boolean expressions. The K-maps for *C*+, *B*+, and *A*+ are shown in Figure 7.5. This leads to the circuit design of Figure 7.6. Note that this isn't obvious immediately from the K-maps. You may want to verify that the logic in this circuit is equivalent to the logic equations suggested by the K-map method. Our objective in transforming them into the form of Figure 7.6 was to make each of the three bits look like they have similar logic, a module that can be replicated. The XOR gate is used to complement the toggle the output of its respective flip-flop or not

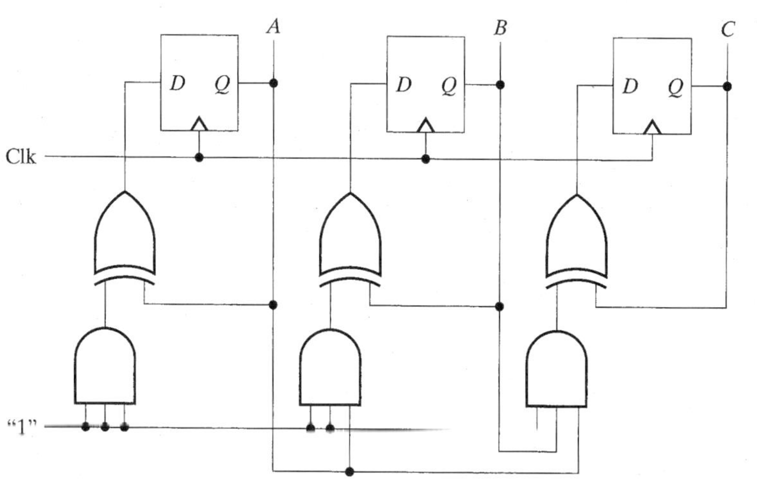

Figure 7.6 **Circuit diagram of three-bit binary up-counter.**

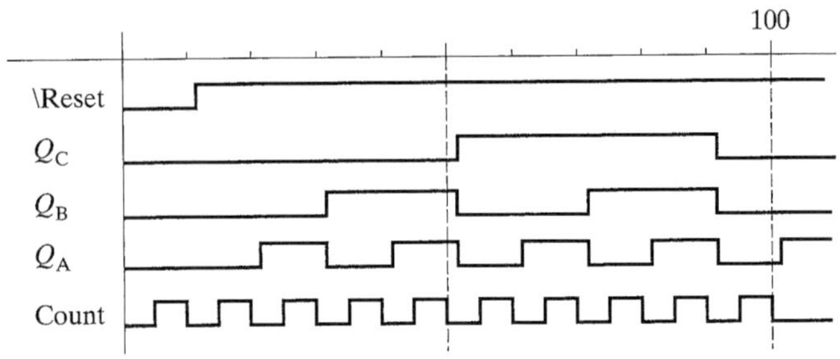

Figure 7.7 Timing waveform of three-bit binary up-counter.

depending on the value of its other input. It toggles if all the lower-order bits are asserted. For the lowest-order bit, this is all the time, hence the three 1s as input to the AND gate. For the *B* its whenever *A* is asserted. For *C* whenever both *A* and *B* are asserted. If we had a fourth bit, *D*, its AND gate would have *A*, *B*, and *C* as inputs. We could have implemented the logic as prescribed by the K-maps, but this implementation is more regular even if it may waste some gates and gate inputs. That is, we can use a similar module containing a flip-flop, an XOR gate, and an AND gate for each *bit slice*. The timing waveform of this implementation is given in Figure 7.7.

7.1.2 Counters with More Complex Sequencing

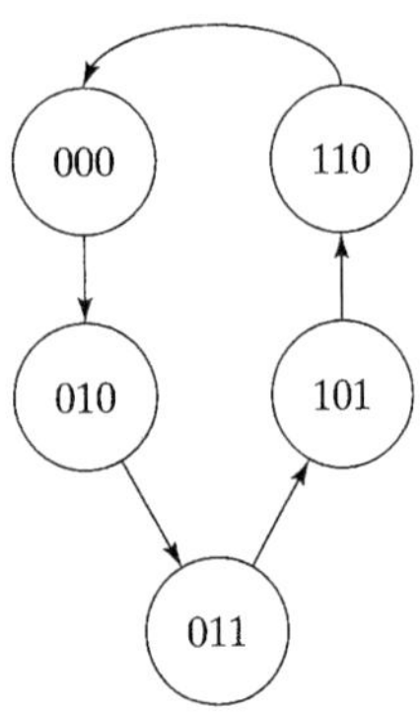

Figure 7.8 State transition diagram.

The generalized design process consists of the following three steps:

STEP 1 From the written specification of the counter, we first draw a state transition diagram that shows the counter's desired sequence of states.

STEP 2 We next derive the state transition table from the state diagram, tabulating the present state with its corresponding next state in the count sequence. Each state bit is implemented by its own flip-flop.

STEP 3 We express each next-state bit as a combinational logic function of the present state bits. This will be the logic function at the input of the flip-flops.

EXAMPLE 7.2 GENERALIZED COUNTER DESIGN

To see this three-step process in action, let's look at another implementation of a counter. We will design a 3-bit counter that advances through the sequence 000, 010, 011, 101, 110, 000 and so on. Not all of the possible combinations of the 3 bits represent a valid state. The unused states, 001, 100, and 111, can serve as don't-care conditions in helping us simplify the logic.

STEP 1 Draw the state transition diagram. This is shown in Figure 7.8.

STEP 2 Derive the state transition table. This is shown in Figure 7.9.

Present State			Next State		
C	*B*	*A*	*C+*	*B+*	*A+*
0	0	0	0	1	0
0	0	1	–	–	–
0	1	0	0	1	1
0	1	1	1	0	1
1	0	0	–	–	–
1	0	1	1	1	0
1	1	0	0	0	0
1	1	1	–	–	–

Figure 7.9 State transition table.

STEP 3 Express each next-state bit as a combinational logic function of three current-state bits. Figure 7.10 shows the appropriate K-maps.

By making use of the don't cares provided by the unused states, we derive the following equations:

$$C+ = A$$
$$B+ = \bar{A}\bar{C} + \bar{B}$$
$$A+ = B\bar{C}$$

Figure 7.11 shows the component-level implementation, with its associated timing waveform in Figure 7.12. We use the label *Count* for the clock signal. Note that we continue to use *Reset* to put our circuit into the 000 starting state and show it in the timing waveform but not in the circuit schematic to reduce clutter. To further reduce wiring clutter in the diagram, we simply label inputs and outputs rather than draw connecting wires. Two wires with the same label are understood to be connected. The proper sequencing through the states 000, 010, 011, 101, 110, and 000 should be clear from the waveform. Note that we use negative edge-triggered flip-flops in this example.

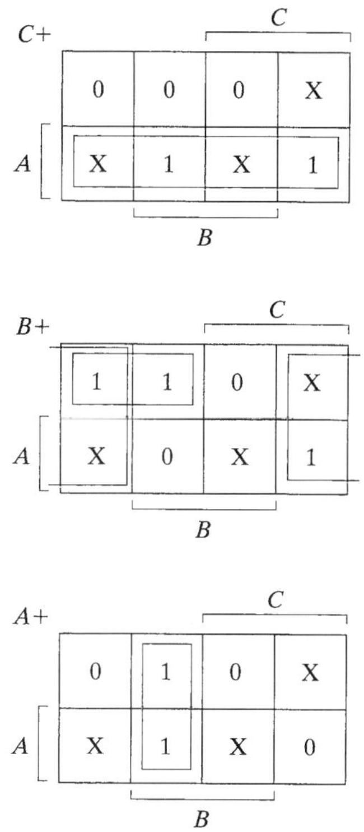

Figure 7.10 K-maps for next-state functions.

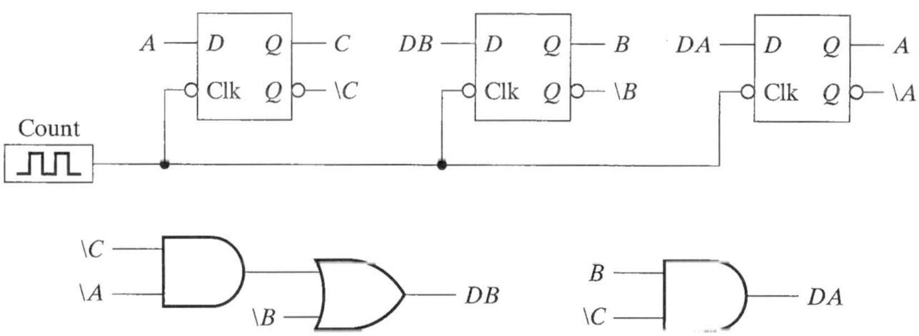

Figure 7.11 Implementation of the 3-bit counter.

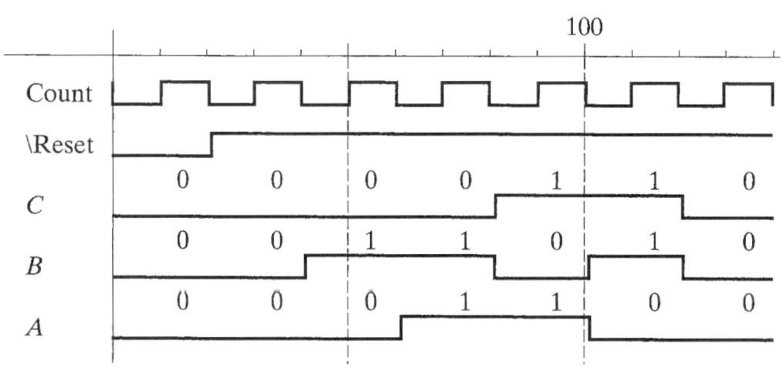

Figure 7.12 Timing waveform of the 3-bit counter.

7.1.3 Self-Starting Counters

In drawing state transition diagrams for the example counters in the previous section, we assumed implicitly that they would begin the sequence in state 000. When working with real hardware, you should never assume that the counter will start in a particular state unless you design it to do so! At the time of power-up, the states of the flip-flops are undefined: they could be 0 or 1 at random. This is why we included the reset signal.

However, if we want to save space, and not use a reset signal or special resettable flip-flops, we have a particularly nasty problem. For counters that use every state this is not a big issue as they will start in one of their states and proceed through their sequence. We may not know where they start but the counting sequence will be the same. However, for counters that do not use all state combinations of the storage elements, such as the last example, it is a much more serious issue. What would happen if, by chance, our example counter had entered state 001 on start-up? Of course, it depends on how don't cares have been mapped into 0s and 1s by the implementation procedure, but the counter could sequence through the noncounter states, 001, 100, and 111, never entering the sequence it was designed for at all as shown in Figure 7.13(a).

A *self-starting* counter is one in which every possible state, even those not in the desired count sequence, has a sequence of transitions that eventually leads to a valid counter state. This guarantees that no matter how the counter starts up, it will eventually enter the proper counter sequence. Counters that use all possible states are always self-starting.

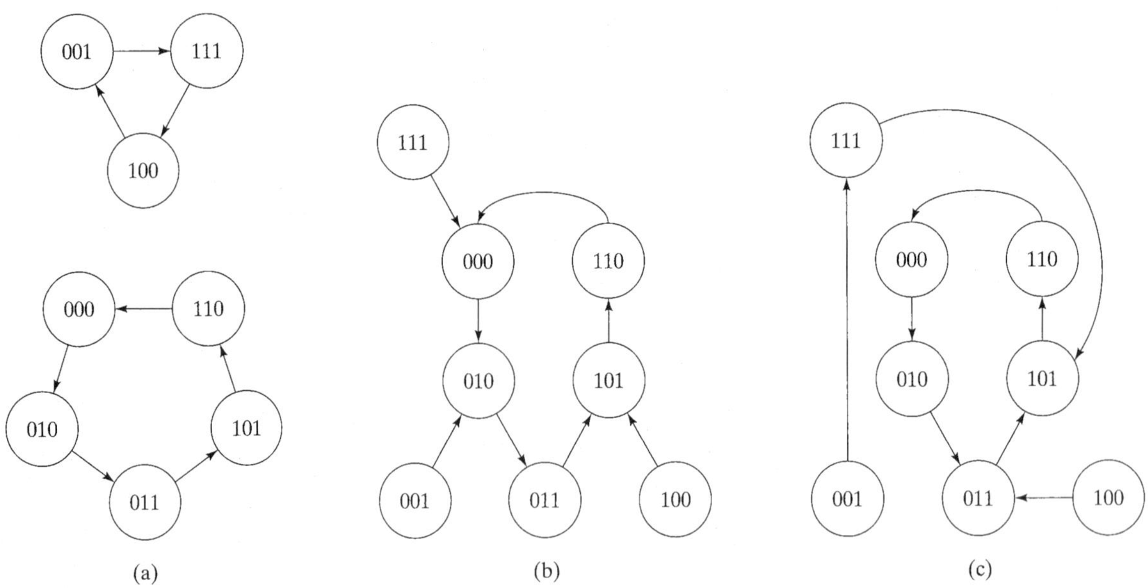

Figure 7.13 **Three counter state diagrams, one non-self-starting (a) and two self-starting (b and c).**

Figure 7.13 shows three different state transition diagrams that meet our counter specification. The first is nonself-starting while the other two are self-starting implementations for the counter of Figure 7.8. In the state diagram of Figure 7.13(b), the counter is guaranteed to be in the correct sequence after one transition. In Figure 7.13(c), the counter may require up to two transitions before it is in the correct sequence.

In general, it is desirable to enter the counter sequence in as few transitions as possible, so we would prefer the state diagram in Figure 7.13(b). However, there may be an advantage in departing from this rule if a particular sequence of noncounter states can lead to reduced hardware.

EXAMPLE 7.3 REVERSE-ENGINEERED COUNTER DESIGN

What is the state diagram for the counter of Figure 7.11? To see how this can be derived, let's reverse-engineer the counter step-by-step as illustrated in Figure 7.14. First, we replace the don't cares in the K-maps of Figure 7.10 with the actually assigned 1s and 0s. By reading across the K-maps, it is easy to see that the current state $CBA = 100$ changes to the new state 010. When we are done filling in the three don't care states, we see that our implementation is self-starting and can take up two cycles to enter the defined five-state periodic sequence.

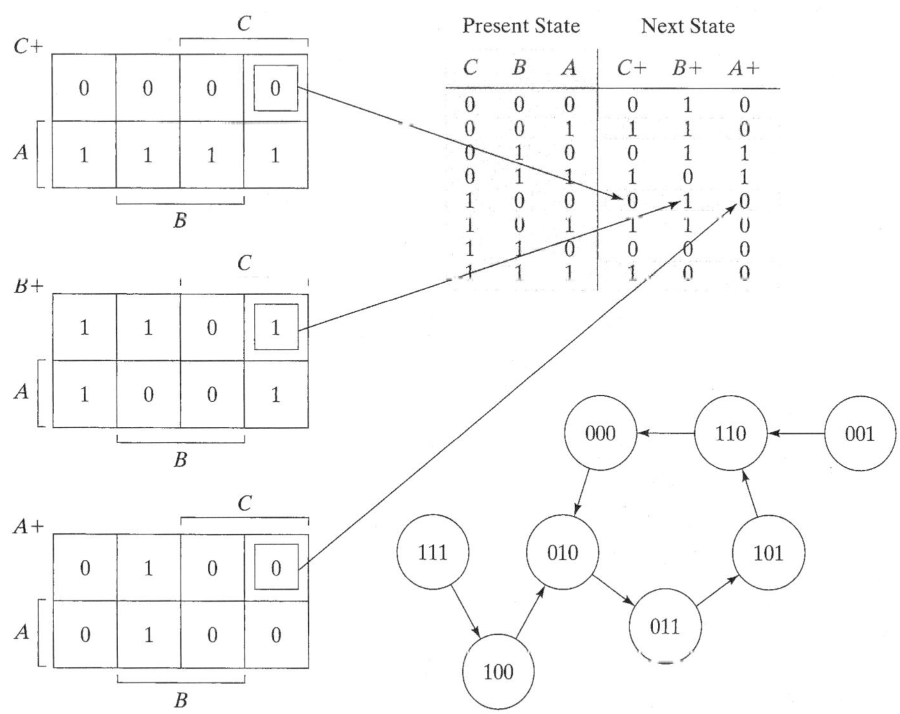

Present State			Next State		
C	B	A	C+	B+	A+
0	0	0	0	1	0
0	0	1	1	1	0
0	1	0	0	1	1
0	1	1	1	0	1
1	0	0	0	1	0
1	0	1	1	1	0
1	1	0	0	0	0
1	1	1	1	0	0

Figure 7.14 **Deriving a state transition table given the don't-care assignment.**

7.1.4 Counter Reset

In the preceding discussion, our primary goal was to enter the count sequence at a valid state. The particular starting state did not matter. Although this may be true in some applications, it is more usual to have a distinguished *starting* state for the counter or finite state machine.

If this is the case, it is desirable to include additional hardware to *reset* the counter to this state. We can accomplish this with an explicit reset button or a special circuit that asserts the reset signal for a short time after power-up. Flip-flops typically come with preset and clear inputs. Thus, we can use an asserted reset signal to place the flip-flops in the desired starting state. The implementation in Figure 7.2 includes such circuitry to reset the flip-flops to state 1000. Of course, by judicious use of preset inputs as well as clear, we can choose any state as the starting state.

7.1.5 Counter Variations

Counters are used in digital systems to count events and find many applications in tracking the passage of time. See the exercises at the end of this chapter for some design problems whose solutions rely on counters. As you might expect, there are many common variations of counter designs that have led to the development of catalog parts. We'll look at the most common of these in detail in this section. This 4-bit counter is designed so that it can be cascaded to build larger counters and create binary sequences that are offset at both start and end.

EXAMPLE 7.4 CATALOG COUNTER

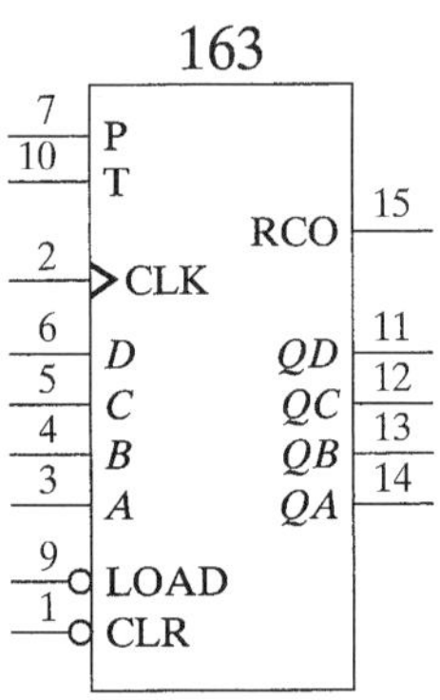

Figure 7.15 **Schematic shape of 4-bit synchronous up-counter.**

Let's examine a counter component from the TTL catalog. Figure 7.15 shows the schematic shape of the 163 synchronous 4-bit counter. The component has four data inputs, four data outputs, four control inputs ($P, T, \overline{LOAD}, \overline{CLR}$), one control output ($RCO$), and the clock.

When the $\overline{LOAD}$ signal is asserted, the data inputs replace the contents of the counter's internal flip-flops. Similarly, when $\overline{CLR}$ is asserted, all the flip-flops are reset to zero. These operations are synchronous. Although the load or clear signal must be asserted before the clock's positive edge, the actual operation occurs at the clocking event. Contrast this with asynchronous operation, in which the load or clear takes place as soon as the appropriate control signal is asserted. We will say more about synchronous versus asynchronous counter operation later.

The P and T inputs are two enable signals. They cause the counter to advance to the next state in the binary sequence when both are asserted. Internally, there is a small amount of logic that ANDs P and T and uses the result to select which is the real next state for the counter: the next state in the sequence (if P AND T is true) or the same state to hold the sequence (if P AND T is false).

Again, the count operation takes place when the clock undergoes a low-to-high transition. RCO is a ripple-carry output that is asserted after the same rising clock edge that advances the counter to its largest value, 1111. This signal can be used as a count enable signal to a second

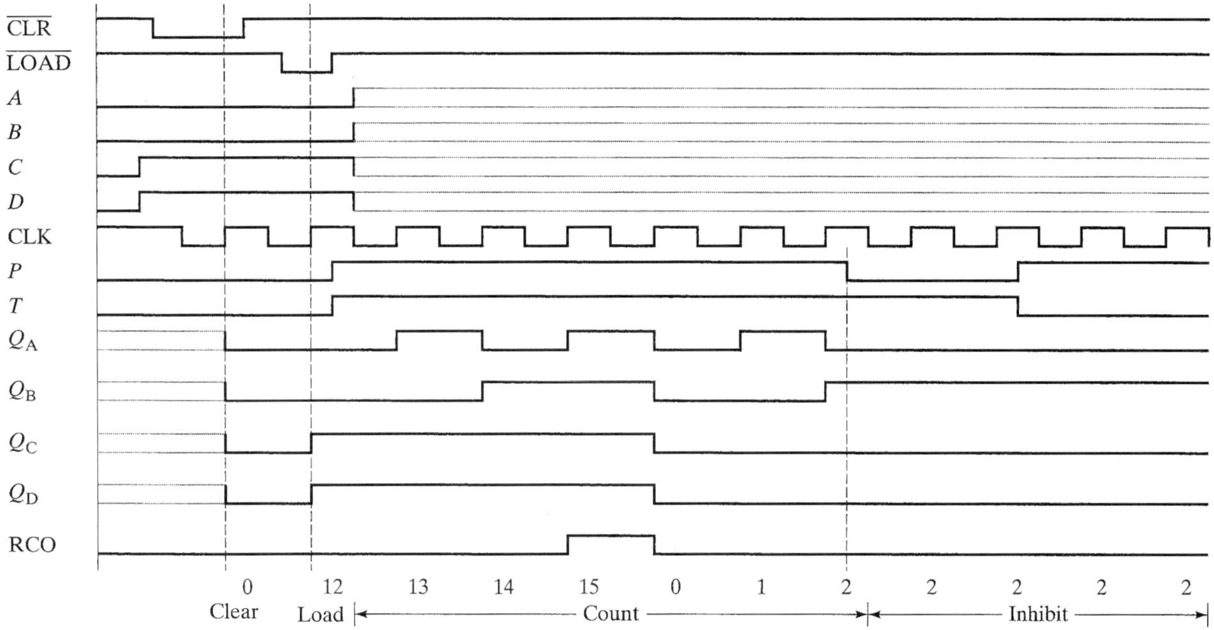

Figure 7.16 **Timing waveform of 163 synchronous 4-bit up-counter.**

cascaded counter. This is why we have two enable signals. One is used to enable counting, the other is connected to the *RCO* output of a previous stage. Thus, subsequent stages are enabled to count every 16th time the previous stage counts making it easy to build larger counters.

The timing waveform of Figure 7.16 illustrates the interaction of these signals. When the clear signal is asserted, the counter's outputs are set to 0 on the next rising clock edge. When load is asserted, the counter is preset to 1100 and the data on the parallel load inputs. When the count enable inputs, *P* and *T*, are enabled simultaneously, the counter counts up on each subsequent rising clock edge. As soon as the counter's output reaches 15, the *RCO* output is asserted. When either of *P* or *T* becomes unasserted, the counter stops counting and holds its current value through subsequent rising edges of the clock (labeled INHIBIT as the counter is *inhibited* from counting). Note that the values of *A*, *B*, *C*, and *D* are immaterial when LOAD is not asserted.

EXAMPLE 7.5 CASCADED COUNTERS

Cascading is the process of combining several smaller-bit-width components into a larger-bit-width function. As you have seen, an 8-bit shift register can be constructed by wiring together two 4-bit registers. You can use a similar strategy to build wider counters from multiple 4-bit slices. The components are wired together in a series ripple fashion; hence, they are referred to as *ripple* or *cascaded counters*.

Figure 7.17 shows an 8-bit cascaded counter. Shortly after the rising edge in which the low-order counter enters the state 1111, the *RCO, ripple carry output,* is asserted. This signal enables the next

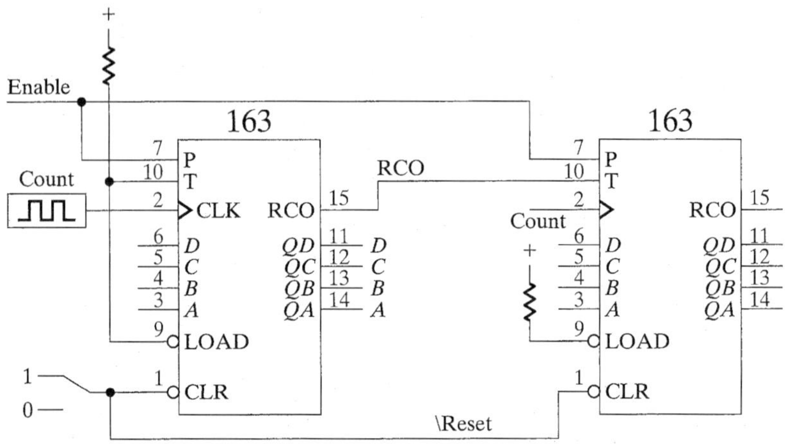

Figure 7.17 **Eight-bit counter constructed from cascaded counters.**

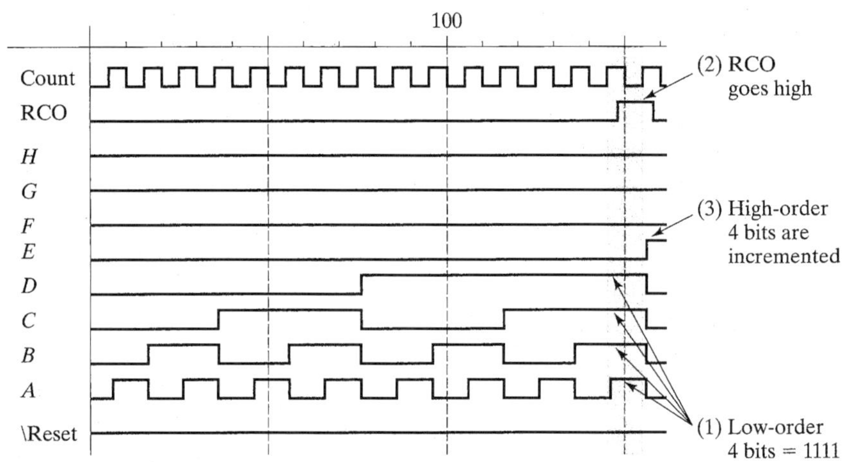

Figure 7.18 **Timing waveform of cascaded counters.**

higher stage for counting. On the next rising edge, the low-order bits enter state 0000, the high-order bits count by 1, and the *RCO* signal of the low-order stage is unasserted. In a full 8-bit binary counter, this causes the high-order stage to only count once for every 16 count cycles of the low-order stage.

The detailed sequence of events is given in the timing waveform of Figure 7.18. First, the low-order counter enters the state 1111. Second, the *RCO* is asserted. Third, the higher-order stage counts by 1.

Note that the cascaded counters of Figure 7.17 are still synchronous. The counter's internal logic computes the *RCO* signal in parallel, simply by ANDing together the current-state bits. The count takes place on the rising edge of the clock. As long as there is enough time between rising edges for *RCO* to be asserted by the low-order stage and recognized by the high-order stage, the cascaded counters function properly. The *Enable* input is used to enable the entire counter.

However, the higher-order 4-bits also requires *RCO* from the first stage to be high.

The ubiquitous 163 counter illustrates the power of synchronous inputs. In addition to its internal synchronous implementation, it provides synchronous clear and load signals. This makes it ideal for implementing more complex count sequences with a beginning offset or limiting cutoff.

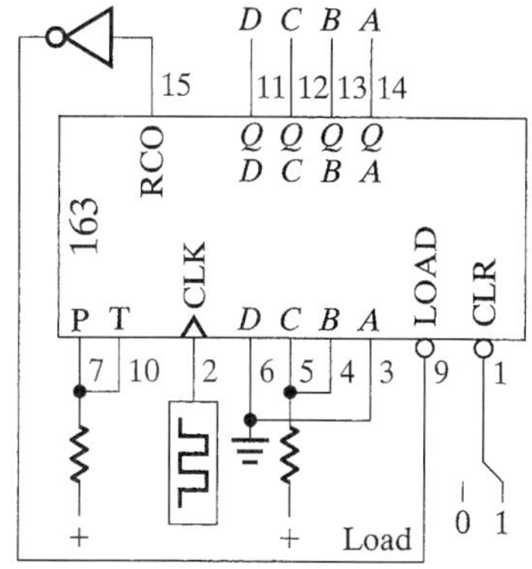

Figure 7.19 Counter with beginning offset.

EXAMPLE 7.6 STARTING OFFSET COUNTER

Suppose we need to implement a counter that follows the sequence 0110 through 1111 and then repeats. The *RCO* signal, in conjunction with a synchronous load input, can implement this function easily.

The logic is shown in Figure 7.19. The *RCO* signal is complemented and used to drive the $\overline{LOAD}$ input. If the desired starting state is placed at the parallel load inputs, it will be entered on the next rising edge of the clock after 1111 instead of 0000. The timing waveform is given in Figure 7.20. The active low control signal $\overline{LOAD}$ (the complement of *RCO*) is asserted when the counter enters state 1111. On the next rising clock edge, the counter enters state 0110.

This counter is self-starting. Resetting clears the counter to 0000. Although it starts off in an invalid state, it reaches the desired sequence within six clock periods.

EXAMPLE 7.7 CUTOFF LIMIT COUNTER

Similarly, we can construct a counter with a cutoff limit by using the synchronous clear signal. For example, we can use the logic of Figure 7.21 to implement a counter that begins at 0000 and sequences through to 1101 before restarting. When the counter enters the cutoff state, the active low $\overline{CLR}$ signal is asserted. This forces the counter into the zero state on the next rising edge.

Figure 7.22 shows the timing diagram. When the counter reaches 1101, it enters the cutoff state, $\overline{CLR}$ goes low, and the next state is 0000.

Dangers of Asynchronous Inputs The designs of Figure 7.19 and Figure 7.21 would not be possible if the clear and load inputs were asynchronous.

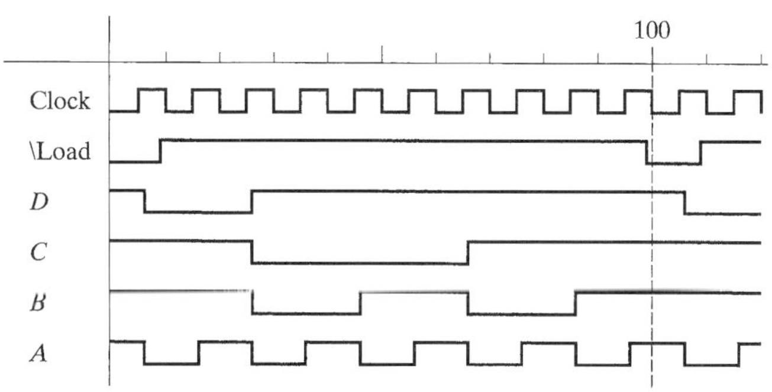

Figure 7.20 Count sequence 0110 through 1111 and repeat.

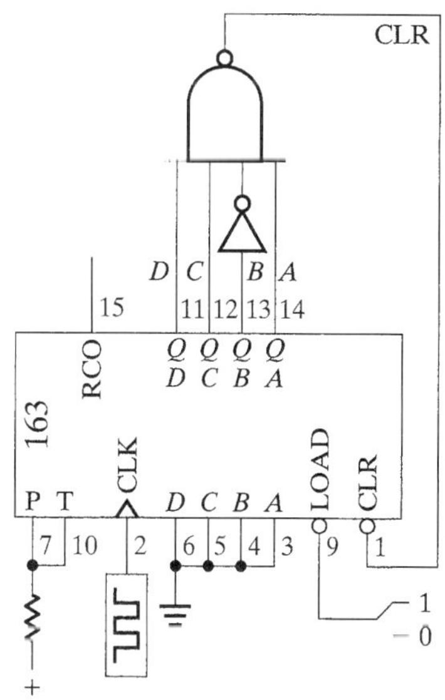

Figure 7.21 Counter with limit.

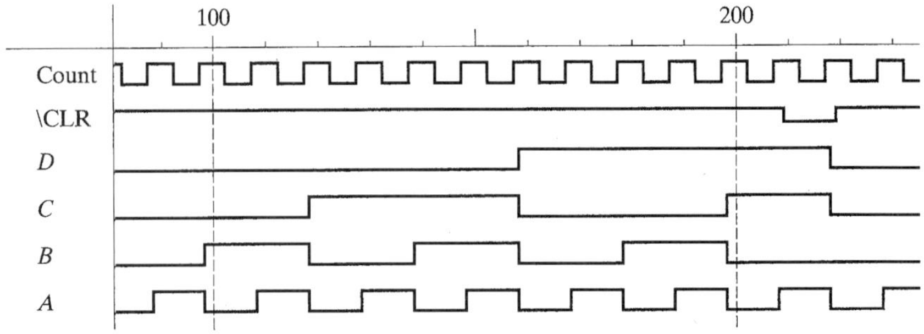

Figure 7.22 Timing waveform for offset limited counter.

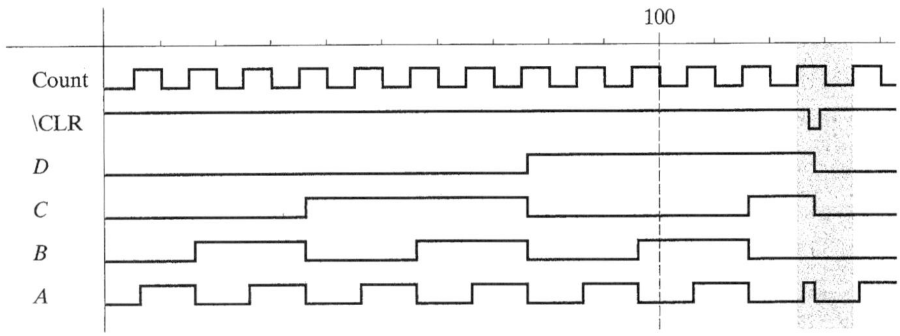

Figure 7.23 Sequence implemented with the 161 with asynchronous clear.

We use external logic to recognize the offset or cutoff state and assert a control input that leads to an out-of-sequence new state. In these synchronous designs, the state does not change until the next rising clock edge.

If the control inputs were asynchronous, the state change would have happened as soon as the control input was asserted, independent of the clock. This violates a fundamental assumption of synchronous systems—that the state changes once and only once at clocking events and not at other times.

Let's reexamine the counter of Figure 7.21, but this time implemented with the 161 counter. This device is identical to the 163, except that the clear signal is asynchronous.

Compare the timing diagram of Figure 7.23 with that of Figure 7.22. The $\overline{CLR}$ signal is obviously a short-duration pulse, and state 1101 is held for a much shorter time than any other state. In effect, if we consider the counter to advance from one state to the next on every rising clock edge, then the behavior of Figure 7.23 actually shows the counter advancing from 1100 to 0000, passing through 1101 in a brief instant (much shorter than a full clock period). This is obviously not the behavior we desired.

Asynchronous inputs should be used only for situations such as power-on reset. Never use them to implement state transitions. As a designer, you should always be aware of the behavior of the catalog parts you select. Remember to choose the right parts for the job at hand.

7.2 The Concept of the State Machine

Counters are very limited finite state machines. They do not have any external inputs besides a clock and reset signal. Now it is time to begin our study of general finite state machines with an example sequential logic function that depends on its history of inputs to determine its output and not just on its past state. We will see how our specification method can be extended to handle inputs as well as more general outputs, those that also depend on current input values in addition to the current state. We will see the complete process of transforming a specification of the function, through a variety of equivalent representations, resulting in an actual implementation using gates and flip-flops.

7.2.1 Odd or Even Parity Checker

Consider the design of a logic circuit that counts the number of 1s in a bit-serial input stream. If the circuit asserts its output when the input stream contains an odd number of 1s, it is called an *odd parity checker.* If it asserts its output when it has seen an even number of 1s, it is an *even parity checker.* The circuit is clearly sequential: the current output depends on the complete history of inputs.

State Diagram The first step of our design process is to develop a state diagram that describes the behavior of the circuit. It's not too hard to see that the circuit can be in one of two different states: either an even or an odd number of 1s has been seen since reset. Whenever the input contains a 1, we switch to the opposite state. For example, if an odd number of 1s has already been seen and the current input is 1, we now have seen an even number of 1s. If the input is 0, we stay in the current state.

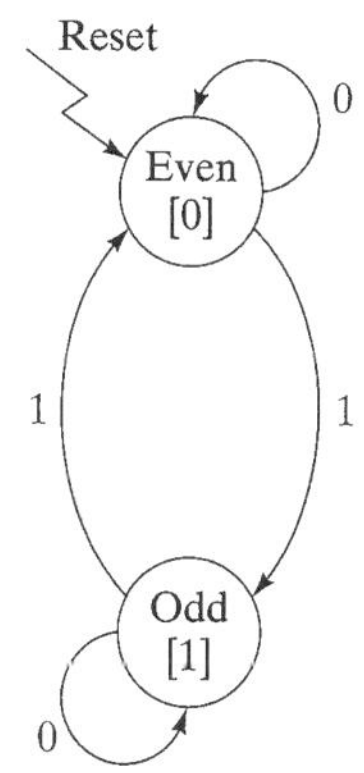

Figure 7.24 Example state diagram.

The state diagram is shown in Figure 7.24. We name the two unique configurations of the circuit *Even* and *Odd.* The outputs are explicitly associated with the states and are shown in square brackets. When an odd number of 1s has been seen, the output is 1 (an odd parity checker). Otherwise it is 0. For this circuit, the output will only be dependent on the state. We associate the corresponding input values that cause a transition from one state to another with the arcs in the state diagram. Note that there are exactly two arcs starting in each state that, together, cover all possible combinations of inputs, in this case, just 0 and 1. We could also written these conditions as $\overline{Input}$ and *Input*, respectively. Some of the arcs are *self-arcs* in that the finite state machine is instructed to stay in the same state for that input condition.

There is one special arc in the state diagram that appears to be coming from nowhere and is labeled with another input, *Reset.* This is just a short hand notation for a set of arcs from every state indicating that whenever this input is asserted, the next state should be *Even.* It also saves a bit more clutter in the diagram in that every arc would have to have a more complex input condition: namely, *Reset* unasserted and the input is 0 ($\overline{Reset} \cdot \overline{Input}$), or *Reset* is unasserted and the input is 1 ($\overline{Reset} \cdot Input$). We usually treat *Reset* this way if it will be connected to the set/reset inputs of flip-flops. If we don't use these

Present State	Input	Next State	Output
Even	0	Even	0
Even	1	Odd	0
Odd	0	Odd	1
Odd	1	Even	1

Figure 7.25 Symbolic state transition table.

Present State	Input	Next State	Output
0	0	0	0
0	1	1	0
1	0	1	1
1	1	0	1

Figure 7.26 Encoded state transition table.

inputs or do not have them available, then we have to treat *Reset* as just another input signal.

State Transition Table A reformulation of the state diagram is the *symbolic state transition table*. This is shown in Figure 7.25. We give meaningful symbolic names to the inputs, outputs, and present and next states. We cannot implement the circuit just yet. We must first assign binary encodings to all the state, input, and output symbols in the transition table.

Figure 7.26 shows the revised representation, called the *encoded state table*. We have assigned the encoding 0 to state *Even* and 1 to state *Odd*. The table now looks more like a truth table.

Next-State and Output Functions At this point, we have the next state (NS) and output (*Output*) expressed as logic functions of the present state (PS) and present input (*Input*). Based on a quick examination of the encoded state table, we write the functions as

```
    NS = PS ⊕ Input
Output = PS
```

Implementation Now we are ready to implement the circuit. The state of the finite state machine is held by a flip-flops. Since we have only two states, we can implement the circuit with a single flip-flop. The next-state function determines the input to this flip-flop.

We show an implementation using a *D* flip-flop in Figure 7.27. The *D* input of the flip-flop will be the value of NS and the XOR gate directly computes the function of the present state and the input.

Figure 7.28 shows the abstract timing behavior of the finite state machine for the input stream 100110101110. Each input bit is sampled on the rising edge of the clock because the state register is implemented by a positive edge-triggered flip-flop. The output changes soon after the rising edge. You should be able to convince yourself that the output is 1 whenever the input stream has presented an odd number of 1s, and is 0 otherwise.

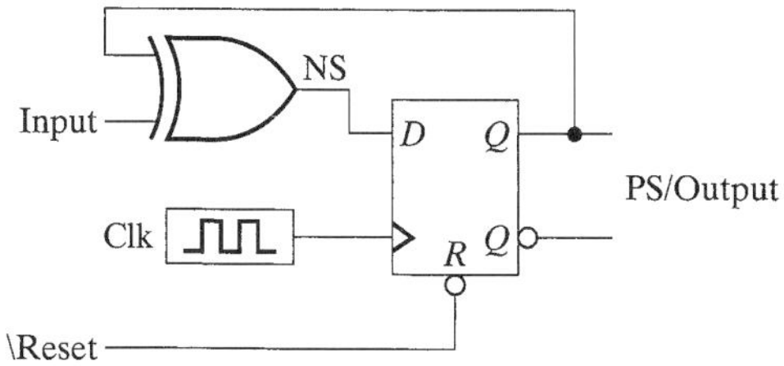

Figure 7.27 **Parity checker hardware.**

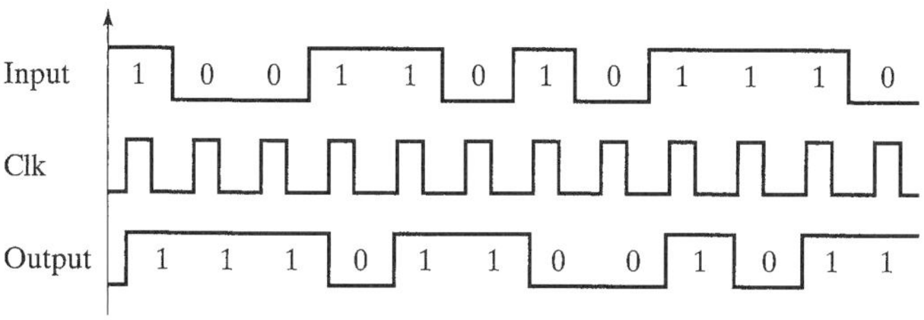

Figure 7.28 **Waveforms for input/output behavior.**

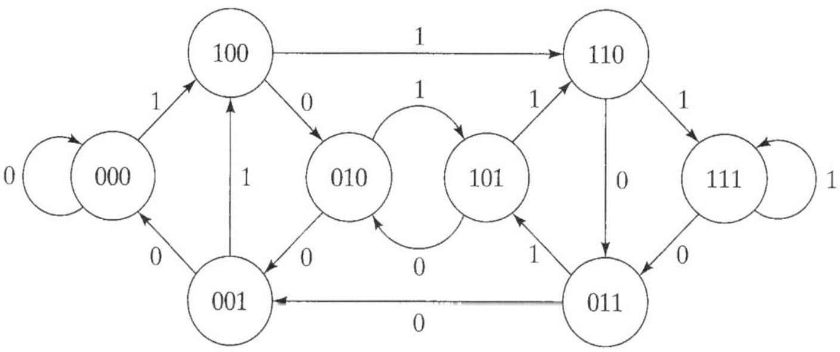

Figure 7.29 **State diagram for a 3-bit shift register.**

EXAMPLE 7.8 SHIFT REGISTER AS FSM

Can any sequential circuit be represented as a state machine? Will our design procedure always work? The answer to both questions is an emphatic: Yes! Let's try an example we've already used. A shift register is a simple sequential circuit that we didn't previously think of as a finite state machine. Figure 7.29 shows the state diagram for a 3-bit shift register. Note that there are exactly eight states for each possible combination of values in the three flip-flops of the register. Depending on the value of the single input, our finite state machine transitions to two different states. Verify for yourself that this state diagram is accurate by going through an example input sequence as we did for the parity checker in Figure 7.28.

The corresponding state transition table for the shift register is shown in Figure 7.30. The current state bits are labeled as C_1, C_2, and C_3 and represent the input of 1, 2, and 3 cycles previously, respectively.

In	C_1	C_2	C_3	N_1	N_2	N_3
0	0	0	0	0	0	0
0	0	0	1	0	0	0
0	0	1	0	0	0	1
0	0	1	1	0	0	1
0	1	0	0	0	1	0
0	1	0	1	0	1	0
0	1	1	0	0	1	1
0	1	1	1	0	1	1
1	0	0	0	1	0	0
1	0	0	1	1	0	0
1	0	1	0	1	0	1
1	0	1	1	1	0	1
1	1	0	0	1	1	0
1	1	0	1	1	1	0
1	1	1	0	1	1	1
1	1	1	1	1	1	1

Figure 7.30 State transition table for shift register.

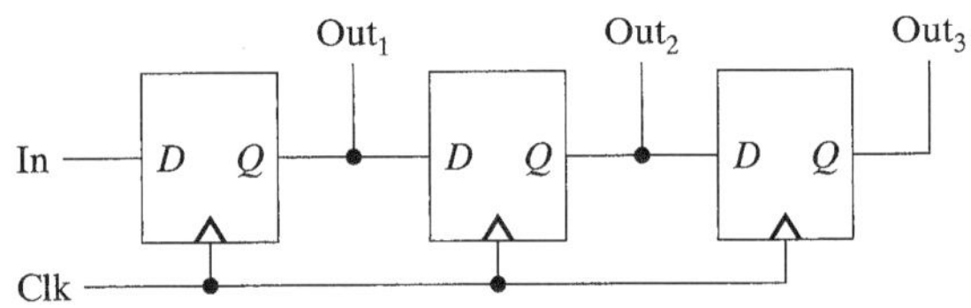

Figure 7.31 Implementation of the state diagram of Figure 7.29 to yield a 3-bit shift register.

N_1, N_2, and N_3 represent the next state to which the flip-flops will change based on the value of *In*. Note that there are two entries for each current state (one for each arc).

Interestingly, we don't need to go to K-maps to realize our circuit. The column for N_1 is identical to that for In, the column for N_2 is identical to that for C_1, and the column for N_3 is identical to that for C_2. Exactly what we would have expected for our shift register: the input to the first flip-flop (N_1) is just the input (*In*), the input to the second flip-flop (N_2) is the output of the first (C_1), and the input to the third flip-flop (N_3) is the output of the second (C_2).

Who would have thought that the relatively complex state diagram of Figure 7.29 would have led to the relatively simple circuit implementation of Figure 7.31? You can't judge a finite state machine by its state diagram. The very simple state diagram of the 3-bit binary counter in Figure 7.4 led to much more complex logic in Figure 7.6.

7.2.2 Timing in State Machines

In designing our finite state machines, we will follow a rigorous *synchronous design methodology*. This means that we will trigger the state changes with a global reference signal, the *clock*. It is important for you to understand when inputs are sampled, the next state is computed, and the outputs are asserted with respect to the clock signal.

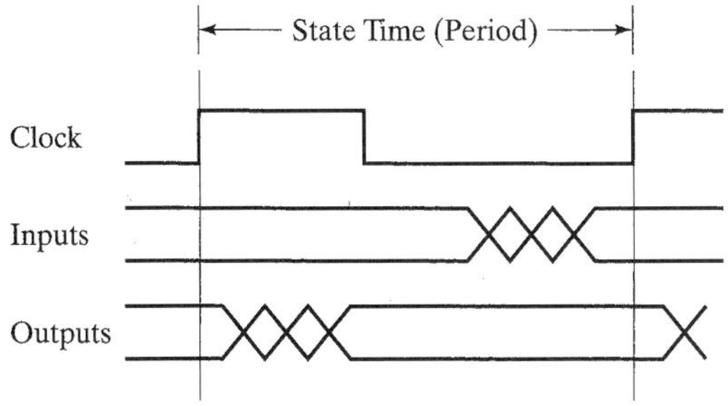

Figure 7.32 Input/output behavior of a positive edge-triggered state machine.

State Time We define *state time* as the time between related *clocking events.* For edge-triggered systems, the clocking events are the low-to-high (positive edge) or high-to-low (negative edge) transitions on the clock. In a positive edge-triggered system, the state time is measured from one rising clock edge to the next. In negative edge-triggered systems, the state time is measured between subsequent falling edges.

In response to a clocking event, the state and the outputs change, based on the current state and inputs. To be safe, and to meet propagation delays and setup times in the next-state logic, the inputs should be stable before the clocking event. After a suitable propagation delay, the finite state machine enters its next state and its new outputs become stable.

Figure 7.32 illustrates the state change, input sampling, and output changes for a positive edge-triggered synchronous system. Waveforms showing both 0 and 1 levels simultaneously are used to indicate a stable value while the criss-crossing lines are used to indicate where signals may be changing. On the rising edge, the inputs and current state are stable and are sampled to compute the next state and new outputs. Note that the output does not change to its new value until a propagation delay after the clock edge. These changes to the outputs and state will work their way through the combinational logic, eventually causing some inputs to change. This propagation delay, from the output of flip-flops through combinational logic to the inputs of flip-flops, must complete in time for sampling at the next positive clock edge while respecting setup time constraints. In mathematical terms,

State Time or *Period* > *Flip-Flop Output Propagation Delay*
+ *Combinational Logic Delay*
+ *Flip-Flop Setup Time*

Or, more concisely:

$$T_{\text{period}} > T_{\text{p}} + T_{\text{delay}} + T_{\text{su}}$$

This clearly shows how increasing combinational logic delay slows down the clock by increasing its period to keep the constraint from being violated.

High-speed designs seek to keep T_{delay} as small as possible.

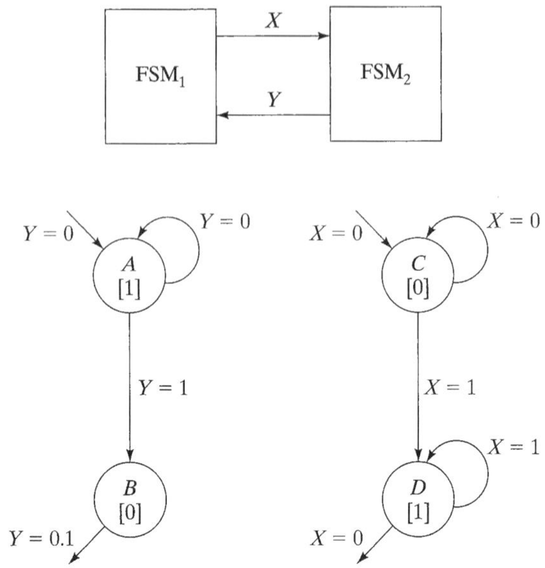

Figure 7.33 **Communicating FSM fragments.**

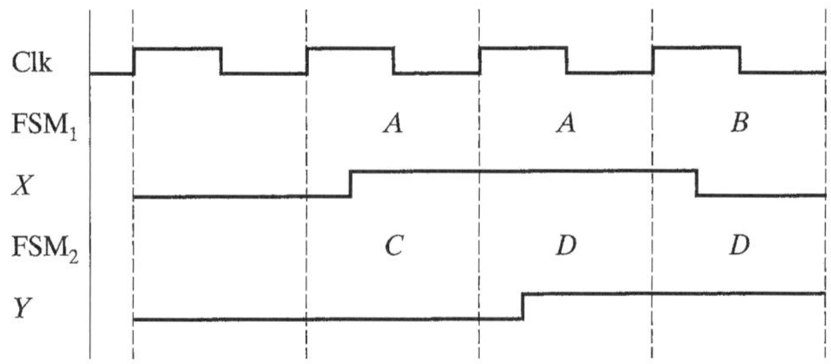

Figure 7.34 **State and output changes associated with the FSM fragments of Figure 7.33.**

Communicating FSMs As an example of detailed, state-machine timing behavior, Figure 7.33 gives fragment state diagrams for two communicating finite state machines (FSMs). We assume that both are positive edge-triggered synchronous systems and the output from each state machine is the input to the other. The interaction between these machines is illustrated by the timing diagram of Figure 7.34.

To start, the clock is in the first period with FSM$_1$ about to enter state A with its output $X = 0$. FSM$_2$ is entering state C with its output $Y = 0$. In the second clock period, FSM$_1$ is in state A and asserts its output X. FSM$_2$ is in state C with its output Y unasserted. On the rising edge that starts the third clock period, FSM$_1$ stays in state A since its input is 0. FSM$_2$ advances to state D, asserting Y, but too late to affect a state change in FSM$_1$. The input value before the clock edge is the one that matters.

Now that Y is 1, FSM$_1$ goes to state B on the next rising edge. In this state, it will output a 0, but this is too late to affect a state change in FSM$_2$. It remains in state D for one more cycle before moving on.

7.3 Basic FSM Design Approach

The counter design procedure forms the core of a more general procedure for arbitrary finite state machines. You will discover that the procedure must be significantly extended for the general case.

7.3.1 Finite State Machine Design Procedure

STEP 1 *Understand the problem.* A finite state machine is often described in terms of an English-language specification of its behavior. It is important that you interpret this description in an unambiguous manner. For counters, it is sufficient simply to enumerate the sequence. For finite state machines, try some input sequences to be sure you understand the conditions under which the FSM transitions between states and the various outputs are asserted.

STEP 2 *Obtain an abstract representation of the FSM.* Once you understand the problem, you must place it in a form that is easy to manipulate by the procedures for implementing the finite state machine. A state diagram is one possibility. Another representation, to be introduced in the next chapter, is the specifications in hardware description languages.

STEP 3 *Perform state minimization.* Step 2, deriving the abstract representation, often results in a description that has many states. Certain paths through the state machine can be eliminated because their input/output behavior is duplicated by other functionally equivalent paths. This is a new step, not needed in the simpler counter design process. We will see an example of this shortly.

STEP 4 *Perform state assignment.* In counters, the state and the output were identical, and we didn't need to worry about encoding a particular state. In general finite state machines, this is not the case. Outputs are derived from the bits stored in the state flip-flops (and possibly also the values of some of the inputs), and a good choice of how to encode the state often leads to a simpler implementation.

STEP 5 *Implement the finite state machine.* The final step is also found in the counter design procedure. Using Boolean equations or K-maps for the next-state and output combinational functions, produce the minimized two-level or multilevel implementation.

In this chapter, we concentrate on Steps 1, 2, and 5 of the design process. We will motivate the optimizations alluded to in Steps 3 and 4 but leave a more detailed treatment to the next chapter.

EXAMPLE 7.9 A SIMPLE VENDING MACHINE

To illustrate the basic design procedure, we will advance through the implementation of a simple finite state machine that controls a vending machine.

Here is how the control is supposed to work. The vending machine delivers an item after it has received 15 cents in coins. The machine has a single coin slot that accepts nickels and dimes, one coin at a

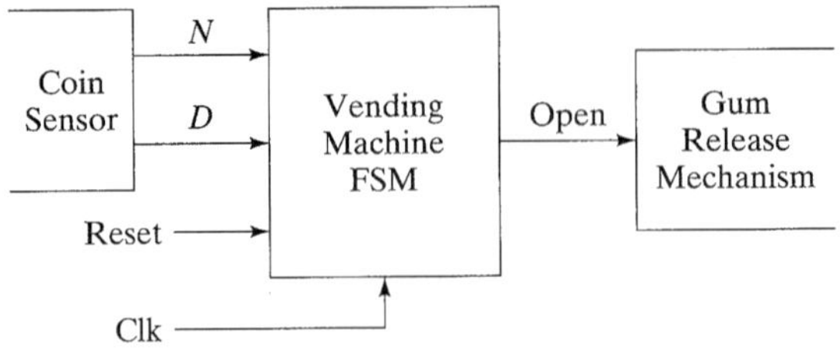

Figure 7.35 **Vending machine block diagram.**

time. A mechanical sensor indicates whether a dime or a nickel has been inserted into the coin slot. The controller's output causes a single item to be released down a chute to the customer.

We will make two additional assumptions to further simplify our specification. First, we will design our machine so it does not give change. A customer who pays with two dimes will lose the five extra cents! Second, we will expect our machine to be reset before each new use. This might be accomplished through a separate mechanism the customer uses to select the item to be purchased.

Understanding the Problem

The first step in the finite state machine design process is to *understand the problem*. Start by drawing a block diagram to understand the inputs and outputs. Figure 7.35 is a good example. Let's assume that N is asserted for one clock period when a nickel is inserted into the coin slot and that D is asserted when a dime has been deposited. Furthermore, we'll postulate that it is enough if the machine asserts *Open* for one clock period to release an item after 15 cents (or more) has been deposited since the last reset. It is important to list these assumptions as they will need to be verified later with other members of the vending machine design team.

The specification may not completely define the behavior of the finite state machine. For example, what happens if someone inserts a penny into the coin slot? Sometimes we have to make reasonable assumptions. We'll assume that the coin sensor is designed so that it returns any coins it does not recognize, leaving N and D unasserted. As you can see, these assumptions can have significant implications for how other parts of the vending machine will have to be designed and implemented. In a design team, it is crucial to have a process that gets these assumptions aired early and keeps track of them as the design progresses.

Abstract Representations

Once you understand the behavior reasonably well, it is time to *map the specification into a more suitable abstract representation*. A good way to begin is by enumerating the possible unique sequences of inputs or configurations of the system. These will help define the states of the finite state machine.

For this problem, it is not too difficult to enumerate all the possible input sequences that lead to releasing an item:

Three nickels in sequence: N, N, N

Two nickels followed by a dime: N, N, D

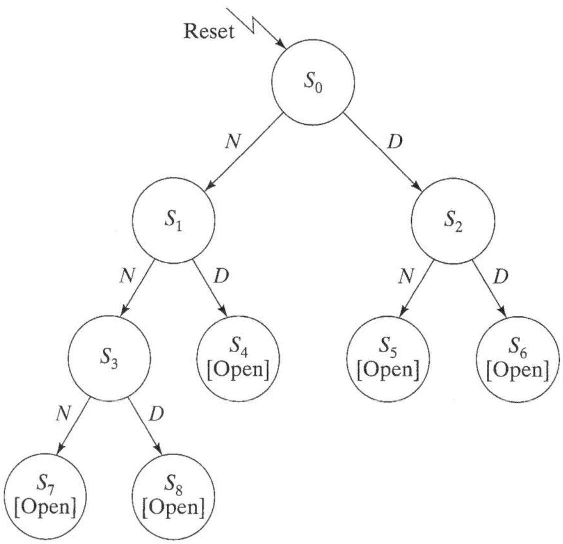

Figure 7.36 Initial vending-machine state diagram.

A nickel followed by a dime: *N*, *D*

A dime followed by a nickel: *D*, *N*

Two dimes in sequence: *D*, *D*

This can be represented as a state diagram, as shown in Figure 7.36. For example, the machine will pass through the states S_0, S_1, S_3, S_7 if the input sequence is three nickels.

To keep the state diagram simple and readable, we include only transitions that explicitly cause a state change (in other words, we leave out the self-arcs). For example, in state S_0, if neither input *N* or *D* is asserted, we assume the machine remains in state S_0 (the specification allows us to assume that *N* and *D* are never asserted at the same time). Also, we include the output *Open* only in states in which it is asserted. *Open* is implicitly unasserted in any other state.

State Minimization

This nine-state description isn't the "best" possible. For one thing, since states S_4, S_5, S_6, S_7, and S_8 have identical behavior, they can be combined into a single state.

To reduce the number of states even further, we can think of each state as representing the amount of money received so far. For example, it shouldn't matter whether the state representing 10 cents was reached through two nickels or one dime.

A state diagram derived in this way is shown in Figure 7.37. We capture the behavior in only four states, compared with nine in Figure 7.36. Also, as another illustration of a useful shorthand, notice the transition from state 10¢ to 15¢. We interpret the notation "*N*, *D*" associated with this transition as "go to state 15¢ if *N* is asserted OR *D* is asserted" and could have also written it as $N + D$.

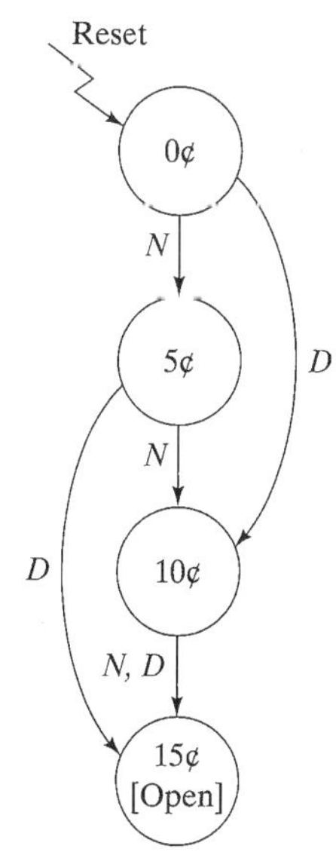

Figure 7.37 Minimized vending-machine state diagram.

Present State	Inputs D	N	Next State	Output Open
0¢	0	0	0¢	0
	0	1	5¢	0
	1	0	10¢	0
	1	1	X	X
5¢	0	0	5¢	0
	0	1	10¢	0
	1	0	15¢	0
	1	1	X	X
10¢	0	0	10¢	0
	0	1	15¢	0
	1	0	15¢	0
	1	1	X	X
15¢	X	X	15¢	1

Figure 7.38 **Minimized vending-machine symbolic state transition table.**

Present State Q_1	Q_0	Inputs D	N	Next State D_1	D_0	Output Open
0	0	0	0	0	0	0
		0	1	0	1	0
		1	0	1	0	0
		1	1	X	X	X
0	1	0	0	0	1	0
		0	1	1	0	0
		1	0	1	1	0
		1	1	X	X	X
1	0	0	0	1	0	0
		0	1	1	1	0
		1	0	1	1	0
		1	1	X	X	X
1	1	0	0	1	1	1
		0	1	1	1	1
		1	0	1	1	1
		1	1	X	X	X

Figure 7.39 **Encoded vending-machine state transition table.**

In the next chapter, we will examine formal methods for finding a state diagram with the minimum number of states. The process of minimizing the states in a finite state machine description is called *state minimization*.

State Encoding

At this point, we have a finite state machine with a minimum number of states, but it is still symbolic. See Figure 7.38 for the symbolic state transition table. The next step is *state encoding*.

The way states are encoded can have a major effect on the amount of combinational logic we will need to implement the machine. A natural state assignment would encode the states in 2 bits: state 0¢ as 00, state 5¢ as 01, state 10¢ as 10, and state 15¢ as 11. A less obvious assignment could lead to reduced hardware. The *encoded* state transition table for this simple assignment is shown in Figure 7.39.

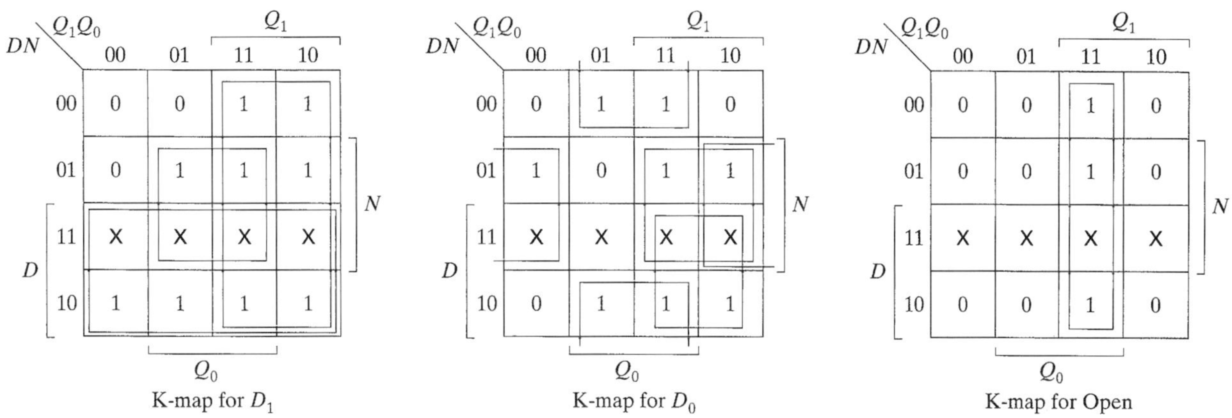

Figure 7.40 K-maps for D flip-flop implementation of vending machine.

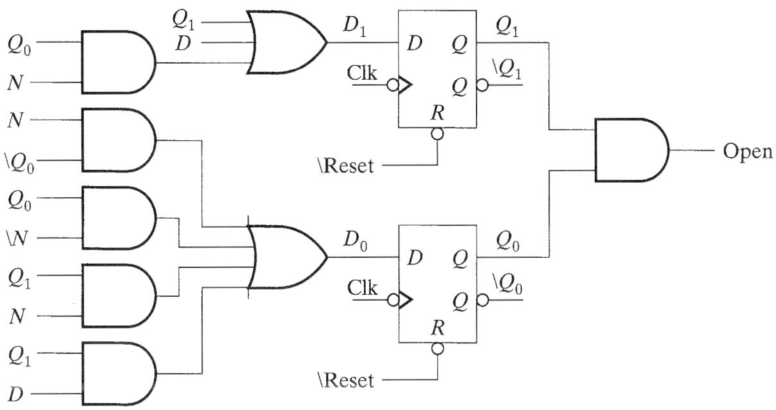

Figure 7.41 Vending machine FSM implementation.

In the next chapter we present a variety of approaches to finding an effective state encoding.

Implementation

The next step is to implement the state transition table. The K-maps for the D flip-flop inputs are shown in Figure 7.40. We filled these in directly from the encoded state transition table. The minimized equations for the flip-flop inputs and the output become

$$D_1 = Q_1 + D + Q_0 \bullet N$$

$$D_0 = N \bullet \bar{Q}_0 + Q_0 \bullet \bar{N} + Q_1 \bullet N + Q_1 \bullet D$$

$$\text{Open} = Q_1 \bullet Q_0$$

The logic implementation is shown in Figure 7.41. It uses eight gates and two flip-flops. Note that the output *Open* is a function of the state bits and will require some additional propagation time after the clock edge (the delay of the AND gate). This is the first state machine we've seen where an output is not simply the output of a state flip-flop.

Discussion

We briefly described the complete finite state machine design process and illustrated it by designing a simple vending machine controller. Starting with an English-language statement of the task, we first described the machine in a more formal representation. In this case, we used state diagrams.

Since more than one state diagram can lead to the same input/output behavior, it is important to find a description with as few states as possible. This usually reduces the implementation complexity of the finite state machine by requiring less flip-flops and, consequently, less logic functions. For example, the state diagram of Figure 7.36 contains nine states and requires four flip-flops for its implementation with four corresponding input functions. The minimized state diagram of Figure 7.37 has four states and can be implemented with only two flip-flops and two next state functions.

Once we have obtained a minimum finite state description, the next step is to choose a good encoding of the states. The right choice can further reduce the logic for the next-state and output functions. In this example, we used only the most obvious state assignment.

7.3.2 Moore and Mealy Machines

There are two basic ways to organize a clocked sequential network:

- *Moore machine:* The outputs depend only on the present state. See the block diagram in Figure 7.42. A combinational logic block maps the inputs and the current state into the appropriate next state which is the input to the flip-flops. The state will change at the next clocking event. The outputs are computed by a combinational logic block whose only inputs are the flip-flops' state outputs. The outputs change *synchronously* with the state transition and the clock edge. The finite state machines you have seen so far have all been Moore machines. In fact, they've had very simple output logic. The counters just had wires connecting the state bits to the outputs—therefore, no real output logic at all. The vending machine example had a single gate ANDing two state bits to form the single output.
- *Mealy machine:* The outputs depend on the present state as well as the present value of the inputs. See Figure 7.43. The outputs can

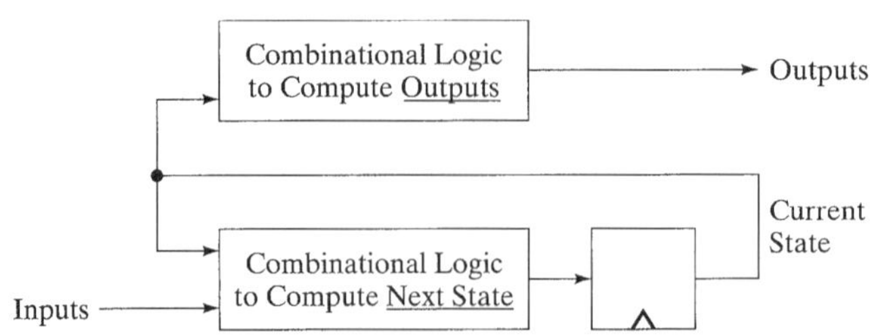

Figure 7.42 Moore machine block diagram (note outputs are a function of current state only).

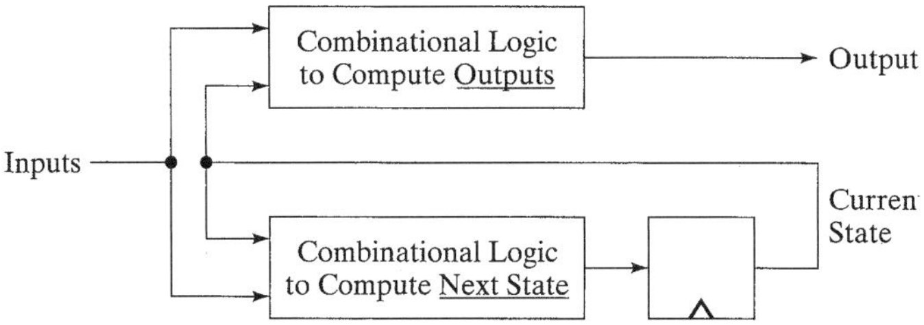

Figure 7.43 **Mealy machine block diagram (note outputs are a function of both inputs and current state).**

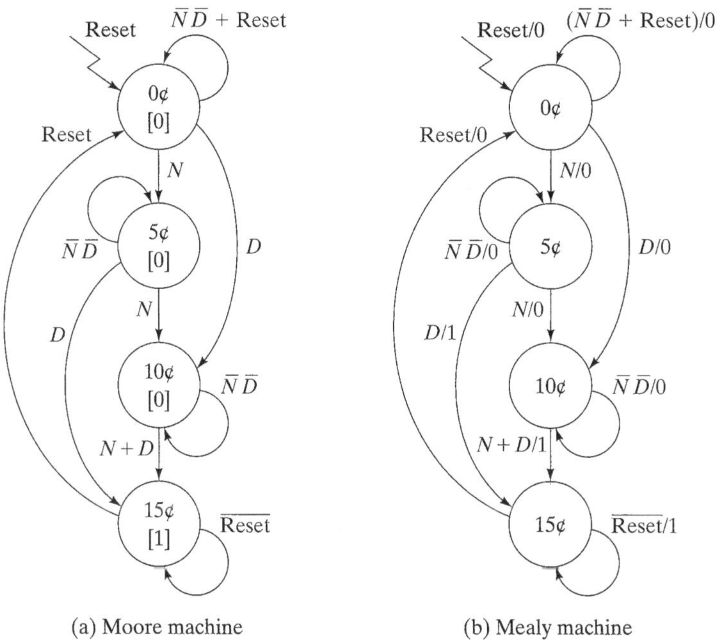

Figure 7.44 **Moore and Mealy machine state diagrams for the vending machine FSM.**

change immediately after a change at the inputs, independently of the clock. A Mealy machine constructed in this fashion has *asynchronous* outputs. We'll see that the Mealy model, being more general, usually leads to implementations with less states.

Moore outputs are synchronous with the clock, while Mealy outputs are asynchronous because they can change in response to any changes in the inputs. This gives Moore machines an advantage in that they have a disciplined timing methodology. However, there is a synchronous variation of the Mealy machine, which we will describe later.

7.3.3 State Diagram Representation

Figure 7.44 shows the notations for Mealy and Moore state diagrams using the vending machine example. For Moore machines, the outputs are associated with the state in which they are asserted. Arcs are labeled with the input conditions that cause the transition from the state at the tail of the arc to the state at its head. Combinational logic functions are perfectly acceptable as arc labels.

In Mealy machines, the outputs are associated with the transition arcs rather than the state bubble. A slash separates the inputs from the outputs. For example, if we are in state 10¢ and either *N* or *D* is asserted, *Open* will be asserted (indicated by the 1 following the /). Any glitch on *N* or *D* could cause the item to be delivered by mistake.

The state diagrams in this figure are labeled more completely than our previous ones for this example. Here, we explicitly show the transitions that cause the machine to stay in the same state. We often do not include such transitions to simplify the presentation of the state diagram. We also associate explicit output values with each transition in the Mealy state diagram and each state in the Moore state diagram. A common simplification places the output on the transition or in the state only when it is asserted. You should state your own conventions whenever you draw state diagrams to minimize misunderstandings.

7.3.4 Comparison of the Two Machine Types

Because it can associate outputs with transitions, a Mealy machine can often generate the same output sequence in fewer states than a Moore machine.

Consider a finite state machine that asserts its single output whenever its input string has at least two 1s in sequence. The minimum Moore and Mealy state diagrams are shown in Figure 7.45.

To represent the 1s sequence, the Moore machine requires two states to distinguish between the first and subsequent 1s. The first state has output 0, while the second has output 1. The Mealy machine accomplishes this with a single state reached by two different transitions. For

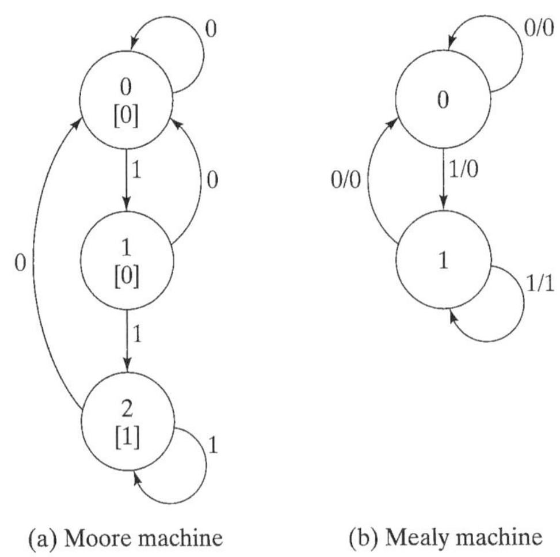

Figure 7.45 Two state diagrams with the same I/O behavior but different number of states.

the first 1, the transition has output 0. For the second and subsequent 1s, the transition has output 1. Despite the Mealy machine's asynchronous timing complexities, designers are quite fond of its reduced state count.

States, Transitions, and Outputs in Mealy and Moore Machines Suppose that a given state machine has M inputs and N outputs and is being implemented using L flip-flops. You might ask a number of questions to bound the complexity of this state machine. For example, what are the minimum and maximum numbers of states that such a machine might have? With L flip-flops, the implementation has the power to represent 2^L states. But for a specific FSM, as few as 1 and as many as all 2^L of these might be valid states.

What are the minimum and maximum numbers of state transitions that can begin in a given state? Since there must be an exit transition for each possible input combination, the minimum and the maximum are the same: 2^M transitions.

A similar question involves the minimum and maximum numbers of state transitions that can end in a given state. Because we can have start-up states reachable only on power-up, the minimum number of input transitions is 0. Since a single state could conceivably be the target of all the transitions of the finite state machine, the maximum number of input transitions is $2^M * 2^L$, the number of possible input combinations multiplied by the number of states.

A final question is the minimum and maximum numbers of patterns that can be observed on the machine's outputs. The minimum number of unique output patterns is 1, of course. Every state and every transition can be associated with the same pattern.

The maximum number depends on the kind of machine. For a Mealy machine, the maximum number of output patterns is the smaller of the number of transitions, $2^M * 2^L$, or the number of possible output patterns, 2^N. If the number of transitions exceeds the number of possible output patterns, then some must be repeated. In the Moore machine, the maximum is the smaller of the number of states, 2^L, and the number of possible output patterns, 2^N. If the number of states exceeds the number of output patterns, then some patterns will also need to be repeated.

As an example, consider a Moore machine with two inputs, one flip-flop, and three outputs. The state, transition, and output bounds are:

Minimum number of states: 1
Maximum number of states: 2
Minimum number of output transitions (per state): 4
Maximum number of output transitions (per state): 4
Minimum number of input transitions (per state): 0
Maximum number of input transitions (per state): 8
Minimum number of observed output patterns: 1
Maximum number of observed output patterns: 2

In this case, the output patterns are limited by the number of states.

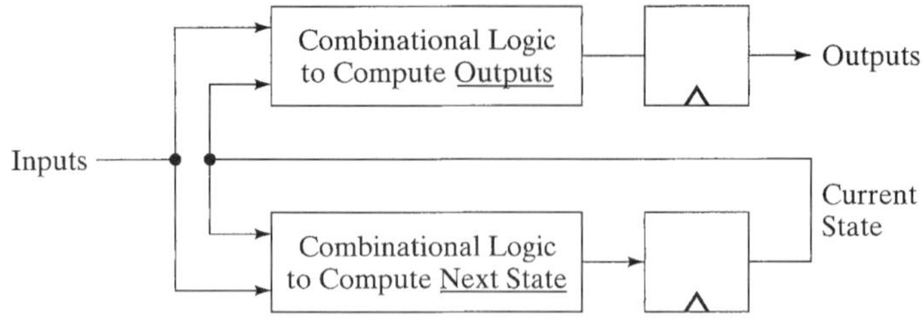

Figure 7.46 **Synchronous Mealy machine block diagram.**

Synchronous Mealy Machines There are two major issues with the structure of the Mealy machine. Firstly, the output logic of Mealy machines can cause glitches on the output. As you have already seen, glitches are undesirable in real hardware controllers. Secondly, the output can change asynchronously to the clock. This can sometimes be an advantage as the circuit can react more quickly to an input change rather than having to wait for the next clock edge to advance the state. However, if we imagine two communicating Mealy machines, it is possible to have a situation where a complete loop between the two machines does not pass through a register. This is exactly the kind of uncontrolled logic feedback loop we worked to tame in Chapter 6. But because Mealy machines encode control in fewer states, saving on flip-flops and the logic of their next-state equations, it is still desirable to use them.

Dealing with these issues leads to alternative *synchronous* design styles for Mealy machines. Simply stated, the way to construct a synchronous Mealy machine is to break the direct connection between inputs and outputs by introducing storage elements.

One way to do this is to synchronize the Mealy machine outputs with output flip-flops. See Figure 7.46. The flip-flops are clocked with the same edge as the state register. This has the effect of converting the Mealy machine into a Moore machine, by making the outputs part of the state encoding! However, this machine does not have exactly the same input/output behavior as the original Mealy machine (can you figure out why?).

In general, fully synchronous finite state machines are much easier to implement and debug than asynchronous machines. Output changes are well-behaved and only occur after a clock edge making it easier to reason about what is happening when and placing less concerns on the delay of different logic paths. Most of the time you will find it easiest to use Moore or synchronous Mealy implementations. You should use asynchronous Mealy machines only after very careful analysis of the input/output timing behavior of the finite state machine and the other logic circuitry from which it gets inputs and/or provides outputs.

EXAMPLE 7.10 THE VENDING MACHINE IN MOORE, MEALY, AND SYNCHRONOUS MEALY IMPLEMENTATIONS

To better illustrate the relationship between these different types of finite state machines, let's return to the vending machine example. We

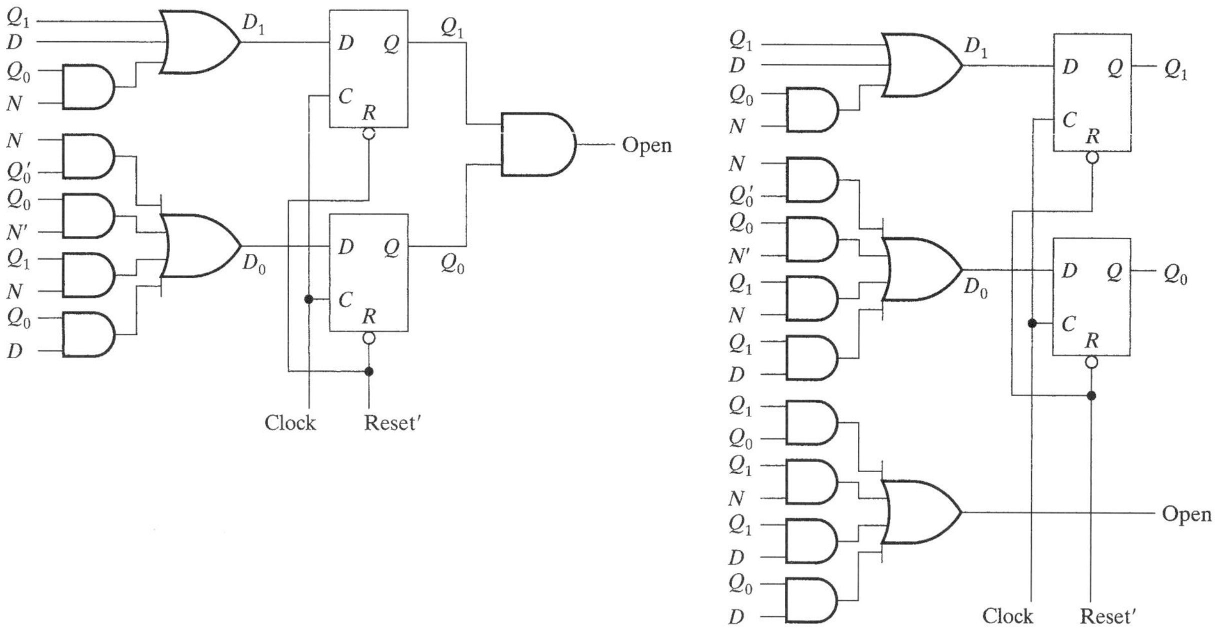

Figure 7.47 Moore and asynchronous Mealy machine implementations of the vending machine.

already have Moore and Mealy state diagrams for the vending machine in Figure 7.44. The straightforward Moore and asynchronous Mealy implementations of the vending machine are shown in Figure 7.47. The Moore implementation is just a slightly redrawn version of Figure 7.41. Note that the output, *Open,* will be slightly delayed after the state changes by as much as the propagation delay of the AND gate. For the Mealy machine, *Open* will change as soon as the last coin in inserted and will not wait for the next clock edge. Thus, *Open* will be asserted while the machine is still in a state other than the one representing 15¢. You can see that already, that asynchronous Mealy machines make it a bit harder to reason about FSM behavior because of this.

Let's now address these issues by making some transformations. First, we'll move the logic for *Open* from one side of the flip-flops to the other in the Moore implementation. We can do this my moving the inputs of the AND gate from the output of the flip-flops to their inputs and adding a flip-flop between the output of the gate and the *Open* output of the FSM. You can think of this as "pushing" the gate through the flip-flops. This process is formally called *retiming* as we are changing the time at which signals are computed and their values seen at the outputs of the FSM. This is equivalent to computing *Open* in the previous state and then waiting until the clock ticks to load it into a flip-flop, advancing both the state and its corresponding output. Now the output will change at the clock edge just as the state does. The result of this transformation for our Moore machine is shown on the left-hand side of Figure 7.48.

For the asynchronous Mealy implementation, we'll transform it into a synchronous Mealy machine by simply adding a flip-flop to the

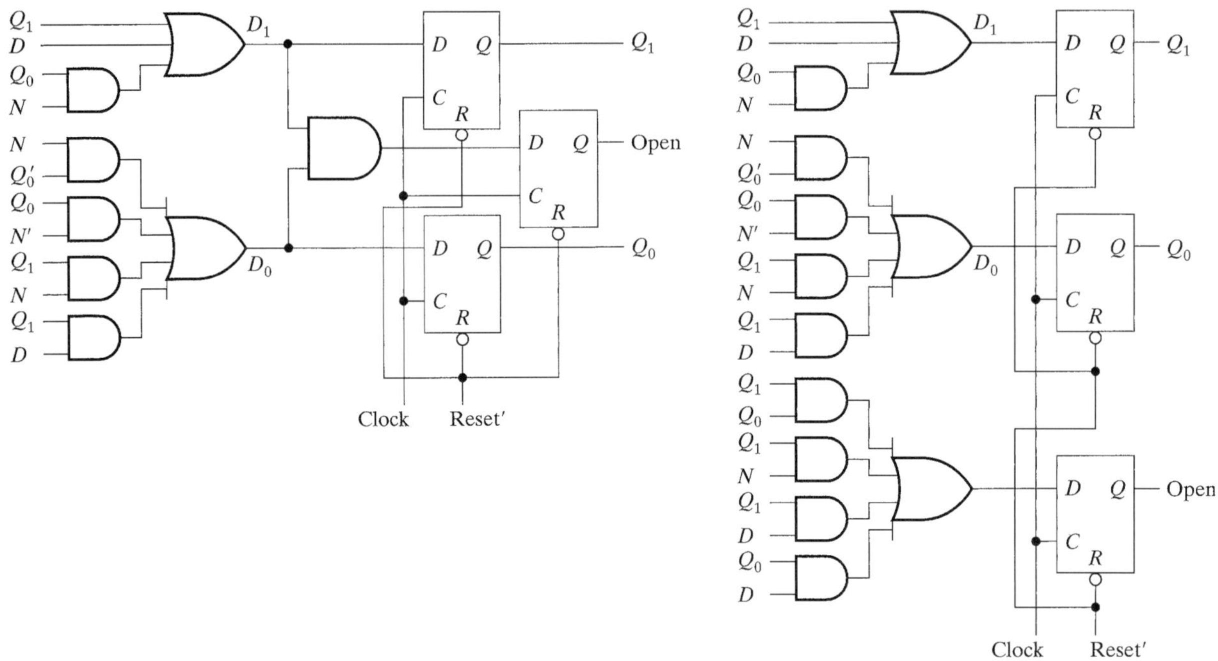

Figure 7.48 **Retimed Moore and synchronous Mealy machine implementations of the vending machine.**

Open output. The resulting circuit is shown on the right-hand side of Figure 7.48. As in the Moore case, now both the state bits and output change with the clock edge. Thus, the vending machine will release an item at the clock edge after the last required coin is inserted. This is exactly how the Moore machine operates. In fact, these are now both Moore machines. We can think of them as having three state bits where one of them is also an output.

If they have identical output behavior, why does the logic for *Open* look so different? We definitely need to look at this in more detail. For the Moore machine, the equation for the input to the *Open* flip-flop is, by inspection:

$$(Q_1 + D + Q_0N)(Q_0'N + Q_0N' + Q_1N + Q_1D)$$
$$= Q_1Q_0N' + Q_1N + Q_1D + Q_0'ND + Q_0N'D$$

On the other hand, for the synchronous Mealy machine, the equation for the input to the *Open* flip-flop is:

$$Q_1Q_0 + Q_1N + Q_1D + Q_0D$$

Unfortunately, these are not the same, or, are they? To really get to the bottom of this, we need to map these equations onto a K-map and see if they really are the same function. The K-maps for each of these is shown in Figure 7.49. It turns out there is a difference! How can that be? Our design process should work whether we implement our FSM as a Moore or a Mealy machine. Was there something wrong with one of our transformations?

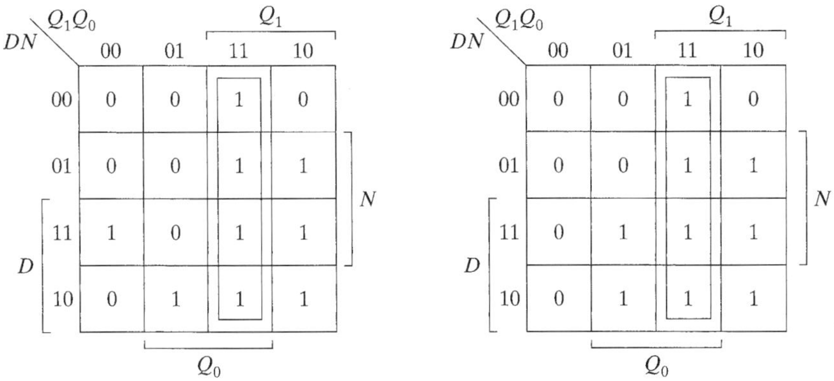

K-map for Open
FF Input in Retimed
Moore Implementation

K-map for Open
FF Input in Synchronous
Mealy Implementation

Figure 7.49 K-maps for the two implementation of *Open* in Figure 7.8.

On closer inspection, it turns out that our process is sound. The only differences in the K-maps occur when both the D and N inputs are asserted—an input don't care condition—we predicated our design on those two signals never being asserted at the same time. The difference in the two logic functions turns out to be a different assignment for the input don't-care conditions. They are, in fact, identical for the conditions we do care about.

You can start to get a sense of why it is important to write down all these assumptions (N and D never both asserted) and carry them along throughout the design. Many verification issues end up being traced back to how don't cares where used and what assumptions were or were not included in the design process.

Hopefully, this example also served to illustrate that Moore and synchronous Mealy styles are just different ways of getting to an implementation. Their respective implementations can be transformed one into the other (if don't-care conditions are taken into account).

7.4 Motivation for Optimization

To review, the finite state machine design process consists of (1) understanding the problem, (2) obtaining a formal description (ultimately, a *symbolic state transition table*), (3) minimizing the number of states, (4) encoding the states, and finally (5) implementing the finite state machine's next-state and output functions. Now that you have seen the process through a few examples, you may be starting to get a sense for the optimization opportunities. In this section, we'll review these and use another example to motivate a few others.

7.4.1 Two State Diagrams, Same I/O Behavior

In the age of very large scale integrated circuits, why should we bother to minimize a finite state machine implementation? After all, as long

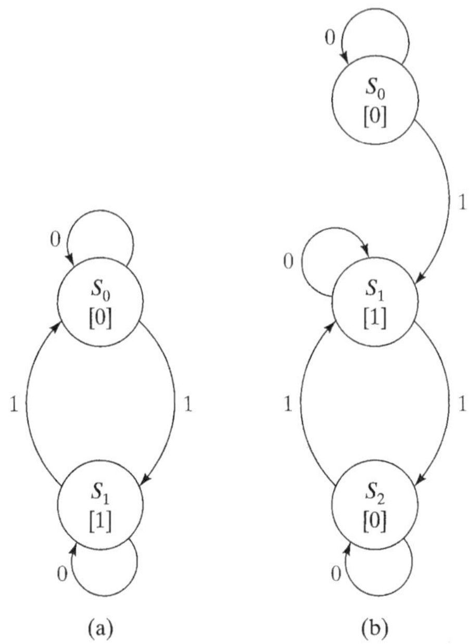

Figure 7.50 **Two equivalent state diagrams for the odd parity checker.**

as the input/output behavior of the machine is correct, it really doesn't matter how it is implemented. Or does it?

Figure 7.50 shows two different state diagrams for the odd parity checker of Section 7.2.1. They have identical output behavior for all input strings. You should try some inputs to convince yourself. We define *equivalence* of finite state machines as follows. Two machines are equivalent if their input/output behavior is identical for all possible input strings.

For a particular finite state machine, there are many equivalent forms. Rather than reusing states while deriving the state diagram, you could simply introduce a new state whenever you need one (to keep the number of states finite, you will need to reuse some of them, of course).

The two implementations of the state diagrams of Figure 7.50 certainly are not the same. The machine with more states requires more flip-flops and more complex next-state logic. For this simple case, the state diagram on the right will require at least 2 flip-flops instead of 1 and, similarly, one more next-state equation.

7.4.2 Advantages of Minimum States

In general, you will find it is worthwhile to implement the finite state machine in as few states as possible. This usually reduces the number of logic gates and flip-flops you need for the machine's implementation. Similarly, judicious mapping between symbolic and encoded states can reduce the implementation logic.

A state diagram with n states must be implemented with at least k flip-flops, where $2^{k-1} < n <= 2^k$. By reducing the number of states to

2^{k-1} or less, you can save a flip-flop. For example, suppose you are given a finite state machine with five state flip-flops. This machine can represent up to 32 states. If you can reduce the number of states to 16 or less, you save a flip-flop.

Even when reducing the number of states is not enough to eliminate a flip-flop, it still has advantages. With fewer states, you introduce more don't-care conditions into the next-state and output functions, making their implementation simpler. Less logic usually means shorter critical timing paths and a higher clock rate for the system.

More importantly, today's programmable logic provides limited gate and flip-flop counts on a single programmable logic chip. A typical programmable logic part might have "10,000 gate equivalents" (rarely approached in practice) yet provide only 100s of flip-flops! An important goal of state reduction is to make the implementation "fit" in as few components as possible. The fewer components you use, the shorter the design time and the lower the manufacturing cost.

State reduction techniques also allow you to be sloppy in obtaining the initial finite state machine description. If you introduce a few redundant states, you will find and eliminate them by using the state reduction techniques introduced in the next chapter.

7.4.3 State, Input, and Output Encoding

Deciding the pattern of bits to assign to each symbolic state is an important optimization opportunity. We first have to decide how many bits to use to represent the state. In synchronous Mealy machines, we even have the option of using some of the outputs as they also have a flip-flop and can be thought of as part of the *state vector,* the set of flip-flops that will hold the patterns for each state.

Optimizations can get quite elaborate when we consider that we can have more state bits than the minimum if they help us derive simpler logic equations. Also, there is no reason why each state can only have a single unique pattern, it can have more than one if that helps makes the logic simpler (as long as each pattern is unique to one state).

In addition to coming up with patterns for each of the states, we often have the liberty to decide how outputs will be encoded. Unless, of course, the outputs are precisely fixed by other parts of the design. Again, the right encoding can lead to smaller logic realizations.

Finally, if inputs are symbolic—because the logic that will be supplying them has yet to be fully realized—it may be possible to make some interesting encoding choices there as well. An example of this is our choice for representing the number of days in a month in the calendar subsystem example of Chapter 1.

Unfortunately, there are no known methods for guaranteeing the best state assignment or output encoding. We will present some heuristics in the next chapter and then discuss some tradeoffs for different implementation technologies in the subsequent chapter.

7.4.4 Factoring State Machines

Often a state machine has recurring elements—a common subsequence of states that occurs more than once in the state diagram. If this set of states can be factored into a separate communicating state machine, then it can greatly reduce the number of states in the original machine. This helps make for smaller, more compact machines that can often be implemented using a smaller number of resources. This is analogous to subroutines in a programming language where a common function is separated out to make the code simpler and put the recurring functionality in one place so that modifications can be done only once rather than having to track down every occurrence in the code.

We'll review several approaches to factoring and extracting subroutines in the next chapter. Before we get there, however, let's look at a finite state machine that helps to better motivate all the optimizations we've discussed by providing a concrete example.

7.4.5 A Traffic-Light Controller

The following description of a traffic-light controller represents a relatively complex control function: "A busy highway is intersected by a little-used farm road, as shown in Figure 7.51. Detectors are placed along the farm road to raise a signal C as long as a vehicle is waiting to cross the highway. The traffic light controller should operate as follows. As long as no vehicle is detected on the farm road, the lights should remain green in the highway direction. If a vehicle is detected on the farm road, the highway lights should change from green to yellow to red, allowing the farm road lights to become green. The farm road lights stay green only as long as a vehicle is detected on the farm road and never longer than a set interval so as not to block traffic flow along the highway for too long. If these conditions are met, the farm road lights change from green to yellow to red, allowing the highway lights to return to green. Even if vehicles are waiting to cross the highway, the highway should remain green for a minimum amount of time.

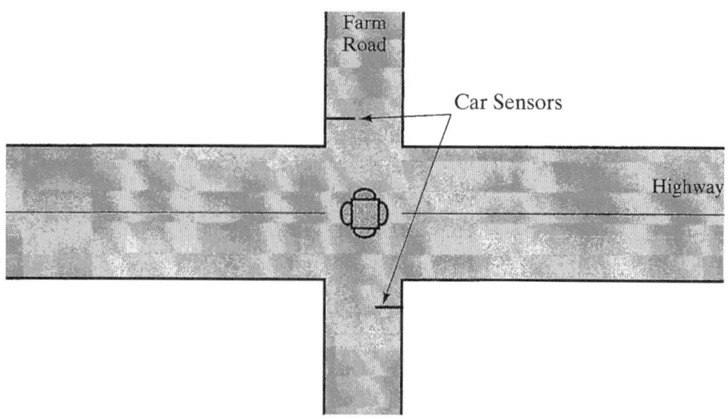

Figure 7.51 **Highway/farm road intersection: placement of lights and sensors.**

It is also important to ensure that the lights stay yellow for a predetermined amount of time. Thus, we'll need to measure a few intervals: one for the amount of time the highway light is yellow, one for the amount of time the farm road light is yellow, one for the minimum time the highway light is green, and one for the longest amount of time the farm road light can stay green, even if there are still cars waiting to cross the highway. We'll set the two yellow light durations to be the same and also make the two longer time intervals for minimum highway green and maximum farm road green be the same. With these assumptions, we'll need only a timer with the ability to signal two time intervals: one short and one long. These assumptions are important to document and keep together with the design so that others can see the reasoning that went behind the decisions.

Understanding the Specification To understand the problem statement, a good way to begin is to identify the inputs and outputs and the unique configurations of the controller. The required inputs and outputs are as follows:

Input Signal	**Description**
Reset	Place controller in initial state
C	Detects vehicle on farm road in either direction

Output Signal	**Description**
HG, HY, HR	Assert green, yellow, red highway lights
FG, FY, FR	Assert green, yellow, red farm road lights

In terms of the unique output combinations, there are only four. If the highway light is green or yellow, the farm road light must be red, and similarly, when the farm road light is green or yellow, the highway light must be red. Thus, there are four major states for the controller has that may need to be further refined into individual states.

State	**Description**
HG	Highway green (farm road red)
HY	Highway yellow (farm road red)
FG	Farm road green (highway red)
FY	Farm road yellow (highway red)

We'll make an assumption that the short time interval for the duration of yellow lights is five cycles of our clock, while the longer maximum time for green on the farm road and minimum green on the highway is 20 cycles.

Formal Representation In developing a state diagram for an FSM, is it often useful to work through a typical execution sequence. Let's start in the state *HG*, highway green. We'll stay in this state, a self-arc, if there is no

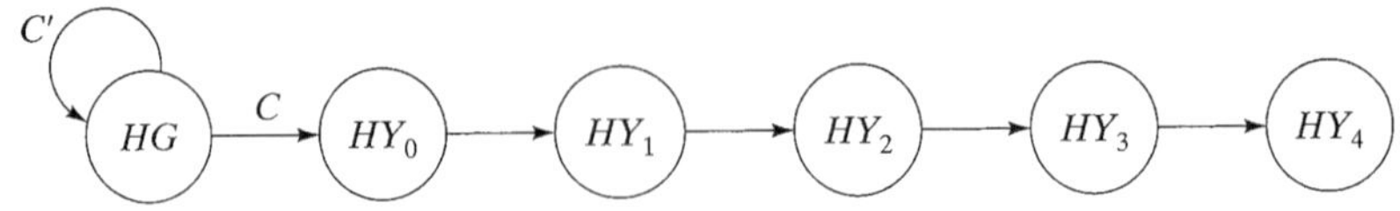

Figure 7.52 **Initial attempt at a state diagram (partial) for the traffic-light controller.**

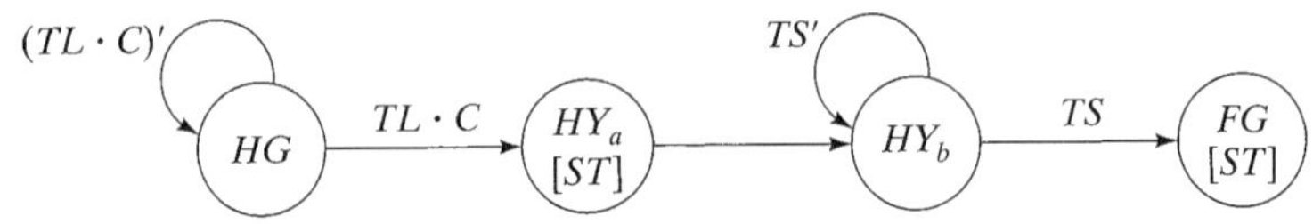

Figure 7.53 **Initial attempt at a Moore state diagram (partial) for the traffic-light controller utilizing an external counter.**

car on the farm road. If a car arrives on the farm road, *C* is asserted, we'll transition to the *HY*, or highway yellow, state. We'll need to stay in the this state for five cycles. We can do this my having a sequence of five states that all keep the highway light yellow as shown in Figure 7.52.

This starts to point us at considering a factoring of this state machine. If we count in this fashion, we'll need five states for each of the "yellow" states, and 20 for the maximum time for green on the farm road. That is already 30 states! Maybe we can do better if we utilize an external counter—a subroutine—for our FSM. We'll design a counter that has a reset signal and counts every cycle. It will have two outputs, one that is asserted after the count reaches 4 (allowing one cycle for communication back to our FSM) and another output that is asserted after 19 cycles.

Let's go back to the previous step and add these inputs and outputs to our specification (two to our input table and one to our output table).

Input Signal	Description
Reset	Place controller in initial state
C	Detects vehicle on farm road in either direction
TS	Short timer interval has expired
TL	Long timer interval has expired

Output Signal	Description
HG, HY, HR	Assert green, yellow, red highway lights
FG, FY, FR	Assert green, yellow, red farm road lights
ST	Reset timer and start timing long and short intervals

Now that we have our external counter, we can make another try at a more compact state diagram. We'll start by considering a Moore machine. This seems appropriate, as the outputs that will control the traffic lights are associated with the states *HG*, *HY*, *FG*, and *FY*. Figure 7.53 shows a

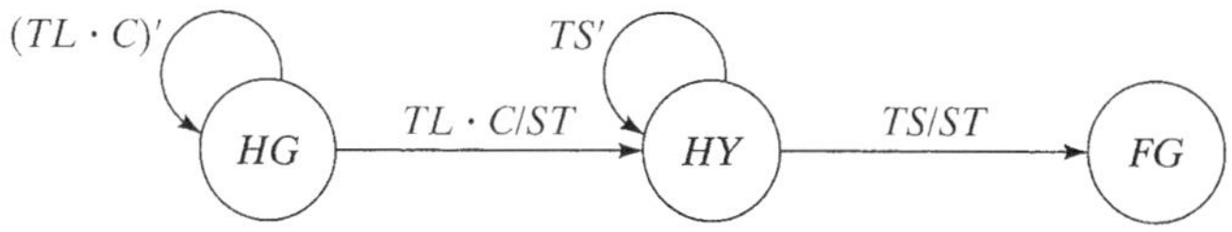

Figure 7.54 Initial attempt at a Mealy/Moore hybrid state diagram (partial) for the traffic light controller utilizing an external counter.

fragment of our state diagram. The outputs for each state are shown in square brackets. We do not show the traffic-light outputs as those are exactly what you would expect given the name of the state: namely, highway green and farm road red in state *HG*. The transition from *HG* to *HY* is now made when the long timer interval has expired, and there is a car on the farm road (recall we wait for the long time interval to ensure that the highway is green for a reasonable minimum amount of time: in this case, 20 cycles).

When we enter the first of the *HY* states shown, HY_a, we need to assert the output that will reset our external counter. However, we can't stay in this state and wait for the counter as it will be reset every cycle. This is why we have a second *HY* state, HY_b, that does not reset the counter. In this state, we can wait until the short time interval expires before moving on to the *FG* (farm road green) state.

This is where we should rethink our use of a Moore machine model. If we had been able to have outputs on transitions, we wouldn't have needed two separate states. Figure 7.54 shows the same fragment using a Mealy machine model for the interaction with the external counter. Note that we have kept the Moore outputs associated with each state that set the traffic lights to the appropriate colors. Therefore, we really have a mixed Mealy/Moore model. Mixing the two styles can be quite reasonable, especially if we intend to implement the Mealy machine as a synchronous one.

We have now seen the motivation for partitioning state machines, in this case, creating an external counter to greatly reduce the number of states in our machine. Furthermore, we've seen the reasoning that may go behind choosing a Moore, Mealy, or hybrid model for the state diagram. In this case, we would have likely needed eight states for a pure Moore model but need only four states in a hybrid model. Figure 7.55 shows the complete state diagram for our hybrid FSM.

Encoding Our next step is to encode the states and outputs of our FSM. With four states we can easily choose a simple assignment of 00, 01, 10, and 11 for the two state bits. Or, we could choose a *one-hot encoding* with four state bits. The states would have the codes 1000, 0100, 0010, 0001. Note that only one bit is set to 1 in each code, hence, the term one-hot.

Yet another possibility presents itself. Why not encode the states so that they are identical to the outputs we'll need for the traffic lights? For example, *HG* could be set to 001100 where the first 3 bits represent the red, yellow, and green bulbs of the highway traffic light and the last 3 bits represent the red, yellow, and green of the farm road light. If we do that, we'll have unique codes for each state and will have no additional

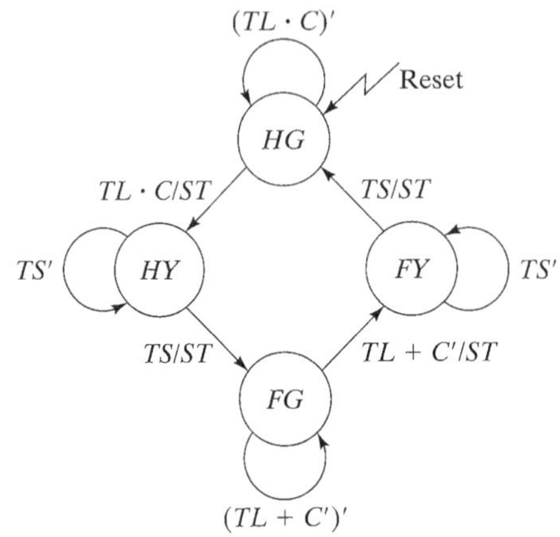

Figure 7.55 Complete hybrid Moore/Mealy state diagram for the traffic-light controller.

logic equations to implement. We'll simply have six state bits doing double duty as outputs, one for each light bulb. However, they may not lead to the six smallest logic implementations. It might have led to less logic if we had only two state bits with additional decoding logic for turning the two state bits into the six outputs needed for the lights. Or, possibly, the one-hot encoding would have been best.

This example demonstrates that output and state encoding can be intertwined very tightly, and there are quite a range of options available, even in a small FSM such as this. We'll come back to this traffic-light controller in the next chapter as we look at these optimization opportunities in more detail.

CHAPTER REVIEW

In this chapter, we have built on the basic sequential logic elements of the previous chapter to show how they can be used to implement counters and finite state machines.

Counters are an important class of sequential circuits, not only because they count events but also because they can be used to generate a periodic sequence. This has important implications for controller design, which we will examine in more detail in the following chapters.

We introduced a design procedure for mapping a state transition diagram, describing the count sequence, into actual hardware. The steps involved deriving the state diagram from the written specification for the counter, deducing a state table from the diagram which tabulates current and next states, expressing each next-state bit as a combinational logic function of the current-state bits. Because real hardware does not necessarily come up in a known state, we described the concept of self-starting counters. Self-starting design methods can be used to get a counter (eventually) into a known, valid state. We also demonstrated the advantages of counters with synchronous load and clear inputs, especially for the design of counters with starting or ending offsets.

We then moved on to develop a more general model of *finite state machines*. Counters are a very simple subclass of finite state machines that have a simple state sequence and limited inputs to control that sequence. Also, their outputs were simply the state of their flip-flops.

We developed the finite-state machine model of computation by understanding the basic timing behavior of a finite state machine: when inputs are sampled, when the next state and outputs undergo transition, and when these become stable. We presented the Moore and Mealy machine organizations, the latter in both asynchronous and synchronous forms. The basic idea is that you can represent a finite state machine by combinational logic functions of the current state (both Moore and Mealy machines) and inputs (Mealy machine) for the next state and output. Flip-flop storage elements hold the current state.

We introduced a five-step process for finite-state-machine design, and the main part of this chapter concentrated on the first two of these steps: *understanding the problem* and *obtaining an abstract representation of the finite state machine from an imprecise description of its behavior.* We concluded the chapter with motivation for the optimization opportunities that present themselves in the remaining three steps of realizing a finite state machine. We'll cover these in more detail in the next chapter along with the use of hardware description languages for describing finite state machines. After that we'll look at technologies for realizing sequential logic with a particular focus on finite state machines. Chapter 10 will provide a collection of examples taken from start to finish.

FURTHER READING

The seminal work in defining models of finite state machines are E. F. Moore's *Gedanken-experiments on Sequential Machines,* pp. 129–153, *Automata Studies, Annals of Mathematical Studies, No. 34,* Princeton University Press, Princeton, N. J., 1956 and G. H. Mealy's *A Method for Synthesizing Sequential Circuits,* Bell System Technical Journal, Vol. 34, pp. 1045–1079, September 1955.

M. Sipser's *Introduction to the Theory of Computation,* PWS Publishing, Boston, 1997 is an excellent introduction into the theoretical foundations of finite state machines and their various representations, including state transition tables. Some other important works that formed the basis for the modern theory of finite automata can be in S.C. Kleene's *Representation of Events in Nerve Nets and Finite Automata,* Automata Studies, Princeton, N. J., 1956 and Hopcroft & Ullman's important text *Introduction to Automata Theory, Languages and Computations*, Addison-Wesley, 1979.

Just about any textbook on digital design will describe counters, counter components, and design strategies that employ them. The concept of finite state machines (FSMs) is also integral to just about every book on digital logic design. You will find many additional examples and alternate explanations there.

EXERCISES

7.1 *(Counter Design)* Design a 2-bit counter that behaves according to the two control inputs I_0 and I_1 as follows: I_0, $I_1 = 0$, 0: stop counting; I_0, $I_1 = 0$, 1: count up by one; I_0, $I_1 = 1$, 0: count down by one; I_0, $I_1 = 1$, 1: count by two.

- **(a)** Draw the state diagram and state transition table.
- **(b)** Implement the counter using *D* flip-flops.
- **(c)** Assume that only 2-input NAND, NOR, XOR, and XNOR gates are available. Draw the schematic for your minimum gate count implementation.

7.2 *(Counter Design)* Design a three flip-flop counter that counts in the following sequence: 000, 010, 111, 100, 110, 011, 001, and repeat. Verify that your implementation is self-starting.

7.3 *(Counter Design)* Consider the design of a 4-bit BCD counter that counts in the following sequence: 0000, 0001, 0010, 0011, 0100, 0101, 0110, 0111, 1000, 1001, and then back to 0000, 0001, etc.

- **(a)** Draw the state diagram and next-state table.
- **(b)** Implement the counter using *D* flip-flops.
- **(c)** Implement the counter making it self-starting.

7.4 *(Counter Design)* Consider the design of a 4-bit Gray-code counter (that is, only one of the state bits changes for each transition) that counts in the following sequence: 0000, 0001, 0011, 0010, 0110, 0111, 0101, 0100, 1100, 1101, 1111, 1110, 1010, 1011, 1001, 1000, and then back to 0000, 0001, 0011, etc.

- **(a)** Draw a state diagram and next-state table.
- **(b)** Implement the counter using *D* flip-flops.
- **(c)** Do you have to worry about self-starting? Why or why not?

7.5 *(Counter Design)* The 4-bit Johnson counter advances through the sequence 0000, 1000, 1100, 1110, 1111, 0111, 0011, 0001, and repeats. Using the standard counter design process, show that the procedure yields the same implementation as in Figure 7.2.

7.6 *(Self-Starting Counters)* Analyze your solutions to Exercise 7.5 to check whether or not it is self-starting. Draw a complete state diagram and show all states and transitions implied by the implementation.

7.7 *(Asynchronous Inputs)* Can you think of any cases in which an asynchronous load or reset signal is acceptable in a digital design?

7.8 *(Offset Counters)* Figure 7.19 shows a possible implementation for a counter with a beginning offset. The counter will eventually enter the correct sequence, but in some applications this may not be acceptable. How might you add a reset signal that will force the counter into a valid state when it is asserted? What assumptions must be made about the duration of the reset signal?

7.9 *(Offset Counters)* Use 163 counter components and only NAND gates (with any number of inputs) and inverters to implement a

BCD counter. Show how you would construct a one BCD digit counter. Show how to build a two BCD digit counter. Does your design generalize to more digits?

7.10 *(Offset Counters)* Use two cascaded synchronous up-counters (for example, a 163 component) to implement a 6-bit offset counter that counts from 000010 to 110011 and repeats. Make sure the counter begins in state 000010 when the external reset signal is asserted.

7.11 *(Counter/Register Applications)* Consider the design of a bit-serial adder. This circuit uses a single full adder to add two binary numbers presented in serial fashion, 1 bit at a time.

(a) Draw the schematic for a 4-bit version of this circuit. Two 4-bit shift registers are loaded with the data to be added in parallel. These are shifted out a bit at a time, starting with the lowest-order bit, into the A and B inputs of the full adder. The partial sum is shifted into a third register. How should the carry out be handled between subsequent bits?

(b) Define your control signals for the bit-serial adder subsystem. Draw a timing diagram that illustrates the sequencing of these signals to implement the 4-bit addition. What happens on reset?

7.12 *(Counter/Register Applications)* Flip-flops, registers, and counters can be used to implement a variety of useful clocking functions.

(a) Design a system that will generate a single clock pulse one period long each time a push-button is pressed (you may assume that an external reference clock is available).

(b) Design a system using a 163 counter that will assert a signal for exactly 13 clock pulses each time a push-button is pressed (once again, an external reference clock is available).

(c) In many applications, it is desirable to single step the clock as well as to halt a free-running clock signal and later restart it. Design a circuit that will generate a single clock pulse each time a STEP push-button is pressed. Provide a separate circuit, independent of STEP, that will cleanly turn the clock on when a RUN switch is in the ON position and off when RUN is in its off position. No marginal/partial clock outputs are allowed.

7.13 *(Counter/Register Applications)* Design a 2-bit binary up-counter to the following specification. The counter has five inputs (not including the clock) and three outputs. The inputs are CLR, LOAD, COUNT, L_B, and L_A. CLR takes precedence over LOAD, which in turn takes precedence over COUNT. The outputs are B, A, and RCO.

When CLR is asserted, the flip-flops of the counter are reset to 0. If LOAD is asserted (when CLR is not asserted), the flip-flops' contents are replaced by the L_B and L_A inputs. If COUNT is asserted (when CLR and LOAD are not asserted), the flip-flops' contents are incremented: 00 becomes 01, 01 becomes 10, 10 becomes 11, and 11 becomes 00. When the counter is in state 11, RCO is asserted.

Implement the counter using *D* flip-flops and as few logic components as possible. *Hint:* Use 4-to-1 multiplexers to implement your next-state function. You may also use XOR gates, as well as standard AND, OR, and NOT gates.

7.14 *(Timing Methodology)* In this chapter, we have encouraged you to think of implementing all state registers of a finite state machine with flip-flops that are clocked in the same way. Consider what (if anything) could go wrong if an FSM was constructed using a combination of positive edge-triggered *D* flip-flops and negative edge-triggered *D* flip-flops to implement the FSM state register.

7.15 *(Timing Methodology)* What impact does the user of both positive and negative edge-triggered flip-flops have on timing constraints that must be satisfied by FSM logic? Consider a 50% duty-cycle clock first, then consider a 90% duty-cycle clock? Are there any advantages to using both types of flip-flops?

7.16 *(Parity Checker)* Redesign the odd parity checker FSM of Section 7.2.1 to make it check for even parity (that is, assert the output whenever the input contains an even number of 1s). Show your state diagram and implement the machine using either a *D* flip-flop.

7.17 *(Parity Checker Subsystem)* The odd parity checker of Section 7.2.1 generates a 1 whenever a bit stream of serial inputs contains an odd number of 1s. This is useful in a data communication subsystem for checking that transmitted data has been sent correctly. Data is transmitted as 8 data bits appended with a ninth parity bit. The 9-bit sequence must be in odd parity. That is, if the data bits have an odd number of 1s, the parity bit is 0. If the data bits have an even number of 1s, the parity bit is 1. You are to design a parity checker that asserts OK if the 9-bit sequence is correct in odd parity and ERROR otherwise.

(a) Is it possible to write a state diagram with a small number of states to describe the behavior of this finite state machine? Does your state diagram need to track all possible sequences of 9 bits?

(b) Consider implementing the subsystem using the parity checker FSM of Figure 7.27 in conjunction with a synchronous 4-bit counter like the 163. Draw a schematic using logic gates, a single flip-flop, and the counter. Draw a timing diagram including a bit sequence that leads to ERROR and one that leads to OK.

7.18 *(Mealy Machines)* Suppose you are told that a Mealy machine is implemented with three flip-flops, two inputs, and six asynchronous outputs. Consider the *complete* state diagram for this machine (that is, there are no don't cares). Answer the following questions:

(a) What are the minimum and maximum numbers of states in the state diagram?
(b) What are the minimum and maximum numbers of transition arrows starting at a particular state?
(c) What are the minimum and maximum numbers of transition arrows that can end in a particular state?
(d) What are the minimum and maximum numbers of different binary patterns that can be displayed on the outputs?

7.19 *(Moore Machines)* Suppose you are told that a Moore machine has five flip-flops, three inputs, and nine outputs. Answer the following questions:

(a) What are the minimum and maximum numbers of states in the state diagram?
(b) What are the minimum and maximum numbers of transition arrows starting at a particular state?
(c) What are the minimum and maximum numbers of transition arrows that can end in a particular state?
(d) What are the minimum and maximum numbers of different binary patterns that can be displayed on the outputs?

7.20 *(Reverse Engineering)* What is the counter state diagram implied by the flip-flop implementation of Figure Ex. 7.20? Note that there are two inputs to this counter. C is asserted to enable counting. D is used to change count direction, that is, go through the sequence in reverse order.

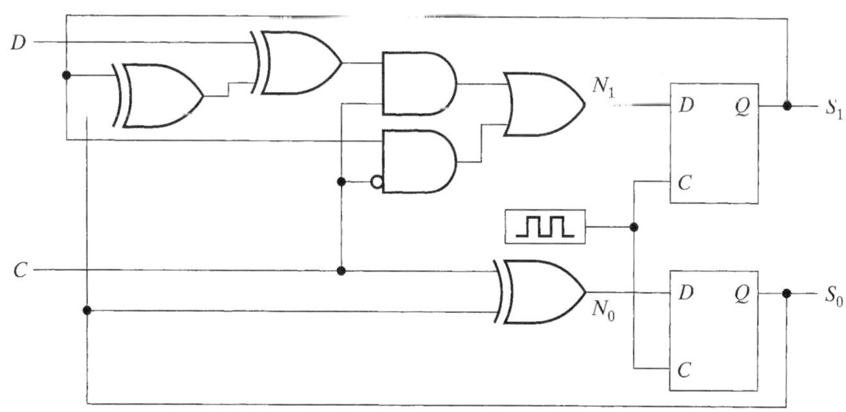

Figure Ex. 7.20 Counter implementation to reverse engineer in Exercise 7.20.

7.21 *(Synchronous Mealy Machine)* Section 7.3.4 describes the implementation of the vending machine in several styles. Draw a timing diagram that shows a difference in the detailed timing behavior of the original (unretimed) Moore, asynchronous Mealy, and synchronous Mealy implementations.

7.22 *(Word Problem)* A finite state machine has one input and one output. The output becomes 1 and remains 1 thereafter when at least two 0s and at least two 1s have occurred as inputs, regardless of the order of occurrence. Assuming this is to be implemented as a Moore machine, draw a state diagram for the machine. (*Hint:* You can do this in nine states.)

7.23 *(Word Problem)* A finite state machine has one input (X) and two outputs (Z_1 and Z_2). An output $Z_1 = 1$ occurs every time the input sequence 101 is observed, provided the sequence 011 has never been seen. An output $Z_2 = 1$ occurs every time the input 011 is observed. Note that once $Z_2 = 1$, $Z_1 = 1$ can never occur. Assuming the machine is to be implemented in the Mealy design style, draw the corresponding state diagram. (*Hint:* The minimum number of states is eight.)

7.24 *(Word Problem)* A Moore machine has two inputs (X_1, X_2) and one output (Z). Produce the state diagram for the machine, given the following specification. The output remains a constant value unless one of the following input sequences occurs:

(a) The input sequence $X_1X_2 = 00$, 11 causes the output to become 0.
(b) The input sequence $X_1X_2 = 01$, 11 causes the output to become 1.
(c) The input sequence $X_1X_2 = 10$, 11 causes the output to toggle value.

7.25 *(Word Problem)* A sequential circuit has one input (X) and one output (Z). Draw a Mealy state diagram for each of the following cases:

(a) The output is $Z = 1$ if and only if the total number of 1s received is divisible by 3 (for example, 0, 3, 6, ...).
(b) The output is $Z = 1$ if and only if the total number of 1s received is divisible by 3 and the total number of 0s received is an even number greater than zero (*Hint:* Nine states are sufficient).

7.26 *(Word Problem)* A sequential circuit has two inputs and two outputs. The inputs (X_1X_2) represent a 2-bit binary number, N. If the present value of N is greater than the previous value, then $Z_1 = 1$. If the present value of N is less than the previous value, then $Z_2 = 1$. Otherwise, Z_1 and Z_2 are 0.

(a) Derive a Mealy machine state diagram. (*Hint:* The machine needs only five states.)
(b) Derive a Moore machine state diagram. (*Hint:* The machine needs at least 11 states.)

7.27 *(Word Problem)* A Moore machine has one input and one output. The output should be 1 if the total number of 0s received at the input is odd and the total number of 1s received is an even number greater than 0. This machine can be implemented in exactly six states. Draw a complete state diagram. Indicate what each state is meant to represent.

7.28 *(Word Problem)* Two two-way streets meet at an intersection controlled by a four-way traffic light. In the east and west directions, the lights cycle from green to yellow to red. The south-facing lights do the same thing, except that they are red when the east-west lights are green or yellow, and vice versa. However, the north-facing lights are augmented with a green left-turn arrow. They cycle red–green arrow–yellow arrow–green–yellow–red. Consider the following additional problem specifications:

(a) When the green or yellow left-turn arrows are illuminated, the lights in the other three directions are red.
(b) The timings for the north-facing lights are as follows: red, 60 seconds; green arrow, 20 seconds; yellow arrow, 10 seconds; green, 45 seconds; and yellow, 15 seconds.
(c) The timings for the other lights can be derived from specifications (a) and (b). Assume you have as many programmable timers as you need. These can be loaded with a time constant (in seconds) and assert an output when they count down to zero.

Construct a chart that shows the timing behavior of the lights in each of the four directions (*Y*-axis). List the illuminated lights for east, west, south, and north along the *Y*-axis. The *X*-axis is calibrated in the elapsed time in seconds. Show what happens in one complete cycle of the lights. How many unique configurations of the lights are there? Derive a chart, explicitly listing all input and output control signals needed to implement the traffic light system.

7.29 *(Word Problem)* You are to develop a state diagram for a washing machine. The machine starts when a coin is deposited. It then sequences through the following stages: soak, wash, rinse, and spin. There is a "double wash" switch, which, if turned on, causes a second wash and rinse to occur. There is one timer—you may assume that each stage should take the same amount of time. The timer begins ticking as soon as the coin is deposited, generates a T signal at the end of the time period, and then resets itself and starts again. If the lid is raised during the spin cycle, the machine stops spinning until the lid is closed. You may assume that the timer suspends ticking while the lid is raised. Identify your inputs and outputs, and draw a state diagram that implements this finite state machine.

7.30 *(Word Problem)* You are to design the control for an automatic candy vending machine. The candy bars inside the machine cost 25 cents, and the machine accepts nickels, dimes, and quarters only. The inputs to the control are a set of three signals that indicate what kind of coin has been deposited, as well as a reset signal. The control should generate an output signal that causes the candy to be delivered whenever the amount of money received is 25 cents or more (no change is given). Once the candy has been delivered, some external circuitry will generate a reset signal to put the control back into its initial state. Identify your inputs and outputs, and draw the state diagram that implements this finite state machine.

7.31 *(Word Problem)* Your task is to design the control for a newspaper vending machine to the following specification. The newspaper costs 50 cents. The vending machine accepts nickels, dimes, and quarters. The customer presses a START button and then begins entering coins. Coin sorter logic indicates to the FSM whether a nickel (N), dime (D), or quarter (Q) has been deposited. (Assume that the FSM advances from one state to the next when a coin is deposited.) If exact change is entered, a latch is released so the customer can get the paper. If the amount of money deposited exceeds 50 cents, the deposited coins are refunded to the customer by asserting a refund (REF) signal. Assume that the money just deposited is kept separated from previously accepted coins. The latter are held in a coin repository. Otherwise, the deposited coins join the repository as the FSM asserts a release (REL) signal. The block diagram for the FSM is shown in Figure Ex. 7.31.

Consider for a moment the signals that indicate the number of nickels and dimes available to make change. What is the maximum number of nickels needed at any time? What is the maximum number of dimes needed? Understanding the answers to these questions may help to simplify your state diagram.

Complete a Mealy machine state diagram for the vending machine's control.

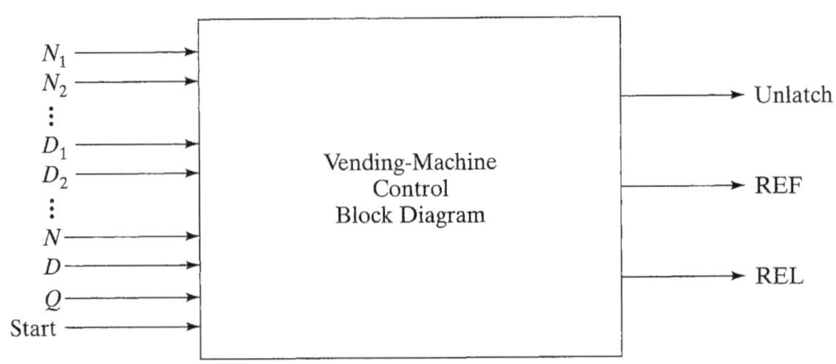

Figure Ex. 7.31 **Inputs/outputs for newspaper vending machine.**

CHAPTER 8

Working with Finite State Machines

Introduction

Now that we have developed a methodology for implementing finite state machines, it is time to turn our attention to how we optimize the implementation process to yield more efficient circuits. We will also see how we can describe them using hardware description languages.

In the previous chapter, we focused on deriving a state diagram from an English language description. In this chapter, we'll look at methods of reducing the number of states in our specifications. We'll also take a longer look at the problem of state encoding and how encodings affect the number of flip-flops and the amount of logic needed to implement a finite state machine.

Often, it is easier and cheaper to implement a finite state machine by braking it up into smaller pieces that communicate. We'll see how and when it makes sense to partition state machines and look at methods for doing so even after we've already derived a state diagram. You may be forced to partition your state machine because it cannot fit into the available logic components or just because the machine is too big and requires complex multilevel logic to implement monolithically that may be too slow for the desired clock period.

The chapter will end with how to describe finite state machines using a hardware description language. We'll revisit all the basic concepts and see how they are represented in Verilog.

In this chapter, we emphasize the following techniques and concepts:

- *Procedures for optimizing a finite state machine.* You will learn methods for state minimization and state assignment.
- *Partitioning methods.* You will learn the techniques for breaking finite state machines into smaller, communicating state machines

that may be faster and/or better suited for implementation with available logic.

- *Describing finite state machine in HDLs.* You will learn how to describe finite state machines and their structure using hardware description languages.

8.1 State Minimization/Reduction

State reduction identifies and combines states that have *equivalent behavior.* Two states have equivalent behavior if, for all input combinations, their outputs are the same and they change to the same or equivalent next states.

For example, in Figure 7.50(b), states S_0 and S_2 are equivalent. Both states output a 0; both change to S_1 on a 1 and self-loop on a 0. Combining these into a single state leads to Figure 7.50(a). On all input strings, the output sequence of either state diagram is exactly the same.

Algorithms for state reduction begin with the symbolic state transition table. First, we group together states that have the same state outputs (Moore machine) or transition outputs (Mealy machine). These are potentially equivalent, since states cannot be equivalent if their outputs differ.

Next, we examine the transitions to see if they go to the same next state for every input combination. If they do, the states are equivalent and we can combine them into a renamed new state. We then change all transitions into the newly combined states and repeat the process until no additional states can be combined.

In the following two subsections, we examine alternative algorithms for state reduction: *row matching* and *implication charts*. The former is a good pencil-and-paper method, but does not always obtain the best reduced state table. Implication charts are more complex to use by hand, but they are easy to implement in software and do find the best possible solution.

We can always combine the two approaches. Row matching quickly reduces the number of states. The implication chart method, now working with fewer states, finds the equivalent states missed by row matching more rapidly.

8.1.1 Row-Matching Method

Let's begin our investigation of the row-matching method with a detailed example. We will see how to transform an initial state diagram for a simple sequence detector into a minimized, equivalent state diagram.

Four-Bit Sequence Detector We'll start with an example finite state machine that, if implemented straighforwardly, generates many equivalent states. A common sequential circuit is a sequence detector that looks for a particular pattern of bits on a single input.

Specification and Initial State Diagram Let's consider a sequence-detecting finite state machine with the following specification. The machine has a single input X and output Z. The output is asserted after each 4-bit input sequence if it consists of one of the binary strings 0110 or 1010. The machine returns to the reset state after each and every 4-bit sequence.

We will assume a Mealy implementation. Some sample behavior of the finite state machine is

$$X = 0010\ 0110\ 1100\ 1010\ 0011\ \ldots$$

$$Z = 0000\ 0001\ 0000\ 0001\ 0000\ \ldots$$

The output is asserted only after the previous four serial inputs match one of the specified strings. Also, the input patterns do not overlap: the machine makes a decision to assert its output after each group of 4 bits.

Because this finite state machine recognizes finite length strings, we can place an upper bound on the number of states needed to recognize any particular binary string of length four. Figure 8.1 shows the Mealy state diagram. There are 16 unique paths through the state diagram, one for each possible 4-bit pattern. This adds up to 15 states and 30 transitions. We highlight the paths leading to recognition of the strings 0110 and 1010 in the figure. Only two of the transitions have a 1 output, representing the accepted strings.

State Table and Row-Matching Method We can combine many of the states in Figure 8.1 without changing the input/output behavior of the finite state machine. For example, it is clear that once we have the input 00 that there will be no way that we will detect one of our two target strings. But how do we find these equivalent states in a systematic fashion?

First, we look at the state transition table, as shown in Figure 8.2. This table is in a slightly different format than we have seen so far. It contains one row per state, with multiple next-state and output columns

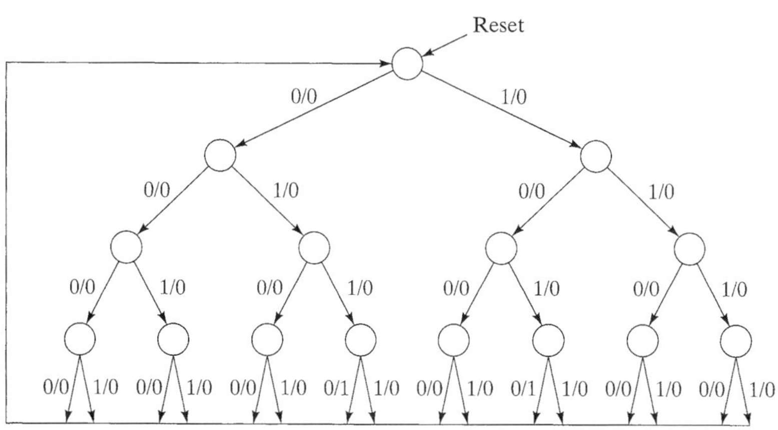

Figure 8.1 **Original state diagram for 4-bit string recognizer.**

Input Sequence	Present State	Next State $X=0$	Next State $X=1$	Output $X=0$	Output $X=1$
Reset	S_0	S_1	S_2	0	0
0	S_1	S_3	S_4	0	0
1	S_2	S_5	S_6	0	0
00	S_3	S_7	S_8	0	0
01	S_4	S_9	S_{10}	0	0
10	S_5	S_{11}	S_{12}	0	0
11	S_6	S_{13}	S_{14}	0	0
000	S_7	S_0	S_0	0	0
001	S_8	S_0	S_0	0	0
010	S_9	S_0	S_0	0	0
011	S_{10}	S_0	S_0	1	0
100	S_{11}	S_0	S_0	0	0
101	S_{12}	S_0	S_0	1	0
110	S_{13}	S_0	S_0	0	0
111	S_{14}	S_0	S_0	0	0

Figure 8.2 **Initial state transition table for the 0110 or 1010 sequence detector.**

Input Sequence	Present State	Next State $X=0$	Next State $X=1$	Output $X=0$	Output $X=1$
Reset	S_0	S_1	S_2	0	0
0	S_1	S_3	S_4	0	0
1	S_2	S_5	S_6	0	0
00	S_3	S_7	S_8	0	0
01	S_4	S_9	S'_{10}	0	0
10	S_5	S_{11}	S'_{10}	0	0
11	S_6	S_{13}	S_{14}	0	0
000	S_7	S_0	S_0	0	0
001	S_8	S_0	S_0	0	0
010	S_9	S_0	S_0	0	0
011 or 101	S'_{10}	S_0	S_0	1	0
100	S_{11}	S_0	S_0	0	0
110	S_{13}	S_0	S_0	0	0
111	S_{14}	S_0	S_0	0	0

Figure 8.3 **Revised state transition table after S_{10} and S_{12} are combined.**

based on the input combinations. It gives exactly the same information as a table with separate rows for each state and input combination.

The input sequence column is a documentation aid, describing the partial string as seen so far. When read from left to right, it describes the sequence of input bits that lead to the given state.

Next we examine the rows of the state transition table to find any with identical next-state and output values (hence the term "row matching"). For this finite state machine, we can combine S_{10} and S_{12}. Let's call the new state S'_{10} and use it to rename all transitions to S_{10} or S_{12}. The revised state table is shown in Figure 8.3.

Input Sequence	Present State	Next State $X=0$	Next State $X=1$	Output $X=0$	Output $X=1$
Reset	S_0	S_1	S_2	0	0
0	S_1	S_3	S_4	0	0
1	S_2	S_5	S_6	0	0
00	S_3	S_7'	S_7'	0	0
01	S_4	S_7'	S_{10}'	0	0
10	S_5	S_7'	S_{10}'	0	0
11	S_6	S_7'	S_7'	0	0
Not (011 or 101)	S_7'	S_0	S_0	0	0
011 or 101	S_{10}'	S_0	S_0	1	0

Figure 8.4 Revised state transition table after S_7, S_8, S_9, S_{11}, S_{13}, and S_{14} are combined.

Input Sequence	Present State	Next State $X=0$	Next State $X=1$	Output $X=0$	Output $X=1$
Reset	S_0	S_1	S_2	0	0
0	S_1	S_3'	S_4'	0	0
1	S_2	S_4'	S_3'	0	0
00 or 11	S_3'	S_7'	S_7'	0	0
01 or 10	S_4'	S_7'	S_{10}'	0	0
Not (011 or 101)	S_7'	S_0	S_0	0	0
011 or 101	S_{10}'	S_0	S_0	1	0

Figure 8.5 Final reduced state transition table after S_3, S_6 and S_4, S_5 are combined.

Row-Matching Iteration We continue matching rows until we can no longer combine any. In Figure 8.3, S_7, S_8, S_9, S_{11}, S_{13}, and S_{14} all have the same next states and outputs. We combine them into a renamed state S_7'. The table, with renamed transitions, is shown in Figure 8.4.

Now states S_3 and S_6 can be combined, as can S_4 and S_5. We call the combined states S_3' and S_4', respectively. The final reduced state transition table is shown in Figure 8.5. In the process, we have reduced 15 states to just 7 states. This allows us encode the state in 3 bits rather than 4. The reduced state diagram is given in Figure 8.6.

Limitations of the Row-Matching Method Unfortunately, row matching does not always yield the most-reduced state table. We can prove this with a simple counterexample. Figure 8.7 shows the state table for the three-state odd parity checker of Figure 7.50. Although states S_0 and S_2 have the same output, they do not have the same next state. Thus, they cannot be combined by simple row matching. The problem is the self-loop transitions on input 0. If we combined these two states, the self-loop would be maintained, but this is not found by row matching. We need another, more rigorous method for state reduction.

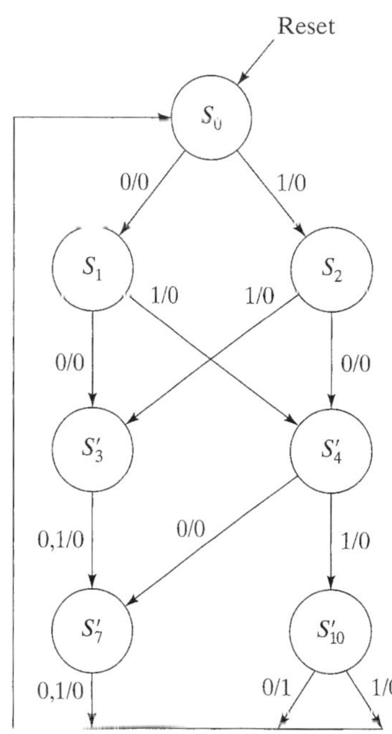

Figure 8.6 Reduced state diagram for 4-bit string recognizer.

Present State	Next State $X=0$	Next State $X=1$	Output
S_0	S_0	S_1	0
S_1	S_1	S_2	1
S_2	S_2	S_1	0

Figure 8.7 **State table for three-state odd parity checker.**

Input Sequence	Present State	Next State $X=0$	Next State $X=1$	Output $X=0$	Output $X=1$
Reset	S_0	S_1	S_2	0	0
0	S_1	S_3	S_4	0	0
1	S_2	S_5	S_6	0	0
00	S_3	S_0	S_0	0	0
01	S_4	S_0	S_0	1	0
10	S_5	S_0	S_0	0	0
11	S_6	S_0	S_0	1	0

Figure 8.8 **Initial state transition table for the 3-bit sequence detector.**

8.1.2 Implication Chart Method

The implication chart method is a more systematic approach to finding the states that can be combined into a single reduced state. As you might suspect, the method is more complex and is better suited for machine implementation than hand use.

Three-Bit Sequence Detector We'll use a slightly smaller sequence detector that looks at only 3 bits at a time to help us highlight this method. It starts off with 7 states, instead of 15.

Specification and Initial State Table Your goal is to design a binary sequence detector that will output a 1 whenever the machine has observed the serial sequence 010 or 110 at the inputs. Figure 8.8 shows its initial state table.

Data Structure: The Implication Chart The method operates on a data structure that enumerates all possible combinations of states taken two at a time, called an *implication chart.* Figure 8.9(a) shows the chart with an entry for every pair of states. This form of the chart is more complicated than it needs to be. For example, the diagonal entries are not needed: it does not reduce states to combine a state with itself! And the upper and lower triangles of entries are symmetric. The chart entry for S_i and S_j contains the same information as that for S_j and S_i. Thus, we work with the reduced structure of Figure 8.9(b). You can now see that if we had done a 4-bit sequence detector, we would have had a chart with 105 cells instead of the 21 of Figure 8.9. In general, for n starting states, our implication chart data structure will require $(n^2 - n)/2$ cells.

We fill in the implication chart as follows. Let X_{ij} be the entry whose row is labeled by state S_i and whose column is labeled by state S_j.

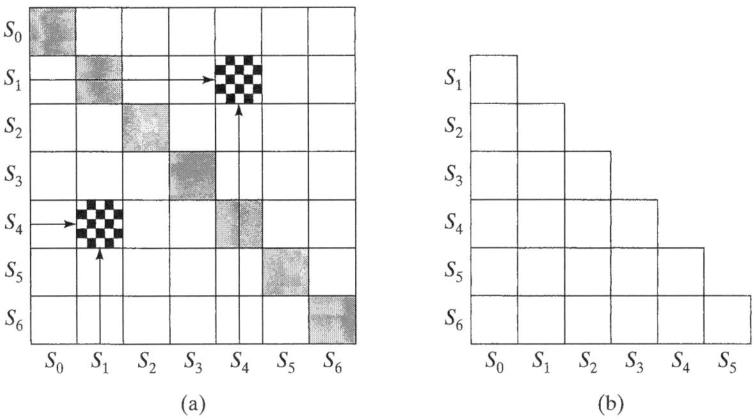

Figure 8.9 **Matrix for state combinations and the corresponding implication chart.**

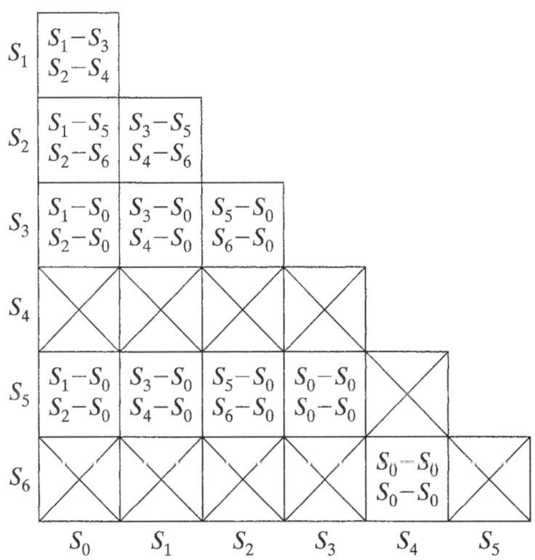

Figure 8.10 **Initial implication chart for the 3-bit sequence detector.**

If we were able to combine states S_i and S_j, it would imply that their next-state transitions for each input combination must also be equivalent. The chart entry contains the next-state combinations that must be equivalent for the row and column states to be equivalent. If S_i and S_j have different outputs or next-state behavior, an X is placed in the entry. This indicates that the two states can never be equivalent.

Initial Implication Chart The implication chart for the example state table is shown in Figure 8.10. S_0, S_1, S_2, S_3, and S_5 have the same outputs and are candidates for being combined. Similarly, states S_4 and S_6 might also be combined. Any combination of states across the two groups, such as S_1 and S_4, is labeled by an X in the chart. Since their outputs are different, they can never be equivalent.

To fill in the chart entry for (row) S_1 and (column) S_0, we look at the next-state transitions. S_0 goes to S_1 on 0 and S_2 on 1, while S_1 goes

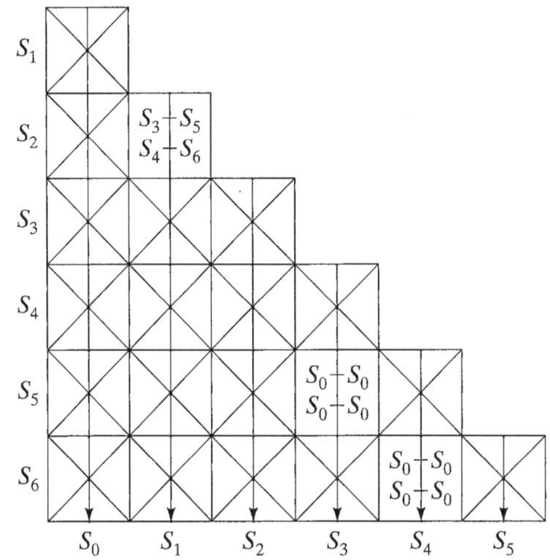

Figure 8.11 Results of first marking pass.

to S_3 and S_4, respectively. We fill the chart in with S_1–S_3, the transitions on 0, and S_2–S_4, the transitions on 1. We call these groupings *implied state pairs*. The entry means that S_0 and S_1 cannot be equivalent unless S_1 is equivalent to S_3 and S_2 is equivalent to S_4. The rest of the entries are filled in similarly.

At this point, the chart already contains enough information to eliminate many impossible equivalent pairs. For example, we already know that S_2 and S_4 cannot be equivalent: they have different output behavior. Thus, there is no way that S_0 can be equivalent to S_1.

Finding these cases is straightforward. We visit the entries in sequence. For example, start with the top square in the first column and advance from top to bottom and left to right. If square S_i,S_j contains the implied state pair S_m–S_n and square S_m,S_n contains an X, then mark S_i,S_j with an X as well.

First Marking Pass Figure 8.11 contains the results of this first marking pass. Entry S_2,S_0 is marked with an X because the chart entry for the implied state pair S_2–S_6 is already marked with an X. Entry S_3,S_0 is also marked, because entry S_1,S_0 (as well as S_2,S_0) has just been marked. The same is true for S_5,S_0. By the end of the pass, the only entries not marked are S_2,S_1; S_5,S_3; and S_6,S_4.

Second Marking Pass We now make a second pass through the chart to see if we can add any new markings. Entry S_2,S_1 remains unmarked. Nothing in the chart refutes that S_3 and S_5 are equivalent. The same is true of S_4 and S_6.

Continuing, S_3,S_5 and S_4,S_6 are now obviously equivalent. They have identical outputs and transfer to the same next state (S_0) for all input combinations.

Since no new markings have been added, the algorithm stops. The unmarked entries represent equivalences between the row and column

Input Sequence	Present State	Next State $X=0$	Next State $X=1$	Output $X=0$	Output $X=1$
Reset	S_0	S_1'	S_1'	0	0
0 or 1	S_1'	S_3'	S_4'	0	0
00 or 10	S_3'	S_0	S_0	0	0
01 or 11	S_4'	S_0	S_0	1	0

Figure 8.12 **Final reduced state transition table for the 3-bit sequence detector.**

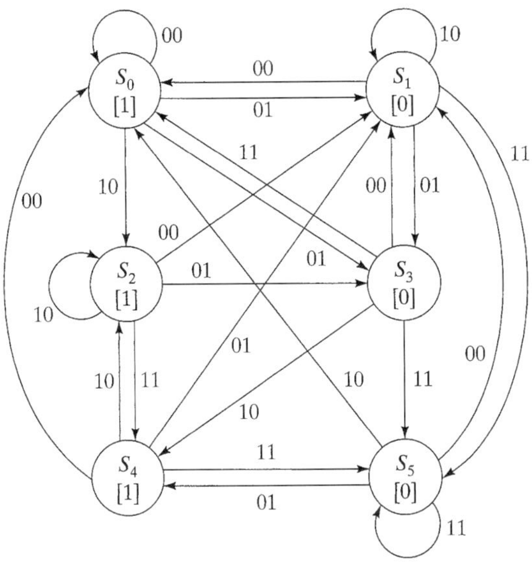

Figure 8.13 **Multiple-input state diagram.**

Present State	Next State 00	Next State 01	Next State 10	Next State 11	Output
S_0	S_0	S_1	S_2	S_3	1
S_1	S_0	S_3	S_1	S_5	0
S_2	S_1	S_3	S_2	S_4	1
S_3	S_1	S_0	S_4	S_5	0
S_4	S_0	S_1	S_2	S_5	1
S_5	S_1	S_4	S_0	S_5	0

Figure 8.14 **Multiple-input state transition table.**

indices: S_1 is equivalent to S_2, S_3 to S_5, and S_4 to S_6. The final reduced state table is shown in Figure 8.12.

Multi-Input State Diagram and Transition Table We can generalize the procedure for finite state machines with more than one input. The only difference is that there are more implied state pairs: one for each input combination.

Let's consider the state diagram for a two-input Moore machine shown in Figure 8.13. Each state has four next-state transitions, one for each possible input condition. The derived state transition table is given in Figure 8.14.

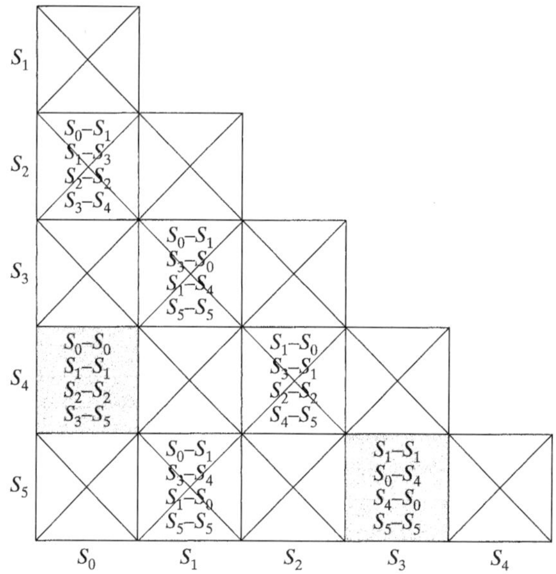

Figure 8.15 Multiple-input implication chart.

Present State	Next State 00	01	10	11	Output
S_0'	S_0'	S_1	S_2	S_3'	1
S_1	S_0'	S_3'	S_1	S_3'	0
S_2	S_1	S_3'	S_2	S_0'	1
S_3'	S_1	S_0'	S_0'	S_3'	0

Figure 8.16 Multiple-input reduced state table.

Implication Chart Processing Figure 8.15 shows the implication chart derived from the state transition table. Let's see how some of the entries are filled in. Since S_1 and S_0 have different state outputs, we place X in entry S_1,S_0. For the S_2,S_0 entry, we list the implied state pairs under the input conditions 00, 01, 10, 11. Because S_0 stays in S_0 on input 00, while S_2 goes to S_1 on 00, we add the implied state pair $S_0–S_1$ to the entry. On input 01, S_0 goes to S_1, S_2 goes to S_3, and we add $S_1–S_3$ to the entry. Similarly, we add the pairs $S_2–S_2$ on 10 and $S_3–S_4$ on 11 to the entry and fill in the rest of the entries.

Now we begin the marking pass. Working down the columns, we cross out entry $S_2–S_0$ because S_0,S_1 is already crossed out. The same thing happens to the entries S_3,S_1; S_5,S_1; and S_4,S_2. This leaves S_4,S_0 and S_5,S_3 unmarked (these are highlighted in the figure). Their being unmarked implies that S_4 is equivalent to S_0 (renamed S_0') and S_3 is equivalent to S_5 (S_3'). The reduced state table is given in Figure 8.16.

EXAMPLE 8.1 STATE REDUCTION OF PARITY CHECKER FINITE STATE MACHINE

The row-matching method could not combine states S_0 and S_2 in the three-state parity checker of Figure 7.50. Can the implication chart method do the job?

The implication chart for the state transition table of Figure 8.7 is given in Figure 8.17. S_1,S_0 and S_2,S_1 are marked immediately because their outputs differ. The remaining square is left unmarked, implying that S_0 and S_2 are equivalent. This is the correct reduced state transition table.

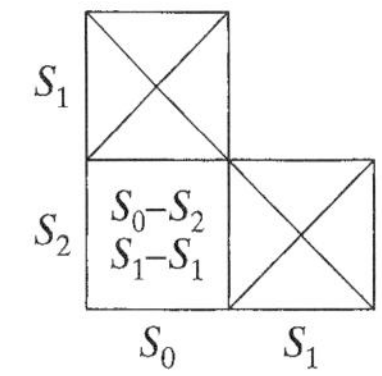

Figure 8.17 **Implication chart for three-state odd parity checker.**

Implication Chart Summary The algorithm for state reduction using the implication chart method consists of the following steps:

1. Construct the implication chart, consisting of one square for each possible combination of states taken two at a time.
2. For each square labeled by states S_i and S_j, if the outputs of the states differ, mark the square with an X; these states are *not* equivalent. Otherwise, they may be equivalent. Within the square, write implied pairs of equivalent next states for all input combinations.
3. Systematically advance through the squares of the implication chart. If the square labeled by states S_i, S_j contains an implied pair S_m–S_n and square S_m,S_n is marked with an X, then mark S_i, S_j with an X. Since S_m and S_n are not equivalent, neither are S_i and S_j.
4. Continue executing Step 3 until no new squares are marked with an X.
5. For each remaining unmarked square S_i, S_j, we can conclude that states S_i and S_j are equivalent.

8.1.3 Equivalent States in the Presence of Don't Cares

Finding equivalent states using the implication chart method is straightforward and can be efficiently implemented in software. However, the problem becomes much more difficult if we consider that we could have don't cares in our state transition tables. Input don't cares allow us to choose the state to transition to for an input combination that is impossible. This expands our ability to merge states.

Output don't cares are much more problematic as both the state minimization methods presented above rely on output patterns as a central part of determining state equivalence. The issue with output don't cares and why they make state minimization much more difficult is best illustrated with a small example.

EXAMPLE 8.2 STATE REDUCTION WITH OUTPUT DON'T CARES

Figure 8.18 shows a simple table of states and their outputs. These are Moore outputs associated with the state and not input values but show some don't cares. The difficulty arises in that state equivalence is

State	Output
S_0	– 0
S_1	1 –
S_2	– 1

Figure 8.18 **States with output "don't cares."**

longer transitive. States S_0 and S_1 can be compatible if we set the output don't cares to appropriate values, namely 10 for both states. Also, S_1 and S_2 can be made compatible by setting their output don't cares so that they both have 11 as output. Notice, however, that it is impossible to come up with an assignment of don't care values that will make all three states compatible. Our row-matching and implication chart methods both rely on the transitive property and here, in this simple example, we have a violation of that property, S_0 can be equivalent to S_1 and S_1 can be equivalent to S_2 but there is no way that S_0 and S_2 can be equivalent. Of course, in this small example it doesn't matter which pair we decide to make equivalent. In a larger state table, however, it is much harder to make local choices that will have the best global results.

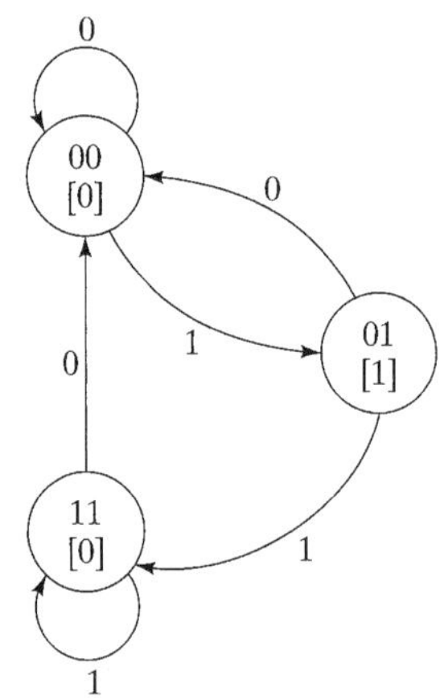

Figure 8.19 Moore state diagram for a 0 to 1 edge detector.

X	Q_1	Q_0	Q_1^+	Q_0^+
0	0	0	0	0
0	0	1	0	0
0	1	1	0	0
1	0	0	0	1
1	0	1	1	1
1	1	1	1	1
–	1	0	0	0

Figure 8.20 Encoded state table for edge detector of Figure 8.19.

8.1.4 When State Minimization Doesn't Help

State minimization can be very useful. It allows us to build smaller finite state machines with less flip-flops and less logic as we need one less logic equation implemented for every flip-flop we save. However, the logic equations themselves can become more complex as we fit more states into a smaller number of bits. Again, this is best illustrated with a small example.

EXAMPLE 8.3 MINIMIZATION LEADING TO A LESS EFFICIENT REALIZATION

Let's implement an edge detector finite state machine. We'll use a Moore configuration to detect when consecutive inputs change from 0 to 1. We will output a one whenever we detect this "rising edge" in our input stream.

Figure 8.19 shows the state diagram for our edge detector. It uses three states. Note that whenever a 0 is seen on the input, the finite state machine goes to state 00 and outputs a 0. State 01 is the only state to output a 1 and the finite state machine enters this state after a 1 is seen following some 0s (the transition from state 00 to state 01). State 11 is used to handle the situation when we see multiple 1s in a sequence but need to change the output to a 0 after the first 1 is seen. None of the three states are equivalent, thus, this is a minimized state diagram.

The encoded state table for this state diagram is shown in Figure 8.20. The equations for the next state Q_1^+ and Q_0^+ are (where X represents the single input and OUT, the single output):

$$Q_1^+ = X(Q_1 \text{ XOR } Q_0)$$
$$Q_0^+ = XQ_1'Q_0'$$
$$\text{Out} = Q_1'Q_0$$

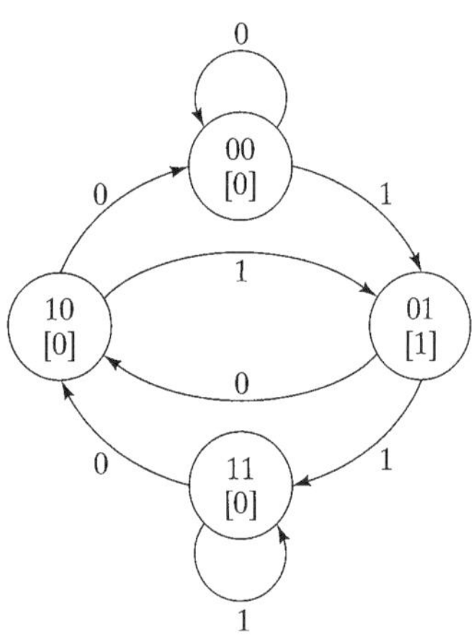

Figure 8.21 Non-minimal state diagram for edge detector.

Now let's consider a different state diagram with four states instead of three, as shown in Figure 8.21. Clearly, one of the four states is

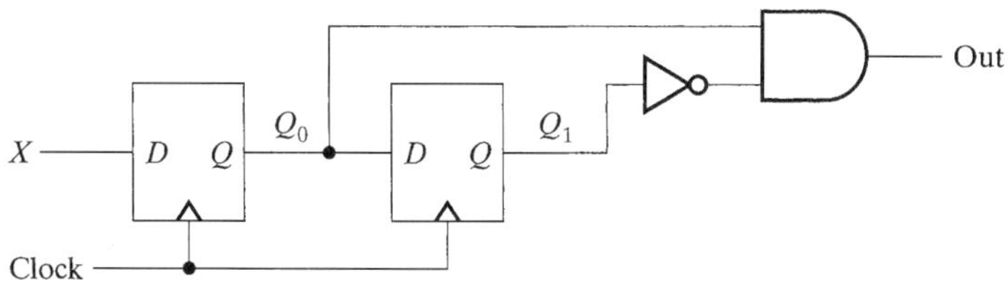

Figure 8.22 Realization of edge detector using the non-minimal state diagram of Figure 8.21.

redundant. However, if we just go ahead an implement this state diagram directly, we end up with much simpler equations:

$$\begin{aligned} Q_1^+ &= Q_0 \\ Q_0^+ &= X \\ \text{Out} &= Q_1' Q_0 \end{aligned}$$

In fact, this is a simple 2-bit register with an AND gate to look for a 0 followed by a 1. The circuit realization is shown in Figure 8.22. This is clearly an easy way to design an edge or sequence detector—just shift in enough bits so that a gate can be used to detect the sequence of interest. The fact that there are redundant states actually helped achieve the smallest implementation.

Minimizing states isn't always the best thing to do. It can lead to a larger and non-obvious implementation. This, coupled with the problems of figuring out how best to assign output don't cares so as to make more states similar, should demonstrate that state minimization is still an art and not anywhere near as straightforward in practice as our neat row-matching and implication chart methods might have led us to believe.

8.2 State Assignment

The number of gates needed to implement a sequential logic network depends strongly on how we assign *encoded* Boolean values to *symbolic* states. Unfortunately, state assignment is an even more difficult problem than state minimization in that we do not have simple techniques that can get us to a good solution. In fact, the only way to obtain the best possible assignment is to try every choice for the encoding, an extremely large number for real state machines. For example, even a simple four-state finite state machine, such as the traffic-light controller of the last chapter (see Figure 7.55), has 4! (4 factorial) = 4 * 3 * 2 * 1 = 24 different encodings (see Figure 8.23 for a list of all 24 possibilities) if we choose two state bits in which to encode them. We can choose any one of the four possible encodings for the first state, on of the three left for second state, and so on. If we use more state bits than the minimum, then there are even more possibilities. In general, if we have *m* states and *n* bits in which to encode them (2^n must be

HG	*HY*	*FG*	*FY*		*HG*	*HY*	*FG*	*FY*
00	01	10	11		10	00	01	11
00	01	11	10		10	00	11	01
00	10	01	11		10	01	00	11
00	10	11	01		10	01	11	00
00	11	01	10		10	11	00	01
00	11	10	01		10	11	01	00
01	00	10	11		11	00	01	10
01	00	11	10		11	00	10	01
01	10	00	11		11	01	00	10
01	10	11	00		11	01	10	00
01	11	00	10		11	10	00	01
01	11	10	00		11	10	01	00

Figure 8.23 Alternative state encodings of the traffic light controller.

greater than or equal to m) then the number of possible state assignments is $(2^n)!/(2^n - m)!$.

There are many possible strategies to choosing the codes to assign to each state. One can imagine some very straightforward ones. For example, a sequential state assignment where we simply assign codes in numerical order, namely, 0..00 to the first state, 0..01 to the second, 0..10 to the third, and so on. An even simpler strategy is to just pick the codes at random from the list of those available. We only have to ensure that each state gets a unique code so that combinational logic can tell them apart.

Other state assignment possibilities include one-hot encoding, where we use m state bits for m states. The code for each state has a 1 in the corresponding state bit and 0 for all the others. For example, for a four state finite state machine, a one-hot encoding could be 0001, 0010, 0100, and 1000. If there is an all-zero reset state, then this could be a modified one-hot encoding of 000, 001, 010, 100, which uses one less state bit.

If we are implementing a Moore machine (or a synchronous Mealy machine), then we may be able to use some of the outputs to encode our states. This follows naturally as we should be able to distinguish states if their outputs differ. Since their outputs already come from flip-flops, it is perfectly appropriate to consider those outputs as if they were state bits. We may need to still have explicit state bits if some states have similar outputs—as we will need other bits to tell them apart.

Of course, designers have also developed some heuristics, or rules of thumb, to help them generate good state assignments. These are not guaranteed to always generate a good circuit but generally do a pretty good job.

We'll now turn to the traffic-light controller example of the previous chapter to help us compare all of these methods and illustrate the impact of state encoding on the next-state and output logic. Let's start with the symbolic state transition table for the traffic-light controller, shown in Figure 8.24. The input combinations that cause the state transitions are shown at the left of the table. The symbolic state names *HG*, *HY*, *FG*, *FY* represent the states highway green/farm road red, highway

Inputs			Present State	Next State	Outputs		
C	TL	TS	$Q_1\ Q_0$	$P_1\ P_0$	ST	$H_1\ H_0$	$F_1\ F_0$
0	X	X	HG	HG	0	00	10
X	0	X	HG	HG	0	00	10
1	1	X	HG	HY	1	00	10
X	X	0	HY	HY	0	01	10
X	X	1	HY	FG	1	01	10
1	0	X	FG	FG	0	10	00
0	X	X	FG	FY	1	10	00
X	1	X	FG	FY	1	10	00
X	X	0	FY	FY	0	10	01
X	X	1	FY	HG	1	10	01

Figure 8.24 Traffic light controller symbolic state transition table.

yellow/farm road red, highway red/farm road green, and highway red/farm road yellow. We have encoded the traffic light outputs a bit differently this time: 00 = Green, 01 = Yellow, and 10 = Red.

8.2.1 Sequential Encoding

In sequential state assignment, we simply replace the symbolic state names HG, HY, FG, and FY with a regular encoding sequence. Rather than just using 00, 01, 10, 11, let's use a Gray-code sequence where $HG = 00$, $HY = 01$, $FG = 11$, $FY = 10$. Note that this is the second encoding listed in Figure 8.23. The finite state machine requires 10 unique product terms (one for each row of Figure 8.24 after substituting our state encoding for the symbolic states). The equations using this state assignment are:

$$\begin{aligned}
P_1 &= TS\ Q_1'\ Q_0 + C\ TL'\ Q_1\ Q_0 + C'\ Q_1\ Q_0 + TL\ Q_1\ Q_0 + TS'\ Q_1\ Q_0' \\
P_0 &= C\ TL\ Q_1'\ Q_0' + TS'\ Q_1'\ Q_0 + TS\ Q_1'\ Q_0 + C\ TL'\ Q_1\ Q_0 \\
ST &= C\ TL\ Q_1'\ Q_0' + TS\ Q_1'\ Q_0 + C'\ Q_1\ Q_0 + TL\ Q_1\ Q_0 + TS\ Q_1\ Q_0' \\
H_1 &= C\ TL'\ Q_1\ Q_0 + C'\ Q_1\ Q_0 + TL\ Q_1\ Q_0 + TS'\ Q_1\ Q_0' + TS\ Q_1\ Q_0' \\
H_0 &= TS'\ Q_1'\ Q_0 + TS\ Q_1'\ Q_0 \\
F_1 &= C'\ Q_1'\ Q_0' + TL'\ Q_1'\ Q_0' + C\ TL\ Q_1'\ Q_0' + TS'\ Q_1'\ Q_0 + TS\ Q_1'\ Q_0 \\
F_0 &= TS'\ Q_1\ Q_0' + TS\ Q_1\ Q_0'
\end{aligned}$$

These equations have a maximum fan-in of five (the three inputs plus the two state bits) and a maximum number of product terms of five (with many terms shared between equations). They are relatively easy to implement on a programmable logic device or field-programmable gate array. If we try to minimize the equations to yield fewer and smaller gates for a multilevel logic implementation, we obtain the following equations:

$$\begin{aligned}
P_1 &= TS\ HY + FG + TS'\ FY \\
P_0 &= X\ HG + HY + Y\ FG \\
ST &= X\ HG + TS\ HY + Y'\ FG + TS\ FY
\end{aligned}$$

$$H_1 = FG + FY$$
$$H_0 = HY$$
$$F_1 = HG + HY$$
$$F_0 = FY$$
$$HG = Q_1' Q_0'$$
$$HY = Q_1' Q_0$$
$$FG = Q_1 Q_0$$
$$FY = Q_1 Q_0'$$
$$X = C\,TL$$
$$Y = C\,TL'$$

Note that the output equations for the traffic lights only depend on the state bits. This is what we would expect for Moore outputs. The *ST* output is a Mealy output and it requires the inputs *C*, *TL*, and *TS*, as well as the state bits. The gates required for this minimized realization have mostly two inputs. Three gates with more than two inputs are required, and these are used to form P_1, P_0, and *ST*. These three logic equations also have three levels of logic to try to keep the number of high fan-in gates to a minimum (H_1, H_0, F_1, and F_0 only require two levels).

8.2.2 Random Encoding

Let's take a look at another encoding to see if there is a difference in the relative complexity of the equations. Randomly, we'll select $HG = 00$, $HY = 10$, $FG = 01$, $FY = 11$ which happens to be the third row in the table of state encodings in Figure 8.23.

Without minimization, the equations for this state assignment are going to look very similar to those of the Gray-code assignment of the previous section. The only difference will be in the use of the state bits and their complements—we are only changing some literals in the equations. The equations' structure will remain the same.

After minimization, however, the difference is more pronounced. The equations for the "random" state assignment are:

$$P_1 = C\,TL\,Q_1' + TS'\,Q_1 + C'\,Q_1'\,Q_0$$
$$P_0 = TS\,Q_1\,Q_0' + Q_1'\,Q_0 + TS'\,Q_1\,Q_0$$
$$ST = C\,TL\,Q_1' + C'\,Q_1'\,Q_0 + TS\,Q_1$$
$$H_1 = Q_0$$
$$H_0 = Q_1\,Q_0'$$
$$F_1 = Q_0'$$
$$F_0 = Q_1\,Q_0$$

With this encoding, we can realize the logic using only two levels rather than three and with no gates of more than 3 inputs. Not only are smaller gates likely to lead to a faster implementation, but lower fan-in also reduces overall wiring and is one reason why it is often more useful to count literals than gates in comparing circuit complexity.

8.2.3 One-Hot Encoding

So far, our goal has been *dense* encodings: state encodings in as few bits as possible. An alternative approach introduces additional flip-flops, in the hope of reducing the next-state and output logic.

One form of this method is called *one-hot encoding.* A machine with n states is encoded using exactly n flip-flops. Each state is represented by an n-bit binary code in which exactly 1 bit is asserted. This is the origin of the term "one-hot."

Let's again turn to our traffic-light controller. The following would be a possible one hot encoding of the machine's state: $HG = 0001$, $HY = 0010$, $FG = 0100$, $FY = 1000$. The state is encoded in four flip-flops rather than two, and only 1 bit is asserted in each of the states. This presents a lot of don't-care opportunities. Two or more of the state bits will never be asserted at the same time.

The encoded state table for this state assignment is shown if Figure 8.25. The logic equations for this state encoding are:

$$P_3 = (C' \cdot Q_2) + (TL \cdot Q_2) + (TS' \cdot Q_3)$$
$$P_2 = (TS \cdot Q_1) + (C \cdot TL' \cdot Q_2)$$
$$P_1 = (C \cdot TL \cdot Q_0) + (TS' \cdot Q_1)$$
$$P_0 = (C' \cdot Q_0) + (TL' \cdot Q_0) + (TS \cdot Q_3)$$
$$ST = (C \cdot TL \cdot Q_0) + (TS \cdot Q_1) + (C' \cdot Q_2) + (TL \cdot Q_2) + (TS \cdot Q_3)$$
$$H_1 = Q_3 + Q_2$$
$$H_0 = Q_1$$
$$F_1 - Q_1 + Q_0$$
$$F_0 = Q_3$$

Inputs			Present State	Next State		Outputs	
C	TL	TS	$Q_3\ Q_2\ Q_1\ Q_0$	$P_3\ P_2\ P_1\ P_0$	ST	$H_1\ H_0$	$F_1\ F_0$
0	X	X	0001	0001	0	00	10
X	0	X	0001	0001	0	00	10
1	1	X	0001	0010	1	00	10
X	X	0	0010	0010	0	01	10
X	X	1	0010	0100	1	01	10
1	0	X	0100	0100	0	10	00
0	X	X	0100	1000	1	10	00
X	1	X	0100	1000	1	10	00
X	X	0	1000	1000	0	10	01
X	X	1	1000	0001	1	10	01

Figure 8.25 Traffic-light controller encoded state transition table for a one-hot encoding.

The realization for a one-hot state assignment is aligned very closely with the state diagram. Note that the equation for each state bit has the same number of terms as there are terms on the arcs of the state diagram that point into that state. For example, P_2 has two terms corresponding to the two arcs that point into state *FG* in the state diagram of Figure 7.55. P_3 has three product terms even though it has only two arcs because one arc has an expression with two terms. The Moore outputs, H_1, H_0, F_1, and F_0 are very straightforward expressions, they are just the simple OR of all the states in which that output is asserted. Because each state has an associated state bit, the number of terms is exactly the number of states in which the output is asserted. The most complex expression is *ST*. It is asserted on four arcs representing a total of five product terms, which we see represented in the equation above.

One-hot encodings are often popular in programmable logic if the number of terms per equation in the PAL is greater than the number of incident arcs (and product terms) on each state. If it is, then the logic is guaranteed to fit. We don't need to concern ourselves with minimization. One-hot encodings are also popular in FPGAs but for a different reason. In this case, we are concerned with the states the transitions come from rather than the complexity of their product terms. For a given state bit, the logic equation will need one input for every state that can transition to the one in question. Thus, sparse state diagrams with only a few transitions to each state lead to equations that easily fit into one logic block of an FPGA. The fan-in is determined by the number of states that can transition to that state plus the inputs that also need to be considered for those transitions. In the case of our traffic-light controller, every state has transitions from only two others and there are, at most, three inputs to consider. This guarantees that no state equation will have a fan-in greater than 5.

Of course, one-hot encodings are not very efficient if the finite state machine has a large number of states. As we'll see in the next chapter, for PALs, and even more so for FPGAs, where flip-flops are relatively cheap, we can easily afford a one-hot encoding. Its also very useful because it makes debugging and modifications easy. Its easy to tell what state the machine is in and a new state can be added easily with another flip-flop. For large state machines, one-hot encodings are usually too inefficient for fixed logic implementations.

In practice, one-hot encoding are often mixed with other types. For example, a group of states may have a particular state-bit asserted for each state and then several other state bits that are encoded sequentially and used to tell apart each state in the group. We often end up with hybrid encoding schemes.

8.2.4 Output-Oriented Encoding

Outputs present an interesting opportunity for state assignment. Moore outputs are derived from the state bits. However, if we implement the Moore output directly (so that the output of a flip-flop is an output) then we can use this "state-bit" as an input to our next state equations.

Inputs			Present State	Outputs		
C	TL	TS	ST H_1 H_0 F_1 F_0	ST	H_1 H_0	F_1 F_0
0	X	X	HG: 00010 + 11001	0	00	10
X	0	X	HG: 00010 + 11001	0	00	10
1	1	X	HG: 00010 + 11001	1	00	10
X	X	0	HY: 10010 + 00110	0	01	10
X	X	1	HY: 10010 + 00110	1	01	10
1	0	X	FG: 10110 + 01000	0	10	00
0	X	X	FG: 10110 + 01000	1	10	00
X	1	X	FG: 10110 + 01000	1	10	00
X	X	0	FY: 11000 + 01001	0	10	01
X	X	1	FY: 11000 + 01001	1	10	01

Figure 8.26 Traffic light controller encoded state transition table for an output-oriented encoding.

Synchronous Mealy outputs, since they are implemented directly as the output of a flip-flop, can also be used this way.

The traffic-light controller is again useful to illustrate the concept of *output-directed* state encoding. If we look carefully at the state transition tables of Figure 8.24 and Figure 8.25, we'll see that the patterns of our five output signals are different for the transitions to each state. When the next state is HG, there are two possible output patterns for ST, H_1, H_0, F_1, and F_0, namely 00010 and 11001. At no other time do we see the same pattern of outputs. Why not use this for our state encoding?

This approach leads to a very different encoded state table where the outputs also appear as inputs (see Figure 8.26). Note the present state is now represented by "present outputs" instead. Let's see how we derived these. For example, when we transition to state HG (0001) in the table of Figure 8.25, we set the outputs to either 00010 (the top two rows) or 11001 (the bottom row). At no other time do we set the outputs to the same values. Therefore, we can tell we are in state HG if we look for the outputs to be 00010 or 11001.

For each state there are two possible output patterns. The following equations represent our states:

$$HG = ST' H_1' H_0' F_1 F_0' + ST\, H_1 H_0' F_1' F_0$$
$$HY = ST\, H_1' H_0' F_1 F_0' + ST' H_1' H_0 F_1 F_0'$$
$$FG = ST\, H_1' H_0 F_1 F_0' + ST' H_1 H_0' F_1' F_0'$$
$$HY = ST\, H_1 H_0' F_1' F_0' + ST' H_1 H_0' F_1' F_0$$

The output equations are very similar to those for our one-hot encoding can be written as follows:

$$ST = (C \cdot TL \cdot HG) + (TS \cdot HY) + (C' \cdot FG) + (TL \cdot FG) + (TS \cdot FY)$$
$$H_1 = FG + FY$$
$$H_0 = HY$$
$$F_1 = HG + HY$$
$$F_0 = FY$$

We still use five flip-flops. For the one-hot encoding we had four state bits and one synchronous Mealy output (the other outputs were functions of the state bits). For this output-directed encoding we also have five flip-flops, one for each output. The equations for *HG, HY, FG,* and *FY* add some more complexity, but you'll note that all four have a fan-in of 5, making them easy to implement in a single FPGA logic block. Thus, H_1, H_0, F_1, and F_0 can be implemented in one logic block as well. *ST* is a bit more problematic as it has a complete fan-in of 8 (three inputs and five state bits) and will likely need to realized as multilevel logic. In a PAL, *ST* would require 10 product terms (as each of *HG, HY, FG,* and *FY* is two terms).

Again, as with one-hot encoding, it may not be the best strategy to use output-directed encoding for the entire state machine. However, the judicious utilization of some of the outputs (those that are synchronous outputs of a flip-flop) may help decrease the number of state bits. In this case, we ended up with no state bits at all, just five synchronous outputs.

8.2.5 Heuristic Methods

As with logic minimization, computer-aided design tools have been developed to help designers generate good state encodings. Hand enumeration using trial and error becomes tedious even for a relatively small number of states. After all, even an n-state finite state machine has $n!$ different encodings. And this is only the lower bound. If the state is not densely encoded in the fewest number of bits, even more encodings are possible.

To make the problem more tractable, designers have developed a collection of heuristic "guidelines." These try to reduce the distance in Boolean n-space between related states. For example, if state Y is reached by a transition from state X, then their encodings should differ by as few bits as possible. The next-state logic will be minimized if you follow such guidelines. We examine these guidelines in this section.

State Maps *State maps,* similar in concept to K-maps, provide a means of observing adjacencies in state assignments. The squares of the state map are indexed by the binary values of state bits; the state given that encoding is placed in the map square. Obviously the technique is limited to situations in which a K-map can be used, that is, up to six variables.

Figure 8.27 presents a state diagram for a five-state finite state machine. Figure 8.28 gives two alternative state assignments and their representations in state maps.

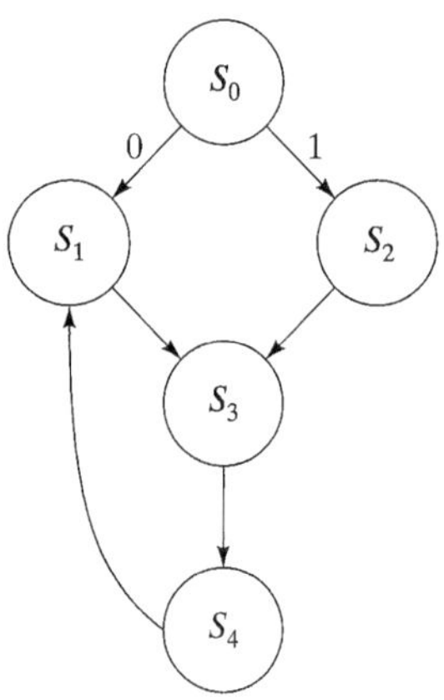

Figure 8.27 Five-state finite state machine.

Minimum Bit-Change Heuristic One heuristic strategy assigns states so that the number of bit changes for all state transitions is minimized.

State Name	Assignment Q_2	Q_1	Q_0
S_0	0	0	0
S_1	1	0	1
S_2	1	1	1
S_3	0	1	0
S_4	0	1	1

Assignment

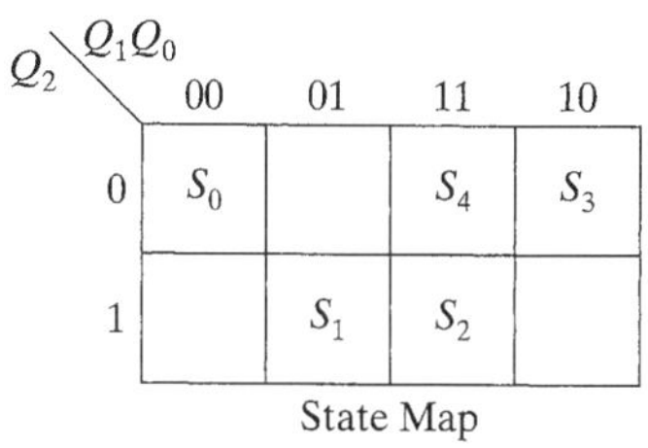

State Map

(a) First state assignment and map

State Name	Assignment Q_2	Q_1	Q_0
S_0	0	0	0
S_1	0	0	1
S_2	0	1	0
S_3	0	1	1
S_4	1	1	1

Assignment

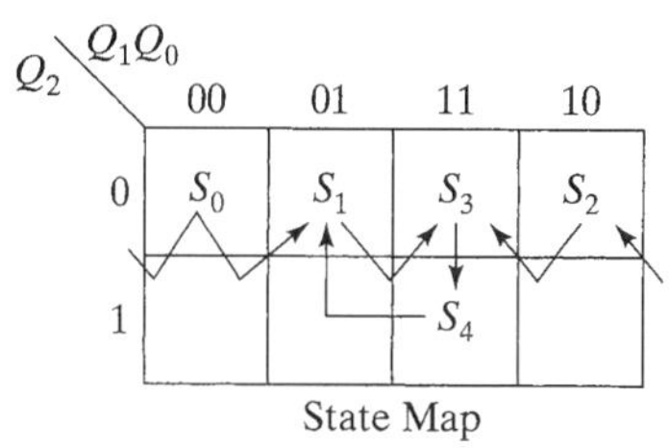

State Map

(b) Second state assignment and map

Figure 8.28 Two state assignments for the five-state finite state machine and their state maps.

For example, the assignment of Figure 8.28(a) is not as good as the one in Figure 8.28(b) under this criterion:

Transition	*First Assignment Bit Changes*	*Second Assignment Bit Changes*
S_0 to S_1	2	1
S_0 to S_2	3	1
S_1 to S_3	3	1
S_2 to S_3	2	1
S_3 to S_4	1	1
S_4 to S_1	2	2

The first assignment leads to 13 different bit changes in the next-state function, the second only seven bit changes.

We derived the first assignment completely at random and the second assignment with minimum transition distance in mind. Here is how we did it. We made the assignment for S_0 first. Because of the way reset logic works, it usually makes sense to assign all zeros to the starting state. We make assignments for S_1 and S_2 next, placing them next to S_0, because they are targets of transitions out of the starting state. Note how we used the edge adjacency of the state map. We can then place S_3 between the assignments for S_1 and S_2, since it is the target of transitions from both of these states.

Finally, we place S_4 adjacent to S_3, since it is the destination of S_3s only transition. It would be perfect if S_4 could also be placed distance 1 from S_0, but it is not possible to do this and satisfy the other desired adjacencies.

The resulting assignment exhibits only seven bit transitions. There may be many other assignments with the same number of bit transitions, and perhaps an assignment that needs fewer.

The minimum bit-change heuristic, although simple, is not likely to achieve the best assignment. For a finite state machine like the traffic light controller, cycling through its regular sequence of states, the minimum transition distance is obtained by a Gray-code assignment: $HG = 00$, $HY = 01$, $FG = 11$, $FY = 10$. This was the sequential state assignment we tried in the previous subsection, and it was not as good as the random assignment, even though the latter did not involve a minimum number of bit changes.

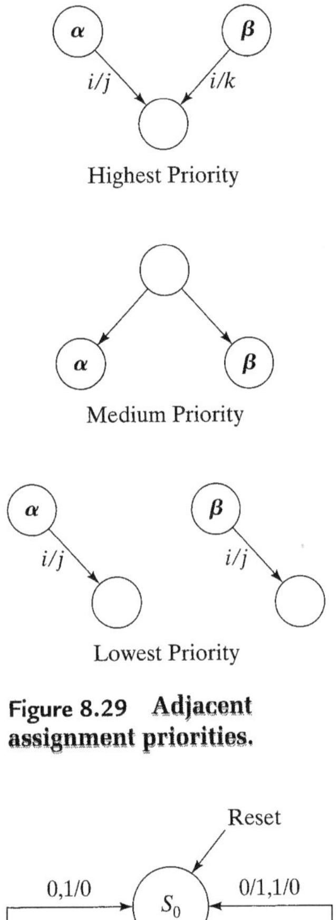

Figure 8.29 Adjacent assignment priorities.

Guidelines Based on Next State and Input/Outputs Although the criterion of minimum transition distance is simple, it suffers by not considering the input and output values in determining the next state. A second set of heuristic guidelines makes an effort to consider this in the assignment of states:

- Highest priority: States with the same next state for a given input transition should be given adjacent assignments in the state map.
- Medium priority: Next states of the same state should be given adjacent assignments in the state map.
- Lowest priority: States with the same output for a given input should be given adjacent assignments in the state map.

The guidelines, illustrated in Figure 8.29 for the candidate states α and β, are ranked from highest to lowest priority. The first two rules attempt to group together 1s in the next-state maps, while the third rule performs a similar grouping function for the output maps. We do a state assignment by listing all state adjacencies implied by the guidelines and then satisfying as many of these as possible.

EXAMPLE 8.4 APPLYING THE GUIDELINES

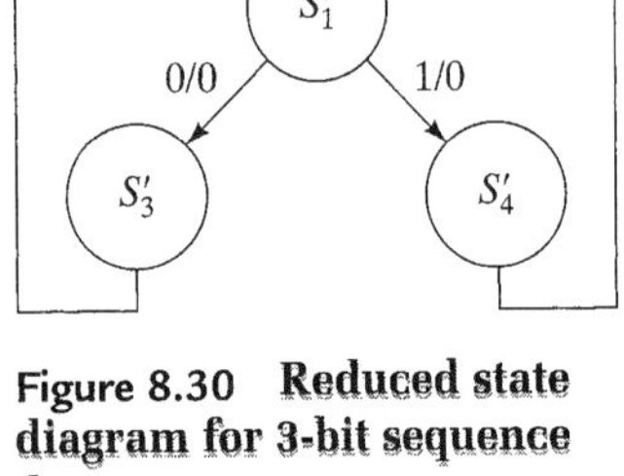

Figure 8.30 Reduced state diagram for 3-bit sequence detector.

Consider the state transition table of Figure 8.12 for the 3-bit sequence detector. The corresponding state diagram is shown in Figure 8.30. Let's apply the state assignment guidelines to this state diagram.

The highest-priority constraint for adjacent assignment applies to states that share a common next state on the same input. In this case, states S'_3 and S'_4 both have S_0 as their next state. No other states share a common next state.

The medium-priority assignment is for states that have a common ancestor state. Again, S'_3 and S'_4 are the only states that fit this description.

The lowest-priority assignments are made for states that have the same output behavior for a given input. S_0, S'_1, and S'_3 all output 0 when the input is 0. Similarly, S_0, S'_1, S'_3, and S'_4 output 0 when the input is 1.

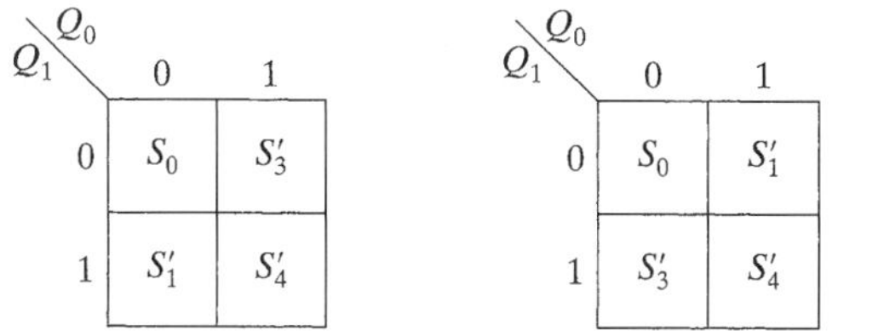

(a) First state assignment (b) Second state assignment

Figure 8.31 **Two possible state assignments.**

The constraints on the assignments can be summarized as follows:

Highest priority:	(S'_3, S'_4)
Medium priority:	(S'_3, S'_4)
Lowest priority:	0/0: (S_0, S'_1, S'_3)
	1/0: (S_0, S'_1, S'_3, S'_4)

Since the finite state machine has four states, we can make the assignment onto two state bits. In general, it is a good idea to assign the reset state to state map square 0. Figure 8.31 shows two possible assignments. Both assign S_0 to 00 and place S'_3 and S'_4 adjacent to each other.

EXAMPLE 8.5 APPLYING THE GUIDELINES IN A MORE COMPLEX CASE

As another example, let's consider the slightly larger state diagram of the 4-bit string recognizer of Figure 8.6. Applying the guidelines yields the following set of assignment constraints:

Highest priority:	(S'_3, S'_4), (S'_7, S'_{10})
Medium priority:	(S_1, S_2), $2 \times (S'_3, S'_4)$, (S'_7, S'_{10})
Lowest priority:	0/0: $(S_0, S_1, S_2, S'_3, S'_4, S'_7)$
	1/0: $(S_0, S_1, S_2, S'_3, S'_4, S'_7, S'_{10})$

Figure 8.32 shows two alternative assignments that meet most of these constraints. We start with Figure 8.32(a) and first assign the reset state to the encoding for 0. Since (S'_3, S'_4) is both a high-priority and medium-priority adjacency, we make their assignments next. S'_3 is assigned 011, and S'_4 is assigned 111.

We assign (S'_7, S'_{10}) next because this pair also appears in the high- and medium-priority lists. We assign them the encodings 010 and 110, respectively. Besides giving them adjacent assignments, this places S_7 near S_0, S'_3, and S'_4, which satisfies some of the lower-priority adjacencies.

The final adjacency is (S_1, S_2). We give them the assignments 001 and 101. This satisfies a medium-priority placement as well as the lowest-priority placements.

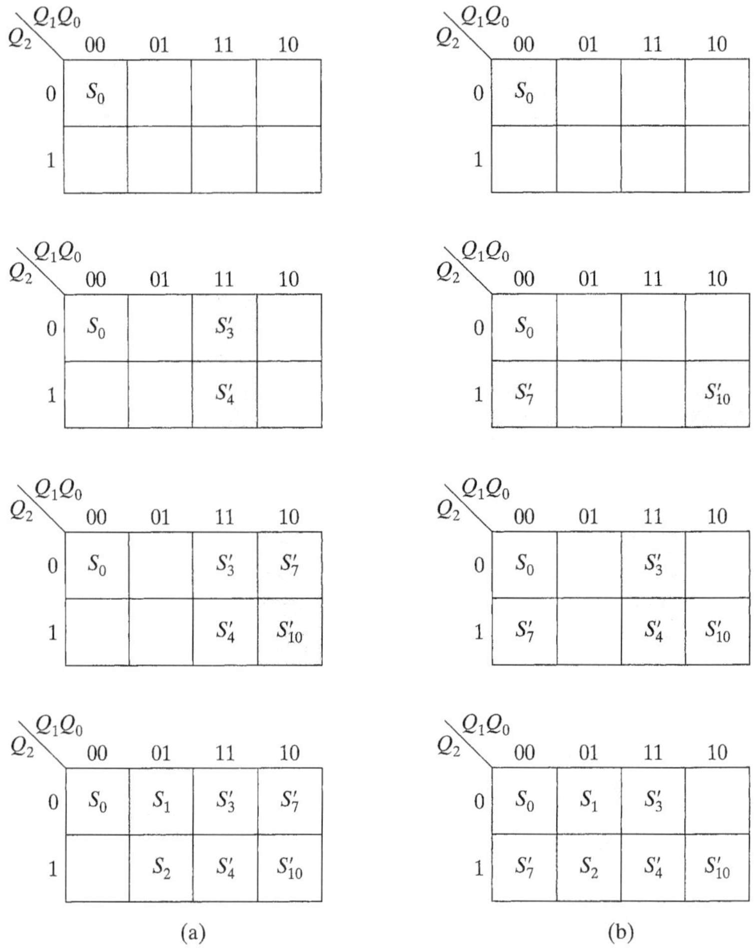

Figure 8.32 **State maps for state assignment example.**

The second assignment is shown in Figure 8.32(b). We arrived at it by a similar line of reasoning, except that we assigned S_7' and S_{10}' the states 100 and 110. The second assignment does about as good a job as the first, satisfying all of the high- and medium-priority guidelines, as well as most of the lowest-priority ones.

Why They Work The state assignment guidelines attempt to maximize the adjacent groupings of 1s in the next-state and output functions. Let P_2, P_1, and P_0 be the next-state functions, expressed in terms of the current state Q_2, Q_1, Q_0 and the input X. To see how effective the guidelines were, let's compare the assignment of Figure 8.32(a) with a more naive assignment: $S_0 = 000$, $S_1 = 001$, $S_2 = 010$, $S_3' = 011$, $S_4' = 100$, $S_7' = 101$, $S_{10}' = 110$.

Figure 8.33 compares the encoded next-state tables and K-maps for the two encodings. The 1s are nicely clustered in the next state K-maps

Current State	Next State $X=0$	Next State $X=1$
(S_0) 000	001	101
(S_1) 001	011	111
(S_2) 101	111	011
(S_3') 011	010	010
(S_4') 111	010	110
(S_7') 010	000	000
(S_{10}') 110	000	000

P_2

Q_0X \ Q_2Q_1	00	01	11	10
00	0	0	0	X
01	1	0	0	X
11	1	0	1	0
10	0	0	0	1

P_1

Q_0X \ Q_2Q_1	00	01	11	10
00	0	0	0	X
01	0	0	0	X
11	1	1	1	1
10	1	1	1	1

P_0

Q_0X \ Q_2Q_1	00	01	11	10
00	1	0	0	X
01	1	0	0	X
11	1	0	0	1
10	1	0	0	1

Current State	Next State $X=0$	Next State $X=1$
(S_0) 000	001	010
(S_1) 001	011	100
(S_2) 010	100	011
(S_3') 011	101	101
(S_4') 100	101	110
(S_7') 101	000	000
(S_{10}') 110	000	000

P_2

Q_0X \ Q_2Q_1	00	01	11	10
00	0	1	0	1
01	0	0	0	1
11	1	1	X	0
10	0	1	X	0

P_1

Q_0X \ Q_2Q_1	00	01	11	10
00	0	0	0	0
01	1	1	0	1
11	0	0	X	0
10	1	0	X	0

P_0

Q_0X \ Q_2Q_1	00	01	11	10
00	1	0	0	1
01	0	1	0	0
11	0	1	X	0
10	1	1	X	0

Figure 8.33 K-maps for state assignment example.

for the assignment derived from the guidelines. We can implement P_2 with three product terms and P_1 and P_0 with one each.

In the second assignment, the 1s are spread throughout the K-maps, since we made the assignment with no attempt to cluster the 1s usefully. In this implementation, the next-state functions P_2, P_1, and P_0 each require three product terms, with a considerably larger number of literals overall.

These heuristics have formed the basis of many computer aids for the state assignment problem. Given the difficulty of the problem, most modern tools include many control parameters that designers can use to guide the solution in a particular direction they care about more. For example, if they are focusing on an FPGA implementation, they may not care as much about the number of state bits and may prefer a solution closer to a one-hot encoding. If they are using PLDs, they may want to stress the heuristics of this section and have the tools try to minimize the number of product terms that will be required.

None of these tools are able to guarantee the best result. In fact, often, just trying a set of random state assignments and choosing the best of the implementations that result does as well or better than some of the most sophisticated tools. In the end, computer tools simply help the designer explore the design space more efficiently than would be possible with pencil-and-paper methods. They are not limited to six state bits, they can generate many alternative assignments rapidly, and they can evaluate the derived assignments by invoking logic minimization tools automatically. In many cases, the designer may be more interested in reducing design time rather than obtaining the most optimized circuit. The tools are simply used to generate a reasonable implementation very quickly.

8.3 Finite State Machine Partitioning

In the preceding sections, we described the design process for a single, monolithic finite state machine. The approach is reasonable for many strategies for implementing a finite state machine, such as using discrete gates.

However, when using some forms of programmable logic, we may need to partition the machine. In some cases, we cannot implement a complex finite state machine with a single programmable logic component. The machine might require too many inputs or outputs, or the number of terms to describe the next-state or output functions might be too large, even after state reduction and Boolean minimization.

EXAMPLE 8.6 FSM PARTITIONING

To illustrate the value of state machine partitioning, suppose we have a finite state machine with 20 inputs and 10 outputs (including next-state outputs). But we only have programmable logic components with 15 inputs and 5 outputs. We cannot implement this finite state machine with a single component.

Suppose we can arrange the outputs in two sets of five, each of which can be computed from different 15-element subsets of the original 20 inputs. Then we could partition the output functions among two programmable logic components, as shown in Figure 8.34. Of course, it isn't always possible to find such a fortuitous partitioning. For example, every output might be a function of 16 inputs.

If we cannot reduce the complexity of the finite state machine by simple input/output partitioning, another way to "make it fit" is to partition the single finite state machine into smaller, individually less complex, communicating finite state machines. We examine this approach in the next subsection.

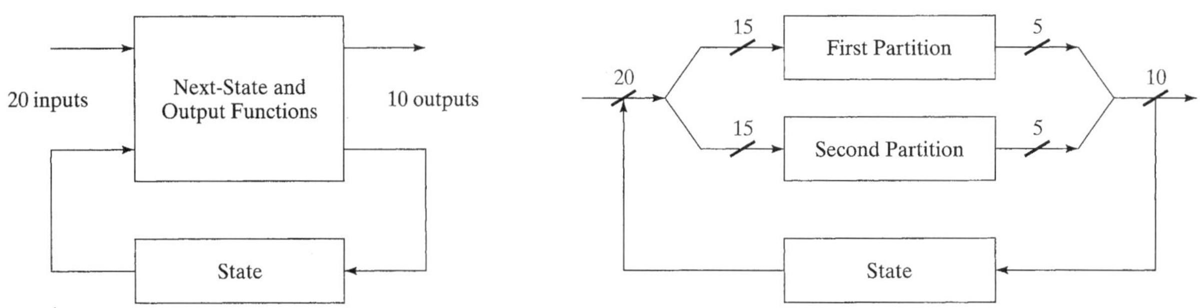

Figure 8.34 Finite state machine partitioning on inputs and outputs.

8.3.1 Finite State Machine Partitioning by Introducing Idle States

Partitioning the finite state machine makes sense if the next-state logic is too complex to implement with the programmable logic components at hand. The problem is that PALs provide a fixed number of product terms per output function. We can make a trade-off between the number of flip-flops needed to encode the state and the complexity of their next-state functions. Our idea is to introduce additional *idle* or *coordination states* into the finite state machine in the hope of reducing the number of terms in the next-state functions.

EXAMPLE 8.7 FSM PARTITIONING: IDLE STATES

For example, Figure 8.35 shows a subset of a state diagram. We have chosen to partition the state diagram into two separate machines, containing states S_1, S_2, S_3 and S_4, S_5, S_6, respectively. The symbols C_i, associated with the transitions, represent the Boolean conditions under which the transition takes place.

What happens if we partition the state diagram, but a transition must take place between the two pieces? We need to introduce idle states to synchronize the activity between the two finite state machines. In essence, the machine at the left hands control off to the machine at the right when a transition from S_1 to S_6 takes place. The left machine must idle in some new state until it regains control, such as when there is a transition from S_6 back to S_1. In this event, the machine on the right must remain idle until it regains control.

The revised state diagrams are shown in Figure 8.36. We have introduced two new states, S_A and S_B, to synchronize the transitions across the partition boundary. Here is how it works for the state sequence S_1 to S_6 and back to S_1. Initially, the machines are in states S_1 and S_B. If condition C_1 is true, then the left-hand state machine exits S_1 and enters its idle state, S_A. At the same time, the right-hand machine exits S_B and enters S_6.

Suppose that the right-hand machine sequences through some states, eventually returning to S_6. Throughout this time, the left-hand

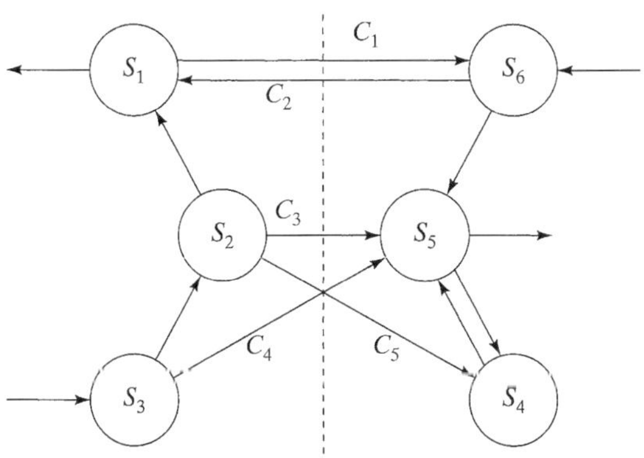

Figure 8.35 **State diagram fragment before partitioning.**

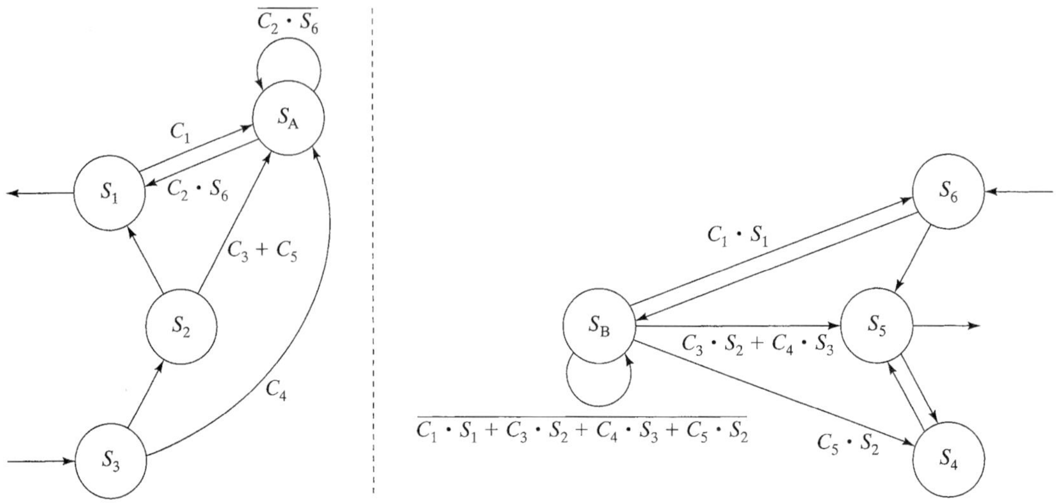

Figure 8.36 **State diagram fragment after partitioning.**

machine remains in its idle state. If the right-hand machine is in S_6 and C_2 is true, it next enters its idle state, S_B. At the same instant, the left-hand machine exits S_A, returning to S_1. While the left-hand machine sequences through states, the right-hand machine idles in S_B. Note that the two finite state machines also need to communicate to each what state the other is in.

Rules for Partitioning We are ready to describe the rules for introducing idle states into a partitioned finite state machine. We illustrate each rule with an example from the partitioned state machine of the previous subsection. All of the rules involve transitions that cross the partition boundary.

The first rule applies for a state that is the source of a transition that crosses the boundary. The case is shown in Figure 8.37(a). The cross-boundary transition is replaced by a transition to the idle state and labeled by the same exit condition as the original transition. For example, the S_1-to-S_6 transition is replaced by a transition with the same condition to S_A.

The second rule applies to the destination of a transition that crosses the partition boundary. This is shown in Figure 8.37(b). The transition is replaced with an exit transition from the idle state and labeled with the original condition ANDed with the source state. For example, the transition from S_6 to S_1 is replaced with a transition from S_A. We exit the idle state when both C_2 is true and the right-hand state machine is in S_6. Hence, the transition is labeled with the condition $C_2 \bullet S_6$.

The third rule applies when multiple transitions share the same source or destination. This case is illustrated in Figure 8.37(c). If a state is the source of multiple transitions across the partition boundary, all of these are collapsed into a single transition to the idle state. The exit conditions are ORed together to label the new transition. For example, S_2 has transitions to states S_5 and S_4. These are replaced with a single transition to S_A and labeled $C_3 + C_5$.

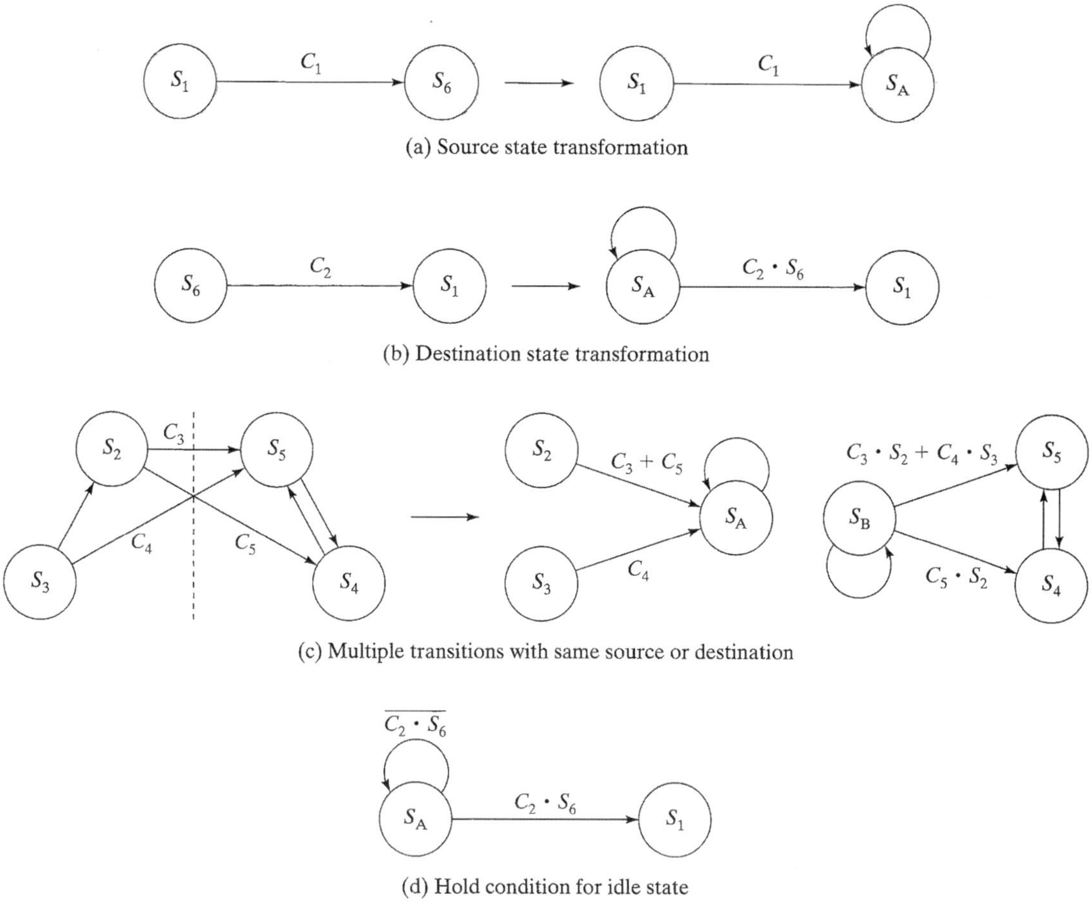

Figure 8.37 Rules for partitioning.

If a state is the target of multiple transitions across the boundary, a single transition is added from the idle state to this state. The transition is labeled with the OR of the conditions associated with the individual transitions in the original state machine. This case is also illustrated in Figure 8.37(c) by the transitions from S_2 and S_3 to S_5. These are replaced by a single transition from S_B to S_5 and labeled $C_3 \bullet S_2 + C_4 \bullet S_3$.

When all of these rules have been applied, the final rule describes the self-loop ("hold") condition for the idle states. Simply form the OR of all of the exit conditions and invert it. This is shown in Figure 8.37(d). Consider the idle state S_A. Its only exit condition is $C_2 \bullet S_6$. So, its hold condition is the inverse of this: namely, $\overline{C_2 \bullet S_6}$.

EXAMPLE 8.8 FSM PARTITIONING

Consider the six-state finite state machine of Figure 8.38(a). The machine implements a simple up/down counter. When the input U is asserted, the machine counts up. When D is asserted, it counts down. Otherwise the machine stays in its current state (self arcs are not shown).

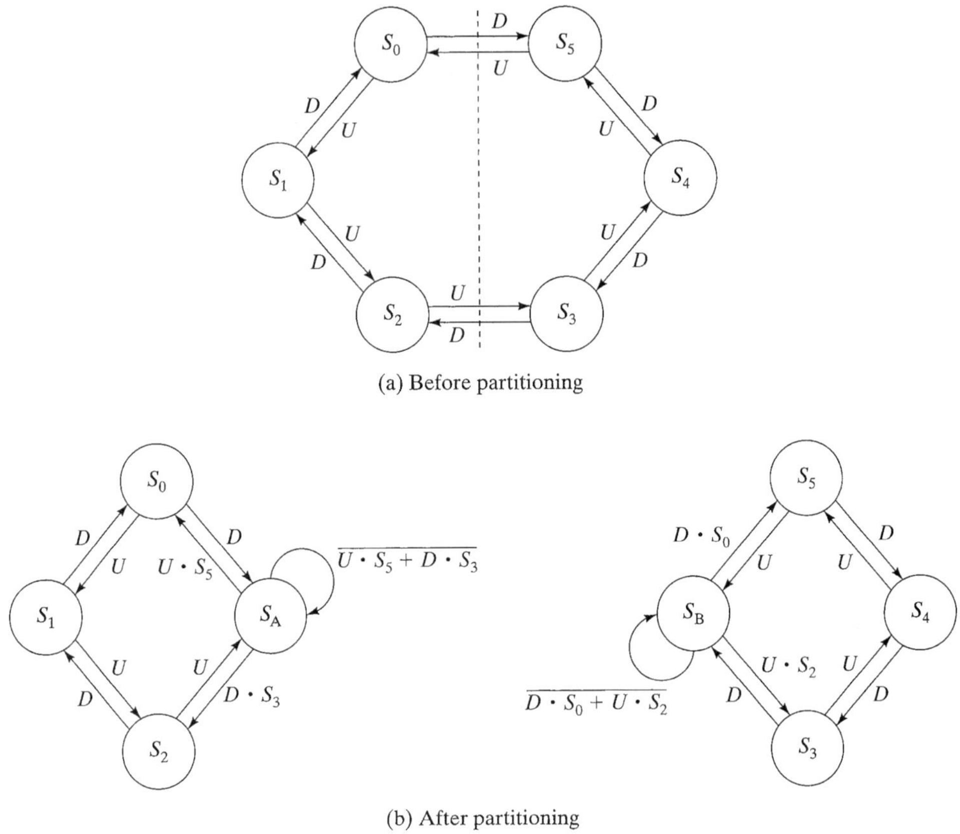

Figure 8.38 **Partitioning example.**

The goal is to partition the machine into two communicating four-state finite state machines. We might need to do this because the underlying logic primitives provide support for two flip-flops within the logic block, as in the Xilinx CLB to be introduced in the next chapter.

Figure 8.38(b) shows the result of the partitioning. States S_0, S_1, and S_2 form the core of one machine and S_3, S_4, and S_6 form the other. We also introduce the two idle states, S_A and S_B.

The machine at the left enters its idle state S_A when it is in S_0 and D is asserted or when it is in S_2 and U is asserted. It exits the idle state when the machine at the right is in S_5 with U asserted or in S_3 with D asserted. Otherwise it stays in its idle state. The machine at the right works similarly.

To see how the machines communicate, let's consider an up-count sequence from S_0 to S_5 and back to S_0. On reset, the machine on the left enters S_0, while the machine on the right enters S_B. With U asserted, the left machine advances from S_0 to S_1 to S_2 to S_A. It will idle in this state until the right machine is ready to exit S_5.

Meanwhile, the right machine holds in S_B until the left machine enters S_2. At the same time that the left machine changes to S_A, the right one exits S_B to S_3. On subsequent clock transitions, it advances from S_3 to S_4 to S_5 to S_B, where it holds. When the right machine changes from S_5 to S_B, the left machine exits S_A to S_0, and the process repeats itself. Down-count sequences work in an analogous way.

EXAMPLE 8.9 PARTITIONING FOR THE TRAFFIC-LIGHT CONTROLLER

As we saw in the previous chapter, the traffic-light controller required some partitioning in order to achieve a practical implementation. In that case, we partitioned a single finite state machine into a main controller and a straightforward counter/timer.

We began with a large state machine for which we showed only a small fragment in Figure 7.52. We then incorporated an idle state in Figure 7.53 and, finally, were able to merge the idle state into an already existing state in Figure 7.54. This example demonstrates that we do not always add extra states by introducing the idle states needed for coordination between the two machines. Often, an already existing state can do the job. However, we do have to be careful about timing. In this case, *ST* must be an asynchronous reset so that the counter will count the very next state. The complete state diagrams for the two state machines of the traffic light controller are shown in Figure 8.39.

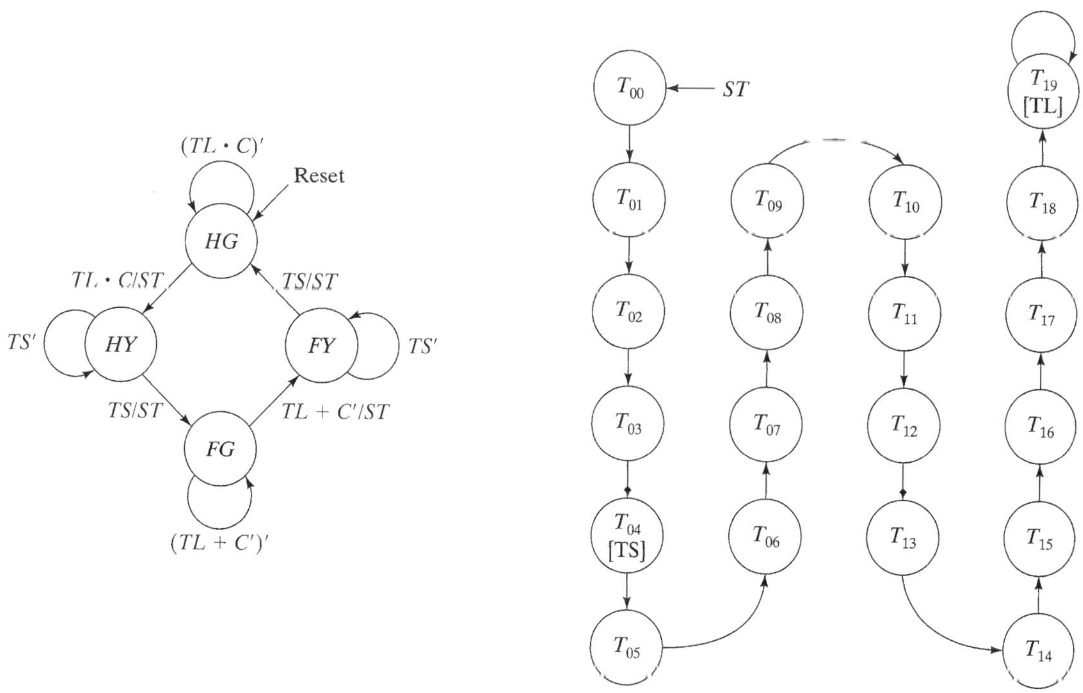

(a) Main controller (b) Counter/timer

Figure 8.39 State diagrams for the two partitioned parts of traffic light controller.

Note that if we had implemented these state machines as a single controller we would have had many more states than the 24 shown in Figure 8.39. The number would have been closer to the product of 4 and 20 rather than their sum. This is why finite state partitioning is often referred to as *finite state machine factoring* and the large single finite state machine is referred to as a *product machine* of its two *factor machines*.

8.4 Hardware Description Languages

It is now time to return to hardware description languages and see how they can be used to describe finite state machines in addition to the basic sequential logic of Chapter 6. In this text, we'll discuss only the representation of the structure of a finite state machine. In other words, in our descriptions we'll see explicitly the state registers and each transition. There are also ways of describing finite state machines more abstractly, but those are outside our scope.

We'll start with the simple pair of Moore and Mealy machines from Figure 7.45. The code for the Moore machine with three states is as follows:

```
module reduce (clk, reset, in, out);
input clk, reset, in;
output out;
reg out;

// start by defining the three states of machine so we can
// refer to them
// by name and easily change the state assignment
// here we just use a sequential encoding
parameter zero = 0;
parameter one1 = 1;
parameter two1s = 2;

// state variables, state is used for the current state,
// next_state for the next state
reg [2:1] state;
reg [2:1] next_state;

always @(posedge clk)
// this first always block is just here to handle the reset case
if (reset) state = zero;
else state = next_state;

always @(in, state) // this block specifies the next
                    // state logic
   case (state)
         zero:    // last input was a zero
            begin
               if (in) next_state = one1;
               else    next_state = zero;
            end
```

```
        one1:    // we've seen one 1
           begin
              if (in) next_state = two1s;
              else    next_state = zero;
           end
        two1s:    // we've seen at least 2 ones
           begin
              if (in) next_state = two1s;
              else    next_state = zero;
           end
    endcase

always @(state) // this block specifies the outputs which
                // are based on the current state
    case (state)
         zero: out = 0;
         one1: out = 0;
         two1s: out = 1;
    endcase

endmodule
```

The specification has three main blocks: one for reset, one for the next state logic, and one for the output logic. Note the use of case statements to capture the different behavior of each state. State transitions are represented by the conditional statements in each case. The first *always* block is the only one triggered by the clocking event and, as such, is the only one making an assignment to a sequential logic element, *state*. The next state logic block is triggered whenever an input or the current state changes, so it can compute the value of `next_state`, finally, the output logic block is triggered by any change in the current state (a Moore machine). Thus, the last two *always* blocks are purely combinational logic.

Now let's turn to the Mealy machine of Figure 7.45. Its Verilog description is:

```
module reduce (clk, reset, in, out);
input clk, reset, in;
output out;
reg out;

parameter zero  = 0;
parameter one = 1;

// state variables
reg state;
reg next_state;

always @(posedge clk)
   if (reset) state = zero;
   else       state = next_state;

always @(in, state)
   case (state)
```

```
            zero: // last input was a zero
               begin
                  out = 0;
                  if (in) next_state = one;
                  else    next_state = zero;
               end
            one: // we've seen one 1
               if (in) begin
                  out = 1;
                  next_state = one;
               end
               else begin
                  out = 0;
                  next_state = zero;
               end
            endcase

endmodule
```

The first block is identical to that for the Moore machine. However, the next-state and output logic are merged into a single block. Note that an input change can cause an immediate change to an output—this is not a synchronous Mealy machine. We could have separated the next state and output logic as we did for the Moore machine but that would have caused the entire case statement to be duplicated for little advantage. It's usually just easier to keep it all together for the case of a basic Mealy machine.

The last case we'll consider here is for the synchronous Mealy machine. Its Verilog specification is:

```
module reduce (clk, reset, in, out);
input clk, reset, in;
output out;
reg out;

parameter zero  = 0;
parameter one = 1;

// state variables
reg state;

always @(posedge clk)
   if (reset) state = zero;
   else case (state)
         zero: // last input was a zero
            begin
               out = 0;
               if (in) state = one;
               else    state = `zero;
            end
         one: // we've seen one 1
            if (in) begin
               out = 1;
```

```
            state = one;
        end
        else begin
            out = 0;
            state = zero;
        end
    endcase
endmodule
```

In this case, there is only one *always* block as all of the outputs are synchronous to the clock. We no longer even need the intermediate variable next_state that held the value of the next state until the next clock edge arrived. This is another reason designers often favor a synchronous Mealy implementation. The next_state and output logic can all be kept together in a more compact form.

We couldn't complete this section without returning to our traffic-light controller example. The Verilog for the main controller is:

```
module FSM(HR, HY, HG, FR, FY, FG, ST, TS, TL, C, reset, Clk);
output HR, HY, HG, FR, FY, FG, ST;
input TS, TL, C, reset;
input Clk;

parameter highwaygreen = 6'b001100;
parameter highwayyellow = 6'b010100;
parameter farmroadgreen = 6'b100001;
parameter farmroadyellow = 6'b100010;

reg [6:1] state;
reg ST;

assign HR = state[6];
assign HY = state[5];
assign HG = state[4];
assign FR = state[3];
assign FY = state[2];
assign FG = state[1];

initial begin state = highwaygreen; ST = 0; end

always @(posedge Clk) begin
   if (reset) begin state = highwaygreen; ST = 1; end
   else begin ST = 0;
      case (state)
         highwaygreen: if (TL & C) begin state =
           highwayyellow; ST = 1; end
         highwayyellow: if (TS) begin state =
           farmroadgreen; ST = 1; end
         farmroadgreen: if (TL | !C) begin state =
           farmroadyellow; ST = 1; end
         farmroadyellow: if (TS) begin state =
           highwaygreen; ST = 1; end
      endcase
```

```
    end
end

endmodule
```

This is a description of a hybrid Mealy/Moore machine. The single synchronous Mealy output, ST, is right in the middle of the next state logic. There are now six Moore outputs, as one output for each light bulb of the traffic light seemed more appropriate an encoding than what we had previously, and these are set using assign statements. Note that the state vector is exactly these six outputs. This machine is using an output-directed state encoding. The codes for each state are defined at the top. This is a very compact representation that leads to a very efficient implementation whose equations are simpler than those earlier in the chapter.

The counter/timer of the traffic light controller has an even simpler description:

```
module Timer(TS, TL, ST, Clk);
output TS, TL;
input ST;
input Clk;
integer value;

assign TS = (value >=  4); // 5 cycles after reset
assign TL = (value >= 19); // 20 cycles after reset

always @(posedge ST) value = 0; // async reset

always @(posedge Clk)
   if (value < 19) value = value + 1;
   else value = 19;

endmodule
```

There are several elements to note in this description. First, the counter is implemented directly using an integer variable that is incremented on every clock tick. The Verilog language provides arithmetic operators. Of course, in any implementation, we would need to specify the number of bits in our counter (Verilog typically defaults to 32 bits for an integer variable). Second, we use assign statements for the two outputs and the value of these outputs is the result of a comparison with an integer. The amount of logic these comparisons represent is also based on how many bits we allow for our counter. Third, we use an *always* block triggered on the positive edge of an input to asynchronously reset the timer. Finally, we only increment the counter up to 19 and then hold that value until it is reset. This is the precise behavior as specified in Figure 8.39.

The only thing left to do now is to connect the two modules of our traffic-light controller together. This is accomplished with a bit of Verilog that describes the schematic of Figure 8.40 (and adds a

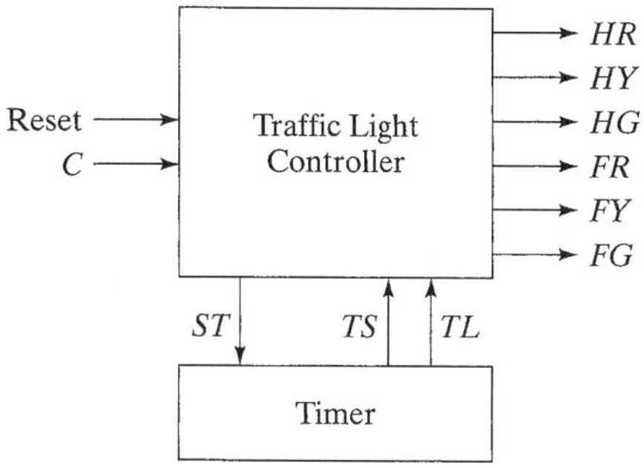

Figure 8.40 Schematic for the two modules of the traffic light controller.

clock input to both modules—not shown in the figure):

```
module main(HR, HY, HG, FR, FY, FG, reset, C, Clk);
output HR, HY, HG, FR, FY, FG;
input  reset, C, Clk;

Timer part1(TS, TL, ST, Clk);
FSM part2(HR, HY, HG, FR, FY, FG, ST, TS, TL, C, reset, Clk);

endmodule
```

These last few lines of Verilog should be self-explanatory. Identically named signals are connected together to form the combined state machine that implements our entire traffic light controller.

CHAPTER REVIEW

This chapter has concentrated on the optimization of finite state machines. We have emphasized the methods for state reduction, state assignment, and state machine partitioning.

For state reduction, we introduced the row-matching and implication chart methods. These can be used to identify and eliminate redundant states, thus reducing the number of flip-flops needed to implement a particular finite state machine.

We then examined various approaches to state assignment. These included simple methods, such as sequential, random, and one-hot encoding strategies, as well as more complex output-oriented methods. We concluded by looking at the heuristics used by computer-aided design tools to obtain good state encodings.

We provided a set of rules for state machine partitioning, including partitioning based on inputs and outputs and partitioning by introducing idle states. These techniques are needed when we cannot implement a finite state machine within a single programmable logic component.

The chapter concluded with a return to hardware description languages and how they are used to represent the structure of Moore, Mealy, and synchronous Mealy machines.

In the next chapter, we will examine implementation technologies in more detail. In particular, we will look at the methods for implementing finite state machines based on structured logic methods, such as ROMs and programmable logic, as well as approaches based on fixed logic components.

FURTHER READING

An excellent pair of books by T. Kam and T. Villa address virtually all of the issues involved in synthesizing the logic of finite state machines. The first, by T. Kam, T. Villa, R. Brayton, and A. Sangiovanni-Vincentelli, is titled *Synthesis of FSMs: Functional Optimization,* Kluwer Academic Publishers, 1997 and focuses on state reduction and effective use of don't cares. The second, by T. Villa, T. Kam, R. Brayton, and A. Sangiovanni-Vincentelli whose title is *Synthesis of FSMs: Logic Optimization,* Kluwer Academic Publishers, 1997, focuses on the state assignment problem and its effect on the size of resulting logic.

The traffic-light controller example used extensively in this chapter is borrowed from the famous text by C. Mead and L. Conway, *Introduction to VLSI Systems,* Addison-Wesley, Reading, MA, 1979. It has been a principal example in many logic-design books over the years. C. Roth's book, *Fundamentals of Logic Design,* West Publishing, St. Paul, MN, 1985, has an extensive discussion of state assignment guidelines. *Modern Logic Design* by D. Green, Addison-Wesley, 1986, has a highly readable, short, direct description of state assignment (pp. 40–43).

Finite state machine partitioning is a topic that waxes and wanes in importance. The original work was done in the late 1950s, became less interesting during the era of VLSI, and is becoming more important again with pervasive use of programmable logic in digital designs. The topic is not well covered by most of today's textbooks. One exception is M. Bolton's book, *Digital System Design with Programmable Logic,* Addison-Wesley, Wokingham, England, 1990, which offers a section on the topic. The partitioning rules we introduced in this chapter were obtained from an applications note in the *Altera Applications Handbook,* Altera Corporation, Santa Clara, CA, 1988.

EXERCISES

8.1 *(State Reduction)* Use the implication chart method to reduce the 4-bit string recognizer state diagram of Figure 8.1.

8.2 *(State Reduction)* Given the state diagram in Figure Ex. 8.2, obtain an equivalent reduced-state diagram containing a minimum number of states. You may use row-matching or implication charts. Put your final answer in the form of a state diagram rather than a state table. Make it clear which states have been combined.

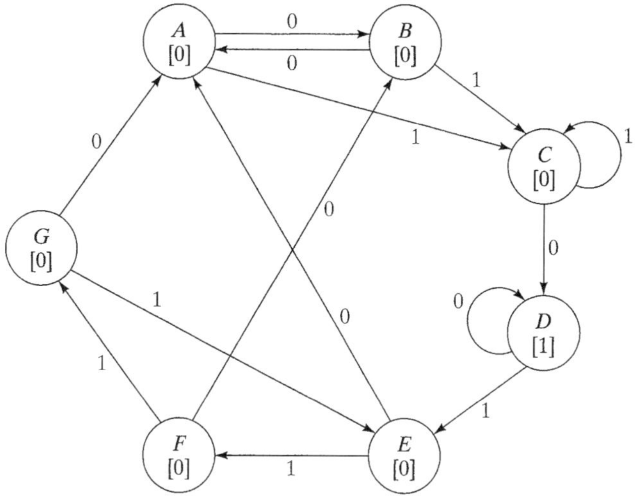

Figure Ex. 8.2 **State diagram for Exercise 8.2.**

8.3 *(State reduction)* Create a Moore state diagram for the 4-bit string recognizer of Figure 8.1. Does is have more states than the Mealy version? Use the implication chart method to reduce the number of states. Do you end with more, less, or the same number of states? Why?

8.4 *(State Reduction)* Given the state diagram in Figure Ex. 8.4, determine which states should be combined to determine the reduced state diagram. You may use row-matching or implication charts.

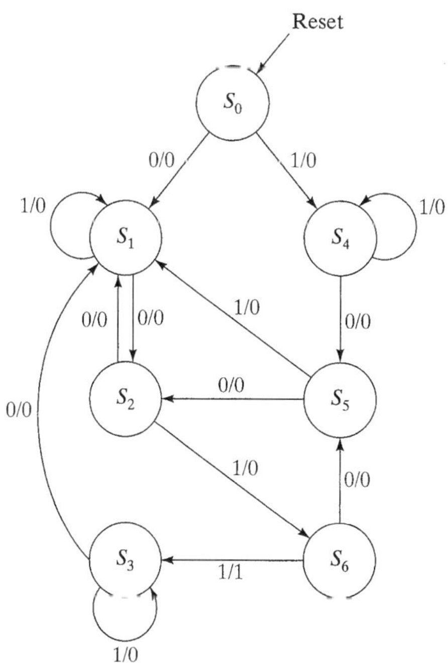

Figure Ex. 8.4 **State diagram for Exercise 8.4.**

8.5 *(State Reduction)* Given the state diagram in Figure Ex. 8.5, draw the fully reduced state diagram. State succinctly what strings cause the recognizer to output a 1.

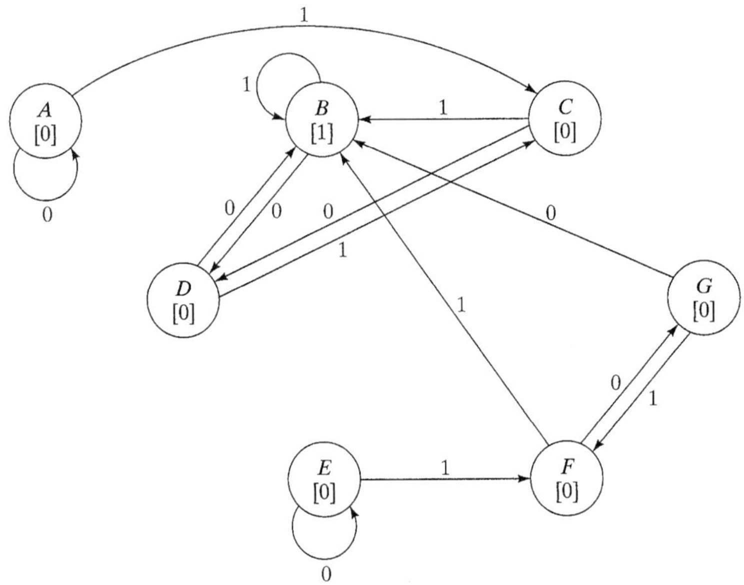

Figure Ex. 8.5 **State diagram for Exercise 8.5.**

8.6 *(State Reduction)* Starting with the state diagram of Figure Ex. 8.6, use the implication chart method to find the minimum state diagram. Which of the original states are combined?

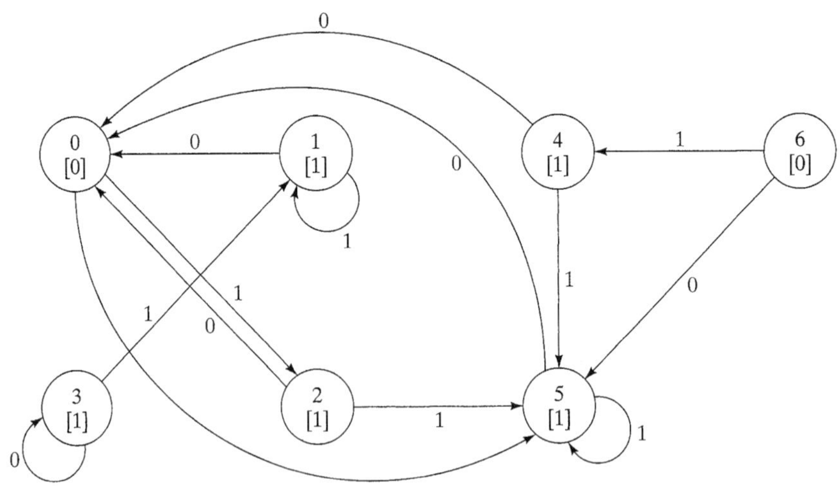

Figure Ex. 8.6 **Original state diagram for Exercise 8.6.**

8.7 *(State Assignment)* Given the state diagram in Figure Ex. 8.7, implement a state assignment using the following techniques:

(a) Minimum bit-change heuristic

(b) State assignment guidelines

Show your assignment in a state map. Explain the rationale for your state assignment.

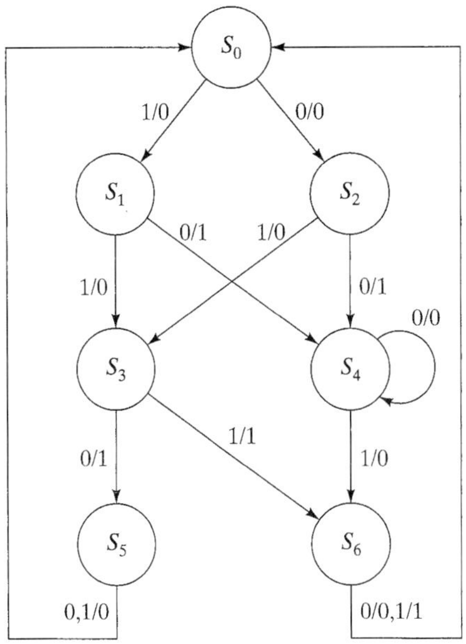

Figure Ex. 8.7 State diagram for Exercise 8.7.

8.8 *(State Assignment)* Given the state diagram in Figure Ex. 8.8, select a good state assignment, justifying your answer in terms of the state assignment guidelines.

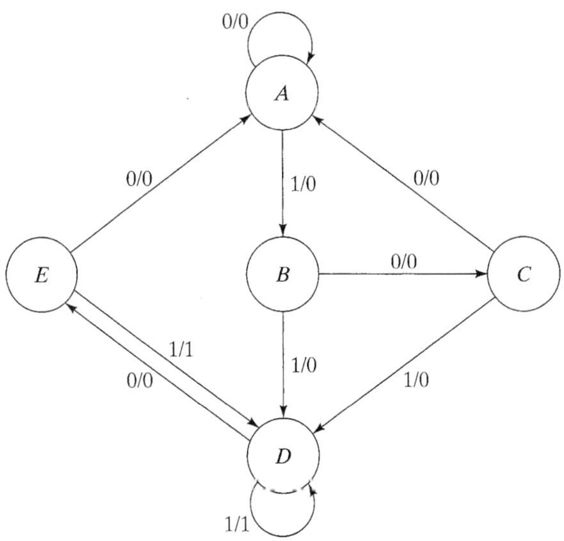

Figure Ex. 8.8 State diagram for Exercise 8.8.

8.9 *(State Assignment)* One method for state assignment is to exhaustively enumerate all the possible state assignments. Given the traffic-light controller symbolic state table of Figure 8.24, use a logic minimization tool to evaluate all 24 possible 2-bit encodings of the states (alternatively, choose two to four random assignments and derive the logic by hand). Using literal count as your metric, which is the best encoding?

8.10 *(State Assignment)* Try some random encodings of the traffic-light controller that are *non-dense*. That is, map the four states into eight (3 bits) or sixteen (4 bits) possible states. How do they compare in terms of literal count to the encodings found in the previous question?

8.11 *(State Assignment)* Given the next-state function of the finite state machine shown in Figure Ex. 8.11, use the implication chart method to find the most reduced state diagram.

Q_3	Q_2	Q_1	Q_0	P_3	P_2	P_1	P_0
0	0	0	0	0	0	0	1
0	0	0	1	0	0	0	0
0	0	1	0	1	0	0	1
0	0	1	1	1	0	1	0
0	1	0	0	0	1	0	1
0	1	0	1	0	1	0	0
0	1	1	0	0	0	0	1
0	1	1	1	0	0	1	0
1	0	0	0	0	0	1	0
1	0	0	1	0	0	0	1
1	0	1	0	1	0	1	0
1	0	1	1	1	0	1	0
1	1	0	0	0	1	1	0
1	1	0	1	0	1	0	1
1	1	1	0	1	1	1	0
1	1	1	1	1	1	1	0

Figure Ex. 8.11 **Next-state functions for Exercises 8.10, 8.11.**

8.12 *(State Partitioning)* Show how to partition the next-state functions of Figure Ex. 8.11, P_3, P_2, P_1, P_0, into two groups, each of which depends on only three of the four possible current state bits, Q_3, Q_2, Q_1, Q_0.

8.13 *(State Partitioning)* Given a 3-bit, eight-state Gray-code up/down counter (similar to the state machine in Figure 8.38 but with eight states instead of six), show how the state diagram can be partitioned into two communicating finite state machines with five states each, including idle states.

8.14 *(State Partitioning)* Given the state diagram in Figure Ex. 8.14, partition the state machine into two communicating finite state machines, one containing the states: S_0, S_1, S_2, S_3, and the other containing: S_4, S_5, S_6.

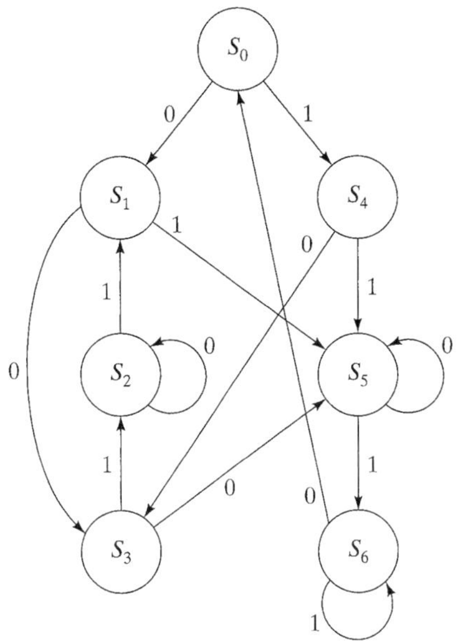

Figure Ex. 8.14 State diagram for Exercise 8.14.

8.15 *(State Partitioning)* Figure Ex. 8.15 gives a state diagram with nine states. Show how to partition the state diagram into three communicating state machines, consisting of the state groups: S_0, S_1, S_2; S_3, S_4, S_5; and S_6, S_7, S_8.

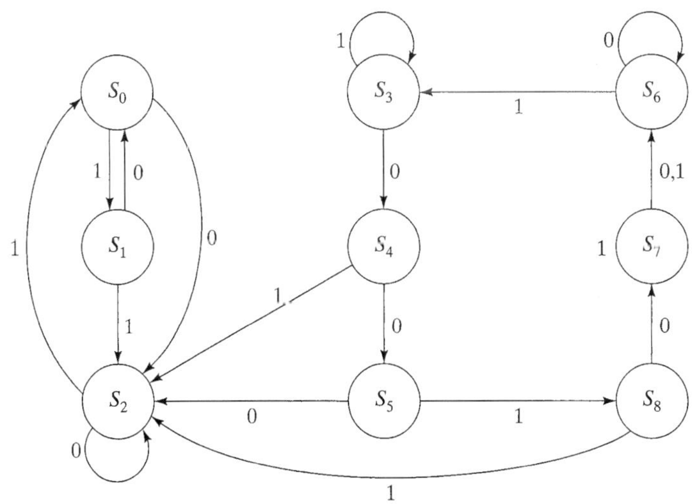

Figure Ex. 8.15 State diagram for Exercise 8.15.

8.16 *(State Partitioning)* The partitioning rules presented in Figure 8.37 describe only the transformations on states and transition conditions. Outputs are not considered.

(a) Describe how the partitioning rules should be modified to handle Mealy outputs. How are the transfers into the idle states affected?

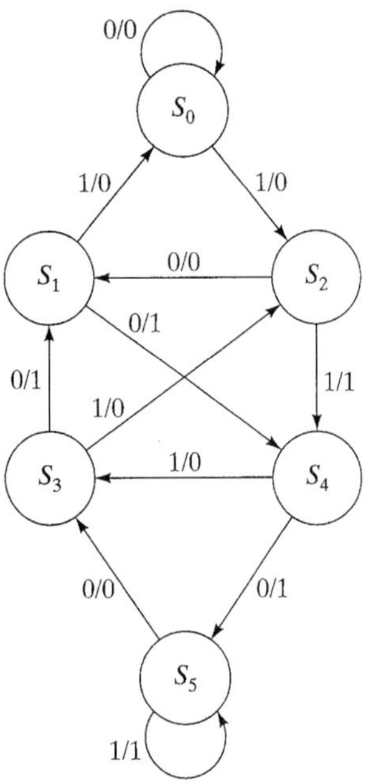

Figure Ex. 8.17 **State diagram for Exercise 8.17.**

(b) Describe how the partitioning rules should be modified to handle Moore outputs. How might the outputs from the partitioned machines be combined?

8.17 *(State Assignment)* Given the state diagram in Figure Ex. 8.17 and the state assignment $S_0 = 000$, $S_1 = 001$, $S_2 = 010$, $S_3 = 011$, $S_4 = 100$, and $S_5 = 101$, write the encoded state table, and derive the minimized Boolean equations for implementing the next-state and output functions.

8.18 *(State Assignment)* Given the state diagram in Figure Ex. 8.18 and the state assignment $A = 000$, $B = 001$, $C = 011$, $D = 111$, $E = 101$, implement the state machine using a minimum number of gates and flip-flops. You may assume that an external RESET signal places the machine in state A (000).

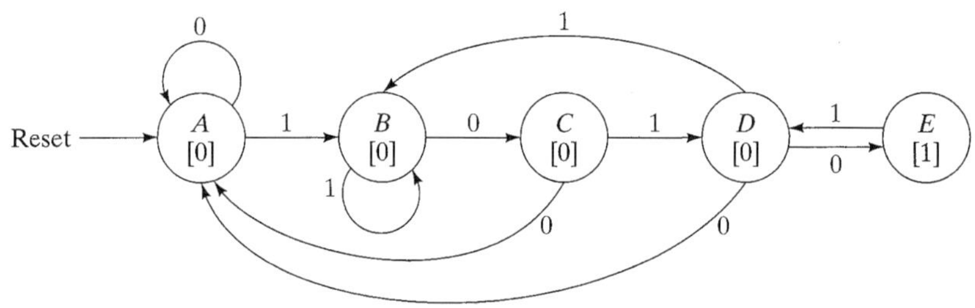

Figure Ex. 8.18 **State diagram for Exercise 8.18.**

8.19 *(Hardware Description Languages)* Derive the equations for the traffic-light controller as given in the Verilog description at the end of Section 8.4.

8.20 *(Hardware Description Languages)* Write a Verilog description of the 4-bit string recognizer of Figure 8.6.

8.21 *(Hardware Description Languages)* Write a Verilog description of the state diagram of Figure 8.13.

8.22 *(Hardware Description Languages)* Write a Verilog description for the partitioned communicating finite state machines of Figure 8.38.

8.23 *(Design Process)* In the traffic light controller we implemented in the state diagrams of Figure 8.39, if the timer of the traffic light goes past state four without the main controller seeing the TS signal, our traffic light will be stuck on yellow forever. You have been asked to make the design more robust to this unlikely failure. Add other states and/or outputs to rectify the problem. Do you need to add more states to either state machine or can the problem be corrected without adding new states?

8.24 *(Design Process)* Implement the following finite state machine description using a minimum number of states and a good state assignment. The machine has a single input X, a single output Z, and will assert $Z = 1$ for every input sequence ending in the string 0010 or 100.

8.25 *(Design Process)* Your task is to implement a finite state machine with the following state diagram. The finite state machine is actually a complex Gray-code counter. The counter has two control inputs, I_1 and I_0, which determine the next state. The counter's functional specification is as follows. When $I_1 I_0 = 00$, it is a Gray-code up-counter. When $I_1 I_0 = 01$, it is a Gray-code down-counter. When $I_1 I_0 = 10$, it is a Gray-code count-by-two. Finally, when $I_1 I_0 = 11$, the counter holds it current state. The state diagram is shown in Figure Ex. 8.25.

(a) Complete a state transition table, including the next-state bits (Q_1 and Q_0) and the needed inputs to the two flip-flops (P_1 and P_0) to obtain that next state.

(b) Produce the four-variable K-maps for the next-state functions. Obtain the minimized two-level implementation.

(c) Draw an implementation schematic, using a minimum number of inverters and two-input NAND, NOR, XOR, and XNOR gates. Assume that complements are not available.

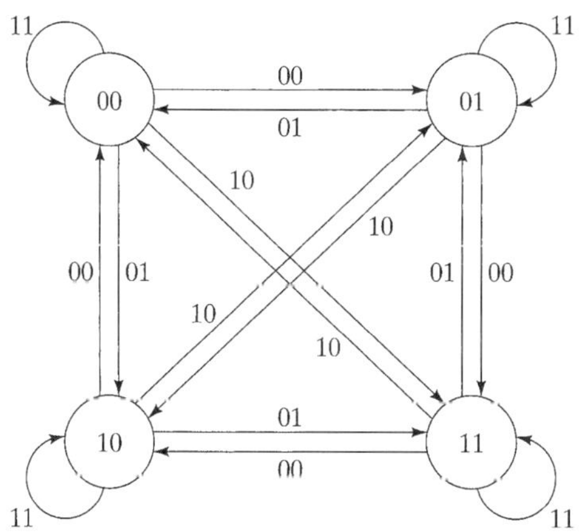

Figure Ex. 8.25 **State diagram for Exercise 8.25.**

8.26 *(Design Process)* Design a Mealy finite state machine with input X and output Z. The output Z should be asserted for one clock cycle whenever the sequence . . . 0111 or . . . 1000 has been input on X. The patterns may overlap. For example, $X =$ . . . 0000111000 . . . should generate the output stream $Y =$ . . . 0000001001 . . . .

(a) Complete the state diagram for the sequence detector, without concern for state minimization.

(b) Complete the state table for the state diagram derived in part (a).

(c) Use row matching or implication charts to minimize the state table derived in part (b).

(d) Use the state assignment guidelines to obtain a good state assignment for the reduced state machine of part (c). Justify your method in terms of the high-, medium-, and low-priority assignment guidelines.

(e) Implement your encoded, reduced state table and show all the minimized logic equations for the next state and outputs.

(f) Describe your finite state machine in a Verilog description.

9

Sequential Logic Technologies

Introduction

All sequential logic is implemented using the basic flip-flops and registers we have already seen in Chapter 6. However, these sequential logic elements always need to be combined with combinational logic that will implement the next-state and output equations. In Chapter 7, we have already seen how sequential logic can be combined with combinational logic to accomplish this.

In this chapter, we will examine the full range of sequential logic technologies. We will see how, for each corresponding combinational logic technology, designers have developed flexible and efficient sequential logic elements. Along the way, we will use design examples to highlight the choices made in each technology. As in Chapter 4, we'll review both fixed and programmable logic technologies. Specifically, we will:

- *Implement state machines with discrete flip-flops.* Flip-flops have been packaged to have many useful features to make it easier to build finite state machines. We will also see how MSI components such as shift registers and counters can further reduce component counts.
- *Implement state machines with ROMs or PALs/PLAs.* The next-state and output functions are implemented with programmable logic connected to state flip-flops. These flip-flops have many of the same programmable features that evolved with discrete logic.
- *Implement state machines using common programmable devices such as PLDs and FPGAs.* These devices are now the most commonly used to implement sequential logic and you will get a sense for the balance that has to be struck between the amount of

chip area devoted to combinational logic versus sequential logic so that neither is wasted unnecessarily. Different design choices in this balance lead to different programmable devices appropriate to different types of finite state machines.

9.1 Basic Sequential Logic Components

Over the years, the basic flip-flops we studied in Chapter 6 have diversified into many varieties fueled by the need to decrease implementation costs. Some of the variations are quite simple. For example, some flip-flops have a reset (and/or set) input. Others have a control input that puts them in a new state determined by another data input (load enable and load input, respectively). Historically, flip-flops with two inputs called JK-flip-flops were also common. The two inputs implemented set, reset, toggle, and hold functions for the state bit. You can learn more about these in Appendix C. Let's see how simple additions such as set and reset to the logic of a flip-flop can simplify the next-state equations needed to realize a finite state machine.

EXAMPLE 9.1 ADVANTAGE OF FLIP-FLOPS WITH RESET AND SET INPUTS

The logic for a finite state machine that includes a reset input takes the following form for each next-state equation:

$$NS_i = \text{Reset}'(PS_a\, IC_a + PS_b\, IC_b + \cdots) = \text{Reset}'\ PS_a\, IC_a + R'\ PS_b\, IC_b + \cdots$$

Where NS_i is a next-state bit, PS_a is a term describing a present state and IC_a is a term describing the input conditions under which a state transition from the present state PS_a is to occur and set the NS_i bit to true. By using *Reset* in the same way for each next-state equation, we can force the machine into the all-zero state whenever *Reset* is asserted.

If we use a flip-flop with a reset input then the equation can be simplified to

$$NS_i = PS_a\, IC_a + PS_b\, IC_b + \cdots$$

This allows us to use simpler logic. If the reset condition was more complex than a simple *Reset* input being asserted, then our savings can be even greater. For example, we may have a state machine that has many transitions to one particular state. It is common to set this *most common target state* to be all zero. Typically, the reset state is the most common target state as a state machine has a transition from each of its states to the reset state.

Let's look at an example of a state machine that benefits from the flip-flop reset and set inputs. Figure 9.1 shows the state diagram of such an FSM. Although the state diagram may look complex at first glance, the FSM has a very simple function. It has a single input (besides the Reset input) and a single Moore output, its rightmost state

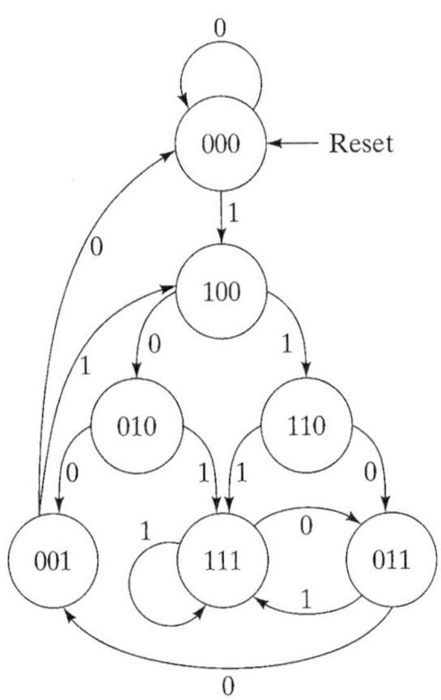

Figure 9.1 Median filter FSM.

I	PS_1	PS_2	PS_3	NS_1	NS_2	NS_3
0	0	0	0	0	0	0
0	0	0	1	0	0	0
0	0	1	0	0	0	1
0	0	1	1	0	0	1
0	1	0	0	0	1	0
0	1	0	1	X	X	X
0	1	1	0	0	1	1
0	1	1	1	0	1	1
1	0	0	0	1	0	0
1	0	0	1	1	0	0
1	0	1	0	1	1	1
1	0	1	1	1	1	1
1	1	0	0	1	1	0
1	1	0	1	X	X	X
1	1	1	0	1	1	1
1	1	1	1	1	1	1

Figure 9.2 State transition table for median filter FSM.

bit. The function it performs is to remove any 0s in its input stream that are between two 1s. This is called a median filter in that it filters out lone 0s by changing them to a 1 on the output. Thus, the output stream of the state machine (the sequence of values on its rightmost state bit) will always have 0s (at least in pairs). Take a few minutes to convince yourself that this is in fact the behavior of the FSM by trying out a sample input stream and checking the output.

The state table for this state diagram is shown in Figure 9.2. Note that it is similar to Figure 7.29, which represented a simple shift register. In fact, our FSM is very close to being a 3-bit shift register, with a slight difference in states 011 and 010 that cause the lone 0s to be changed to 1s. State 101 does not appear in the state diagram and is treated as a don't-care condition. The equations derived from this table, by inspection or using a K-map, are:

$$NS_1 = \text{Reset}' I$$
$$NS_2 = \text{Reset}'(PS_1 I' + (PS_1 + PS_2) I) = \text{Reset}' PS_1 + \text{Reset}' PS_2 I$$
$$NS_3 = \text{Reset}' PS_2$$
$$O = PS_3$$

This does, in fact, look a lot like a shift register, except for NS_2. We can make this logic simpler if we employ flip-flops with set and reset inputs. The equations now become

$$R = \text{Reset}$$
$$S = PS_2 I$$
$$NS_1 = I$$
$$NS_2 - PS_1$$
$$NS_3 = PS_2$$
$$O = PS_3$$

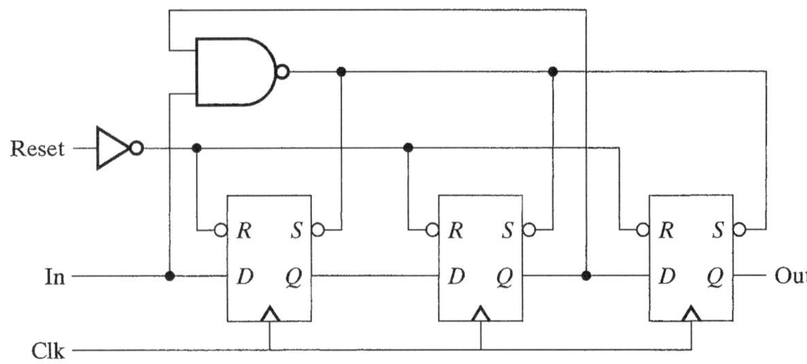

Figure 9.3 **Realization of the median filter FSM.**

Note that the reset and set inputs of all the flip-flops are the same. The set function is now invoked whenever there is a 1 on the input and the next-to-last input was also 1. This does exactly the filtering function by setting the state to all 1s and thus eliminating the 0 that may or may not have been in between the two 1s.

We now need only a single AND gate to implement our logic (rather than the four AND gates and one OR gate) if we used basic flip-flops without set and reset capabilities. In Figure 9.3, we show an implementation using flip-flops with active-low reset and set inputs.

Flip-flops have been packaged together in many ways. The most common of these, that we have already seen in previous chapters, are shift registers and counters. Often, designers use these components to help them implement the state equations of their FSMs if there is a good match between the state diagram and the sequential function that are already embodied in these components. Let's again turn to some examples to see how this is done.

EXAMPLE 9.2 USING A SHIFT REGISTER TO IMPLEMENT AN FSM

String recognizers are perfect candidates for implementation using shift registers. Let's return to the example 4-bit string recognizer of Figure 8.1. Rather than implementing this 15-state Mealy machine using the typical design procedure of state minimization, state assignment, and logic-equation generation, we can implement it directly using a shift register. The circuit of Figure 9.4 implements the 4-bit string recognizer using a packaged 4-bit shift register and a few logic gates to implement the output function. Recall that the output of the string recognizer is asserted if we have seen the strings 0110 and 1010 in the input stream. Since it is a Mealy machine, the output is a function of the input. The shift register comes with a parallel load input to get the shift-register bits into a known state. We do not use this input and unassert it by setting it to a logic one since it is active low. The serial input of the shift register is labeled *SI* and the output bit of the first flip-flop in the shift chain is Q_0. Note that we only use the first three

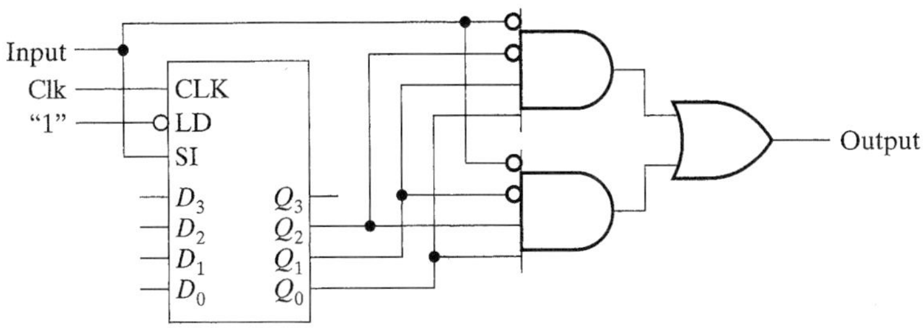

Figure 9.4 Initial realization of the 4-bit string recognizer of Figure 8.1 using a shift register.

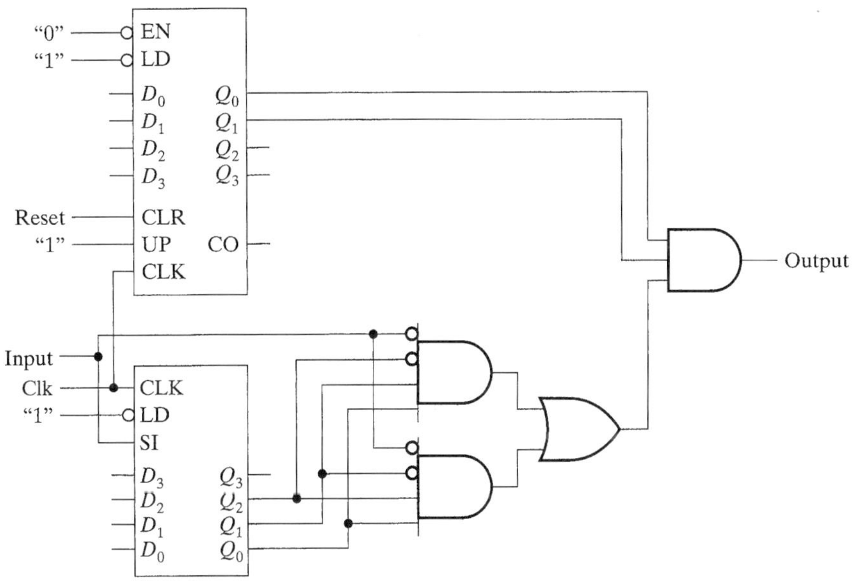

Figure 9.5 Corrected realization of the 4-bit string recognizer of Figure 9.4 using a shift register and counter.

shift-register flip-flop outputs because we look at the fourth input directly (Mealy machine) rather than wait until it is shifted in.

There is a problem with our realization, however. The state machine of Figure 8.1 worked on 4-bit groups of input values. Our shift-register implementation looks for the two patterns anywhere in the input stream and not just in 4-bit groups. To correct our implementation, we need to make sure that our output is only asserted on 4-bit boundaries. We can do this by adding a 2-bit counter to count out the groups. We can then enable our output with the fourth state of the counter (an 11 in the low-order two bits which occurs every four cycles) using an AND gate. Figure 9.5 shows this alteration to our original circuit. Now, only patterns discovered when the value of the counter is . . . 11 can cause the output to be asserted.

We have realized our state machine using two parallel machines: one to look for a pattern and the other to align it to 4-bit groups of inputs. An interesting exercise would be to see if this realization could be derived using the FSM optimization and transformation techniques of the previous chapter.

Effect on the Design Process Most designers design such simple FSMs by hand and easily can map them to components like shift registers, counters, and flip-flops with set and reset inputs. The process entails keeping in mind the state diagram of these components and looking for similarities in the state diagram of the FSM to be implemented. State assignment can have a big impact on this, especially when we consider the use of set and reset inputs. Most often, designers will either set or reset all flip-flops in the same way. However, it is also possible to mix sets and resets, but that usually makes the mental mapping much more difficult.

9.2 FSM Design with Counters

As already hinted with the previous example, synchronous counters are potentially attractive for implementing finite state machines. In a single package they provide a state register with mechanisms for advancing the state (CNT), resetting it to zero (CLR), and "jumping" to a new state (LD). Rather than encode the next-state function as actual state bits, it can be implemented by asserting the counter control signals CNT, LD, and CLR under the right conditions. We examine a counter-based implementation strategy in this section.

For correct operation in state sequencing, the counter must be implemented with synchronous LD and CLR signals. As pointed out in Chapter 6, asynchronous control signals lead to invalid behavior of the state machine.

Figure 9.6 shows the kinds of state transitions supported by a counter. You should choose the state assignment to exploit the special sequencing capabilities of a counter. The state encodings you assign to sequential states should follow the binary sequence supported by the counter. Let's look at a counter-based implementation of a BCD-to-excess-3 converter next.

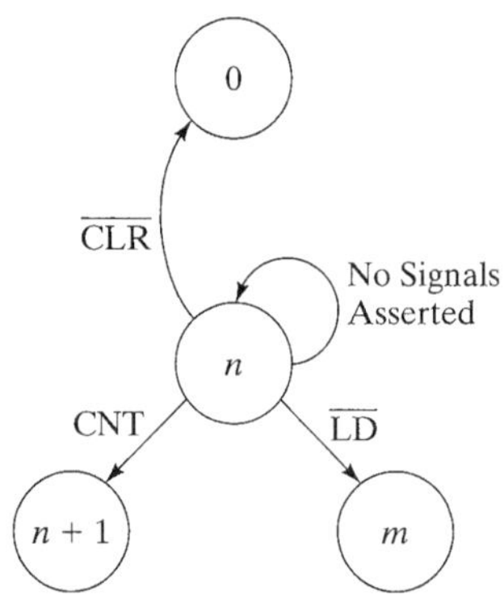

Figure 9.6 State transitions of a counter-based finite state machine.

BCD	Excess-3 Code
0000	0011
0001	0100
0010	0101
0011	0110
0100	0111
0101	1000
0110	1001
0111	1010
1000	1011
1001	1100

Figure 9.7 BCD and excess-3 code.

EXAMPLE 9.3 BCD-TO-EXCESS-3 CODE CONVERTER USING A COUNTER

The function to be implemented is a BCD-to-excess-3 code converter. We obtain the excess-3 code by adding 11_2 to the BCD number, as shown in Figure 9.7. We easily could implement this as a 4-bit combinational logic function. However, our machine will be designed to accept a bit-serial BCD number, starting with the least significant bit. The machine has one input X (the BCD bit) and one output Z (the corresponding excess-3 bit). We will implement the finite state machine as a synchronous Mealy machine.

Reduced State Diagram/Symbolic State Transition Table
The state transition table and state diagram are shown in Figures 9.8 and 9.9, respectively. You should verify that the state diagram actually

Present State	Next State X=0	Next State X=1	Output X=0	Output X=1
S_0	S_1	S_2	1	0
S_1	S_3	S_4	1	0
S_2	S_4	S_4	0	1
S_3	S_5	S_5	0	1
S_4	S_5	S_6	1	0
S_5	S_0	S_0	0	1
S_6	S_0	–	1	–

Figure 9.8 Symbolic state transition table for the code converter.

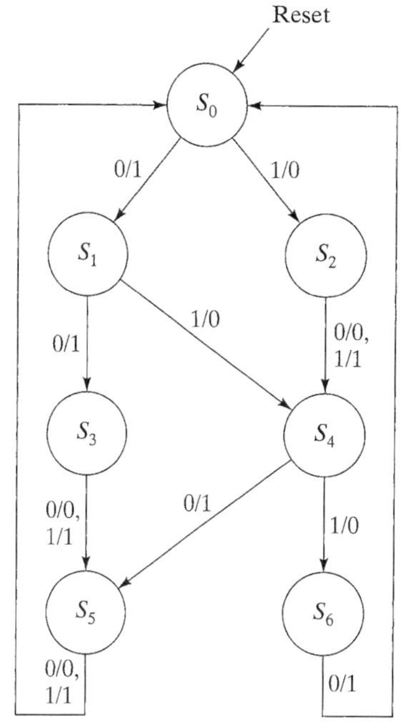

Figure 9.9 State diagram corresponding to Figure 9.8.

implements the conversion. By the way, you may wonder why state S_6 has only one transition. Since the input is always a BCD digit, if you get to this state, the next input can never be 1.

Figure 9.10 gives a state assignment for the state diagram of the code converter. This assignment is reasonably well suited for a counter-based finite state machine implementation. As much as possible, we assign sequential state encodings to flows in the state diagram. We also try to reduce the number of transitions in which a jump (LD) to a state (other than 0) takes place. The figure shows three such jumps: from state 0 to 4, from 1 to 5, and from 5 to 3.

State Transition Table

To implement a finite state machine with a counter, we replace the next-state bits with logic to generate the counter control signals. However, we must still specify directly the next-state bits for the jump cases, and they must be provided at the parallel inputs whenever $\overline{LD}$ is asserted.

Figure 9.11 shows the finite state machine state transition table, extended with output columns for the counter control signals $\overline{CLR}$, $\overline{LD}$, EN (count), as well as the counter load inputs C, B, and A. Our basic procedure obtains minimized logic expressions for these signals rather than the next-state bits.

For our catalog counter, the $\overline{CLR}$ signal has precedence over $\overline{LD}$, which in turn has precedence over EN (count). This makes it possible to exploit many more don't-care conditions within the K-maps.

We derive the state transition table in the following manner. We examine each state transition in turn. Consider the first transition, from state 0 to 1. If the machine is currently in state 0 with input 0, the finite state machine changes to state 1 with output 1. Thus, $\overline{CLR}$ and $\overline{LD}$ should be unasserted (left high) and EN asserted to cause the state register to count up from 0 to 1. These settings fill out the rest of the row in the transition table.

If the input is 1 in state 0, the next state is 4 with an output 0. This calls for a jump: $\overline{CLR}$ is left unasserted high, $\overline{LD}$ is asserted low, and EN can be a don't care. The value to be loaded, 1 0 0, must be available on the C, B, A signals. The collection of values fill out the ninth row of the state transition table.

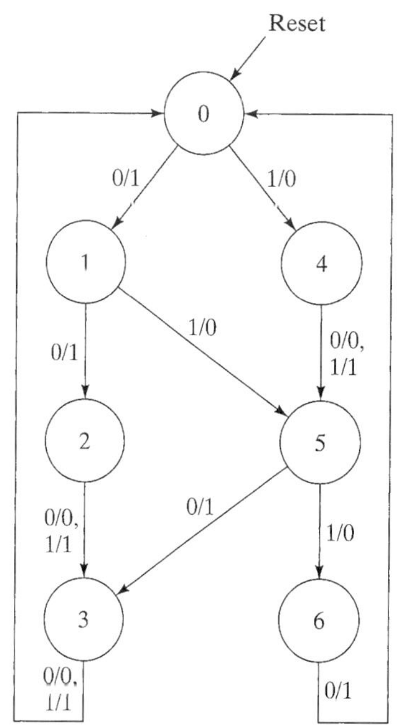

Figure 9.10 Code-converter state diagram with counter-based state assignment.

Inputs/Current State				Next State			Outputs						
X	Q_2	Q_1	Q_0	Q_2^+	Q_1^+	Q_0^+	Z	$\overline{\text{CLR}}$	$\overline{\text{LD}}$	EN	C	B	A
0	0	0	0	0	0	1	1	1	1	1	X	X	X
0	0	0	1	0	1	0	1	1	1	1	X	X	X
0	0	1	0	0	1	1	0	1	1	1	X	X	X
0	0	1	1	0	0	0	0	0	X	X	X	X	X
0	1	0	0	1	0	1	0	1	1	1	X	X	X
0	1	0	1	0	1	1	1	1	0	X	0	1	1
0	1	1	0	0	0	0	1	0	X	X	X	X	X
0	1	1	1	X	X	X	X	X	X	X	X	X	X
1	0	0	0	1	0	0	0	1	0	X	1	0	0
1	0	0	1	1	0	1	0	1	0	X	1	0	1
1	0	1	0	0	1	1	1	1	1	1	X	X	X
1	0	1	1	0	0	0	1	0	X	X	X	X	X
1	1	0	0	1	0	1	1	1	1	1	X	X	X
1	1	0	1	1	1	0	0	1	1	1	X	X	X
1	1	1	0	X	X	X	X	X	X	X	X	X	X
1	1	1	1	X	X	X	X	X	X	X	X	X	X

Figure 9.11 Transition table for code converter implemented with a counter-based finite state machine.

An example of a return to state 0 is the 0 transition from state 3. In this case, the $\overline{\text{CLR}}$ signal is asserted low, while $\overline{\text{LD}}$ and EN can be don't cares. Rows 7 and 12 of the transition table provide other examples of these "return to 0" transitions.

Boolean Minimization of the Counter Control Signals Our next step is to minimize the functions for Z, $\overline{\text{CLR}}$, $\overline{\text{LD}}$, EN, C, B, and A. Because we are already using the $\overline{\text{CLR}}$ signal for a control function, we need a separate *RESET* input to place the finite state machine in state 0 initially. Note that this is a synchronous reset; but let's assume that the reset on the quad registers is asynchronous. Another way to implement a synchronous reset is to include the reset signal as an input to the next-state logic.

Discussion For this implementation of the BCD-to-excess-3 code converter, the logic is quite complex: 10 product terms, an additional input, and three additional outputs. The schematic is shown in Figure 9.12. Since we are designing synchronous Mealy machines, we need a flip-flop between the combinational logic and the output Z, adding yet another component.

The counter-based implementation was a bad choice for this particular state machine. However, when the state diagram has fewer out-of-sequence jumps, a counter-based implementation can be very effective.

As we have seen with this last example, it can be quite difficult to map large FSMs to fixed sequential-logic components. It is more difficult to find matching patterns and usually impossible to implement a large FSM using only one or two components. For these reasons, designers have developed programmable logic approaches to building FSMs.

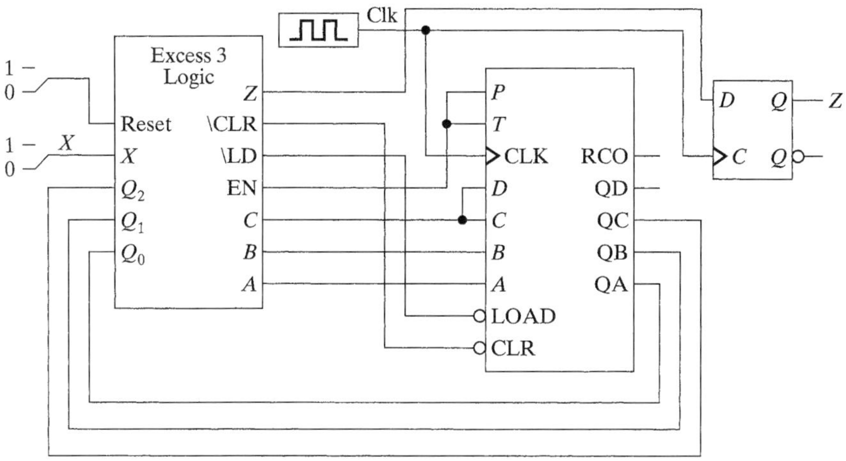

Figure 9.12 Counter-based implementation of code converter.

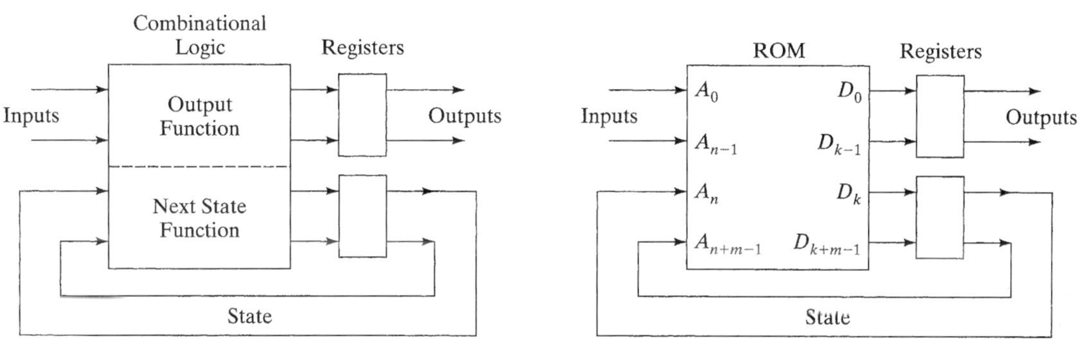

Figure 9.13 ROM implementation of a synchronous Mealy finite state machine.

9.3 FSM Design with Programmable Logic

In this section, we examine alternative schemes for implementing finite state machines using programmable logic components.

9.3.1 Mapping a State Machine into a ROM Implementation

A ROM or PAL/PLA is a convenient way to implement the combinational logic of a finite state machine. The state is stored in an external register whose outputs are fed back as addresses to the ROM or inputs to the PAL/PLA. Figure 9.13 shows the general block diagram of a synchronous Mealy finite state machine and how this is mapped onto a ROM implementation. It is particularly convenient to implement a Mealy finite state machine this way as all of the inputs go into the same combinational-logic block and all of the outputs are treated identically, whether they are state bits or outputs.

In the figure, the finite state machine has n inputs, k outputs, and m state bits. The ROM address bits are at the left of the ROM block, with the data bits at the right. The $n + m$ input bits and next-state bits form the ROM's address lines. The $k + m$ output and state bits form the ROM's output lines. Thus, we need a ROM of 2^{n+m} words of $k + m$ bits each to implement this finite state machine.

It is easy for you to fill in the ROM contents, given an encoded state transition table organized as a truth table. Each row of the table, identified by the input and current-state bits, selects a unique word of the ROM. The bit pattern for the next state and outputs is simply stored in this word.

Effect on the Design Process Using programmable logic for the finite state machine implementation has important implications for state assignment and state machine partitioning. Unlike discrete logic, which has few external constraints, programmable logic has many. For example, programmable logic devices have a fixed number of pins, preassigned to be inputs or outputs. The internal resources, be they product terms or ROM words, are severely limited. This leads to more complexity when mapping a design onto programmable logic.

It simplifies your design if the next-state and output functions can be made to "fit" within a single component. Unfortunately, this is not always possible for complex finite state machines. You may need the partitioning strategies of Chapter 8 to map the state machine onto as few components as possible.

The choice of programmable logic also affects the state-assignment process. Since the complexity of a ROM is determined solely by the number of inputs and outputs, the kind of ROM you need is the same, regardless of the state assignment. However, a PAL/PLA, with its limited product terms, requires a good state assignment to keep manageable the number of terms to implement the next-state and output logic.

9.3.2 ROM Versus PLA-Based Design

Let's compare the implementation of the same finite state machine using a ROM and a PLA and we'll reuse the BCD-to-excess-3 code converter to do so.

State Assignment for ROM-Based Implementation We assume that the seven states of the finite state machine are encoded in 3 bits. Thus, the ROM-based implementation requires a 16-word ROM (four address bits: input X and three current-state bits, Q_2, Q_1, Q_0) with four data outputs (output Z and three next-state bits: D_2, D_1, D_0). Since the size of the ROM is independent of the state assignment, a straightforward assignment will be used (that is, $S_0 = 000$, $S_1 = 001$, etc.). This leads us to the truth table of Figure 9.14. The inputs at the left form the ROM's address and the outputs at the right are the contents of the ROM word at that address.

Hardware Implementation for the ROM-Based Design We choose a synchronous Mealy machine implementation, as shown in Figure 9.15. The

ROM Address				ROM Outputs			
X	Q_2	Q_1	Q_0	Z	D_2	D_1	D_0
0	0	0	0	1	0	0	1
0	0	0	1	1	0	1	1
0	0	1	0	0	1	0	0
0	0	1	1	0	1	0	1
0	1	0	0	1	1	0	1
0	1	0	1	0	0	0	0
0	1	1	0	1	0	0	0
0	1	1	1	X	X	X	X
1	0	0	0	0	0	1	0
1	0	0	1	0	1	0	0
1	0	1	0	1	1	0	0
1	0	1	1	1	1	0	1
1	1	0	0	0	1	1	0
1	1	0	1	1	0	0	0
1	1	1	0	X	X	X	X
1	1	1	1	X	X	X	X

Figure 9.14 ROM truth table.

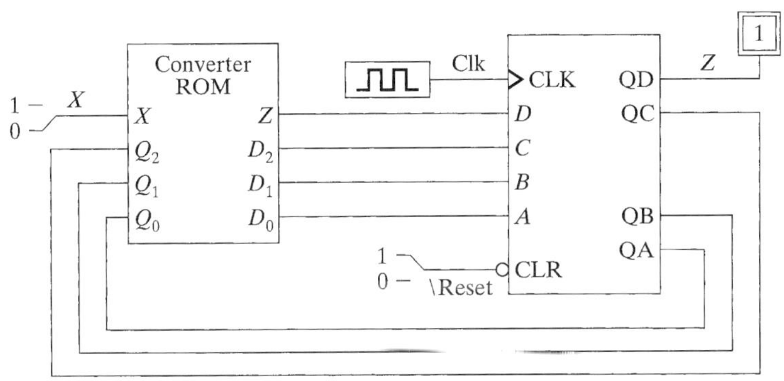

Figure 9.15 Excess-3 synchronous Mealy ROM-based implementation.

ROM addresses are at the left of the converter ROM, with the outputs at the right. We used a flip-flop component that contains quad positive edge-triggered D flip-flops with common clear and clock signals. This is much simpler than our counter-based implementation. There are fewer equations since we do not have to generate logic for each of the counter's control signals. Furthermore, the ROM implementation is indifferent to state assignment so we didn't have to spend time finding counter-like flows among the states. These factors make ROM-based FSM design much more attractive in most cases.

It is important to understand the timing behavior of our implementation. Figure 9.16 gives the timing waveforms for the converter's output behavior, for the input strings 0000 (0) and 1110 (7). We show these values backwards as they arrive from least significant bit to most significant bit (the leftmost bit arrives first). By the way, this timing behavior is identical to that of our counter-based implementation in Figure 9.12.

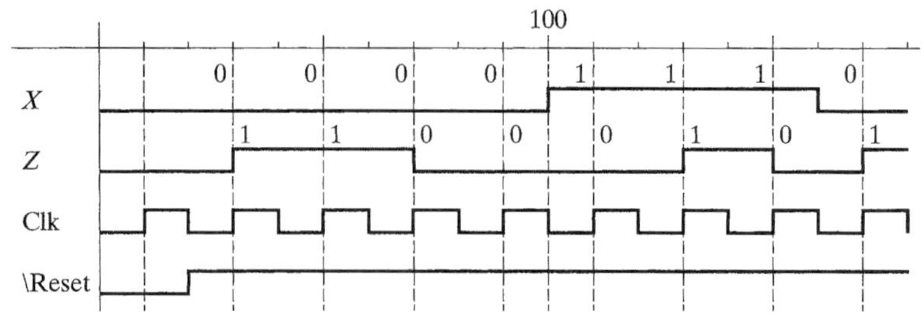

Figure 9.16 **Timing behavior of ROM converter.**

Let's start with the first string, 0000. Inputs are sampled on the rising edge, with the output becoming valid just after the clock edge. After reset, a 0 at the input results in a 1 at the output after the clock edge. At the next rising edge, the input is still 0 and the output stays 1. The output falls and stays low for the next two positive clock edges. Thus, the string 0000 is converted to 1100, representing a binary three when read backward.

Similarly, the string 1110 generates the output 0101. On the first rising edge, the input is 1 and the output stays 0. On the next positive edge, the input is 1 but the output rises. The input stays 1 for the third positive edge, but the output goes low. Before the fourth and final edge, the input goes low, causing the output to go high. Thus, a backward seven has been converted to a backward ten.

State Assignment for PAL/PLA-Based Implementation For a PAL/PLA-based implementation, we must first derive a state assignment and then follow up with two-level logic minimization.

Suppose that our state-assignment heuristics yield the following state assignment (other assignments might be equally good):

$$S_0 = 000$$
$$S_1 = 001$$
$$S_2 = 011$$
$$S_3 = 110$$
$$S_4 = 100$$
$$S_5 = 111$$
$$S_6 = 101$$

To find the minimized two-level implementation, we derive K-maps or use logic-minimization tools and make sure to include the don't-care transitions in this state assignment to obtain the best possible reductions. We derive the following reduced logic:

$$D_2 = \bar{Q}_2Q_0 + Q_2\bar{Q}_0$$
$$D_1 = \bar{X}\bar{Q}_2\bar{Q}_1Q_0 + X\bar{Q}_2\bar{Q}_0 + \bar{X}Q_2\bar{Q}_0 + Q_1\bar{Q}_0$$
$$D_0 = \bar{Q}_0$$
$$Z = XQ_1 + \bar{X}\bar{Q}_1$$

If we select a PLA to implement the logic, it must provide four inputs, four outputs, and at least nine product terms. The implementation schematic is identical to that of Figure 9.15, except that we replace the converter ROM with a converter PLA.

PAL versus PLA-Based Implementations As mentioned in Chapter 4, the delay in a PLA tends to be worse than that in a comparable PAL. Although PLAs are very popular in full, custom IC designs as they can be sized as needed, PALs are more likely to be used in conventional board-level designs.

If you select a PAL to implement the code-converter finite state machine, then (at least) one of the gates of the OR plane must have a four-input fan-in. Not every PAL has this kind of internal architecture. For example, a very small 10H8 PAL (10 inputs, eight outputs, two inputs per OR gate) could not implement the next-state logic without some modification to the equations, but an only slightly larger 12H6 PAL (12 inputs, six outputs, two OR gates with four inputs each, four OR gates with two inputs each) could implement it without modification.

Suppose that as the designer, you have only the simpler 10H8 PAL available. It is not difficult to get the equations into the right form. You simply rewrite them using subexpressions that do not exceed the maximum product-term count for the PAL. Here is a way to write the next-state function D_1 as expressions with no more than two product terms per function:

$$D_1 = D_{11} + D_{12}$$
$$D_{11} = \bar{X}\bar{Q}_2\bar{Q}_1Q_0 + X\bar{Q}_2\bar{Q}_0$$
$$D_{12} = \bar{X}Q_2\bar{Q}_0 + Q_1\bar{Q}_0$$

Figure 9.17 shows the partial 10H8 PAL programming map for this logic.

The programming map includes indices for the vertical input lines and the horizontal product term lines, to let you program down to the individual crosspoints in the device. The horizontal rows correspond to the following product terms:

Vertical	**Horizontal**
0. $\bar{Q}_2Q_0$	**24.** D_{11}
1. $Q_2\bar{Q}_0$	**25.** D_{12}
8. $\bar{X}\bar{Q}_2\bar{Q}_1Q_0$	**32.** $\bar{Q}_0$
9. $X\bar{Q}_2\bar{Q}_0$	**33.** Not used
16. $\bar{X}Q_2\bar{Q}_0$	**40.** XQ_1
17. $Q_1\bar{Q}_0$	**41.** $\bar{X}\bar{Q}_1$

The D_{11} and D_{12} outputs are simply fed back as inputs, which are ORed within the PAL to form the expression for D_1. Figure 9.18 shows the

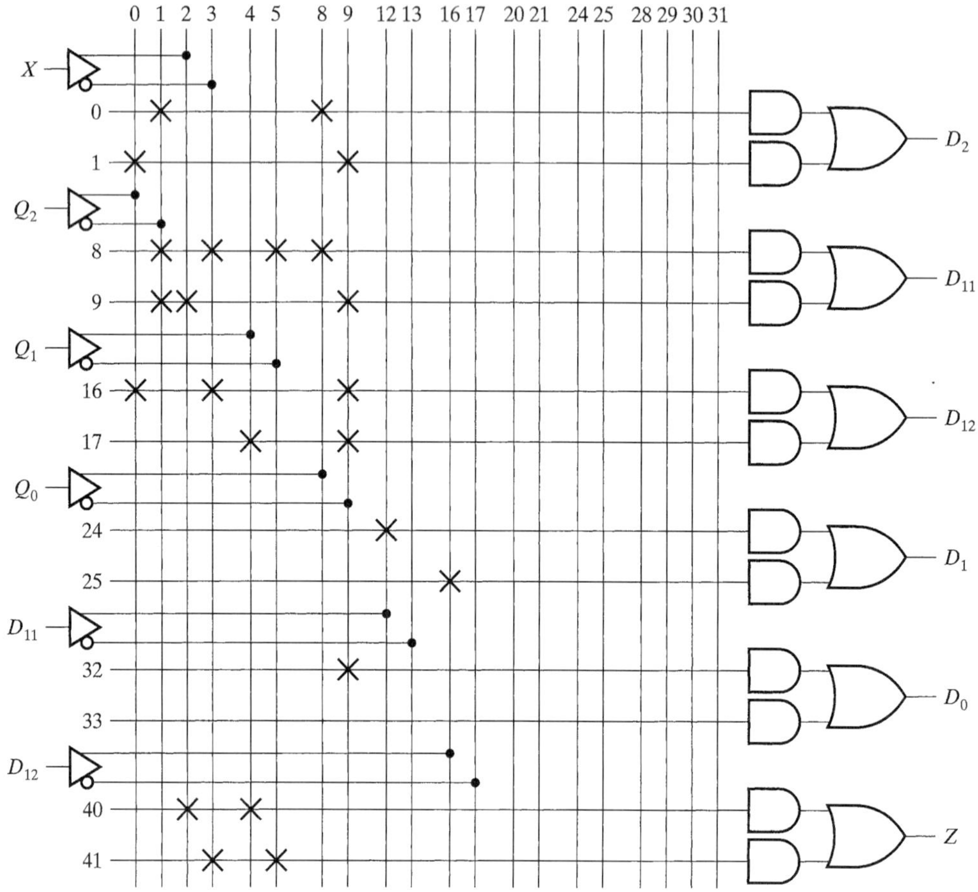

Figure 9.17 **Partial PAL programming map for the code converter.**

pin-out and wiring. The PAL still has three inputs and two outputs that are uncommitted. These are available to implement other logic functions.

Partitioning functions in this way has some performance implications. The propagation delay to compute D_1 is essentially twice that of the other outputs, since it passes through the AND-OR network twice while they pass through only once. Thus, D_1 could easily determine the performance-limiting critical-delay path that sets the minimum clock period possible.

One alternative is to use a more sophisticated PAL that provides interconnect for outputs to be fed back as inputs in the AND array. This is a little faster than the implementation suggested by Figure 9.18, because the signals do not have to exit and re-enter the chip (a transit that is much slower than any connection that is within the same chip package), but still results in a fundamentally multilevel-logic circuit. If this is a serious problem for the performance of the system, a PAL containing OR gates with a sufficient number of product term fan-ins to guarantee a two-level implementation will be needed.

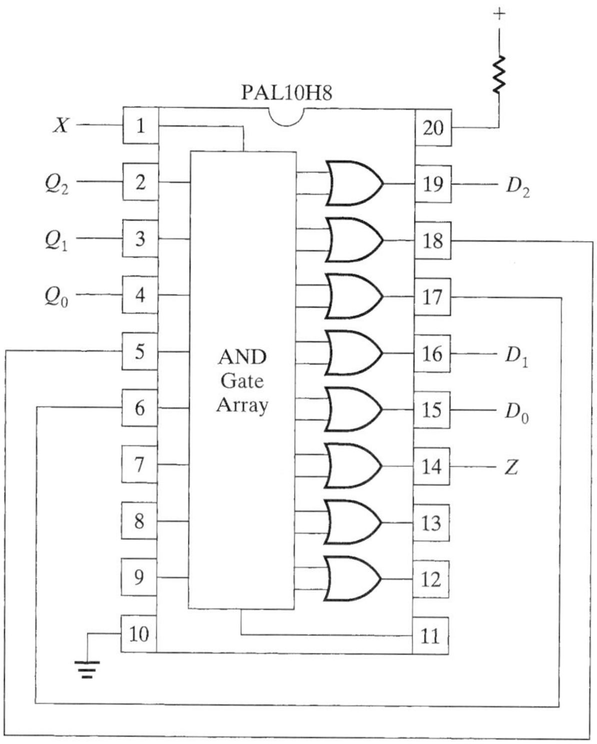

Figure 9.18 **10H8 PAL pin-outs and wiring.**

9.3.3 Alternative PAL Architectures

So far, we have examined PALs in their simplest form: relatively small fan-in structures suitable for implementing simple combinational logic. PAL architectures are actually much richer than this. For example, some members of the PAL family contain on-chip flip-flops: they can generate either positive- or negative-logic outputs, they can allow a pin to be selectively programmed for input or output, or they include XOR gates. Let's examine these alternatives.

Registered PALs A *registered PAL* is a programmable AND-array device with on-chip flip-flops associated with the output pins. These devices particularly are useful for implementing synchronous Mealy machines or Moore machines.

Part of the programming map for a registered PAL is shown in Figure 9.19. Positive edge-triggered *D* flip-flops latch the OR-plane outputs. These are gated to the output pins through tri-state inverting buffers when the $\overline{OE}$ signal is asserted low.

Depending on the details of the PAL architecture, you may supply the $\overline{OE}$ signal by a dedicated input pin or drive it from a product term computed within the array. Similarly, you can provide the clock signal from off-chip or compute it within the PAL.

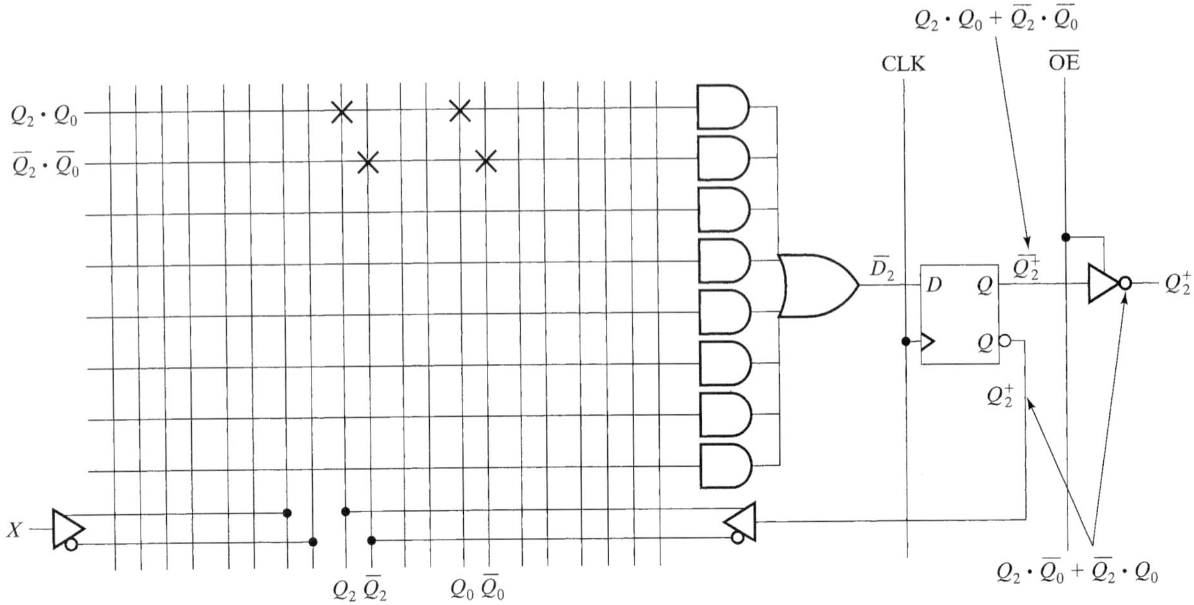

Figure 9.19 Partial registered PAL programming map.

Figure 9.19 shows the output of the flip-flop fed back to the AND array. If this is a next-state function, the signal fed back serves as the current-state input.

The feedback line is the negative-logic output of the flip-flop. This isn't as confusing as it may seem. It is common practice to implement the negative-logic form of the function within the AND array. The inverting tri-state buffer complements this function to get back to the positive-logic form. The feedback buffer drives the function and its complement back into the AND array.

The indispensable CAD software used in programming PALs usually hides the issues of signal polarities from you. Typically, each PAL on the market has a description within a library for the CAD software package that itemized all of its capabilities and idiosyncrasies, including the number of product terms and options for signal polarities. The CAD software implements sophisticated *fitting algorithms* to transform logic so that it can be implemented with the widest variety of devices as possible.

EXAMPLE 9.4 CODE CONVERTER IMPLEMENTED WITH A REGISTERED PAL

The function implemented at the input of the flip-flop in the PAL fragment of Figure 9.19 is the negative logic for the next-state function D_2. CAD tools either handle the translation to negative logic directly; or apply minimization tools that find the optimized form of a function's complement.

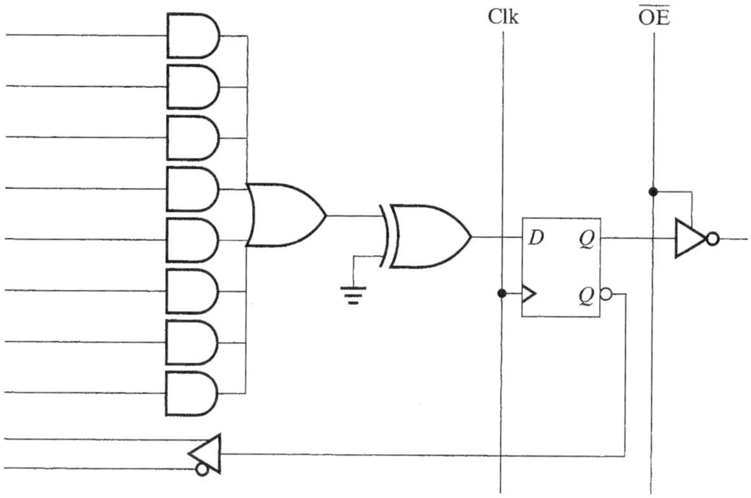

Figure 9.20 **PAL architecture with programmable output polarity.**

In negative logic, the code-converter next-state equations become

$$\bar{D}_2 = Q_2Q_0 + \bar{Q}_2\bar{Q}_0$$
$$\bar{D}_1 = \bar{X}\bar{Q}_2\bar{Q}_1\bar{Q}_0 + XQ_2\bar{Q}_1 + XQ_0 + Q_2Q_0 + Q_1Q_0$$
$$\bar{D}_0 = Q_0$$
$$\bar{Z} = X\bar{Q}_1 + \bar{X}Q_1$$

Since both the positive and negative forms of the current-state bits are available to the AND plane, the detailed programming of the functions is much as before.

PALs with Programmable Outputs More sophisticated PALs allow you to program the output's polarity, giving you the option of positive- or negative-logic outputs. Figure 9.20 shows how this is done in the PAL. An XOR gate is placed between the output of the OR gate and the input to the flip-flop. One of the XOR's inputs can be programmed with a connection to ground. When this fused connection is blown, the input is treated by the circuity as though it were high, and the XOR behaves like an inverter. It passes the inverted signal to the flip-flop, which inverts the signal one more time on the way to the output pin and the feedback path. When the connection is left intact, the input is 0 and the XOR behaves like a non-inverting buffer.

Programmable polarity can also help you overcome the limited product-term inputs to the OR gate. The complement of a function in sum of products form may use fewer terms than the function itself. You can implement the function in its complemented form, using the XOR gate to return the function to its true sense.

Additional Variations on PAL Architectures PAL architectures continue to evolve, giving the basic PAL structure ever more function and flexibility. We mention some of the newer variations here.

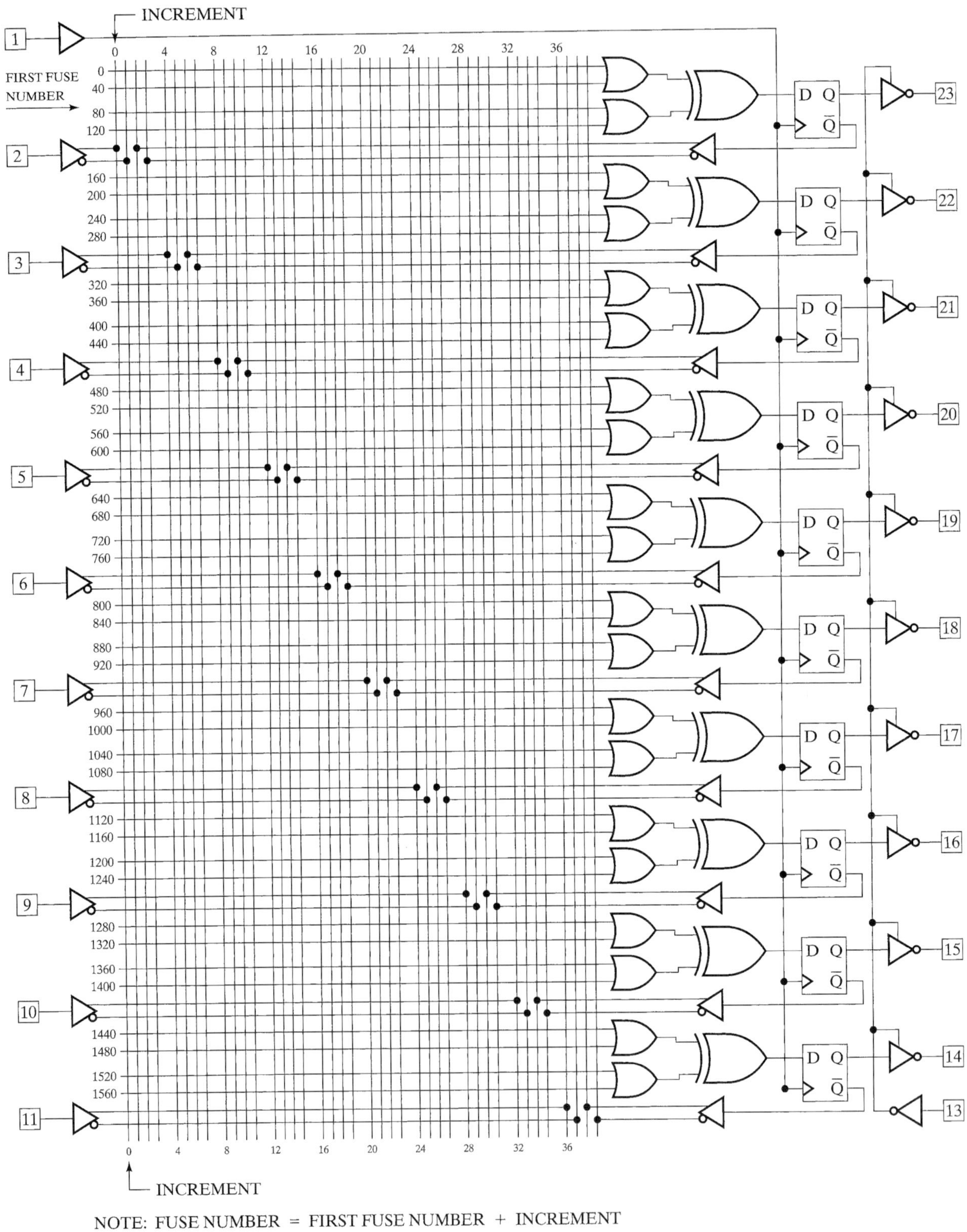

Figure 9.21 **P20X10 PAL.**

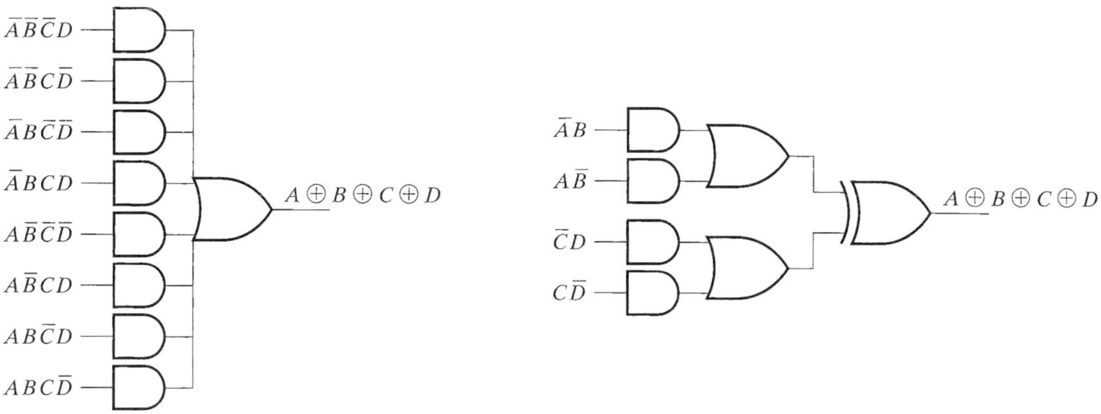

Figure 9.22 Alternative PAL-based implementations of a 4-variable XOR function.

XOR PALs (as distinct from PALs with programmable polarity) contain internal XOR gates whose inputs are fed from AND-OR array structures. They are well suited for computing certain arithmetic functions that would otherwise generate many product terms.

An example is the P20X10 PAL of Figure 9.21. It also illustrates some of the complexities of the detailed structure of a typical PAL. The 20X10 contains 10 data-input pins and 10 output pins with internal flip-flops. The outputs are fed back into the AND plane for a total of 20 possible AND-array inputs. Pin 1 is a dedicated input that drives the flip-flop clocks. Pin 13 controls the tri-state enables of the inverting output buffers of the registered outputs. Two-input XOR gates at the inputs to the flip-flops are driven by OR gates with two product terms each.

As an example, let's look at alternative PAL-based implementations of the expression $A \oplus B \oplus C \oplus D$. Using a conventional AND-OR PAL, we need eight four-literal product terms. But we could also implement this in only four terms in an AND-OR-XOR structure. The two alternatives are shown in Figure 9.22.

Typical registered PALs have several limitations on their inputs and outputs. For example, an output pin is dedicated to every flip-flop output even if it is used to hold state information that is never provided off-chip. One solution is to have *buried registers,* with the output pins decoupled from the registers. The flip-flop can feed back a signal to the AND array, while the output pin carries a different signal. Or the outputs may be multiplexed, allowing the output pin to be connected to the combinational output from the array or the registered output from the flip-flop. An example of this more sophisticated and quite common PAL structure is the P22V10 (to be shown in Figure 9.28). We will discuss it in Section 9.4.

EXAMPLE 9.5 VENDING MACHINE IMPLEMENTED WITH A REGISTERED PAL

Let's see how we can realize Moore and Mealy machines using programmable logic devices. A good example is the vending machine

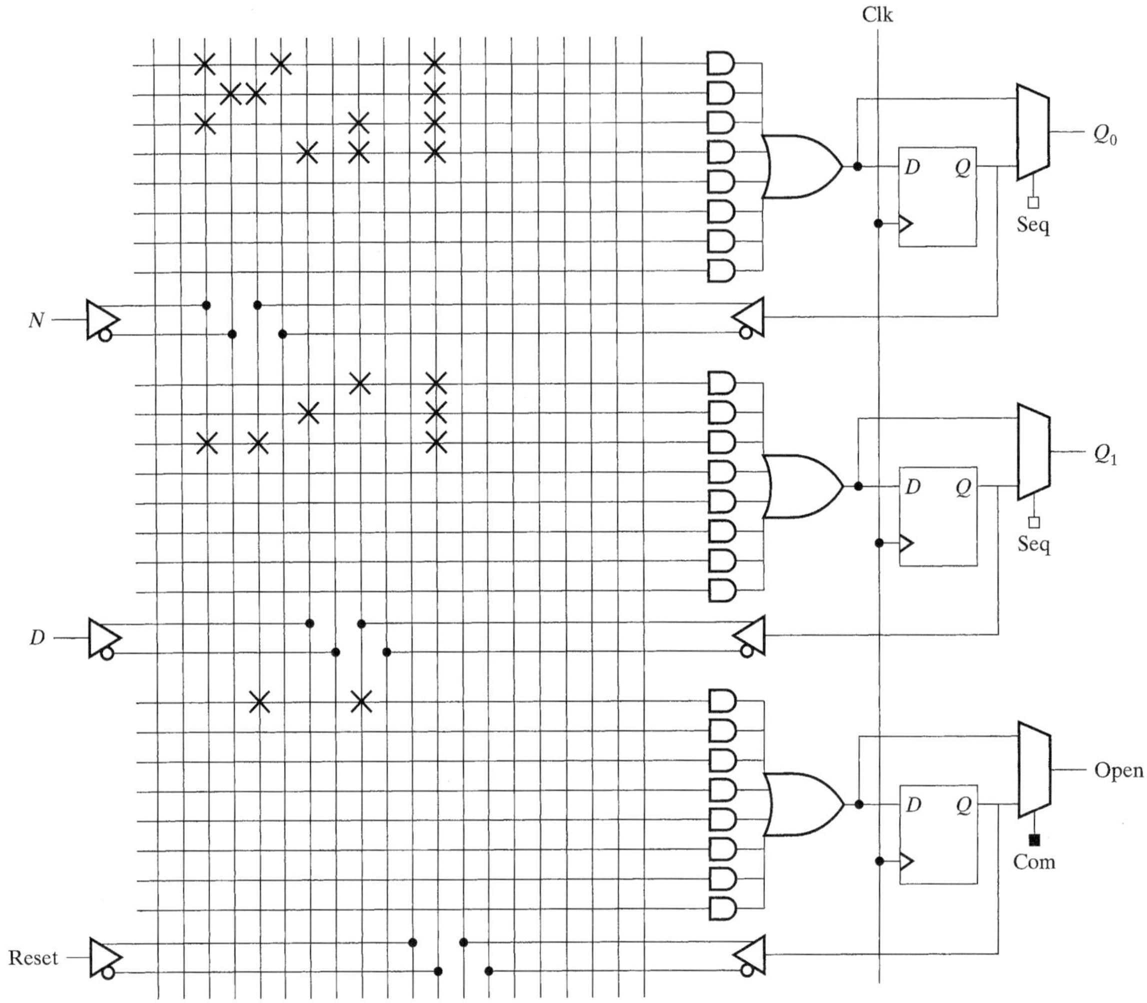

Figure 9.23 **Moore machine mapping of vending machine to a PLD.**

logic of Figures 7.47 and 7.48. The Moore machine of Figure 7.47 can be implemented readily in a 3-output PLD with at least four product terms per function. Figure 9.23 shows such a realization. In this PLD, the outputs can come either from the flip-flop output or the combinational logic block output. The small box near each flip-flop is used to indicate the status of the programming for its output multiplexer (set either to a combinational or a sequential output). Note the way the state bits are fed back into the AND plane so that they can be used as input-to-logic functions. We implement OPEN as a function of the two fed-back state bits and set its output multiplexer to output the combinational-logic function and not use the flip-flop.

Figure 9.24 shows a realization of the synchronous Mealy machine implementation of the same state machine. In this case, the largest function has five product terms. As expected for a synchronous Mealy machine, all of the outputs are from flip-flops.

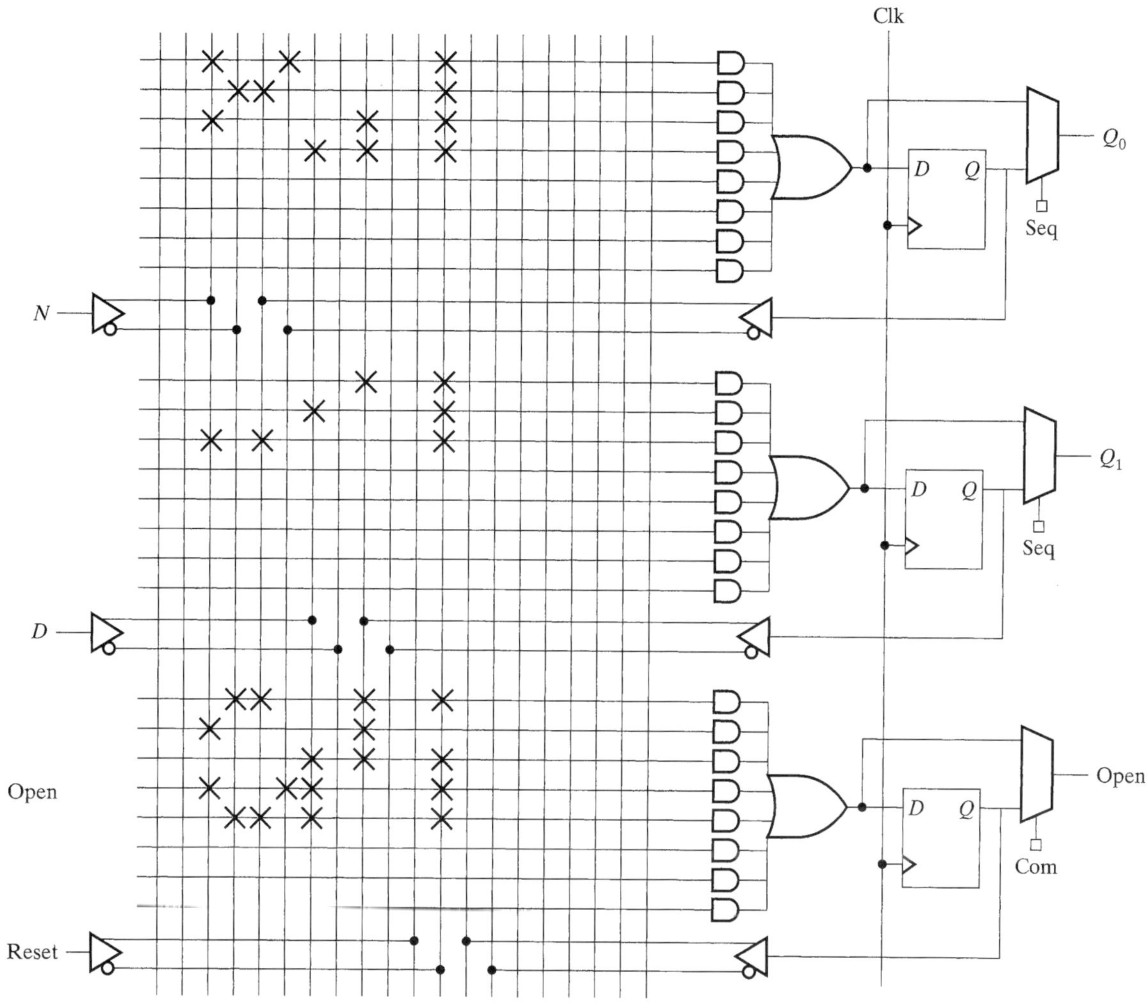

Figure 9.24 **Synchronous Mealy machine mapping of vending machine to a PLD.**

9.4 FSM Design with More Sophisticated Programmable Logic Devices

The PAL concept was pioneered in the 1970s by Monolithic Memories (which has since merged with Advanced Micro Devices, also known as AMD). It was based on bipolar fuse technology developed for programmable ROMs (all connections are available initially; *blow* the connections you do not want). The primary goal of PAL-based designs was to reduce parts count by replacing conventional TTL logic with more highly integrated programmable logic. Designers frequently report that four TTL packages (10 to 100 "gate equivalents") can be replaced by a single small PAL.

PAL-based designs also have the advantage of reduced parts inventory, since a PAL is a "universal" device. You don't need a large stock of miscellaneous fixed-logic components. In addition, PALs support rapid prototyping, because they reduce the number of component-to-component interconnections. Designers can implement bug fixes and new functions within the PALs, often without making changes at the printed circuit board level.

9.4.1 PLDs: Programmable Logic Devices

A number of companies have extended the PAL concept by changing the underlying technology, as well as the component's array of gates and interconnections. Generically, these components are called *programmable logic devices* (PLDs), with the more sophisticated devices called *field-programmable gate arrays* (FPGAs). We will examine three representative PLD architectures in this section: Altera MAX, Actel programmable gate array, and Xilinx logic cell array.

9.4.2 Altera Erasable Programmable Logic Devices

Except for very high speed programmable logic, the general trend has been toward CMOS implementation, with its much higher levels of circuit integration and lower power demands than bipolar technologies. PALs initially were based on the same "program once" technology as bipolar PROMs. Altera pioneered the development of erasable programmable logic devices (EPLDs) based on CMOS erasable ROM technology. The EPLD can be erased simply by exposing it to ultraviolet (UV) light and then reprogrammed at a later time. Today, there are also electrically erasable PLDs (EEPLDs) that can have their programming cleared for reuse using an input wire rather than requiring exposure to UV light. This allows them to be packaged more compactly and cheaply, and they do not require additional equipment for programming. In fact, they can even be programmed after being soldered to a printed circuit board (in-circuit programming). Altera MAX7000 Series EEPLDs are equivalent to 600 to 10000 conventional two-input gates, depending on the model selected.

EEPLD Macrocell Architecture The basic element of the EEPLD is the macrocell, containing an eight-product-term AND-OR array and several programmable multiplexers. Multiplexers are particularly easy to implement in MOS technology, so it is no surprise that they are pervasive in CMOS-based programmable logic.

Figure 9.25 gives a block diagram/schematic view of the macrocell's contents. Its elements include a programmable AND array, a multiple fan-in OR gate to compute the logic function with programmable output polarity (via the XOR gate), a tri-state buffer driving an I/O pin, a programmable sequential-logic block, and a programmable feedback section. Depending on the component, an EEPLD may contain from 32 to 512 such macrocells, each of which can be programmed independently.

Let's look at each of the programmable elements of the macrocell. As you will see, it offers more flexibility than any of the PAL architectures we have seen so far.

The macrocell's AND array is crossed by the true and complement of the EEPLD's dedicated input and clock pin signals and the internal

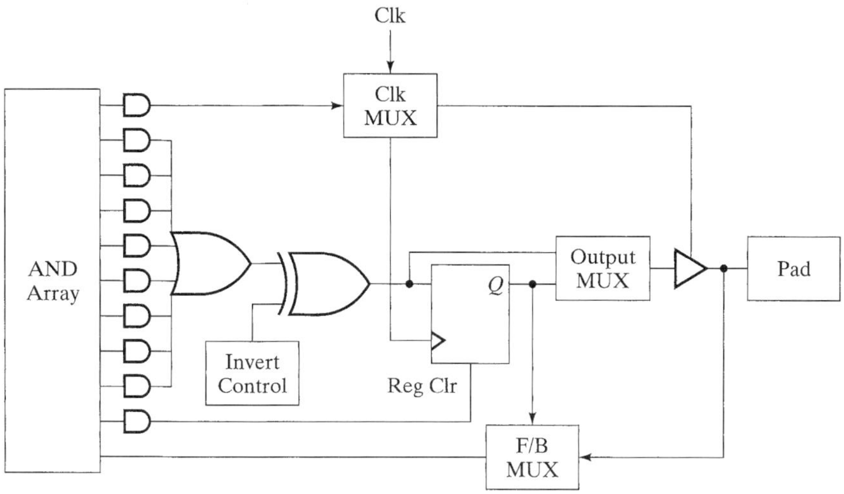

Figure 9.25 Altera macrocell schematic.

feedbacks from each of the component's outputs. Crosspoints can be implemented by EPROM connections that are connected initially. Unwanted connections are broken as needed by a high current during *programming*. In electrically erasable devices, these connections are controlled via a switch that is in turn controlled by an internal-memory bit that can be re-programmed as desired.

The multiplexers allow the feedback, output, and clock sections to be programmed independently. The MUX selection lines are controlled by their own EPROM bits. Under MUX control, the combinational function can bypass the flip-flop on the way to the output. Thus, you can program any output to be either combinational or registered.

Similarly, macrocell feedback into the AND arrays can come from the registered output or from the external pin. You can program many of the pins to be either output or input.

Some variations on the macrocell architecture support dual feedback, making it possible to use the register for internal state while the pin is used as an independent input. This is an application of the concept of buried registers mentioned previously.

The programmable clock section allows you to (1) clock the registers synchronously in groups by a dedicated clock input or (2) clock them by a local signal within the macrocell. Since the latter is a product term, the register's clock signal can be any combination of inputs or external clocks.

The two possible configurations of the clock multiplexer are shown in Figure 9.26. Depending on the value of the bit programmed within the EPROM cell, the flip-flop is controlled with the global clock while the AND array's product term selectively enables the output. This is called *synchronous mode* because the output register is clocked by a global clock signal, shared among all macrocells. This signal can cause all outputs to change at the same time.

Alternatively, the clock multiplexer can be configured so a local clock, computed from a distinguished AND array product term, controls

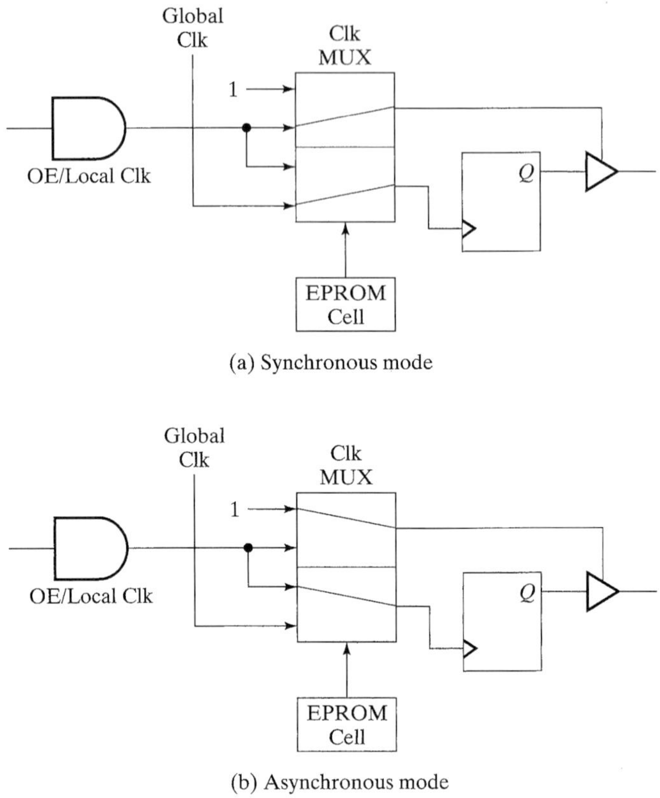

Figure 9.26 Clock MUX modes.

the output register. In this mode, the output is always enabled, driving the output pin. Since every macrocell can generate its own local clock, the output can change at any time. This is called *asynchronous mode.*

The final programmable element of the macrocell is the register clear signal. One of the AND array's product terms is dedicated to provide this function.

Altera MAX Architecture The major problem with all AND-OR structures is the difficulty of sharing product terms among macrocells. In a conventional PAL, you cannot share the same product term across different OR gates. The term must be repeated for each output. This can lower the efficiency of the PAL, reducing the number of equivalent discrete gates it can replace.

As programmable logic devices become even more highly integrated, the architectures must evolve to provide more area for global routing of signals. It must be possible to share terms and outputs between macrocells more easily. Altera has addressed these problems in their *multiple array matrix* (MAX) family of parts. We describe the structure of MAX components in this subsection.

Macrocells, similar in structure to Figure 9.25, are grouped into *Logic Array Blocks* (LABs). Associated with each LAB is a group of

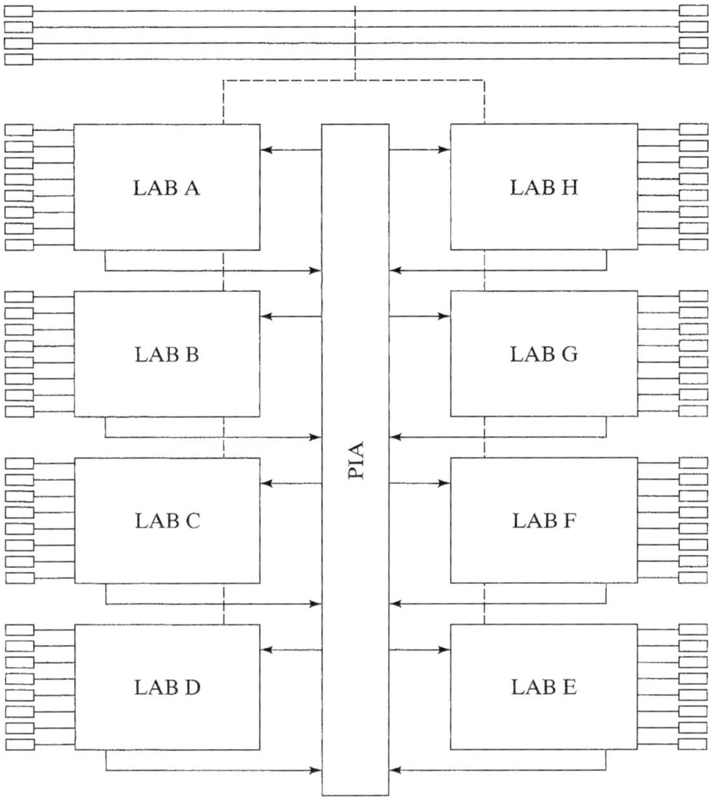

Figure 9.27 Altera EPLD block diagram.

additional product terms, usable by any of the macrocells within the LAB. These *Expander Product Terms* make it possible to implement a function with up to 35 product terms inside a single macrocell. This compares to only eight terms per function in most PAL families.

In addition, a *Programmable Interconnect Array* (PIA) can route the LAB's macrocell outputs globally throughout the device. Some lower-density devices also use the PIA to route the product term expanders.

Figure 9.27 shows the generic architecture of a MAX component. This simplified device has eight dedicated inputs (including the clock), 64 programmable I/O pins, eight LABs, 16 macrocells per LAB (128 macrocells total—not all macrocells are connected to an output pin), and 32 product term expanders per LAB (256 total). The dedicated input pins come in along the top and are distributed to each of the eight LABs. The PIA routes global signals. All on-chip signals have a connection path to the PIA. Only the signals needed by a particular LAB are connected to it under EPROM programming.

Altera claims that a single 192 macrocell device organized into 12 LABs can replace up to 100 SSI and MSI components or 20 P22V10 PALs (for the implementation power of this kind of PAL, still one of the most commonly used, see Figure 9.28). Altera's MAX7000 series

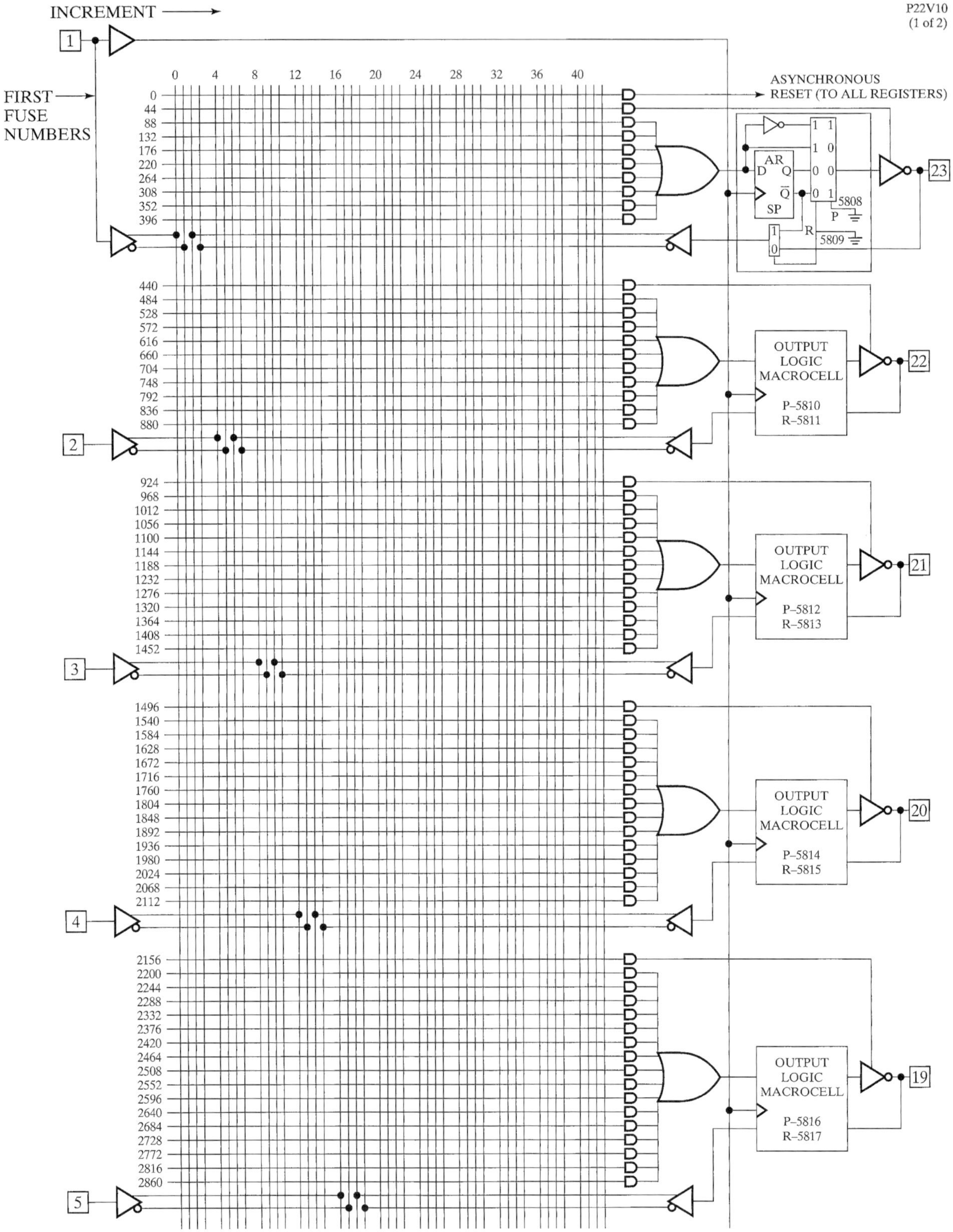

Figure 9.28 P22V10 PAL.

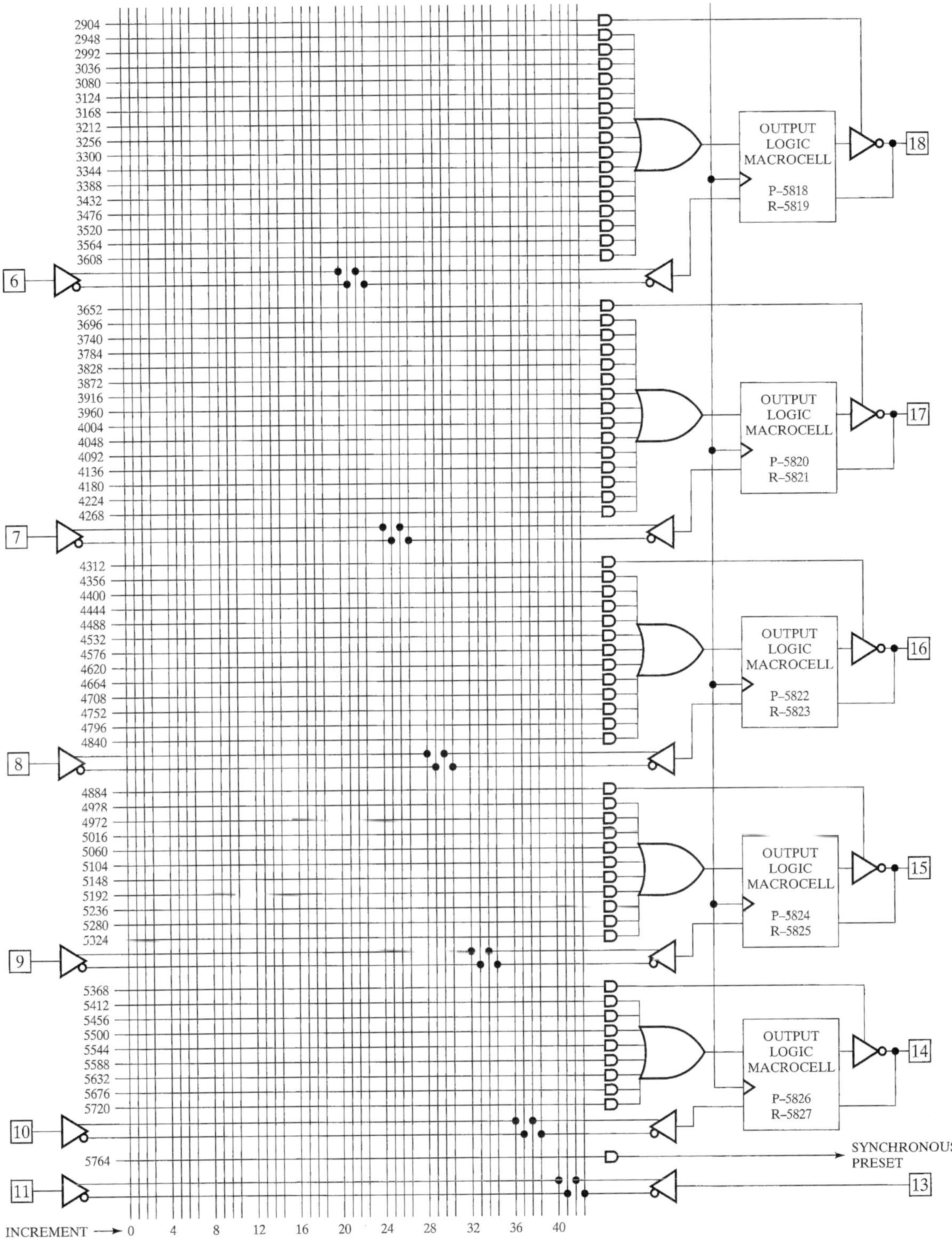

NOTES:
1. INSIDE EACH MACROCELL, THE "P" FUSE NUMBER IS THE POLARITY FUSE, AND THE "R" FUSE IS THE REGISTER FUSE
2. FUSE NUMBER = FIRST FUSE NUMBER + INCREMENT

Figure 9.28 ***(Continued)***

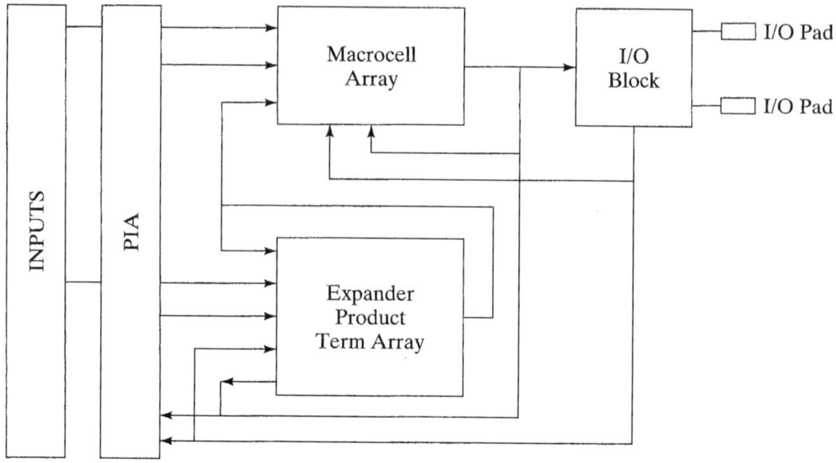

Figure 9.29 **Internal architecture of the LAB.**

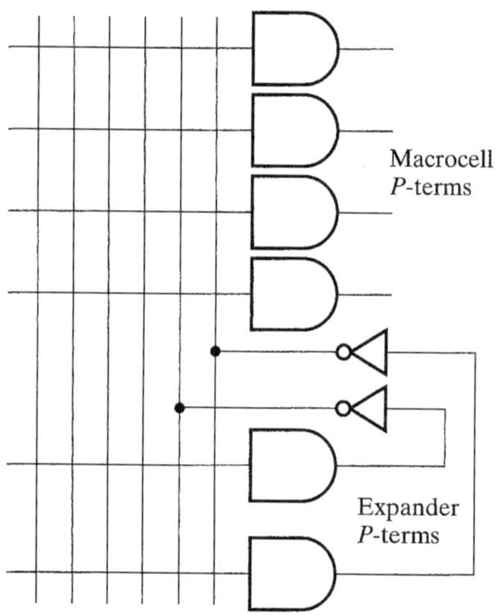

Figure 9.30 **Expander product terms internal organization.**

of parts organizes the LABs into multiple rows and columns around the programmable interconnect and can reach up to 10,000 gate equivalents.

Figure 9.29 shows more details of the LAB's internal organization. It consists of an array of macrocells sharing a product-term expander array. All macrocell outputs and I/O block inputs are connected to the PIA. Selected signals from the PIA are input to the macrocells.

Figure 9.30 gives more details of the implementation of the expander product terms. The AND array is crossed by the dedicated inputs and the feedback signals from the macrocells. The expander terms also form some of the columns of the array. In other words, they appear like inputs to the macrocells. Any expander term can be shared by all of the macrocells in the LAB.

If you use expander terms you quickly get into multilevel logic structures. Optimization techniques such as those in Chapter 3 are absolutely necessary to do a good job of mapping logic onto these structures.

Once you reach devices as complex as the advanced MAX family, you would be unlikely to try to generate the personality map by hand. Altera provides an extensive tool set for mapping logic schematics onto the primitives supported by their PLD structures.

9.4.3 Actel Field-Programmable Gate Arrays

Actel programmable logic chips provide what amounts to a field-programmable gate array structure. The chip contains rows of personalizable-logic building blocks separated by horizontal routing channels. The programming method is a proprietary "anti-fuse" technology, so called because the connector's resistance changes from high to low when a high voltage is placed across it. This is the opposite of a conventional PALs or PROMs that are based on fuse technology. The anti-fuses require a very small area, so the chip can have more connections than with other technologies. Unlike the EPLDs, Actel anti-fuse parts can be programmed only once.

The elements of the Actel architecture are I/O buffers, logic modules, and interconnect. Figure 9.31 shows a chip *floorplan* or block diagram. Programmable I/O buffers, and special programming and test logic are along the chip's edges. The I/O pins can be configured as input, output, tri-state, or bidirectional buffers, with or without internal latches.

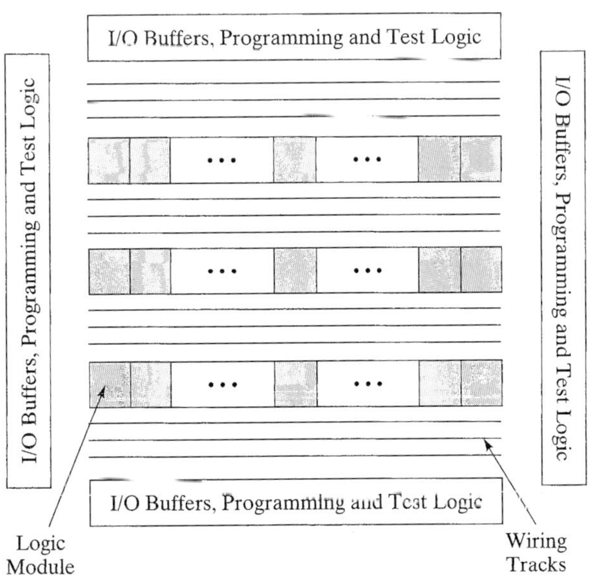

Figure 9.31 Chip floorplan of Actel FPGAs.

Internally, the building blocks are organized into multiple rows of *logic modules* separated by wiring tracks. Each logic module consists of two basic-logic cells: one combinational and one sequential. There are two identical combinational-logic cells per logic module. Each can be used to implement a ten-input, one-output configurable combinational-logic function (the internal structure is described in the next subsection). You can program the module to implement a large number of logic functions of all the inputs or subsets. There is also one sequential-logic cell per logic module. It consists of one D flip-flop with reset and set inputs. In addition to the logic cells, the logic module of the Actel Axcelerator FPGA family also has resources dedicated to routing signals including switches (to connect available wires) and repeaters (to boost signal strength for signals that have to travel a further distance across the chip). A further optimization includes an additional input to the basic logic module that can be used to implement fast carry chains for arithmetic logic.

Horizontal wiring tracks provide the main interconnection. Although the tracks run across the length of the chip, a given wire can be partitioned into segments so it can make several interconnections instead of only one. In addition, vertical wires pass through the logic modules and span multiple wiring channels. Inputs come from the tracks both above and below the logic module.

Other resources on-chip include large memory arrays that can be used to incorporate data storage directly on-chip. These provide a total of 16 K to 256 K bits of available storage across the FPGA family. Further storage is available at I/O pins in the form of first-in/first-out storage queues.

The Axcelerator component family is organized around a basic "tile" consisting of 336 logic modules and some memory. These are then arranged into larger hierarchical tiles ranging from a single tile to a 16-tile arrangement with the current largest member of the family. Capacity in equivalent gates ranges from 125 K to 2 M gates. Although in practice, designers are able to effectively utilize about half.

Actel Axcelerator Combinational Logic Cell The *C*-cell is shown in Figure 9.32. It is very similar at its core to the cell of Figure 4.41 but includes several additional features, including fast carry-chain logic and additional inputs. A carry-chain stretching through adjacent cells is connected directly to make arithmetic circuits faster by not requiring them to use programmable interconnect on their critical path. The core of the module is still a 4:1 multiplexer as in the early Actel ACT family of programmable gate arrays. See Chapter 4 for details on how to use these logic modules.

Actel Axcelerator Sequential Logic Cell The *R*-cell is shown in Figure 9.33. At its core is a D flip-flop. Surrounding the flip-flop are capabilities for altering the clocking signal and selecting an input. The input select logic includes the capability to simply feed back the current value of the flip-flop so that it not necessary to route the current state through combinational logic in implementing a simple hold function. To make

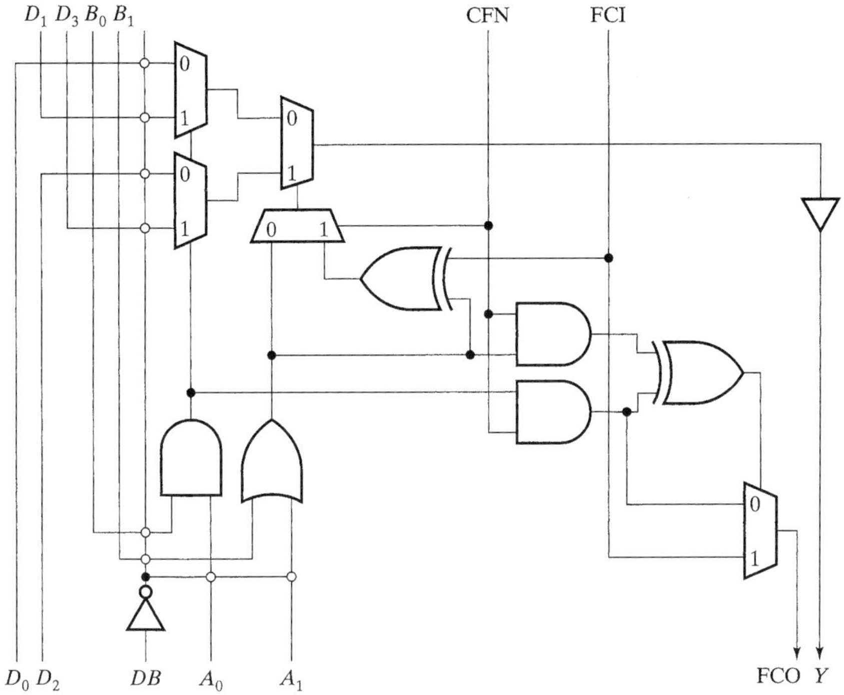

Figure 9.32 **Actel Axcelerator *C*-cell combinational logic cell.**

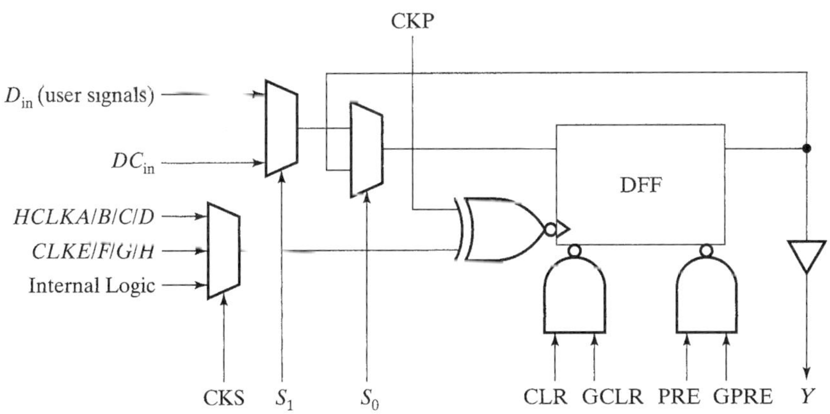

Figure 9.33 **Actel Axcelerator *R*-cell sequential logic cell.**

sequential logic circuits faster, there is a direct connection from one of the *C*-cell outputs of a logic module to an *R*-cell input. This eliminates the need for the signal to use the programmable interconnect to get to the flip-flop.

Actel Axcelerator Interconnection Fabric The interconnection "fabric" surrounds the logic modules. Special receivers and transmitters are part of the logic module and enable signals to pass in and out of the routing structures. Connections are made between wires using anti-fuses. To increase wire utilization, it is possible to "cut" long wires into segments so that they can be used for different signals. However, this has a cost in that each potential cut point must include an anti-fuse and this makes a signal propagate slower. Therefore, there are also longer range wires that can not be segmented but offer faster transmission speeds.

The logic modules must be placed carefully and then wired by routing interconnections through the network of anti-fuses. Because of the resistance and capacitance associated with crossing an anti-fuse, speed-critical signals must minimize the number of anti-fuses they pass through. Most connections can be performed in two or three hops. Worst cases may require many more.

9.4.4 Xilinx Field Programmable Gate Arrays

Xilinx takes another approach to bringing the PLD concept to higher levels of integration. Their programming method is based on CMOS static RAM technology: RAM cells sprinkled throughout the chip determine the personality of logic blocks and define the connectivity of signal paths. Static RAM circuits are important in integrated circuit technology and will continue to get denser and faster. ROM-based programmable logic can take advantage of this trend.

The RAM cells are linked into a long shift register, and the programming involves shifting in strings of ones and zeros to personalize the function of the chip. The devices come with an on-board hardwired finite state machine that allows the program to be downloaded from a standard ROM part. The Xilinx approach has the advantage of fast reprogrammability, although the chip loses its program each time it is powered down.

Figure 9.34 shows a portion of the chip architecture of the Xilinx logic cell array. The major components are I/O blocks (IOBs) and

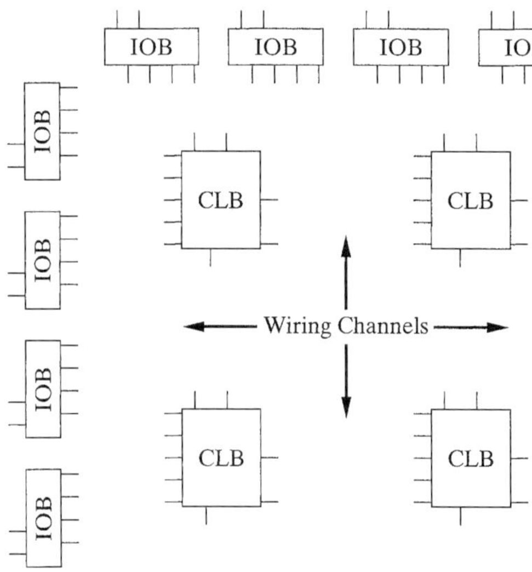

Figure 9.34 Segment of the Xilinx chip architecture.

configurable logic blocks (CLBs). The programmable I/O blocks are placed around the periphery, while the CLBs are arrayed in the central part of the chip. Horizontal and vertical wiring channels separate the various components. We will examine these components in the following subsections.

Xilinx currently supports many component families, the older 4000 series, and the more recent Spartan, and Virtex-II are among them. We will discuss the Virtex-II series, the one most commonly encountered in new designs. The Virtex-II family of FPGAs ranges in size from 88 to over 1,000 I/O pins. The logic arrays vary from 64 to over 10,000 CLBs. Each CLB consists of four slices that can be utilized in several different ways: combinational and sequential logic using look-up tables and flip-flops (our focus), random-access memory, or shift registers for use as buffer storage. In addition, Virtex-II FPGAs include a set of multipliers regularly placed throughout the CLB array to accelerate digital signal-processing applications. The largest member of the family, XC2V8000, contains 11,648 CLBs and 1,108 IOBs, over 90,000 available flip-flops, 3 M-bits of RAM organized in 168 18 K-bit blocks, 168 multipliers, and is claimed to be equivalent to 8 million two-input gates. Quite a formidable amount of computing power.

Xilinx Virtex-II I/O Block Figure 9.35 shows the internal architecture of the I/O block. The basic components of the I/O block are a tri-state driver so that the pin can be bi-directional and a set of registers for each of three signals involved: input, output, and tri-state enable. Each signal has two registers to latch its values with separate clocks. For large pin-out designs, separate clocks help to stagger when signals change to avoid large current spikes if they all changed together. The input flip-flops can be used for synchronization as well as latching. Many details

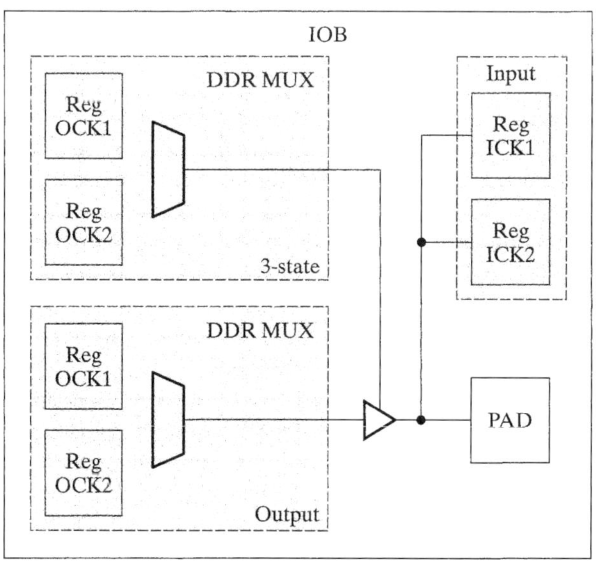

Figure 9.35 Xilinx Virtex-II I/O Block.

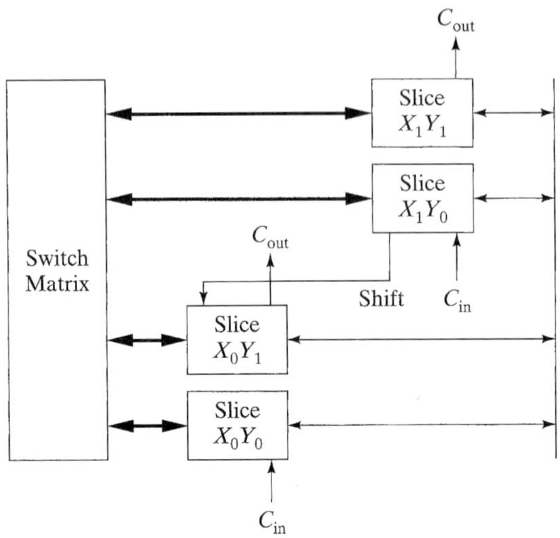

Figure 9.36 Xilinx Virtex-II Configurable Logic Block.

are not shown in the diagram, but each of the signals can be multiplexed from different sources (including the clock signals), all three signals may pass through flip-flops or be connected directly to the tristate driver, and there are global reset signals to all of the flip-flops.

First, let's consider how the block can be used when it is associated with an output pad. The active-high or -low sense of the OUT signal and the output enable can be set by internal options, stored in RAM cells within the block. The output signal can be direct (combinational) or from the dedicated output register (registered). Let's now consider the block with the pad used as an input. The input signal can come from the dedicated input register or directly from the input pad. The input register can be an edge-triggered flip-flop or a transparent latch.

Xilinx Virtex-II Configurable Logic Block (CLB) Figure 9.36 gives the internal view of the CLB. It has four basic slices organized into two groups. Each group is optimized to have a fast and efficient carry-chain. We have seen this optimization for arithmetic circuits in the Actel FPGAs as well. For large FPGAs, arithmetic circuits are quite common features, so it is worth the extra effort to make them run fast and be compact in space. Each CLB also has local interconnect to wire together the logic implemented by each slice and connect to the CLB array. The switch matrix is a large collection of programmable switches.

The internal structure of only one half of one of the four slices is shown in Figure 9.37. Thus, the logic shown in Figure 9.37 represents only one-eighth of the logic in a single CLB! A slice is organized around two main components: a 4-input look-up table (in inputs A_1, A_2, A_3, and A_4) and a flip-flop. The output of the logic can be combinational or sequential. There is a programmable connection to route the output of the look-up table directly to the flip-flop input. There are

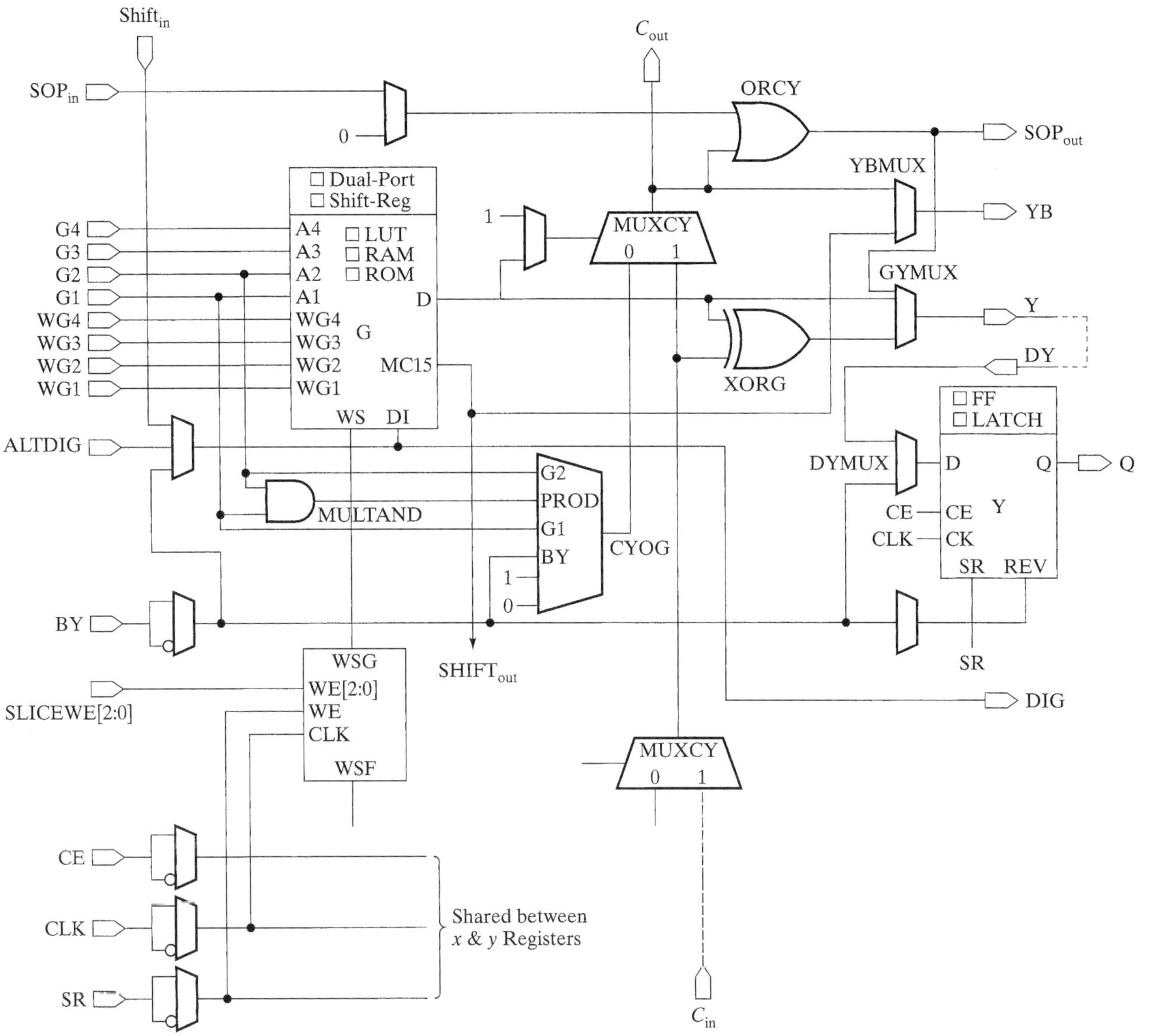

Figure 9.37 **One half of a Xilinx Virtex-II Configurable Logic Block Slice (there are four slices in a CLB).**

many programmable elements within the logic block to select the appropriate signals to pass through each of the multiplexers.

A CLB slice is a very powerful combinational-logic element. Not only can it implement two arbitrary 4-input logic functions, but through the use of the MUXCY and ORCY logic elements it can implement very large sum-of-products functions. Note the SOPIN and SOPOUT signals in the slice. They make direct connections between slices so that multiple terms can be ORed together across slices. Note how the look-up table output controls MUXCY, whose inputs can in turn be set to be the fixed-logic values 0 and 1 or the MUXCY output of another look-up table.

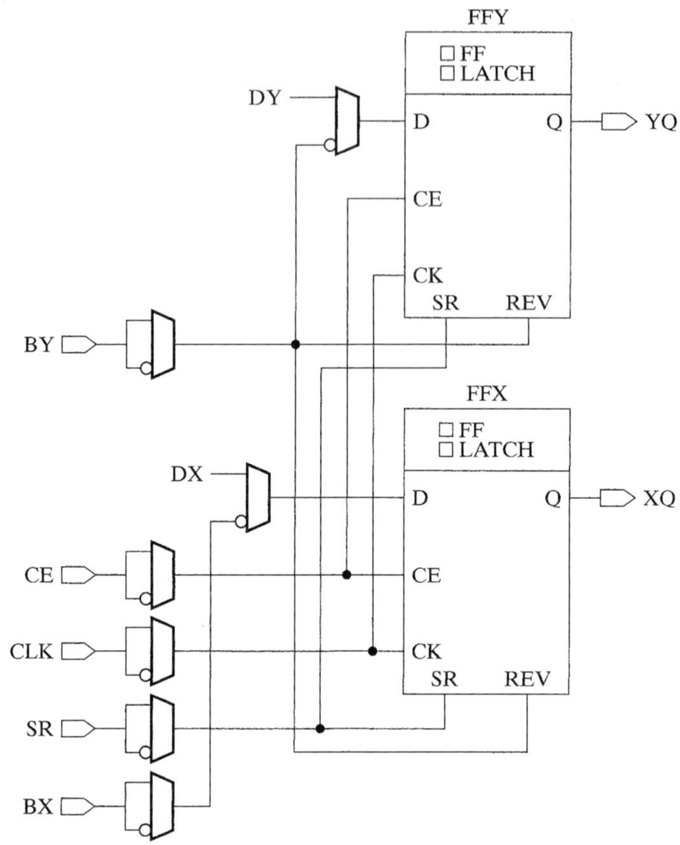

Figure 9.38 **The sequential portion of a Xilinx Virtex-II Configurable Logic Block Slice.**

The sequential portion of a slice is shown in more detail in Figure 9.38. The two positive edge-triggered flip-flops can be configured in a variety of ways. They can act as transparent latches as well as flip-flops. They can have asynchronous or synchronous sets and resets. And, they can be set to initialize to different values at power-up. There is also some programmability around the flip-flops themselves, including clocks and load enables that can be complemented or not.

Slices have multiple uses. Figure 9.39 shows a complete slice and three different uses for its lookup table. Personality RAM bits, the programmable bits that underlie the CLBs, let us configure the function block as: a 4-input function generator, 16 bits of dual-ported random-access memory (with separate address inputs for read (G1 to G4) and write (WG1 to WG4), and a 16-bit variable-tap shift register. Of course, only one of these uses is possible at any one time, but a slice is made to be quite versatile so that it can accommodate a variety of designs.

Through additional multiplexers that interconnect the outputs of the two look-up tables, a single slice can have all of its resources used to generate any function of up to 8-inputs and even some functions of 9-inputs.

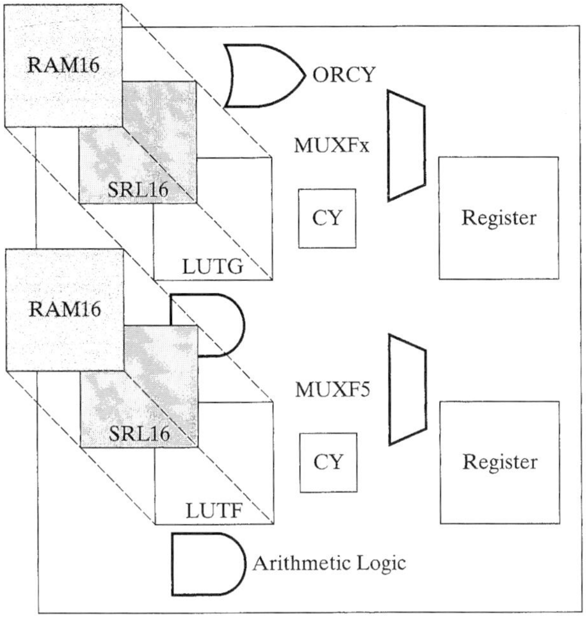

Figure 9.39 Xilinx Virtex-II Configurable Logic Block Slice and the three possible uses of its look-up tables.

The slice can output both the registered and unregistered versions of the function-block outputs. The slice and overall CLB organization is reasonably symmetric, making it possible to easily interchange slices or functions within a slice to reduce the complexity of inter-block routing.

The wealth of flip-flops available in modern FPGAs causes designers to approach optimization of sequential logic quite differently than when they tried to minimize the number of flip-flop packages. One way this manifests itself, is that many finite state machine designs tend not to optimize state assignment for the minimum number of bits. The most common state assignment in many FPGAs is the one-hot assignment. This is true because the complexity of next-state equations in one-hot encodings is related directly to the number of transitions into a state (approximately one term for each arc incident on a state). This tends to lead to functions with a smaller number of inputs that are more easily mapped into lookup tables.

Xilinx Virtex-II Interconnect Fabric The Xilinx chip architecture supports many methods of interconnecting the CLBs and IOBs: (1) direct fast connections within a CLB, (2) direct-connections between adjacent CLBs, (3) double-lines to fanout signals to CLBs one or two away in the array, (4) hex lines to connect to CLBs three or six away, and (5) long lines that span the entire chip. Figure 9.40 provides a schematic of these wiring resources. The switch matrices in each CLB connect internal CLB signals to all these wires.

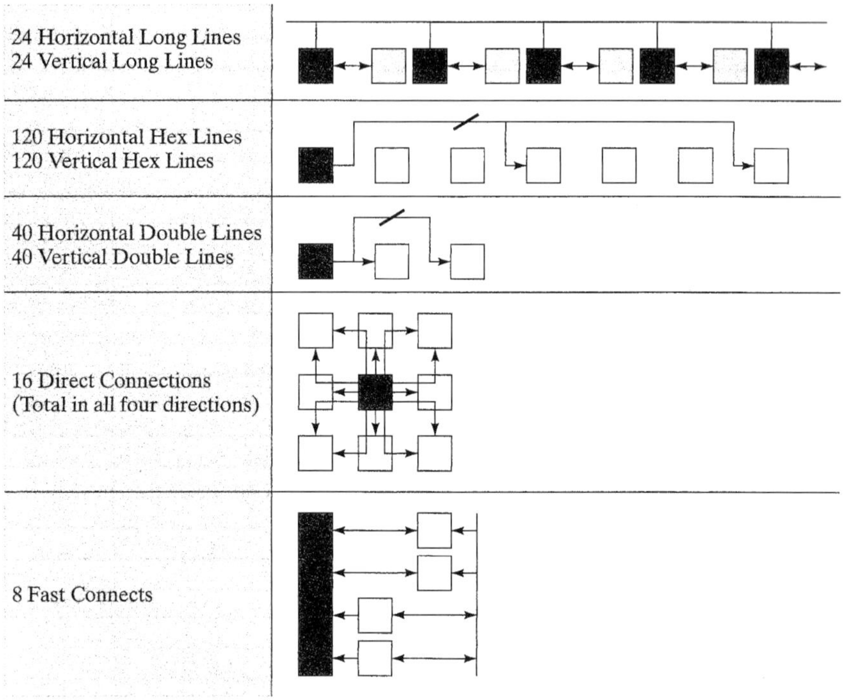

Figure 9.40 **Xilinx Virtex-II wiring resources.**

With direct connections, adjacent CLBs get fast access to neighbors vertically and horizontally. Double and hex lines provide a slightly larger range. The long lines are saved for time-critical signals that must be distributed to many CLBs with minimum signal skew, such as clock signals or common control signals. It is important for the software that assigns logic functions to available CLBs to place related logic within the same or adjacent CLBs.

Interleaved among the checkerboard of CLBs are the horizontal and vertical wiring channels of the Xilinx general-purpose interconnect. Also interleaved every few CLBs is a multiplier unit.

EXAMPLE 9.6 IMPLEMENTING THE BCD-TO-EXCESS-3 CONVERTER WITH A XILINX LOGIC CELL ARRAY

In this subsection, we examine the implementation of the BCD-to-excess-3 converter finite state machine using the LCA structure. The next-state and output equations are:

$$D_2 = \bar{Q}_2 Q_0 + Q_2 \bar{Q}_0$$
$$D_1 = \bar{X}\bar{Q}_2\bar{Q}_1 Q_0 + X\bar{Q}_2\bar{Q}_0 + \bar{X} Q_2 \bar{Q}_0 + Q_1\bar{Q}_0$$
$$D_0 = \bar{Q}_0$$
$$Z = XQ_1 + \bar{X}\bar{Q}_1$$

Suppose we use a synchronous Mealy implementation. Then each of the four functions requires its own flip-flop. Since no function is more complex than four variables, we can implement each one in one-half of a CLB slice, comfortably fitting all four functions within half of a single CLB.

To give you a feeling for the size of functions that can be implemented in a Xilinx chip, the smallest configuration contains 64 CLBs. The example finite state machine uses only 1/128 of the CLB resources of the array.

Xilinx provides software to map a logic schematic into a placed and routed collection of CLBs, so designers rarely have to deal with the array at this level of detail. Still, it is instructive to understand some of the routing details, because they have a critical effect on performance. Xilinx permits you to hand route critical nets to tune circuit performance (or to help the automatic router complete a difficult routing), if desired. As FPGAs get larger and larger, this approach is quickly becoming obsolete and designers are ever more dependent on powerful CAD tools to realize their designs.

9.5 Case Study: Traffic-Light Controller

In this section, we will examine several alternative implementations for the traffic light controller finite state machine using the sequential logic technologies we've learned about in this chapter. We start by decomposing the basic machine into its constituent subsystems. Besides the next-state and output functions, we also need logic for the timing of the lights and for detecting the presence of a car at the intersection.

9.5.1 Problem Decomposition: Traffic-Light State Machine

As we've already seen, there are many possible ways to organize the components of the traffic-light system. Here is the decomposition we will use for this case study:

- Controller finite state machine
  next state/output combinational functions
  state register (n state bits, where n is determined by the state encoding)
- Short time/long time interval counter
- Car sensor
- Output decoders and traffic lights

System Block Diagram A block diagram description for this decomposition is shown in Figure 9.41. The controller finite state machine takes as input the *Reset*, *Clk*, *TL*, and *TS* signals, as well as a synchronized *C* signal, and generates the *ST* signal and encoded light signals (00 = green, 01 = yellow, 10 = red). The interval counter subsystem

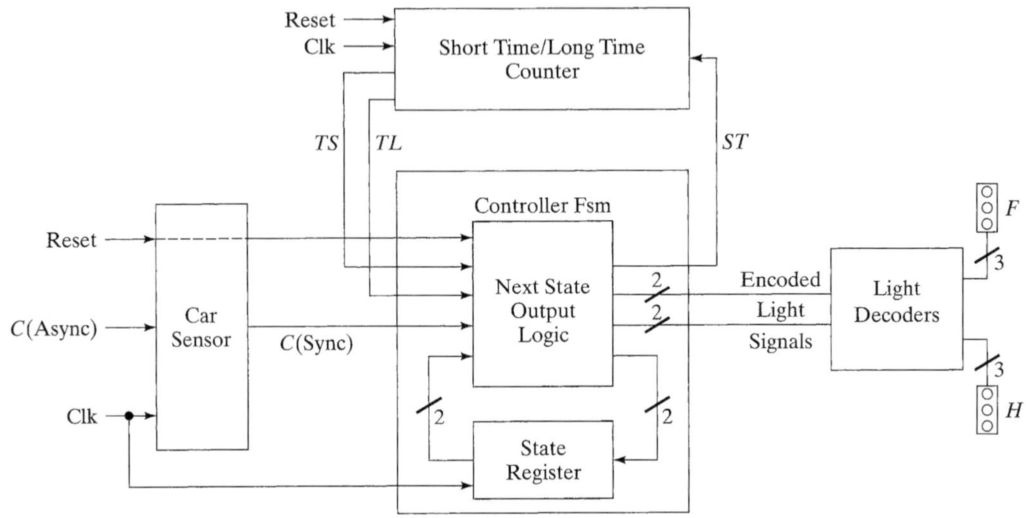

Figure 9.41 Block diagram of complete traffic light system.

takes *Clk*, *Reset*, and *ST* as inputs, generating *TL* and *TS* as outputs. The car sensor subsystem has an asynchronous sensor input, *C*, which it outputs as a synchronized signal. The light decoders translate the encoded light-control signals into signals to drive the individual lights. All of the inputs—*C* (sync), *TS*, and *TL*—change with the clock. To be thorough, *Reset* should also be synchronized.

It is reasonable to generate outputs that directly control the lights, rather than the encoded scheme we have chosen. Since the actual traffic lights are probably relatively far from the traffic light controller hardware, the encoded scheme has the advantage of fewer wires that need to be routed that distance. But the approach requires additional logic to do the decoding near the lights.

Next-State Logic and Outputs The finite state machine has six inputs: *Reset*, *C*, *TL*, *TS*, and the 2 bits of the current state (Q_1 and Q_0); and seven outputs: the 2 bits of the next state (P_1 and P_0), *ST*, $H_{1,0}$ (encoded highway lights), and $F_{1,0}$ (encoded farm road lights).

We'll use the symbolic state table of Figure 8.24 and the random state encoding of Section 8.2.2 ($HG = 00$, $HY = 10$, $FG = 01$, and $FY = 11$). We can fill in these codes for the states into the state table and derive an encoded truth table that we can use to generate discrete gate, ROM, PAL, or PLA-style logic for the next state and other outputs.

Car Detector The car detector logic is a simple debounced switch (see Appendix B for details). A two-position switch embedded in the road determines whether a car is present. This signal should be stable during the transition from one setting to the other, and this calls for a debouncing circuit.

Since a car can arrive at any time, the car detector is asynchronous with respect to the rest of the traffic light system. To synchronize the

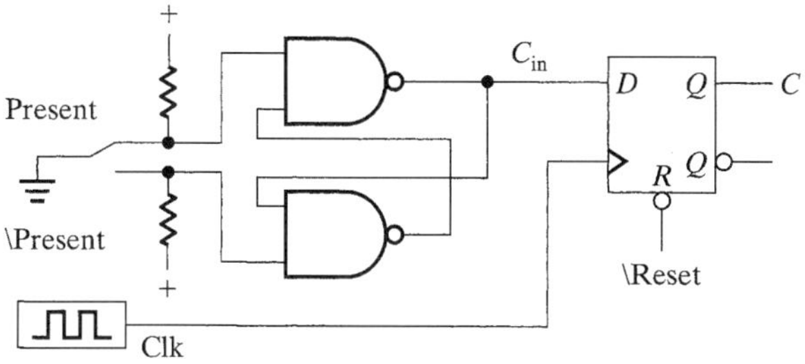

Figure 9.42 Car detector circuit.

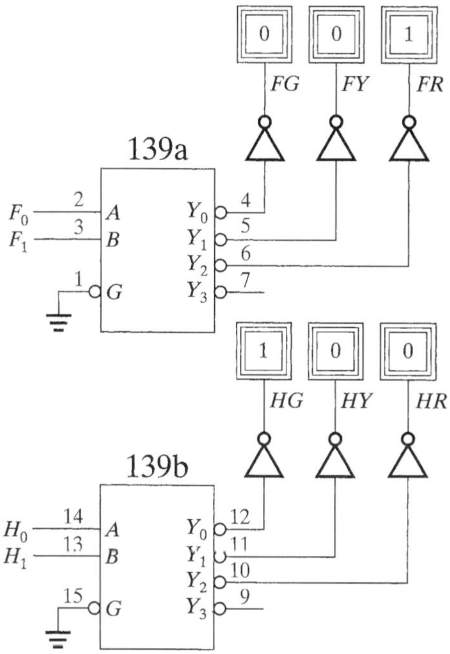

Figure 9.43 Light decoder circuitry.

car sense signal, we must pass it through a synchronizer flip-flop, clocked by the system clock. The circuit is given in Figure 9.42.

Light Decoders The light decoder circuitry is reasonably straightforward. We can use 2:4 decoders, such as the TTL 139. Figure 9.43 contains the necessary logic.

Interval Timer The last major component is the interval timer, designed to generate the signals *TL* and *TS* after being set by *ST*. We could implement this in many ways, perhaps the simplest being to use a counter and external decode logic. The counter is cleared when *ST* is asserted, and *TL* and *TS* are asserted by the external logic when the counter counts up to the appropriate threshold value. For this discussion,

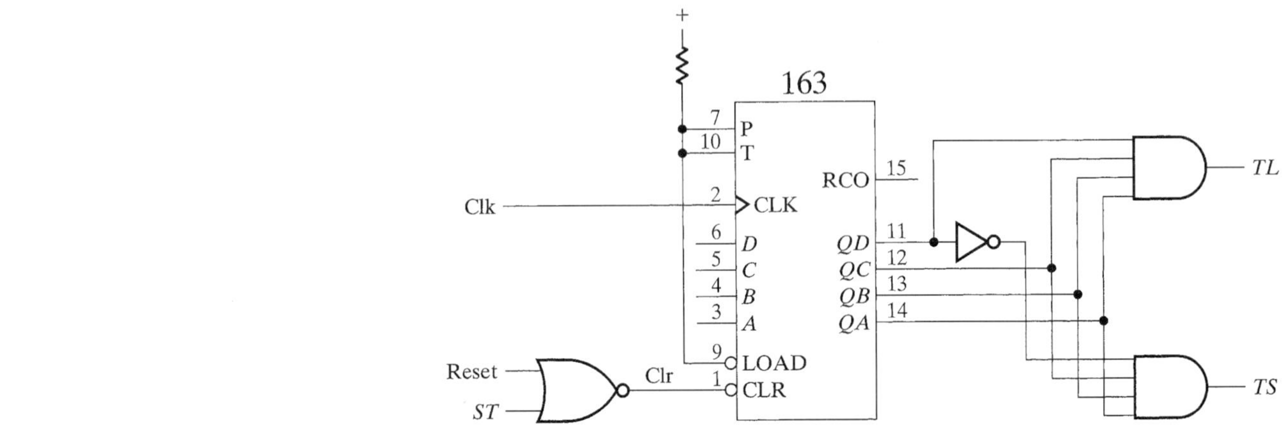

Figure 9.44 **Simple interval timing.**

we will assume that *TS* is asserted when the 4-bit counter reaches 0111_2 and *TL* is asserted when it reaches 1111_2. Wider counters can be used for more realistic interval timings.

Figure 9.44 shows how the logic could be implemented using a 163 synchronous up-counter. In the figure, the OR of *ST* and *Reset* is complemented to reset the counter. When either *ST* or *RESET* is asserted, the $\overline{CLR}$ input is asserted, and the counter is set to zero. This strictly is not necessary: whatever state the counter comes up in when powered on, it will cycle eventually through the states that cause *TS* and *TL* to be asserted.

9.5.2 PLA/PAL/ROM-Based Implementation

In this subsection, we will use the best 2-bit state encoding we've found so far: $HG = 00$, $HY = 10$, $FG = 01$, and $FY = 11$. This yields an implementation for the next-state and output functions that requires eight unique product terms:

$$
\begin{aligned}
P_1 &= CTLQ_1' + TS'Q_1Q_0' + C'Q_1'Q_0 + TS'Q_1Q_0 \\
P_0 &= TSQ_1Q_0' + Q_1'Q_0 + TS'Q_1Q_0 \\
ST &= CTLQ_1' + C'Q_1'Q_0 + TSQ_1Q_0' + TSQ_1Q_0 \\
H_1 &= TSQ_1Q_0 + Q_1'Q_0 + TS'Q_1Q_0 \\
H_0 &= TS'Q_1Q_0' + TSQ_1Q_0' \\
F_1 &= Q_0' \\
F_0 &= TS'Q_1Q_0 + TSQ_1Q_0
\end{aligned}
$$

These equations are the same as those of Section 8.2.2, except they've been optimized to share as many product terms as possible. Any PLA component with five inputs, seven outputs, and eight product terms could implement these functions.

PLA/PAL Implementation Because no function is more complex than four product terms, they can be implemented easily by many of the available sequential PALs. For example, we could use the P22V10 PAL (see Figure 9.28). This device has 11 dedicated inputs and 10 programmable input/outputs. When the latter are programmed as outputs, they can be either registered or combinational. The OR array varies from 8 to 14 product-term inputs, sufficient to implement any of the functions above. The embedded registers can be reset through a dedicated reset line that is routed to each output register, so it is not necessary to include a Reset input signal in the equations.

ROM Implementation A ROM-based implementation requires a complete tabulation of the state transition table. With five inputs and seven outputs, this is a 32-word by 8-bit ROM. If *Reset* is to be handled by the next-state logic directly, this should be included as one of the inputs to the ROM, thus doubling its size.

9.5.3 Counter-Based Implementation

Although the two-level implementation just described is appropriate for a PAL- or PLA-based approach, it is not necessarily the best strategy when using packaged components. The equations in the preceding subsection require eight 3-input gates (three packages), two 4-input gates (one package), three 2-input gates (one package), four inverters (one package), and many wires. An MSI-based implementation could lead to fewer components and certainly fewer interconnections.

Counters, Multiplexers, Decoders If you examine the traffic light finite state machine carefully, you should see that a counter could be used to implement the state register. After all, the machine either holds in its current state or advances to the next state in a well-defined sequence. This is precisely the behavior of a counter. It appears we have a good match.

Let's make the state assignment $HG = 00$, $HY = 01$, $FG = 10$, and $FY = 11$, a simple sequential state assignment that will match the counter. We will implement the state register with a 163 synchronous up-counter. An external reset signal can be wired to the counter's synchronous clear input.

The question now becomes how to implement the counter's count input that will determine when the counter will advance to the next state and when it will hold its current state. Figure 9.45 reproduces the state diagram for the traffic controller but is redrawn to look more like a counter's state diagram. In state *HG*, the exit condition is $TL \bullet C$. In *HY* and *FY*, it is *TS*. In *FG*, it is $TL + \overline{C}$. We could use logic that takes the relevant condition of the current-state bits and ANDs it with the appropriate exit condition to form the count signal. Unfortunately, this would take a fair amount of discrete logic.

A better way is to use a multiplexer (the 153 is a good MSI device for this purpose) to implement the count signal. We drive a 4:1 multiplexer's

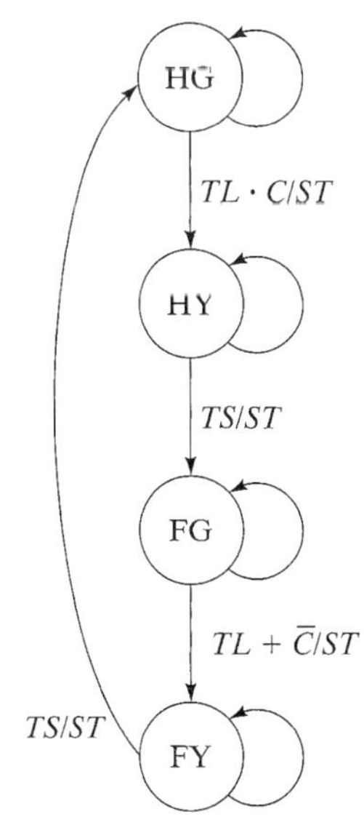

Figure 9.45 Traffic-light controller finite state machine.

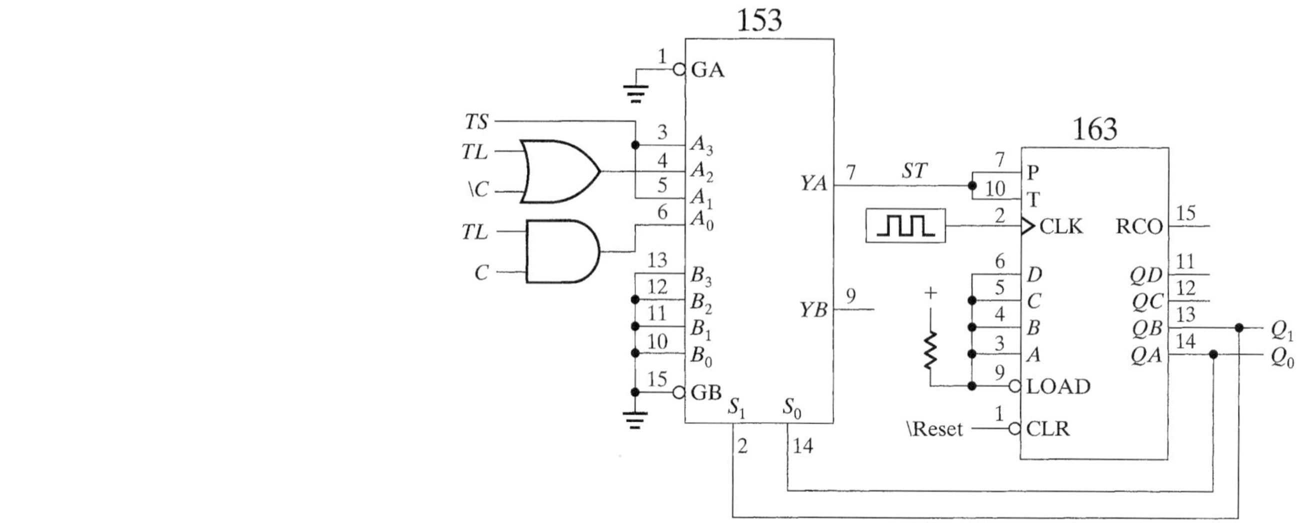

Figure 9.46 **Counter-based implementation of the next-state function.**

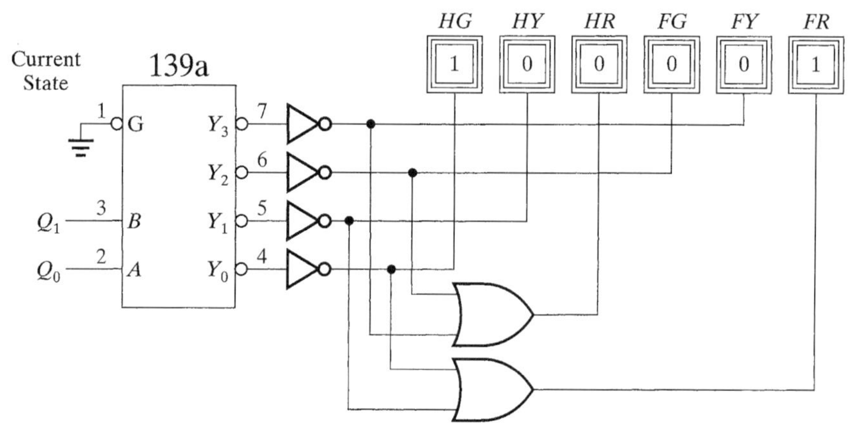

Figure 9.47 **Traffic light decoder.**

selection lines with the current-state bits, Q_1 and Q_0. The inputs are wired for the appropriate exit condition. This is shown in Figure 9.46. The count signal and the start timer signal, *ST*, are identical. As you can see, this approach represents a substantial reduction in package count.

As one final application of MSI components, it is possible to drive the traffic lights from signals that have been directly decoded from the current state. For example, if the machine is in state *HG*, the highway lights are green and the farm road lights are red. Similarly, the highway lights are yellow and the farm road lights are red when the machine is in state *HY*. Thus, the highway green light and yellow light can be decoded directly from state 0 and 1, respectively, while the farm road red light is driven by the OR of these decoded signals. The logic is shown in Figure 9.47.

9.5.4 FPGA-Based Implementation

We'll now map our traffic light controller logic to a Xilinx Virtex-II FPGA. Clearly, even the smallest Virtex-II device is much more than we need for our simple controller. This discussion is only for illustrative purposes.

Let's begin with the same set of equations we used for the PAL/PLA-based implementation. The next-state and output functions have up to five variables: P_1 and ST use five variables; P_0, H_1, H_0, and F_0 use three variables; and F_1 uses only one variable. Thus, we could implement the finite state machine in a total of five slices (one each for P_1 and ST, one for both H_1 and H_0, one for both F_0 and F_1, and one for P_0). The two flip-flops for the state are readily available within the slices that implement P_1 and P_0. Our output decoder generates four functions of only two inputs (the current state). These will require another two slices with two functions implemented in each slice. The logic for our car-detector circuit can be implemented in one more slice where we can cross-couple the two look-up tables to form the two cross-coupled NAND gates. The synchronizing flip-flop is also available in the slice. This makes a total of eight slices or only two CLBs to implement all of this logic. We assume that the traffic-light decoding happens at the traffic lights in a separate logic chip.

The only thing left to implement is the timer. If we stick with a 4-bit counter, we can easily implement its counting functions within a pair of slices (a larger counter would not scale linearly in size as its functions would have increasing numbers of inputs). The TL and TS outputs can be computed in a third slice. Our total circuit will require only 2.75 CLBs.

Xilinx provides software that takes as input a schematic description of a circuit, automatically partitions the logic to the CLB components, chooses specific CLBs in the logic array to implement this partitioning, and selects a routing of the interconnections to complete the implementation. In general, the designer need not know in detail how the schematic is mapped into the array structure. For this very simple circuit, it will be easy for the placement and routing tool to place all the logic in adjacent CLBs and use direct connections to complete the wiring. Some longer wires will be needed to get to the I/O blocks at the edge of the chip.

In larger, and more realistic, FPGA designs, it is not always possible to pack the CLBs tightly as more wiring resources than are available may be needed. In this case, the mapping software will create space by leaving some CLBs unused. Dense designs do sometimes cause these inefficiencies due to limited routing resources. Another way to alleviate local congestion is to duplicate logic. In this approach, rather than having to connect wires from one area to another, the mapping software creates duplicate logic to recompute a function. Thus, the output is available in the two places that is needed without having to run wires in between. Of course, this works in designs where the inputs that logic needs are available in both locations.

In general, the designer need not see this level of detail. So why bother to understand it? Routing decisions can have a serious impact on performance, and no routing software is perfect, especially given the complex constraints imposed on the routing task by the Xilinx architecture. By working at the detailed level of the interconnection fabric, you can do hand routing for critical signals or force them onto the smaller-delay long-line interconnections by directly instructing the mapping software. Of course, these tools are constantly improving along with the technology and manual intervention is becoming less and less necessary.

CHAPTER REVIEW

In this chapter, we have reviewed the many technologies available to implement sequential logic. We focused on alternative strategies for implementing finite state machines. We covered structured logic methods, based on ROMs, PLAs, and PALs, to compute the next-state and output functions. We also examined MSI-based implementation strategies, using components like counters, multiplexers, and decoders to implement finite state machines using a small number of component packages.

If the goal is minimum package count, the best solution is to use some of the more advanced programmable logic components, such as those provided by Altera, Actel, and Xilinx. Altera's architecture is a generalization of the more traditional PAL structures. Their major innovations are the ability to share product terms among multiple output functions, form large functions of many inputs in wide AND arrays, and provide more general method of interconnect.

Actel bases its architecture on a very simple building block, a generalization of the 4:1 multiplexer, and a flexible orthogonal wiring grid for interconnections. Most FPGA technologies are reprogrammable, some even after they've been packaged into a final design, providing extraordinary flexibility. Some FPGAs are one-time programmable such as Actel's anti-fuse technology. Few designers map circuits to any of these technologies by hand. Computer-aided design tools provide automatic mapping of a schematic to the available resources. Many tools also include the ability to synthesize the circuit from an HDL description.

We concentrated a bit more on Xilinx technology, because it was among the first in the programmable gate array business and is one of the most prevalent in industry today. The architecture is based on complex building blocks, CLBs, containing considerable logic and flip-flops. CLBs are surrounded by interconnect resources that include a network of vertical and horizontal tracks of various lengths joined by switching matrices.

In the final section of the chapter, we looked at the detailed implementation of the traffic light controller, a running example throughout this book. In particular, we examined PAL/PLA/ROM implementations, counter/multiplexer/decoder implementations, and, lastly, how it might be implemented within a Xilinx FPGA. As expected, the MSI-inspired implementation yielded a much more elegant solution than the one obtained from discrete gates. Many MSI building blocks, or suitably modified versions of them, can map nicely onto the primitives provided by many FPGAs. In fact, many FPGA providers have libraries of pre-mapped, commonly used functions very much like the entries in the old catalogs.

FURTHER READING

The best place to learn about special PAL/PLD architectures is in the data sheets available from the manufacturers. These usually are filled with excellent tutorial materials and detailed design examples in the form of supplementary applications notes. Data sheets are readily available on the manufacturers' web sites. Here is a partial list of the most relevant sites:

Actel: www.actel.com

Altera: www.altera.com

Cypress Semiconductor: www.cypress.com

Xilinx: www.xilinx.com

Some books have recently been published that focus exclusively on digital design techniques for programmable logic. D. Van Den Bout's book, *The Xilinx Practical Designer,* Prentice Hall, 1999, is one of the first textbooks specifically based on Xilinx FPGAs. M. Cilletti's book, *Modeling Synthesis and Rapid Prototyping with the Verilog HDL,* makes extensive use of programmable logic. R. Sandige's book, *Modern Logic Design,* McGraw-Hill, 1990, focuses on PROM/PAL/PLA-based design. P. K. Lala, *PLD: Digital System Design Using Programmable Logic Devices,* Prentice Hall, 1990, gives a very detailed description of the various PAL and PLD families and many excellent design examples, as does D. Pellerin and M. Holley's *Practical Design Using Programmable Logic,* Prentice Hall, 1991. M. Boulton, *Digital Systems Design with Programmable Logic,* Addison-Wesley, 1990, covers similar ground.

Some seminar papers describing the underlying technology of Actel's anti-fuse technology in are K. A. El-Ayat, *et al.,* "A CMOS Electrically Configurable Gate Array," originally published in the *IEEE Journal of Solid State Circuits,* 24 (3), pp. 752–762 (June 1989), and A. El Gamal, *et al.,* "An Architecture for Electrically Configurable Gate Arrays," *IEEE Journal of Solid State Circuits,* 24 (2), pp. 394–398 (April 1989).

EXERCISES

9.1 *(State Reduction)* Verify that the BCD-to-excess-3 code-converter problem statement of Section 9.2 yields the reduced state transition table and state diagram of Figure 9.9. Start with a general state diagram derived directly from the problem statement, and show how it is reduced.

9.2 *(Reset in Finite State Machines)* Consider the BCD-to-excess-3 code-converter state machine. In various implementations of this machine, *Reset* was used to control the state register's *Clear* signal directly. How does the next-state logic change when *Reset* is treated as an explicit input, for the following implementations?

(a) A ROM-based approach
(b) A PLA-based approach

9.3 *(Finite State Machine Transformations)* Consider the counter-based implementation of the 4-bit string recognizer in Figure 9.5. Show whether the FSM partitioning and transformation guidelines are adequate for deriving this implementation from the original state diagram. If the guidelines are not sufficient, can you suggest some new ones to add? (*Hint:* Try to extract a counter-like FSM from the original state diagram.)

9.4 *(Counter-based FSMs)* Construct a finite-state machine with nine states and two inputs in addition to reset. On reset, the machine starts in the middle state of the nine. When the left input is asserted, it transitions one state to the left; when the right input is asserted, it transitions one state to the right. If it reaches a state at either end of the chain, it stays in that state until the next reset. The leftmost state will assert the Left-LED output, the rightmost state will assert the Right-LED output. Use the up/down counter of Figure 9.5 to implement this state machine with as little external logic as possible. (*Hint:* Consider using the load signal to implement the reset behavior.)

9.5 *(PALs)* Re-implement the FSM of the previous problem using a 22V10 PAL (see Figure 9.28) and a one-hot encoding for the states. Show the details of the PAL programming directly on a copy of the PAL data sheet.

9.6 *(PALs)* Implement an 8-bit register using the 20X10 PAL of Figure 9.20 to the following specifications:

(a) The register has three control inputs, $\overline{OE}$ (output enable), $\overline{LD}$ (load), and CLK (clock), and eight data inputs D_7–D_0. It has eight registered outputs Q_7–Q_0. When $\overline{OE}$ is unasserted, the outputs are in high impedance. When $\overline{LD}$ is unasserted, the register holds its current value. When $\overline{LD}$ is asserted, the register's contents are replaced by the data inputs on the next rising clock edge. What are the Boolean equations for each register input? Draw a wiring diagram, similar to Figure 9.18, for the PAL's pin inputs and outputs.

(b) The register has three control inputs: $\overline{OE}$ (output enable), $CLR/\overline{INC}$ (clear/increment), and CLK (clock), and eight data outputs, Q_7–Q_0. When $\overline{OE}$ is unasserted, the outputs are in high impedance. When *CLR* is asserted, the register is set to 0 on the next rising clock edge. When *CLR* is unasserted, the register is incremented by 1 on the next clock edge. In essence, the register is a free-running counter. What are the

Boolean equations for each register input? Draw a wiring diagram like Figure 9.18 for the PAL's pin inputs and outputs. (*Hint:* Consider the carry-lookahead logic described in Chapter 5, except simplified for the specific case where the sum is $A + 1$.)

9.7 *(PALs)* Re-implement the circuit of Figure 9.5 using a 22V10 PAL (see Figure 9.28). Show the details of the PAL programming directly on a copy of the PAL data sheet

9.8 *(PALs)* Re-map the vending machine FSM implementation given in Figure 9.24 onto a 20X10 PAL (see Figure 9.20). You will need to restructure the equations to use the XOR of this different PAL.

9.9 *(PALs)* Implement a '163 counter using the 22V10 PAL (see Figure 9.28).

9.10 *(PALs)* Extend your solution to the previous exercise so that the counter is as wide as possible and still fits in a single PAL. Make sure to always consider the inputs and outputs you need to chain counters together.

9.11 *(PALs)* Implement the up/down counter of Figure 8.38 using a 20X10 PAL (see Figure 9.20) without partitioning it. Implement it again using its partitioned form into two 4-state FSMs. What implications does the partitioning have for the minimum clock period?

9.12 *(Actel Logic Module)* Figure 9.32 showed a basic Actel combinational logic *C*-cell. Show how a single logic module can be used to implement the following functions:

(a) A two-input AND gate
(b) A two-input NAND gate
(c) A two-input NOR gate with one input inverted
(d) A two-input XOR gate
(e) The logic function $Y = (A \bullet B) + C$
(f) The logic function $Y = (A + B) \bullet C$
(g) The logic function $Y = (A \bullet B) + (A \bullet C) + (B \bullet C)$. This function is also known as the majority function.

9.13 *(Actel Logic Module)* Describe how to implement the following functions using two Actel combinational logic *C*-cells:

(a) A four-input AND-OR-Invert structure with two 2-input AND gates at the first level
(b) $Y = \overline{A} + \overline{B} + C + D$

9.14 *(Actel Logic Module)* Show how to implement a half adder in terms of an Actel logic module (more than one module may be needed). How would you implement the full-adder circuit? (*Hint:* This can be done in two logic modules if the carry in and carry-out are designed to be active-low.)

9.15 *(Actel Logic Module)* Implement the state machine of Figure 9.5 using Actel *C*-cells and *R*-cells. How many of each will you need?

9.16 *(Actel Logic Module)* Re-implement the FSM of Exercise 9.4 using Actel *R*-cells and *C*-cells.

9.17 *(Xilinx CLB)* The basic function generator within the Xilinx CLB can implement any combinational logic function of four inputs. This is equivalent to how many 2-input gates? Explain how you derived your answer.

9.18 *(Xilinx CLB)* A single Xilinx CLB can implement up to eight 4-input functions (two for each of the four slices in a CLB).

(a) Show how a two-level 4-input LUT structure can implement a 25-input parity function.
(b) How many input parity functions can be implemented by a three-level LUT structure?

9.19 *(Xilinx CLB)* The Xilinx look-up table can implement any single combinational logic function $F(A,B,C,D)$ of four variables.

(a) Show how this might be implemented by wiring up a 16-input function generator (multiplexer/selector) using two 8:1 and a single 2:1 multiplexers. Assume that input_0 is selected when $ABCD = 00000$, input_{15} when $ABCD = 1111$, and input_8 when $ABCD = 1000$.
(b) What must the input settings be to implement the four-variable function $F = A \oplus B \oplus C \oplus D$?
(c) What must the input settings be to implement the two 3-variable functions $F(A,B,C) = A \oplus B \oplus C$ (full-adder sum) and $G(A,B,C) = A \bullet B + B \bullet C + A \bullet C$ (full-adder carry)?

9.20 *(Xilinx Logic Module)* Implement the state machine of Figure 9.5 using Xilinx CLBs. How many CLB slices will you need?

9.21 *(Xilinx Logic Module)* Re-implement the FSM of Exercise 9.4 using Xilinx CLBs.

9.22 *(Traffic Light Controller)* The description of the traffic-light controller in Section 9.5 assumes that the traffic-light outputs from the state machine are encoded. This saves only two output lines yet introduces the extra complexity (and parts count) of external decoders.

(a) Re-derive the equations for the traffic light control signals, assuming fully decoded outputs from the state machine: *HG*, *HY*, *HR*, *FG*, *FY*, and *FR*. Also show the revised state transition table.
(b) How does this affect the PAL-based implementation? In particular, will you require a PAL with more product terms per output or fewer terms per output?
(c) How does the change affect the ROM-based implementation?

9.23 *(Traffic Light Controller)* The traffic light controller, as presented in Section 9.5, is a Mealy machine. Modify the description to make it part Moore/part Mealy by decoding the traffic-light control signals directly from the state, as we did in Section 9.5.3.

(a) What are the implications for a PAL-based implementation?
(b) How does this affect the ROM-based implementation?

9.24 *(Traffic Light Controller)* The end of Section 9.5.4 described how to implement the traffic-light-controller finite state machine in terms of MSI-style components mapped onto Xilinx CLBs. Show how the ST function can be implemented by a single CLB slice by giving the truth table for the 5-variable function $ST(TL, TS, C, Q_1, Q_0)$ and show how it maps to the 4-input function generators in the CBL slice.

CHAPTER 10

Case Studies in Sequential Logic Design

Introduction

Perhaps the most difficult problem the novice hardware designer faces is mapping an imprecise behavioral specification of an FSM into a more precise description (for example, a state diagram or a Verilog description). In this chapter, we will illustrate the process by examining several detailed case studies: an FSM that can recognize patterns in its inputs, a complex counter, a digital combination lock, a memory controller, a sequential multiplier, and a serial line transmitter/receiver that connects a telephone keypad and small display.

Each example will highlight different aspects of sequential-logic design. We will apply our design methodology and show where trade-offs can be made. Different implementation technologies will be used for the examples with an emphasis on programmable logic devices. We will also consider FPGA, ROM, and fixed-logic technologies when they are appropriate for illustrative purposes.

After absorbing this chapter, and implementing some of these designs yourself, you should have a solid understanding of the process of sequential logic design. These case studies were chosen for the variety of issues they bring to light. Of course, there are many ways to design and realize these examples. In no way should you take the designs in this chapter to be the best along any of the metrics we've discussed.

10.1 A Finite String Recognizer

Finite state machines are often used to recognize patterns in an input sequence.

Problem Specification Consider the following finite state machine specification: "A finite state recognizer has one input (X) and one output (Z). The output is asserted whenever the input sequence ...010... has been observed, as long as the sequence ...100... has not been seen since the last reset."

Understanding the Specification For problems of this type, it is a good idea to write down some sample input and output behavior to make sure you understand the specification. Here are some input and output strings:

X: 00101010010 ...

Z: 00010101000 ...

X: 11011010010 ...

Z: 00000001000 ...

In the first pair of input/output strings, we find three overlapping instances of 010 (...0101010...) before detecting the termination string (...100). Once this is found, additional 010 strings in the input cause no changes in the output. We have written the outputs so they lag behind the inputs. This is the kind of timing we would expect to see in the real machine.

Similar behavior is illustrated in the second pair of strings. The detected sequence ...010 is immediately followed by another 0. Since this is the terminating string, the output stays at 0 despite further 010 strings in the input.

Formal Representation Now that we understand the desired behavior of the finite state machine, it is time to describe its function by a state diagram. Suppose we choose to represent this example FSM with a state diagram for a Moore machine. It is a good idea to start by drawing state diagram fragments for the strings the machine must recognize: 010 and 100.

Figure 10.1(a) shows the initial Moore state diagram, assuming state 0 is reached on an external *Reset* signal. One path in the diagram leads to a state with the output asserted when the string 010 has been encountered. The other path leads to a looping state for the string 100.

Given that there is only one input, each state should have two exit arcs: when the input is 0 and when it is 1. To refine the state diagram, we add the remaining arcs, and perhaps additional states, to make sure the machine recognizes all valid strings.

For example, what happens when we exit state S_3? To get to S_3, we must have recognized the string 010. If the next input is 0, then the machine has seen ...0100, the termination string. The correct next state is therefore S_6, our termination looping state.

What if the input had been a 1 in state S_3? Then we have seen the string ...0101. This is a prefix of ...010 if the next input turns out to be a 0. We could introduce a new state to represent this case. However, if we carefully examine the state diagram, we find that an existing state, S_2, serves the purpose of representing all prefix strings of the form ...01. The new transition from S_3 to S_2 is shown in Figure 10.1(b).

Continuing with this approach, let's examine S_1. You should realize that any number of zeros before the first 1 is a possible prefix of ...010. So we can loop in this state as long as the input is 0. We define S_1 to represent strings of the form ...0 before a 1 has been seen.

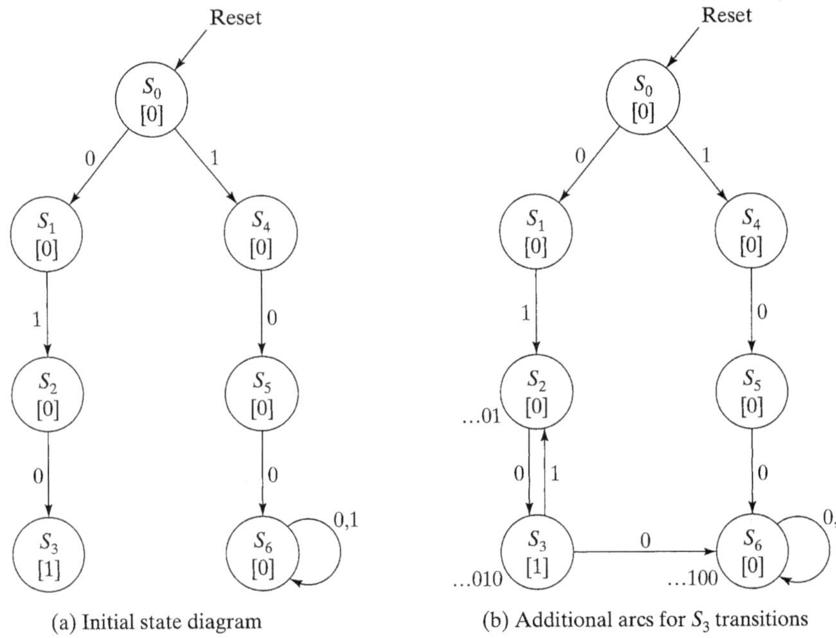

Figure 10.1 State diagrams for the finite string recognizer.

State S_4 plays a similar role for strings of 1s, which may represent a prefix of the terminating string 100. So we can loop in this state as long as the input is 1. The next refinement of the state diagram, incorporating these changes, is shown in Figure 10.2(a).

We still have two states with incomplete transitions: S_2 and S_5. S_2 represents strings of the form ...01, a prefix of the string 010. If the next input is a 1, it can no longer be a prefix of 010 but instead is a prefix of the terminating string 100. Fortunately, we already have a state that deals with this case: S_4. It stands for strings whose last input was a 1 which may be a prefix of 100. So all we need to do is add a transition arc between S_2 and S_4 when the input is a 1.

The final state to examine is S_5. It represents strings consisting of a 1 followed by a 0. If the next input is a 1, the observed string is of the form ...101. This could be a prefix for 010. S_2 already represents strings of the form ...01. So we add the transition between S_5 and S_2 when the input is a 1.

We show the completed state diagram in Figure 10.2(b). You should run through the sample input strings presented at the beginning of this section to make sure you obtain the same output behavior. It is always a good strategy to check your final state diagram for proper operation.

Verilog Description It is straightforward to map the state diagram of Figure 10.2(b) into a finite state machine description. We will use a sequential state encoding into three state bits for the seven states—one code will be unused. Since we are dealing with a Moore state machine, the output is a function of the state bits, namely, that the

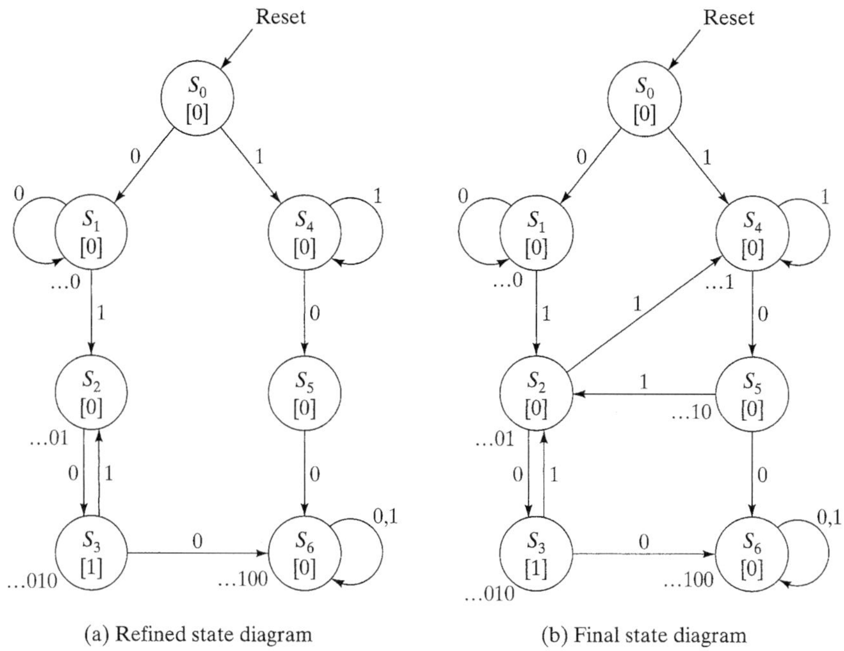

Figure 10.2 Refinement of the state diagrams.

FSM is in state S_3. The description becomes:

```
module string_recognizer (Clk, X, Reset, Z);
input Clk, X, Reset;
output Z;

reg state[0:2];
parameter S0 = [0,0,0]; //reset state
parameter S1 = [0,0,1]; //strings ending in    ...0
parameter S2 = [0,1,0]; //strings ending in   ...01
parameter S3 = [0,1,1]; //strings ending in ...010
parameter S4 = [1,0,0]; //strings ending in    ...1
parameter S5 = [1,0,1]; //strings ending in   ...10
parameter S6 = [1,1,0]; //strings ending in ...100

assign Z = (state == S3);

always @(posedge Clk) begin
    if Reset
        state = S0;
    else
        case (state)
            S0: if (X) state = S4 else state = S1;
            S1: if (X) state = S2 else state = S1;
            S2: if (X) state = S4 else state = S3;
            S3: if (X) state = S2 else state = S6;
            S4: if (X) state = S4 else state = S5;
            S5: if (X) state = S2 else state = S6;
            S6: state = S6;
```

```
                default: begin
                    $display ("invalid state reached");
                    state = 3'bxxx;
                end
            endcase
        end
    endmodule
```

The Verilog description uses a 3-bit state register. The output Z is derived directly from these state registers.

The state transitions are described in the *always* block, using simple conditional and case statements. For example, if the clock ticks a positive edge and the *Reset* input is asserted then the FSM is reset into state S_0. In most of the other states, the FSM decided on the next state based on the value of the input X. For example, if the FSM is in state S_0 and the input X is 1, the machine's next state is S_4. If the input is 0, the next state is S_1. We include a default state just in case the FSM should, for some reason, have its state register set to the unused code. If this should occur, we print a message that an invalid state was reached and set the state to unknown. The state will stay unknown until the next reset.

Discussion Let's briefly review the steps we followed in arriving at the final state diagram:

STEP 1 Write sample inputs and outputs to make sure you understand the statement of the problem.

STEP 2 Next, write sequences of states and transitions for the distinguished strings your FSM is expected to recognize.

STEP 3 Most likely, Step 2 will not cover all transitions. You then add the missing transitions, taking advantage of states you already have introduced wherever possible. You should view the states as "remembering" certain input string sequences. In this example, when the FSM is in state S_1 it means that a string of all zeros has been seen so far; S_2 represents all strings in which the last two inputs are a 0 followed by a 1; and so on.

STEP 4 Finally, verify that the input/output behavior of your state diagram matches the specified behavior of the FSM. You may need to juggle some transitions or introduce additional states when you encounter a "counterexample" input string that does not yield the expected outputs.

Implementation This is a small example. Let's use fixed logic to implement it. We'll start with an encoded state table (see Figure 10.3) from which we can derive K-maps for each of the state bits. The minimization process yields the following equations:

$$\begin{aligned}
NS_0 &= CS_0\,Q_2' + CS_2'\,X + CS_1\,CS_2\,X' + CS_0\,X' \\
NS_1 &= CS_1\,X' + CS_0\,CS_2 + CS_0\,CS_1 + CS_2\,X \\
NS_2 &= CS_0'\,CS_2'\,X' + CS_1'\,CS_2'\,X' + CS_0'\,CS_1'\,X' \\
Z &= CS_0'\,CS_1\,CS_2
\end{aligned}$$

CS_0	CS_1	CS_2	X	NS_0	NS_1	NS_2
0	0	0	0	0	0	1
0	0	0	1	1	0	0
0	0	1	0	0	0	1
0	0	1	1	0	1	0
0	1	0	0	0	1	1
0	1	0	1	1	0	0
0	1	1	0	1	1	0
0	1	1	1	0	1	0
1	0	0	0	1	0	1
1	0	0	1	1	0	0
1	0	1	0	1	1	0
1	0	1	1	0	1	0
1	1	0	0	1	1	0
1	1	0	1	1	1	0
1	1	1	0	X	X	X
1	1	1	1	X	X	X

Figure 10.3 **Encoded state table for the string recognizer.**

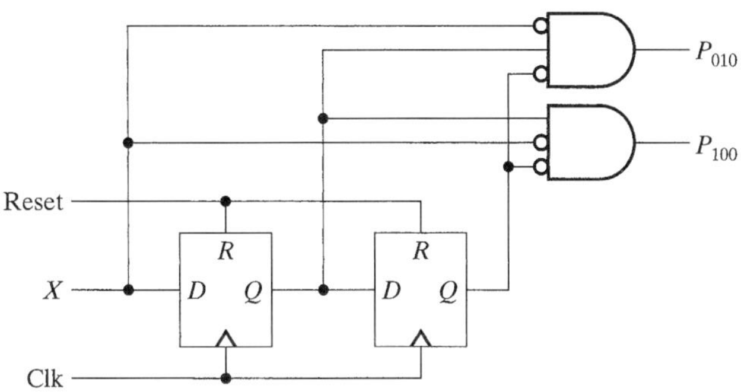

Figure 10.4 **Pattern recognition logic for the new string recognizer.**

These are fairly complex equations for such a small FSM. This will require 12 gates with up to 4 inputs. Could we do better? We used a sequential state assignment. Surely, there must be a better choice of codes for the states. Instead of running a state assignment tool, let's try to re-think our problem.

We are trying to build a string recognizer that looks at the last three values of the input. A shift register is perfect for this function. We only need a 2-bit shift register to hold the two previous inputs. Figure 10.4 shows how we can add two 3-input gates that will check for the two patterns we are interested in, namely, 010 and 100.

We can now design a completely different state machine to generate the output we want. It will use the output of the two-pattern AND gates as inputs (P_{010} and P_{100}). It will have three states: a zero-output state S_0 that will await one of the patterns, a one-output state S_1 that will signal that the pattern 010 has been seen, and a zero-output state S_2 which will trap our FSM after it has seen the pattern 100. Figure 10.5 shows the state diagram for this 3-state FSM. Note that only a Reset will allow our FSM to transition from S_2.

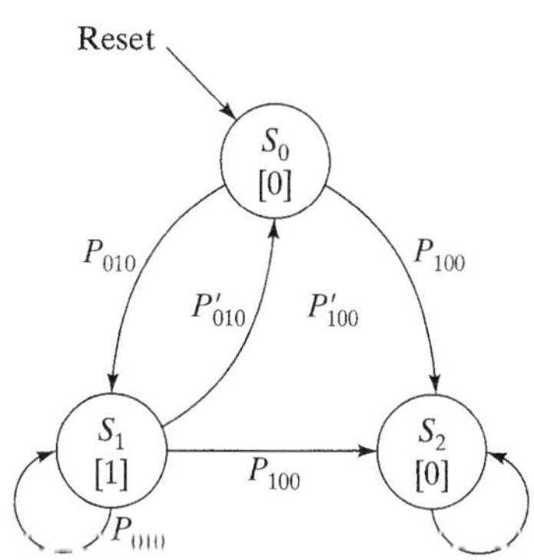

Figure 10.5 **FSM for the new string recognizer.**

CS_1	CS_2	P_{010}	P_{100}	NS_0	NS_1
0	0	0	0	0	0
0	0	0	1	1	0
0	0	1	0	0	1
0	0	1	1	X	X
0	1	0	0	0	0
0	1	0	1	1	0
0	1	1	0	0	1
0	1	1	1	X	X
1	0	0	0	1	0
1	0	0	1	1	0
1	0	1	0	1	0
1	0	1	1	X	X
1	1	0	0	X	X
1	1	0	1	X	X
1	1	1	0	X	X
1	1	1	1	X	X

Figure 10.6 Encoded state table for the new string recognizer.

We'll use a sequential state encoding for this machine as well. However, this time it will work to our advantage. Note that the output of the FSM simply will be the first bit of the 2-bit state register. There are a couple of don't care conditions in this new FSM. First, the P_{010} and P_{100} inputs can't ever be true at the same time. Second, the state code 11 is unused. The encoded state table is provided in Figure 10.6 and leads to the following equations after minimization:

$$NS_0 = CS_0 + P_{100}$$

$$NS_1 = CS_0' \, P_{010} + CS_1 \, P_{010} = CS_0' \, P_{010}$$

$$Z = CS_1$$

Note that we don't even need the second term of the second equation as there is a *sequential don't care* that we did not use. A sequential don't care is a *sequence* of inputs that is impossible. After we enter state S_1 by seeing the pattern 010, the P_{010} signal can't be asserted again in the very next clock cycle—the next pattern will be either 100 or 101, it can't be 010 again. That means whenever we enter state S_1, our FSM will transition to either S_0 or S_1 in the next cycle. Thus, the term CS_1 P_{010} will never be true.

With this simplification, our implementation will require only four gates with, at most, 3 inputs. There may be a need for some inverters on inputs and outputs of these gates, but the implementation undoubtedly is simpler and easier to understand than our original design. Figure 10.7 shows the complete circuit. Note that the FSM is implemented using two 2-input gates and two flip-flops.

The Verilog for this more compact version of the string recognizer is:

```
module string_recognizer_2 (Clk, X, Reset, Z);
input Clk, X, Reset;
output Z;
```

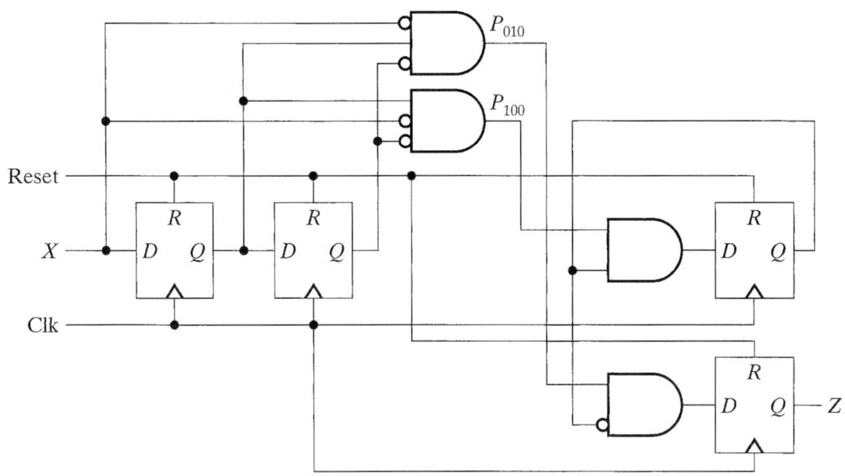

Figure 10.7 Complete circuit for the new string recognizer.

```
wire P010, P100;
reg shift[0:1];

reg state[0:1];
parameter S0 = [0,0]; // reset state
parameter S1 = [0,1]; // the pattern 010 was seen
parameter S2 = [1,0]; // the pattern 100 was seen

assign Z = state[1];
assign P010 = !X && shift[0] && !shift[1];
assign P100 = X && !shift[0] && !shift[1];

always @(posedge Clk) begin
    shift[1] <= shift[0];
    shift[0] <= X;
    if Reset
        state = S0;
    else
        case (state)
            S0: if (P010) state = S1 else if (P100) state
                = S2 else state = S0;
            S1: if (P010) state = S1 else if (P100) state
                = S2 else state = S1;
            S2: state = S2;
            default: begin
                $display ("invalid state reached");
                state = 2'bxx;
            end
        endcase
    end
endmodule
```

Note that the conditional statements for the state transitions do not have to consider the case when P_{010} and P_{100} are both true. This allow us to use simple nesting in the `if` statements.

The shift register is implemented with the first two statements in the `always` block. We must use blocking (or delayed assignment, <=) to indicate that the value should not change until the `always` block completes its computation. The new values in the shift register will be used the next time the clock ticks and this `always` block is triggered again.

10.2 A Complex Counter

As we saw in Chapter 7, counters are a special case of finite state machines: their state and their outputs are always identical. In this case study, we will combine a simple control function with basic sequencing.

Problem Specification The task is to create a complex counter that can count in binary or in Gray code, depending on the value of a mode input: "A synchronous 3-bit counter has a mode control input M. When $M = 0$, the counter steps through the binary sequence 000, 001, 010, 011, 100, 101, 110, 111, and *Repeat*. When $M = 1$, the counter advances through the Gray-code sequence 000, 001, 011, 010, 110, 111, 101, 100, and *Repeat*."

Understanding the Specification Start by making sure you understand the interaction of the control input and the counter's sequence. Let's label the control input as signal M and the outputs as signals Z_2, Z_1, and Z_0. To check our understanding, let's write down a sample counter sequence as the M input varies. The following is one example of a valid count sequence:

Mode Input (M)	*Current State*	*Next State ($Z_2Z_1Z_0$)*
0	000	001
0	001	010
1	010	110
1	110	111
1	111	101
0	101	110
0	110	111

Formal Representation Since all of the eight possible states can be reached by some count sequence, you might start by tabulating the eight states in a state diagram. Then you simply connect the states with the appropriate transitions. These are straightforward. On a mode input of 0, the transitions follow the normal, binary count sequence. When the input is 1, the machine follows the more complex Gray-code sequence.

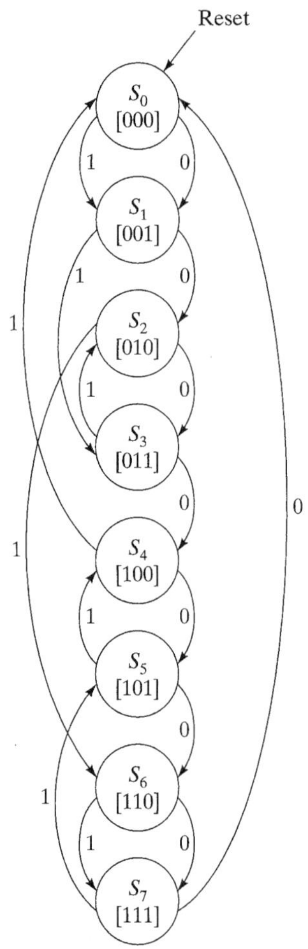

Figure 10.8 State diagram for complex counter.

Figure 10.8 shows the state diagram. The states are named according to the binary encoding of the state's output. State S_0 has output 000, state S_2 has output 010, and so on. *Reset* places the finite state machine into state S_0. The state transitions for Gray-code counting are on the left while the transitions for binary counting are on the right.

Verilog Description The machine's state sequencing is again quite easy to capture in terms of conditional and case statements. As with the previous example, the FSM state bits are directly connected to the corresponding output. The Verilog description follows:

```
module complex_counter (Clk, M, Reset, Z0, Z1, Z2);
input Clk, M, Reset;
output Z0, Z1, Z2;

reg state[0:2];
parameter S0 = [0,0,0];
parameter S1 = [0,0,1];
parameter S2 = [0,1,0];
parameter S3 = [0,1,1];
parameter S4 = [1,0,0];
parameter S5 = [1,0,1];
parameter S6 = [1,1,0];
parameter S7 = [1,1,1];

assign Z0 = state[0];
assign Z1 = state[1];
assign Z2 = state[2];

always @(posedge Clk) begin
    if Reset
        state = S0;
    else
        case (state)
            S0: state = S1;
            S1: if (M) state = S3 else state = S2;
            S2: if (M) state = S6 else state = S3;
            S3: if (M) state = S2 else state = S4;
            S4: if (M) state = S0 else state = S5;
            S5: if (M) state = S4 else state = S6;
            S6: if (M) state = S7 else state = S7;
            S7: if (M) state = S5 else state = S0;
        endcase
end

endmodule
```

Implementation Since we are dealing with an FSM that is at least partly a binary counter, let's try to use a counter component to implement this more complex counter. We'll start with a 4-bit synchronous counter with *Clear* and *Load*. These are readily available parts in most catalogs of MSI parts or other fixed logic. Our default behavior will be a binary counter, and this will be indicated by *M* being unasserted. On the other hand, when *M* is asserted, we will want our counter to follow the Gray-code sequence. We'll do this by having the next Gray-code value computed from the current state and ready at the data-load inputs of the counter.

Z_2	Z_1	Z_0	D_2	D_1	D_0
0	0	0	0	0	1
0	0	1	0	1	1
0	1	0	1	1	0
0	1	1	0	1	0
1	0	0	0	0	0
1	0	1	1	0	0
1	1	0	1	1	1
1	1	1	1	0	1

Figure 10.9 Truth table for computing the next Gray-code value.

We'll load this new value into the counter only if M is asserted. Thus, our circuit is a binary counter if M is low, and a Gray-code counter if M is high.

The truth-table for computing the next Gray-code value is given in Figure 10.9. It is a simple combinational function of three inputs yielding three outputs. The equations after minimization are:

$$D_2 = Z_2 Z_0 + Z_1 Z_0'$$
$$D_1 = Z_2' Z_0 + Z_1 Z_0'$$
$$D_0 = Z_2' Z_1' + Z_2 Z_1$$

This is a very simple circuit that can be implemented in a few 2-input gates. Figure 10.10 shows the complete schematic for our complex counter. The counter component we used has an active-low load input (connected to $\bar{M}$, Gray mode), an active-low enable input for the binary counter (connected to M, binary mode), an UP-direction control input (connected to $\bar{M}$, binary mode), and an active-high clear input (connected to *Reset*).

The schematic was purposely drawn to look like a PAL. With a PAL that has only two product terms per output we can implement our complex counter with two chips. It would be nice to include the binary counter logic in the PAL as well, but this is not always possible as there may not be enough terms per output.

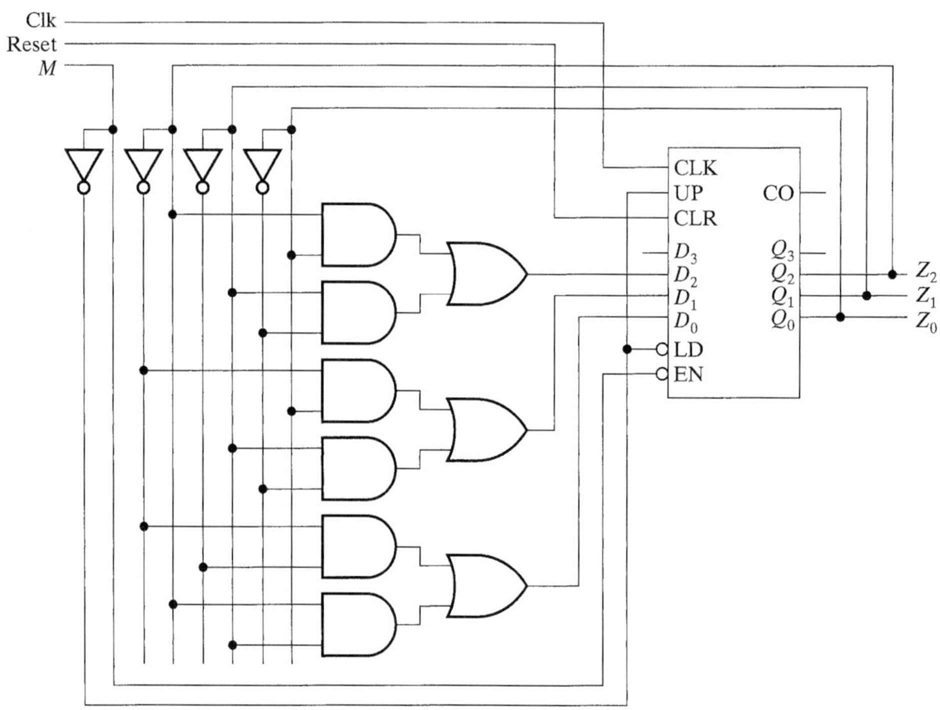

Figure 10.10 Circuit for complex counter using a binary counter component.

Discussion When working with circuits that follow periodic sequences, like counters, it is always a good strategy to begin by writing down the states and connecting them in the order of the sequence. You can add unusual/exceptional sequencing among the states later. In implementation, the more closely the behavior of the FSM matches a well-known common component, the more attractive it will be to utilize that part in the implementation. In this case, the binary counter was a very good match and its load capability made combining the binary with Gray-code counting behavior quite straightforward.

10.3 A Digital Combination Lock

For this case study, we will go back to the combinational lock example used way back in Section 1.4.2. We were asked to design a three-digit combination lock. But before we do, let's review what we've learned since Chapter 1. Combinational logic operates on a set of logic inputs and produces a set of logic outputs. It reacts to any change in the inputs and always delivers the same output values for the same combination of input values. Combinational logic can be used to compare values, select values from a set (multiplexer), and decode signals. Sequential logic is used to remember values so that future outputs can depend on the history of the inputs and not just their current values. Let's see how these two types of logic appear in our combination lock.

Problem Specification We are given the following description: "A 3-digit combination lock is used to allow entry to a locked room. The lock has a *RESET* button, an *ENTER* button, and a 16-key keypad on which the user can enter one value at a time. When the signal *UNLOCK* is asserted, an electromechanical relay is released, allowing the door to open. The unlock process begins when the operator presses *RESET*. He or she then presses a key, followed by pressing the *ENTER* button. This is repeated for the second and third key digits. After entering the three 4-bit digits, the door will be unlocked or not depending on whether the sequence of key presses was correct. The process can be retried at any time simply by pressing *RESET* again."

Understanding the Specification Recall from Figure 1.23 the basic structure of the implementation. There is a data path, a collection of combinational logic that operates on data values, and a controller or sequencer that keeps track of where we are in checking the combination. Figure 10.11 gives a block diagram of the entire combination lock showing these two principal components: data path and controller. It is a slightly different diagram than Figure 1.23. We have added three load signals to be able to change the combination of the lock. The three registers that hold the combination (C_1, C_2, and C_3) all take their input from the same 4-bit value. You can imagine that the load signals are in a protected part of the lock accessible only to authorized persons. One can use the keypad to enter a new value and assert a load signal to change a digit of the combination. The schematic uses a slash across a wire to indicate multiple wires and a number to indicate how many wires there are in the group—this is a

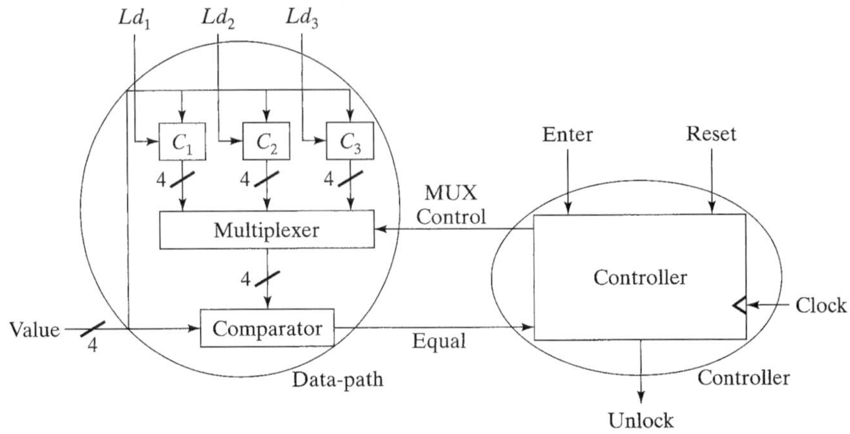

Figure 10.11 Block diagram of a combination-lock design showing data path and controller components.

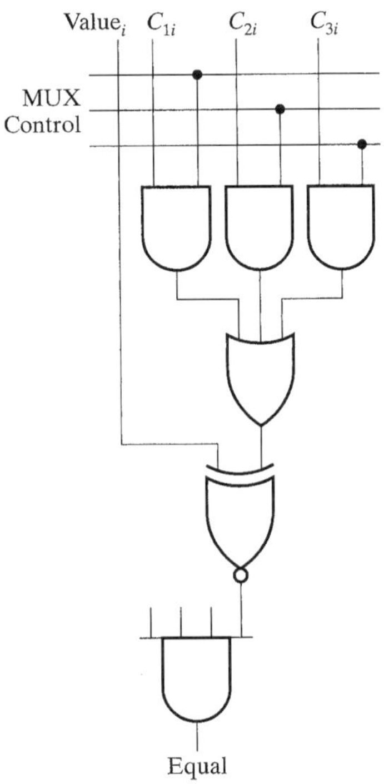

Figure 10.12 Bit-slice of the data path for the combination lock (not showing combination registers).

useful shorthand rather than drawing multiple wires that go between the same points and cluttering the diagram unnecessarily.

Implementation Let's begin our implementation with the data-path component. The data path has three registers, a multiplexer, and a comparator. We can design the data path by creating a one *bit-slice* and then replicating it four times. An initial design for the data-path bit-slice is shown in Figure 10.12. This design shows the multiplexer and comparator and does not show the combination registers and their load control. As in Chapter 1, the mux control consists of three signals using a one-hot encoding to choose one of the three combination values to compare with the key entered on the value input. The comparator consists of an XNOR gate and a 4-input AND gate to combine the comparison results across the 4 bits.

The 4-input AND gate needs wires from each of the four bit-slices. An alternative to the AND gate is to use open-collector logic to form a wired AND. Figure 10.13 shows this alternative. By using an open-collector XNOR gate to form the wired AND we can now just run a single wire across all four bit-slices rather than routing four signals to an AND gate. Of course, we can't forget the pull-up resistor at the end of the wire to ensure the signal will be pulled-up to 1 when none of the open-collector gates are pulling down. In addition, to the wired AND, we also changed the design of the multiplexer to use tri-state drivers. As long as one of the MUX control wires is asserted, one of the combination values will be used to drive the input to the XNOR gate. It is very uncommon to find programmable logic devices with open collector logic. Some devices support open-collector outputs at their chip pins but not internally. Choosing wired-AND connections usually makes sense only at the board level when multiple chips are being interconnected. Many FPGAs do have tri-state drivers available for the purpose of interconnecting many outputs to long wires that span a distance on the chip. In this simple example, they are not that advantageous, as the wires didn't have far to go. In general, however, tri-state drivers can be very useful in

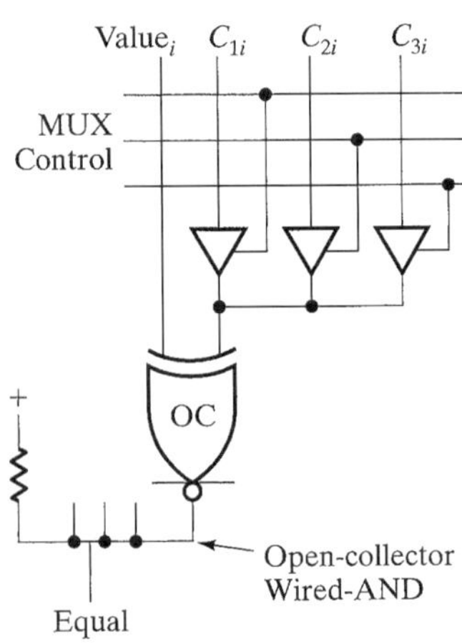

Figure 10.13 Bit-slice of the data-path for the combination lock using wired-AND and tri-state logic.

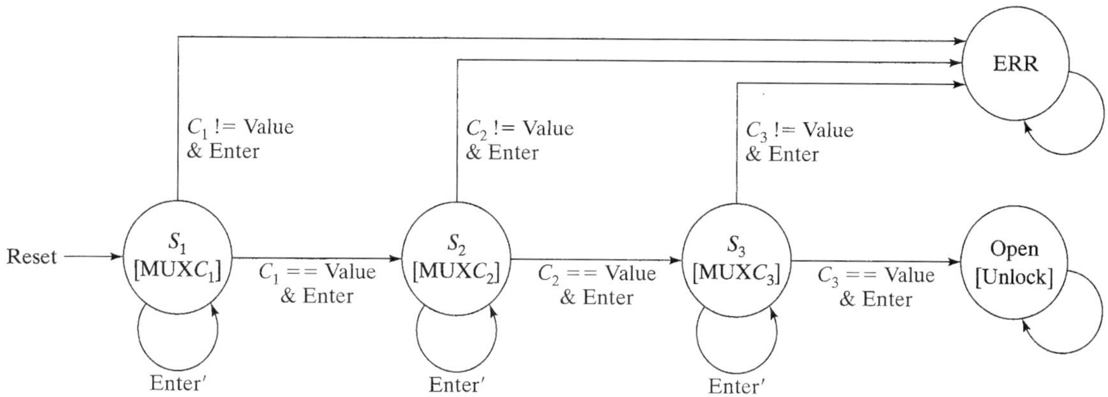

Figure 10.14 **State diagram for the combination lock.**

connecting multiple outputs that are distributed widely by permitting the use of only one wire to connect them all.

Now, turning to the controller, we have the same state diagram as in Figure 1.21. We reproduce it here in Figure 10.14 with updated notation for convenience. The encoded state table is also identical to that of Figure 1.25, and we use it to write the following Verilog description of our Moore controller:

```
module combination_lock_controller (Clk, Reset, Equal,
  Enter, Unlock, Mux);
input Clk, Reset, Equal, Enter;
output Unlock, Mux[3:1];

reg state[4:1];
parameter S1 = [0,0,0,1];
parameter S2 = [0,0,1,0];
parameter S3 = [0,1,0,0];
parameter OPEN = [1,0,0,0];
parameter ERR = [0,0,0,0];

assign Unlock = state[4];
assign Mux[3:1] = state[3:1];

always @(posedge Clk) begin
    if Reset
        state = S1;
    else
        case (state)
            S1: if (Enter) begin if (Equal) state = S2;
                else state = ERR; end else state = S1;
            S2: if (Enter) begin if (Equal) state = S3;
                else state = ERR; end else state = S2;
            S3: if (Enter) begin if (Equal) state = OPEN;
                else state = ERR; end else state = S3;
            OPEN: state = OPEN;
```

```
            ERR: state = ERR;
        endcase
end

endmodule
```

The Verilog description of our data-path component will also have Clk as an input. There will not be a state register, but there will be three registers to hold the combination digits. The Verilog for our data-path component is:

```
module combination_lock_datapath (Clk, Ld1, Ld2, Ld3,
  Value, Mux, Equal);
input Clk, Ld1, Ld2, Ld3, Value [3:0], Mux [3:1];
output Equal;

reg C1, C2, C3 [3:0];
wire MuxOutput [3:0];

assign Equal = (Value == MuxOutput);

always @(C1, C2, C3, Mux) begin
    if (Mux[3]) MuxOutput = C3;
    if (Mux[2]) MuxOutput = C2;
    if (Mux[1]) MuxOutput = C1;
end

always @(posedge Clk) begin
    if (Ld1) C1 = Value;
    if (Ld2) C2 = Value;
    if (Ld3) C3 = Value;
end

endmodule
```

We used two always blocks in our description. The first is to implement the 3:1 multiplexer. It is sensitive to any change in its three inputs (the combination digit registers) and the multiplexer control signals. Whenever any of these change, the *always* block will fire and compute a new output. The second block implements the load control for the combination digit registers. It is just three statements that determine if a load signal is asserted and store the value from the keypad into that register. In both cases, we've made an assumption that the control signals are one-hot. That is, only one of the three multiplexer control lines will ever be asserted in any one clock cycle and, similarly, for the three load control signals. The third parallel block is an assign statement that sets *Equal* to true if *Value* matches the output of the multiplexer, *MuxOutput*. This implements our comparator.

Discussion These two Verilog modules together implement our combination-lock design. Sophisticated synthesis tools will typically be used to turn these module descriptions into actual logic. Whether the tools will decide to use tri-state, open-collector, or gate logic for the data path is an open question. Some tools are able to discern higher-level functions (e.g., a multiplexer). Others need more guidance from the designer.

We have left a few details out from our design. Can we assume the *Reset* and *Enter* inputs to our controller are synchronous? Probably not, and, thus, to be complete, we should make sure to include a synchronization flip-flop on each signal before completing the design. Is there a way we can derive the Enter input from a key press on the keypad rather than requiring a separate button? This would be another module that would look at the Value coming from the keypad and determine when it changes and only trigger an *Enter* assertion at that point. When we use the phrase "determine when an input changes," we are implying sequential logic, as we need to have the history of the input to determine if it has changed. Of course, we also need to consider if we want to recognize every single change on the input, even if it is transitory. We may decide to only consider changes that are stable for some number of clock cycles and disregard the rest as being too fleeting (due to noise or slight movement on the button). This is called debouncing and it can be done with sequential logic or hardware components (see Appendix B for details).

10.4 A Memory Controller

In this section, we will examine memory components in more detail. In particular, we will focus on two important aspects of memory system design: the detailed timing waveforms for a static RAM component, and the design of the register and control logic that surrounds the memory subsystem, making it possible to interface the memory to the rest of a digital system. But first, we must begin with the basics.

10.4.1 RAM Basics: A 1024 × 4-Bit Static RAM

We begin with a relatively simple memory component: a 1024 × 4-bit static RAM. The basic storage element of the static RAM is a six-transistor circuit, shown in Figure 10.15. The *static* storage element is provided by the cross-coupled inverters. This circuit configuration will hold a 1 or 0 as long as the system continues to receive power.

The transistors provide access to the storage element from two buses, denoted Data_j and $\overline{\text{Data}_j}$. To write the memory element, special circuitry in the RAM drives the data bit and its complement onto these lines while the word enable line is asserted. When driven in this fashion, the data bit can overwrite the previous state of the element.

To read the contents of the storage element, the word enable line is once again enabled. The data lines (also called *bit lines*), are tri-stated so that the data contents of the cell can be "sensed" by a different collection of special circuits. These circuits, called *sense amplifiers,* can detect small voltage differences between the data line and its complement. If Data_j is at higher voltage than $\overline{\text{Data}_j}$, the cell contained a logic 1. If the situation is reversed, the cell contained a logic 0.

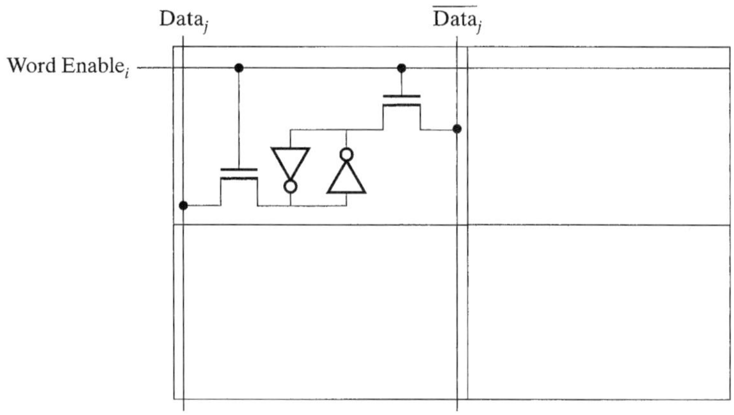

Figure 10.15 **Static RAM storage elements.**

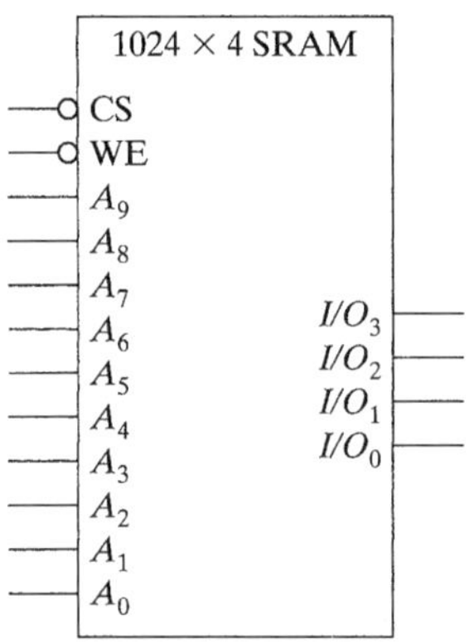

Figure 10.16 **Schematic shape of the 1024 × 4-bit SRAM.**

RAMs are efficient in packing many bits into a circuit package for two reasons. First, only a small number of transistors are needed to implement the storage elements. And second, it is easy to arrange these elements into rows and columns. Each row of memory cells shares a common word enable line. Each column shares common bit lines. The number of columns determines the bit width of each word. Thus, you can find memory components that are 1, 4, 8 or more bits wide and that read or write the bits of a single word in parallel.

Figure 10.16 shows the pin-out for a 1024 × 4-bit SRAM (static RAM). This is a very small SRAM and is somewhat dated, but it serves our purposes as an illustrative example with less pins than is common on today's memories. The pins can be characterized as address lines, data lines, and control lines. Since this RAM has 1024 words, there must be 10 lines to address them. Since each word is 4-bits wide, there are four data lines. The same *bidirectional* pins are used for reading or writing. The value on the active low control signal *Write Enable* ($\overline{\text{WE}}$) determines their direction.

The final signal on the chip is the chip select control line ($\overline{\text{CS}}$). When this signal goes low, a read or write cycle commences, depending on the value of $\overline{\text{WE}}$. If $\overline{\text{WE}}$ is also low, the data lines provide new values to be written into the addressed word within the RAM. If it is high, the data lines are driven by the memory itself with the contents of the addressed word.

Internal Block Diagram From the preceding discussion, you might infer that the RAM is organized as an array with 1024 words and four columns. In terms of performance and packaging, this is not the best internal organization. A long, thin array leads to long wires, which take more time to drive to a given logic voltage. Also, long thin integrated circuits are difficult to fabricate. A square configuration is much more desirable.

Figure 10.17 gives a more realistic block diagram of the internal structure of a typical 1024 × 4-bit SRAM. The RAM array consists of four banks of 64 words by 16 bits each. This makes the array square.

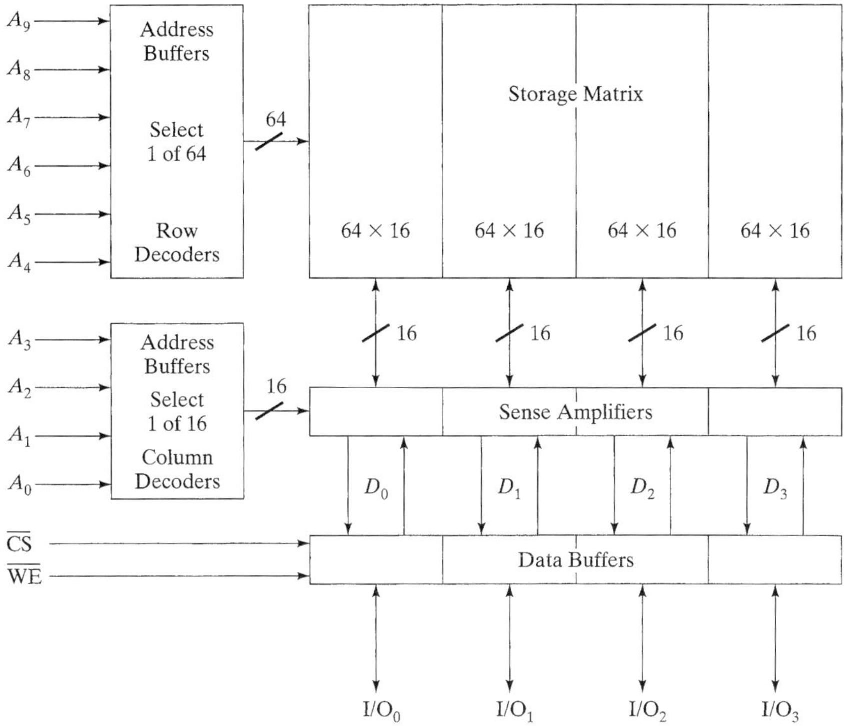

Figure 10.17 Internal block diagram of 1024 × 4-bit SRAM.

Let's consider a read operation. The high-order 6 bits of the address select one of 64 words. Four groups of 16 bits each emerge from the storage array: one group for each of the possible data bits. The four low-order address bits select one of 16 bits from each of the four groups to form the 4-bit data word. Writes are similar, except with data flowing in the opposite direction. Note that a word's bits are not stored next to each other but are inter-leaved across the four banks. This arrangement limits the effect of errors and defects and is amenable to *error-correcting codes*.

This form of 2-dimensional decode, with row and column decoders, is used universally in memory components. Not only does it keep the memory array square, it also limits the longest lines in the decoders.

Simplified Read-Cycle and Write-Cycle Timing Controlling the function of a RAM chip requires precise sequencing of the address pins and control signals. Figure 10.18 gives a simplified-logic timing diagram for the RAM read cycle (we defer a more precise description of RAM timing to Section 10.4.4). First, a valid address must be set up on the address lines. Then the Chip Select ($\overline{CS}$) line is taken low while the Write Enable ($\overline{WE}$) stays high. The *memory access time* is the time it takes for new data to be ready to appear at the output. It is measured from the last change in the address lines, although the output is not visible off-chip unless the chip select is low. Once the chip-select line goes high again (deselecting the chip) the output on the data lines will no longer be valid.

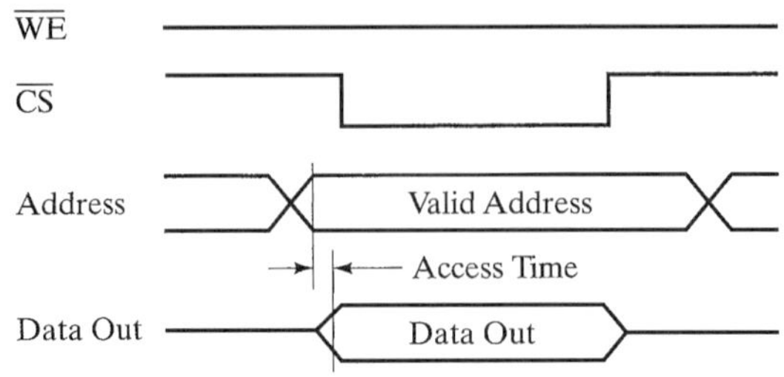

Figure 10.18 **Read cycle logical sequencing.**

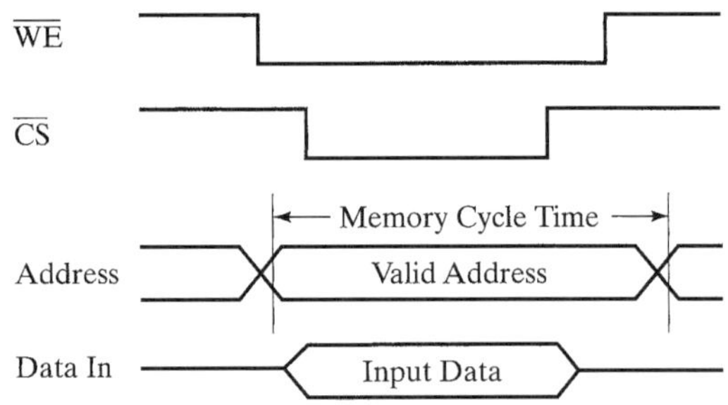

Figure 10.19 **Write cycle logical sequencing.**

Figure 10.19 gives the write cycle sequencing. Because an erroneous Write could have destructive consequences, we especially must be careful during the sequencing of the write signals. To be conservative, $\overline{WE}$ should be brought low and the address and data lines should be stable before $\overline{CS}$ goes low (similar to setup time constraints on flip-flops). A similar sequence occurs in reverse to end the write cycle.

While conceptually correct, this specification is more restrictive than it needs to be. Technically speaking, the write cycle begins when both $\overline{WE}$ and $\overline{CS}$ go low. It ends when $\overline{WE}$ goes high. The only absolute requirement is that, when the address is stable, a setup time before both signals goes low and satisfies a hold time constraint after the first of $\overline{WE}$ and $\overline{CS}$ goes high. The data setup and hold times are also measured from the first control signal to rise.

Another important metric for RAMs is the *memory cycle time*. This is the time between subsequent memory operations. In general, the access time is less than or equal to the memory cycle time.

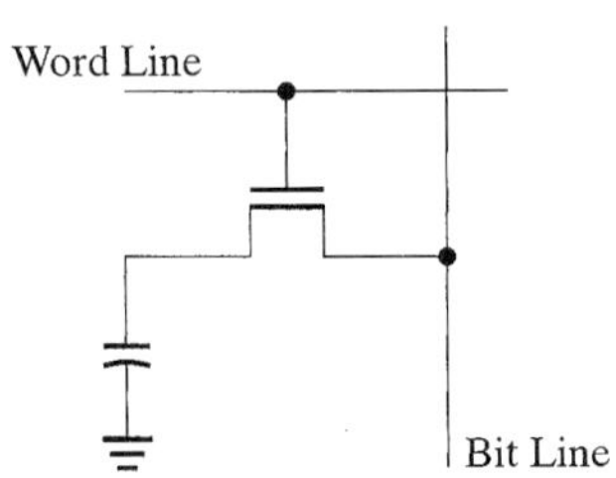

Figure 10.20 **One-transistor memory cell.**

10.4.2 Dynamic RAM

Static RAMs are the fastest memories (and the easiest to interface with), but the densest memories are dynamic RAMs (DRAMs). Their high capacity is due to an extremely efficient memory element: the one-transistor (1-T) memory cell.

The 1-T memory cell, consisting of a single access transistor and a capacitor, works as follows (see Figure 10.20). The word line and bit line provide exactly the same function as in the SRAM. To write the memory cell, the bit line is charged to a logic 1 or 0 voltage while the word line is asserted. This enables the access transistor, thereby charging the storage capacitor with the desired logic voltage.

The read operation takes place by asserting the word line. The access transistor is turned on, sharing the voltage on the capacitor with the bit line. Very sensitive amplifier circuits detect small changes on the voltage of the bit line to determine whether a 1 or 0 was stored in the selected memory element. However, by sharing the charge previously in the cell with the bit line, we lost most of it and with it the memory of the data value.

This *destructive read* operation makes DRAMs complex. Thus, external circuitry in the DRAM must buffer the values that have been read out and then write them right back immediately or they will be lost.

The second problem with DRAMs, and the most significant one from the viewpoint of a system designer, is that their contents decay over time. Usually, within a few milliseconds, the charge on the storage capacitors leaks off. To counteract this, the DRAM must be refreshed. Periodically, the memory elements must be read and written back to their storage locations. The process of writing the value back restores the charge on the capacitor to reasonable levels.

To make this operation reasonably efficient, the DRAM's memory array is a 2-dimensional matrix organized along the lines of the SRAM block diagram of Figure 10.17. Figure 10.21 shows the block diagram of a 4096×1-bit DRAM. Rather than refresh individual bits or even a word, a refresh cycle reads out and writes back an entire row. This happens about once every few microseconds. Just as in the SRAM, the row is typically a multiple of the DRAM's word size. In this case, it is 64-bits wide. The refresh cycles are generated by an external memory controller that manipulates the inputs to the control logic of the DRAM.

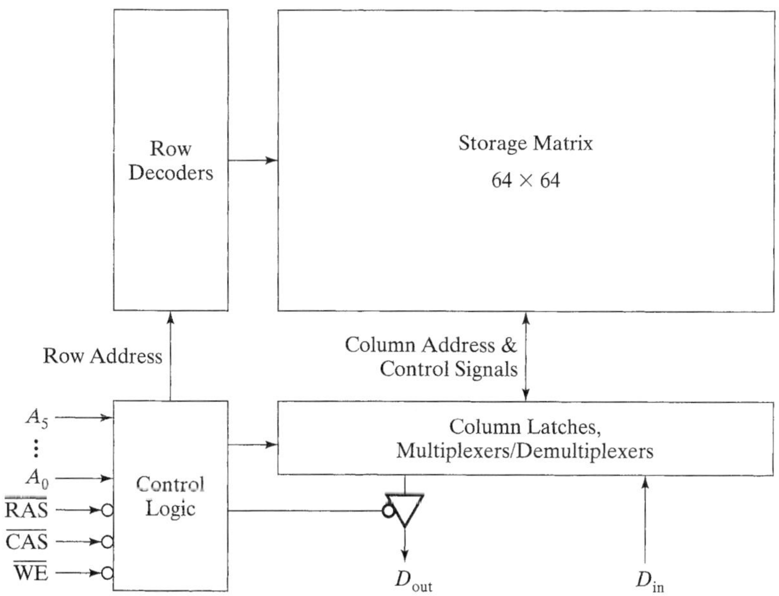

Figure 10.21 **4096×1-bit DRAM block diagram.**

DRAM Access with Row and Column Address Strobes Every time a single-bit word is accessed within the memory array of Figure 10.21, the DRAM actually accesses an entire row of 64 bits. The column latches select one of the 64 bits for reading or writing. The DRAM often accesses adjacent words in sequence, so it is advantageous if access is rapid.

Memory chip designers have developed clever methods to provide rapid access to the DRAM. The key is to provide separate control lines for DRAM row access, RAS (*row address strobe*), and column access, CAS (*column address strobe*). Normally, access involves specifying a row address followed by a sequence of column addresses that choose different entries in the single row that was read. This can speed accesses dramatically in applications where memory is read sequentially—such as screen pixels in a graphics board. This scheme has the extra advantage of reducing the number of address pins needed. A single (smaller) set of address lines are multiplexed over time to specify the row and column to be accessed. This becomes a critical issue as memory chips exceed 1 million bits (20 address pins).

In Figure 10.21, we have done away with the chip-select signal and replaced it with two signals: $\overline{\text{RAS}}$ and $\overline{\text{CAS}}$. The address lines, normally 12 for a 4096×1-bit memory, can now be reduced to 6. Memory access consists of a RAS cycle followed by a CAS cycle. During the RAS cycle, the six address lines specify which of the 64 rows to access. In the following CAS cycle, the address lines select the column to access.

Figure 10.22 shows the RAS/CAS timing for a memory read. Throughout this sequencing, the $\overline{\text{WE}}$ line is held high. First, the row address is provided on the address lines. When the $\overline{\text{RAS}}$ line is brought low, the row address is saved in a latch within the DRAM and the memory access begins. Meanwhile, the address lines are replaced by a column address. When $\overline{\text{CAS}}$ goes low, the column address is latched. At this point, the output is enabled, although it is valid only after a propagation delay. When $\overline{\text{RAS}}$ goes high again, the accessed row is written back to the memory array, restoring its values. When $\overline{\text{CAS}}$ goes high, the output returns to the high-impedance state.

Figure 10.23 shows the RAS/CAS sequencing for a memory write. The signaling begins as before: The row address appears on the address lines and is internally latched when $\overline{\text{RAS}}$ goes low. The row is now read out from the memory array. While the address lines are changing to the column address, valid data is placed on the data-input

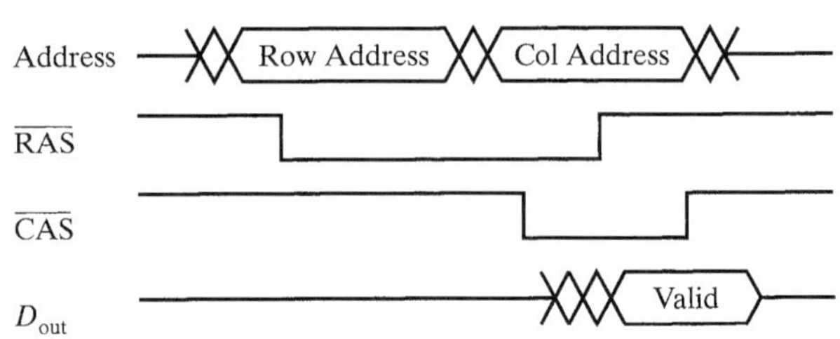

Figure 10.22 **RAS/CAS sequencing for a memory read.**

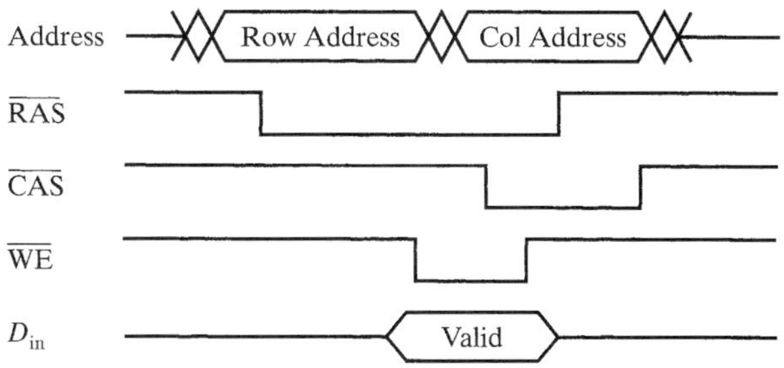

Figure 10.23 RAS/CAS sequencing for a memory write.

line and $\overline{\text{WE}}$ is taken low. Once the address lines are stable, $\overline{\text{CAS}}$ can also be taken low. This latches the column address and also replaces the selected bit with the value of D_{in} within the column latches. At this point, $\overline{\text{WE}}$ can be driven high. When $\overline{\text{RAS}}$ goes high again, the entire row, including the replaced bit, is written back to the memory array. Finally, $\overline{\text{CAS}}$ can be driven high, and another memory cycle can commence.

Refresh Cycle The storage capacitor at the heart of a DRAM memory cell is not perfect. Over time, it leaks away the charge it is meant to hold. Thus DRAMs must undergo periodic *refresh cycles* to maintain their state. In its simplest form, a refresh cycle looks like an abbreviated read cycle: data is extracted from the storage matrix and then immediately written back without appearing at the output pin.

Suppose every DRAM word must be refreshed once every 4 ms. This means that the 4096-word RAM would require a refresh cycle once every 976 ns. Assuming the cycle time is 120 ns, approximately one in every eight DRAM accesses would be a refresh cycle!

Fortunately, the 2-dimensional organization of the storage matrix makes it possible to refresh an entire row at a time. Since the DRAM of Figure 10.21 has 64 rows, we can refresh the rows in sequence once every 62.5 μs, still meeting the overall 4-ms requirement. This is approximately one refresh cycle every 500 accesses.

The *RAS-only* refresh cycle provides a simple form of refresh. It looks very much like the read timing with the CAS phase deleted. The row address is placed on the address lines and $\overline{\text{RAS}}$ is taken low. This causes the row to be read out of the storage matrix into the column latches. When $\overline{\text{RAS}}$ goes high again, the column latches are written back, refreshing the row's contents.

This refresh cycle requires an external memory controller to keep track of the last row to be refreshed. To simplify the memory controller design, some DRAMs have a refresh row pointer in the memory chip. A special *CAS-before-RAS* signaling convention implements the refresh. If $\overline{\text{CAS}}$ goes low before $\overline{\text{RAS}}$, the chip recognizes this as a refresh cycle. The indicated row is read and written back, and the internal indicator is updated to point to the next row to be refreshed. This usually is implemented as a simple counter.

10.4.3 DRAM Variations

We have described the basic internal functioning of a DRAM with RAS/CAS addressing, but have not shown how such an organization can improve DRAM performance. Variations on the basic DRAM model take advantage of the row-wide access to the storage matrix to reduce the time to access bits in the same DRAM row. These are called page mode, static column mode, and nibble mode DRAMs. All three support conventional RAS/CAS addressing. They differ in how they specify accesses to additional bits in the same row.

Page mode DRAMs can read or write a bit within the last accessed row without repeating the RAS cycle. The first time a bit within a row is accessed, the controller sequences through a RAS followed by a CAS cycle, as described earlier. To access a subsequent bit in the row, the controller simply changes the column address and pulses the CAS strobe ($\overline{\text{RAS}}$ is held low throughout). CAS pulsing can be repeated several times to access a sequence of bits in the row. The result is much faster access than is possible with complete RAS/CAS cycling.

Static column mode DRAMs provide a similar function but present a slightly simpler interface to the memory controller. Changing the column address bits accomplishes a static column read, eliminating multiple strobes on the $\overline{\text{CAS}}$ line altogether. Writes are a little more complicated. To protect against accidentally writing the wrong memory location, either $\overline{\text{CAS}}$ or $\overline{\text{WE}}$ must be driven high before the column address can be changed.

Nibble mode DRAMs are yet another variation on page mode. Most memory locations are accessed in sequence, and the DRAM can take advantage of this to reduce the complexity of the control sequencing. After the first RAS/CAS cycle, a subsequent CAS pulse accesses the next bit in sequence. This can be done three times, yielding 4 bits in sequence, before a RAS/CAS cycle is needed again. Thus, the sequence is RAS/CAS, CAS, CAS, CAS, RAS/CAS, CAS, CAS, CAS, etc.

Video RAMs (*VRAMs*) are DRAMs that can be used as frame buffers for computer displays. A *frame buffer* is a display memory that allows new data to be written to storage without affecting how the screen is being refreshed from the old data. A VRAM has a conventional DRAM storage matrix and four serial-access memories (SAMs). Its signaling convention allows a data row to be transferred from the storage matrix to the SAMs. Once in a SAM, the data can be read out a bit at a time at a high rate even while new data is being written into the storage matrix. In this way, VRAMs support a kind of dual-port access to memory: one from the standard read/write interface and one from the serial memories.

In addition, some VRAMs support logical operations, such as XOR, between the current contents of the memory and the bit that is overwriting it. This is useful for certain graphics-oriented operations, such as moving items around smoothly on the display.

10.4.4 Detailed SRAM Timing

Here we expand on the discussion of SRAM components begun in the previous section. We will describe the detailed timing of a 1024×4-bit static RAM. We have already shown the basic pin-out in Figure 10.16: 10 address lines, 4 data input/output lines, and active-low chip select ($\overline{CS}$) and write enable ($\overline{WE}$) control signals. The generic read and write cycle sequences were shown in Figures 10.18 and 10.19.

Read Cycle Let's reexamine the read cycle timing in more detail. The following discussion assumes that $\overline{WE}$ is held high throughout the read operation. Any change on the address lines causes new data to be extracted from the storage matrix, independent of the condition of the chip select. Once $\overline{CS}$ goes low, the output buffers become enabled, latching the data from the storage array and driving the output pins.

An important metric of a memory component is the *access time*, t_A. This is the time it takes for an address change to cause new data to appear at the output pins (the memory has already been selected by driving $\overline{CS}$ low). Our memory component has an access time of 200 ns, which is very slow by today's standards. High-speed static RAMs now have an access time of 8 to 10 ns and have surpassed sizes of 64 Mb (8 M words $\times$ 8 bits).

An equally important metric is the *read cycle time*, t_{RC}. This is the time between the start of one read operation and a subsequent read operation. In the case of our memory chip, it is also 200 ns. In modern SRAM components, the access time and cycle time are usually the same. However, this is not the case for DRAMs, where cycle times are often longer than access times.

Figure 10.24 shows the read-cycle timing waveform. It shows two back-to-back read cycles, the first commencing before $\overline{CS}$ goes low and the second while $\overline{CS}$ is low.

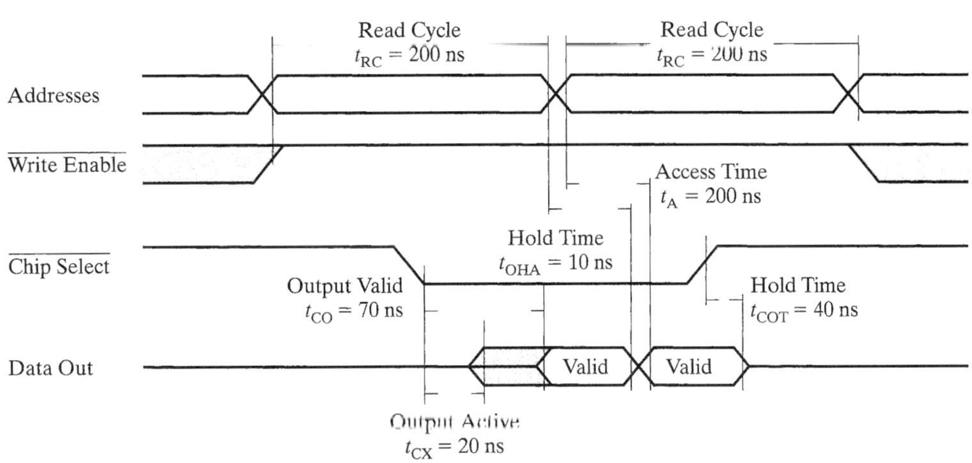

Figure 10.24 **Read cycle timing for an SRAM chip.**

The first cycle begins with a change on the address lines. Valid data does not appear on the output lines before an access time, 200 ns, has expired. The timing waveform assumes that the limiting condition is not the access time, but rather the time from when the chip is selected. T_{CX}, the time from chip select to output active, is 20 ns. This means that the memory chip will begin to drive its outputs no sooner than 20 ns after chip select goes low, although the data may not yet be valid. Any other components driving the same wires as the RAM's output lines must be tri-stated within this time.

T_{CO}, the time from chip select to output valid, is 70 ns. Thus, the time to data valid is determined by which is longer: 200 ns from the last address line change or 70 ns from when the chip is selected.

Once the address lines begin to change for the next read cycle, the outputs are guaranteed to remain valid for another 10 ns, the *output-hold-from-address-change time, t_{OHA}*. External logic latching the output of the RAM must factor this into its timing before it can allow the addresses to change.

Once the address lines are stable, it takes another 200 ns for valid data to appear at the outputs. If a third read cycle were to commence now, by changing the address lines, the output data would remain valid for 10 ns. However, the read cycle is terminated in the waveform by deselecting the chip. In this case, the hold time is determined by t_{COT}, the *time from chip select to output tri-state*. For our memory chip, t_{COT} is a minimum of 0 ns and a maximum of 40 ns. Thus external logic must wait at least 40 ns before it can begin to drive the wires at the RAM's I/O pins.

Write Cycle As in all RAMs, the write cycle timing is more complex and requires more careful design. The write operation is enabled whenever $\overline{CS}$ and $\overline{WE}$ are simultaneously driven low. This is called the *write pulse*. To guard against incorrect writes, one or both of the signals must be driven high before the address lines can change.

Figure 10.25 gives the timing waveform for the write operation. Once the address lines are stable, we must wait an address setup time before driving the last of the chip select and write enable signals low. This is the *address-to-write setup time, t_{AW}*, and is 20 ns. The *write cycle time, t_{WC}*, is defined from address change to address change and must be at least 200 ns for this component. The *write pulse width time,* t_{WP}, is at least 100 ns.

The write cycle ends when the first of the two control signals goes high. Thus, data setup and hold times are measured with respect to this event. The *data setup time, t_{DS}*, is at least 100 ns. The *data hold time, t_{DH}*, is 0 ns.

The *write recovery time*—the time between the end of the write pulse and when the address lines are allowed to change—is denoted by t_{WR}. For this RAM component, the recovery time is 0 ns.

The bottommost waveform in Figure 10.25, labeled Data Out, illustrates the interaction between read and write cycles. The writing circuitry must realize that the outputs could take as long as 40 ns to be

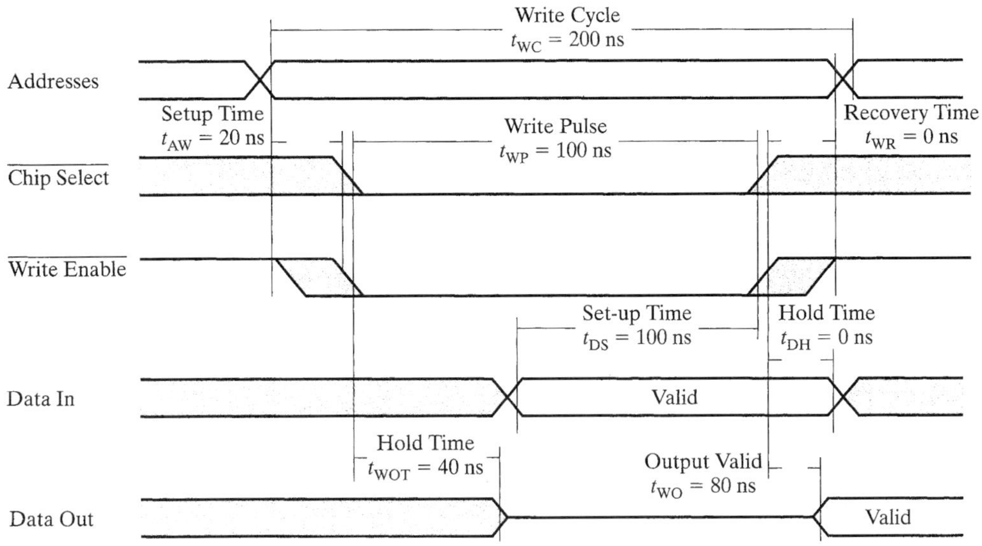

Figure 10.25 **Write cycle timing for an SRAM chip.**

tri-stated, so this amount of time must pass before the data to be written can be placed on the I/O pins. This is indicated by t_{WOT}, the *time from write enable to output tri-state.*

The final timing specification, t_{WO}, the *write-enable-to-output-valid time,* is the time between the end of the write cycle and when the RAM turns around to drive the output lines with the data just written (write enable high). In this case, it is 80 ns.

10.4.5 Design of a Simple Memory Controller

At the heart of most digital systems is a data path consisting of one or more interconnection pathways (called *buses*) and several registers, arithmetic circuits, and memory attached to some or all of these interconnections. This is the "switchyard," which routes data items from memory to a unit that executes some operation on them and then back into memory.

In this subsection, we will design a simple memory controller, using some basic fixed logic (TTL) register and counter components, as well as our static RAM chip. We will develop control circuitry for sequencing through the write enable and chip select signals to implement read and write operations.

Memory Subsystem Data Path Figure 10.26 gives a block diagram view of the data and address paths of a simple memory subsystem. To keep it simple, the address and data paths are just 4-bits wide. We can read data from four input switches and store them in the static RAM when the tri-state buffer is enabled. Data stored in the RAM can be read out, latched into a register, and then displayed on LEDs attached to the register's outputs (inverter drivers are used to buffer the LEDs). We use a

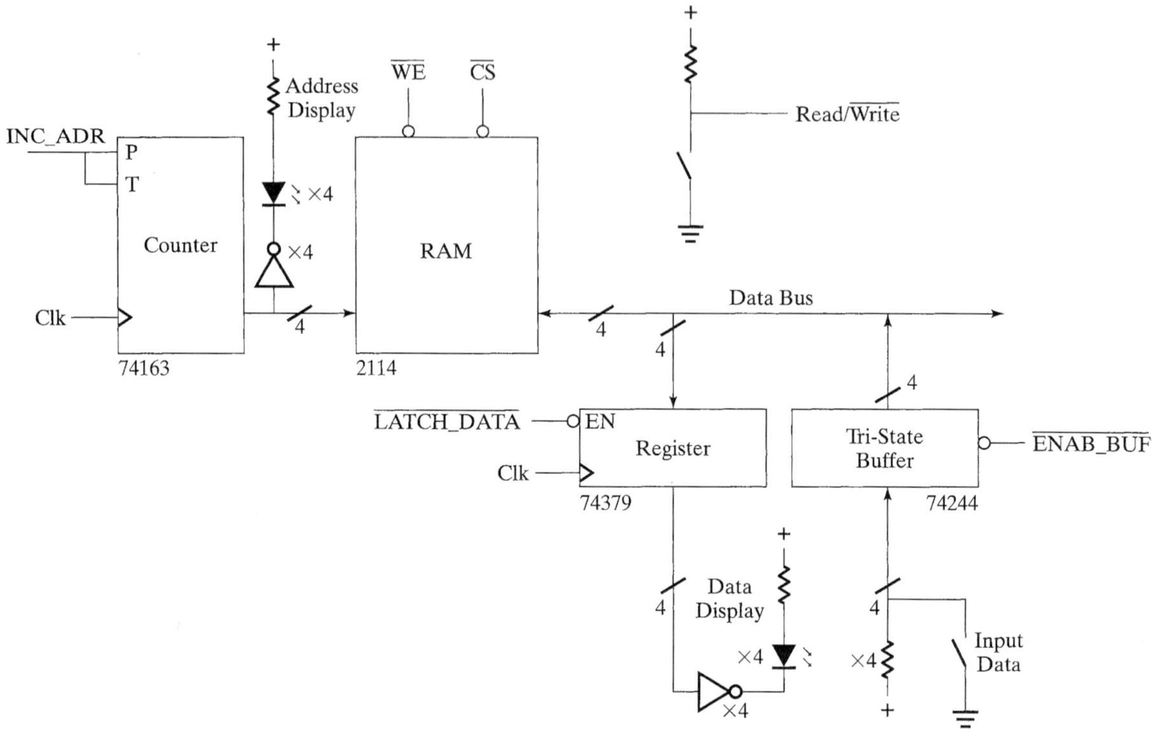

Figure 10.26 **Data path of simple memory subsystem.**

4-bit binary counter to access locations in memory sequentially for reading or writing. We also use LEDs to monitor memory addresses.

Figure 10.27 provides the schematic representation for the data path, using TTL components to implement the logic blocks of Figure 10.26. We implement the address counter with a 163 4-bit binary up-counter. For the tri-state buffers and output latches we use one-half of a 244 and a 379 component, respectively. The 379 is a 4-bit register with complementary outputs available. For the purposes of the schematic, we have replaced the output LEDs by symbols for hexadecimal displays and the input switches by a hex keypad.

Memory Controller The following signals control the data path:

INC_ADR	Add one to address
$\overline{WE}$	Write Enable on SRAM
$\overline{CS}$	Chip Enable on SRAM
$\overline{LATCH\_DATA}$	Latch valid data on data bus in display register during read cycle
$\overline{ENAB\_BUF}$	Enable buffer to put switch data on data bus during write cycle
READ/$\overline{WRITE}$	User input to select read or write mode

In addition, we need a global reset signal to force the counter to the 0000 state.

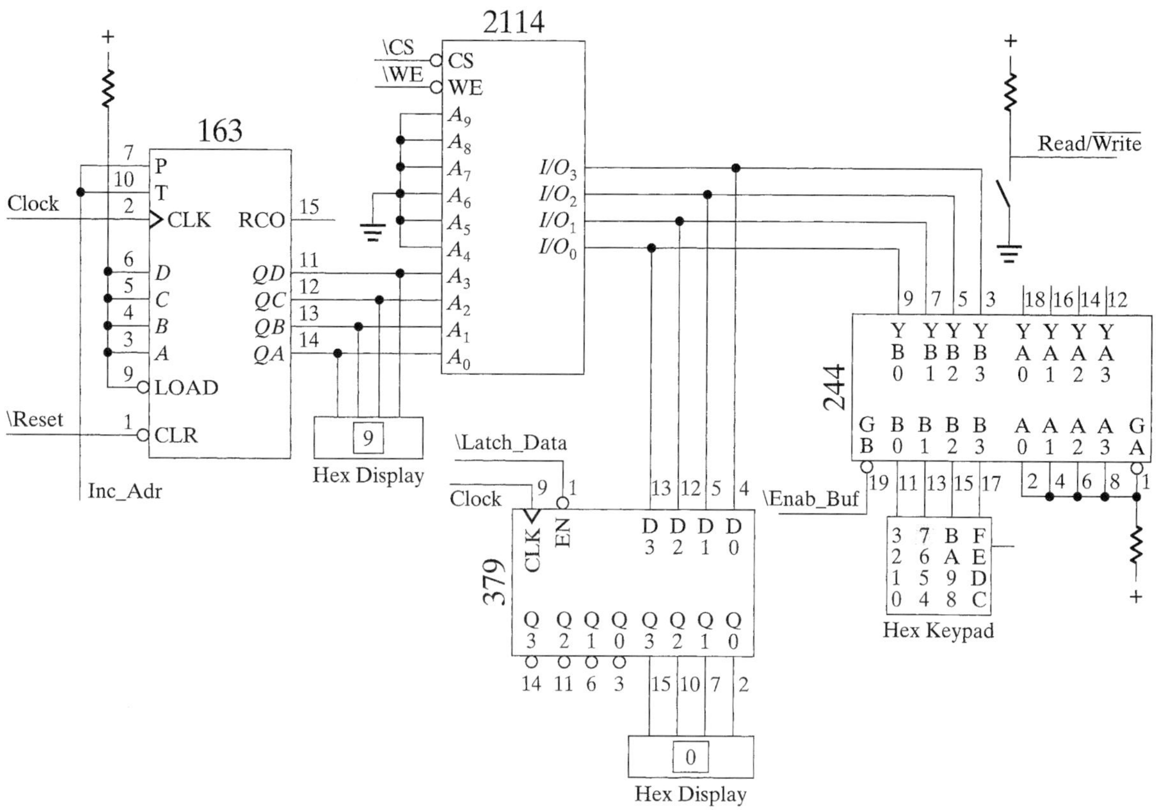

Figure 10.27 **Schematic for memory subsystem data path.**

The memory controller reads from or writes to the current address, then increments the counter to point to the next address. First, a sequence of write cycles fills the RAM with data. Then the controller is reset, setting the address counter back to 0. A sequence of read cycles then views the data that has been stored in the RAM.

Figure 10.28 shows a skeletal sequencer circuit diagram. The timing waveform it generates is given in Figure 10.29. Pressing the momentary push-button switch generates a $\overline{\text{GO}}$ signal that lasts for one clock cycle. The $\overline{\text{GO}}$ signal enables the 194 shift register, which shifts right, generating overlapping clock signals Φ_1, Φ_2, Φ_3, Φ_4. The circuit halts when all of the clock signals return to 0. This sequencer could be driven with a slow clock (such as the 555 timer described in Appendix B).

We can use simple combinational logic to derive pulses of the correct start time and length for the various control signals from the multiphase clock. Figure 10.30 shows how this is accomplished.

We start by partitioning the overlapping clocks into seven distinct periods, each of which is defined as a unique function of two of the clock phases. These seven distinct periods divide the memory-access cycle into seven states. By combining these functions, we can obtain equations for the individual control signals in each state. For example,

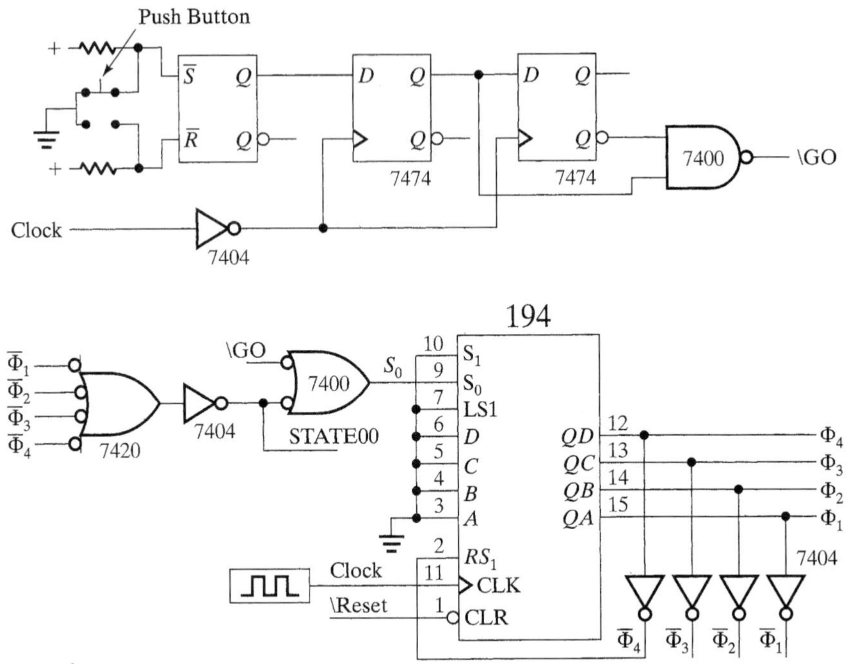

Figure 10.28 **Multi-phase clock for sequencer.**

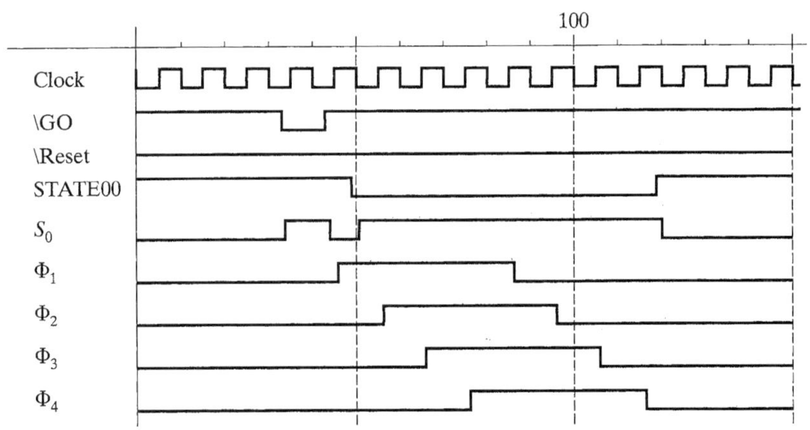

Figure 10.29 **Timing waveforms for sequencer.**

we choose to implement $\overline{CS}$ simply as the inversion of clock phase Φ_2. To be safe, we will design the high-to-low-to-high transitions on the $\overline{WE}$ signal to be properly nested within the $\overline{CS}$ transitions.

$\overline{WE}$ should be low exactly during the time that clock phases Φ_1 and Φ_3 overlap (periods 3 and 4 in Figure 10.30). This is easy to generate with combinational logic:

$$WE = \Phi_1 \bullet \Phi_3 \bullet \overline{R/\overline{W}} \qquad \overline{WE} = (\overline{\Phi_1 \bullet \Phi_3}) + R/\overline{W}$$

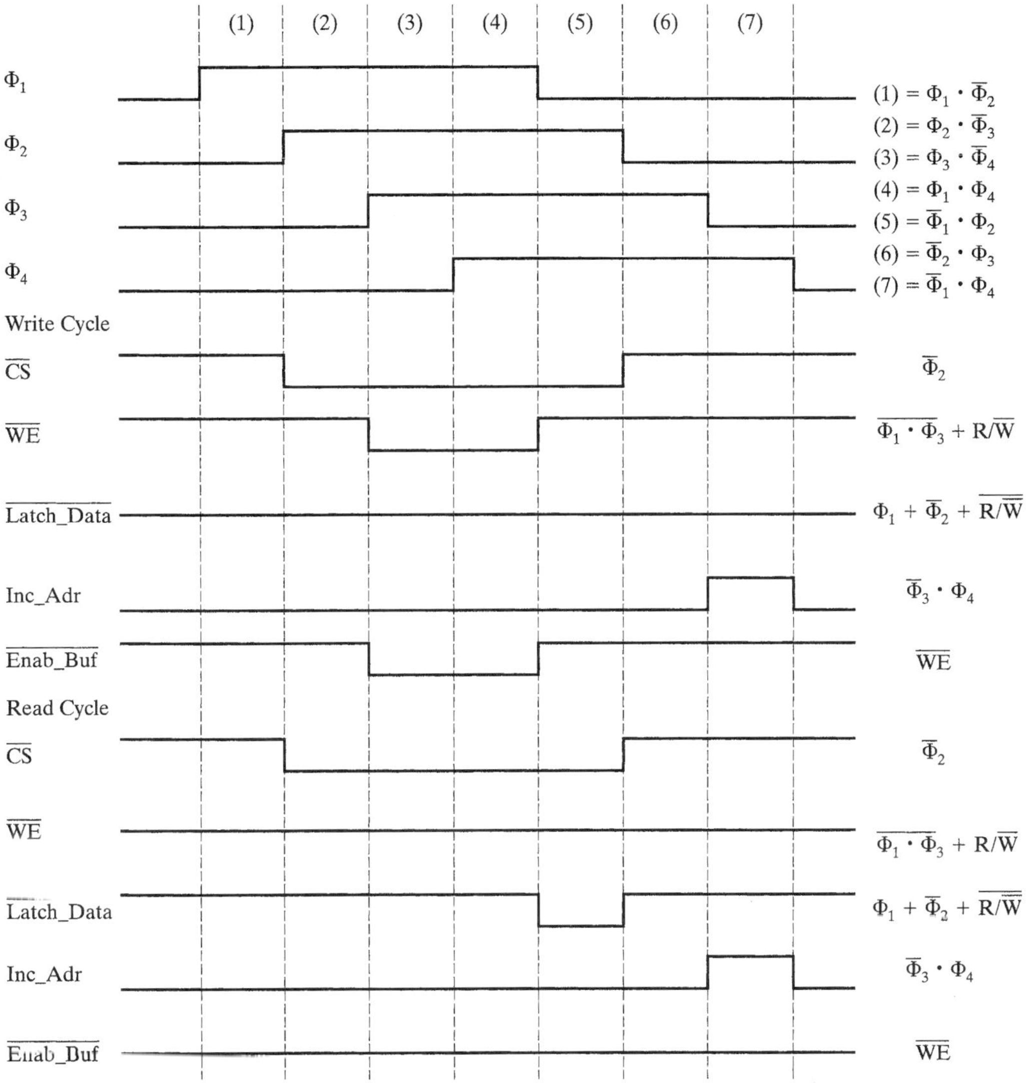

Figure 10.30 **Generation of control signals from overlapping clocks.**

The write enable signal should be asserted whenever clock phases Φ_1 and Φ_3 are asserted simultaneously and the READ/$\overline{\text{WRITE}}$ signal is low. Since $\overline{\text{WE}}$ is active-low, we invert this logic to obtain the current sense of the signal.

We assert *LATCH_DATA* during period 5, when Φ_1 is low and Φ_2 is high. Since the signal is active low, the implementation for the control signal becomes

$$\overline{\overline{\Phi_1} \cdot \Phi_2 \cdot R/\overline{W}} = \Phi_1 + \overline{\Phi_2} + \overline{R/\overline{W}}$$

INC_ADR is active during period 7, so its implementation becomes $\overline{\Phi_3} \cdot \Phi_4$. Finally, $\overline{\text{ENAB\_BUF}}$ is identical to the signal $\overline{\text{WE}}$.

Of course, alternative implementations are possible as long as they lead to valid sequencing of the control signals. Also, the logic must be designed so that the sequence meets the setup and hold time requirements for the SRAM.

Discussion The memory controller presented a new issue in sequential-logic design. Namely, how to create a state machine that can generate the appropriate timing on a set of signals so that their rising and falling edges are positioned to meet all timing constraints. In this implementation, we accomplished this by dividing up the memory read and memory write cycles into seven distinct steps, we then used a counter to step our logic through these seven steps, and, finally, we showed how to derive the waveforms we need from the state bits of the sequencer. This approach is often employed when we need to connect one circuit to another and need to generate carefully timed interface signals.

10.5 A Sequential Multiplier

In Section 5.8, we designed a combinational multiplier. This case study will redesign that multiplier so that it can handle two 8-bit quantities rather than two 4-bit ones. Unlike the combinational multiplier that computes the product in one clock cycle (a long one, to be sure, as the signals need to propagate through quite a bit of logic), a sequential multiplier computes the product in steps much as we would if we were to do a multiplication with pencil and paper. Our sequential multiplier will take 8 clock cycles to compute a product but each cycle can be relatively short as we'll only need to do one 16-bit addition and a shift of our multiplier.

Problem Specification We have been asked to design an 8×8-bit multiplier. The two 8-bit inputs, the multiplier and multiplicand, will be already loaded into two registers by other logic. Our multiplier is to compute the 16-bit product within approximately 8 clock cycles and leave the result in another register where it should stay until the next multiply operation.

Understanding the Specification Let's begin by clearing up two issues. First, we'll assume that we are dealing with unsigned binary values. Second, we'll assume that other logic will generate the load signals that will get the two 8-bit numbers into two registers. Finally, a start signal will be generated to initiate the 8-step multiplication. The result will end up in a separate 16-bit register 8 cycles later. Figure 10.31 shows a block diagram of the multiplier's data path and controller. Note that the controller has two inputs (besides the clock signal) and three outputs. The inputs are the start signal and the value of the rightmost (least significant) binary digit of the multiplier value. The outputs are a signal to clear the 16-bit accumulator register, a signal to load that same register, and a shift signal to shift the multiplier value one bit to the right (divide-by-2). The are three registers for the multiplier, multiplicand,

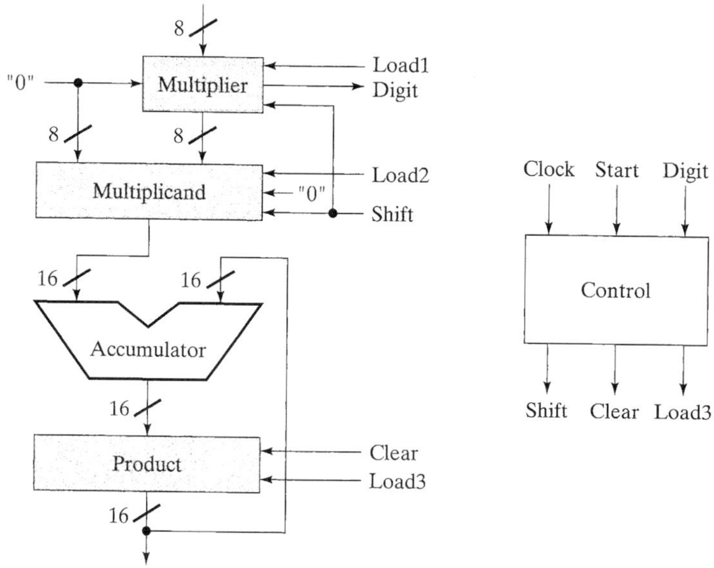

Figure 10.31 Sequential multiplier block diagram.

and product. The multiplier and multiplicand registers are loaded from the same eight data wires, with an additional eight set to zero for the multiplicand register. Of course, there is also a 16-bit adder to perform the additions of the partial products.

Implementation Since we must be able to shift the multiplier value to the right for each of eight clock cycles, the multiplier register must also implement a right shift in addition to a load capability. The multiplicand register is extended with with an extra 8-bits initially loaded with zeroes. This is so that we can multiply-by-2 via a left shift of the multiplicand register and generate the partial products we need to accumulate. Thus the multiplicand register is very similar to the multiplier register except it has to shift left rather than right (with zeroes filling in from the right rather than from the left as in the multiplier register) and is twice the size.

The accumulator is a simple 16-bit adder that could be designed using any of the techniques we learned in Chapter 5 for designing fast arithmetic circuits. It adds the contents of the product and multiplicand registers and makes the sum available at the input of the product register.

The operation of our sequential multiplier is straightforward. For eight cycles, our controller will look at a bit of the multiplier register and if it is 0, will do nothing as the partial product is 0, and if it is 1, it will add the multiplicand to the partial product and put the result in the product register. The multiplicand will have been shifted left one bit at a time in each cycle so as to adjust the value of the multiplicand to correspond to the position of the bit of the multiplier currently being used.

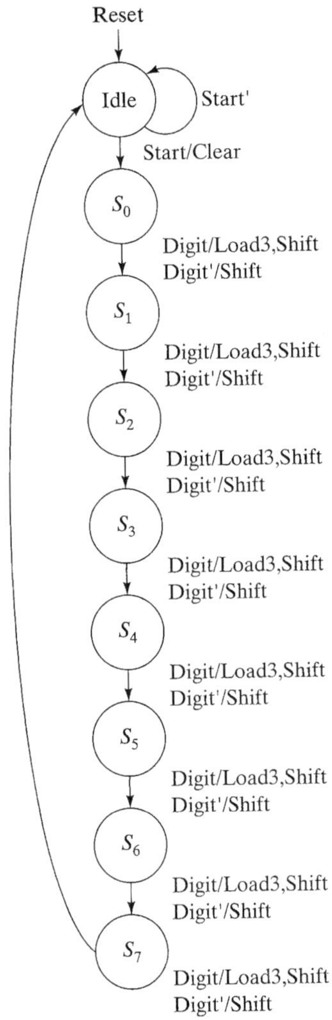

Figure 10.32 Sequential multiplier controller state diagram.

At every clock edge the following actions take place in parallel:

- The multiplier shifts one bit to the right (dividing the multiplier by 2) so that its next bit can be observed
- The multiplicand is shifted one bit to the left (multiplying the multiplicand by 2) so that it can be added conditionally to the partial product through the accumulator adder
- The accumulator adder adds the current multiplicand to the partial product from the product register
- The product register loads the next accumulated partial product

Because these actions happen in parallel, the product register loads the addition result prior to the multiplicand's shift. Let's take a look at the Mealy state diagram for the controller FSM in Figure 10.32. You'll note that in reality our multiplier takes nine cycles to compute a result as it uses the first cycle to clear the product register when it first receives a start signal. Because this controller is a Mealy machine, *Clear* is asserted as soon as *Start* is and is able to have the clearing effect on the product register on the same clock edge that takes our FSM into state S_0. In each of the states S_0 through S_7, the FSM observes the value of the low-order multiplier digit and based on that value either loads and shifts (the multiplier and multiplicand registers right and left, respectively) or it simply shifts. The load and shift happens at the next clock edge so that by the time the FSM returns to the idle state it has observed all eight bits of the multiplier and performed up to eight partial product updates (depending on the number of bits that were equal to 1 in the multiplier register).

A timing diagram for the behavior of this FSM should help make things clearer. Figure 10.33 shows the FSM starting in the *Idle* state and then advancing through the eight state sequence of the multiplication after *Start* is asserted. Note how *Clear* is asserted soon after *Start* and then is unasserted when the machine enters the S_0 state even before *Start* is deasserted. *Shift* is asserted through all eight states, *Load3* is only asserted if the digit is 1. The first partial product may be

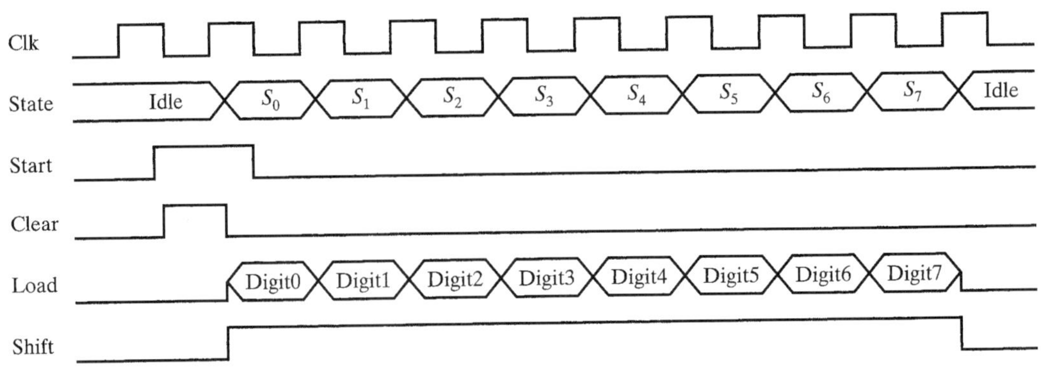

Figure 10.33 Sequential multiplier controller timing.

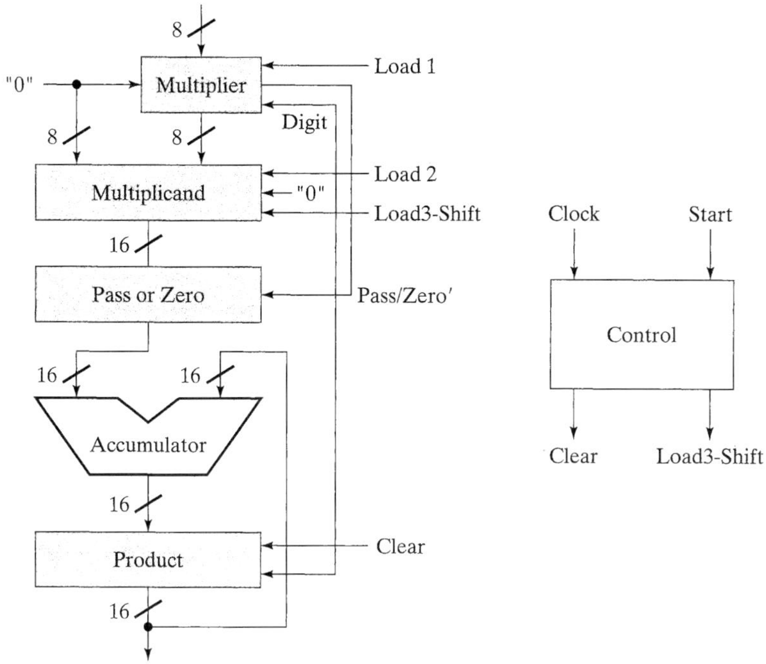

Figure 10.34 Modified sequential multiplier block diagram.

loaded into the product register as the FSM transitions from state S_0 to state S_1. The last may be loaded as it transitions from S_7 back to Idle.

Given that the value of the multiplier digit has such a direct effect on whether we add in a partial product or not, we can make the design of our multiplior FSM much simpler if we add just a bit more combinational logic to our implementation. Figure 10.34 shows a slightly modified multiplier design that includes "pass" logic. This is a very simple set of 16 AND gates, one for each bit, where one input of every AND gate is the *Pass* signal. If *Pass* is 1 then the output of the AND is just the value of its other input ("pass"). If *Pass* is 0 then the output of the AND is 0 as well ("zero"). *Pass* is connected directly to the low-order bit of the multiplier register (the previous *Digit* input to the controller).

The controller no longer uses the *Digit* input and the *Load3* and *Shift* outputs have been combined into a single signal used for both purposes. We now load the product register on each of the eight clock cycles of the multiplication. However, the value accumulated is either the multiplicand (appropriately shifted) or 0. This accomplishes the same effect as the selective loading we were doing before. Using combinational logic to implement what could have been a conditional in the state machine is a common optimization employed when a data path and controller are designed together. Designers can trade-off the complexity of the FSM for a bit more combinational logic. Often, this additional combinational logic can have multiple uses. In the case of our multiplier, it makes the

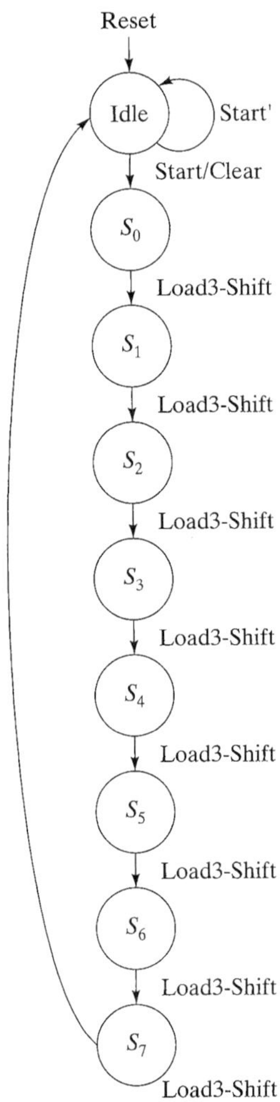

Figure 10.35 Modified sequential multiplier-controller state diagram.

state diagram for the controller considerably simpler as shown in Figure 10.35.

The encoded state table for this state diagram is shown in Figure 10.36. The state assignment is fairly straightforward. Since the FSM has two basic parts: the idle state and the eight states that actually do the multiplication. We assign 0000 to the Idle state and 1000 through 1111 to the eight other states so that they easily can be implemented with a counter. By assigning 0000 to the idle state, we make it easy to implement with resettable flip-flops.

Discussion This case study helps to show why field-programmable logic arrays (FPGAs) are so popular with designers. If we had to build our multiplier with fixed logic, we would need many components to implement the three registers (likely, five 8-bit registers), the 16-bit adders (four 4-bit adders combined together into one larger one), the controller FSM (a 4-bit counter), and miscellaneous logic for the remainder of the FSM logic and the pass logic. A PAL is not very appropriate technology for this design as there are many flip-flops (8 + 16 + 16 for the registers and four more for the controller state) and this would require several chips. The more complex PLDs of the previous chapter make a good choice as the equations for each flip-flop input are likely to have a small number of terms. However, the accumulator logic may be cumbersome in any PAL-based device, as we've seen adders can be difficult to implement with a two-level logic arrangement.

On the other hand, FPGAs seem very appropriate in this case. As we've seen, many modern FPGAs include support for building fast adders, and they have plenty of flip-flops (one for every block of combinational logic). With the 44 flip-flops in this design, it is highly likely we would be able to fit it into six or seven Xilinx Virtex-II CLBs, especially if we can arrange for the accumulator logic to be implemented in slices that have a fast carry-chain between them. Of course, if we need a faster adder, we may want to use more CLBs to implement carry-lookahead logic. Today, large FPGAs often have adders and multipliers built-in because they are so common in many large designs.

Current State		Start	Next State		Clear	Load3-Shift
Idle	0000	0	0000	Idle	0	0
Idle	0000	1	1000	S_0	1	0
S_0	1000	–	1001	S_1	0	1
S_1	1001	–	1010	S_2	0	1
S_2	1010	–	1011	S_3	0	1
S_3	1011	–	1100	S_4	0	1
S_4	1100	–	1101	S_5	0	1
S_5	1101	–	1110	S_6	0	1
S_6	1110	–	1111	S_7	0	1
S_7	1111	–	0000	Idle	0	1

Figure 10.36 Encoded state table for multiplier controller.

10.6 A Serial Line Transmitter/Receiver

Our last case study is the most substantial and will touch upon almost every aspect of digital design we have covered so far in this text. We'll consider the problem of designing hardware to communicate over a single wire. The data will, of course, be sent serially, and we'll have to make our circuit work in the general case when the clocks on the two ends are completely asynchronous. Rather than designing our own communication scheme, we'll use one of the most heavily used standards ever developed for computing equipment, the RS-232 serial line protocol.

Problem Specification Our task is to design two circuits. The first circuit will take input from a telephone-like keypad and send a byte corresponding to the key that was pressed over a single wire one bit at a time. The second circuit, at the other end of the wire, will receive the serial data sent by the first and display it on a small LCD screen.

We will be using the RS-232 protocol for formatting the data on the wire. Figure 10.37 shows how a byte of data is represented in this standard. Note that it is assumed that the wire is normally high. This is called the *quiescent* value. The byte of a data begins with a *start bit* that consists of the wire being low for one bit time. A bit time is the duration of a single bit's value on the wire. The start bit is then followed by the eight data bits with the most significant bit of the byte being sent first. After the eight bits have been sent, the wire must be high for at least one bit time before the next byte's start bit can be sent. This is called a *stop bit*. The transmission of a byte, therefore, has a duration of 10 bit times in all.

Our input device needs to support a telephone keypad consisting of 10 number keys and two more for * and #. Our display will also be similar to a simple telephone handset LCD screen that has 2 lines of up to 16 characters each. These are relatively standard parts that easily can be obtained.

Understanding the Specification An important issue to clear up is that the two circuits will be completely asynchronous to each other. The single wire we are using for communication will carry only data. The clock rate for the sending side will need to be have a period corresponding to a bit time on the wire so that our circuit will be able to switch to the next bit at the right time. On the receiving side, the situation is quite different. We need our circuit to be able to determine when a start bit is starting. If we sample the wire once every bit time

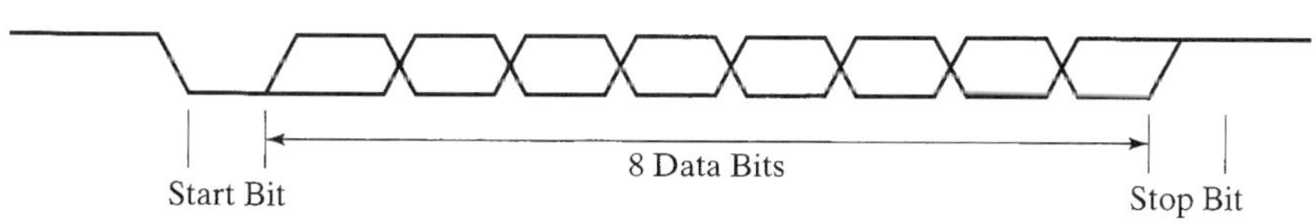

Figure 10.37 **RS-232 serial line protocol for one byte of data.**

we may just miss the start bit between two high values. Therefore, we need a faster clock that will sample the wire faster than once every bit cycle. Let's choose to sample four times during each bit time to make sure that we pick out the starting falling edge of each byte's transmission. This is called *oversampling* and is a common technique in communication circuits.

Finally, we need to understand our input and output devices. The keyboard we've found has a wire for each row and each column. When a button is pressed, the signals corresponding to the row and column of the key both go low while all the others stay high. You can imagine that the wires under every button are pressed into a connection with a common ground wire. Our LCD has an 8-bit data input along with a few control inputs. One of these control inputs is an enable input. Whenever there is a rising edge on the enable input, the character corresponding to the byte on the data inputs is displayed on the screen. In addition, the screen has some special command inputs that clear the screen, set the cursor in position, etc.

Implementation Figure 10.38 gives a block diagram of our two circuits. Our sender is on the left. It consists of two major blocks we will need to design in addition to the keyboard input device. The first block will decode the signals from the keypad and turn them in to the appropriate character code for the button that was pressed. The second block takes this byte and serially transmits it over the single wire. Our receiver is on the right. It also has two major blocks in addition to the LCD screen. The first block observes the signal wire coming from the sender and determines when a byte has been received. The second block takes this data and controls the LCD appropriately to get the corresponding character to show up on the screen. Let's take a look at each of our six components in turn. We'll start at the input to our system, the keypad, and work our way around to the output device, the LCD.

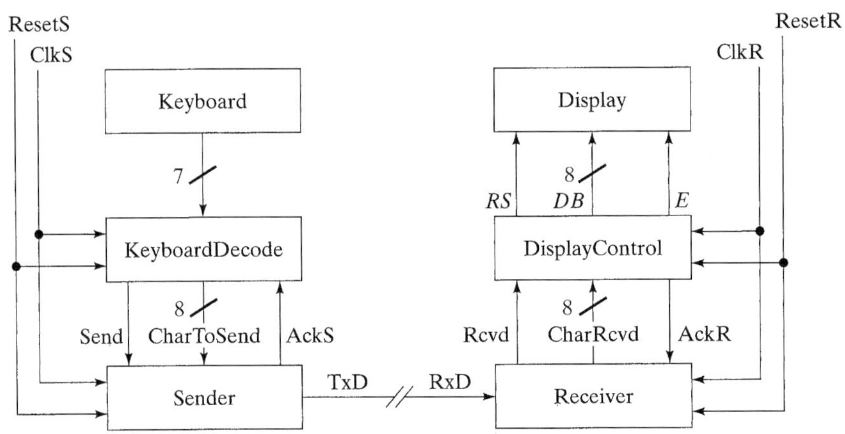

Figure 10.38 **Block diagram of serial line transmitter/receiver.**

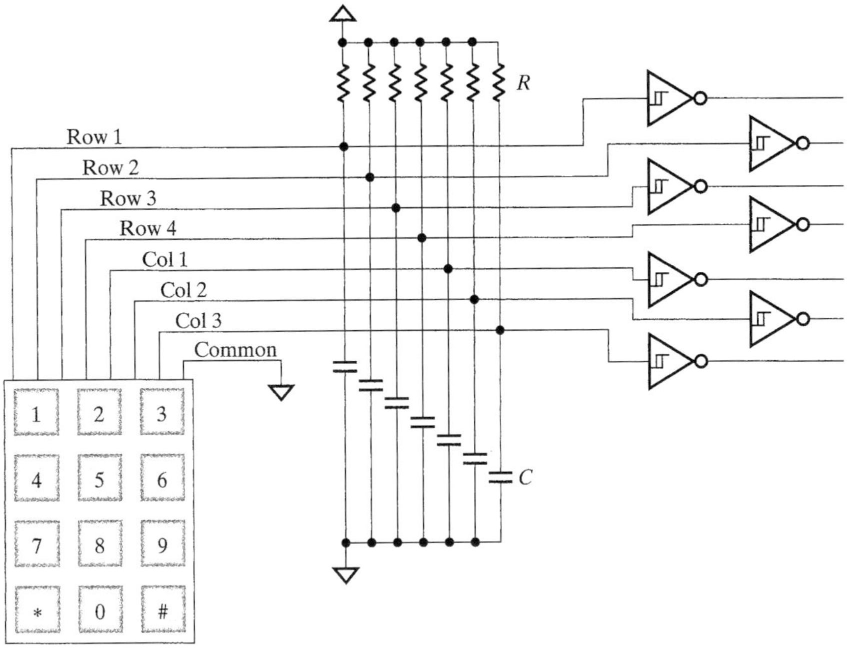

Figure 10.39 **Schematic diagram of a keyboard block with debouncing circuitry.**

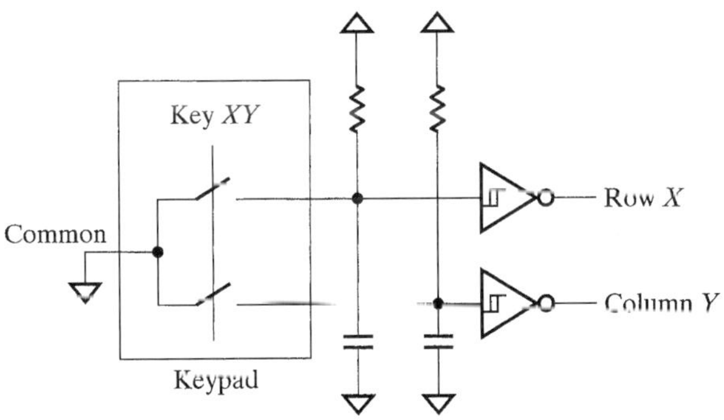

Figure 10.40 **Schematic diagram of a single keyboard key.**

Keyboard The keypad has seven wires and an eighth common wire. When a key is pressed, one of the four row wires and one of the three column wires will be low. We'll want to debounce all of these signals (details on debouncing can be found in Appendix B).

Figure 10.39 shows the circuitry we'll place around the keypad. The *Rs* and *Cs* are used to debounce each of the row and column wires along with a Schmitt trigger inverter that, in addition to helping with the debouncing, also inverts the signals to positive logic. With this arrangement, we'll see the corresponding row and column wires at the inverter outputs go high when a key is pressed. Figure 10.40 shows the detail of a single key.

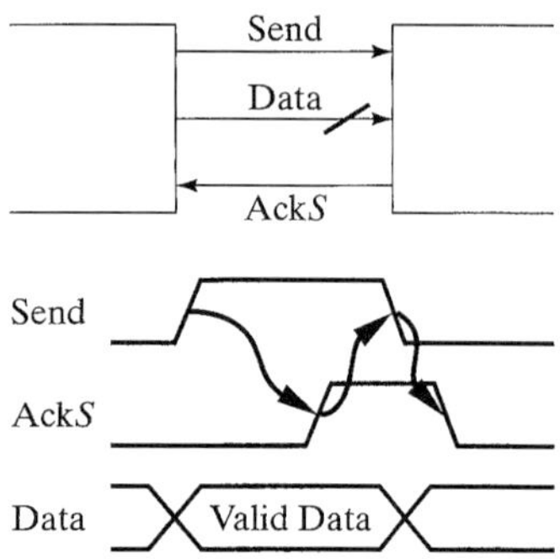

Figure 10.41 Four-cycle handshake with data.

Keyboard Decode Our next task is to decode key presses into the 8-bit character code that will need to be sent over the wire. The *KeyboardDecode* block will perform this function. To make our system more robust and modular we'll also include a four-cycle handshake between the *KeyboardDecode* block and the Sender block. Remember also that a key press is an asynchronous event and the extra handshaking will help with this as well. Figure 10.41 shows a simple four-cycle handshake using two signals, *Send* and *AckS* (for "acknowledge the send"). *Send* is raised first and will be seen by the other block that will then raise *AckS* in response. This in turn will be seen by the original block that is now assured its raising of the *Send* output was observed. It then lowers *Send* which allows the other block to also lower *AckS* and return to the original state and be ready for the next handshake cycle. Either block can take more time to do what it needs to do by delaying when it raises or lowers its signal. If data is being sent along with the handshake, then the data should be held constant from when *Send* is raised to when the acknowledgement, via *AckS*, is received.

The Verilog description of the *KeyboardDecode* block is shown next. It has two major blocks: one to determine if a key had been pressed and another to communicate the corresponding data to the next block.

```
module KeyboardDecode (ClkS, ResetS, R1, R2, R3, R4, C1,
  C2, C3, AckS, Send, CharToSend);

input     ClkS, ResetS, R1, R2, R3, R4, C1, C2, C3, AckS;
output    Send, CharToSend[7:0];

reg [7:0] CharToSend;
reg       Send;
wire      KeyPress;

assign KeyPress = (R1 || R2 || R3 || R4) && (C1 || C2 || C3);

always @(posedge ClkS) begin
    if (ResetS) begin Send <=0; end
    else begin
        if (KeyPress && !AckS) begin
            if      (R1 && C1) CharToSend <= 8'b00110001;
             // code for 1
            else if  (R1 && C2) CharToSend <= 8'b00110010;
             // code for 2
            else if  (R1 && C3) CharToSend <= 8'b00110011;
             // code for 3
            else if  (R2 && C1) CharToSend <= 8'b00110100;
             // code for 4
            else if  (R2 && C2) CharToSend <= 8'b00110101;
             // code for 5
            else if  (R2 && C3) CharToSend <= 8'b00110110;
             // code for 6
            else if  (R3 && C1) CharToSend <= 8'b00110111;
             // code for 7
```

```
            else if  (R3 && C2) CharToSend <= 8'b00111000;
            // code for 8
            else if  (R3 && C3) CharToSend <= 8'b00111001;
            // code for 9
            else if  (R4 && C1) CharToSend <= 8'b00101010;
            // code for *
            else if  (R4 && C2) CharToSend <= 8'b00110000;
            // code for 0
            else if  (R4 && C3) CharToSend <= 8'b00100011;
            // code for #
            Send <=1;
        end
    if (AckS) Send <=0;
end

endmodule
```

KeyPress is a simple combinational signal that is true if one row and one column are high. On every positive edge of the clock, the module checks if *KeyPress* is high and that *AckS* from the next block is low (so that it knows it can start another four-cycle handshake). It then computes the ASCII character code corresponding to the key that was pressed through a long chain of `IF` statements and raises the *Send* signal. We use ASCII character codes because that is what our display will use. This way we won't need to do any further translation on the receive side. Nothing changes until *AckS* is asserted by the other block. At that point, *Send* can be lowered and the cycle can begin anew.

Note that if the user takes his finger off the key while the handshake is happening, the data will not change, as the byte to be sent is stored in the *CharToSend* register. Also, if a second key is pressed, it will be ignored until the handshake is complete. In fact, if it is a quick key press that is no longer there when the handshake completes (that is, *KeyPress* is no longer true), then it will be missed entirely. This is probably a reasonable assumption as our circuitry will be working at much faster-than-human speeds.

Sender Our next module is the Sender that serializes the data it receives from the *KeyboardDecoder* into the RS-232 format. It will implement the other half of the four-cycle handshake with the *KeyboardDecode* module and will need to include a counter to help it step through the 10 bits that will need to be sent over the serial line for each key pressed on the keyboard. The Verilog for this module is as follows:

```
module Sender (ClkS, ResetS, Send, CharToSend, AckS, TxD);

input  ClkS, ResetS, Send, CharToSend [7:0];
output AckS, TxD;

reg    TxD; // register alias for RS232 line
reg    AckS;// register alias for acknowledgement line
```

```
reg [7:0] charToSend; // holds character to be sent over
                      // the serial line
reg [3:0] bitCounter; // keeps count of which bit is
                      // being sent
reg       go;         // indicator to start sending character

always @(posedge ClkS) begin
    if (ResetS) begin TxD <= 1; go <=0; end
    else begin
        if (Send) begin
            // load character sent from keypad and hold
            // it just in case
            // reset bitCounter to zero for start of new
            // transmission and set the go bit
            charToSend <= CharToSend;
            bitCounter <= 0; go <= 1;
        end
        else begin
            bitCounter <= bitCounter + 1;
            // if go is set, its time to send the
            // next bit of the 10 to be sent -
            // start bit, 8 bits of character to send
            // (most significant bit first),
            // and a stop bit at the end - reset go when done
            if (go) begin
                if (bitCounter == 0) TxD <= 0;
                else if (bitCounter > 0 && bitCounter <= 8)
                TxD <= charToSend[8 - bitCounter];
                else if (bitCounter > 8) begin TxD <= 1;
                  go <= 0; end
            end
        end
    end
end

assign AckS = go;

endmodule
```

This module keeps the serial line (*TxD*) high until *Send* is asserted. When that occurs, it stores the data byte to be sent in its own register for safe keeping, resets its bit counter to 0, and raises the *go* signal. On the next clock cycles, it will send out bits of data and increment the bit counter. Note that delayed assignment is used for the bit counter so that its value does not change until after the always completes execution. The `IF` statement is comparing the value of *bitCounter* before the increment takes effect. Once we reach the last bit (the stop bit), *go* is lowered.

The handshake with the *KeyboardDecode* block is implemented by having *go* directly assigned to *AckS*. This will keep the *KeyboardDecode* module from telling the *Sender* module about another key being pressed until the character has been sent and the *Sender* module is ready again.

Receiver We now turn to the second block that will implement our receiving circuitry on the other side of the wire and display the data received on the LCD. The *Receiver* module is the most complex in this case study. It will have to sample the input to determine when a start bit occurs and then store the value of the 8 bits after the start bit and pass them on to the *DisplayControl* module. Its Verilog description is shown next.

```
module Receiver (ClkR, ResetR, RxD, AckR, CharRcvd, Rcvd);
input    ClkR, ResetR, RxD, AckR;
output   CharRcvd [7:0]; Rcvd;

reg [4:0]  bitCounter;         // keep count of number of
                               // bits received
reg [1:0]  cycleCounter;       // used to divide input
                               // clock by 4
reg [8:0]  characterReceived;  // start bit plus 8 data bits
reg        go;                 // flag indicating a character
                               // is arriving
reg        Rcvd;               // register alias for output
                               // telling DisplayControl
                               // module a character has
                               // been received
reg [7:0]  CharRcvd;           // register alias for
                               // character value received

// finite state machine to receive characters
always @(posedge ClkR) begin
    if (ResetR) begin go <= 0; Rcvd <= 0; end
    else begin
        if (!go && !Rcvd && !AckR && RxD) ;
        // serial line is still quiet, nothing to do
        if (!go && !Rcvd && !AckR && !RxD) begin
        //found a start bit, reset the counters
            go <=1;
            bitCounter <= 0;
            cyclcCounter <= 0;
        end
        if (go) begin
            // every four cycles . . .
            if (cycleCounter == 0) begin
                // copy sample into character bit array
                characterReceived[8 - bitCounter] <= RxD;
                // if the counter has not reached 8,
                // increment it
                if (bitCounter < 8) bitCounter
                  <= bitCounter + 1;
                else begin
                    // otherwise, signal that the entire
                    // character has been received by
                    // starting handshake and resetting
                    // counter and
        flag
                    go <= 0;
```

```
                        Rcvd <= 1;
                    end
                end
                cycleCounter <= cycleCounter + 1;
                // increment counter
            end
            if (AckR) Rcvd <= 0;
// complete four cycle
//handshake with DisplayControl
    end
end

// connect the CharRcvd lines to the character received
// ignoring the start bit
assign CharRcvd[7:0] = characterReceived[7:0];

endmodule
```

The important control signal in this module is the *go* signal. When it is high, the module is busily receiving a character. When it is low, the serial line is quiet. The *go* signal is raised when we see a 0 on the serial line signifying a start bit. In the very next cycle, we'll store that start bit in the *characterReceived* register that will hold the start bit and the eight data bits.

We use an internal counter to count off four cycles of the clock between samples of the serial-line input. Recall that we decided to oversample at four times the sender's clock rate so that we'd be sure to catch the start bit of a character. Now, we have to pay the price and divide the clock down by a factor of 4 using a 2-bit counter called *cycleCounter*. We only increment the counter when a character is being received (*go* is high). We sample the serial line every four cycles or, in other words, whenever the *cycleCounter* is 0 again.

We also keep another counter, the *bitCounter*, to keep track of how many bits we've sampled. Once we reach 8 on this counter (corresponding to 9 bits sampled including the start bit), we know we have received a character completely and can signal the next module, the *DisplayControl* module, using a four-cycle handshake consisting of the *Rcvd* output signal and the *AckR* input signal coming back. Note that we ensure we won't set *go* high unless we've finished any handshake we may have started and both *Rcvd* and *AckR* are low.

Before we turn to the *DisplayControl* module, we'll need to understand how the LCD screen's interface operates.

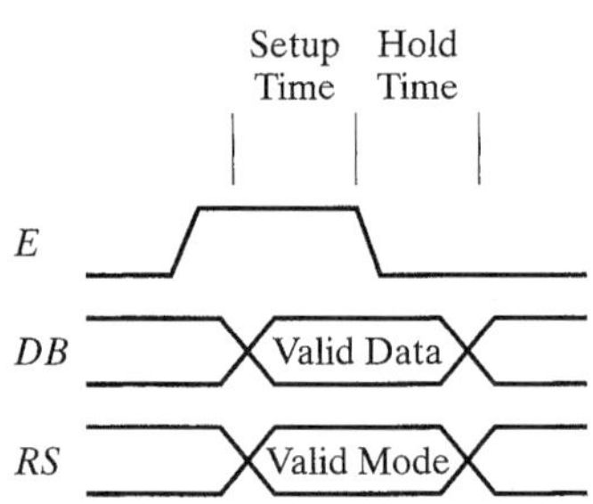

Figure 10.42 Timing of LCD screen interface.

Display As with many similar devices, the LCD has a simple enable-driven interface. There are a set of data lines and mode inputs that must be stable when a distinguished signal (the enable signal) has a falling edge on it. When this occurs, the LCD interprets the mode and data inputs and takes the appropriate action. The timing for the LCD's interface is shown in Figure 10.42. Think of the enable signal (*E*) as acting very similarly to a clock signal with setup and hold times for the corresponding data. Our clock will be very slow compared to the

Operation	*RS*	DB7...DB0
Clear display	0	0000 0001
Function set	0	0011 0011
Display on	0	0000 1100
Entry mode set	0	0000 0110
Write character	1	DDDD DDDD

Figure 10.43 LCD screen operations.

timing constraints of the LCD, and since we'll only be changing signal values at clock edges, we can be confident in knowing that we'll meet the display's timing constraints. If we had a much faster clock, we would have had to examine the situation more carefully, possibly taking multiple clock cycles to make changes to the LCD inputs.

The display module we've chosen supports many operations. A sampling of them, the ones we will use, are shown in Figure 10.43. The *RS* input is used to indicate how to interpret the data bits, as a command if it is low and as a character to be displayed if it is high. To properly initialize and clear the display, we'll execute the operations in the first four rows of Figure 10.43 whenever we reset our circuit.

We can now move on to the last module of our implementation, the *DisplayControl* module.

DisplayControl The controller for the display needs to accomplish two main tasks: initializing the display and displaying the characters received by the receiver module. It has to use the signalling convention of the LCD module and implement the four-cycle handshake with the *Receiver* module using the *Rcvd* and *AckR* signals. It is by far the longest Verilog description of all the modules in this case study, but it is not particularly complex.

The Verilog description of this module is shown next and consists of three principal parts: an initialization state machine that steps sequentially through 10 states, a simple two-state machine that sends characters to the display, and logic to implement the four-cycle handshake with the *Receiver* module.

```
module DisplayControl (ClkR, ResetR, Rcvd, CharRcvd,
  AckR, DB, RS, Enable);

input         ClkR, ResetR, Rcvd, CharRcvd[7:0];
output        AckR, DB[7:0], RS, Enable;

reg [7:0] databus;     // internal databus that can
                       // be connected to DB lines
reg [7:0] DB;          // register aliasing for LCD
                       // screen DB lines
reg       Enable;      // register aliasing for
                       // LCD screen E signal
reg       RS;          // register aliasing for LCD
                       // screen RS signal
```

```
reg       AckR;        // signal to Rcvr module to
                       // clear its Rcvd output
                       // this implements a four-cycle
                       // handshake with Rcvd
reg       initMode;    // indicates whether the initialize
                       // sequence is in progress
reg[4:0]  initState;   // states for initialization sequence

parameter start = 0;
parameter step1 = 1;
parameter endstep1 = 2;
parameter step2 = 3;
parameter endstep2 = 4;
parameter step3 = 5;
parameter endstep3 = 6;
parameter step4 = 7;
parameter endstep4 = 8;
parameter final = 9;

// drive DB lines from internal databus during
// initialization sequence
// or from CharRcvd register when displaying characters
assign DB = (initMode) ? databus : CharRcvd;

always @(posedge Clk) begin
  if (ResetR) begin initMode <= 1; initState <= start;
    Enable <= 0; AckR <= 0; end
  else if (initMode) begin
       // do initialization sequence
       case(initState)
         start: // set display to receive commands
              begin RS <= 0; initState <= step1; end
         step1: // 1st step of initialization
                // sequence - clear display
              begin Enable <= 1; databus <= 8'b00000001;
                initState <= endstep1; end
         endstep1:
              begin Enable <= 0; initState <= step2; end
         step2: // 2nd step of initialization sequence -
                // function set
              begin Enable <= 1; databus <= 8'b00110011;
                initState <= endstep2; end
         endstep2:
              begin Enable <= 0; initState <= step3; end
         step3: // 3rd step of initialization sequence -
                // display on
              begin Enable <= 1; databus <= 8'b00001100;
                initState <= endstep3; end
         endstep3:
              begin Enable <= 0; initState <= step4; end
         step4: // 4th step of initialization sequence -
                // entry mode set
```

```
            begin Enable <= 1; databus <= 8'b00000110;
               initState <= endstep4; end
         endstep4:
            begin Enable <= 0; initState <= final; end
         final:
            begin
            RS <= 1;            // set display to
                                // receive characters
            initMode <= 0;      // indicate initialize
                                // completed
            initState <= final; // stay in this state
                                // until next reset
            end
      endcase
   end
   else begin
      // or see if character has been received
      if (Rcvd && !AckR) begin
         // character has been received by Receiver module
         // display character on display and
         // begin handshake with Receiver module by
         // raising AckR
         case(Enable)
               0: begin Enable <= 1; end
               1: begin Enable <= 0; AckR <= 1; end
            endcase
         end
         else begin
            // complete handshake with Receiver module by
            // lowering AckR
            // after Rcvd is lowered
            if (!Rcvd) AckR <= 0;
         end
      end
   end

   endmodule
```

The initialization sequence consists of issuing the four commands at the top of the table in Figure 10.43. Each command requires two states. In the first state, the data is put on the internal data bus and the enable signal is raised. In the second state, the enable signal is lowered triggering execution of the command by the LCD. When the initialization sequence ends, the mode input *RS* is set high and the state machine stays in its final state until the next reset.

When a character is received (indicated by *Rcvd* being high and *AckR* as low to be consistent with our four-cycle handshaking protocol), the module cycles the enable signal in the same way as during the initialization sequence, only this time with the character received on the data bus to the LCD. This multiplexing of data to the LCD is accomplished in a separate assign statement and the multiplexor control is the internal

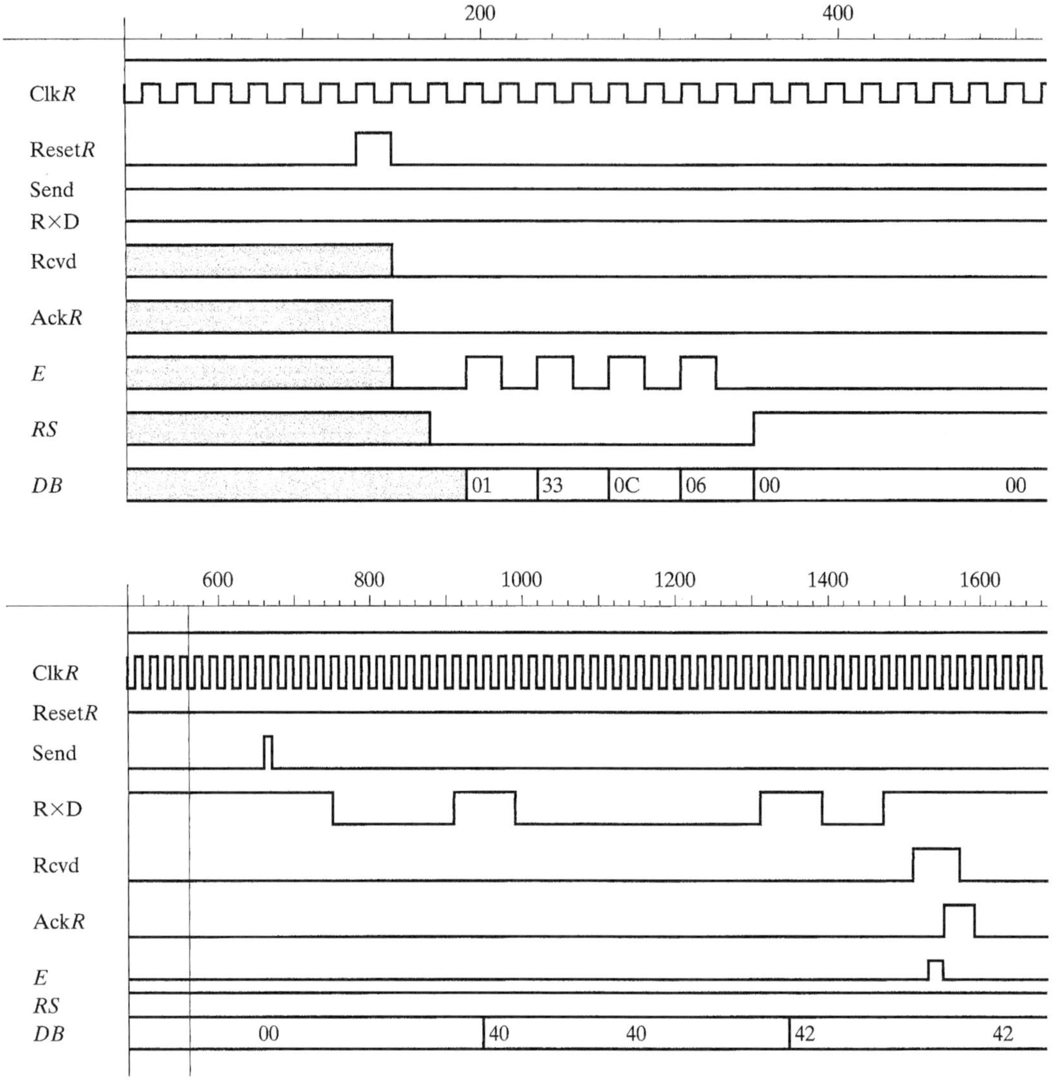

Figure 10.44 Timing diagram for the receiver half of the transmitter/receiver.

initMode signal. *AckR* is raised as the character is written to the LCD. The last clause lowers *AckR* once *Rcvd* is lowered by the *Sender* module.

Detailed timing for the receiver part of our circuit is shown in Figure 10.44. The top timing diagram shows the initialization sequence of the *DisplayControl* module. The bottom timing diagram shows the *Receiver* receiving a character and the handshake with the *DisplayControl* module to get it on the LCD.

Discussion The two parts of our serial-line controller could be implemented with many of the technologies we've discussed in this and earlier chapters. We've already seen all of the components of this example mapped to these technologies so we will not discuss it again here. The point of this case study was to show how to breakdown a larger problem into components and how to define the interfaces between those components. There are many ways to accomplish this, the design presented here is merely one of these. Most of the design choices were made so as to make the implementation easy to understand.

CHAPTER REVIEW

This chapter consisted of a sequence of case studies, progressively more complex, with each highlighting different aspects of the sequential-logic design techniques we have mastered in these last few chapters. The objective was to get familiar with the process of understanding a problem, breaking it down into its components, realizing each component in a circuit or hardware description language description, understanding how the components communicate, and, finally, realizing the circuit using the technologies from the last chapter.

The six case studies began with a simple string recognizer and moved on to a complex counter and the combination lock from Chapter 1. The last three case studies were much more substantial than the first three with many more design choices available. Each of these three highlighted a different aspect of digital design. The memory controller focused on timing, the sequential multiplier on the split between control and datapath (that we'll revisit in the next chapter), and the serial line transmitter/receiver that emphasized creating components and their interfaces to solve a larger problem.

All the technologies of Chapter 9 can be used to realize any of these case studies. However, by now it should be obvious, that as the size and complexity of the circuits increases programmable logic solutions are the most productive way to go. Of these, field-programmable gate arrays have the most promise and are, in large part, the dominant technology for logic design today.

FURTHER READING

Logic design problems are found all around us. Every textbook for logic design includes numerous design problems with solutions to varying levels of detail. In addition to textbook resources, however, many programmable logic manufacturers include many application notes on their web sites that detail solutions to common problems encountered by their customers and the solutions they generated. Finally, there are a large number of popular and hobbyist magazines, such as *Circuit Cellar*, that have many articles per issue on interesting and timely logic-design solutions.

EXERCISES

10.1 *(Word Problem)* You are to design a finite state machine to control the position of a mechanical arm. Your inputs include two registers, R_0 and R_1, that contain the current position of the arm and the target position, respectively, encoded as 32-bit twos-complement numbers. R_0 automatically is updated by external logic on every clock pulse.

The machine should operate as follows. The FSM commences operation when a *Start* pulse is asserted. If the current position is less than the target, the machine should assert a *Forward* signal. If the position is greater than the target, a *Reverse* signal should be asserted. If the position is already correct, return to the initial off state. When the arm has moved seven eighths of the way to the target from its initial position, an additional *Slow* output should be activated to brake the motion of the arm. You may assume that the arm moves slowly enough with respect to your FSM's clock rate that you need not worry about overshooting the target.

You will probably need additional data-path objects besides registers R_0 and R_1. Draw a register diagram of your data path, showing the elements and how they are interconnected. Then draw the controller's state diagram, showing the high-level register transfer operations that are asserted in each state or transition (you may choose Moore or Mealy implementation, at your own discretion).

10.2 *(Word Problem)* Consider the following variation on the classical traffic-light controller. The intersection is shown in Figure Ex. 10.2. A Street runs north to south, B Street runs east to west, and C Street enters the intersection from the southeast. A Street is quite busy, and it frequently is difficult for cars heading south on A to make the left turn onto either B or C. In

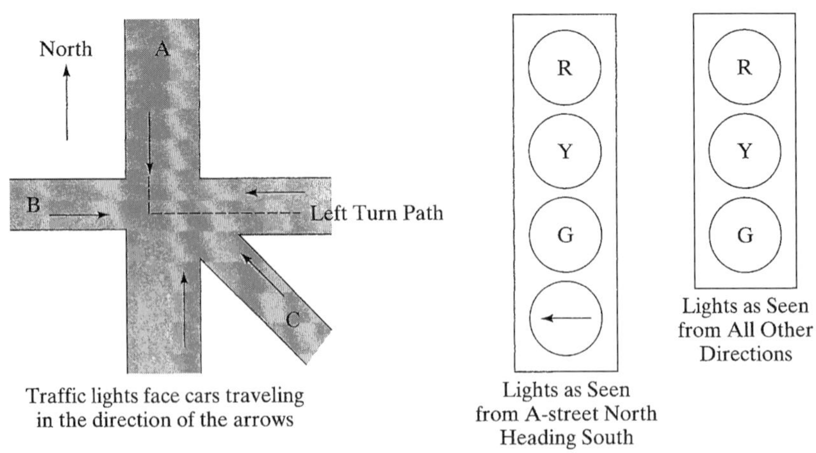

Figure Ex. 10.2 **Traffic light problem specification.**

addition, cars rarely enter the intersection from C Street. Design a traffic light state diagram for this three-way intersection to the following specifications:

(a) There are five sets of traffic lights facing cars coming from A north, A south, B east, B west, and C southeast, respectively.
(b) The red, yellow, and green lights facing cars from A Street north are augmented with a left-turn arrow that can be lit up as either green or yellow or not lit up at all.
(c) The normal sequencing of lights facing the cars coming from A Street north is arrow green, arrow yellow, traffic light green, traffic light yellow, traffic light red, and repeat. In other words, the left-arrow light is illuminated in every complete cycle of the lights.
(d) However, it should be possible for traffic going from north to south on A Street to cross the intersection even when the left turn arrow is illuminated. Therefore, the traffic light green should also be illuminated while the turn arrow is lit up.
(e) Cars traveling from south to north on A Street (and all directions on B and C Streets) must see a red light while the left turn arrow is illuminated for the traffic heading south.
(f) A car sensor *C* is embedded in C Street to detect whether a car is waiting to enter the intersection from the southeast.
(g) A timer generates a long interval signal *TL* and a short interval signal *TS* when set by an *ST* signal.
(h) Red and green lights are lit up for at least a *TL* unit of time. Yellow lights, the green arrow, and the yellow arrow are lit up for exactly a *TS* unit of time.
(i) The C Street lights cycle from red to green only if the embedded car sensor indicates that a car is waiting. The lights cycle to yellow and then red as soon as no cars are waiting. Under no circumstances is the C Street green light to be lit for longer than a *TL* unit of time.

Draw a state diagram for the traffic-light controller. Indicate the logical conditions for remaining in the current state and for exiting it to the next state. Also, create a table that indicates precisely which lights are illuminated for each of your states.

10.3 *(Word Problem)* You are to design a Mealy state diagram for a digital lock. Assume that two debounced push-buttons, A and B, are available to enter the combination. An electromechanical interlock guarantees that the buttons cannot be activated simultaneously. The lock should have the following features:

(a) The combination is A-A-B-A-B-A. If this sequence is correctly entered, an output signal is asserted that causes the lock to open.
(b) For any state, three B pulses in a row should guarantee to reset the control to its initial state.

(c) When any out-of-sequence use of the A push-button occurs, an output is asserted that rings a bell to warn that the lock is being tampered with.

Once the lock is open, pressing either A or B will cause the lock to close without signaling an error. Draw a Mealy state diagram for this finite state machine. Indicate what each state represents and what input conditions cause state and output changes. Not everything may have been specified, so write down any assumptions you make.

10.4 *(Word Problem)* Design a state diagram to perform the following function. There are two data inputs A and B, a check input C, and an output D. The FSM takes as input two continuous, synchronous streams of 4-bit twos-complement numbers in a bit-serial form with the most significant (sign) bit first. The least significant bit is marked by a 1 on the check line (C). During the time slot in which C is asserted, the output D should go to a 1 if the twos-complement number on A is larger than the twos-complement number on B.

(a) Complete the timing diagram in Figure Ex. 10.4 to make sure you fully understand the statement of the problem.

(b) Draw a state diagram that implements this specification using as few states as possible. (*Hint:* It is possible to implement this machine in six or fewer states.)

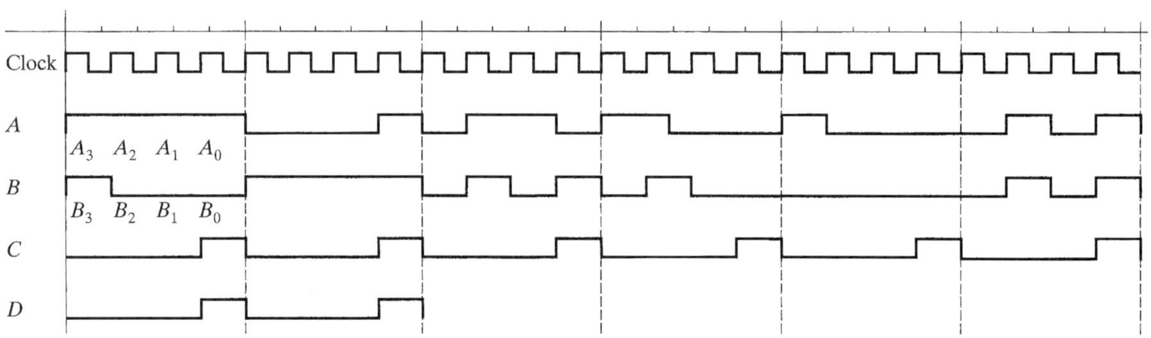

Figure Ex. 10.4 **Timing diagram for serial number comparator.**

10.5 *(Word Problem)* Your task is to design the control for a sequential 4-bit multiplier. The data path is shown in Figure Ex. 10.5. It consists of a 4-bit adder, a 4-bit register, and a 9-bit shift register. The latter shifts right when its *Sh* input is asserted (assume that zeros are entered at the left for this operation). A new value is loaded into the high-order 5 bits of the shift register when *Ld* is asserted. The same 5 bits are zeroed when *Cl* is asserted. These signals are synchronous.

(*Hint:* As a simple example, consider the 2-bit version of the device forming the product of 11_2 and 10_2.)

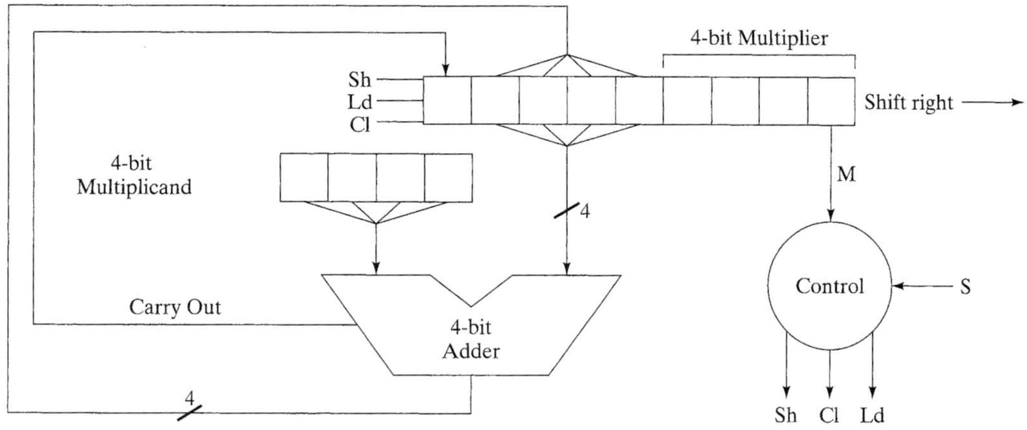

Figure Ex. 10.5 Multiplier data path for Exercise 10.5.

Draw a Mealy machine state diagram for a 4-bit multiplier. The inputs are *S* (a multiply start signal) and *M* (the low-order bit of the multiplier). The outputs are the *Sh*, *Ld*, and *Cl* signals.

10.6 *(Word Problem)* You are to design the state diagram for a simple controller that turns a lamp on and off at preset times. This is a *timed light switch.* The finite state machine has six inputs: *Reset, SetTime, SetLiteOn, SetLiteOff, Run,* and *Advance.* The first five inputs are generated by a five-position rotary switch that advances through *Reset, SetTime, SetLiteOn, SetLiteOff,* and *Run* (the inputs are mutually exclusive and are encountered in the specified order). The *Advance* input is a push-button. See Figure Ex. 10.6(a). When you hold the *Advance* button down (asserted), the displayed time rapidly advances through 24 hours, a minute at a time.

The typical operation of the timed light switch works as follows. It is normally in *Run* mode. The lamp is turned on whenever the internal clock matches an internal register (*LiteOn*) that holds the time to turn the light on. The lamp is turned off whenever the internal clock matches an internal register (*LiteOff*) that holds the time to turn the light off.

To operate the timed light switch, you must set the current time, then the time on, and finally the time off. This is accomplished as follows. The mode switch is moved from *Run* to *Reset.* This causes an internal timer register to be loaded with the time 08:00. Next, the mode switch is moved to the *SetTime* position. Whenever *Advance* is pushed and held down, the timer register rapidly cycles through the minutes and hours. You "pulse" or single step *Advance* as it gets close to the current time. When you move the switch to *SetLiteOn,* the current value in the timer register overwrites the value in the internal clock register. At the same time, the internal timer register is reset to 08:00.

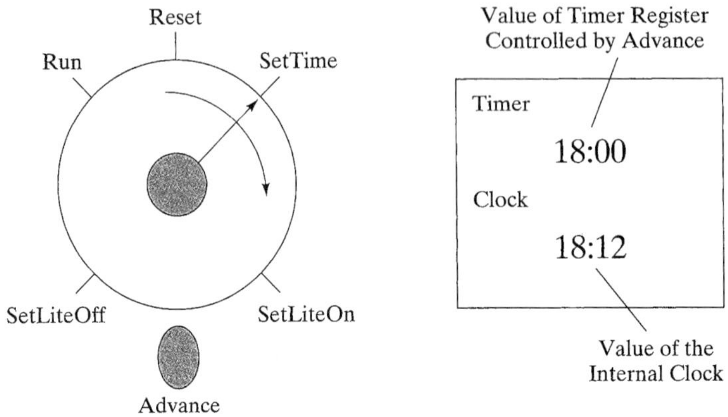

(a) Buttons and displays

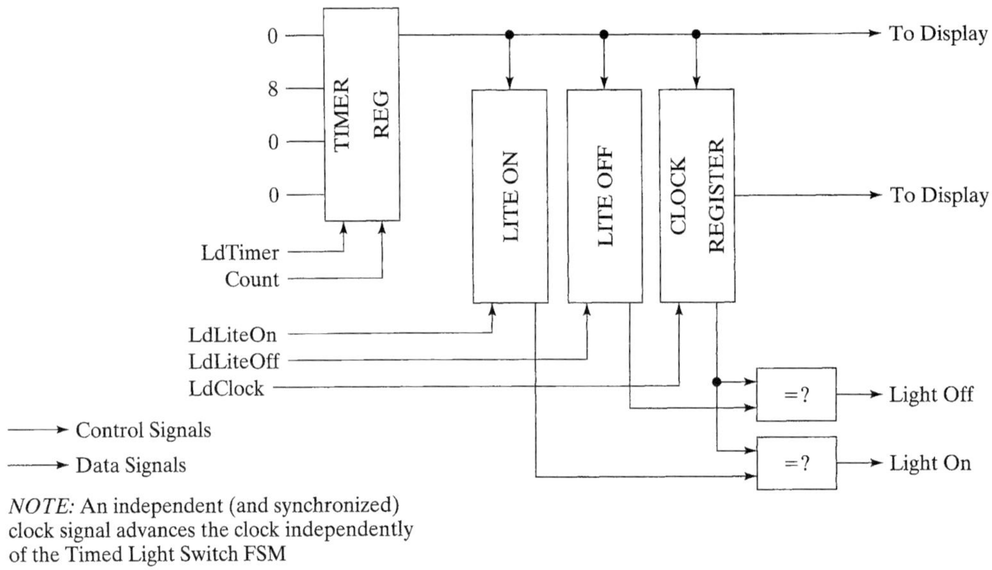

(b) Data path

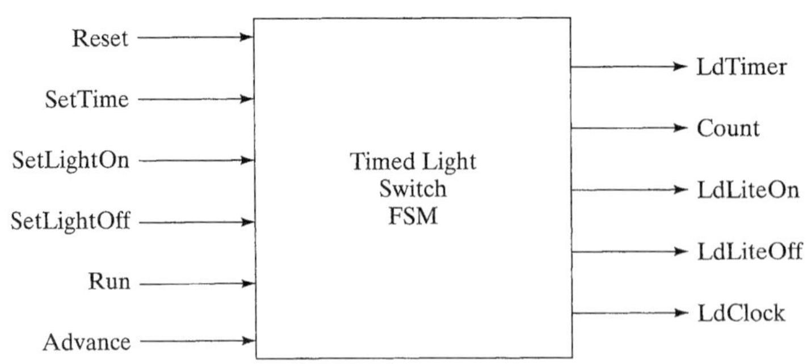

(c) Control Input/Output

Figure Ex. 10.6 The timed light switch.

By working with the *Advance* button, you set a new time at which the light is to be turned on. Moving the mode switch to *SetLiteOff* causes the *LiteOn* register to be overwritten by the timer register.

Using the *Advance* button once again, you advance the timer from its last value (the "lights on" time) to the desired time to turn the lights off. Once the mode switch is set to *Run,* the timed light switch goes into its running mode.

The data path associated with the timed light switch is shown in Figure Ex. 10.6(b). The block diagram is given in Figure Ex. 10.6(c).

Complete a Moore state diagram for the timed light-switch controller.

10.7 *(Word Problem)* Given the combination lock of Section 8.5.4, describe how you would design a combination lock with a variable number of values in the key.

Hint: How could you use a counter to assist in this?

Draw a block diagram showing signals between the finite state machine and the counter. Draw a revised state diagram for the combination lock incorporating your design changes.

10.8 *(Random-Access Memories)* A microprocessor with an 8-bit-wide data bus uses RAM chips of 4096 × 1-bit capacity. How many chips are needed and how should their address lines be connected to provide a memory capacity of 16 K-bytes (1 byte = 8 bits).

10.9 *(Random-Access Memories)* Consider a 1-megabit dynamic memory component. The memory is organized into 512 rows of 2048 bits each. Assume that every bit must be refreshed within 4 ms. How frequently should a row refresh operation be scheduled? If the memory has an 80 ns access time, approximately what fraction of memory accesses must be dedicated for refresh?

10.10 *(Random-Access Memories)* Consider the memory controller design described in Section 7.6.5. Show an alternative implementation of the control signals *INC_ADR*, *LATCH_DATA*, and *ENAB_BUF* that leads to a different detailed sequencing of the signals that is still logically correct.

10.11 *(Random-Access Memories)* Consider the read and write timing of the 2114 memory component in Figures 7.55 and 7.56. What is the minimum clock width for the overlapping clocks generated by the memory controller that will still meet the memory's timing specification? Justify your answer.

10.12 *(Interfacing Memories)* In this exercise, you will design a memory-controller finite state machine that implements a processor-memory handshake to the following specification. The processor initiates a transfer request by asserting REQ (request) while specifying a Read or Write (RW) operation. During a Read operation (RW asserted), the processor waits for the memory controller to

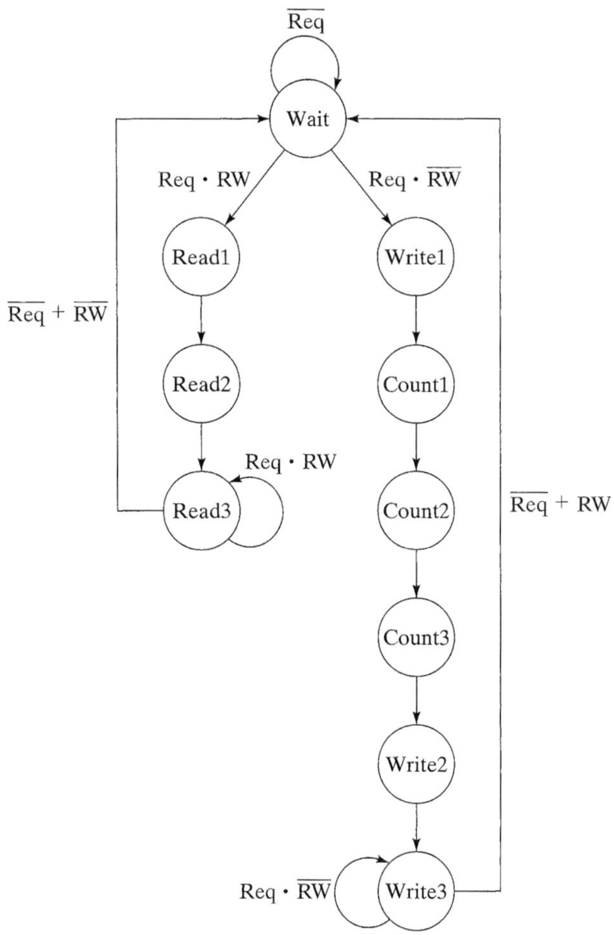

(a) State diagram

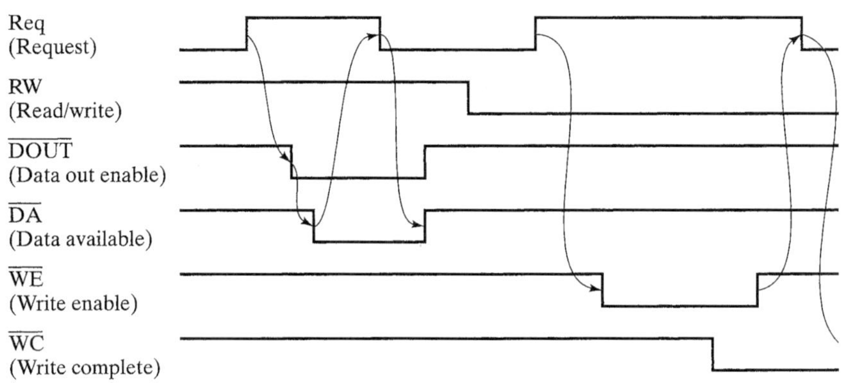

(b) Timing diagram

Figure Ex. 10.12 Memory controller.

State	$\overline{DOUT}$	$\overline{DA}$	$\overline{WE}$	$\overline{WC}$	C_0	C_1
Wait	1	1	1	1	0	0
Read1	0	1	1	1	0	0
Read2	0	0	1	1	0	0
Read3	1	1	1	1	0	0
Count1	1	1	0	1	1	0
Count2	1	1	0	1	0	1
Count3	1	1	0	1	1	1
Write1	1	1	0	1	0	0
Write2	1	1	0	0	0	0
Write3	1	1	1	0	0	0

(c) State output behavior

Figure Ex. 10.12 *(Continued)*

assert $\overline{DA}$ (Data Available). The processor can then sample the data. It unasserts the REQ line to end the memory cycle. During a write operation (RW unasserted), the processor drives data to the memory system, waiting for the memory controller to assert $\overline{WC}$ (Write Complete). When the processor sees this, it unasserts REQ to end the cycle. This is a variation of the four-cycle handshake described in Chapter 4.

The Moore state diagram for the memory controller is shown in Figure Ex. 10.12a). Note that the read and write require multiple states for their execution. A timing diagram, showing the relationships between the critical control signals for a read and a write cycle, is given in Figure Ex. 10.12(b). The state control-signal outputs are listed in Figure Ex. 10.12(c). Several of the memory controller's signals listed here are used to control the memory components. Their detailed meaning is not important, except that the appropriate signals should be asserted in the listed states.

(a) Choose a good state assignment and implement using discrete gates and flip-flops. What kind of PAL would you need to implement this machine in a single chip (in particular, number of inputs/outputs, flip-flops, product terms per output, etc.)?

(b) How many Xilinx CLBs would it take to implement your solution to part (a)? Justify your answer.

(c) Implement the state machine using a four-bit counter as a state register. Show your counter-based state assignment, and your implementation of the next-state function in terms of clear, count, and load. What kind of PAL would you need to implement the next state control in a single chip. (*Hint:* Assume the state register is implemented externally with a 163 synchronous up-counter).

(d) How many Xilinx CLBs would it take to implement your solution to part (c), including the counter state register and the output logic?

10.13 *(Sequential Multiplier)* The multiplier of Section 10.5 computes partial products by looking at one bit of the multiplier during each clock cycle. Construct a sequential multiplier that operates on two bits of the multiplier in every clock cycle. You will now need to accumulate 0, 1, 2, or 3 times the multiplicand. Add circuitry around the accumulator to implement these four functions. Modify the shift registers to shift two bits at a time instead of one. Now your multiplier will compute a result in only four clock cycles? But will the logic you added have an effect on the length of a clock period? Do you expect the cost (more logic) to outweigh the benefits (fewer clock cycles)?

10.14 *(Sequential Multiplier)* The sequential multiplier of Section 10.5 only operates on unsigned numbers. Construct a multiplier that operates on twos-complement numbers. Consider carefully how multiplication works in twos-complement form. In the worst case you will need circuitry to convert one or both operands to positive values and then adjust the result's sign after it has been computed.

10.15 *(Word Problem)* The telephone keypad decoder of Section 10.6 is severely limited in that it only allows the user to enter numbers. Construct a sequential keypad decoder that supports alphanumeric characters in the same way as many cell phones do by entering different characters through multiple key presses. For example, pressing "2" once enters an "*A*," pressing it twice in a row enters a "*B*," three times is a "*C*," four is a "2," and more key presses can be used for other special characters. Copy a cell phone with which you are familiar and create this more complex decoder. (*Hint:* you will need to consider what constitutes a multiple press or a new press of the same key. For example, how will you tell apart a single press for an "*A*" followed by a double press to enter a "*B*" and not misinterpret it as a "*C*?")

10.16 *(Word Problem)* A different kind of character encoding on transmission lines is called "Manchester encoding." A 1 is composed of low period followed by a high period. A 0 is composed of a high period followed by a low period. The wire is typically quiescent at a high value and the first bit of any transmission is always a 0.

(a) Re-implement the *Sender* module of Section 10.6 to use Manchester encoding in transmitting its byte of data. Assume that you have a clock that runs at twice the bit rate of your data so that you can use a single clock to generate the two complementary parts of each (one high and one low, or vice versa).

(b) Re-implement the *Receive* module of Section 10.6 to receive a Manchester encode byte. Assume that you have a clock that can sample each bit four times (twice while it is high, twice while it is low, or vice versa).

Epilogue

It has been a long way since Chapter 1. We sincerely hope that we've delivered on the promises we made in the first pages of the book. Our hope is that you now have a solid understanding of the fundamental concepts of logic design—Boolean algebra and finite state machines—but also a sense for the practical. After all, this book describes an engineering discipline and the engineer's objective is to find solutions to problems that meet a set of complex constraints. We emphasized this aspect by including many examples throughout the text. The examples show the approaches to designing solutions and how the choice of technology affects that solution. Design is, in large part, an art. Practicing the art, developing the intuitions that distinguish good designers, is very important to any engineering discipline. We hope you found the examples interesting and thoroughly explained so that they helped you develop those intuitions.

This book does not have everything you will need to become an effective digital logic designer. It is important to use the tools of the trade. The schematic editors, hardware description languages, compilers, simulators, and synthesis tools we only briefly described form the toolbox every designer needs close at hand. We only provided a general introduction to these tools. Fortunately, the tools are developing rapidly—improving with each generation—and are quite varied in their details of operation. Unfortunately, this makes it difficult to choose one set and weave it into the text. It would be very unlikely you would have the same tools at your disposal. This is why we made the conscious decision to provide the concepts but not the details. Hopefully, the course work that accompanied reading this text filled in this gap.

This is also the case for the implementation technologies. Many students start with older technologies available in their laboratories. Newer technologies, for larger scale designs, are used in later courses on more advanced topics of the logic designer's art. Again, we chose to give only general information in the text and left it to the particular laboratory you had available to you to provide the components and data sheets. Our goal was to ensure that you would understand any component you will encounter and be able to read its documentation.

You are likely to encounter logic design in several courses in your curriculum. We hope we have provided a solid foundation for learning computer architecture—the organization of machines to perform complex calculations efficiently. It was too large an area to do justice in the limited space available to us. Instead, we have provided some material to bridge the transition on our web site. We hope you'll find it useful.

Thank you for making it this far. Our satisfaction will come from you applying the knowledge you've gained and, hopefully, continuing in this discipline. We wish you the best.

A

Number Systems

Introduction

In this appendix, we briefly review the concept of positional number systems, the methods for conversion between alternative number systems, and the basic elements of binary addition and subtraction. If you are not familiar with these concepts, it is probably a good idea to read this appendix before even starting out with Chapter 1.

Throughout much of our lives, we have been exposed to the base-10 number system. The preference for 10-digit number systems is no surprise: We have 10 fingers! However, this is not natural for digital hardware systems, where arithmetic is based on the binary digits 0 and 1. We will also discuss number systems that are variations on the binary system: octal (the digits 0 through 7) and hexadecimal. The latter is a base-16 system, with 0 through 9 extended by the additional digits A (10), B (11), C (12), D (13), E (14), and F (15).

A.1 Positional Number Notation

In this section, we cover the main positional number systems used in digital hardware: decimal, binary, octal, and hexadecimal.

A.1.1 Decimal Numbers

The decimal number system represents quantities using the digits 0 through 9, arranged in a positional notation. For example, in base 10, the number 154 can be represented as

$$\begin{aligned} 154_{10} &= 1 \times 100 + 5 \times 10 + 4 \\ &= 1 \times 10^2 + 5 \times 10^1 + 4 \times 10^0 \end{aligned}$$

This is called *positional* because a digit's "place" in the sequence determines its weight. The least significant digit, in the rightmost position, has a weight of 1. The next digit to the left has a weight of 10. The most significant digit, in the leftmost position, has a weight of 100.

Each additional position to the left has a weight 10 times as much as the position to its immediate right. This is why the decimal number system is called a base 10 system. You should also notice that numbers are represented by sequences consisting of the ten digits 0 through 9.

A.1.2 Binary, Octal, and Hexadecimal Numbers

Digital hardware systems almost universally use the binary number system rather than base 10. However the basic concepts of positional number systems still apply. A number is written from the most significant digit at the left to the least significant digit at the right.

Binary Numbers A binary number can be represented only by using the two digits 0 and 1. These are called *binary digits,* or simply *bits.* As the number is written down, each bit has twice the weight of its neighbor to its immediate right.

For example, consider the 8-bit binary number 10011010_2. The subscripted 2 reminds us that the number is in base 2. When a number is represented without a subscript, it usually means that it is a base-10 number.

What is the value of 10011010_2? Let's rewrite it in positional notation:

$$\begin{aligned} 10011010_2 &= 1 \times 2^7 + 0 \times 2^6 + 0 \times 2^5 + 1 \times 2^4 + 1 \times 2^3 \\ &\quad + 0 \times 2^2 + 1 \times 2^1 + 0 \times 2^0 \\ &= 1 \times 128 + 0 \times 64 + 0 \times 32 + 1 \times 16 + 1 \times 8 \\ &\quad + 0 \times 4 + 1 \times 2 + 0 \times 1 \\ &= 128 + 16 + 8 + 2 = 154_{10} \end{aligned}$$

The binary number 10011010_2 denotes the same quantity as the decimal number 154_{10}. We can always place a binary number into base 10 by expanding it using positional notation.

Octal and Hexadecimal Numbers Writing down even relatively small quantities in base 2 requires a large number of bits. To simplify the chore, designers have introduced alternative octal and hexadecimal number systems, based on 8 and 16 digits, respectively. It is easy to convert between binary and these systems, because the base in each case is a power of 2.

An octal number is represented by a sequence of digits drawn from 0 through 7. For example, the number 232_8 denotes the same quantity as 154_{10}. We can verify this by expanding the positional notation:

$$\begin{aligned} 232_8 &= 2 \times 8^2 + 3 \times 8^1 + 2 \times 8^0 \\ &= 128 + 24 + 2 = 154_{10} \end{aligned}$$

Converting from base 16 is very similar. Remember that the 16 digits used in the hexadecimal system are 0 through 9 and *A* through *F.*

Thus, the hexadecimal number $9A_{16}$ can be expanded as follows:

$$\begin{aligned} 9A_{16} &= 9 \times 16^1 + 10 \times 16^0 \\ &= 144 + 10 = 154_{10} \end{aligned}$$

Once again, the hexadecimal represents the same quantity as 154_{10}.

A.2 Conversion Between Binary, Octal, and Hexadecimal Systems

We cover the methods for converting between the various binary-based systems in this subsection.

A.2.1 Conversion from Binary to Octal or Hexadecimal

It is always easy to rewrite a binary number in the octal or hexadecimal system. All we have to do is group the binary digits into 3-bit groupings (octal) or 4-bit groupings (hexadecimal), starting at the right of the number. For example, starting with the binary number 10011010_2, we derive its octal and hexadecimal equivalents as follows:

$$\underbrace{10}_{2}\,\underbrace{011}_{3}\,\underbrace{010}_{2_8}{}_2 \qquad \underbrace{1001}_{9}\,\underbrace{1010}_{A_{16}}{}_2$$

To see why this grouping strategy is correct, let's look at a generalized representation of a binary number and how we convert it to a hexadecimal number. We assume that the binary number always has a multiple of 4 bits. Of course, for any binary number, we can prepend additional zeros to the most significant bits to make this assumption true.

The generalized representation is shown below. The binary number starts out with n bits and will be converted to a hexadecimal number with $k = n/4$ hex digits:

$$\begin{aligned} &a_{n-1}2^{n-1} + a_{n-2}2^{n-2} + a_{n-3}2^{n-3} + a_{n-4}2^{n-4} + \cdots \\ &\quad + a_{4i+3}2^{4i+3} + a_{4i+2}2^{4i+2} + a_{4i+1}2^{4i+1} + a_{4i}2^{4i} + \cdots \\ &\quad + a_3 2^3 + a_2 2^2 + a_1 2^1 + a_0 2^0 \end{aligned}$$

The process of putting the bits into groups of four is equivalent to factoring powers of 16 within the expression. By doing the grouping, we can rewrite the generalized expression as follows:

$$\begin{aligned} &(a_{n-1}2^3 + a_{n-2}2^2 + a_{n-3}2^1 + a_{n-4}2^0)2^{4(k-1)} + \cdots \\ &\quad + (a_{4i+3}2^3 + a_{4i+2}2^2 + a_{4i+1}2^1 + a_{4i}2^0)2^{4(i)} + \cdots \\ &\quad + (a_3 2^3 + a_2 2^2 + a_1 2^1 + a_0 2^0)2^{4(0)} \end{aligned}$$

We can now rewrite each of the 4-bit numbers in the parentheses as a single hexadecimal digit, raised to an appropriate power of 2^4. We could apply a similar method to demonstrate why grouping by threes performs the conversion from binary to octal.

A.2.2 Conversion from Octal to Hexadecimal and Vice Versa

To convert between octal and hexadecimal, you should first convert the number into binary. By appropriate grouping of bits, you then form the number in the target base. Let's see how this works for the conversion between 232_8 and $9A_{16}$:

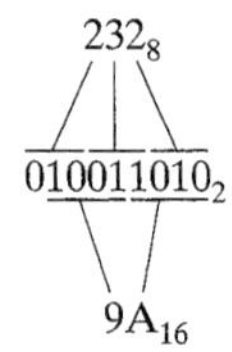

232_8 maps into the binary number 010 011 010_2, which can be regrouped to form the number $9A_{16}$. The mapping from hex to octal works analogously.

A.2.3 Conversion from Base 10 to Base 2: Successive Division

So far, you have seen how to map binary, octal, and hexadecimal numbers into base 10, by expanding the positional notation. You have also seen how to map between the binary system and the octal and hex systems. The missing link is how to get from base 10 to base 2.

The method we use is called *successive division*. It works by successively dividing the base-10 number by the base to which it is to be converted. For example, we successively divide by 2 if the target is base 2. We collect the remainders to form the number in the target base. The first division yields the least significant bit as its remainder. The process continues until the quotient is 0.

Let's look at an example: the conversion of 154_{10} to base 2. The process is as follows:

$154 \div 2 = 77$ Remainder 0 → 10011010_2
$77 \div 2 = 38$ Remainder 1
$38 \div 2 = 19$ Remainder 0
$19 \div 2 = 9$ Remainder 1
$9 \div 2 = 4$ Remainder 1
$4 \div 2 = 2$ Remainder 0
$2 \div 2 = 1$ Remainder 0
$1 \div 2 = \boxed{0}$ Remainder 1

The same method works for conversion from base 10 to other bases. For example, to convert 154_{10} to base 8, we proceed as follows:

$154 \div 8 = 19$ Remainder 2 → 232_8
$19 \div 8 = 2$ Remainder 3
$2 \div 8 = \boxed{0}$ Remainder 2

The conversion to base 16 works like this:

$$154 \div 16 = 9 \text{ Remainder } 10 \longrightarrow 9A_{16}$$
$$9 \div 16 = \boxed{0} \text{ Remainder } 9$$

The Theory Behind Successive Division Given that the number N in base 10 is equivalent to an n-digit number in a different base R, we should be able to rewrite N in base R as $(a_{n-1}a_{n-2} \ldots a_0)_R$. As we successively divide by R, each subsequent division liberates the next lower-order digit, a_i, from the base-R representation.

To see this, consider the following description of successive division. Since N is represented as an n-digit number in base R, the first division yields the quotient, Q_0, and the remainder, a_0. The latter is exactly the value of the lowest-order digit in base R. Dividing R into Q_0 results in a new quotient, Q_1, and a new remainder, a_1, the next lowest-order digit. If we repeat the process n times, we obtain a quotient of zero and a remainder that is equal to the highest-order digit in base R. The process of successive division is shown below:

$$\begin{aligned} N &= (a_{n-1}a_{n-2}\ldots a_0)_R \\ &= a_{n-1}R^{n-1} + a_{n-2}R^{n-2} + \cdots + a_1R^1 + a_0R^0 \end{aligned}$$

$$Q_0 = \frac{N}{R} = a_{n-1}R^{n-2} + a_{n-2}R^{n-3} + \cdots + a_1R^0 \qquad \text{Remainder } a_0$$

$$Q_1 = \frac{Q_0}{R} = a_{n-1}R^{n-3} + a_{n-2}R^{n-4} + \cdots + a_2R^0 \qquad \text{Remainder } a_1$$

$$\vdots$$

$$Q_{n-1} = \frac{Q_{n-2}}{R} = 0 \qquad \text{Remainder } a_{n-1}$$

To summarize, Figure A.1 shows the various conversion methods that we have covered. We use successive division to convert from base 10

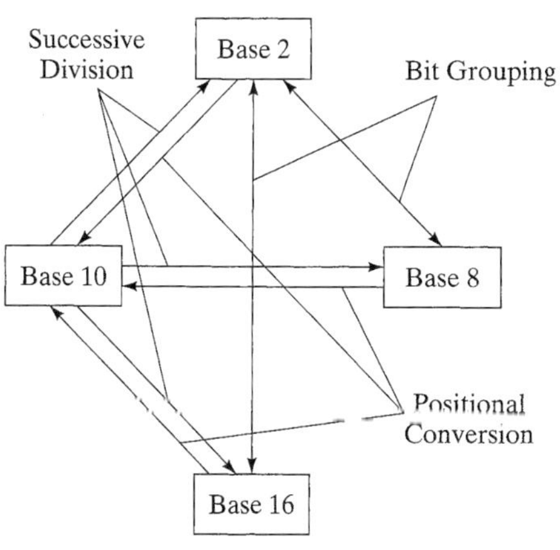

Figure A.1 Conversion methods.

to any other base. By expanding the positional notation, we can convert from any base to base 10. We perform conversions between base 2 and other bases that are powers of two, such as base 8 and base 16, through the process of bit grouping.

A.3 Binary Arithmetic Operations

We examine the arithmetic operations in the binary number system next.

A.3.1 Addition in Positional Notation

Let's begin by reviewing how addition works in the base-10 positional number system. Consider the addition of 95 and 16 to yield 111:

$$\begin{array}{r} 95_{10} \\ +\ 16_{10} \\ \hline \downarrow \\ 111_{10} \end{array} \qquad \begin{array}{r} 9 \times 10^1 + 5 \times 10^0 \\ +\ 1 \times 10^1 + 6 \times 10^0 \\ \hline 10 \times 10^1 + 11 \times 10^0 \\ \downarrow \\ 1 \times 10^2 + (0+1) \times 10^1 + 1 \times 10^0 \end{array}$$

The numbers are added column by column, one position at a time. Should the column sum exceed 9, the largest digit that can be represented in base 10, we must generate a carry-out to the next higher-order position.

You are familiar with a shorthand method for addition in positional notation. You proceed column by column, from right to left. 5 + 6 = 11, which is written as a sum of 1 with a carry of 1 to the next column. The *carry* represents an excess quantity that is too big to be represented in a single column. It is then added to the column sum of the next higher position. In the example, 1 + 9 + 1 = 11, which again is represented as a column sum of 1 with a carry of 1. Since there are no additional column sums to be formed, the final carry-out is written as part of the sum. This process, as illustrated by the example, is summarized below:

$$\begin{array}{r} 11 \leftarrow \text{Column carries} \\ 95_{10} \\ 16_{10} \\ \hline 111_{10} \end{array}$$

The same process applies to addition in any base. Next, we will see how addition is applied to base 2.

Addition in Base 2 In the binary number system, we have the following addition table:

$$\begin{aligned} 0 + 0 &= \boxed{0} \\ 0 + 1 &= \boxed{1} \\ 1 + 0 &= \boxed{1} \\ 1 + 1 &= \boxed{0} \text{ with a carry of 1} \end{aligned}$$

Let's illustrate addition in binary with an example. Consider the addition of 101_2 and 11_2 to yield 1000_2 $(5_{10} + 3_{10} = 8_{10})$:

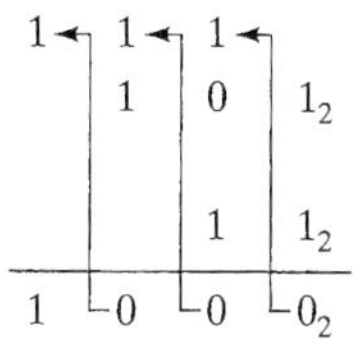

$$
\begin{array}{r}
1 \times 2^2 + 0 \times 2^1 + 1 \times 2^0 \\
1 \times 2^1 + 1 \times 2^0 \\
\hline
1 \times 2^2 + 1 \times 2^1 + 10 \times 2^0 \\
\downarrow \\
1 \times 2^2 + (1+1) \times 2^1 + 0 \times 2^0 \\
\downarrow \\
(1+1) \times 2^2 + 0 \times 2^1 + 0 \times 2^0 \\
\downarrow \\
1 \times 2^3 + 0 \times 2^2 + 0 \times 2^1 + 0 \times 2^0
\end{array}
$$

The step-by-step process is shown at the right, the shorthand method at the left. In longhand, the columns are added up individually. If the column quantity cannot be represented, it must "carry out" a power of two to the next higher column.

Since 10_2 in the sum's ones column cannot be represented directly, we must add 1×2^1 to the next higher column, leaving the 0 behind. The twos column undergoes a similar carry-out: 1 + 1 yields 0 with a carry of 1 to the fours column. This, in turn, cascades to the eights column: 1 + 1 again yields 0 with a carry-out of 1.

It is much easier to work with the shorthand method, of course. We proceed from right to left, adding one column at a time. Beginning with the ones column, 1 + 1 = 0 with a carry of 1 to the twos column. In the twos column, we add the carry of 1 + 0 + 1 = 0 with a carry of 1 to the fours column. Continuing with the fours column, the carry-in of 1 plus 1 is 0 with a carry-out of 1 to the eights column. This results in the final sum: 1000_2.

As another example, let us look at the addition of 95_{10} and 16_{10}, but this time in binary:

		1 ←					
95_{10} =	1	0	1	1	1	1	1
+ 16_{10} =	0	0	1	0	0	0	0
	1	1	0	1	1	1	1 = 111_{10}

95_{10} maps into 1011111_2, while 16_{10} is equivalent to 10000_2. In adding, we work from the rightmost column to the left. In the first column, the sum is 1, as it is in columns two, three, and four. In the fifth column, two 1s are being summed, yielding 0 with a carry of 1. This leads to a sum in the sixth column of 1. The final column also sums to 1. You should verify that $1101111_2 = 111_{10}$.

A.3.2 Subtraction in Positional Notation

Let's review the process of subtraction in base 10. Then we will show how the process applies in base 2.

To illustrate, consider the subtraction of 16_{10} from 95_{10} to yield 79_{10}:

$$\begin{array}{r|l} & \rightarrow 10 \\ 9 & 5_{10} \\ -1 & 6_{10} \\ -1 & \\ \hline 7 & 9_{10} \end{array}$$

$$\begin{aligned} 95 &= 9 \times 10^1 + 5 \times 10^0 = 8 \times 10^1 + 15 \times 10^0 \\ -16 &= 1 \times 10^1 + 6 \times 10^0 = 1 \times 10^1 + 6 \times 10^0 \\ & \qquad\qquad\qquad\qquad\quad \overline{7 \times 10^1 + 9 \times 10^0} \end{aligned}$$

The longhand method is shown at the right, the shorthand at the left. At the right, the numbers are rearranged into the positional notation. If the digit to be subtracted is larger than the digit it is being subtracted from, we must *borrow* from the next higher position to the left. A 10 is subtracted from this column and added back to the original column. This borrowing guarantees that the subtraction can now proceed in the ones column.

Let's start by looking at the example in its long form at the right. Starting with the ones column, 5 is smaller than 6, so we must move 10 units from the tens position to the ones position. Now we can subtract 6 from the resulting 15 to yield 9. For the next column, we subtract 1 from the remaining 8 to obtain 7 in the tens position.

Looking at the shorthand form, since 5 is less than 6, we borrow from the tens column to form 15. Then, $15 - 6 = 9$. For the tens column, we subtract 1 from 8 to get 7. So the result is 79. Obviously, the process of borrowing can cascade from right to left in much the same way that carries can propagate.

Subtraction in Base 2 Subtraction in base 2 is much like subtraction in base 10, except that the borrowing process moves two from a higher-order column to a lower-order column. If a column subtraction cannot proceed without yielding a negative result, borrowing causes 1×2^1 to be subtracted from the adjacent column to the left, which is then added back as 10_2 to the original column.

To see how subtraction proceeds in base 2, we begin with the example of subtracting 11_2 from 101_2 to get 10_2 ($5_{10} - 3_{10} = 2_{10}$):

$$\begin{aligned} 101_2 &= 1 \times 2^2 + 0 \times 2^1 + 1 + 2^0 = 0 \times 2^2 + 10 \times 2^1 + 1 \times 2^0 \\ \frac{-11_2}{10_2} &= \qquad\qquad 1 \times 2^1 + 1 \times 2^0 = \frac{1 \times 2^1 + 1 \times 2^0}{1 \times 2^1 + 0 \times 2^0} \end{aligned}$$

Starting with the ones column, we subtract 1 from 1 to obtain 0. In the twos column, 0 is smaller than 1, so we must borrow from the fours column. In essence, we borrow one from the fours column to add two to the twos column. $2 - 1 = 1$, which is the result in the middle column. The high-order column is left with a 0.

As another example of borrowing, consider the subtraction of 111_2 from 10001_2 to yield 1010_2. To see borrowing in the longhand form, let's look at the step-by-step process:

$$
\begin{array}{lr}
\text{(i)} & 1 \times 2^4 + 0 \times 2^3 + 0 \times 2^2 + 0 \times 2^1 + 1 \times 2^0 \\
\text{(ii)} & 1 \times 2^4 + 0 \times 2^3 + (0-1) \times 2^2 + (0+10) \times 2^1 + 1 \times 2^0 \\
\text{(iii)} & 1 \times 2^4 + (0-1) \times 2^3 + (0-1+10) \times 2^2 + (0+10) \times 2^1 + 1 \times 2^0 \\
\text{(iv)} & (1-1) \times 2^4 + (0-1+10) \times 2^3 + (0-1+10) \times 2^2 + (0+10) \times 2^1 + 1 \times 2^0 \\
\text{(v)} & 1 \times 2^3 + 1 \times 2^2 + 10 \times 2^1 + 1 \times 2^0 \\
 & - \quad 1 \times 2^2 + 1 \times 2^1 + 1 \times 2^0 \\
\hline
 & 1 \times 2^3 + 0 \times 2^2 + 1 \times 2^1 + 0 \times 2^0
\end{array}
$$

Steps (i) through (v) show the transformations to the positional notation to get a form suitable for subtraction of 111_2. (i) is 10001_2 in the standard form. Since the twos position is smaller than the digit being subtracted from it, we must borrow from the fours position. This is shown in (ii). Now the fours column is negative, so the borrowing process must cascade. The result is shown in (iii). Now the eights column is negative, so the process continues, as (iv) shows. The final form, suitable for subtraction, is shown in (v). Now we can perform the subtraction column by column to obtain the final result of 1010_2.

In general, the strategy for subtraction is summarized by

$$
\begin{array}{ll}
0 - 0 = 0 & \\
0 - 1 = 1 & \text{With a borrow of 1} \\
1 - 0 = 1 & \\
1 - 1 = 0 &
\end{array}
$$

Using this table, let's subtract 10000_2 from 1011111_2 $(95_{10} - 16_{10})$:

$$
\begin{array}{rcrrrrrrrl}
95_{10} & = & 1 & 0 & 1 & 1 & 1 & 1 & 1 & \\
-16_{10} & = & -0 & 0 & 1 & 0 & 0 & 0 & 0 & \\
\cline{3-9}
 & & 1 & 0 & 0 & 1 & 1 & 1 & 1 & = 79_{10}
\end{array}
$$

In this example, there is no need to borrow. You should verify that $1001111_2 = 79_{10}$.

As a final example, let's consider the subtraction of 1 from 10000_2 to yield 1111_2:

$$
\begin{array}{rccccc}
 & & \to 10 & \to 10 & \to 10 & \to 10 \\
16_{10} = & 1 & 0 & 0 & 0 & 0 \\
-\ 1_{10} = & 0 & 0 & 0 & 0 & 1 \\
 & -1 & -1 & -1 & -1 & \\
\hline
 & 0 & 1 & 1 & 1 & 1 = 15_{10}
\end{array}
$$

This example exhibits cascaded borrows. We begin with the rightmost column. 0 is smaller than 1, so the result is 1 with a borrow of 1 from the column to the left.

Let's look at the second column. It has a deficit of 1, because it has lent a 1 to the column to its right. This is equivalent to subtracting 1 from this column. We must borrow from the next column to the left to make good on this deficit. Thus, the column computation is the same as $10_2 - 1 = 1$ with another borrow from the left.

The process repeats for the third column. The deficit of 1 used as the borrow into the second column must be made good. Since the current column has 0 in it, we must borrow from the fourth column. After the borrow, the calculation becomes $10_2 - 1 = 1$ with yet another borrow from the left.

The reasoning for the fourth column is the same as for the third. The borrow comes from the fifth column to make good on the deficit for the fourth column. $10_2 - 1 = 1$. In the fifth column, the calculation becomes $1 - 1$, based on the borrow deficit. The final result is 01111_2.

A.4 Representation of Negative Numbers

Within digital systems, all data, whether characters or numbers, are represented by strings of binary digits. This is fine as long as you never have negative numbers. Unfortunately, this is not normally the case.

Over the years, hardware designers have developed three different schemes for representing negative numbers: *sign and magnitude, ones complement,* and *twos complement.* In this section, we will examine these schemes and their implications for addition and subtraction of signed binary numbers.

In mathematics, there are infinitely many positive and negative integers. However, in a practical hardware system, only a fixed number of integers can be represented based on the number of bits allocated to the representation. In most modern computer systems, numbers are represented in 32 bits. This means that over 4 billion unique numbers can be represented—quite a few, but certainly not infinite! An *overflow* occurs when an arithmetic operation results in a number outside the range of those that can be represented.

Throughout this section we will assume that our system operates on 4-bit binary quantities. Thus we can represent 16 unique binary numbers. Roughly half of these will represent positive numbers and zero, while the remainder will be negative numbers. Each of the three representation schemes handles negative numbers slightly differently, as we now examine.

A.4.1 Sign and Magnitude

In *sign and magnitude* systems, the most significant bit represents the number's sign, while the remaining bits represent its absolute value as an unsigned binary magnitude. If the *sign bit* is a 0, the number is

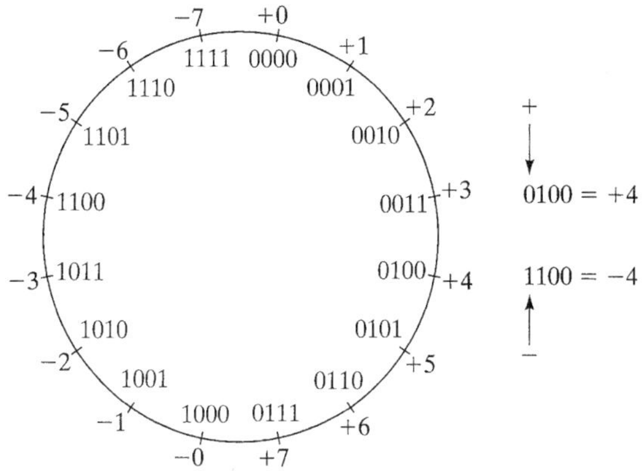

Figure A.2 **Sign and magnitude number representation.**

positive. If the sign bit is a 1, the number is negative. We negate a number simply by replacing the sign bit with its complement.

Figure A.2 depicts a "number wheel" representation of our 4-bit number system. The figure shows the binary numbers and their decimal integer equivalents, assuming that the numbers are interpreted as sign and magnitude. The largest positive number we can represent in three data bits is $+7 = 2^3 - 1$. By a similar calculation, the smallest negative number is –7. Zero has two different representations (0000 and 1000), even though +0 and –0 don't make much sense mathematically.

Adding two positive or two negative numbers is straightforward. We simply perform the addition and assign the result the same sign as the original operands. When the signs of the two operands are not the same, addition becomes more complex. In this case, we should subtract the smaller magnitude from the larger. The resulting sign is the same as that of the number with the larger magnitude.

This is what makes arithmetic operations with sign and magnitude numbers so cumbersome—any adder circuit must also include a subtractor and a comparator. The number wheel gives us a hint of this difficulty, because it does not show a progression through –2, –1, 0, 1, and 2 as we would expect on a number line. Instead, –1 is closer to +7 and +1 is closer to –7. Because of this burdensome complexity, hardware designers have proposed other schemes for representing negative numbers.

A.4.2 Ones-Complement Numbers

A *ones complement* approach represents the positive numbers just as in the sign and magnitude system. The only difference is in how it represents negative numbers.

We use the following procedure to derive a negative ones-complement integer, denoted $\bar{N}$, from a positive integer, denoted N. If

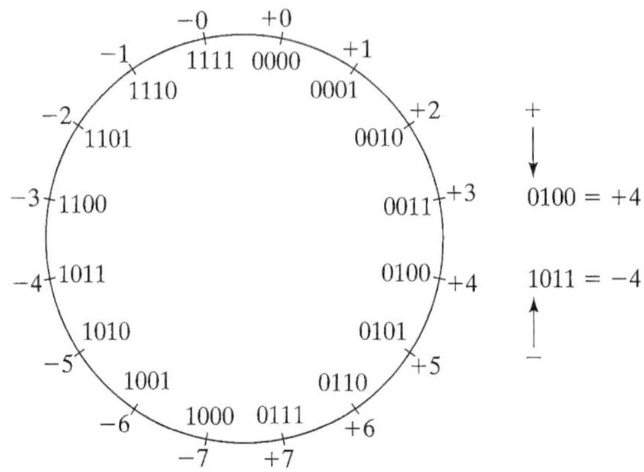

Figure A.3 **Ones complement number representation.**

the word length is n bits ($n = 4$ in our case), then $\bar{N} = (2^n - 1) - N$. For example, in a 4-bit system, +7 is represented as 0111. We compute −7 as

$$\begin{array}{lrl} 2^4 = & 10000 & \\ \text{Subtract 1} & -\ \underline{0001} & \\ & 01111 & \\ \text{Subtract 7} & -\ \underline{0111} & \\ & 0\underline{1000} & \text{Representation of } -7 \end{array}$$

This rather complicated method is just one way to compute the negative of a ones-complement number. A simpler method forms the ones complement by taking the number's bit-wise complement. Thus, +7 = 0111 and −7 = 1000, similarly +4 = 0100 and −4 = 1011, and so on.

The number wheel representation of the 4-bit ones-complement number system is shown in Figure A.3. All negative numbers have a 1 in their sign bit, making it easy to distinguish between positive and negative numbers. Note that we still have two different representations of zero (0000 and 1111).

Computing negative numbers is just as easy for ones-complement numbers as it was for the sign and magnitude representation. A big advantage is that subtraction can be easily implemented by a combination of addition and negation: $A - B = A + (-B)$. Thus, we don't need a separate subtractor circuit. However, addition is still complicated by the two zeros. We must skip two steps along the number wheel in moving from +1 to −1 and vice versa. For example, if we add −2 and +3 our result should be +1, but if we move three to the right on the number wheel, we end up at +0. Therefore, we will need a way to detect when we need to cross over the two-zeros. This complexity leads us to twos-complement numbers.

A.4.3 Twos-Complement Numbers

The *twos complement* scheme is similar to ones-complement, except that there is only one representation for zero making addition and subtraction much easier. Figure A.4 shows how the twos-complement

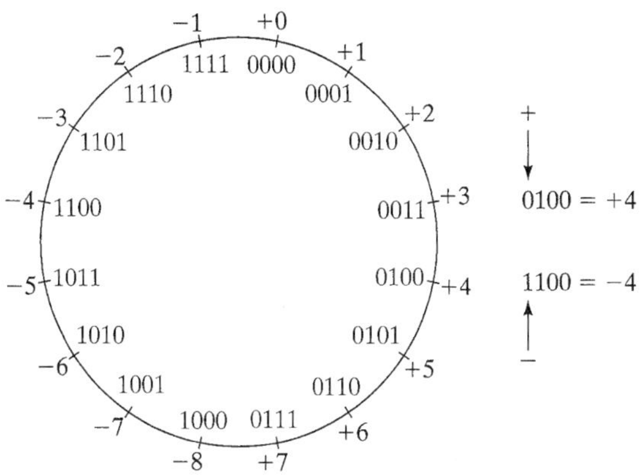

Figure A.4 Twos-complement number representation.

numbers are derived from the ones complement representation. We've taken the negative numbers and shifted them one position in the clockwise direction. This allows us to represent one more negative number, –8, than we were able to represent in ones complement. The negative numbers still have a 1 in their highest-order bit, the sign bit.

More formally, a twos-complement negative number, denoted $N*$, is derived from its positive number, N, by the equation $N* = 2^n - N$, where n is the number of bits in the representation. This equation omits the ones-complement step that subtracts 1 from 2^n.

Let's compute the twos complement of +7, represented as 0111_2:

2^4 =	10000	
Subtract 7	– 0111	
	0 1001	Representation of –7

Note that the calculation works equally well in deriving the twos complement of –7:

2^4 =	10000	
Subtract – 7	– 1001	
	0 0111	Representation of 7

It should come as no surprise that the same shortcut we used to find ones-complement numbers also applies for the twos-complement system, but with a twist. The adjustment we made to the number wheel suggests the scheme. Simply form the bit-wise complement of the number and then add 1 to form its twos complement. For example, +7 = 0111_2; its bit-wise complement is 1000_2, and plus 1 is 1001_2. This is the same twos-complement representation of –7 we derived by the last calculation. For the number –4, represented as 1100_2, its bit-wise complement is 0011_2, and plus 1 is 0100_2. This is exactly the twos-complement representation of +4.

As with ones complement, twos complement allows subtraction to be implemented with addition and negation. Even though it is a little harder to compute the twos complement, this is the form used almost universally in today's digital systems. In the next subsection, we will see how easily we can compute addition and subtraction by using twos-complement numbers.

A.4.4 Addition and Subtraction of Numbers

In this subsection, we will examine how addition and subtraction are performed in the three different number systems.

Sign and Magnitude Calculations Continuing with our 4-bit number scheme, let's look at some examples of addition and subtraction with sign and magnitude numbers. We need only consider addition, because subtraction is implemented by adding the negative of the subtracted number.

Here are some examples adding and subtracting 3 and 4:

(a)
$$\begin{array}{r} 4 \\ +3 \\ \hline 7 \end{array} \qquad \begin{array}{r} 0100 \\ +0011 \\ \hline 0111 \end{array}$$

(b)
$$\begin{array}{r} -4 \\ +(-3) \\ \hline -7 \end{array} \qquad \begin{array}{r} 1100 \\ +1011 \\ \hline 1111 \end{array}$$

(c)
$$\begin{array}{r} 4 \\ -3 \\ \hline 1 \end{array} \qquad \begin{array}{r} 0100 \\ -0011 \end{array} = \begin{array}{r} 0100 \\ +1011 \\ \hline 0001 \end{array}$$

(d)
$$\begin{array}{r} -4 \\ +3 \\ \hline -1 \end{array} \qquad \begin{array}{r} -0100 \\ +0011 \end{array} = \begin{array}{r} 1100 \\ +0011 \\ \hline 1001 \end{array}$$

Examples (a) and (b) are cases in which the signs are the same. The result is simply the sum of the magnitudes, and the sign of the result is the same as the signs of the operands.

Examples (c) and (d) represent the more complex situations in which the signs of the two operands differ. In (c), we have converted $4 - 3$ to $4 + (-3)$. The smaller magnitude, 3, is subtracted from the larger magnitude, 4, to obtain the magnitude of the result, 1. Since 4 is greater than 3, its sign is given to the result. This makes the result +1.

Case (d) looks at the operation $(-4) + 3$. The subtraction of the smaller magnitude from the larger yields a result of −1. The larger magnitude is negative, so the result must also be negative.

Sign and magnitude calculations are complicated because we need both an adder and a subtractor even to implement addition. The adder is used when the signs are the same, the subtractor when they differ. Subtraction is just as complicated.

Ones-Complement Calculations Let's repeat the examples with ones complement arithmetic:

(e)
$$\begin{array}{r} 4 \\ +3 \\ \hline 7 \end{array} \qquad \begin{array}{r} 0100 \\ +0011 \\ \hline 0111 \end{array}$$

(f)
$$\begin{array}{r} -4 \\ +(-3) \\ \hline -7 \end{array} \qquad \begin{array}{r} 1011 \\ +1100 \\ \hline 1\,0111 \\ \hookrightarrow 1 \\ \hline 1000 \end{array}$$

(g)
$$\begin{array}{r} 4 \\ -3 \\ \hline 1 \end{array} \qquad \begin{array}{r} 0100 \\ -0011 \end{array} = \begin{array}{r} 0100 \\ +1100 \\ \hline 1\,0000 \\ \hookrightarrow 1 \\ \hline 0001 \end{array}$$

(h)
$$\begin{array}{r} -4 \\ +3 \\ \hline -1 \end{array} \qquad \begin{array}{r} -0100 \\ +0011 \end{array} = \begin{array}{r} 1011 \\ +0011 \\ \hline 1110 \end{array}$$

Adding two positive numbers, case (e), gives the same result as before. This should not be too surprising, since positive numbers are represented in the same fashion in all three systems.

Example (f) introduces one considerable difference between sign and magnitude addition and ones-complement addition: the concept of *end-around* carry. In ones complement, -4 is represented as 1011_2 and -3 as 1100_2. When we add these two numbers, we get a carry-out of the high-order bit position. Whenever this occurs, we must add the carry bit to the result of the sum. $1011_2 + 1100_2$ yields $1\ 0111_2$. When the carry-out is added to the 4-bit result, we get $0111_2 + 1 = 1000_2$. This is the representation of -7 in ones complement.

The end-around carry also happens in example (g). The sum of 4 (0100_2) and -3 (1100_2) yields $1\ 0000_2$. Adding in the carry gives 0001_2, the ones-complement representation of 1.

The last example, (h), obtains the sum 1110_2. This is precisely the ones-complement representation of -1.

Why does the end-around carry scheme work? Intuitively, the carry-out of 1 means that the resulting addition advances through the origin of the number wheel. In effect, we need to advance the result by 1 to avoid counting zero twice.

More formally, the operation of the end-around carry is the equivalent of subtracting 2^n and adding 1. Consider the case in which we compute the sum $M + (-N)$ where $M > N$:

$$M - N = M + \bar{N} = M + (2^n - 1 - N) = (M - N) + 2^n - 1$$

This is exactly the situation of example (g). The end-around carry subtracts off 2^n and adds 1, yielding the desired result of $M - N$.

Now consider the case shown in the second example. The sum to be formed is $-M + -N$, where $M + N$ is less than 2^{n-1}. This results in the following sequence of equations:

$$\begin{aligned} -M + (-N) = \bar{M} + \bar{N} &= (2^n - M - 1) + (2^n - N - 1) \\ &= 2^n + [2^n - 1 - (M + N)] - 1 \end{aligned}$$

After the end-around carry, the result of the sum becomes $[2^n - 1 - (M + N)]$. This is the correct form for representing $-(M + N)$ in ones-complement form.

Twos-Complement Calculations Twos-complement calculations behave very much like the ones-complement method, but without the end-around carry. Let's revisit the four examples:

(i)
$$\begin{array}{r} 4 \\ +3 \\ \hline 7 \end{array} \qquad \begin{array}{r} 0100 \\ +0011 \\ \hline 0111 \end{array}$$

(j)
$$\begin{array}{r} -4 \\ +(-3) \\ \hline -7 \end{array} \qquad \begin{array}{r} 1100 \\ +1101 \\ \hline 1|1001 \end{array}$$

(k)
$$\begin{array}{r} 4 \\ -3 \\ \hline 1 \end{array} \qquad \begin{array}{r} 0100 \\ -0011 \end{array} = \begin{array}{r} 0100 \\ +1101 \\ \hline 1|0001 \end{array}$$

(l)
$$\begin{array}{r} -4 \\ +3 \\ \hline -1 \end{array} \qquad \begin{array}{r} -0100 \\ +0011 \end{array} = \begin{array}{r} 1100 \\ +0011 \\ \hline 1111 \end{array}$$

Subtraction is handled as before: we negate the operand and perform addition. Carry-outs can still occur, but in twos-complement arithmetic we ignore them.

Example (i), summing two positive numbers, is identical to the two previous representation schemes. Summing two negative numbers is also straightforward. We simply perform binary addition, ignoring any carry-outs. Since we no longer have two representations for zero, there is no need to worry about correcting the summation. Mixed addition of positive and negative numbers is handled exactly like the other cases.

Why is it all right to ignore the carry-out? The same kind of analysis we used in the ones-complement case can be applied here. Consider the sum $-M + N$ when $N > M$. This can be rewritten as

$$M* + N = (2^n - M) + N = 2^n + (N - M)$$

Ignoring the carry-out is equivalent to subtracting 2^n. Doing this to the foregoing expression yields the result $N - M$, which is exactly what we desire. Consider another case: $-M + -N$, where $M + N$ is less than or equal to 2^{n-1}. This can be rewritten as

$$\begin{aligned} -M + (-N) = M* + N* &= (2^n - M) + (2^n - N) \\ &= 2^n - (M + N) + 2^n \end{aligned}$$

By subtracting 2^n, the resulting form is exactly the representation of $(M + N)$, the desired twos-complement representation of $-(M + N)$.

The trade-off between twos-complement and ones-complement arithmetic should now be a little clearer. In the twos-complement case, addition is simple but negation is more complex. For the ones-complement system, it is easy to perform negation, but addition becomes more complicated. Because twos complement only has one representation for zero, it is preferred for most digital systems.

A.4.5 Overflow Conditions

Overflow occurs whenever the sum of two positive numbers yields a negative result or when two negative numbers are summed and the result is positive. We can use the number wheel to illustrate overflow. Think of addition as moving clockwise around the number wheel. Subtraction moves counterclockwise. Using the twos-complement number representation, we can divide the number wheel into two halves: one representing positive numbers (and zero), the other representing the negative numbers. Whenever addition or subtraction crosses the positive/negative line, an overflow has occurred.

This concept is illustrated in Figure A.5, with the two example calculations 5 + 3 and −7 − 2. On the number wheel, starting with the representation for +5, we advance three numbers in the clockwise direction. This yields −8: an overflow has occurred. Similarly for

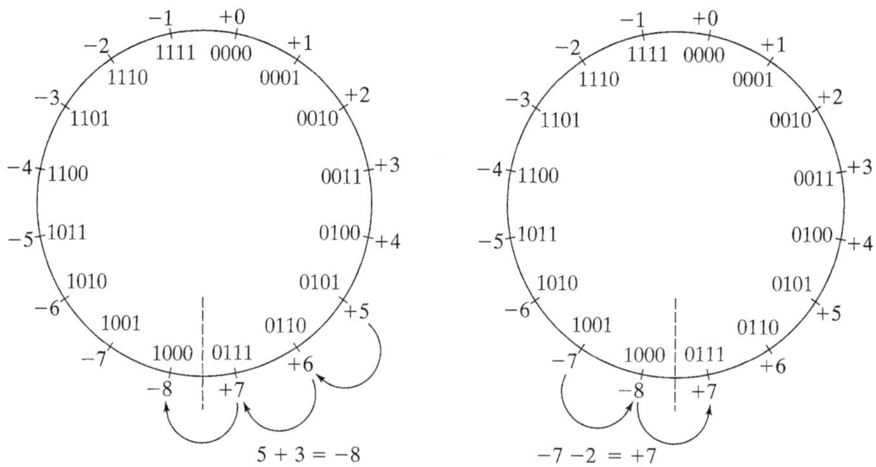

Figure A.5 Illustration of overflow conditions.

subtraction. Starting with the representation for −7, we move two numbers in the counterclockwise direction, obtaining the representation for +7. Once again, we have an overflow.

There is another way to detect when overflow has taken place. Let's look at the detailed calculations:

rry-in ≠ carry-out, overflow   Carry-in = carry-out, no over

	0111		1000		0000		1111
5	0101	−7	1001	5	0101	−3	110
3	+0011	−2	+1110	2	+0010	−5	+101
8	1000	7	10111	7	0111	−8	110

The carry-ins are shown at the top of each column of bits. In the first calculation, 5 + 3, the carry-in to the high-order bit is 1 while the carry-out is 0. In the second calculation, −7 + −2, once again, the carry-in to the final bit is different from the carry-out. In two cases in which overflow does not occur, where 5 + 2 and −3 + −5, the carry-in and the carry-out of the final stage are identical.

In general, overflow occurs when the carry-in and carry-out of the sign bit are different.

A.5 BCD Number Representation

BCD, or *binary-coded decimal,* represents the 10 decimal digits in terms of binary numbers. It is possible to build digital hardware that manipulates BCD directly, and such hardware could be found in early computers and many hand-held calculators. The BCD system was chosen for the internal number system in these machines because it is easy to convert it to alphanumeric representations for printouts and displays. The compelling advantages of BCD have waned over time, and these digits are supported by more modern hardware simply to provide backward compatibility with earlier generations of machines. In this section, we briefly examine the BCD representation. Details on constructing a BCD adder can be found in Section 5.6.4.

In the BCD representation, the decimal digits 0 through 9 are represented by the 4-bit binary strings 0000 through 1001. The remaining 4-bit encodings, 1010_2 through 1111_2, are not used and will be usually be treated as don't cares.

Just as in conventional decimal addition, BCD addition is performed one decimal digit at a time. The question is, what happens when the sum exceeds what can be represented in 4 bits? Stated differently, what are the conditions under which a carry is generated to the next highest-order BCD digit?

For example, let's consider the addition of the two BCD digits, 5 and 3:

$$\begin{array}{rr} 5 = & 0101 \\ 3 = & \underline{0011} \\ & 1000 \;=\; 8 \end{array}$$

Now consider the sum of 5 and 8:

$$\begin{array}{rr} 5 = & 0101 \\ 8 = & \underline{1000} \\ & 1101 \;=\; 13! \end{array}$$

The sum is $1101_2 = 13$, but this result should be represented correctly as 0001 0011 in BCD notation. Fortunately, there is a simple way to find the correct result. We add 6 (0110_2) to the digit sum if it exceeds 9. Let's examine the following cases:

$$\begin{array}{rrl} 5 = & 0101 & \\ 8 = & \underline{1000} & \\ & 1101 & = 13 \text{ in decimal} \\ & \underline{+\,0110} & \\ & 1\,0011 & = 1\,3 \text{ in BCD} \end{array} \qquad \begin{array}{rrl} 9 = & 1001 & \\ 7 = & \underline{0111} & \\ & 1\ 0000 & = 16 \text{ in decimal} \\ & \underline{+\,0110} & \\ & 1\,0110 & = 1\,6 \text{ in BCD} \end{array}$$

In both cases, by adding 6 we obtain the correct answer in BCD. Deciding when 6 needs to be added is the basis for designing BCD arithmetic elements (see Section 5.6.4).

APPENDIX REVIEW

In this appendix, we have examined the methods for representing numbers in positional notation. We use positional notation to represent numbers in a variety of different bases, including base 10 (decimal), base 2 (binary), base 8 (octal), and base 16 (hexadecimal). Of course, it is also possible to represent numbers in bases other than these four.

We then presented the methods for converting numbers in one base to the other. A number can be converted to base 10 simply by expanding the positional notation. Base 10 can be converted to another base by successive division. Since 8 and 16 are powers of 2, there is a simple method for mapping between these bases. All we need do is group the binary representation into adjacent groups of three bits for base 8 and four bits for base 16.

We next turned to the mechanics of binary addition and subtraction, using our intuition about how these operations work for base 10. For the addition operation, when the column sum exceeds the quantity that can be represented by a single digit, the "overflow" amount is carried over to the next higher-order column. Similarly for subtraction, when the column difference is negative, we must borrow an amount equal to the underlying base from the adjacent higher-order column. This guarantees that the column difference yields a non-negative result. Both carries and borrows can cascade from the rightmost columns toward the left columns.

Our next topic was the representation of negative numbers. We compared and contrasted three different representations: sign and magnitude, ones complement, and twos complement. The evolution of these representations was explained in terms of the complexity of implementing arithmetic elements. Twos-complement notation is used universally today, because it makes addition and subtraction easy to implement. Section 5.6 provides the details of constructing this circuitry and how to make it fast.

Finally, we concluded with a short description of the binary-coded decimal representation. Although no longer common in the internals of computers, this representation is closer to the base 10 system humans use and is always a part of user interfaces.

EXERCISES

A.1 *(Conversion to Base 10)* Convert each of the following numbers into its equivalent form in base 10:

(a) 1000_2
(b) 1101100_2
(c) 0101011_2
(d) 757_8
(e) 1000_8
(f) 123_8
(g) FFA_{16}
(h) 1000_{16}
(i) $3AE_{16}$

A.2 *(Conversion to Target Base)* Convert each of the following base-10 numbers into its equivalent form in the indicated base:

(a) 53 to base 2
(b) 500 to base 2

(c) 129 to base 2
(d) 127 to base 8
(e) 74 to base 8
(f) 798 to base 8
(g) 1023 to base 16
(h) 4000 to base 16
(i) 240 to base 16

A.3 *(Conversion Between Base 2 and Base 8 or Base 16)* Convert each of the following binary numbers to the indicated base:

(a) 1110011_2 to base 8
(b) 1011011_2 to base 8
(c) 1001000111000101_2 to base 8
(d) 10011100_2 to base 16
(e) 11110011_2 to base 16
(f) 11100011001100011000_2 to base 16

A.4 *(Convert Between Base 8 or Base 16 and Binary)* Convert each of the numbers in the indicated base to binary:

(a) 252_8
(b) 4077_8
(c) 101_8
(d) $AFE0_{16}$
(e) 4077_{16}
(f) $8FC_{16}$

A.5 *(Convert Between Base 8 and Base 16)* Using the same numbers as in Exercise A.4, convert the numbers in base 8 to base 16 (a, b, c) and the numbers in base 16 to base 8 (d, e, f).

A.6 *(Alternative Number Systems)* People on the planet Mars use a quarternary number system, consisting of the digits 0, 1, 2, and 3 (that is, base 4). Perform the following conversions to and from base 4:

(a) Convert 597_{10} to base 4
(b) Convert 32021_4 to base 10
(c) Convert 110010011_2 to base 4
(d) Convert 223_4 to base 2
(e) Convert 771_8 to base 4
(f) Convert 32210_4 to base 8
(g) Convert $AB0_{16}$ to base 4
(h) Convert 1230322_4 to base 16

A.7 *(Binary Addition)* Perform the following additions in binary:

(a) $100110_2 + 111_2$
(b) $110111_2 + 101_2$
(c) $111110_2 + 10111_2$
(d) $111001_2 + 10001_2$
(e) $11011100110_2 + 10011001_2$
(f) $10101010_2 + 1111111_2$

A.8 *(Binary Subtraction)* Perform the following subtractions in binary:

(a) $100110_2 - 111_2$
(b) $110111_2 - 101_2$
(c) $111110_2 - 10111_2$
(d) $111001_2 - 10001_2$
(e) $11011100110_2 - 10011001_2$
(f) $10101010_2 - 1111111_2$

A.9 *(Base Conversions)* Perform the following base conversions. Use successive division where necessary.

(a) 200_{10} to base 7
(b) 200_{16} to base 3
(c) 356_7 to base 2
(d) $A9DE_{16}$ to base 3
(e) 591_{10} to base 5
(f) 1001011_2 to base 4

A.10 *(Addition/Subtraction in Different Bases)* Perform the indicated arithmetic operations:

(a) $200_3 + 22_3$ in base 3
(b) $43_5 - 24_5$ in base 5
(c) $10A_{16} + 201_8$ in base 16
(d) $77_8 - 25_{10}$ in base 8
(e) $95_{10} + 211_3$ in base 3
(f) $71_8 - 32_4$ in base 8

A.11 *(Negative Numbers)* Express the following negative numbers in sign and magnitude representation (using 6 bits):

(a) −13
(b) −27
(c) −5

A.12 *(Negative Numbers)* Express the following negative numbers in ones-complement representation (using 6 bits):

(a) −13
(b) −27
(c) −5

A.13 *(Negative Numbers)* Express the following negative numbers in twos-complement representation (using 6 bits):

(a) −13
(b) −27
(c) −5

A.14 *(Addition with Twos Complement Numbers)* Perform the following additions by first converting the two base-10 number to twos-complement notation (using 6 bits):

(a) 15 + 1
(b) −1 + 0

(c) −5 + −23
(d) −16 + 15
(e) −14 + −2
(f) −27 + 12

A.15 *(BCD Numbers)* Perform the following addition of binary-coded decimal numbers:

(a) 0000 1001 + 0000 0001
(b) 0010 0111 + 0110 0001
(c) 0001 0101 + 0000 0101

B

Basic Electronics

Introduction

Digital hardware systems can be viewed from many alternative perspectives, including Boolean logic, logic gates, and behavioral specifications. For most of this book, the most primitive abstraction we use is the logic gate. In this appendix, we peek beneath the sheets to get some idea of how logic gates are actually implemented by more primitive electrical components.

We begin by reviewing briefly the concepts of electricity that influence the fundamental operation of logic gates. We will also examine the basic implementation technologies from which logic gates are constructed, such as diodes, capacitors, resistors, bipolar transistors, and MOS transistors.

B.1 Basic Electricity

In this section, we introduce the basic terminology, fundamental quantities, and laws of electricity.

B.1.1 Terminology

Electricity is energy that can be transported. An *electric circuit* consists of an energy source (such as a battery or power supply) and interconnected electrical components implementing a useful function. The connections are formed by wires, also known as *conductors,* which are made of materials such as copper or some other metal that can conduct electricity. Electrical charge transported across a conductor is called electric *current.* Charge is carried by *electrons,* which are negatively charged, or by positively charged *ions* in the conductor. Current is the intensity of the flow of charge. Between two points in a circuit, electrons flow from the more negatively charged point toward the one that is more positively charged. Positive charges, sometimes called *holes,* move in the opposite direction. By convention, current flows in the direction of holes, which is opposite to the direction of electron flow.

Voltage is associated with any two points in the circuit, and represents the difference in electrical potential between those points. Stated differently, voltage is the electric force that causes electrons to flow in a circuit. Voltage is defined as a relative quantity. In a typical digital circuit, the lowest possible voltage is called *ground* and is arbitrarily assigned 0 volts. In most existing digital circuits, the highest possible voltage value is defined not to exceed 3.6 volts. In digital systems, we assign logic 1 to "high" voltages and 0 to "low" voltages, but these assignments are somewhat arbitrary. Today's CMOS technology operates in a wide voltage range (1–6 V) with most circuits operating with a 3.3 V supply although 3.0 V is also common. In contrast, older TTL technology operated with a narrow range around 5.0 V. For the popular Advanced Low-Voltage CMOS (ALVC) technology, a voltage in the range of 0 to 0.8 volts is interpreted as logic 0, while 2.0 volts or higher is interpreted as a logic 1. Voltages outside these ranges are not guaranteed to be interpreted as either a 0 or a 1.

The fundamental concepts of electricity can often be described by analogy with water. The greater the electrical potential, the larger the voltage, and the greater the force on the flow of the charge-carrying electrons. Think of a waterfall. A large voltage corresponds to a waterfall of great height. As a water molecule flows "downhill," a good deal of pressure is exerted on it by gravity and the force of water behind it. By analogy, water molecules correspond to electrons, and electrical current corresponds to the speed of the water flow.

Suppose that the voltage difference is 0, so that both points in the circuit are at the same potential. In this case, the water is a stagnant pool with no water flow, and there is no current. Given a waterfall of only modest height, the water trickles slowly downhill. This is analogous to a small current. But if the waterfall is of a great height, the flow of water will be forceful and the current is large.

Whereas conductors transport electricity, other materials (called *insulators*) are impervious to electricity. An important class of materials is the *semiconductors;* materials that can change from being conductors at one moment to being insulators at the next. This makes it possible to form electrically controlled switches, which are at the heart of all digital logic circuits.

B.1.2 Fundamental Quantities and Laws

The key electrical quantities and their units are shown in Figure B.1. Charge measures the number of positive or negative charges at a given point in the circuit and is described in units of coulombs (C). One coulomb is equivalent to the charge on 6.24×10^{18} electrons. Voltage is

Quantity	Symbol	Units
Charge	Q	Coulomb
Voltage	V	Volt
Current	I	Ampere
Resistance	R	Ohm
Capacitance	C	Farad

Figure B.1 **Electrical quantities and their units.**

the difference in electrical potential between two points in a circuit and is measured in volts (V). At a given point in a circuit, current is the change in charge as a function of time. Thus, one ampere is defined to be one coulomb per second.

Resistance Figure B.1 introduces two important new quantities: resistance and capacitance. *Resistance* is the "friction" that limits current. Doubling resistance cuts the current in half. When two resistors are connected in series, one immediately following the other: their resistances add. Placing two resistors in parallel results in a resistance that is less than the component resistances. To be more precise, if the resistances, measured in ohms, Ω are R_1 and R_2, the parallel resistance will be $(R_1 \times R_2)/(R_1 + R_2)$.

A *short circuit* is a path of conductors with no (or very low) resistance. An *open circuit* is a conductive path with infinite resistance. Semiconductor materials make it possible to construct connections between two points that can be varied between low and high resistance.

One of the most important expressions for analyzing electrical circuits is Ohm's law (for the German scientist Georg Simon Ohm). It describes the relationship between voltage (V), current (I), and resistance (R) as follows:

$$V = I \times R$$

To understand this relationship, let's consider the water analogy again. V is the height of the waterfall. We can think of R as inversely proportional to the diameter of a water pipe: a high resistance corresponds to a narrow pipe, a low resistance to a wide pipe. A narrow pipe restricts the flow more than a wide pipe. With a high resistance, current is reduced, because fewer electrons can move through the conductor per unit time. By reducing the resistance (that is, increasing the cross section of the pipe), we increase the flow of electrons.

An alternative formulation of Ohm's law allows us to describe resistance as a function of voltage and current: $R = V/I$. Thus, if a power supply provides voltage V and the current is measured as I, then the resistance of the circuit being driven by the power supply is R.

Capacitance Capacitance is the ability to store charge and is measured in units of farads (named for the great 19th-century British scientist Michael Faraday). A capacitor is a device with two parallel conducting plates separated by a nonconducting material. Placing negative charges on one plate will attract positive charges to the other plate. A capacitor uses current to charge the plates up slowly to a new voltage. Once charge is stored, the capacitor can also provide a "discharge" current to the rest of the circuit. Thus, capacitors are often used to smooth out variations in the current provided by the circuit's power supply.

Continuing with our water analogy, a capacitor behaves much like a water holding tank. A hole at the bottom of the tank provides a steady "outflow" of current, even though the inflow may be sporadic.

Charge, voltage, and capacitance are related by

$$Q = C \times V$$

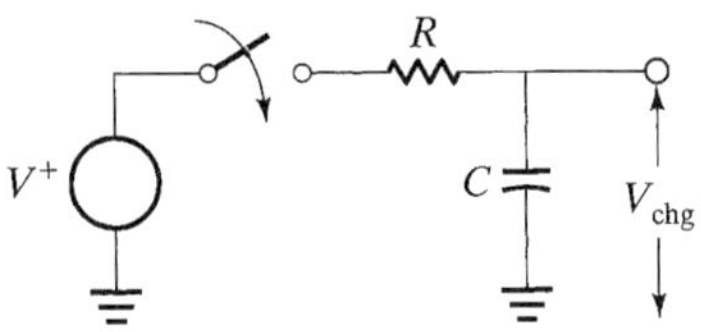

Figure B.2 **Charging a capacitor through a resistor.**

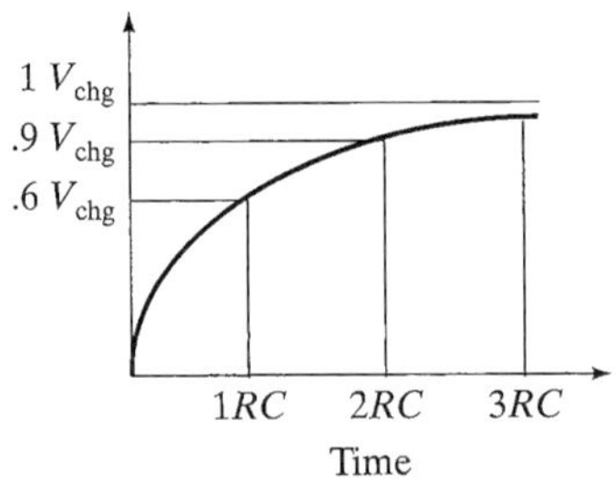

Figure B.3 ***RC* time constant and time to charge a capacitor.**

Charge is equal to capacitance times voltage. By placing a voltage V across a capacitor of C farads, we can store a charge of Q coulombs.

RC Delay There is an interesting relationship between time, resistance, and capacitance. Consider how long it takes to charge up a discharged capacitor. Figure B.2 shows a possible setup. In the schematic, the voltage source is labeled V, the resistor R, and the capacitor C.

We assume that the capacitor is discharged completely and the switch is in the open position. When the switch is closed, the power supply begins to charge up the capacitor toward the voltage V_{chg}. If you measure the voltage across the capacitor with a voltmeter, initially the voltage changes very quickly, but then it slows down.

There is a precise relationship between the resistance and capacity of the circuit and the time it takes to charge the capacitor. It is related directly to $R \times C$, also known as the *RC time constant* (it may seem strange that ohms times farads is seconds, but this is the case). After one *RC* delay, the capacitor is charged up to slightly more than 60% of its final value. After two *RC* delays, it reaches almost 90% of its final value. It takes five *RC* delays before the capacitor reaches 99% of its final value. This is shown in Figure B.3.

RC delays play an important role in determining the true performance of digital circuits. Even though wires are excellent conductors, they do present some resistance to the current flow. But even more important, wires introduce capacitance: a wire forms one plate of a capacitor whose second plate is the circuit board itself. Changes in voltages on wires require this capacitance to be either charged or discharged, and this translates into a significant source of delay in real circuits.

B.2 Logic Gates from Resistors and Diodes

In this section, we investigate how to construct logic functions from fundamental electrical objects like resistors, diodes, and transistors.

B.2.1 Voltage Dividers

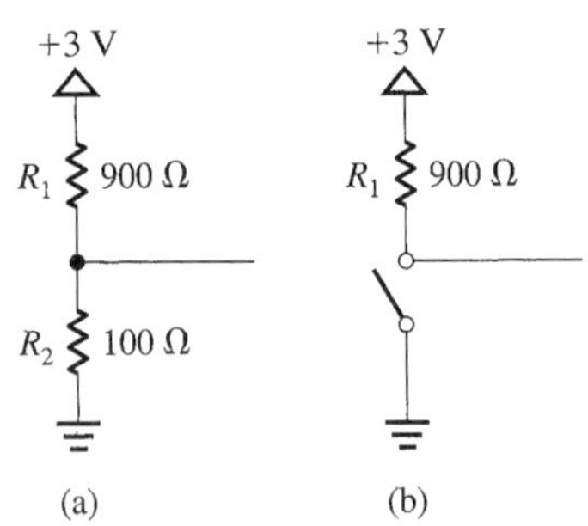

Figure B.4 **Voltage divider view of a logical inversion.**

When a voltage spans two resistors in series, the voltage measured at the point between the resistors is divided in proportion to the ratio of the individual resistors and the sum of the resistances. As an example, see Figure B.4(a). The voltage across R_1 is given by $V \times R_1/(R_1 + R_2)$, where V is the voltage across the series resistors, R_1 is the resistance between the power supply and the output node, and R_2 is the resistance between the ground and the power supply. R_1 is often called the *load* or *pull-up* resistor, while R_2 is the *pull-down* resistor.

In the example, R_1 = 900 ohms, R_2 = 100 ohms, and V = 3 volts. The voltage drop across R_1 is

$$\frac{R_1}{R_1 + R_2} \times V = \frac{900}{1000} \times 3 = 2.7 \text{ V}$$

Thus, the voltage measured between the output node and ground would be 0.3 V. Alternatively, we could have arrived at this directly by calculating the voltage across R_2, namely $V \times R_2/(R_1 + R_2)$. By appropriately sizing the relative resistances, we can choose any desired output voltage.

Now suppose that R_2 is a variable resistor with two basic settings: low resistance or very high resistance. When the pull-down resistor has a sufficiently high resistance, it behaves like an open circuit, as shown in Figure B.4(b). The output node reaches 3 V.

An electrical device that can switch between low resistance and very high resistance is called a transistor, as we will see in Sections B.3 and B.4. A *transistor* is a three-terminal device that establishes a low-resistance path between two terminals when a high voltage is placed on the third terminal. When a low voltage is placed on this control terminal, the remaining two terminals are separated by a high resistance. If R_2 is replaced by a transistor, it's easy to see that we obtain an inverter. When the input voltage is high, the output voltage is low. When it is low, the output voltage is high.

But before we can examine transistors in more detail, we need to take a look at a simpler, two-terminal device: the diode.

B.2.2 Diode Logic

A *diode* is a two-terminal electrical device that allows current to flow in one direction but not the other. It is like a pipe with an internal valve that allows water to flow freely in one direction but shuts down if the water tries to flow backward. The schematic diagram for a diode is shown in Figure B.5. The diode's two terminals are called the *anode* and *cathode*. In the diode symbol, the arrow points from the anode (flat part of triangle) toward the cathode (point of the triangle).

The device operates by allowing current to flow from anode to cathode, basically in the direction of the triangle. Recall that current is defined to flow from the more positive voltage toward the more negative voltage (electrons flow in the opposite direction). If the diode's anode is at a higher voltage than the cathode, the diode is said to be *forward biased:* its resistance is very low, and current flows. The diode is not a perfect conductor, so there is a small voltage drop, between 0.3 and 0.7 V,

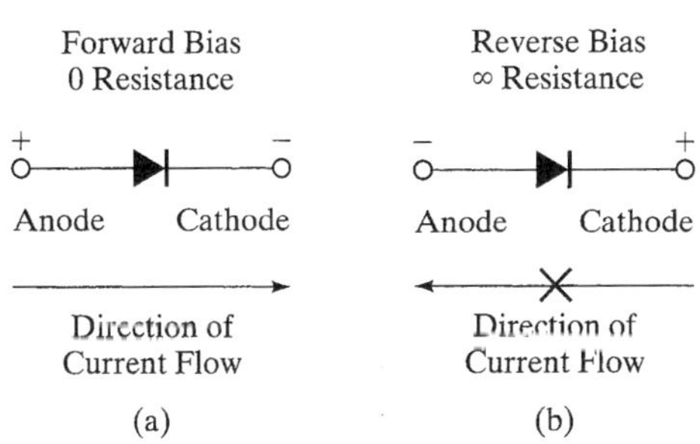

Figure B.5 Diode operation.

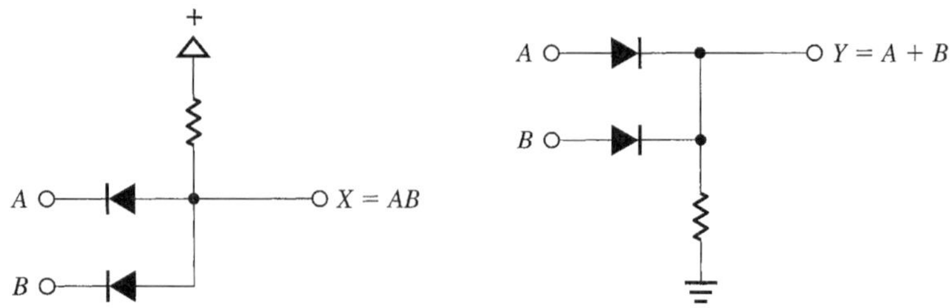

Figure B.6 Simple gates from diodes and resistors.

across it. If the anode is at a lower voltage than the cathode, the diode is *reverse biased:* its resistance is very high, and no current flows.

We can construct simple gates with nothing more than two or more diodes and a resistor (see Figure B.6). At the left of the figure is a diode AND gate, and at the right is a diode OR gate. Let's examine the AND gate first. If one of the inputs A or B is grounded, current flows through the diode and the output node X is at a low voltage. The only way to get a high output is by having both inputs high. This is clearly a logical AND function.

Now we turn to the OR gate. Whenever one or the other of the inputs A and B are high, current flows through the associated diode. This brings the output node Y to a high voltage. This circuit clearly implements a logical OR.

Unfortunately, it is difficult to cascade circuits of this kind into multiple levels of logic gates. The voltage drops across the diodes and adds up as they are cascaded in series, leading to significantly degraded voltage levels.

For example, suppose we wire up five diode-resistor AND gates in series. If a string of inputs are logic 0 and the series diodes are conducting, then the output from the final stage should be recognized as a logic 0 as well. But because each diode adds a voltage drop, the measured output would actually get higher with each stage as the diode offsets add up until the voltage could be higher than that recognized as logic 0.

One solution is to increase the power-supply voltage, redefining the range that is recognized as a logic 0 and logic 1. Of course, the higher the voltage, the higher the power consumed, and the more heat the circuit generates. And no matter what you set the power supply to, there is still a limit to the number of logic levels that can be cascaded. This is hardly an adequate solution.

Also note that it is not possible to construct an inverter with only diodes and resistors. AND and OR functions by themselves are not a complete logic without NOT. Thus, there are some logic functions that cannot be implemented in diode-resistor logic. Fortunately, transistors solve all of these problems.

B.3 Bipolar-Transistor Logic

In this section, we examine how to build logic gates from bipolar transistors, the dominant technology of the 1970s and early 1980s.

B.3.1 Basic Bipolar-Transistor Logic

A *bipolar transistor* is a three-terminal semiconductor device. Under the control of one of the terminals (called the *base*), current can flow selectively from the *collector* terminal to the *emitter* terminal. Using transistors as electronically controlled switches is critical for building modern digital logic.

Using our water analogy, a transistor is like a water spigot. The base is like a screw. When it is tightened down, no water (current) flows through the spigot. The transistor is in its "off" region of operation. As the screw is turned, the flow of water begins to trickle and then increases. This is the "linear operating" region. The current increases linearly as the voltage across the transistor is increased. Eventually, the flow reaches a point where opening the screw further does not increase the flow, being limited by the diameter of the pipe. This is called the saturation region. Despite changes in voltage across the transistors, there is no change in current. Transistors in digital logic pass quickly from the off region to the saturation region.

The basic inverter constructed from transistors and resistors is shown in Figure B.7 (for those in the know, the transistor shown is an NPN transistor). A high voltage at the base turns on the transistor. The output F is discharged to ground, getting close to 0 V but never quite reaching it (it reaches a voltage drop away from 0 V).

When a low voltage is placed on the base, the transistor is turned off. The output node F is charged up toward the power supply voltage through the pull-up/load resistor R_1.

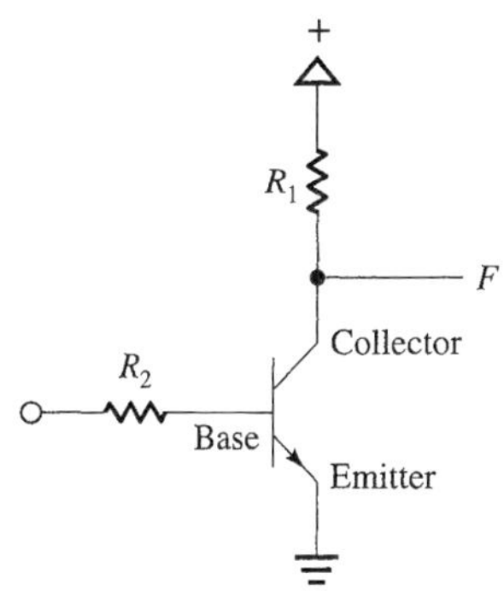

Figure B.7 Transistor–resistor inverter.

B.3.2 Diode-Transistor Logic

Diodes, transistors, and resistors can be used to implemented a wide variety of gates. Basically, we combine the diode logic of Figure B.6 with the transistor inverter of Figure B.7 to form NAND (not AND) and NOR (not OR) functions.

A two-input DTL (diode-transistor logic) NAND gate is shown in Figure B.8. It works like this. The diodes marked D1 and D2, together with resistor R_1, form a two-input AND function. At the input to D3, a logic 0 is represented by approximately 0.7 V, while a logic 1 is close to

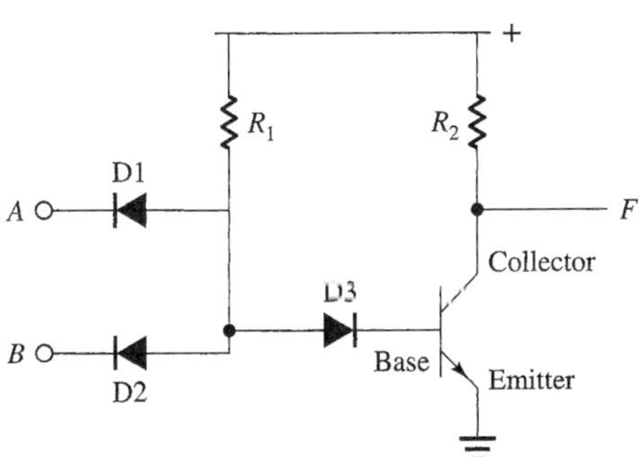

Figure B.8 Diode-transistor logic (DTL) NAND gate.

the supply voltage. D3 increases the voltage required to turn on the transistor. This gives a better separation between the voltage levels recognized as a logic 0 and logic 1.

For the transistor to conduct, D3 must be turned on. This happens when the anode voltage reaches 1.4 V. If the voltage at the anode is much higher than this, the base will be driven to a high voltage. The transistor will be strongly turned on, with low resistance, and thus, F will be discharged toward 0 V. If the anode is at a low voltage, the base will also be low. This keeps the transistor off (essentially infinite resistance), allowing the output node to reach a logic 1 voltage level.

Gates constructed as in Figure B.8 have a limit to the number of gate inputs to which their output can be connected. This is called *fan-out*. The pull-up resistor R_2 is what limits the fan-out. The output F is at Vcc as long as no current is being drawn from the power supply to charge electrical nodes to which F is connected. However, if F is connected to the A or B input of a similar gate, current is drawn through R_2. The voltage at F is reduced according to Ohm's law. If there are too many connections drawing current, the voltage at F may be so reduced that it can no longer be recognized as a logic 1. Thus, the number of fan-outs must be limited carefully in this kind of logic.

DTL has several advantages. NAND and NOR functions, which are easy to build in DTL, are *logically complete*. This means that any logical function can be expressed as a collection of only NAND gates or only NOR gates. NAND and NOR gates form the heart of all logic designs. In addition, this logic family uses lower voltages, less power, and operates at higher speeds, since only small currents are needed to turn on the transistors.

Another feature of DTL is its ability to implement a "wired AND" function. For example, if we wired several DTL NAND gates together, we would observe the following behavior. If any one of the NAND gates had a logic 0 as its output, the whole output function would also be at a logic 0 voltage. All the output functions would have to be at logic 1 for the output to be 1. See Figure B.9, which shows the internal wiring and the equivalent logic schematic for the wired-AND function. The AND gate actually does not exist. The notation is used simply to represent that the interconnection forms the AND function.

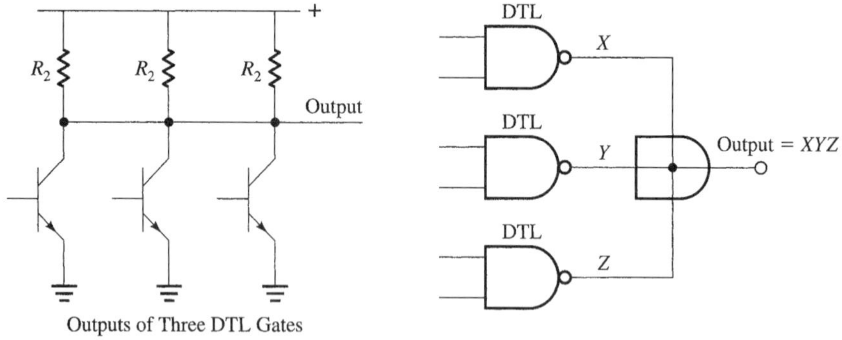

Figure B.9 Wired-AND DTL gates.

B.3.3 Transistor–Transistor Logic

We can think of a bipolar transistor as two diodes placed very close together, with the point between the diodes being the transistor base. Thus, we can use transistors in place of diodes to obtain logic gates that can be implemented with transistors and resistors only. This is called *transistor–transistor logic* (TTL), and it is the most widely used family of components available today.

A smart designer realized that the DTL NAND gate of Figure B.8 could be constructed from a two-emitter transistor connected to a transistor inverter. The multi-emitter transistor is the critical piece of technology that makes TTL logic gates possible.

A *two-input* TTL NAND gate is shown in Figure B.10 (some of the details have been eliminated). This circuit replaces the three diodes of Figure B.8 with a dual-emitter transistor. This configuration has one significant advantage over the diode implementation. Besides being voltage-controlled switches, transistors also act as amplifiers. When the transistor base is undergoing a change in voltage, the transistor can amplify this change, thus speeding up the rate at which the transistor turns on or off. The result is faster gate switching.

In simplified terms, the circuit of Figure B.10 works as follows. When one of the inputs A or B is low, the current available through R_1 at the transistor base is diverted to ground. No current flows from the base to the collector, and therefore no current reaches the base of the output transistor. Thus, the output transistor is off. The pull-up resistor, R_2, charges the output node to the high-voltage state. Only when both inputs are high can the current flow through R_1 from base to collector to turn on the output transistor. In this case, the output path discharges to ground.

A more realistic circuit for a NAND logic gate, such as that found in the TTL 7400 component, is shown in Figure B.11. The output configuration with transistor Q_4 in the pull-up path and transistor Q_3 in the pull-down path is called a *totem pole output*. Q_4 makes it possible to pull up the output faster, using lower power than is possible with just a resistor. In principle, this circuit behaves much like the simplified schematic of Figure B.10.

In Section 4.4, we introduce the concept of *open-collector* gates. These are gates with internal organizations that allow them to participate

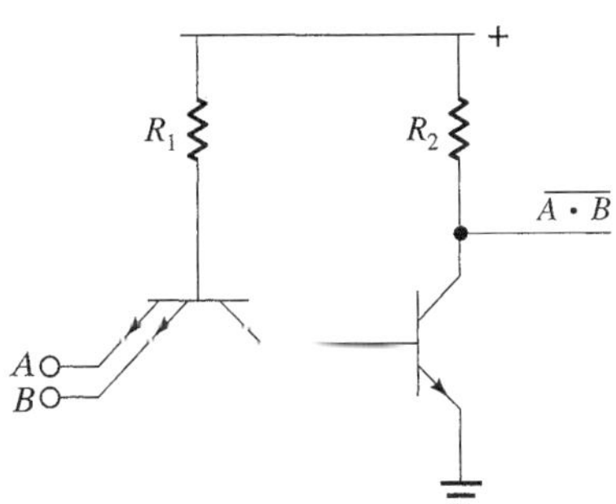

Figure B.10 Two-input TTL NAND gate.

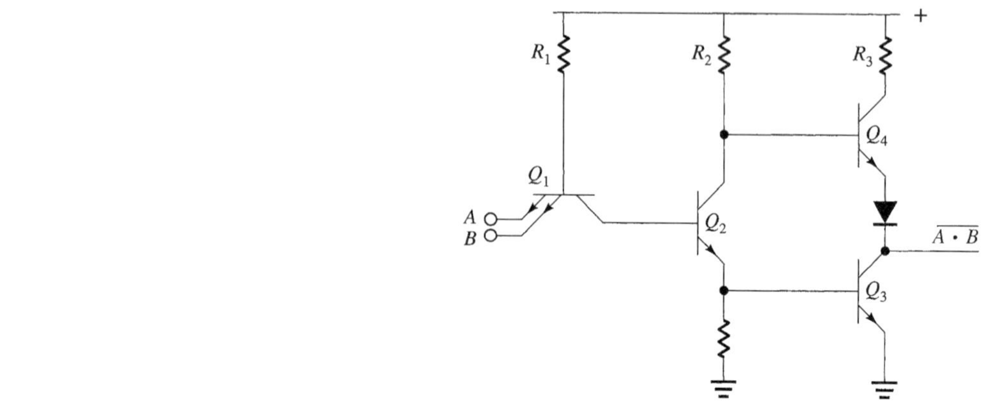

Figure B.11 **Two-input TTL NAND gate with totem pole output.**

in a wired-AND configuration, similar to the DTL gates in Figure B.9. The pull-up path (namely the R_3 resistor, Q_4 transistor, and the diode of the output stage of Figure B.11) is eliminated in an open-collector gate. Thus, an open-collector gate only has the ability to pull down its output node. We need a resistor external to the gate. This pulls up the wired-AND output if none of the attached gates have an input combination that provides a pull-down path to ground.

Section 4.4 also introduces the concept of *tri-state gates*. These are gates with three possible interpretations of their outputs: logic 0, logic 1, and no connection. The latter is called the *high-impedance state* and is denoted by the "value" *Z*. In Figure B.11, the totem-pole-output transistors Q_3 and Q_4 are not designed to be on simultaneously. Q_4 is on when the output is 1, and Q_3 is on when the output is 0. In a tri-state gate, both of these transistors can be off at the same time when a special *enable* input is left unasserted. The output node is disconnected from the supply or ground, making it appear to other logic as an open circuit.

B.3.4 TTL Circuits and Noise Margin

A major achievement of TTL logic is the ease with which different circuits can be interfaced and cascaded to form more complex logic functions. In part, this is due to the concepts of guaranteed voltage levels and noise margins. A *guaranteed voltage* is one at which circuits always detect the correct voltage level, within a specified temperature range (0 to 70°C), voltage range (5 V ± 5%), loading, and the parametric variance of the semiconductor devices themselves.

TTL circuits are characterized by four voltage specifications: V_{oh}, V_{ol}, V_{ih}, and V_{il}. V_{oh} (output high voltage) is the minimum voltage at which the circuit delivers a logic 1. V_{ol} (output low voltage) is the maximum voltage at which the circuit can produce a logic 0.

Similarly, V_{ih} (input high voltage) is the minimum voltage at which a circuit detects a logic 1. V_{il} (input low voltage) is the maximum voltage at which it recognizes a logic 0.

For TTL circuits, $V_{oh} = 2.4$ V, $V_{ol} = 0.4$ V, $V_{ih} = 2$ V, and $V_{il} = 0.8$ V. The input and output voltages differ by 0.4 V. This permits the output signals to be degraded by the wires between circuits but still be recognized as good logic values. The difference between V_{oh} and V_{ih} is called the *high-state DC noise margin.* The difference between V_{ol} and V_{oh} is called the *low-state DC noise margin.*

B.4 MOS Transistors

We now turn our attention to logic functions constructed from MOS transistors, the dominant technology of today.

B.4.1 Voltage-Controlled Switches

The operation of an MOS transistor is considerably easier to explain than that of a bipolar transistor. The voltage-controlled switches introduced in Section 1.3.2 correspond directly to MOS transistors. In this section, we describe the operation of MOS transistors and how they can be used to implement logic gates.

An MOS transistor is nothing more than a voltage-controlled switch. It has three connection points: a *source,* a *drain,* and a *gate* (a transistor gate bears no resemblance to a logic gate, an unfortunately ambiguous use of the term).

A cross section of the metal-oxide-silicon sandwich that forms the transistor is shown in Figure B.12. The bottommost material layer is made of silicon, an insulating oxide layer sits on top of it, and the topmost layer is the metal gate. (More modern integrated circuit processes have replaced the metal layer with a material called polycrystalline silicon, but the older "metal gate" terminology still holds.) The source and drain regions contain silicon material with a large excess of electrons separated by the slightly positively charged bulk silicon. The source and drain are called *diffusion regions* because of the chemical process used to create them. Negatively charged ions (atoms with extra valence electrons) are placed onto the silicon surface and are diffused into the surface by heating the silicon material. The materials of the source and drain are identical. By convention, the source is the electrical node with the lower of the two voltage potentials at either end of the channel.

The electrical behavior of the transistor is generally as follows. When a positive voltage is placed on the gate, electrons from the silicon bulk are attracted to the transistor *channel:* an initially nonconducting region between the source and drain very close to the silicon surface. When the

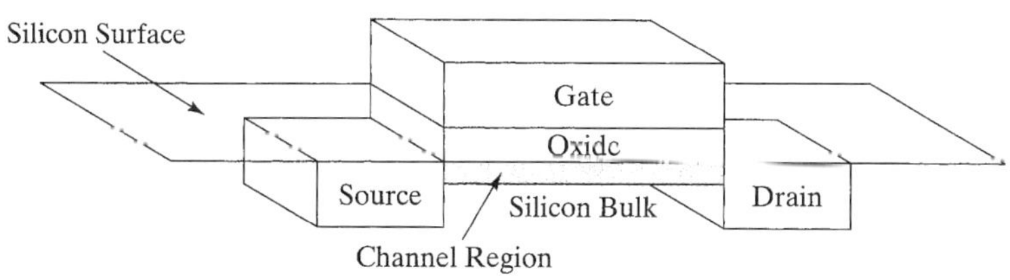

Figure B.12 Cross section of a MOS transistor.

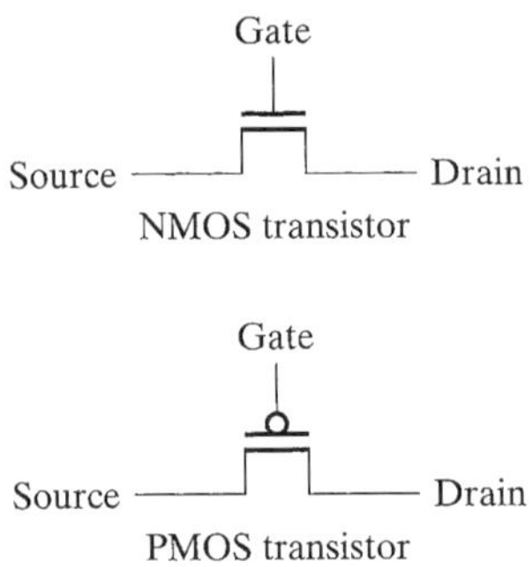

Figure B.13 NMOS and PMOS transistor symbols.

gate voltage becomes sufficiently positively charged, enough electrons are pulled into the channel from the bulk to establish a charged path between the source and the drain. Electrons flow across the transistor channel, and the voltage-controlled switch is conducting. If a 0 or very small voltage is placed on the gate, no electrons (or at least very few) are attracted to the channel. The source and drain are disconnected, no current flows across the channel, and the switch is not conducting.

There are two fundamentally different kinds of MOS transistors, called *n-channel* and *p-channel* transistors, or *NMOS* and *PMOS* for short. Their schematic symbols are shown in Figure B.13. Because they are made from materials with different affinities for electrons, the two transistor types behave quite differently. The transistor operation described earlier is actually for the NMOS transistor. The bulk is positively charged, while the diffusion is negatively charged. The transistor switch is "closed" (conducting) when a logic 1 is placed on its gate and "open" (nonconducting) when the gate is connected to a logic 0. The PMOS transistor is complementary. The diffusion regions are positively charged and the silicon bulk is negatively charged. A PMOS transistor behaves in a complementary way: It is "closed" (conducting) when a logic 0 is placed on the gate and is "open" (nonconducting) when a logic 1 is placed there.

The symbols for the two different kinds of transistors make it easy to remember how they operate. An NMOS transistor conducts when the gate voltage is asserted in positive logic. The PMOS transistor conducts when the gate is asserted in negative logic. This is why there is a polarity bubble on the gate of the PMOS transistor's symbol.

B.4.2 Logic Gates from MOS Switches

Any logic gate can be constructed from a combination of NMOS and PMOS transistors. Figure B.14 shows transistor networks for (a) an inverter, (b) a two-input NAND gate, and (c) a two-input NOR gate. The power supply (+3 V) and ground (0 V) represent logic 1 and logic 0, respectively. The inverter is constructed from an NMOS and a PMOS transistor connected in series between power and ground. It operates as follows (see Figure B.15). When A is a logic 1, the NMOS transistor

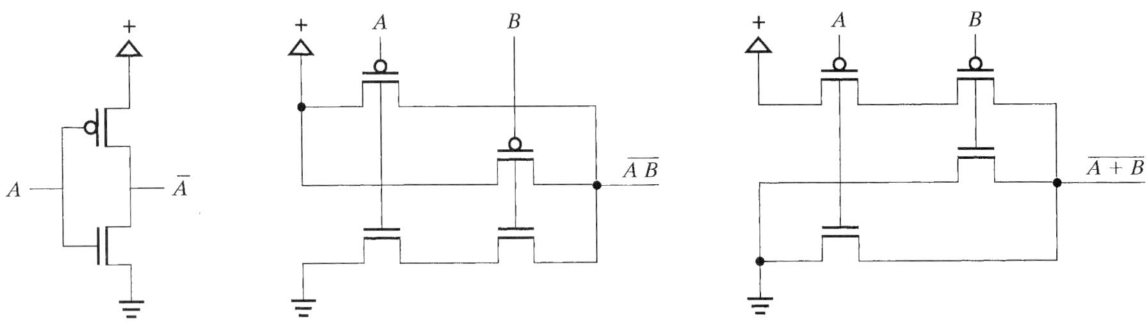

(a) Inverter (b) NAND (c) NOR

Figure B.14 Gates implemented from MOS transistors.

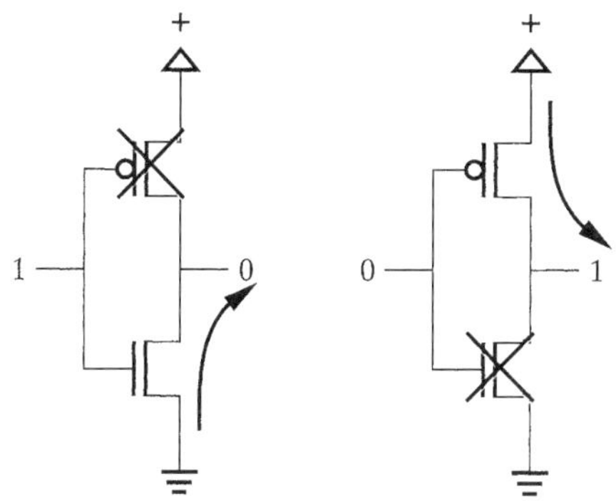

Figure B.15 **Inverter operation.**

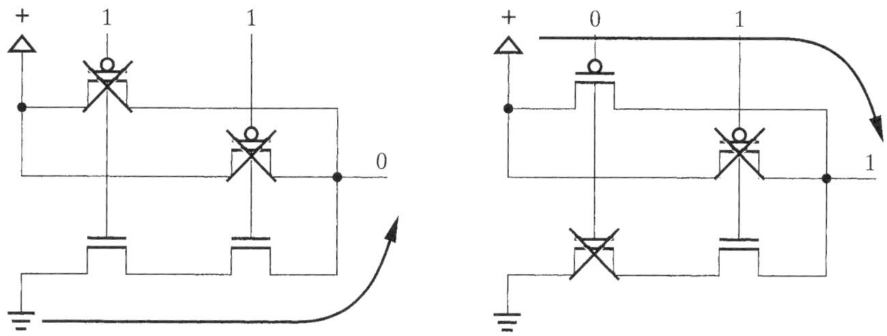

Figure B.16 **NAND-gate operation.**

is conducting and the PMOS transistor is not. The only unbroken connection path is from ground to the output node. Thus, a logic 1 at the input yields a logic 0 at the output.

Now let's look at the case in which A is a logic 0. Now the PMOS transistor conducts while the NMOS transistor does not. The output node is connected to a logic 1. A 0 at the input yielded a 1 at the output. The series transistors implement an inverter.

The transistor-level implementations of the NAND and NOR gate work similarly. Let's start with the NAND logic gate, constructed from two NMOS transistors in series between the output node and ground and two PMOS transistors in parallel between the output node and the power supply. A path between the output node and ground can be established only when both of the NMOS transistors are conducting. This happens only if A and B are both at a logic 1. In this case, the two PMOS transistors are not conducting, breaking all paths between the output node and the logic 1 at the power supply. This is the case when $A = B = 1$ and Output = 0.

Now what happens if one or both of A and B are at a logic 0? Let's take the case $A = 0$ and $B = 1$ (see Figure B.16). The NMOS transistor controlled by A is not conducting, breaking the path from the output to ground. The PMOS transistor it controls is conducting, establishing the path from the power supply to the output. The other path, controlled by B, is broken, but this has no effect on the output node as long as some path exists between it and some voltage source.

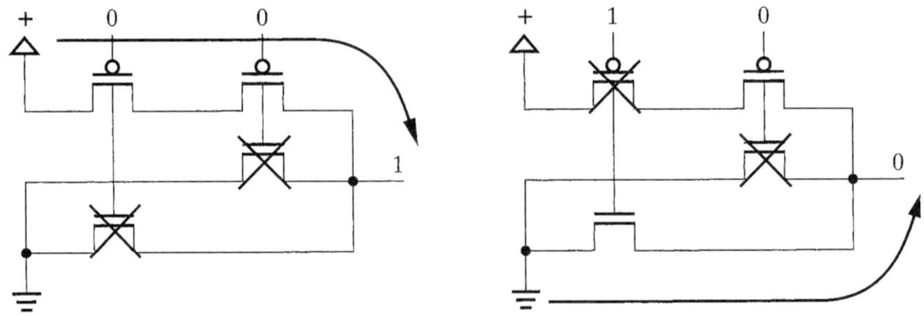

Figure B.17 **NOR-gate operation.**

The case $A = 1$ and $B = 0$ is symmetric. If both inputs are 0, there are now two paths between the power supply's logic 1 and the output node. Of course, this doesn't change the logic signal at the output: it is still a logic 1. From this discussion, it should be obvious that the circuit configuration performs the function of a NAND gate.

The *pull-down network* switches between the output and the ground signal. We define the *pull-up network* similarly between the output and the power supply. The pull-down network is $A \cdot B$, since the transistors are two NMOS transistors in series. The pull-up network consists of two PMOS devices in parallel. This logic function is $\overline{A} + \overline{B}$. Another way to say this is that the function is 0 when $A \cdot B$ is true and is 1 when $\overline{A} + \overline{B}$ is true. Of course, by DeMorgan's law, $\overline{A} + \overline{B}$ is the same function as $\overline{A \cdot B}$.

We can apply the same kind of analysis to the NOR gate implemented as transistors. Using the observation we just made, the pull-up network is $\overline{A} \cdot \overline{B}$, the pull-down network is $A + B$, and the function is 0 when $A + B$ is true and is 1 when $\overline{A} \cdot \overline{B}$ $(\overline{A + B})$ is true. Analyzing the transistor network directly (see Figure B.17), the output node can be 1 only if both transistors between it and the power supply are conducting. Thus, the output is 1 if both inputs are 0. If either or both inputs are 1, then the path to the power supply is broken while at least one path from the output to ground is established. The network does indeed implement the NOR function.

B.4.3 CMOS Transmission Gate

Any gate logic function can be implemented as a pull-up network of PMOS transistors and a pull-down network of NMOS transistors. In fact, this is the standard way to construct digital logic from CMOS transistors. Yet, there is an even richer set of things you can do with switching networks. The key to constructing these networks is a special circuit structure called the transmission gate. We examine it next.

For electrical reasons that are beyond the scope of this discussion, it turns out that PMOS transistors are great at transmitting a logic 1 voltage without signal loss, but the same cannot be said about logic 0 voltages. Having 0 V at one side of a conducting PMOS transistor yields a voltage at the other side somewhat higher than 0 V. NMOS transistors have a complementary problem: they are great at passing logic 0 but awful at passing logic 1. In the circuits we have looked at so far, PMOS transistors in the pull-up network passed only ones while the NMOS transistors in the pull-down network passed only zeros. So everything works out fine.

As you may guess, the best possible transmission behavior can be obtained by combining both kinds of transistors. This yields the *CMOS transmission gate,* which is shown in Figure B.18. The PMOS and NMOS transistors are connected in parallel and are controlled by complementary control signals in the figure. When signal A is asserted, the transmission gate conducts a logic 0 or 1 equally well. Signal A at a logic 1 makes the NMOS transistor conduct, while $\bar{A}$ at a logic 0 makes the parallel PMOS transistor conduct as well. When A is unasserted, the gate no longer conducts. A at logic 0 breaks the connection through the NMOS transistor, while $\bar{A}$ at 1 has the same effect on the PMOS transistor. In circuit diagrams, the transmission gate is often denoted by a "butterfly" or "bow tie" symbol, as shown in the figure.

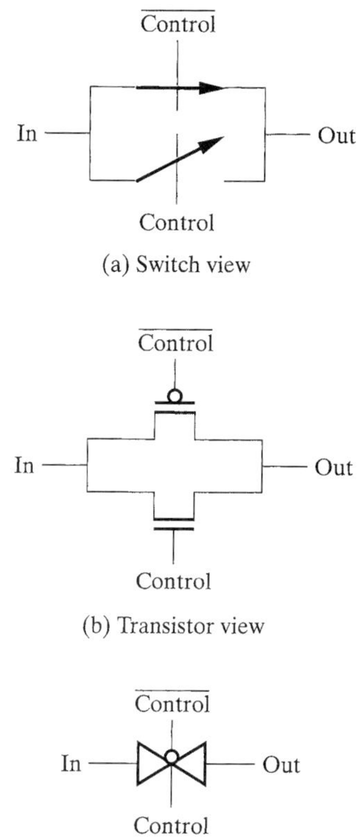

(a) Switch view

(b) Transistor view

(c) Schematic symbol

Figure B.18 CMOS transmission gate.

B.4.4 Switch and Steering Logic

In this subsection, we examine alternative ways to implement logic functions as networks of CMOS switching elements, based on CMOS transmission gates.

Switching networks provide an alternative to discrete gates for constructing digital systems. They operate by steering or directing inputs to outputs through a network of switching paths rather than by computing a Boolean function. The switching networks for standard gates like NAND and NOR route 0 or 1 to the output under the control of the data inputs.

For general-purpose switching networks, the primitive component is the CMOS transmission gate. It is constructed from a normally open switch (NMOS transistor) wired in parallel with a normally closed switch (PMOS transistor), with complementary control signals. Figure B.18 shows the switch, transistor, and schematic representations of the transmission gate. A transmission gate is equally good at passing a 0 or 1 when an external control signal is asserted.

Transmission gates are not equivalent in function to any of the logic gates you have encountered so far. They are not available to TTL logic designers, only to those designing directly in CMOS technologies.

So why bother with transmission gates? They are of interest because some important digital functions can be implemented in many fewer transistors if we use transmission gates. These are usually functions that can be recast in terms of steering networks, such as multiplexer/demultiplexer circuits, which we describe in Section 4.2.

CMOS Transmission Gate Transmission gates provide an efficient way to build steering logic. *Steering logic circuits* are circuits that route data

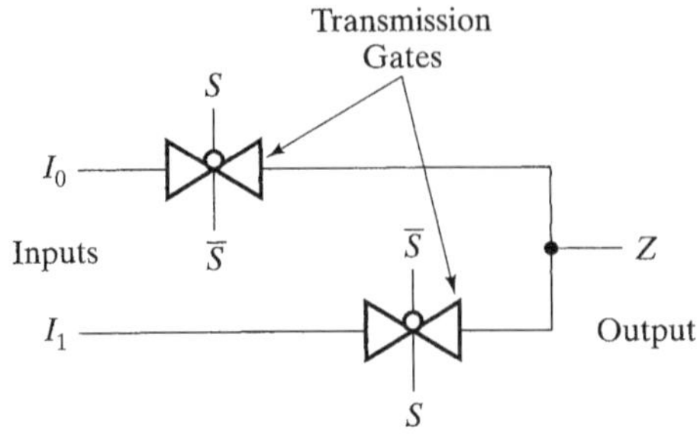

Figure B.19 **Selector function implemented with transmission gates.**

inputs to outputs based on the settings of control signals. As an example of such a circuit, consider the following. The circuit has two data inputs, I_0 and I_1, a single output Z, and a control input S. The function steers I_0 to Z when S is 0 and I_1 to Z when S is 1.

We call this a *selector function* or a *multiplexer.* It operates just like a switch in a railroad yard. Two independent pieces of track have to be merged into a single track. In a similar way, we often need to merge signal paths in digital logic.

Steering logic implements the selector as shown in Figure B.19. When S is asserted, the lower transmission gate conducts and the upper gate does not. Thus I_1 is gated to Z. When S becomes unasserted, the upper transmission gate conducts and the lower gate breaks the connection. Thus, I_0 is steered to Z. This multiplexer or selector function is described in greater detail in Chapter 4.

We make two observations about steering logic. First, for any combinations of the control inputs, there must be at least one conducting path from an input to the output. In other words, the output node should always be driven from some input. It should not be left floating. Second, there should never be more than one conducting path between the inputs and an output. If more than one path could exist, one of the paths might attempt to drive the output node to logic 1, while another drives it to logic 0. The resulting conflict would yield neither a logic 1, nor a logic 0 at the output node.

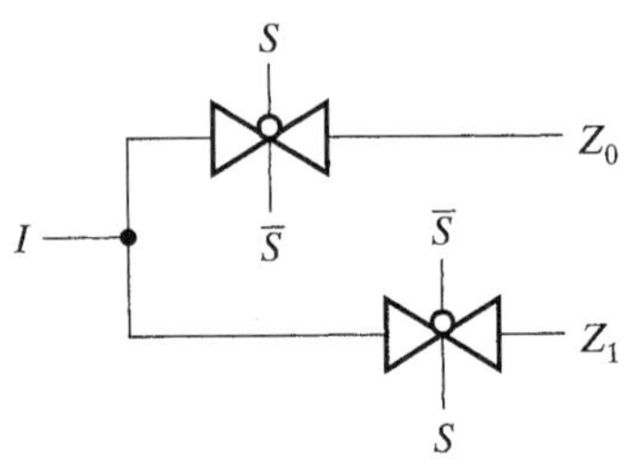

Figure B.20 **Possible implementation of demultiplexer functions.**

We give a second example of steering logic in Figure B.20. This network steers its single input to Z_1 if S is asserted and Z_0 when S is unasserted. This function is called a *demultiplexer* because it performs the reverse operation of a multiplexer. Multiplexers and demultiplexers are used to implement multiple connections between components, as shown in Figure B.21. The figure shows connection paths between A and Z and B and Y. When the control settings are changed, the paths connect A to Y and B to Z. Once again, the structure looks just like the switches in a railroad yard.

There is one problem in the circuit of Figure B.20. It violates one of our conditions for a properly functioning network. When S is asserted, I is steered to Z_1, but what value is placed on the output Z_0? Unfortunately, the output Z_0 is neither 0, nor 1, but floats. The same problem exists for Z_1 when S is unasserted. A solution with the correct

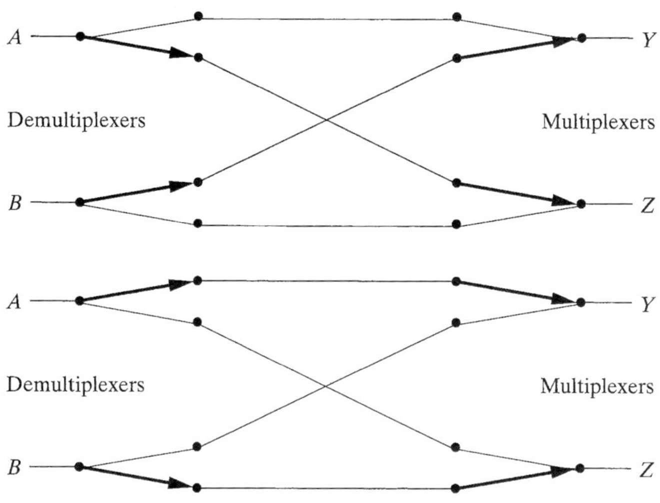

Figure B.21 **Multiplexer/demultiplexer functions.**

behavior is given in Figure B.22. When the input steers to Z_1, an additional transmission gate steers a 0 to Z_0. We do something similar for output Z_1 when the input steers to Z_0.

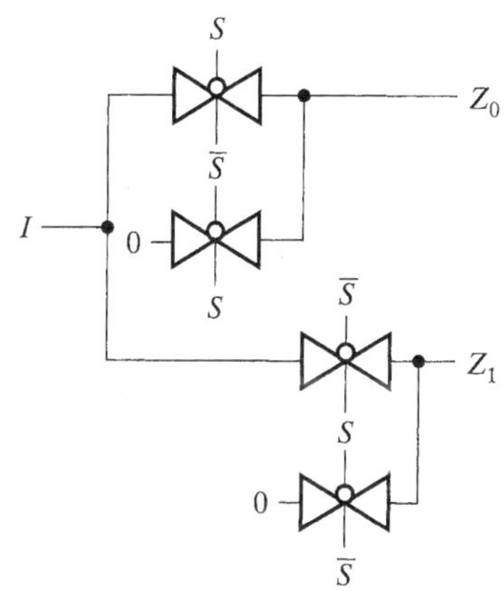

Figure B.22 **Correct implementation of demultiplexer function.**

A Complex Steering Logic Example We are now ready to look at a more complex design example of what steering logic can do. Let's examine a function called the Tally circuit. A Tally circuit has N inputs (I_1, I_2, . . . , I_n) and $N+1$ outputs (Zero, One, . . . , N). The circuit counts the number of its inputs that are at logic 1. Only one of the outputs is asserted at any time: If none of the inputs are 1, the output Zero is asserted; if one is 1, then the output One is asserted, and so on.

Figure B.23 gives the truth table and gate-level implementation for a single-input Tally function. It has the single input I_1 and the two outputs Zero and One. The truth table makes it clear that Zero is simply the complement of I_1 and One is identical to I_1.

How would we implement this function with steering logic? Remember that steering logic is good at routing inputs to outputs under the direction of control signals, so we must rethink the definition of the Tally circuit in these terms. While it may seem counterintuitive at first, in steering logic it is quite acceptable to use data signals to control transmission gates.

Figure B.24(a) shows the switch network for a single-input Tally function implemented from transmission gates. The inputs are 0 and 1: I_1 is used as a control signal, and the outputs are labeled Zero and One. Each output is connected to two transmission gates controlled by complementary signals. This guarantees that the network is well formed: It is not possible for both transmission gates to be conducting at the same time if their control signals are complementary.

We'll call the upper of these the "straight-through" transmission gate, and the lower the "diagonal" transmission gate. Depending on the setting of the control signals, the inputs can be steered straight through

I_1	Zero	One
0	1	0
1	0	1

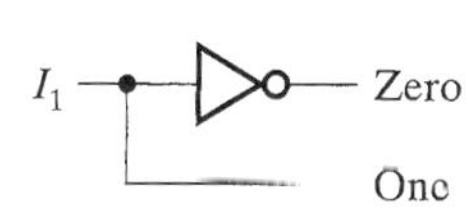

Figure B.23 **Truth table and gate implementation of single-input tally function.**

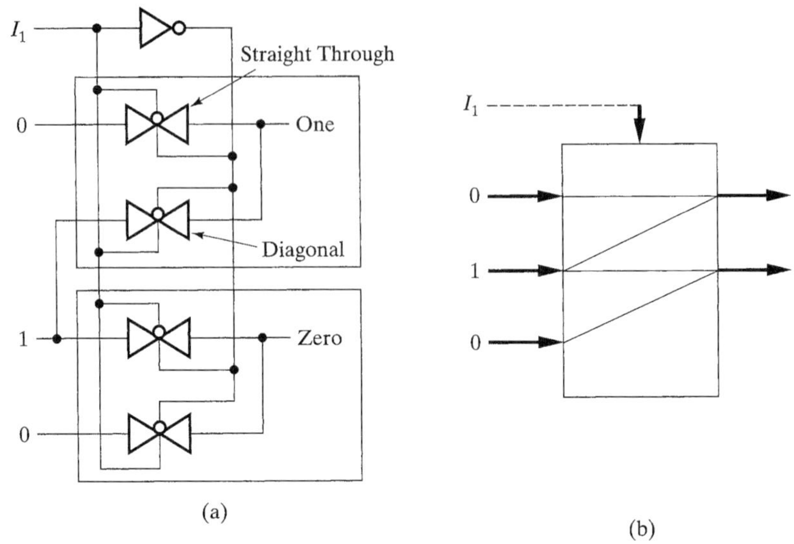

Figure B.24 **Single-input tally circuit implemented via transmission gates.**

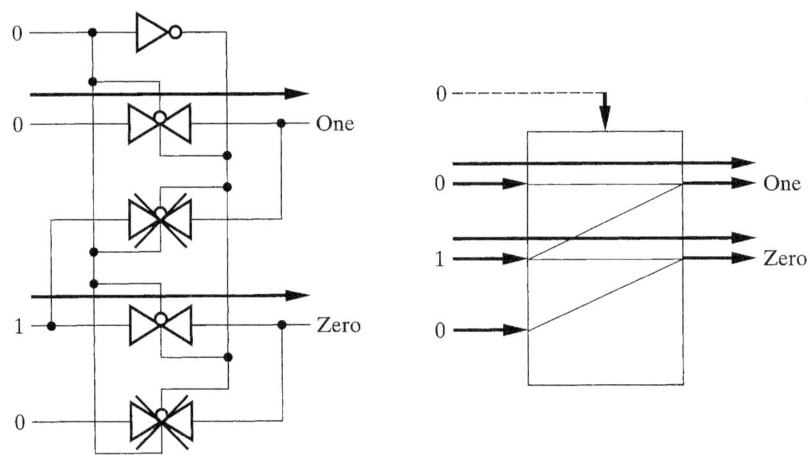

Figure B.25 **Tally input of 0.**

the network or can be shifted diagonally. The steering paths can be seen more clearly in the block diagram of Figure B.24(b).

When I_1 is unasserted, the straight-through gates are conducting and the diagonal gates are not. Thus, the input 0 steers to the output One while 1 steers to Zero. This is shown in Figure B.25.

When I_1 is asserted, the opposite is the case. The straight-through gates are nonconducting while the diagonal gates conduct. This causes a 1 to be steered to the One output and a 0 to the Zero output. This is shown in Figure B.26.

The transmission-gate implementation is significantly more complex than the simple gate implementation, so it offers no real advantage. However, it does become more attractive when we have a larger number

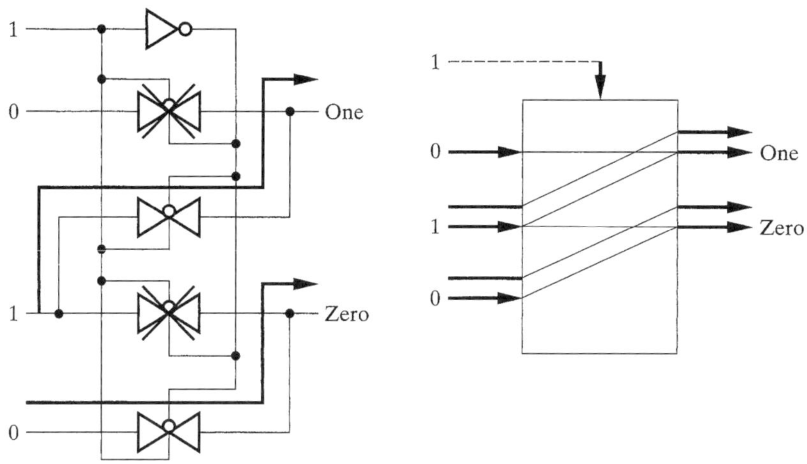

Figure B.26 Tally input of 1.

I_1	I_2	Zero	One	Two
0	0	1	0	0
0	1	0	1	0
1	0	0	1	0
1	1	0	0	1

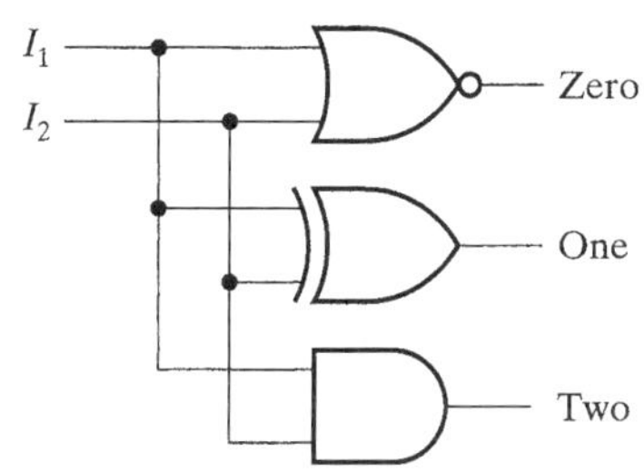

Figure B.27 Truth-table and gate implementation of the two-input tally function.

of inputs. Let's turn our attention to a two-input Tally circuit. The truth-table and logic-gate implementations are given in Figure B.27.

The transmission-gate implementation builds on the design of the single-input case for the two-input circuit. In essence, we decompose a two-input Tally circuit into two cascaded single-input networks. The inputs to the second stage are the Zero and One outputs from the first stage. The transmission-gate implementation is given in Figure B.28(a). Figure B.28(b) gives its block-diagram representation.

The four different cases of the two inputs and the resulting paths through the switching network are shown in Figure B.29. Figure B.29(a) shows what happens when both inputs are 0. The straight-through gates are enabled, setting the Zero output to 1 and the One and Two outputs to 0.

Figure B.29(b) shows the situation when I_1 is 0 and I_2 is 1. The first stage passes through, while the second stage shifts diagonally. The intermediate One and Zero signals are set to 0 and 1, respectively, and these are shifted up by the second stage. The second stage's One output is asserted while the Two and Zero outputs are 0.

Figure B.29(c) illustrates what happens when I_1 is 1 and I_2 is 0. The network behaves symmetrically. The One and Zero intermediate outputs read 1 and 0, respectively; these are passed directly through by the second stage.

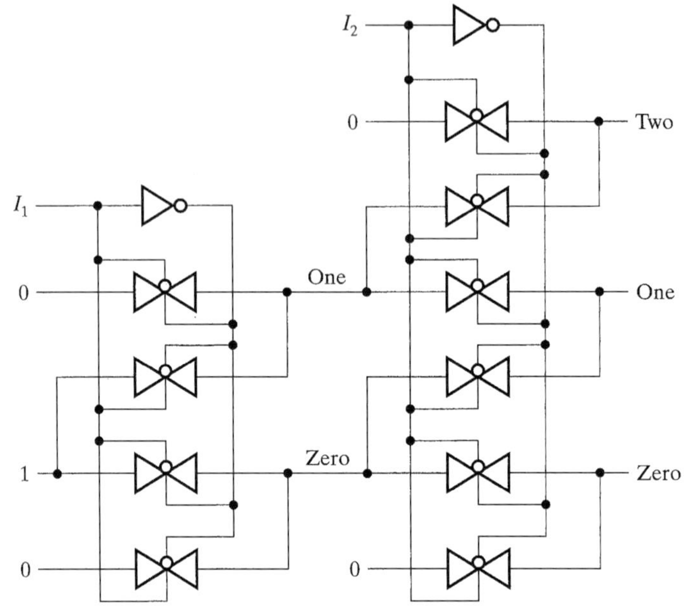

(a) Construction from transmission gates

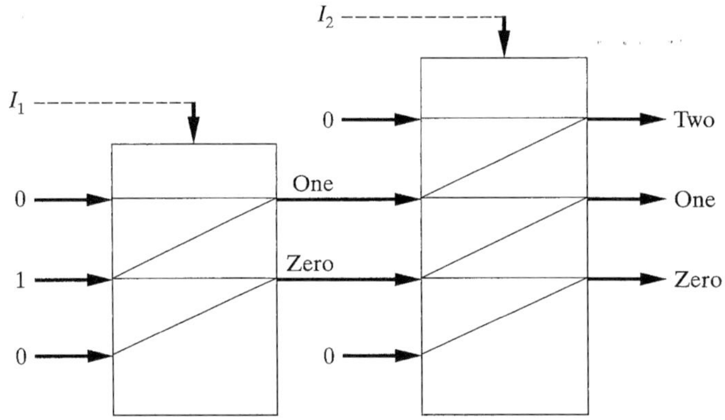

(b) Block-diagram representation

Figure B.28 Two-input tally function.

The final case is shown in Figure B.29(d), when both inputs are 1. All of the diagonal transmission gates are enabled. The One and Zero outputs at the first stage are driven to 1 and 0, respectively. These are shifted once again by the second stage, yielding 1, 0, 0 on the Two, One, and Zero outputs.

The transmission gate implementation still looks more complicated than the approach using discrete gates. But which method is more transistor efficient? The switching network makes use of 2 inverters and 10 transmission gates: a total of 24 transistors. The gate method actually

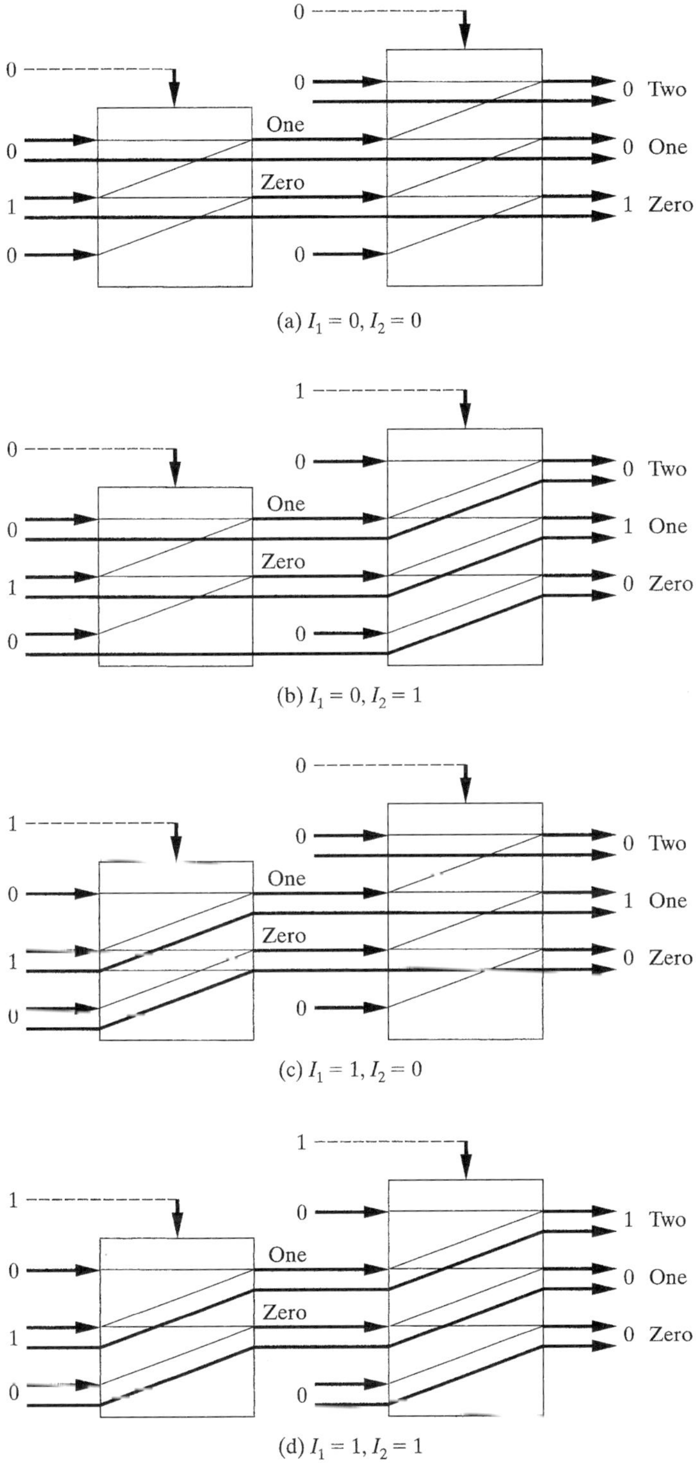

(a) $I_1 = 0, I_2 = 0$

(b) $I_1 = 0, I_2 = 1$

(c) $I_1 = 1, I_2 = 0$

(d) $I_1 = 1, I_2 = 1$

Figure B.29 Operation of the two-input tally circuit.

uses 26 transistors! Here is how we counted them. The two-input NOR gate is implemented with four transistors. We assume that the AND gate is implemented with an inverter and a two-input NAND gate, for six more transistors. The XOR gate actually consists of four interconnected two-input NAND gates. That's 16 more transistors. The total is 26. Thus, the switching network makes use of fewer transistors than the gate network to implement the same function. The advantage would be even more pronounced for a three- or four-input Tally circuit.

B.5 Elements of the Data Sheet

In this section, we will first learn how to read a component data sheet. Then we will see how to perform some simple electrical calculations on the data it presents. We will be using the data sheet for the 7400 quad two-input NAND-gate component as a running example.

A data sheet contains all of the relevant documentation that you need to use the component. The basic elements include (1) an English-language description of the function performed by the component, (2) a function/truth table, (3) a logic schematic with labeled inputs and outputs, (4) Boolean expressions defining the outputs in terms of logic functions of the inputs, (5) alternative package pin-outs, (6) internal transistor schematics, (7) operating specifications for the component, (8) recommended operating conditions, (9) electrical characteristics, and (10) switching characteristics. Not every data sheet will have all of these pieces, but most of these will be present. Let's look at each of these for a very basic component, the ALVC 7400 2-input NAND gate (see logic.ti.com for this and many other data sheets for a wide variety of devices in many logic families.)

English-Language Description "This quadraple 2-input positive-NAND gate is designed for 1.65 V to 3.6 V operation."

The description tells you succinctly what the component does, a NAND gate with non-inverted inputs. In the case of a simple logic gate, the description is quite brief. For more complex MSI components, the description can go on for several paragraphs.

Inputs		Output
A	*B*	*Y*
H	H	L
L	X	H
X	L	H

Figure B.30 Function table for 7400.

Function/Truth Table The function table describes the operation of the component by tabulating all input and output combinations. To avoid confusion over positive and negative logic, the truth table is defined in terms of signal levels, H and L, rather than logic levels, 0 and 1.

Figure B.30 shows the functional table for the NAND gate. When both inputs are high, the output is low. When an input is low, independent of the other input, the output is high. Writing the truth table for the NAND function in this way is a nice shorthand. For more complex functions, it is more common to take this functional approach than to write down a complete truth table.

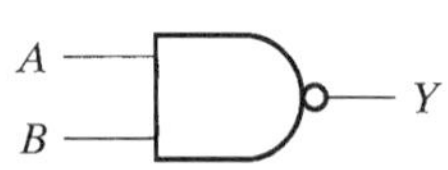

Figure B.31 Logic diagram for 7400.

Logic Schematics Our example is the simple gate shown in Figure B.31. Even in this simple case, the labeling is important because it ties together the functional table and the package pin-out.

Boolean Expression The Boolean expression for the gate is in positive logic. The expression is written in two alternative forms: $Y = \overline{A \bullet B}$ or $Y = \bar{A} + \bar{B}$.

Package Pin-Out Figure B.32 shows the standard package pin-out for the 7400. Note that there are several alternative package types, so it is always important to refer to the one that matches your particular component. The 7400 comes in a 14-pin dip package. The inputs and outputs are labeled by gate number (1 through 4), and the signal names are derived from the function table and the logic schematic.

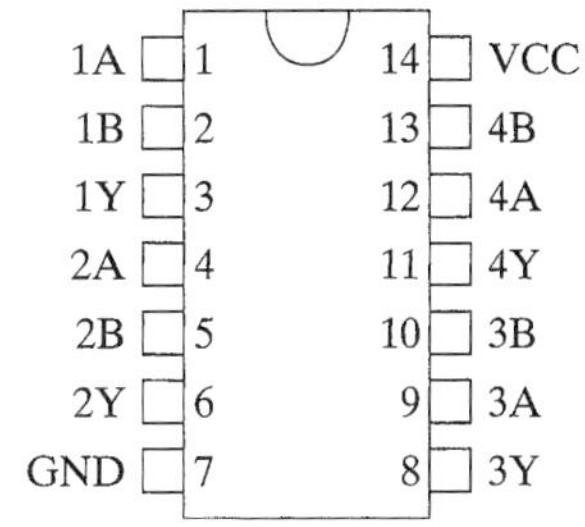

Figure B.32 **Package pin-out for the 7400.**

Absolute Maximum Ratings These ratings are the absolute worst-case conditions under which the component can operate or be stored. They should never be exceeded. The maximum supply voltage for an ALVC component is 4.6 V, the maximum input voltage is also 4.6 V, the operating free-air temperature is from –40 to 85°C, and the storage temperature can range from –65 to 150°C.

Recommended Operating Conditions These specifications describe the normal operating conditions for the supply voltage, input voltages, output currents, and temperature. Supply voltages are described in terms of minimum, nominal (normal), and maximum settings.

The input voltages, V_{IH} and V_{IL}, describe the minimum voltage that is recognized as a logic 1 and the maximum voltage that is recognized as a logic 0, respectively. For ALVC components operating at 3 V, anything above 2 V is a logic 1 and anything below 0.8 V is a logic 0. Voltages between these values will not be recognized as a 1 or a 0.

The output currents, I_{OH} and I_{OL}, describe the maximum currents the gate can supply to maintain the output at a voltage that will be recognized as a logic 1 or 0, respectively. For ALVC, these are –24 mA (current flows from the output pin) and 12 mA (current flows into the output pin).

Electrical Characteristics This section defines several voltages and currents that can be observed at the inputs and outputs of the components. The entries are V_{OH}, the minimum output high voltage, V_{OL}, the maximum output low voltage, I_{CC}, the package's steady-state power supply current when all outputs are low, and ΔI_{CC}, the package's current when the values of outputs are changing. The latter two values are per package and must be divided by the number of gates in the package to obtain the current per gate.

For the ALVC components, an output high voltage is 2.4 V minimum. The output low voltage is 0.4 V maximum. These values determine the noise margin. ALVC enjoys a 0.6 V noise margin on logic 1 (2.4 to 3.0 V) and a 0.4 noise margin on logic 0 (0.4 to 0.0 V).

Switching Characteristics This section gives the typical and maximum gate delays under specified test conditions of output resistance and capacitance. For example, the test conditions for the ALVC 7400 entries use a load resistance of 1000 Ω and a capacitance of 30 pF. With a 3 V supply, the propagation delay is 1 to 3 ns. These are typical values for most ALVC gates.

B.5.1 Simple Performance Calculations

In this subsection we describe some simple calculations you can use to determine the performance of your circuit.

Typical Propagation Delay Data sheets report minimum, maximum, and, sometimes, typical propagation delays. Using typical delays is always somewhat controversial. A truly conservative design would always use the maximum propagation delays.

It is important to note that the delays quoted in the data book are for specific test conditions. Under conditions of increased resistance or capacitance, the delay may vary from that described in the data sheet.

Power Consumption Compute steady-state power consumption by multiplying the gate current in holding outputs high or low by the power supply voltage. The nominal power-supply voltage is 3 V. The current is I_{CC}. For the ALVC 7400, the typical average package power consumption will be

$$I_{CC} \times V_{CC} = 10\ \mu A \times 3\ V = 30\ \mu W$$

This is the typical power consumption per package. It should be divided by 4 to obtain the power consumption per gate.

In CMOS circuits, higher currents are generated only when signal values change. For the 7400, ΔI_{CC} is 750 μA (75 times the steady-state current). Power consumption calculations must take this into account as well. To do this, they need to estimate the frequency of change. If values are changing 10% of the time then the total power consumption is:

$$30\ \mu W + 750\ \mu A * 3\ V * 10\% = 255\ \mu W$$

Of course, 10% is only a guess. To be precise we would have to look at the details of how the gate is used in a circuit.

B.6 Schematic Documentation Standards

An important part of reading data sheets and understanding others' designs are the schematic diagrams that describe a component's logic. There are standard techniques for documenting logic schematics. Although it is frequently ignored, proper documentation is one of the most important skills of a designer. You must be able to describe your design to other designers. This means that documentation is primarily about standard ways of doing things and conventions that make the information clear and easy to grasp. It includes the following:

- *Representing the design on paper.* This involves ways of drawing components and describing their compositions. A project team must develop and adhere to a standard convention for naming components and the signal wires that interconnect them.
- *Representing interfaces.* An interface describes everything you need to know about a subsystem to use it without understanding all

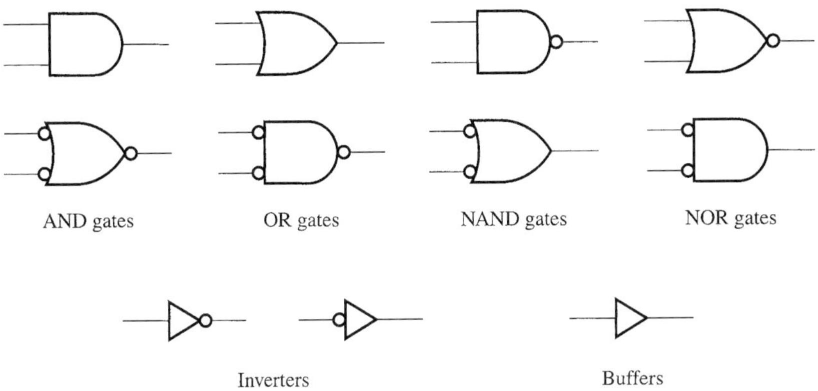

Figure B.33 **Standard SSI gate symbols.**

its internal workings. This is like the description of a subroutine in a computer program: the documentation describes what the subroutines do as well as the names, types, and functions of its parameters. Hardware interfaces are similar to software interfaces, but somewhat more complicated. A hardware interface describes the behavior of the design, as well as the names of signals and how to connect to them.

Standard Schematic Symbols Throughout this section, we have used standard gate symbols to represent common SSI components, such as AND, OR, NAND, NOR, XOR, and inverter gates. Gates always have two representations: normal logic and dual logic. The dual representation is used to conserve inversions when using active-low signals. Some examples are shown in Figure B.33. These are recognized by digital designers everywhere, so you should use them in your work as well. Do not use any other symbols for the same functions.

Conventions for MSI components are less rigid, but the following are typical. Functions are represented by blocks with input/output signals rather than discrete gates. Figure B.34 contains a schematic symbol for the 112 dual *J-K* flip-flop, a 16-pin TTL component. Inputs are drawn on the left, outputs on the right. The general flow of data is from left to right and top to bottom. All signals are labeled with meaningful names. Bubbles on pins or names that end in a slash (/) indicate signals that are active-low. The numbers identify package pin numbers. The connections to ground (pin #8) and the power supply (pin #16) usually are not shown in schematics. Every logic symbol must, without exception, have its part number written inside it.

At times, you will use different instances of the same set of gates in several places in your schematic. The best way to handle this is to draw the gates once, then box them in with a dashed line and label them with a detail letter, as in Figure B.35. When you use these gates in a particular place in your schematic, draw a single symbol for the function to be performed (like the single 4-bit wide inverter in the figure), and label it with the detail letter. The idea is that you expand the detail with its definition.

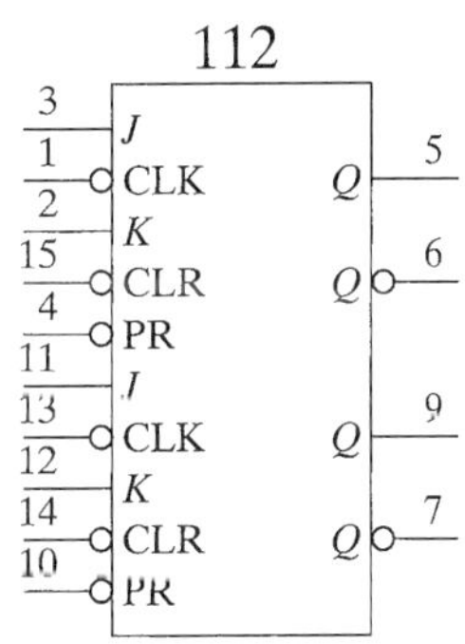

Figure B.34 **MSI component schematic.**

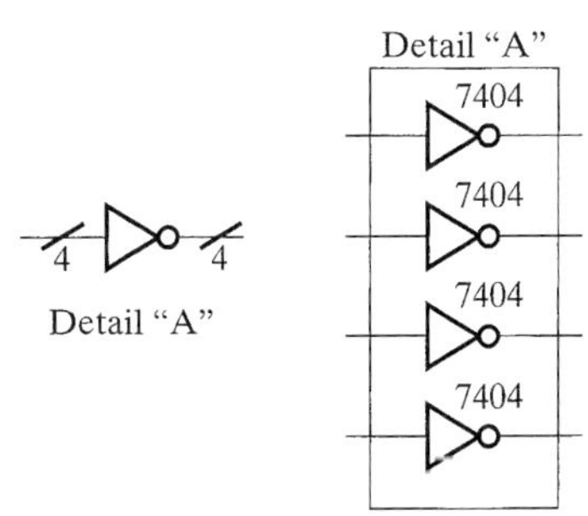

Figure B.35 **Use of a detail letter.**

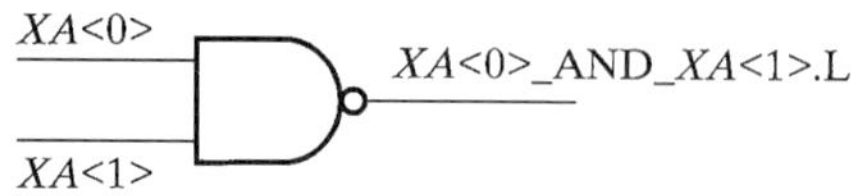

Figure B.36 Example of polarity inversion.

Names All signals that are not entirely local to an individual schematic drawing must be given a name. If a signal connects to many places in one drawing, it is more convenient to name the signal once and label local wires with this name where it is used, rather than draw wires to connect the uses together.

Names are an important form of documentation, so it is a good idea to name any wire whose usage is not trivial. You should use names that are understandable and describe the function performed by the signal. For example, if a signal causes the B Register to be cleared, then name the signal *CLEAR BREG*, not *52* or *CB*. More than one word in a name is fine. Many people capitalize and/or italicize signal names so that they are easy to distinguish from text in documentation. Each signal name must be unique within the project.

Polarization and Bubbles To see an application of practical bubble matching, consider Figure B.36. The designer is trying to AND together two related data bits, identified as XA_0 and XA_1. Unfortunately, the NAND-gate output has the opposite sense from what the designer wants: If both bits are 1s, then the NAND output is a 0. As signals pass through levels of logic, it is quite typical for the polarity to switch back and forth.

To make it easier to deal with inverting logic, we think about signals in two separate ways, both of which are reflected in the convention for signal names. The first element of a signal name indicates its function: `LOAD_PC`, or `XA<0>_AND_XA<1>`. The second gives the signal's polarity as either high (`.H`) or low (`.L`). An `.H` or `.L` polarity indicator is added to every signal name to indicate whether its function occurs when the signal is 1 or 0. For example, in Figure B.36, the signal `XA<0>_AND_XA<1>.L` is at a low voltage (true) when both `XA<0>` and `XA<1>` are at high voltages (also true). Thus, the signal is given an `.L` polarity. If an AND gate had been used instead of NAND, the polarity would be `.H`. If a signal is not marked with a polarity indicator, it defaults to positive logic. To be absolutely clear, it is a good idea to mark all signals explicitly.

A signal with positive (`.H`) polarity is asserted at a high-voltage level, and a signal with negative (`.L`) polarity is asserted at a low-voltage level. A bubble on a logic symbol indicates that an input or output is inverted. An input with a bubble means that the input signal is to be asserted low. A bubbled output is asserted when its voltage is low. A bubbled input should almost always match a bubbled output or another signal that is specified as being asserted active-low. Inputs and outputs without bubbles match `.H` signals or other "bubbleless" inputs and outputs.

Because it makes drawings so much easier to read, you should match bubbles wherever possible. Where they match, you can simply ignore polarity. For example, although Figure B.37 and Figure B.38 are

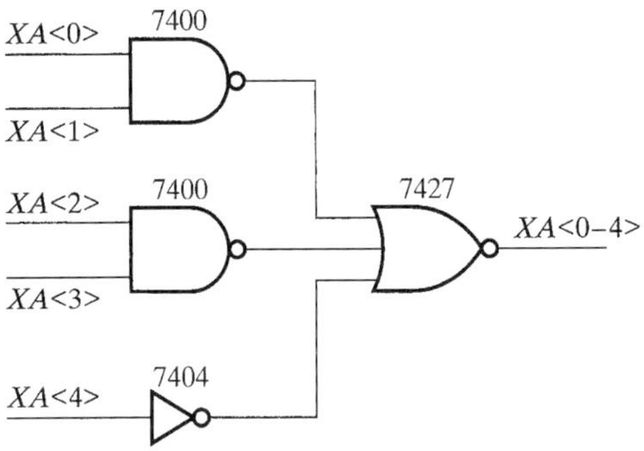

Figure B.37 **Incorrect bubble matching.**

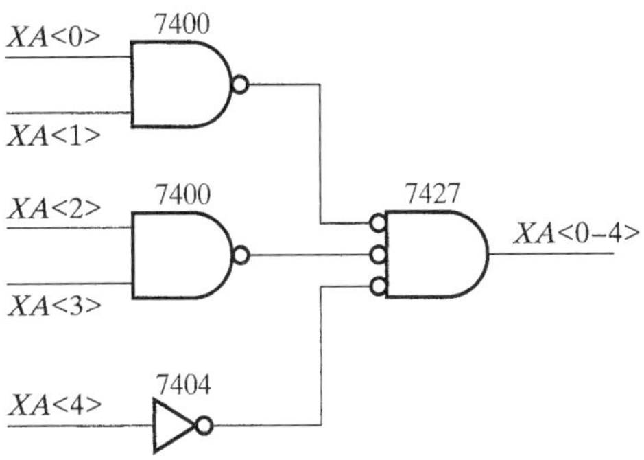

Figure B.38 **Correct bubble matching.**

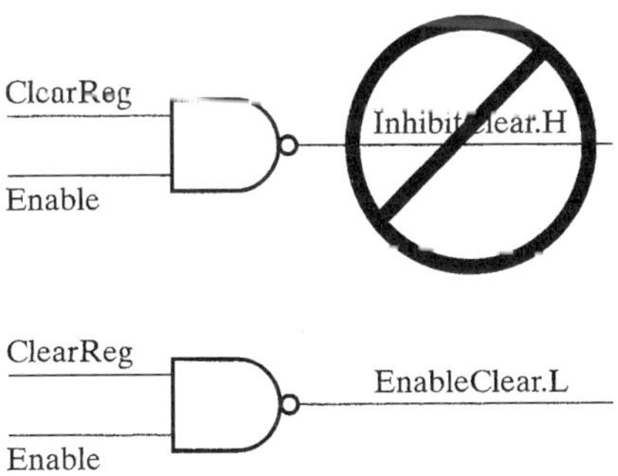

Figure B.39 **A case where polarity and bubbles don't match.**

equivalent, in Figure B.38 it is much easier to see that five active-high input signals are being ANDed together.

In a few cases, bubbles cannot be made to match. Usually these cases involve some form of *inhibition*. That is, when the signal is asserted, something is NOT happening. Figure B.39 shows an example where a clear pulse (*ClearReg*) is being controlled by another signal (*Enable*). The actual clear signal is active-low and should be asserted only when both *Enable* and *ClearReg* are asserted. When the output of the NAND gate is high, the *Clear* signal is inhibited. Note the use of the positive-logic form of the output-signal polarity.

Since the gate's output is active-low but the signal label is active-high, we have a mismatch. It would be clearer if the gate output were labeled with the action that takes effect when the signal is asserted active-low. Careful renaming of a signal can make the mismatch go away. In this case, you simply replace `InhibitClear.H` with `EnableClear.L`.

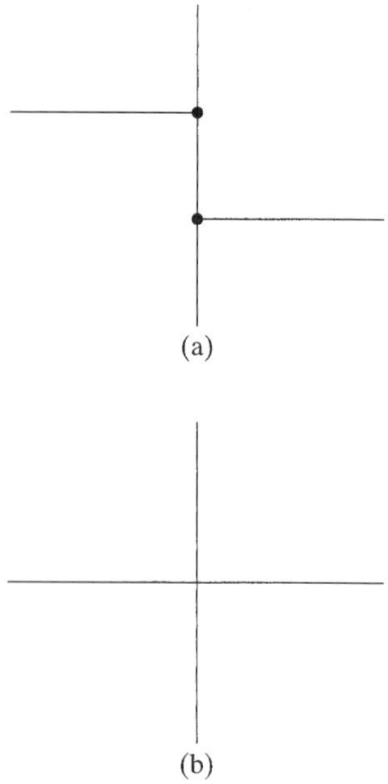

Figure B.40 **Wiring connection (a) and crossover (b).**

Crossovers and Connections All connections between wires must be marked with "blobs," as shown in Figure B.40. Be careful to make it crystal clear when signals cross without connection and when connections are made.

Hierarchical Documentation and Cross References Large designs cannot fit on a single sheet of paper. They must be suitably distributed over many pages, and signals will cross page boundaries. Even if everything did fit on one sheet, it probably wouldn't be practical to draw wires for every connection: the sheet would turn into a rat's nest of lines. Thus, you need to make proper use of hierarchy and abstraction in your presentation of the design.

The first pages should contain a coarse block diagram showing the main group of components. Only signals that leave or enter one of these blocks should be shown. Subsequent pages will expand these blocks hierarchically into more and more details. Thus, the same signal may appear in several disconnected places. It is important to keep track of where signals are used, making it easier to scan the drawings and to verify signal fan-outs.

You are expected to observe several conventions in order to keep track of signal usage. All inputs to a sheet should enter at the top or left edge of the sheet; all outputs from the sheet should terminate at the bottom or right edge. If a signal is both an input and an output, you may take your pick.

Each page of your project should be named in some conspicuous place and should also be numbered. Signals that leave a page should carry a unique label and an indication of other pages on which they can be found entering. Correspondingly, all entering signals should carry the proper name label and an indication of the page from which they come. Ideally, if each page corresponds to some block in a previous page, then that previous page is a block diagram that shows all the interconnections between the blocks. The inputs and outputs to any block should correspond to all of the entering and exiting signals on the detailed page and should carry the same signal names.

B.7 Practical Aspects of Inputs, Outputs, and Clocks

In this section, we look at three important practical issues related to the design of synchronous digital systems: input/output using switches and LEDs, methods for debouncing mechanical switches, and the use of timer generating chips. A combinational-logic circuit is of no use unless you can provide it with inputs and observe its outputs.

We will examine how to use switches to provide inputs to a circuit, while a light-emitting diode (LED) can be used to observe the circuit's output. Although mechanical switches provide a very simple input device, they must be used carefully because a single switch transition could appear as multiple transitions to internal logic. Programmable timer chips provide a simple way to generate clock signals within a system.

B.7.1 Switches and LEDs as Inputs and Outputs

Single-Pole/Single-Throw Switches A single-pole/single-throw (SPST) switch has two point connections to the outside. The switch can make or break the connection between these two points.

To implement a useful source of logic 1s and 0s, the switch can be configured as in Figure B.41. It works as follows. The symbol that looks like a top is a standard way to represent ground. When the switch is open, ground is disconnected from the output node. The resistor is called a "pull-up" because it brings the voltage on the output node up to something that will be recognized as a logic 1. The box with a number inside it represents a logic probe, and we use it to indicate the logic value on the circuit node it is connected to.

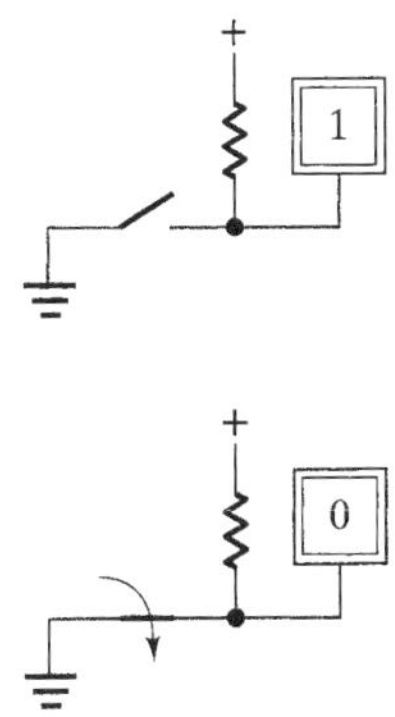

Figure B.41 Single-pole/single-throw switches.

When the switch is closed, the ground potential brings the output node down to a logic-0 voltage. The resistor should be chosen to limit the current between the power supply and ground when the switch is closed. Too small a resistance will cause too much power to be consumed when the switch is closed. If the resistance is set too large, the drop across the resistor may yield a voltage at the circuit input node that is too small to be recognized as a logic 1. A good compromise resistor value for these kinds of pull-ups is 10,000 Ω.

Single-Pole/Double-Throw Switches A single-pole/double-throw (SPDT) switch has three connections to the outside. Internal mechanics make it possible to connect selectively one of two of the connections to the third connection.

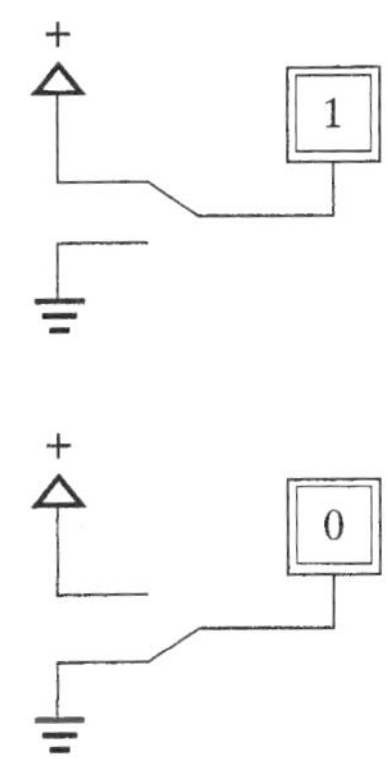

Figure B.42 Single-pole/double-throw switches.

Figure B.42 shows the SPDT switch configuration. One connection point is wired to the power supply and the other is connected to ground. The switch can be placed in one of two positions. When in the up position, the switch connects the power supply to the output node. The output node is driven to a logic-1 voltage. In the other position, the ground is connected to the output. Now the output node is driven to a logic-0 voltage.

Single-Pole/Double-Throw Push-Button A variation on the single-pole/double-throw switch is the momentary contact push-button. This is shown in Figure B.43. A distinguished input pole is connected through to the output when the push-button is at rest. When the push-button is pressed, the switch makes a connection between the second input pole and the output. When you release the button, it immediately returns to its rest position.

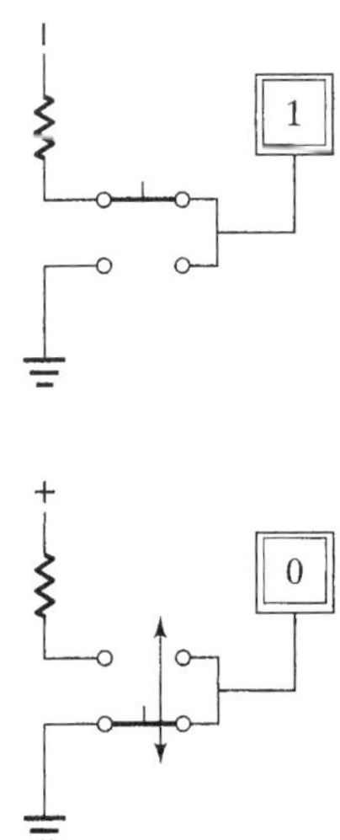

Figure B.43 Single-pole/double-throw push-button.

Light-Emitting Diodes Light-emitting diodes, or LEDs, are electronic elements that emit light whenever a current flows across them. Thus, they are ideal optical output devices. An LED has two connections, called the *anode* and the *cathode*. The device is illuminated whenever the anode voltage exceeds the cathode voltage by a certain threshold. The LED is a unidirectional element, so it is important to be able to distinguish

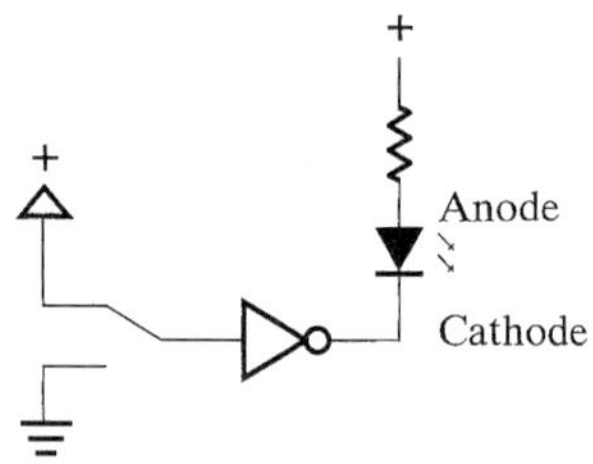

Figure B.44 Light-emitting diode driven by a TTL gate.

between the two connections. The cathode is usually the longer lead or the lead closest to the flat side of the LED's plastic housing.

Figure B.44 shows one of the many ways in which a gate can drive an LED. In a schematic, an LED is represented as a diode.

The LED is illuminated when the switch is set to its upper position. In this case, the inverter's input is a logic 1 and its output is at logic 0. This position ensures that the cathode potential is less than the anode potential.

The LED is dark when the switch is set to its lower position. This position gates a logic 0 to the inverter's input, yielding an output of logic 1. A logic-1 output places too high a voltage on the cathode to allow the LED to light up.

The resistor between the LED and the power supply is called a *current-limiting resistor.* An unprotected LED wired directly between ground and the power supply would burn out and could destroy the output circuitry of the gate.

We can compute the value for the current-limiting resistor as follows. We size the resistor so that the current across it comes close to but does not exceed the I_{OL} value for the gate that will drive the LED. For example, the I_{OL} of the ALVC 7400 gate is 24 mA. The resistance should be minimum of is 130 Ω because across a 3-V power supply, a 130 Ω resistor has a current of 3 V/130 Ω or 23 mA. In practice, we'll use a resistor as large as possible, to minimize current, while still causing a bright light from the LED.

B.7.2 Debouncing Switches

A problem with the use of mechanical switches in digital circuits is a phenomenon known as *bouncing*. When a switch is flipped from one terminal to another, it does not make a clean, solid contact with the new terminal. Instead, it bounces off the terminal into the air several times before finally coming to rest. Because of this and the fact that TTL chips treat floating inputs as logic 1s, a single flip of the switch that should cause a 1-to-0 transition in fact causes several transitions from 1 to 0. This can cause errors in the circuitry that is reading the switch's output.

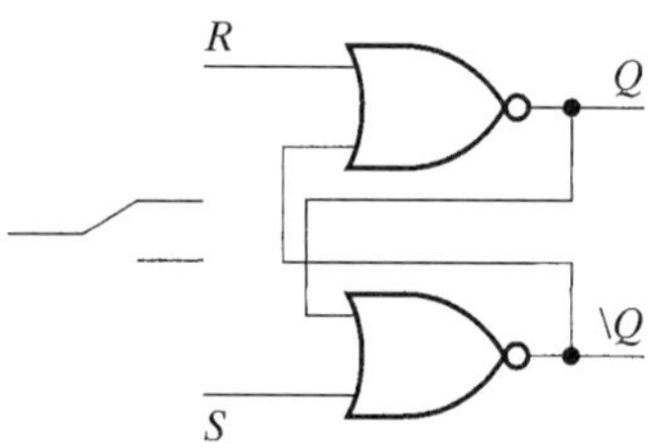

(a) Initial setup

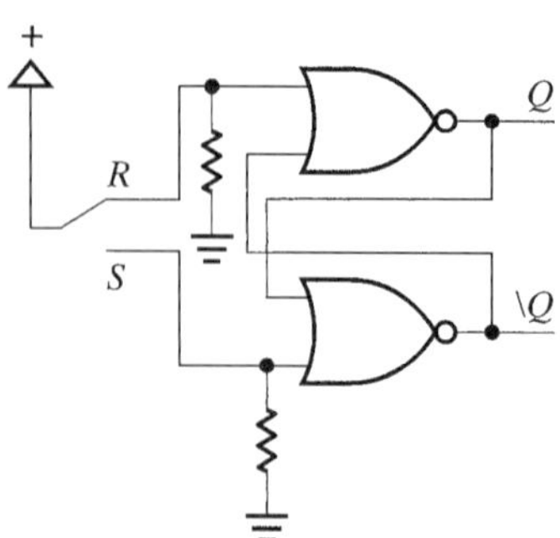

(b) Final setup

Figure B.45 Debouncing a switch using an *R-S* latch.

A Solution The simple *R-S* latch provides the solution for debouncing mechanical switches. We depend on the fact that the latch will hold its current state when both inputs are 0.

Figure B.45(a) shows the initial setup, a single pole/double throw (SPDT) switch and an *R-S* latch. We should get a 0 at the output *Q* when we place the switch in one position and a 1 at *Q* when we place it in the opposite position. Since the holding state occurs when *R* and *S* are both 0, we want these to be the default values while the switch is in transition between its settings. The switch must be able to force one of these inputs to a 1 depending on its current setting.

This setup leads to the wiring diagram of Figure B.45(b). The single contact side of the switch is connected to the power supply, while the double contact side is connected to the *R* and *S* inputs of the latch. When the switch is connected to the top output, *R* is asserted, and *Q* is reset to 0. When it is placed in the lower position, *S* is asserted, and *Q* is set to 1.

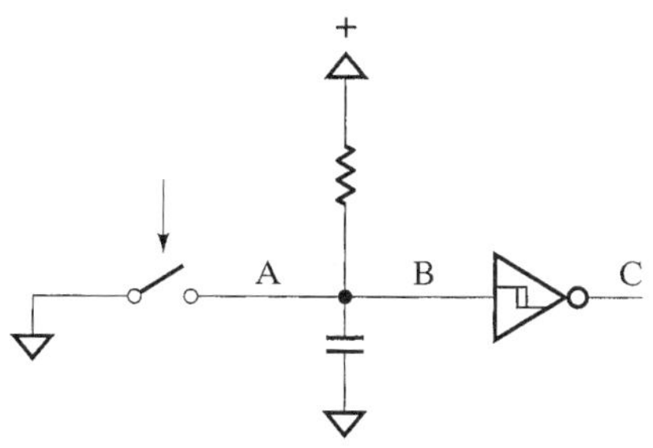

Figure B.46 **Debouncing a switch using a Schmitt trigger.**

This diagram does not yet solve the bounce problem. We must ensure that, when the switch is not connected to one side, the latch input remains steady at 0. To accomplish this, we wire both R and S to ground through resistors. The resistors are sized so that the input can be pulled high when the switch is trying to do so. They also protect the circuit against a short circuit between ground and the power supply.

When the switch is in the reset position, the R input to the latch is set high and Q is low. Now suppose that the switch is in transition from the R to the S position. The switch breaks connection with the R terminal, and the ground connection pulls the latch input low. The latch is now in its holding state because both inputs are 0.

When the switch first touches the S contact, the S input to the latch goes high, and the latch is set with Q equal to 1. If the switch bounces, temporarily breaking the connection, the latch input returns to 0, leaving the latch in the holding state. If the switch bounces back, remaking the S connection, the latch is simply set again and no state change occurs.

As long as the switch does not bounce far enough to remake the R connection, the Q output will remain high as long as the switch is bouncing into its final setting position. The same analysis applies for a switch transition from resting at the S terminal to connecting the R terminal.

Another Solution R-S latches work well for dobouncing single-pole/double-throw switches that have two terminal wires, one usually being "high" while the other is "low". But what about single-pole/single-throw switches? These commonly are found in keypads of various form factors. We used such a keypad in the example of Section 10.6 (see Figures 10.39 and 10.40).

A simpler configuration for a single switch is shown in Figure B.46. Notice that the switch selectively can connect its output to ground. The resistor is used to pull up the signal to "high" when the switch is open. The resistor is sized to limit current to a reasonable value when the switch is closed. The capacitor is added to "slow down" voltage changes in the switches output. Recall from earlier in this appendix that a signal changes at a rate consistent with its RC time constant. For a typical switch, we might set this time constant to 10 to 50 ms. Slowing down the signal make any bouncing less pronounced. Figure B.47(a) shows a typical signal coming directly from a switch (at point A in Figure B.46). Figure B.47(b) shows the effect of the RC circuit in creating a smoother version (at point B in Figure B.46).

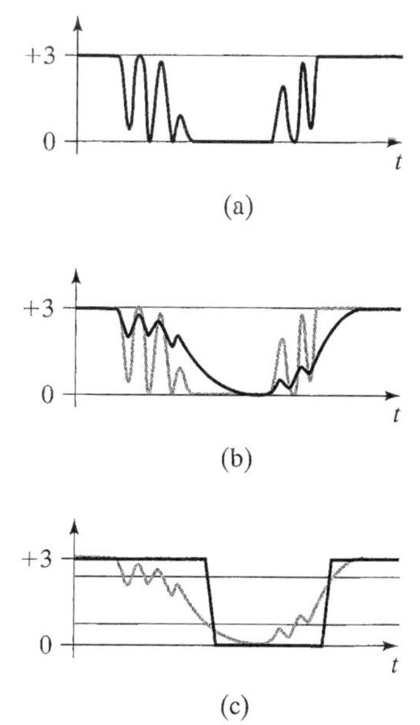

Figure B.47 **Detailed timing of Schmitt trigger debouncing circuit.**

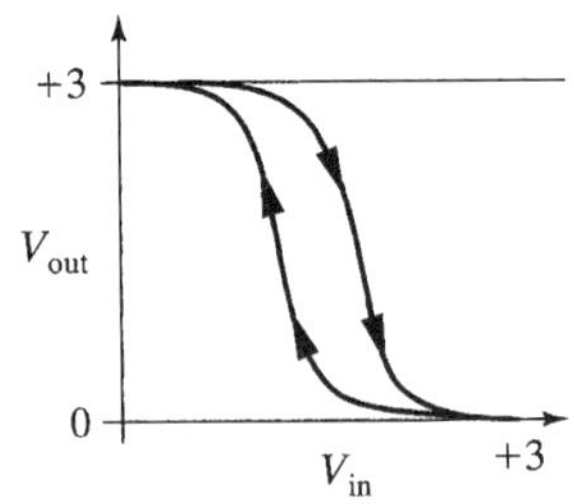

Figure B.48 Schmitt trigger transfer characteristic.

However, by doing this we've now ensured that whatever digital logic will use this signal will have to deal with values that are often between 0 and 1. To make sure that we get sharp transitions between 0 and 1, we use a special inverter called a *Schmitt trigger*. It takes advantage of a *hysteresis effect*. Hysteresis occurs when a value at the input of an inverter has to reach a certain threshold to cause a change in the output value but then needs to go to a much lower level on the input to change the output back. Figure B.48 shows the input output characteristic of a Schmitt trigger inverter such as the TTL '14. Notice that the curve splits in two. The direction arrows explain which curve to use when the input is rising or falling.

Figure B.47(c) shows the final output of our Schmitt trigger based debouncing circuit. Notice the two threshold values and how they determine when the output (at point C in Figure B.46) actually changes. The values now have sharper edges and can be used more safely as input to other gates.

APPENDIX REVIEW

In this appendix, we have examined the basic electrical building blocks of digital logic: resistors, capacitors, diodes, and transistors.

We began with the basic concepts of electricity: voltage, current, resistance, and capacitance. Voltage is electrical force; current is the intensity of the flow of electrical charge; resistance restricts current flow; and capacitance represents an ability to store electrical charge. Ohm's law ($V = IR$) and the charge-capacitance-voltage equation ($Q = CV$) describe the relationships among these electrical quantities.

Next we examined how to build useful logic functions from the primitive electrical components at our disposal: resistors, diodes, and transistors. We started with primitive diode–resistor logic. This has the serious drawbacks that it is not easy to cascade and an inverter cannot be built in the logic. The introduction of the transistor changed all this, and the resulting diode–transistor logic was a popular implementation technology in the 1950s and 1960s. It was replaced by the more efficient transistor–transistor logic that was the mainstay of logic designers through the 1980s.

We also covered an important class of transistor structures, the *field effect MOS (metal-oxide-silicon) transistors*. Logic gates constructed from such transistors are much simpler to analyze than bipolar transistors. We also discussed MOS switching structures that can be much more efficient in implementing multiplexer functions than logic gates. Steering logic devices, such as multiplexers and demultiplexers, are examples of logic of this form. Even complex functions, like the Tally circuit, can be implemented efficiently by casting them in terms of data signals routed through a network of switches. CMOS logic now dominates logic design because of its low power consumption and ease of use.

The final sections dealt with several important practical matters. These were reading component specifications from a data book;

performing simple calculations to determine delays and power consumption; documentation standards for schematics; wiring up switches and LEDs as circuit inputs and outputs; and debouncing switches to yield well-behaved digital signals.

EXERCISES

B.1 *(Basic Electronics)* How long does it take to charge a gate input from 0 to 2.7 V through a switch to 3 V with a resistance of 1000 Ω and a capacitance of 10 pF?

B.2 *(Basic Electronics)* It is never a good idea to allow inputs to a CMOS gate to "float," that is, to be left unconnected. Give four ways to wire up the unused input of a NAND gate to ensure proper operation of the gate.

B.3 *(Tally Circuit)* Examine the truth table for the two-input Tally function (Figure B.27) very carefully. Is there a way to implement the three outputs without using an XOR gate? (*Hint:* Think of a way to implement the One output in terms of the Zero and Two functions.) How does this revised gate-level implementation compare with the switch implementation?

B.4 *(Data Sheets)* Suppose you want to drive two LEDs with the output of an AVLC 7400 gate. The LED require 15 mA of current to emit the amount of light required. Can you connect both LEDs to a single output of the 7400? If not, what could you do instead?

B.5 *(Data Sheets)* TTL logic has a fanout limit not found in CMOS because inputs on TTL require current while CMOS gates do not. Therefore, an output pin can only supply current to so many input pins of other gates. What solutions are there when we have a signal with too high a fanout? List all the solutions you can devise.

B.6 *(Data Sheets)* Compare the power consumption of two technologies: a low-voltage TTL technology that has a steady-state power consumption of 0.5 mW; and a CMOS technology that has power consumption dependent on operating frequency such as 30 mW + 750 mW (% of time switching). Under what operating conditions is the TTL technology more efficient.

B.7 *(Data Sheets)* Can two technologies be interconnected if one has inputs that must be less than 0.4 V to be low and greater than 2.6 V to be high and outputs that are 0 V and 3.0 V, respectively, while the other technology has inputs that must be less than 0.8 V to be low and greater than 2.4 V to be high while its outputs are guaranteed to be less than 0.6 V for a low and greater than 2.4 V for a high? If not, what is the problem?

B.8 *(Switch Debouncing)* How would the debounce circuit of Figure B.45 change if you were to use an $\overline{R}$-$\overline{S}$ latch instead of an R-S latch?

B.9 *(Switch Debouncing)* Suppose you are to design a debouncing circuit using a single pole/single throw (SPST) switch. Can it be done? What problems do you face?

C

Flip-Flop Types

Introduction

We describe the evolution of basic sequential logic components, latches and flip-flops, in Chapter 6. We discuss master–slave arrangements and the role of edge-triggered flip-flops in supporting robust timing methodologies. The remainder of the text stresses the use of edge-triggered *D*-type flip-flops which are now the mainstay of sequential logic design. However, it is important to note that there are other possible types of flip-flops: *R-S*, *J-K*, *D*, and *T*. Chapter 6 discusses *R-S* latches, *J-K* master–slave flip-flops, and edge-triggered *D* flip-flops. In this appendix we will develop a design method for all four types.

C.1 Flip-Flop Components

We'll begin with a brief discussion of the basic components of Chapter 6.

R-S Flip-Flop The level-sensitive *R-S* latch has limited utility as a stand-alone element for holding state. However, it is the principal building block in implementing master–slave or edge-triggered flip-flops, and it serves some special functions for debouncing switches (see Appendix B). Two *R-S* latches can be combined in a master–slave arrangement to form an edge-triggered *R-S* flip-flop.

J-K Flip-Flops *J-K* flip-flops are a natural extension of *R-S* latches where we use the unused input combination ($R = 1$, $S = 1$) to toggle the stored value. This is accomplished by feeding back the outputs to the input AND gates. They are quite versatile in that they can hold their current value ($J = 0$, $K = 0$), toggle it ($J = 1$, $K = 1$), set it ($J = 1$, $K = 0$), or reset it ($J = 0$, $K = 1$). These flip-flops were very popular when packaged logic was dominant, as they tend to lead to the smallest implementations of next-state functions. However, each *J-K* flip-flop requires two inputs, and this could lead to more complex wiring. If the goal is to minimize wires, *D* flip-flops are attractive. In VLSI technologies, where the size of the wiring area is typically a greater concern than gate area, *D* flip-flops are used almost universally.

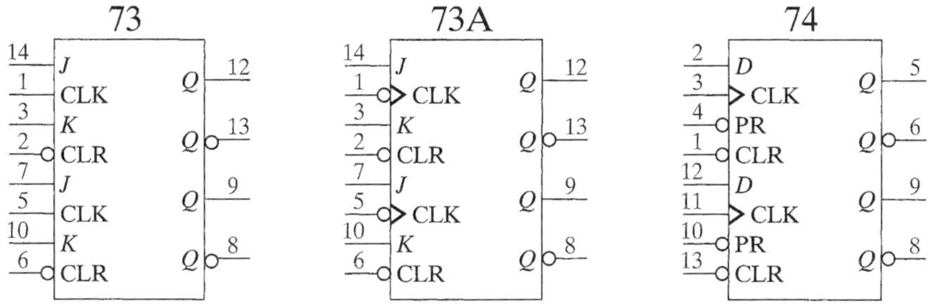

Figure C.1 **TTL flip-flop components.**

T Flip-Flops The *toggle flip-flop,* or *T flip-flop,* either holds its value (if the *T* input is 0) or toggles it (if the *T* input is 1). They rarely are available in packaged logic because they are formed so easily from *J-K* flip-flops by tying the *J* and *K* inputs together. *T* flip-flops turn out to be good building blocks for *counters*. However, you should note that they really require a set or reset control, as there is no other way to load a specific value into the flip-flop.

D Flip-Flops Today, the most frequently encountered flip-flop type is the *D* flip-flop. They are by far the easiest to understand; however, when they need to hold a value, they do require the output to be fed back to the input next-state function. If the goal is minimizing wires rather than logic, as is typical in FPGAs and VLSI technologies, then *D* flip-flops are more attractive than *J-K* flip-flops.

The TTL catalog contains a number of *R-S* latches and *J-K* and *D* flip-flops. Figure C.1 shows three of the most popular of these: the 73 *J-K* flip-flops and the 74 *D* flip-flops. Each package contains dual independent flip-flops. The 73 comes with independent active-low clear signals (CLR/). When they are asserted, the flip-flops' state is set to 0, independent of the current value of the inputs and the clock. The 74 has both preset (PR/) and clear (CLR/). In a similar way, when preset is asserted, the flip-flops' state is set to 1.

The 73 component comes in two versions: the standard version contains master–slave flip-flops, and the "A" version contains negative edge-triggered devices. The notation on the 73As clocks shows you how to distinguish between edge-triggered and level-sensitive operation. Edge triggering is denoted by a small triangle on the clock signal. Negative edge triggering is denoted by the negation bubble on the clock line. Be sure you are selecting the correct component for the job at hand. The detailed timing behaviors of the 73 and 73A are not identical.

The 74 is a positive edge-triggered *D* flip-flop. The edge triggering of the clock is indicated by the triangle on the clock signal. Without the bubble, the element is positive edge triggered.

We usually form *R-S* latches from discrete cross-coupled NOR gates using 02 components. However, you can find four $\overline{R}$-$\overline{S}$ latches in a single package if you use the 279 component.

D flip-flops are available almost universally in positive edge-triggered configurations. What if you need a negative edge-triggered device? One solution is to invert the clock signal on the way into a D flip-flop, like the 74. An alternative is to use a negative edge-triggered J-$\bar{K}$ flip-flop, such as the 276. By simply wiring the J and $\bar{K}$ inputs together, you can construct a D flip-flop.

C.2 Realizing Circuits with Different Kinds of Flip-Flops

In this section, we describe the methods for implementing one kind of flip-flop with another. The procedure is useful because a given flip-flop type may be the best choice to implement a given storage element, but flip-flops of that type may not be available to you.

We already know that a D flip-flop can be formed from a J-K flip-flop: simply tie the set input J to the data input D and the reset input K to $\bar{D}$. In similar fashion, a T flip-flop can be derived from a J-K flip-flop by connecting both the J and K inputs to T. But in this section, we will develop a general design procedure that will serve as the basis for designing synchronous sequential circuits in using any of the flip-flop types.

C.2.1 Conversion of One Flip-Flop Type to Another

We have already introduced characteristic equations as a shorthand for describing flip-flop behavior. They are useful particularly when implementing next-state logic by making it possible to relate the desired flip-flop outputs (*state*) to the inputs that must be generated to obtain the necessary behavior.

If Q is the current state and Q^+ is the next state, the equations for the four flip-flop types are

R-S latch: $Q^+ = S + \bar{R}Q$

J-K flip-flop: $Q^+ = J\bar{Q} + \bar{K}Q$

T flip-flop: $Q^+ = T\bar{Q} + \bar{T}Q$

D flip-flop: $Q^+ = D$

Any flip-flop can be implemented as combinational logic for the next-state function in conjunction with a flip-flop of another type. As an example, Figure C.2 shows how to implement a D flip-flop with a J-K flip-flop and, correspondingly, a J-K flip-flop with a D flip-flop.

Consider the leftmost circuit. If D is 1, we place the J-K flip-flop in its set input configuration ($J = 1$, $K = 0$). If D is 0, J-K's inputs are configured for reset ($J = 0$, $K = 1$). In the case of the rightmost circuit, the D flip-flop's input is driven with logic that implements the characteristic equation for the J-K flip-flop (namely $J\bar{Q} + \bar{K}Q$).

General Procedure We can follow a general procedure to map among the different kinds of flip-flops. It is based on the concept of an *excitation*

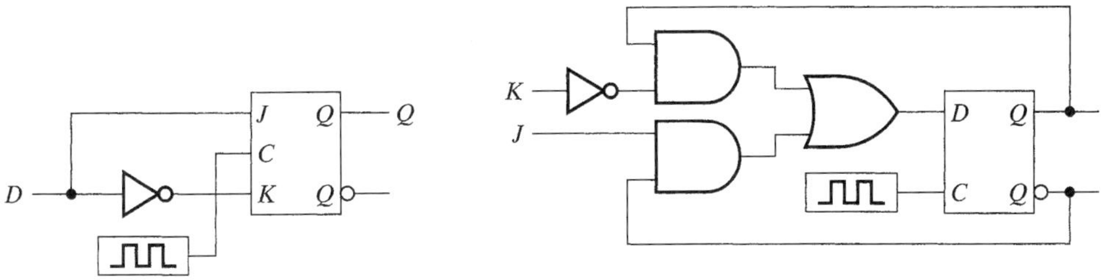

Figure C.2 ***D* flip-flop implemented by *J*-*K* flip-flop, and *J*-*K* flip-flop implemented by *D* flip-flop.**

Q	Q^+	R	S	J	K	T	D
0	0	X	0	0	X	0	0
0	1	0	1	1	X	1	1
1	0	1	0	X	1	1	0
1	1	0	X	X	0	0	1

Figure C.3 **Excitation tables for *R*-*S*, *J*-*K*, *T*, and *D* flip-flops.**

table, that is, a table that lists all possible state transitions and the values of the flip-flop inputs that cause a given transition to take place.

Figure C.3 gives excitation tables for *R-S*, *J-K*, *T*, and *D* flip-flops. If the current state is 0 and the next state is to be 0 too, then the first row of the table describes the flip-flop input to cause that state transition to take place. If an *R-S* latch is being used, it doesn't matter what value is placed on *R* as long as *S* is left unasserted. $R = 0$ and $S = 0$ holds the current state at 0; $R = 1$ and $S = 0$ resets the state to 0. The effect is the same.

If we are using a *J-K* flip-flop, the transition from 0 to 0 is accomplished by ensuring that *J* is left unasserted. The value of *K* does not matter. If $J = 0$ and $K = 0$, the current state is held at 0; if $J = 0$ and $K = 1$, the state is reset to 0.

If we are using a *T* flip-flop, the transition does not change the current state, so the input should be 0. If a *D* flip-flop is used, we set the input to the desired next state, which is 0 in this case. The same kind of analysis can be applied to complete the excitation table for the three other cases.

A flip-flop's next state function can be written as a K-map. For example, the next-state K-map for the *D* flip-flop is shown in Figure C.4(a). To realize a *D* flip-flop in terms of a *J-K* flip-flop, we simply remap the state transitions implied by the *D* flip-flop's K-map into equations for the *J* and *K* inputs. In other words, we express *J* and *K* as functions of the current state and *D*.

The procedure works as follows. First we draw K-maps for *J* and *K*, as in Figure C.4(b). Then we fill them in the following manner. When $D = 0$ and $Q = 0$, the next state is 0. The excitation table tells us that the inputs to *J* and *K* should be 0 and X, respectively, if we desire a 0-to-0 transition. These values are placed into corresponding entries of the

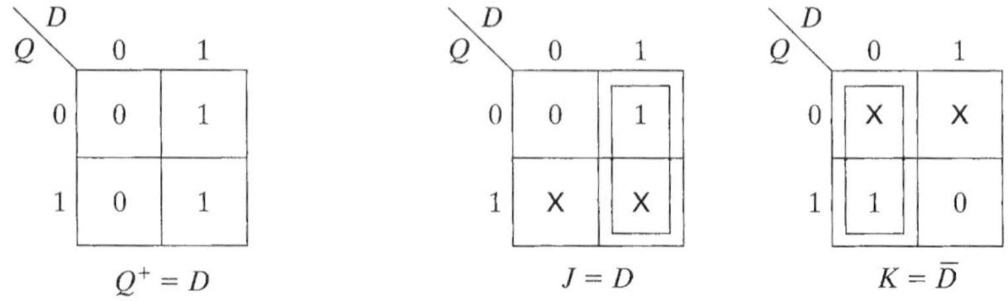

(a) K-map for D flip-flop and characteristic equation

(b) K-maps for J and K inputs

Figure C.4 **Implementing a given flip-flop by one of another type.**

J and K K-maps. The inputs $D = 0$, $Q = 1$ lead to a next state of 0. This is a 1-to-0 transition, and J and K should be X and 1, respectively. For $D = 1$ and $Q = 0$, the transition is from 0 to 1, and J must be 1 and K should be X. The final transition, $D = 1$ and $Q = 1$, is from 1 to 1, and J and K are X and 0. A quick look at the K-maps confirms that $J = D$ and $K = \bar{D}$.

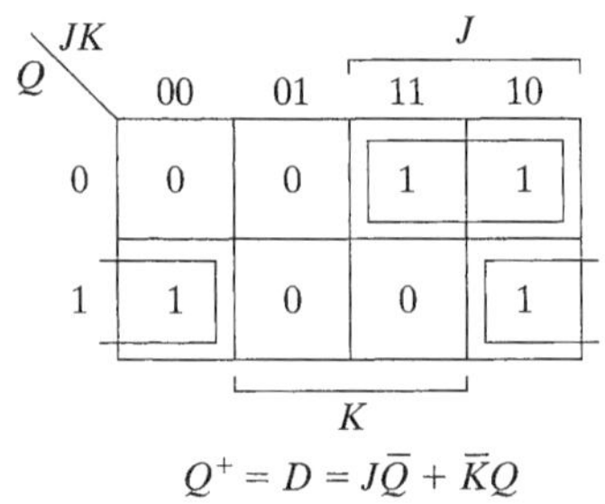

Figure C.5 **K-map for J-K flip-flop.**

The implementation of a J-K flip-flop by a D flip-flop follows the same procedure. We start with a K-map to describe the next state in terms of the three variables: J, K, and the current state Q. To obtain the transition from 0 to 0 or 1 to 0 requires that D be 0; similarly, D must be 1 to implement a 0-to-1 or 1-to-1 transition. In other words, the function for D is identical to the next state. The equation for D can be read directly from the next-state K-map for the J-K flip-flop:

$$D = J\bar{Q} + \bar{K}Q$$

This K-map is shown in Figure C.5.

C.3 Shift Registers and Counters

We'll start the discussion of how to implement sequential logic with different types of flip-flops using two simple examples: a 4-bit shift register and a 3-bit binary counter.

Figure C.6 shows the logic of a simple right-shifting *circular* shift register constructed from master–slave flip-flops. Data moves from left to right. On every shift pulse, the contents of a given flip-flop are replaced by the contents of the flip-flop to its left. The leftmost device receives its inputs from the rightmost. Because flip-flop propagation times far exceed hold times, the values are passed correctly from one stage to the next (cascaded flip-flops are discussed in Chapter 6).

Figure C.7(a) reproduces the state diagram for the 3-bit binary counter of Section 7.1. Figure C.7(b) shows a modified state transition table that includes the required inputs for a T flip-flop implementation of the counter. For this simple example, we can determine the logic just by examining the transition table. Flip-flop A toggles on each state transition, B toggles whenever A is asserted, and C toggles whenever A and B are asserted.

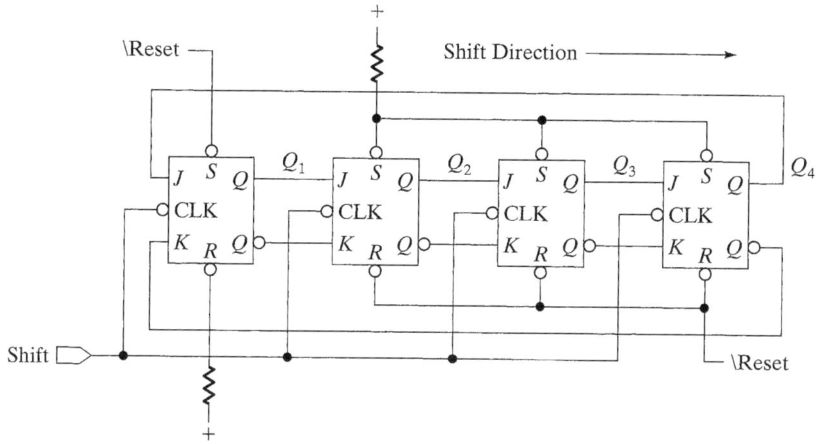

Figure C.6 **Quad right-shifting circular shift register.**

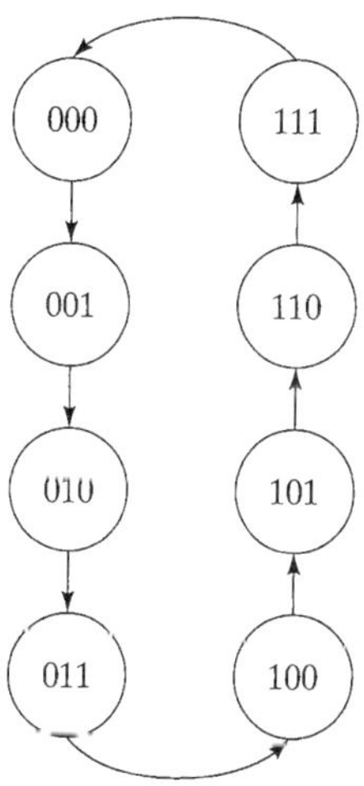

(a) State transition diagram

Present State			Next State			Flip-Flop Inputs		
C	B	A	C^+	B^+	A^+	TC	TB	TA
0	0	0	0	0	1	0	0	1
0	0	1	0	1	0	0	1	1
0	1	0	0	1	1	0	0	1
0	1	1	1	0	0	1	1	1
1	0	0	1	0	1	0	0	1
1	0	1	1	1	0	0	1	1
1	1	0	1	1	1	0	0	1
1	1	1	0	0	0	1	1	1

(b) State transition table

Figure C.7 **State transition diagram and table for a 3-bit binary up-counter.**

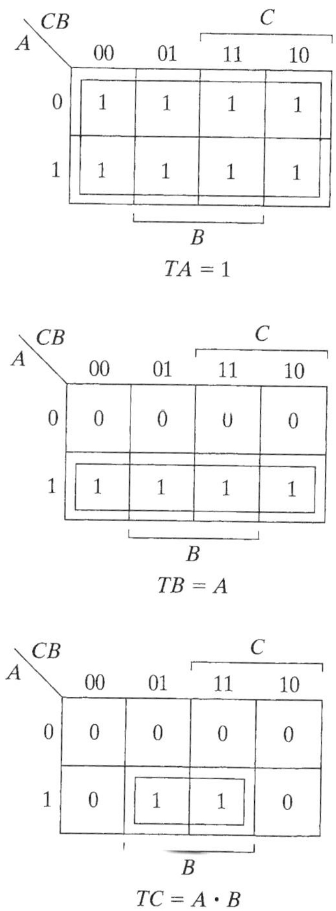

Figure C.8 **K-maps for up-counter toggle flip-flops.**

For more complex examples, we can view the transition table as a truth table that specifies the flip-flops' inputs as a function of *C*, *B*, and *A*. We would use standard K-map methods to obtain the reduced Boolean expressions. The K-maps for *TC*, *TB*, and *TA* are shown in Figure C.8. This leads immediately to the circuit design of Figure C.9.

Toggle flip-flops are a natural choice for implementing binary counters, but other flip-flop types may need less hardware for implementation of other types of state machines. Also, the availability of computer-aided design software may influence our choice of flip-flop. Existing CAD software, favors designs with *D* flip-flops, in part because *D* storage elements are pervasive in programmable logic and VLSI design. Although *J-K* devices often require the fewest gates to implement a

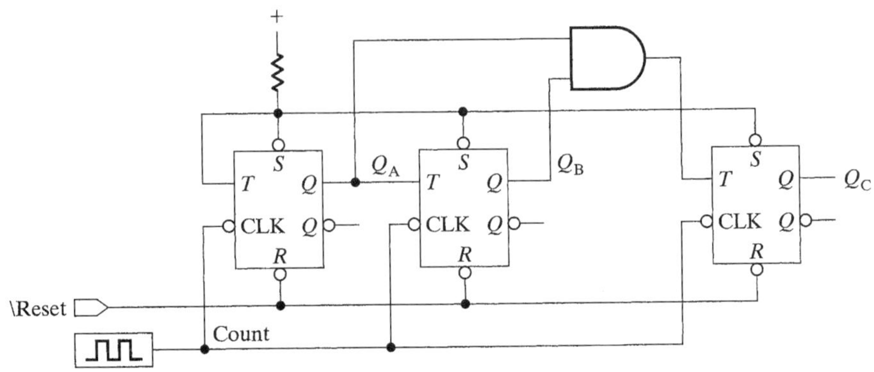

Figure C.9 Circuit diagram of three-bit binary up-counter using *T* flip-flops.

given state diagram, with the ever-increasing level of integration, saving a gate here or there becomes less of an issue.

Section 7.1.2 describes how to implement the five-state up-counter of Figure 7.8 (state transition diagram) and Figure 7.9 (state transition table) using *D* flip-flops. In this section, you will see three more implementations of this counter, using *R-S*, *J-K*, and *T* storage elements.

C.3.1 Implementation with *R-S* Flip-Flops

To implement a finite state machine with a different type of flip-flop, we need to remap the next-state functions into control inputs for the chosen flip-flop.

Q	Q^+	R	S
0	0	X	0
0	1	0	1
1	0	1	0
1	1	0	X

$$Q^+ = S + R\overline{Q}$$

Figure C.10 *R-S* excitation table.

R-S Excitation Table We start with the excitation table for the *R-S* flip-flop, which is given in Figure C.10. The state transitions are now encoded in terms of the *R* and *S* inputs that cause the flip-flops to make those transitions. For example, if the flip-flop's current state is 0 and the next state is also to be 0, we can accomplish this by setting the *S* input to 0 and the *R* input to either 0 (hold current state) or 1 (reset current state). We complete the remaining three entries of the tables in a similar manner.

Remapped State Transition Table The next step is to re-express the state transition table in terms of the *R* and *S* inputs to the three state flip-flops: *C*, *B*, and *A*. This is shown in Figure C.11.

We complete the remapped state transition table as follows. First we examine how the state bit is to change—for example, from a 0 to a 1. We use this information and the *R-S* excitation table to determine the *R* and *S* inputs needed to make that transition happen. We repeat the procedure for each state bit (column) and each state transition (row) of the state transition table.

Let's see how we obtained the table of Figure C.11. In the transition from state 000 to 010, *C* changes from 0 to 0, *B* from 0 to 1, and *A* from 0 to 0. The *A* and *C* transitions are implemented by the *R* and *S* inputs X and 0. The 0-to-1 transition for *B* is accomplished by setting *R* and *S* to 0 and 1, respectively. For the state transition 010 to 011, *C* goes from 0 to 0, *B* from 1 to 1, and *A* from 0 to 1. Based on the excitation table, *C*'s *R*

Present State			Next State			Remapped Next State					
C	B	A	C^+	B^+	A^+	RC	SC	RB	SB	RA	SA
0	0	0	0	1	0	X	0	0	1	X	0
0	0	1	X	X	X	X	X	X	X	X	X
0	1	0	0	1	1	X	0	0	X	0	1
0	1	1	1	0	1	0	1	1	0	0	X
1	0	0	X	X	X	X	X	X	X	X	X
1	0	1	1	1	0	0	X	0	1	1	0
1	1	0	0	0	0	1	0	1	0	X	0
1	1	1	X	X	X	X	X	X	X	X	X

Figure C.11 **State transition table and remapped next-state functions.**

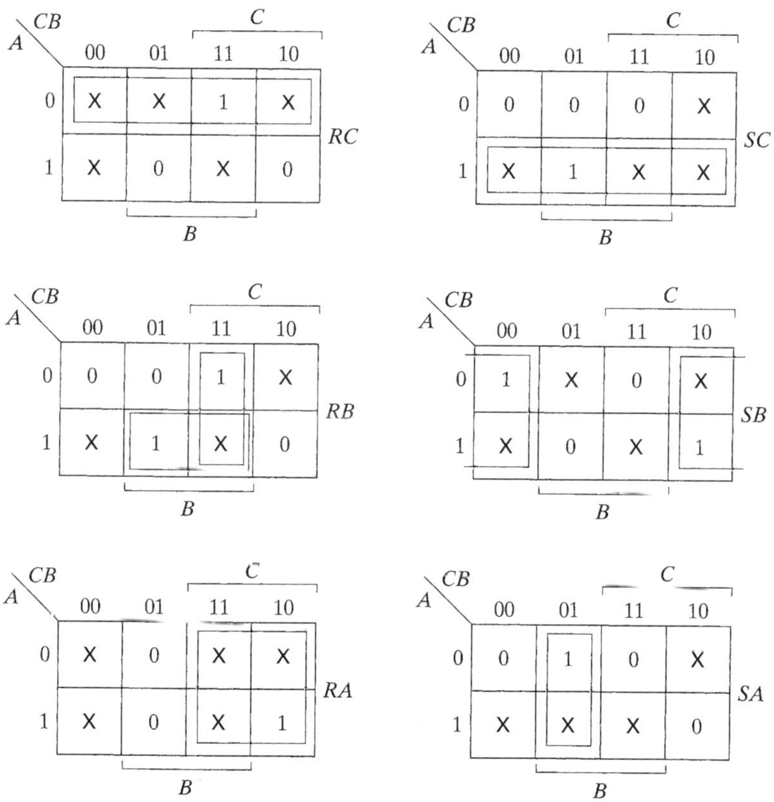

Figure C.12 **Remapped K-maps for *R-S* implementation.**

and S inputs should be X and 0, B's inputs are 0 and X, and A's are 0 and 1. We determine the rest of the transitions in a similar manner.

Remapped K-maps and Counter Schematic Figure C.12 shows the remapped K-maps. The minimized next-state functions are

$$R_C = \bar{A} \qquad S_C = A$$
$$R_B = AB + BC \qquad S_B = \bar{B}$$
$$R_A = C \qquad S_A = B\bar{C}$$

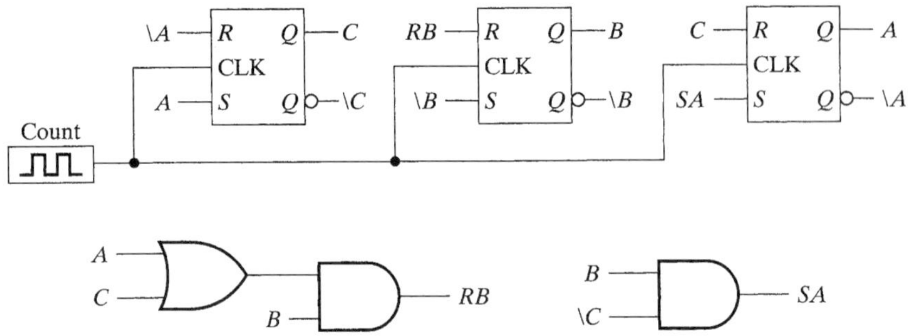

Figure C.13 ***R-S* flip-flop implementation of 3-bit counter.**

This implementation requires four gates and 10 literals. If we factor the expression for R_B, that is, $R_B = B(A + C)$, we can save another gate and a literal. The implementation logic is shown in Figure C.13: three gates, 9 literals, and a total of 12 wires if we consider the flip-flop inputs. We do not count the clock in this tabulation. The figure doesn't show the reset logic.

C.3.2 Implementation with *J-K* Flip-Flops

Q	Q^+	J	K
0	0	0	X
0	1	1	X
1	0	X	1
1	1	X	0

$Q^+ = J\overline{Q} + \overline{K}Q$

Figure C.14 ***J-K* flip-flop excitation table.**

***J-K* Excitation Table** To use *J-K* flip-flops we use the *J-K* excitation table. This is shown in Figure C.14. For a transition from a current state of 0 to a next state of 0, the J input should be set to 0 and the K to 0 (hold) or 1 (reset). In a 0-to-1 transition, J should be set to 1 with K at 0 (set) or 1 (toggle). We handle the 1-to-0 transition and 1-to-1 transition similarly.

State Transition Table and Remapped Next-State Functions The state transition table and the remapped next-state functions are given in Figure C.15. Let's examine the first two state transitions: from 000 to 010 and from 010 to 011. For the first transition, C stays at 0, B toggles from 0 to 1, and A holds at 0. For this transition, C's J and K inputs are 0 and X (0-to-0 transition); B's inputs are 1 and X (0-to-1 transition); and A's inputs are the same as C's. For the second state transition, C holds at 0, B holds at 1, and A toggles from 0 to 1. The excitation table tells us that the J and K inputs should be 0 and X for C; X and 0 for B; and 1 and X for A.

Remapped K-maps and Counter Schematic Because it can use the forbidden state of the *R-S* flip-flop, the *J-K* flip-flop's excitation table contains many more don't-care conditions than in the *R-S* case. As you would expect, this leads to an implementation with a reduced gate and literal count.

Figure C.16 shows the K-maps for the remapped next-state functions. The reduced expressions become

$$J_C = A \qquad K_C = \overline{A}$$
$$J_B = 1 \qquad K_B = A + C$$
$$J_A = B\overline{C} \qquad K_A = C$$

Present State			Next State			Remapped Next State					
C	B	A	C^+	B^+	A^+	JC	KC	JB	KB	JA	KA
0	0	0	0	1	0	0	X	1	X	0	X
0	0	1	X	X	X	X	X	X	X	X	X
0	1	0	0	1	1	0	X	X	0	1	X
0	1	1	1	0	1	1	X	X	1	X	0
1	0	0	X	X	X	X	X	X	X	X	X
1	0	1	1	1	0	X	0	1	X	X	1
1	1	0	0	0	0	X	1	X	1	0	X
1	1	1	X	X	X	X	X	X	X	X	X

Figure C.15 **State transition table and remapped next-state functions.**

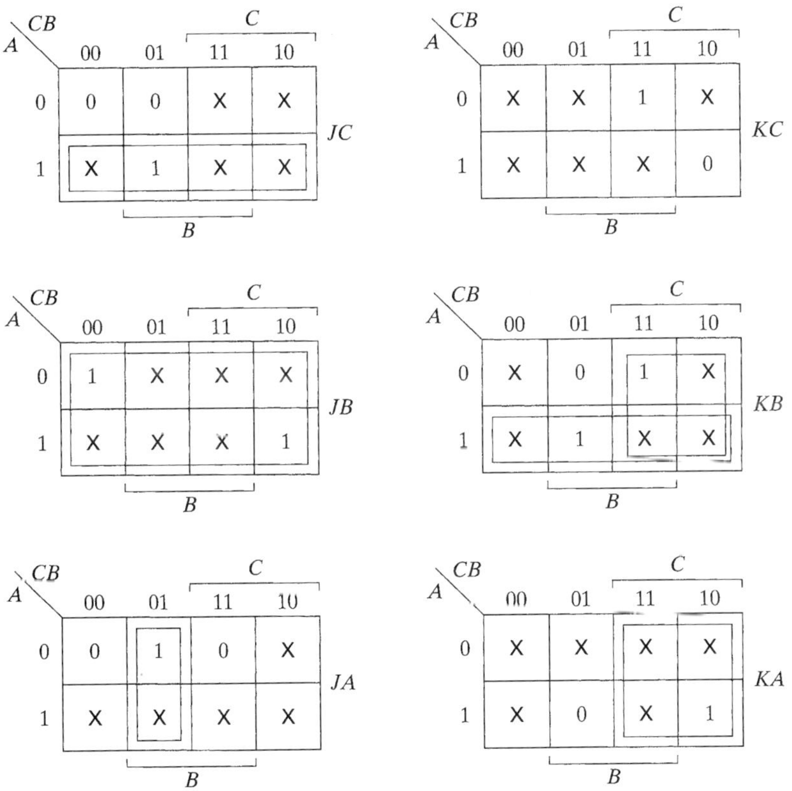

Figure C.16 **Remapped K-maps for *J-K* implementation.**

This implementation reduces the gate count to two and the literals to seven. The implementation logic appears in Figure C.17. The wire count for this implementation is nine (we don't count the hardwired input for J_B and again don't show the reset logic).

There actually is little to distinguish between the *R-S* and *J-K* implementations. This should come as no surprise, given the similar behavior of these two types of devices.

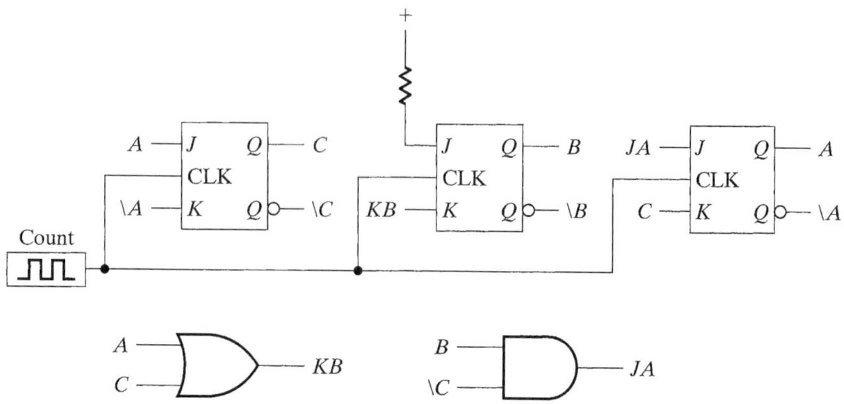

Figure C.17 ***J-K* flip-flop implementation of 3-bit counter.**

Q	Q^+	T
0	0	0
0	1	1
1	0	1
1	1	0

Figure C.18 **Toggle excitation table.**

Present State			Toggle Inputs		
C	*B*	*A*	*TC*	*TB*	*TA*
0	0	0	0	1	0
0	0	1	X	X	X
0	1	0	0	0	1
0	1	1	1	1	0
1	0	0	X	X	X
1	0	1	0	1	1
1	1	0	1	1	0
1	1	1	X	X	X

Figure C.19 **Remapped state transition table.**

C.3.3 Implementation with *T* Flip-Flops

Since this is almost a straight binary sequence, toggle flip-flops seem like a good match. We use the toggle flip-flop excitation table in Figure C.18 to derive new next-state maps. Then, we replace the desired state bits in the K-map with the values needed to control the selected flip-flops to perform the necessary state changes.

Figure C.19 shows the toggle inputs needed to implement the state transitions. For example, counter state 000 advances to 010, so the inputs to the toggle flip-flops should be 0 (don't toggle) for *C*, 1 (toggle) for *B*, and 0 (don't toggle) for *A*. Similarly, state 110 returns to 000. In this case, the control for *C*, *B* and *A* is toggle, toggle, and don't toggle, respectively (or 110).

Reflecting this remapping of functions, the K-maps become those of Figure C.20. The minimized functions become

$$TC = \bar{A}C + A\bar{C} = A \oplus C$$

$$TB = A + \bar{B} + C$$

$$TA = \bar{A}B\bar{C} + \bar{B}C$$

Figure C.21 shows the component-level implementation. To reduce wiring complexity, we simply label input and output nets rather than draw them as wires. Two nets with the same label are understood to be connected. The proper sequencing through the states 000, 010, 011, 101, 110, 000 should be clear from the waveform.

C.3.4 Implementation with *D* Flip-Flops

The choice of *D* flip-flops yields the most straightforward implementation. We dispense with the remapping step altogether. The *D* inputs are identical to the next-state outputs that are already tabulated in the state transition table.

We simply place the next-state outputs into K-maps and find the minimized functions by the usual methods. The K-maps are identical

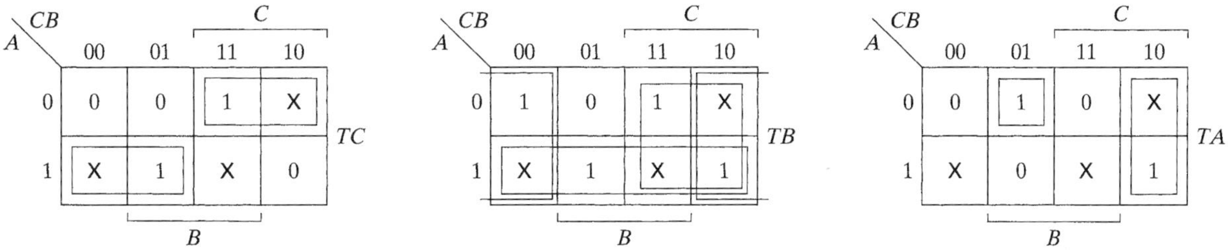

Figure C.20 Remapped K-maps for toggle implementation.

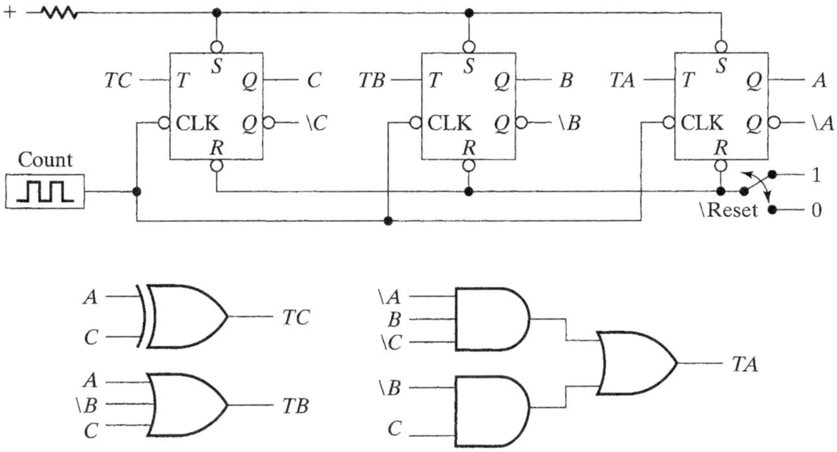

Figure C.21 Toggle flip-flop implementation of 3-bit counter.

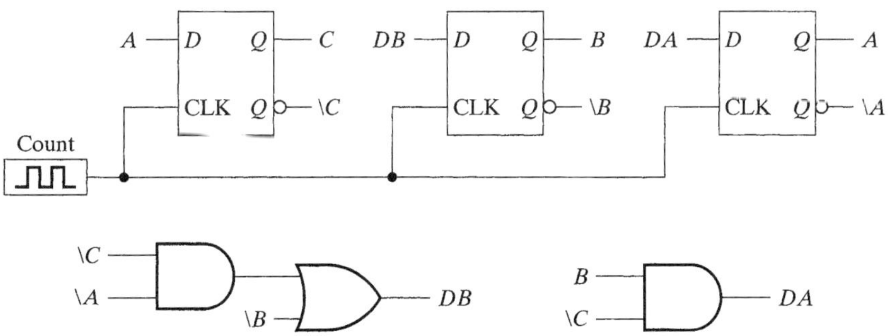

Figure C.22 *D* flip-flop implementation of 3-bit counter.

to those of Figure 7.10, yielding the following minimized functions:

$$D_C = A$$

$$D_B = A\overline{C} + \overline{B}$$

$$D_A = B\overline{C}$$

This implementation requires three gates and six literals and is shown in Figure C.22, which is identical to Figure 7.11 and is reproduced here for easier comparison (reset logic is not shown). Again, the wire count is nine.

C.3.5 Comparison and Summary

Exactly the same state diagram led to somewhat different implementation costs:

R-S flip-flops:	3 gates, 5 literals, 12 wires (Figure C.13)
J-K flip-flops:	2 gates, 4 literals, 9 wires (Figure C.17)
T flip-flops:	5 gates, 10 literals, 15 wires (Figure C.21)
D flip-flops:	3 gates, 5 literals, 9 wires (Figure C.22)

Although it is difficult to generalize from a single example, *J-K* flip-flops usually yield the most gate- and literal-efficient implementations. Since the *R-S* flip-flop behavior is a proper subset of a *J-K*, there is never any advantage in using *R-S* devices as state elements. In fact, you would be hard pressed to find them in the usual parts catalogs (although *R-S* latches are readily available).

T flip-flops are well suited for implementing straightforward binary counters, but their advantage is mitigated when the counter must follow a sequence that is not in direct binary order. In the example of this section, the *T* flip-flop implementation was worst by a wide margin.

Although they do not often yield the most gate-efficient solution, *D* flip-flops have some important advantages. First, they simplify the design procedure, allowing you to skip the next-state remapping step. Tools like *espresso* and *misII* can be applied to the state transition table, treated as a truth table, to obtain a minimized implementation rapidly.

Second, if wiring complexity rather than gate count dominates the design, *D* flip-flops usually have the advantage (in our example, the *D* and *J-K* implementations had the same wiring complexity). Only a single input must be routed to a *D* flip-flop, compared to two inputs in the case of a *J-K* storage element.

Wiring complexity especially is important when using programmable-logic technologies. Often the number of functional outputs supported by a programmable-logic part is more constrained than the complexity of the logic that can be implemented, such as the number of product terms.

Finally, *D* storage elements are particularly transistor efficient in MOS VLSI technologies. *J-K* flip-flops require many more transistors for their implementation.

To summarize, for conventional packaged MSI/SSI TTL design, *J-K* flip-flops usually are preferable, especially when the design criterion is minimum gate and literal count. *D*-type devices are preferred when designing with programmable logic or in highly integrated technologies, where minimum wire count or a simplified design procedure is the goal.

APPENDIX REVIEW

This brief appendix described the different kinds of flip-flops that may be encountered in digital designs. We discussed their respective advantages and disadvantages in the context of a 3-bit counter implementation. We extended the FSM design procedure of Chapters 7 and 8 to include a step that uses the flip-flops excitation table to re-map the state transition table.

In general, *J-K* flip-flops yield counter implementations with the fewest gates. However, we often choose *D* flip-flops because of the simplified design procedure. No remapping step is necessary, and fewer wires are needed to control a *D* flip-flop. *D* flip-flops are most appropriate for modern programmable logic technologies such as PALs and FPGAs.

EXERCISES

C.1 *(Flip-Flops)* Show how to implement a *J-K* flip-flop starting with a *T* flip-flop.

C.2 *(Flip-Flops)* Show how to implement a *D* flip-flop starting with a *T* flip-flop.

C.3 *(Flip-Flops)* Show how to implement a *T* flip-flop starting with a *J-K* flip-flop.

C.4 *(Flip-Flops)* Show how to implement a *T* flip-flop starting with a *D* flip-flop.

C.5 *(Counter Design)* Design a 2-bit counter that behaves according to the two control inputs I_0 and I_1 as follows: $I_0, I_1 = 0, 0$: stop counting; $I_0, I_1 = 0, 1$: count up by one; $I_0, I_1 = 1, 0$: count down by one; $I_0, I_1 = 1, 1$: count by two.

(a) Draw the state diagram and state transition table.
(b) Implement the counter using *T* flip-flops, *D* flip flops, and *J-K* flip-flops.
(c) Which choice of flip-flops leads to the minimum gate count? Assume that only two-input NAND, NOR, XOR, and XNOR gates are available. Draw the schematic for your minimum gate count implementation.
(d) Which choice of flip-flops leads to the minimum wire count? The same kinds of gates are available as in part (c). Draw the schematic for your minimum wire count implementation. Indicate how you have counted the interconnections.

C.6 *(Counter Design)* Design a three flip-flop counter that counts in the following sequence: 000, 010, 111, 100, 110, 011, 001, and repeat. Design the counter using toggle flip-flops. Verify that your implementation is self-starting.

C.7 *(Counter Design)* Consider the design of a 4-bit BCD counter that counts in the following sequence: 0000, 0001, 0010, 0011,

0100, 0101, 0110, 0111, 1000, 1001, and then back to 0000, 0001, etc.

(a) Draw the state diagram and next-state table.
(b) Implement the counter using *D* flip-flops, toggle flip-flops, *S-R* flip-flops, and *J-K* flip-flops.
(c) Implement the counter making it self-starting just for the *D* flip-flop case.

C.8 *(Counter Design)* Consider the design of a 4-bit Gray-code counter (that is, only one of the state bits changes for each transition) that counts in the following sequence: 0000, 0001, 0011, 0010, 0110, 0111, 0101, 0100, 1100, 1101, 1111, 1110, 1010, 1011, 1001, 1000, and then back to 0000, 0001, 0011, etc.

(a) Draw a state diagram and next-state table.
(b) Implement the counter using *D* flip-flops, toggle flip-flops, *R-S* flip-flops, and *J-K* flip-flops.
(c) Do you have to worry about self-starting? Why or why not?

C.9 *(Counter Design)* The 4-bit Johnson counter advances through the sequence 0000, 1000, 1100, 1110, 1111, 0111, 0011, 0001, and repeat. Using the standard counter design process, show how to implement this count sequence using (a) *D* flip-flops and (b) *T* flip-flops. How does your solution compare with the *J-K* implementation of Figure 7.10 in terms of gates and wiring complexity?

Index

D

Q

R

S

T